MOLECULAR PHOTOFITTING

MOLECULAR PHOTOFITTING

PREDICTING ANCESTRY AND PHENOTYPE USING DNA

Tony N. Frudakis

With a Chapter 1 Introduction
by Mark D. Shriver

AMSTERDAM • BOSTON • HEIDELBERG • LONDON
NEW YORK • OXFORD • PARIS • SAN DIEGO
SAN FRANCISCO • SINGAPORE • SYDNEY • TOKYO

Academic Press is an imprint of Elsevier

Acquisitions Editor: Jennifer Soucy
Assoc. Developmental Editor: Kelly Weaver
Project Manager: Christie Jozwiak
Publishing Services Manager: Sarah Hajduk
Cover Designer: Alisa Andreola
Composition: SPi
Printer: China Translation & Printing Services, Ltd.

Academic Press is an imprint of Elsevier

30 Corporate Drive, Suite 400, Burlington, MA 01803, USA

525 B Street, Suite 1900, San Diego, California 92101-4495, USA

84 Theobald's Road, London WC1X 8RR, UK

This book is printed on acid-free paper. ∞

Library of Congress Cataloging-in-Publication Data
Application submitted.

British Library Cataloguing-in-Publication Data
A catalogue record for this book is available from the British Library.

ISBN: 978-0-12-088492-6

For information on all Academic Press publications
visit our Web site at www.books.elsevier.com

Printed and bound by CPI Group (UK) Ltd, Croydon, CR0 4YY
Transferred to Digital Print 2011

TABLE OF CONTENTS

FOREWORD

There is a deep desire within us all to find out who we are as individuals by tracing our ancestors through history, asking where they may have come from and what they may have looked like. Only over a relatively short period of time—at best several generations—can most of us follow our personal genealogies, using family trees drawn from oral histories or public records, before we are quickly lost in the depths of inaccessible ancestors. Another approach, only available to us in recent years, is to peer into our genes to examine the DNA record encoded in the human genome. This can provide a wealth of information about our family ties, the level of relatedness within and between populations, and ultimately even the origin of the human lineage. The linking of our shared genetic ancestries with the geographical distribution of prototypical human populations is one of the keys to finding our own affiliations as well as the distribution of physical traits within present day admixed populations. The concept of Biogeographical Ancestry (BGA) is the term comprehensively defined by Tony Frudakis in this landmark reference work, with the motivation being to correlate ancestry and sequence differences within our DNA to an individual's physical appearance. This whole process is referred to as "molecular photofitting," with downstream applications for forensic identification purposes.

The considerable effort expended in characterizing the frequency distribution of single nucleotide changes within human populations has rewarded the DNAPrint Genomics team with a unique set of Ancestry Informative Markers (AIMS). With these tools and a noticeably pragmatic approach, a detailed description is given of the theoretical basis for choosing a model with four main ancestral continental groups (West African, European, East Asian, and Indigenous American). These are the geographical extremes that can be used to plot the admixture of a present day individual. They also allow indirect methods to predict physical traits such as the degree of pigmentation present within the hair, eye, and skin based on the primordial characteristics of these groups. While not definitive, this is a clear improvement on the current inaccurate means of inferring physical appearance from a DNA sample. Although the DNAWitness™ protocol is currently operational as a molecular photofitting test, the future is in directly correlating physical phenotypes with the genes that are part of the biochemical process determining or modifying these characteristics. The complement of human pigmentation genes is presently being characterized, and this book contains a good description of the polymorphisms that will direct the color traits of hair, eye, and skin. This is a major advancement of a fledgling field; the future will surely be based on an individual's genotype at specific loci. A glimpse of this is seen with correlative genes such as MC1R with hair, OCA2 with eye, and MATP with skin color.

The use of DNA fingerprinting systems such as CODIS are now well accepted in courts of law. This is not yet true for molecular photofitting techniques, and the contentious issue of predicting phenotypes based on ethnic group stereotypes has social connotations beyond forensic analysis. The discussion of the first case reports utilizing DNA for physical profiling shows the need for high levels of accountability with this breakthrough science, testing the limits of understanding of the lay public, police force, and judiciary alike. As a new form of evidence, the ''DNA-witness'' based on accumulating databases of AIMS profiled individuals must be compared with the accepted but not necessarily reliable eye-witness testimony of appearance. The final chapter on the politics and ethics of testing for genetic ancestry, as described in this textbook, is challenging and confrontational to our beliefs about what the idea of race and our own racism represents, and, as such, deserves to be read and considered by a wide audience. *Molecular Photofitting* is a very thoughtful and rigorous treatise on a socially contentious issue, but one that is very likely to help contribute to the policing of our communities.

RICHARD A. STURM, PhD
Melanogenix Group
Institute for Molecular Bioscience
University of Queensland
Brisbane, Qld 4072, Australia

PREFACE

This textbook is meant to serve less as an instructive tool for the classroom and more as a reference for the forensic, clinical, and academic scientist. It is my hope that scientists seeking to develop or use methods for the inference of phenotype from DNA will find some of the ideas presented here useful.

Most of the book focuses on data, results, and observations derived from a small number of Ancestry Informative Marker (AIM) panels, which cynics might point out are the same panels my company (DNAPrint Genomics, Inc., Sarasota, FL) sells to the forensics community. This is no accident, and to a point the cynics are right. I founded DNAPrint and invested a substantial portion of its future into developing the data, databases, and tools described herein. I did this because it had not been done before, and believed there existed a corresponding niche that needed to be filled. The fact is that before this text, to my knowledge, there have been no other descriptions of bona-fide AIM panels—describing their expected accuracy and bias, results across populations, comparison of the results with self-held notions, or social constructs and correlation with elements of physical appearance and so forth. Indeed, the AIM panels developed by my colleagues and me at DNAPrint, including Mark Shriver of Penn State University, is among the first ever to be studied in these ways. The statistical methods for estimating individual ancestry, based on what Fisher referred to in 1912 as "inverse probability," were worked out for the first time only in the 1970s. Bayesian methods in the field of phylogenetics and individual ancestry estimation are even more recent, having been developed only within the past 10 years. It has been possible to construct panels of good AIMs—not those found in genes, but bonafide AIMs of neutral evolutionary character—for the first time only in the past decade, as the sequence for the human genome has been released.

No multifactorial human phenotype has ever before been predicted from an appreciation of polymorphisms in human DNA. My colleagues and I were simply among the first to invest in these types of molecular resources and apply them for the purposes described in this book. As the field of molecular photo-fitting is completely new, having become enabled by the recent completion of the human genome project, it is just a matter of time before other panels and improved methods will be described. Perhaps future editions can incorporate a more diverse collection of panels and databases.

I have done my best to provide a good theoretical background necessary to appreciate the methods and data discussed. I learned much of the theoretical material in the book (see Chapter 2) during my years in the field from such textbooks as *The Genetics of Human Populations*, by L.L. Cavalli-Sforza, and W.F. Bodmer; *The Handbook of Statistical Genetics*, by D.J. Balding, M. Bishop, and C. Cannings; and the various publications of R. Chakraborty. For the student, the theoretical material discussed in this book may serve as stimulation for further reading from a proper population genetics textbook and papers. For the forensics professional, the lay-description provided for these complex ideas may help in better understanding how the machinery of molecular photo-fitting works, and possibly obviate the need to go to a proper population genetics textbook (which to a nongeneticist may not be a pleasant experience).

Parts of the book may seem redundant—for example in a later chapter, you might notice an explanation for an idea that was treated fully in an earlier chapter. When dealing with a new topic that requires background information the approach taken here is to recap the background information learned in an earlier chapter rather than to refer you to that earlier chapter. This also serves many readers better—rather than reading the text from cover to cover, most will probably scan quickly for sections of interest, and if the background information was not recapped in that section, the point could be lost.

There are likely to be mistakes discovered in the first edition, which will be corrected in the next edition. We (my colleagues and I)would be grateful for your feedback. As with most books, this one is intended to stimulate thought, discussion, and ultimately activity in the field. It may be that 20 years from now, the field will have advanced very little due to funding limitations and various sources of resistance. On the other hand it may be that 20 years from now molecular photofitting will be standard practice, and we will be doing amazing things, such as using computers to provide most of the information on a person's drivers license from DNA left at a crime scene, even creating "artist's renderings" from DNA "eyewitness testimony." Either way, we are honored to have the opportunity to share our work and interpretation of other relevant works with you, and we hope you find this book useful.

ACKNOWLEDGMENTS

A very significant THANKS is given to Mark Shriver, who helped write Chapter 1, provided assistance writing the parts of Chapter 6 that covered his work, and edited the first four chapters. It was in collaboration with Mark that the panel of Ancestry Informative Markers (AIMs) discussed in this book were developed, and he was involved in the collection of many of the samples we used. In addition, he and his colleagues wrote the very first version of the MLE program we used, and we later optimized and expanded this program with his help. Much data from his papers appear within these pages and the book would not have been possible without his efforts.

Matt Thomas managed the laboratory that produced most of the data described in this book, and helped develop many of the ideas, algorithms, and figures that are presented. Without his tireless work over the past six years, most of the data discussed within these pages would not have existed and this book would not have been possible. Zach Gaskin, Shannon Boyd, and Sarah Barrow produced most of the genotype data for the 71AIM and 171 AIM panels and iris color work discussed herein. It was through collaboration with Nick Martin and colleagues at the Queensland Medical Research Institute in Brisbane, Australia that the hair color data and discussion was possible.

Lastly, the most significant THANKS needs to be given to the investors of the company that funded much of the research in this book—DNAPrint Genomics, Inc. Even after countless rejected grant applications, average, every-day citizens invested in the commercial viability of the concepts we describe by buying DNAPrint stock. Although the value of DNAPrint continues to sag, and products based on the ideas and data presented here have not yet sold well in the forensic, academic, or clinical world, the field is a very new one and these investors need only look at the outcome of the Louisiana Multiagency Homicide Task Force Investigation to know that their investment has made a positive difference in the world. This book is partial evidence of the value of their investment, and I thank them for making this work possible.

CHAPTER 1

FORENSIC DNA ANALYSIS: FROM MODEST BEGINNINGS TO MOLECULAR PHOTOFITTING, GENICS, GENETICS, GENOMICS, AND THE PERTINENT POPULATION GENETICS PRINCIPLES

With an Introduction by Mark D. Shriver

PART I: INTRODUCTION: BRIEF HISTORY OF DNA IN FORENSIC SCIENCES

The forensic analysis of DNA is one of the clear successes resulting from our rapidly increasing understanding of human genetics. Perhaps much of this success is because this particular application of the molecular genetic revolution is ultimately pragmatic and because the genetic information required for efforts such as the Combined DNA Index System (CODIS) and The Innocence Project (www.innocenceproject.org) are relatively simple. Although the requirements of DNA in these instances, namely individualization, are indeed, relatively simple, they are somewhat technical, especially for the reader unfamiliar with molecular methods or population genetics. They nonetheless provide an important framework for the bulk of the material presented in this book. Though they are important for the rest of our discussion in the book, in this chapter, we provide only a brief summary of the standard forensic DNA methods, because these are well documented in other recent texts (Budowle et al. 2000; Butler 2001; Rudin & Inman 2002).

Modern forensic DNA analysis began with Variable Number of Tandem Repeats (VNTR), or minisatellite techniques. First discovered in 1985 by Sir Alex Jeffreys, these probes, when hybridized to Southern blot membranes (see Box 1-A), produced highly variable banding patterns that are known as DNA fingerprints (Jeffreys et al. 1985). Underlying these complex multi-banded patterns are a number of forms (alleles) of genetic loci that simultaneously appear in a given individual. The particular combinations of alleles in a given individual are highly specific, yet each is visible because they share a common DNA sequence motif that is recognized by the multilocus molecular probe through complementary base pairing. These multilocus probes are clearly very individualizing, but problematic when it comes to quantifying results. Some statistics can be calculated on multilocus data, but certain critical calculations cannot be made unless individual-locus genotype data are available. In answer to this need, a series of single-locus VNTR probe systems were developed, and these became standard in U.S. forensic labs from the late 1980s through the early 1990s.

Box 1-A

The Southern Blot is named after Edwin Southern, who developed this important first method for the analysis of DNA in 1975. This method takes advantage of several fundamental properties of DNA in order to assay genetic variation, generally called *polymorphism*. The first step is to isolate high molecular weight DNA, a process known as genomic DNA extraction. Next, the DNA is digested with a restriction enzyme, which makes double-stranded cuts in the DNA at every position where there is a particular base pair sequence. For example, the restriction enzyme, *EcoRI*, derived from the bacteria, *Escherichia coli* strain RY13, has the recognition sequence, GAATTC, and will cut the DNA at every position where there is a perfect copy of this sequence. Importantly, sequences that are close to this sequence (e.g., GATTTC) will not be recognized and cut by the enzyme. The restriction digestion functions to reduce the size of the genomic DNA in a systematic fashion, and originally evolved in the bacteria as a defense mechanism as the bacteria's own genomic complement was protected at these sequences through the action of other enzymes.

After DNA extraction the DNA is generally a series of large fragments averaging 25,000 to 50,000 bp in length. Because of the immense size and complexity of the genome, the results of a restriction enzyme digestion are a huge mix of fragments from tens of base pairs to tens of thousands of base pairs. When these fragments are separated by size on agarose gels using the process known as electrophoresis, they form a heavy smear. Although it's hard to tell by looking at these smears since all the fragments are running on top of each other, everyone has basically the same smear since all our DNA sequences are 99.9% identical. Places where the restriction patterns differ because of either changes in the sequence of the restriction sites (e.g., GAATTC → GATTTC) or the amount of DNA between two particular restriction sites are called Restriction Fragment Length Polymorphisms (RFLPs).

The key advancement of the Southern Blot was to facilitate the dissection of these restriction enzyme smears through the ability of DNA to denature (become single-stranded) and renature (go back to the double-stranded configuration), and to do so in a sequence-specific fashion such that only DNA fragments that have complementary sequences will hybridize or renature. The DNA in the gel is denatured using a highly basic solution and then transferred by capillary action, using stacks of paper towels onto a thin membrane, usually charged nylon. After binding the digested DNA permanently to the membrane, we can scan it by annealing short fragments of single-stranded DNA, called probes, which are labeled in such a way that we can detect their presence. The probes will anneal with DNA at locations on the genomic smear to which they have complete, or near complete complementarity depending on the stringency of the hybridization and wash conditions. Since the probes are radiolabeled or chemiluminescently labeled, the result is a banding pattern where the location of particular sequences on the genome emerge as blobs called bands. The lengths of the bands can be estimated as a function of the position to which they migrated on the gel relative to size standards which are run in adjacent lanes.

The single-locus forensic VNTR systems are highly informative, with each marker having tens to hundreds of alleles. At every locus each person has only two alleles, which together constitute the genotype, one received from the mother and one from the father. Given such a large number of alleles in the population, most genotypes are very rare. A standard analysis with single-locus VNTRs typically included six such single-locus VNTR markers, each run separately on a Southern blot gel. The data from the separate loci would then be combined into a single result expressing one of two outcomes:

- Exclusion—the suspect and evidence samples do not match
- When the genotypes match, a profile or match probability, which is an expression of the likelihood that the two samples matched by chance alone.

Exclusions are pretty intuitive since the lack of genetic match between the samples eliminates any chance that the suspect could have donated the evidence (baring the very rare occurrences of somatic mutation, chimerism and mosaicism, each cell in our bodies has identical DNA). This of course presumes careful lab procedures and an intact and unquestioned chain of evidentiary custody. Given a match, profile probabilities are also quite intuitive, being expressions of the chances or likelihood that a particular genotype exists in a population. Profile probabilities are essentially a means to express the statistical power of a set of makers to demonstrate exclusion. For example, consider that both the suspect and biological evidence have blood type AB, the least common ABO genotype in most populations. There is no exclusion, but does that mean the suspect left the sample? Since about 4% of people have the AB genotype we say that the profile probability is 0.04 and that given no other information, the chances of having a match by chance alone are 1 in 25. Another way to read this profile probability is to say that 4% of the people match the person who left the sample.

Maybe these are good betting odds in the casino, but in both science and in court where the destiny of human lives are at stake, more stringent criteria are required. For one thing, the frequency of 4% in the population does not necessarily mean there is a 1 in 25 chance that the suspect donated the evidence. When tests of such limited power were used, other forms of evidence that contribute to the prior probability the suspect donated the evidence would have to be taken into account. Generally, genetic markers are not the only evidence against the defendant and other pieces of information can be combined with the genetic data to comprise a preponderance of evidence. With DNA markers commonly used today, profile probabilities are much smaller than 4%, and thus the weight of the evidence is so great that convictions could be and sometimes are made solely on DNA results, without other evidence or prior probabilities taken into account.

Single-locus VNTRs were replaced by newer marker systems that became possible as a result of the Polymerase Chain Reaction (PCR), a process of amplifying DNA *in vitro*, which won a Nobel prize for its inventor, Kary Mullis. These newer markers are most commonly called Short Tandem Repeats (STRs) although they first were referred to as microsatellites since their repeat units are shorter than minisatellites. In many ways STRs are different from VNTRs. For example, STRs generally mutate one or two repeat units at a time and VNTRs mutate in steps of many repeats. There are a number of other differences and similarities in how these markers evolve and how they can be used but these are beyond the scope of this presentation, and interested readers should consult Goldstein and Schlotterer (1999).

Table 1-1 presents a summary of some of the important characteristics of STRs and VNTRs. In terms of how the markers are used in forensic analyses, STRs are quite similar to VNTRs. The most significant difference is that with VNTRs, a process of allelic binning is required to interpret the genotype. Since the range in allele size at forensic VNTRs is large, alleles at a single locus can take the form of both very small and very large fragments, and so a variety of patterns could comprise a given genotype. Gel electrophoresis methods are limited in the resolution of fragments to about 5% of fragment length and as

Table 1-1

Comparison between STRs and VNTRs in forensic DNA analysis.

Characteristic	VNTRs	STRs
Repeat unit size	10–30 base pairs	2–5 base pairs
Repeat array length	10's to 1000's of repeats	10's of repeats
Number of alleles per locus	100's to 1000's	10's
Number of loci in genome	10,000	100,000
Laboratory method	Southern Blot	Polymerase Chain Reaction
Scoring of alleles	Computer-assisted	Automated
Precision of allele size	Binning of alleles	Exact size estimation
Amount of DNA required	1000 ng	0.1 ng

such, a gel that effectively resolves the smaller alleles at a particular locus will be ineffective at resolving with the greatest precision the largest alleles at this locus.

Southern blot gels are also susceptible to other phenomenon (e.g., band smiling, overloading, and DNA degradation) that compromise the precision of allele size estimation. Protocols therefore were developed to score alleles with the highest levels of size precision possible with the understanding that the estimated size of a particular allele contained some degree of error. Allelic bins became the accepted solution to the question of estimating allele frequencies for profile probabilities and allelic identity (National Research Council 1996; Weir 1996). This is not as sloppy as it might seem, since the nature of the polymorphisms and molecular clock and maximum parsimony theories tell us that alleles of similar size are evolutionarily related (i.e., cousins of one another), but it certainly leaves some precision to be desired. The range in allele size for STRs on the other hand is more restricted and because smaller sizes can be scored on sequencing gels accurate to one base pair size intervals they can be scored without question as to the exact length of alleles.

Although STR analysis using the PCR is very sensitive, allowing subnanogram amounts of genomic DNA to be analyzed, there are limitations that led to the development of a subcategory of STRs and increased interest in developing other types of markers, like Single Nucleotide Polymorphisms (SNPs). In particular, degraded remains, like those recovered from the site of the World Trade Center disaster, often are composed of highly fragmented DNA. For PCR to work effectively, the DNA templates spanning the region to be amplified must be present in the sample (that is, the templates must be full-length, and not degraded). The standard STR markers are 100 to 450 base pairs in length after amplification and this size has proven to be too large for highly degraded samples, where DNA can be sheared into small fragments. A special panel of smaller mini-STR markers has been defined, which can replace the standard panel for use on highly degraded samples (Butler et al. 2003).

Given the target for SNPs is only one base pair, these markers are better suited than even the mini-STRs for amplification and analysis in highly degraded samples. SNPs can provide much of the same identity information as STRs, and are a source of information far beyond the simple exclusion and profile probabilities typically gleaned from STR data. This is because most genetic variation among humans is in the form of SNPs, and SNP variation likely underlies most functional variation. That is, SNPs represent the type of polymorphisms one tends to find within or near coding regions of human genes, and since genes cause phenotypes, they are therefore relevant for determining phenotype. In contrast, STRs are of relatively low prevalence and only a few STRs are known to have functional effects, for example, particular STR alleles that have expanded in size result in rare genetic diseases such as Huntington's chorea and Fragile X syndrome. It is notable that we are referring here to the phenotypic variation among individuals that is genetic in origin. Organisms are actually combinations of their environments and their genomes, a point that is important for many traits, but less so for those that are highly genetic.

Since SNPs are usually biallelic (having two alleles, e.g., C and T) there are only three genotypes (CC, CT, and TT). One situation in which STRs provide information that is difficult to get from SNP data is for mixed samples. Having three- or four-allele patterns at an STR or VNTR is clear evidence for more than one contributor since everybody has only one mother and only one father, and so two alleles are the maximum that should be seen in a sample left by one person. If a sample registers even all the possible alleles at a SNP, it is just a heterozygote (the CT genotype from the previous example) and it is not clear from the genotype alone whether it is a mixture of a CC and TT individual, two CT individuals, or a single CT individual.

Showing all STR alleles in the form of an allelic ladder is a useful laboratory technique. Although some SNP analysis protocols can quantitatively assess the relative concentrations of the two alleles helping to recognize mixed samples, it is unlikely that SNPs can replace STRs in the case of a mixed sample when human identification is the goal. STRs are simply better in the case of sample mixtures. That being said, there are usually multiple evidentiary samples at a crime scene and it's unusual for them all to be mixed and sometimes when they are, there are methods for resolving them. For example, in the case of vaginal swab analysis, the male and female DNA fractions can be separated by differential lysis of cells, male sperm cells being hardier than female epithelial cells (Budowle et al. 2000; Gyske & Ivanov 1996).

THE STATISTICS OF FORENSIC DNA ANALYSES

There are a few basic statistics that are fundamental to forensic identity analysis. Earlier we introduced the profile probability as an important measure of the meaningfulness of a genotype match. Calculating the profile probability requires that we are first able to ascertain the genotype of a person. An inability to do this is one of the main limitations for multilocus DNA fingerprinting systems. Even though profile matches using DNA fingerprinting probes are visually stunning and quite compelling as evidence of identity (see Figure 1-1), we do not know which of the multitude of bands on a multilocus gel are allelic; that is, which are the two (maternal and paternal) bands that together comprise the genotype.

Once a genotype is determined, the next step is to determine the expected frequency of the genotype. How many people in the population should we find with the genotype we measured? The expected genotype frequency is dependent on the allele frequencies of the two alleles that make up a particular genotype. The relationship between the allele frequencies and the genotype frequencies was first recognized by Hardy and Weinberg independently in 1908 and so today is called the Hardy-Weinberg Equilibrium (HWE). HWE is the foundation for population genetics and explains how populations continue unchanged in allele frequencies (and thus phenotype) unless acted on by one of the four evolutionary forces (mutation, selection, gene flow (or admixture), and drift). For the AB heterozygote in the previous example, given the genotype is a heterozygote with the frequency of A in the population as 0.1 and the frequency of B as 0.02, we calculate the expected genotype frequency as $2 \times 0.1 \times 0.2 = 0.04$ or 4%. Box 1-B gives more detail on the derivation of the HWE relationship. It is important to recognize that not only can the four evolutionary forces affect HWE, but so can the mating structure and the levels of ancestry stratification in the population. These effects are particularly important for markers of the type that we focus on in much of this book, the Ancestry Informative Markers (AIMs).

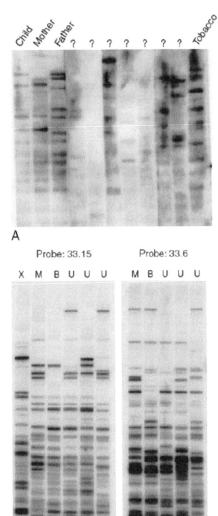

Figure 1-1

Examples of DNA fingerprinting profiles as obtained using the Southern blot method. Shown are: A) The first-ever Southern blot, which used a minisatellite probe. This blot was created by Alec Jeffries and colleagues September 10, 1984. The first three lanes are of a child, the child's mother, and the child's father, respectively, and the remaining lanes are DNA samples from various other species including baboon, lemur, seal, cow, mouse, rat, frog (of order long since forgotten, labeled "?"), and tobacco in the last lane. As the first experiment of its kind, the blot takes a somewhat messy appearance but the DNA fingerprint banding pattern for the child is clearly discernable as a subset of those of the combination of the mother and father. B) The first Southern blot ever used in an immigration case, using two multilocus DNA fingerprinting probes for detecting different minisatellite sequences. In this case, a mother (M) and her three undisputed children (U) are compared with a child of disputed paternity (B). The father's DNA was unavailable but we can see that each band present for the child B that is not present in the mother M can be found in at least one of the other children. The likelihood that the child of disputed paternity is a child of the mother rather than being unrelated happened to be 5×10^8. Both photographs supplied by Alec Jeffries (2005).

Box 1-B

The Hardy-Weinberg Equilibrium formula describes the relationship between observed allele frequencies and expected genotype frequencies in a population. If the population is panmictic (i.e., randomly mating with no population substructure) and of infinite size (so that genetic drift is absent) and there is no admixture, mutation, or natural selection at the locus in question, then given allele frequency for the C allele is p and for T allele q,

$$p^2 + 2pq + q^2 = 1 \qquad (1\text{-}1)$$

where

$$\text{freq}(CC) = p^2 \qquad (1\text{-}2)$$

$$\text{freq}(CT) = 2pq \qquad (1\text{-}3)$$

$$\text{freq}(T\,T) = q^2 \qquad (1\text{-}4)$$

To estimate the frequency of the C allele from the data, we can use allele counting where N_{CC}, N_{CT}, N_{TT}, and N_{TOT} are the counts of the numbers of individuals observed for each of the three genotypes and the combined total number in the sample, respectively.

$$p = \frac{2N_{CC} + N_{CT}}{2N_{TOT}} \qquad (1\text{-}5)$$

$$q = \frac{2N_{TT} + N_{CT}}{2N_{TOT}} \qquad (1\text{-}6)$$

The χ^2-distribution can be used to calculate the significance of deviations from HWE when the numbers of observations per cell exceeds five. There are a number of alternative methods for computing the significance when there are smaller numbers of observations as often is the case with larger numbers of alleles, because some of them are of low frequency and even for large populations of samples, not all alleles are represented.

In calculating profile probabilities we usually are combining genotype data from a number of loci across the genome into a single summary. If the markers are unlinked (either located on different chromosomes or far apart on the same chromosome), they are statistically independent and so the individual locus profile probabilities can be multiplied together to estimate the multi-locus profile probability. However statistical independence of unlinked markers may not always be the case as population structure caused by assortative mating and ancestry stratification can lead to correlations among unlinked markers. Like the deviations from HWE mentioned earlier, which result from population stratification, these unlinked marker correlations exist only for markers that measure ancestry information (i.e., AIMs). Most genetic markers have very little information for ancestry and so this is generally not an important concern.

In the early 1990s there was substantial debate over the importance of population structure in multiplying across loci to calculate the profile probabilities. The conclusion was that although the effects of stratification were minimal, forensic scientists should err on the conservative side, implementing the ceiling principal in lieu of a straight product rule calculation. The ceiling principal and other methods to adjust for low levels of population stratification are covered in detail in the NRC report (National Research Council 1996).

An important development in DNA forensics in the United States was the establishment of the Combined DNA Index System (CODIS) database. CODIS is a national database of DNA profiles from convicted felon and evidentiary samples. The vast majority of the genotype information in CODIS is from STR marker systems, although it was designed to allow for other marker systems (VNTRs and mtDNA) as well. The main purpose of CODIS is to allow criminal investigators to search for matches between convicted felons and evidentiary samples from unsolved cases. Another aim of CODIS was to allow investigators to link different crime scenes for which the same DNA had been donated. Concerns of privacy protections and the logistics were carefully considered in the establishment of the CODIS system (see http://www.fbi.gov/hq/lab/codis/ and Inman & Rudin (2002) for more information). CODIS has proven immensely successful in assisting investigations. Quoting from the CODIS web site:

> As of February 2006, CODIS has produced over 30,000 hits assisting in more than 31,700 investigations.

As of this date there are a total of 3,072,083 DNA profiles in CODIS; 130,877 are evidentiary and 2,941,206 are convicted offenders.

A downside of the CODIS database is that, since it is composed of only previously convicted felons, most crime scene specimens do not provide a hit or match. It is for primarily these types of samples that molecular methods for photofitting are useful. As we will discuss throughout this book, certain elements of physical appearance can now or in the near future be gleaned from the DNA. The recently completed human genome project and its various and sundry databases and tools have accelerated DNA research to the point that it will be possible to paint a rough portrait of an individual from their DNA. However, it is notable that were there a stage in our societal development where everyone was sampled and included in a national database like CODIS, the types of ancestry and phenotype assays we describe in this book would be of little interest or use to forensic scientists. Everyone who could commit a crime, baring emigration from untyped jurisdictions, would already be in the database and therefore linking a crime scene with DNA to an individual would be trivial. The likelihood of such a scenario coming about in the near term in the United States seems very small and the wisdom questionable. For example, it would seem that with such a database DNA matches may take on even extra significance, to the exclusion of other types of evidence, and many might begin to feel as if they were a simple laboratory slip-up, or a corrupted computer file, or program away from facing an aggravated first-degree murder charge.

However, even if such a national database were to be established and everyone entering the United States legally could be compelled to provide a cheek swab, the rate of illegal immigration to the United States would have to be very small for there to no longer be any forensic utility of the methods we propose. Apart from forensics there are other important reasons to pursue molecular photofitting research. Even if molecular photofitting tests are no longer needed forensically because of an exhaustive CODIS database, it is clear that molecular photofitting research will facilitate many important advances in our understandings of human evolution and human genetics, and will serve as model systems for developing efficient methods for studying other phenotypes, like complex diseases.

THE NATURE OF HUMAN GENETIC VARIATION

In order to effectively present the content of this book, it is important to consider a few of the terms used in discussing genetic variation and the presumptions underlying them. We can say, somewhat paradoxically, that as more genetic markers are combined into an analysis, the results become less and less genetic. This is true whether we are meaning that the analysis in question is being done by a scientifically minded observer or by the cells of a developing organism. Genetic is the process of inheritance first described by Gregor Mendel, who observed that characters were the result of the segregation of factors, one coming from each parent (Mendel 1865). Today we call these factors alleles, and know that they are different forms of genes. Such different forms of a gene are ultimately the result of variability in the DNA sequence comprising the gene, sometimes affecting the sequence of amino acids for the protein encoded by the gene, and sometimes the ways in which the genes are transcribed into mRNA and then translated into protein. These variations have distinct effects on the phenotype (physical character or trait), some as described by Mendel being either recessive or dominant effects.

Many traits are like this; the genetic paradigm has been used with tremendous success over the past 100 years. As of April 2006, the molecular basis for 2256 phenotypes transmitted in a monogenic (single gene) Mendelian fashion have been described (Online Mendelian Inheritance in Man database, OMIM: http://www.ncbi.nlm.nih.gov/Omim/mimstats.html). All these traits have distinct genetic effects, meaning they can be observed to segregate in families consistently enough to be mapped onto specific chromosome regions and then the particular genes in those regions identified. Most of these traits that have been mapped to date give rise to disease. It is often the case, however, that not everyone with the risk-allele is affected with the disease; this phenomenon is known as incomplete penetrance. Incomplete penetrance is sometimes the result of modifying loci that interact with the risk-allele, altering either the chances of showing the disease or its severity (formally a related concept called variable expressivity). Despite these caveats, genetic traits are those that are inherited in families and show distinct and dichotomous phenotypes (e.g., normal and affected).

However, variability in many traits is not the result of one gene (with or without modifying genes), but is instead the result of variation in a number of genes. These traits are called polygenic and do not segregate by Mendelian rules, but instead appear to undergo mixing. Charles Darwin, Francis Galton, and others described this as blending inheritance. Of course, genes still are underlying phenotypic variability in polygenetic traits (e.g., skin pigmentation and stature), and it is the combined effect of these multiple genes that results in the continuous nature of trait variation. Thus, polygenic traits are properly *genic*, the result of gene action, but not *genetic* in the formal sense of being inherited simply. This distinction may seem merely semantic, but there are important conceptual implications inherent in these terms and these distinctions are important for the population genomic paradigm on which modern studies of genetic variation and evolution are founded.

POPULATION GENETICS AND POPULATION GENOMICS

The differences between genics, genetics, and genomics are key points in the new paradigm of population genomics. Population genetics is different from Mendelian genetics in that it is primarily concerned with the behavior of genetic markers and trait-affecting alleles in populations, not in families. Population genetics is, fundamentally, the study of evolutionary genetics that arose out of the new synthesis in which evolutionary theory was reinterpreted in the light of the rediscovery of Mendel's work in 1900. Mendelian inheritance is thus part of the foundation of population genetics, akin to an axiom in mathematics, but rarely the object of direct attention. Human population genetics research generally focuses on estimating population parameters that help us model the demographic histories of populations. Questions such as which populations are most closely related, whether particular migrations occurred over short or long periods of time, how admixture has affected contemporary populations, when and how severe were population bottlenecks (or reductions in the size of the breeding population) in the past, are the focus of these efforts.

Much human population genetic research has been carried out using mitochondrial DNA (mtDNA) and nonrecombining Y-chromosomal (NRY) markers. Since the genes contained in mtDNA and the NRY are present in a single (unpaired) form in each cell of the body, they are passed on intact from parent to offspring without undergoing recombination. Thus, all variants are physically connected forever, regardless of the number of base pairs separating them. Since the mtDNA and NRY are very large and highly variable, they are remarkably informative regarding human demographic history. Additionally, since the mtDNA is always inherited through the maternal lineage and the NRY through the paternal lineage, these markers can be used to test hypotheses about differences between female and male migratory patterns and other aspects of mate pairing (Jobling et al. 2004). But despite being very informative, mtDNA and NRY studies do not provide more than one measured locus and at least in individuals, there is no way to calculate confidence intervals about the point estimates of the parameters that are being studied (such as estimates of ancestry, or likelihood to express a disease phenotype). Since there is no recombination within the mtDNA and NRY, and since these contain a number of genes, adaptation in any of the genes will affect all the genetic variation and an assumption of selectively neutral evolution will not hold.

Another branch of human population genetics has focused on averaging together as many unlinked genetic markers as possible to estimate population parameters. Unlinked genetic markers are found on the autosomal chromosomes (chromosomes 1–22) and the X chromosome, and conclusions are drawn considering collections of markers together. The same types of questions that are investigated with mtDNA and NRY markers are addressed with many-marker averages but one procedural difference between the two methods is that for mtDNA and NRY studies, many more individuals are needed in the population samples compared to studies of large numbers of independently segregating markers. For example, consider two population genetics papers appearing in *Science* a few years ago: the paper by Ke and colleagues (2001) focused on NRY analysis in 12,000 Asian men, and the paper by Rosenberg and colleagues (2001) used average population samples of 20 individuals typed for 377 microsatellite markers (a paper that we will discuss in more detail later). Both of these papers were able to draw interesting conclusions, one using a single genetic marker system but very many persons, the other using many markers, but fewer people.

Population genomics is a new branch of population genetics, which is specifically focused on a consideration of both the averages across many markers as well as specific loci individually. A concise summary of this new perspective on genetic variation has been described as, "the process of simultaneously sampling numerous variable loci within a genome and the inference of locus-specific effects from the sample distributions." In other words, population genomics incorporates a model of genome evolution that allows for the analysis of the unique, locus-specific patterns. These patterns result from genetic adaptation in the context of the genomewide averages, and are represented largely by selectively neural loci affected primarily by demographic history. In essence, population genomics focuses stereoscopically on both the forest and the trees; we are interested simultaneously in the evolution of both individual genes and the populations carrying them.

Briefly, there are four primary evolutionary forces: mutation, genetic drift, admixture, and selection. Mutation (i.e., a change in the sequence, number, or position of nucleic acids along the strand of DNA) is the ultimate source of all genetic variation and occurs more or less at random across the genome. Mutation events are unique, occurring in just one chromosome at one particular genomic location and at one point in time, although some nucleotide positions are hypermutable and undergo changes repeatedly. Therefore, although each mutation is unique, each variable position may have had multiple origins. SNPs, for example, are particular nucleotide positions now variable as the result of a base-pair substitution in the recent past that have spread to become prevalent enough that we recognize them as variants in a reasonable sample of the modern-day population. STRs change in repeat number and thus allele size largely through the process of slipped-strand mispairing, a shift of the template and nascent strands during DNA replication. Other types of mutation include unequal recombination, thought to be a primary force in changing VNTR repeat lengths and in causing some insertion deletion polymorphisms and retrotransposition, the process by which Short Interspersed Elements (SINEs) like *Alus* get copied and then inserted into new genomic positions. Mutation is important as an evolutionary force over the long term, such as when making comparisons among species. However, in terms of variation within a species, we are generally more concerned with the frequencies of alleles that already exist, and so the other three evolutionary forces are more important.

Genetic drift occurs as the result of segregation of parental alleles in the generation of the current population cohort. In a population of infinite size, there is no genetic drift; that is, all chromosomes present in the parental generation are transmitted to the next. But in the quite finite population sizes in which actual persons and all other organisms live, the magnitude of the effect of drift on the genetic variation in the population is inversely proportional to the size of the mating population, increasing as population size decreases. Over time, genetic drift leads both to the loss of some alleles and fixation (i.e., a particular variant becomes the only allele in a population) of others, and to genetic differentiation among populations that have become reproductively isolated (i.e., populations drift apart after separation). Both the time since the separation and the sizes of the daughter populations determine the levels of differentiation. As we are most interested in real populations, we also need to recognize that populations are rarely ever completely isolated.

Admixture is the process of gene flow or gene migration of alleles from one population to another that occurs when individuals from reproductively isolated populations interbreed. There is a range of possible scenarios by which admixture could happen. The two extreme models of admixture are Hybrid Isolation (HI), where admixture happens in one generation resulting in the formation of a new population that is then isolated from both of the parental populations, and the Continuous Gene Flow (CGF) model, where admixture happens slowly over a long period of time with one population (or both) exchanging migrants (see Figure 1-2). Admixture counteracts the effects of genetic drift and differential selection, making populations more similar. Admixture also creates Admixture Linkage Disequilibrium (ALD), which can be used for mapping the genes determining variation in traits that distinguish populations. Linkage Disequilibrium (LD) is the nonrandom association of

Figure 1-2

Shown are A) Hybrid Isolation (HI) and B) Continuous Gene Flow (CGF) models of admixture. The equations indicate the amount of Admixture Linkage Disequilibrium that is formed by each of these models. Abbreviations: t = generation; θ = recombination fraction between loci; D_0 = amount of LD present in the admixed population immediately after admixture; δ_A, δ_B = allele frequency differentials between parental populations A and B, respectively; α = the contribution of population 2 in each generation; $1 - \alpha$ = the contribution of the admixed population in each generation; $\delta_{A,t}$ and $\delta_{B,t}$ = the allele frequency differentials at generation t; D_{t-1} = the amount of LD in the previous generation. From Long, Figure 1, Genetics, 1991, vol.127, p.198–207. Reprinted with permission.

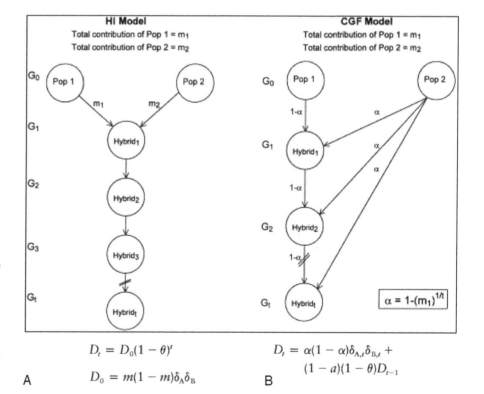

HI Model
Total contribution of Pop 1 = m₁
Total contribution of Pop 2 = m₂

CGF Model
Total contribution of Pop 1 = m₁
Total contribution of Pop 2 = m₂

$$D_t = D_0(1 - \theta)^t$$
$$D_0 = m(1 - m)\delta_A\delta_B$$

A

$$D_t = \alpha(1 - \alpha)\delta_{A,t}\delta_{B,t} + (1 - a)(1 - \theta)D_{t-1}$$

$$\alpha = 1 - (m_1)^{1/t}$$

B

alleles at different loci and an important source of statistical power for genetic association testing. This is because we typically cannot survey every base of DNA for samples as part of a research project and connections between points along the DNA enable us to identify regions and genes of interest through surveys of markers along the chromosomes (like milepost signs on a highway). Additionally, admixture creates a useful type of genetic structure within populations that might best be called Admixture Stratification (AS).

Admixture stratification can be defined as variation in individual ancestry within an admixed population such that, in the simplest case of a dihybrid population—one with two parental populations—some members of the population have more genomic ancestry from one parental population and others have more genomic ancestry from the other parental population. AS has been shown to be an important source of information in testing the extent to which variation in a trait across populations is due to genetic differences between the parental populations. Because there is extensive AS in most admixed populations it is possible to do experiments that test for correlations between phenotypes (common traits or disease traits) and estimates of genomic ancestry (see Chapter 5).

Natural selection is the process of allele frequency change due to differential fitness (survival and reproduction) of individuals of different genotype. Natural selection contrasts with the effects of genetic drift and admixture insofar as only the genomic region around the gene that is under selection is changing, whereas genetic drift and admixture both affect markers across the whole genome. It is notable that, given the random nature of the effects of genetic drift, a substantial amount of variation in the level of divergence is expected even though the evolutionary force of drift is applied evenly to all markers across the genome. In other words, drift alone will lead to a very wide range in outcomes across the genome because of the combined effects of many random events. This wide variance means that alleles that are very different in frequency between populations are not necessarily this way because of the action of natural or sexual selection. Drift will cause some neutral alleles to have changed in frequency across populations, but for the most part, alleles in the far tail of allele frequency distributions, the AIMs, may well be enriched for the action of selection.

Natural selection, which makes populations different at particular genes, often is referred to as directional selection or positive selection. Not all forms of natural selection lead to high levels of allele frequency differentiation among populations. Another major form of natural selection, which does the opposite, is balancing selection. Balancing selection is also called overdominant selection because it acts to favor the heterozygote. If each population is subject to balancing selection, the alleles will remain at similar frequencies over very long periods of time, making for lower levels of genetic differentiation than would be expected if genetic drift only were acting. By considering this range of evolutionary forces in both their global and regional (geographic and genomic) effects, we can use the natural human biodiversity to explore our evolutionary histories and physiologies in a way that has simply not been possible before. For instance, we can now address questions not only regarding whether this particular mountain dwelling population is related to the populations on one side of the mountain range or the other or both, but how might they have changed genetically to adapt to life at high altitude. Which genes changed? Were these passed on to descendants who left the mountains? What can we learn about the physiology of oxygen metabolism from knowing these genes?

THE PROMISE OF MOLECULAR PHOTOFITTING AS A TOOL IN FORENSIC SCIENCE

Our working definition of molecular photofitting is methods to produce forensically (or biomedically) useful predictions of physical features or phenotypes from an analysis of DNA variation. Some useful phenotypes are only very broadly described (such as genetic ancestry or sex), and others are more specific (such as the particular melanin concentration of the skin or the distance between the eyes). In one sense we can think of a molecular photofitting description sort of like a driver's license: a summary list of physical traits like height, weight, hair color, eye color, and race, and a fuzzy or low-resolution photograph. In addition to these classic descriptive traits we should consider that there are ways to make these traits less categorical and thus expressed on a continuous scale where they may be more informative.

There are likely to be other phenotypes that are not included on a driver's license that could be predicted and of use in a forensic setting, such as handedness, body proportions, the distribution of body hair, secondary sexual characteristic form, dermatoglyphic patterns, finger-length ratios, and dental features. As such we are proposing that there are a number of physical traits for which we will be able to make usefully accurate descriptions using an understanding of the genetic variation determining them. We are not proposing that it will be possible to make useful estimates for each of these traits for every person. Presently, these are research questions and we see our current task as an exploration of the structure of the relationships between common traits and genetic variation—an investigation into the genetic architecture of normal variation within and among populations. We suggest that it may not be possible to know how much success can be expected for any particular trait until some level of detailed research has been conducted on the trait in question. More detail on working models for practical molecular photofitting is provided in Chapter 5.

A number of these traits are superficial in that they are expressed on the exterior of the body and are thus immediately evident to observers. This fact combined with the extensive, even superfluous, degree of variation in these traits across world populations argues for the importance of sexual selection in the genesis and maintenance of variation in these traits and the genes they control. There are some very interesting expectations of this hypothesis that are encouraging to both our ability to discover and make use of such traits. First, if these physical variations are the result of sexual selection, then the human neurology and sensorium have coevolved to detect them. In fact, it is only the variations that are evident to the human observer that can be affected by sexual selection. This is not to say that we, even as observers, are always conscious of the variations that affect the attractions underlying mate choice and mate competition, but that we are necessarily able to perceive these variations and act on them accordingly.

Not all obvious external variations require explanations through sexual selection. Since the surface of the body is in direct contact with the environment, it is likely that some of these features are physiological adaptations resulting from natural selection. Alternatively, variations can be the result of either random genetic drift or pleiotropy (i.e., some other phenotype is under either sexual or natural selection, and the gene affecting that trait also affects other traits).

Regardless which evolutionary force or combination of forces created any particular physical trait, that a trait is different among populations supports that the genes directing its development recently have changed in frequency. Such changes increase the likelihood that there may be a small number of alleles at a small number of loci that determine variation in the trait. Traits that are explained by small numbers of loci and only a few alleles at these loci are easier to detect using the standard methods of linkage analysis and genetic association testing, respectively. Additionally, it is likely that when more than one functional allele is responsible for variation among populations and selection has been involved, these alleles will have effects in the same direction. If this is the case, we will find that all these alleles together will contribute to the statistical power for mapping the locus using admixture mapping methods.

An interesting example of a locus like this is the *MC1R* gene, at which alleles are found that determine the classic red hair and freckles phenotype. There are seven well-characterized alleles that have been associated with red hair, freckles, and an inability to tan, all of which are basically restricted to Northwestern Europe (to be discussed in more detail later in Chapter 5; Duffy et al. 2004; Harding et al. 2000).

One can also look to the data on genetic heritability to support that there are genes determining physiological traits of forensic interest. Although the amount of genetic segregation data for facial features, skin pigmentation, and the other such traits currently available is quite limited, we can draw on common knowledge regarding similarities. One of the most powerful tools for segregation analysis is twins studies. Monozygotic twins are basically nature's clones, being identical genetically, and they can be compared against same-sexed siblings or dizygotic twins, both of which have on average 50% genetic identity (see figure 1–3). Across human cultures twins have been a source of amazement in terms of their physical similarity. Indeed it is often possible to recognize particular traits (such as grandma's eyes or dad's nose) in extended families, but nothing is more to the point of demonstrating a genetic influence on facial and other common traits than identical twins. Despite the fact that there is a dearth of data in the scientific literature on twins studies for facial features, it's clear from a cursory examination of the physical similarities between monozygotic twins that variation in these traits are primarily inherited.

Despite these lines of evidence for the primary role of inherited variation in the determination of physical traits like skin color, hair texture, and facial features, we do not conclude that such traits will be necessarily predictable from a DNA analysis. Clearly, the developing organism can interpret the genes encoded in our DNA with such precision and consistency that populations are physically identifiable, children look like their parents, and monozygotic twins

Figure 1-3

Photographs of monozygotic twins. Photographs provided by www.fotosearch.com.

are identical. Therefore it might seem that we should be able to learn to predict the phenotype from the genotype. However, to expect immediate success for all traits and across all populations and individuals is unreasonable given the complexity of the cells, the developmental processes, and both the gene-gene and gene-environment interactions that contribute to these phenotypes.

Population genomics is a powerful framework within which to organize these tools and approaches, and can and should be applied to the study of physiology of normal phenotypic variation. Indeed, we argue that given the much greater role of environmental exposures in complex disease, these genetic studies on the common traits that are less dependent on the environment provide an excellent real-world training set for the optimal application of human genetics and genomics methods to studies of complex disease.

PART II: THE BASIC PRINCIPLES

SAMPLING EFFECTS

Before we discuss the evolutionary histories of human populations it is necessary to explain the theory of population structure, and before we can explain the theory of population structure it is necessary to explain the probabilistic nature of how genes are inherited by offspring from parents. Each eukaryotic organism has a nucleus within each cell, and the structure of every metazoan organism is comprised of multiple such cells. Within the nucleus of human beings exist 46 chromosomes, which are DNA-protein conglomerates (the function of the proteins is to provide functional and constitutional structure for the DNA). The 46 chromosomes are divided into 23 pairs: the numbered chromosomes 1–22 and either two X chromosomes in females or one X and one Y chromosome in males. Normal animals containing two copies of every chromosome are said to be diploid, and in diploid animals, every type of cell except the sex cells contains a pair of each chromosome, one inherited from the mother and one inherited from the father.

In mammals, reproduction occurs via the union of sperm and egg (sex cells) to produce a zygote, or the earliest, single-cell form of an offspring, which divides itself into two, then four, eight, and so on cells to form a multicellular embryo inside of the mother's womb. Over a lifetime each male produces billions of sperm cells, and each female produces thousands of egg cells. During the creation of the sperm and the egg (gametogenesis), which takes place in the progenitor cells for the sperm and the egg, the chromosome pairs are divided into singletons, and one copy of each chromosome is distributed in a random fashion to each gamete (egg or sperm). Although each gamete receives 23 chromosomes, which chromosome of the original pair it obtains is randomly determined. Gametogenesis happens over and over to create many millions of sex cells. Though the sperm and egg contain only single copies of the chromosomes (i.e., they are haploid), the new cell formed from their union has the normal complement of two chromosomes of each type (diploid) rather than the four (polyploid) that would result if gametes received all the progenitor cells' chromosomes.

The distribution of chromosomes to the sex cells is completely random. Considering one chromosome, say chromosome 1, about half of the sperm produced by a male will contain the copy of chromosome 1 contributed by that male's father and half will contain the copy contributed by his mother. The copy of chromosome 1 distributed to a sperm by the progenitor cell is completely independent of the copy of chromosome 2 distributed to that same sperm. This is known as the genetic law of independent assortment, and was first recognized by Gregor Mendel (1865) even though he did not know about chromosomes. In effect, this law states that the inheritance of chromosomes in gametes and therefore diploid offspring is like shuffling a deck of cards, where the nature of one card received is independent of a card previously received. Hence each sibling in a family has half of each of their parent's chromosomes, just a different, unpredictable assortment of those chromosomes.

BINOMIAL SAMPLING

The theory of Mendelian genetics is based on the assumption of random matings between individuals of infinite population size, but in real populations matings are not random and populations are of discrete size. In addition to natural and sexual selection, there is a random sampling effect from generation to generation, know as genetic drift, that causes allele frequencies to change slowly over time. Since there are two alleles for each locus in each person's genome, one from mom and one from dad, the sampling of each genetic locus from one generation to the next is a binary sampling formally called binomial sampling.

If we have two mutually exclusive states that can be used to describe each of many events, such as heads or tails when flipping a coin, male or female in terms of births, or allele frequencies p and q for biallelic loci, then $p + q = 1$, and the probability of a certain combination of states is based on the basic laws of probability. The probability of A or B happening is the sum of the two probabilities $Pr(A) + Pr(B)$. The probability of A *and* B happening is the product of the two probabilities $[Pr(A)][Pr(B)]$. For example, in a coin flip, the probability of heads or tails is $Pr(head) + Pr(tails) = 0.5 + 0.5 = 1$. In five flips, the probability of five heads in a row is $(0.5)(0.5)(0.5)(0.5)(0.5)$ or

$(0.5)^5 = 0.031$ or approximately three in a hundred. In genetics we are interested in the probability of certain allele combinations both in individuals and populations.

Considering alleles p and q for each of n loci, the probability of a gamete receiving r p's and then, the remainder $(n - r)$ q's, in order, is:

$$Pr = (P_p^r)(P_q^{n-r}) \tag{1-7}$$

where P_p is the frequency of allele p and P_q is the frequency of allele q. If we wish to drop the qualification "in order" and calculate the probability that a gamete would receive r p's and $(n - r)$ q's in any order, or of n flips, we would observe r heads in total, we must add a term expressing all the possible ways that r p's and $(n - r)$ q's can be assembled starting with n loci. This happens to be known as the binomial coefficient:

$$[n(n - 1)\ldots(n - r + 1)]/[1 \times 2 \times \ldots \times r] \tag{1-8}$$

and the probability of r events out of n (the rest $n - r$ events) is:

$$P_r = [n(n - 1)\ldots(n - r + 1)]/[1 \times 2 \times \ldots \times r](P_p^r)(P_q^{n-r}) \tag{1-9}$$

Equation (1-3) is known as the binomial distribution. The sum of all possible P_r's works out to be 1,

$$\begin{aligned} S_{r=0-n}P_r &= q^n + npq^{n-1} + \ldots [n(n - 1)\ldots(n - r + 1)] \\ &\quad [1 \times 2 \times \ldots \times r]p^r q^{n-r} + \ldots + p^n \\ &= (p + q)^n = 1 \end{aligned} \tag{1-10}$$

and the mean and variance of this distribution can be shown to be $m = np$ and $v = np(1 - p)$, respectively (Kimura & Crow 1964; Wright 1967; also see Cavalli-Sforza & Bodmer 1999).

GENETIC DRIFT

Accumulation in Random Fluctuations in Allele Frequency from Generation to Generation

Genetic drift is the process by which deviations in expected allele frequencies develop in finite populations over time as a function of statistical sampling of genes from one generation to the next (as opposed to deviations that may develop in finite populations due to selection, mutation, or admixture). To illustrate genetic drift, let us assume a simple population of finite size, with no overlapping generations and for which all individuals of a given generation are born, reproduce, and die at the same time. Let us also assume that there is no mutation, selection, or admixture and two alleles, A and B with frequencies p and q, respectively. We have a population of N individuals, 2N alleles, and before the first generation we have p_0, q_0 and $p_0 + q_0 = 1$, where the subscript of p and q represent the generation.

We are interested in the allele frequencies $p_1, p_2 \ldots p_n$ over time, each of which is expected to be similar to, but not identical to, the value for the preceding generation. Each gamete receives an allele A or B from the parent as a function of probability and the alleles A or B in the population. Although each male produces billions of sperm and each female produces thousands of eggs, only a small fraction of sperm and egg are united during conception to form a new individual, so the alleles that successfully make it from one generation to the next are a sample of the preceding generation. In fact, if we assume that each genetic position (locus) is characterized by two states (such as the case with biallelic loci), the sample is a binomial sample of the preceding generation. If the loci are more complex, the sample is a multinomial sample of the preceding generation (for a treatment of the multinomial distribution in genetic terms, see Cavalli-Sforza & Bodmer 1999). Proceeding from generation to generation, where the allele frequency of one generation is a function of binomial sampling and the frequency in the preceding generation, it is easy to visualize how the allele frequencies would begin to change or drift slightly over time.

Genetic drift occurs when fluctuations in allele frequency from generation to generation begin to accumulate over many generations. Based on the rules of binomial sampling the smaller the sample from each preceding generation that contributes to the next, the greater is the magnitude of the change, or drift. To illustrate this, let us consider a biallelic locus with one of two possible states—A and B of frequency p_0 and q_0, respectively, in a population of $N = 100$ individuals. If we were to flip a coin an infinite number of times, we would find that precisely 50% of the flips would be heads and 50% tails, but if we flip only 60 times we would find that sometimes we would obtain 30 heads (most likely), 40 (quite possible), 50 (less likely), or even 60 heads (least likely). So too, if we were to take 10 samples of $n = 20$ from the initial 100 individuals, we find that for some of these samples p_1 is slightly less than, but similar to p_0, and for others, p_1 is a bit more than, but similar to p_0. If we took one sample of $n = 20$ from the initial 100 individuals, where the p_1 in this sample was a bit less than p_0, and used it to generate a second generation, the gametes from this second $n = 20$ are again a random sample of the $2n = 40$ sites of the second, which were a random sample of those from the first. Though the expected p_2 would also be a bit less than p_0, by the rules of binomial sampling, it could be a bit more than a bit less or a bit less than a bit less.

The rules of the binomial sampling process from one generation to the next remain the same, but the results accumulate from generation to generation. In other words, allele frequencies change slightly from generation to generation and the result is a directional fluctuation, or drift in allele frequencies. Why is the fluctuation directional—wouldn't we expect it to meander aimlessly up and down for ever? No; if $p_1 < p_0$, p_2 might be greater than p_1, and thus return to p_0 or close to it, but it is equally likely to be less than p_1 (the odds are 50% either way). If p_2 happens to be less than p_1, p_3 could be greater than p_2 or equally likely (50% chance) less than p_2. About a quarter of the time, therefore, $p_1 < p_0$, and $p_2 < p_1$. Over the next two generations, only a quarter of the time will $p_3 > p_2$ and $p_4 > p_3$ return to p_0 or close to it, and three-quarters of the time, p_4 will remain less than p_2 because a quarter of the time $p_4 > p_3$ and $p_3 < p_2$, a quarter of the time $p_4 < p_3$ and $p_3 > p_1$, and a quarter of the time $p_4 < p_3$ and $p_3 < p_2$ ($1/4 + 1/4 + 1/4 = 3/4$). Thus, after the allele frequency drifts lower in the initial generations, it tends to stay lower rather than return to its initial level. Indeed, it will tend to get lower and lower. If it drifts even lower still, which it will in some of the cases, it will tend to stay at this new lower level. The same is true for allele frequencies drifting higher than the initial lower level, and though the allele frequencies may meander up and down over the generations, there is bias or inertia toward a continuation of the initial trend until the allele becomes either fixed or eliminated.

Over the long term, the probability that a particular allele will be fixed in the population is equivalent to its frequency. The smaller the sample at each generation, the stronger the bias or inertia, and the more quickly the allele becomes fixed or eliminated. Also, the greater or lower the p_o, the greater or lower the probability that drift will lead to fixation or elimination, respectively, in the next generation.

Computer simulations are the best way to study the theoretical dynamics of genetic drift. Cavalli-Sforza and Bodmer (1999) described a very simple random number generator and transformation for illustrating genetic drift, and we will describe the simulation he presented here. To simulate genetic drift:

1. Select a starting group of N simulated individuals with 2N sites, of a given allele frequency.
2. Simulate (i.e., create) X_i new individuals with two X_i sites by making random picks of alleles from the preceding generation with respect to the allele frequency in that generation (i represents the generation).
3. Determine the allele frequency for generation i.
4. Return to 1 and loop for as many generations as desired.

Executing these four steps over and over, it is clear that if the process of selecting alleles at each generation is random, then the expected frequency of one generation is determined by that of the preceding generation. Due to the accumulation of binomial sampling fluctuations over many generations, we expect allele frequencies to drift higher or lower than the starting allele frequencies. Of course the magnitude of the sampling fluctuations depends on the size of N and X_i's.

To illustrate the point, consider N individuals with one allele each. We create a rule for transforming an N-digit random number into alleles. If the starting allele frequency is 50% (half a and half A), and N = 10, then this rule might be [if 0, 1, 2, 3 or 4 then "a" and if 5, 6, 7, 8 or 9 then "A"]. Using this rule, any X digit number can be transformed into a population of X loci. For example, with X = 10, 2536645632 becomes aAaAAaAAaa and the frequency of the A allele in this particular population is also 50%. To generate the next generation, we pick another X digit number. If X is 10 again, an example is 1633546887, and using the transformation rule this corresponds to aAaaAaAAAA. In this new population, the frequency of the A allele happens to now be 60%, not 50% like it was in the first generation. We thus have observed a fluctuation in the allele frequency from one generation to the next.

To make a third-generation we randomly pick a new 10-digit number such as 1250825234, but in order to make the selection of alleles random with respect to the allele frequency of the preceding generation, the transformation rule must now be changed to reflect the fact that there is now a 60% probability of randomly selecting an A allele and a 40% chance of randomly selecting an a allele, such as [0,1,2,3=a;4,5,6,7,8,9=A]. Using this new rule, the random 10-digit number becomes aaAaAaAaaA and we can see that the frequency of the A allele is now 40% by chance.

An example of such a simulation is shown in Table 1-2, which shows a population of N = 10 loci giving rise to new populations of 10 loci 12 times (generations). We can see in this table that the frequency of the A allele drops initially but then climbs up to 90%, only to drop back down to 10% and eventually 0% where it becomes permanently fixed. In Figure 1-4, we have plotted the data from Table 1-2, as well as 10 other simulated populations, all starting with the same allele frequency ($p_0 = 50\%$). Each population has 10 alleles and the frequency of the A allele is tracked over 12 generations. Most of the populations become fixed for the A or a allele—only two (populations 1 and 9) retain both alleles after 12 generations. Some become fixed very soon after the initial generation (e.g., population 7), but most meander from higher to lower allele frequencies until relatively extreme frequencies are reached, which leads rapidly, in most cases, to fixation (100%) or elimination (0%).

Generation	Random number	Transformation rule	Population of alleles	Frequency of "A" allele
0				50
1	8532217936	0,1,2,3,4=a; 5,6,7,8,9=A	AAaaaaAAaA	50
2	2082369431	0,1,2,3,4=a; 5,6,7,8,9=A	aaAaaAAaaa	30
3	8292072470	0,1,2,3,4,5,6=a; 7,8,9=A	AaAaaAaaAa	40
4	6627876530	0,1,2,3,4,5=a; 6,7,8,9=A	AAaAAAAaaa	60
5	8984438409	0,1,2,3=a; 4,5,6,7,8,9=A	AAAAAaAAAA	90
6	9100800854	0=a; 1,2,3,4,5,6,7,8,9=A	AAaaAaaAAA	60
7	7806023910	0,1,2,3=a; 4,5,6,7,8,9=A	AAaAaaaAaa	40
8	9832622115	0,1,2,3,4,5=a; 6,7,8,9=A	AAaaAaaaaa	30
9	2562052591	0,1,2,3,4,5,6=a; 7,8,9=A	aaaaaaaaAa	10
10	3792128378	0,1,2,3,4,5,6,7,8=a; 9=A	aaAaaaaaaa	10
11	3254906232	0,1,2,3,4,5,6,7,8=a; 9=A	aaaaAaaaaa	10
12	2177445238	0,1,2,3,4,5,6,7,8=a; 9=A	aaaaaaaaaa	0

Table 1-2

Computer simulation of genetic drift.

Figure 1-4

Simulated Genetic Drift. Population allele frequencies for a biallelic locus were simulated over 12 generations in 10 different populations of finite size as described in the text. The symbols for the populations are shown in the right-hand key. Though each of the populations start with a frequency of each allele of 50%, the frequencies immediately change in the second generation, and we can see a tendency toward momentum, where increases tend to beget further increases and decreases lead to continued decline. Eventually the frequency becomes fixed or lost in many of the populations by the tenth generation, and appears well on the way toward fixation or elimination in the remainder.

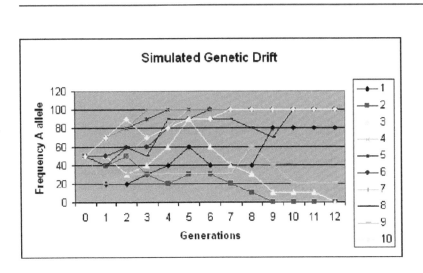

Population 3 in Figure 1-4 is the sample shown in Table 1-2. Its progress was an exception, and shows that the inertia toward fixation/elimination is not absolute; the frequency of the A allele drifted up to 90% only to come back down all the way to 50%, then 10% and soon after, elimination (0%).

From simulations such as this one, the distribution of allele frequencies after t generations was shown by Kimura (1955, 1964) to be an expansion of a very long and complex function, part of which is shown here:

$$f(p,x,t) = 6p(1 - p)e^{-t/2N} + 30p(1 - p)(1 - 2p)(1 - 2x)e^{-3t/2N} + \ldots \quad (1\text{-}11)$$

where p is p_0, the initial allele frequency, x is the allele frequency after t generations and f is the proportion of populations with allele frequency x at time t, given p. When the number of generations is high, the equation simplifies to the probability of fixation or elimination,

$$P(t,p_0) = 1 - 6p_0(1 - p_0)e^{-t/2N} \quad (1\text{-}12)$$

which is 1 minus the first term of (1-11). From (1-12) we can see that, as we expect, when p_0 is very low the probability that at time t a population of size N will have reached elimination, or fixation is greater than when p_0 is higher. From (1-11), a mean time for fixation or elimination in generations t can be computed for any given p and N (Ewens 1969). It happens to be:

$$T = 4N[p_0 \log_e p_0 + (1 - p_0) \log_e (1 - p_0)] \quad (1\text{-}13)$$

and performing this calculation for many values of p_0 it's easy to see that the expected value over all p_0 (i.e., the mean value of T) is simply 2. Thus, for the average allele, it will take twice as many generations as there are individuals for the allele to become fixed or eliminated as long as there is no selection, no new mutation or migration from other populations to disturb the binomial sampling process.

Mean Allele Frequency from Generation to Generation

In a population of infinite size for each generation (i.e., theoretical populations), the expected allele frequency from generation to generation is the mean of the initial generation. The same happens to be true for populations of finite size (i.e., real populations) and for subpopulations of larger initial populations. That is, for subpopulations differentiating under drift, the expected value of the allele frequency among all of them remains the initial allele frequency. Of course, the frequency in any one of the populations will fluctuate, and heterogeneity in frequency among the populations would naturally increase as discussed so far, eventually reaching fixation or elimination. However, even after the mean time to fixation or elimination has been reached, the ratio of those that are fixed to those eliminated is still a function of the starting allele frequency, and the average among all of them would remain that starting frequency.

Considering the influence of selection rather than just drift is more complex, because we have to consider genotypes (i.e., pairs of alleles in an individual) as well as alleles. This is because selection acts on phenotypes, which are determined by genotypes, not alleles. However, when molecular photofitting from a consideration of population structure, we are by definition primarily concerned with alleles of frequencies molded by drift, not selection, and when photofitting in a more direct manner, where we draw inferences on traits based on the genotype for the underlying loci, we are not so much concerned with the distribution of these loci among populations as we are the association between the alleles and phenotype. Thus, we shall refer the interested reader to other sources for more detail (e.g., Cavalli-Sforza & Bodmer 1999).

Heterogeneity Between Populations under HWE and Wahlund's Variance of the Standardized Variance in Gene Frequencies Between Populations

As we have seen, genetic drift is the random fluctuation of allele frequencies generation after generation, and in the absence of mutation, migration, and selection, the ultimate outcome of genetic drift is always fixation of one allele and loss of the other. However, for species found in very large numbers and globally distributed such as the human metapopulation, the time until fixation or elimination for most alleles is a rather long time, probably longer than our species' lifetime, and it so happens that most polymorphic alleles are somewhere along the process of becoming less heterozygous and more homozygous but still nowhere near fixation/elimination. Heterogeneity in allele frequency distribution among subgroups of a larger population provide much of the fuel (information) we need for defining subpopulations so that we can reconstruct our history as a species, and harness natural population diversity for practical aims such as molecular photofitting. We shall therefore briefly outline the nature of population-to-population divergence in genetic character under random genetic drift.

Although the allele frequency among subpopulations remains the same as drift proceeds, the frequencies of each fluctuate, drifting toward fixation or elimination. Thus, in the total population there is a net progression toward a

Table 1-3
Wahlund effect.

	Alleles		Genotypes		
	Freq A	**Freq a**	**Freq AA**	**Freq Aa**	**Freq aa**
Population 1	p_1	q_1	p_1^2	$2p_1q_1$	q_1^2
Population 2	p_2	q_2	p_2^2	$2p_2q_2$	q_2^2
Population 3	p_3	q_3	p_3^2	$2p_3q_3$	q_3^2
Population k	p_k	q_k	p_k^2	$2p_kq_k$	q_k^2

loss of heterozygosity and a gain of homozygosity. The rate at which the loss of heterozygosity proceeds is described by Wahlund's theorem.

Consider a large population of k populations pooled together, each of the k populations of unique allele frequencies pk and qk, where $p_k = 1 - q_k$ as shown in Table 1-3.

The mean allele frequency across all the k populations is

$$p_{avg} = (\Sigma_k p_i)/k \tag{1-14}$$

for all populations i = 1 to k.

$$
\begin{aligned}
H_{avg}^S &= \left[\Sigma_k(2p_iq_i)\right]/k = \Sigma_k\left[2p_i(1-p_i)\right]/k = \Sigma_k(2p_i - 2p_i^2)/k \\
&= \Sigma_k(2p_i)/k - \Sigma 2p_i^2/k = 2\left[p_{avg} - \Sigma_k(p_i^2)/k\right] \\
&= 2\left[p_{avg} - p_{avg}^2 + p_{avg}^2 - \Sigma_k(p_i^2)/k\right] \tag{1-15} \\
&= 2\left[p_{avg}(1 - p_{avg}) - (\Sigma_k(p_i^2)/k - p_{avg}^2)\right] \\
&= 2\left[p_{avg}q_{avg} - (\Sigma_k(p_i^2)/k - p_{avg}^2)\right] \tag{1-16}
\end{aligned}
$$

where in (1-15) the two added terms $-p_{avg}^2 + p_{avg}^2$ equate to zero. As always, the variance is the square of the standard deviation or expected deviation of values from the mean, and so here, the variance of the allele frequencies over the subpopulations is the square of the expected deviation of p_i from the mean, p_{avg}

$$
\begin{aligned}
\sigma^2 &= E(p_{avg} - p_i)^2 \\
&= E(p_{avg}^2 - 2p_{avg}p_i + p_i^2) \\
&= E\left[\Sigma(p_i)^2/k - 2\Sigma(p_i)/kp_i + p_i^2\right]
\end{aligned}
$$

$$= E[p_i^2 - \Sigma(p_i)^2/k]$$

$$= E(p_i^2) - p_{avg}^2$$

$$= \Sigma(p_i)^2/k - p_{avg}^2 \qquad (1\text{-}17)$$

Substituting equation (1–17) into equation (1–16), we obtain

$$H_{avg}^S = 2[p_{avg}q_{avg} - \sigma^2]$$

$$= 2p_{avg}q_{avg} - 2\sigma^2 \qquad (1\text{-}18)$$

$$= H_{avg}^T - 2\sigma^2 \qquad (1\text{-}19)$$

where H_{avg}^S is the mean heterozygosity among the subpopulations, $H_{avg}^T = 2p_{avg}q_{avg}$ is the heterozygosity of the entire population by definition, and σ^2 is the variance in allele frequencies among the subpopulations (in your mind you can exchange σ^2 with σ_{BG}^2 where BG means Between Groups, if it makes it easier to remember this). Note that the average heterozygosity among the subpopulations is $H_{avg}^S = [\Sigma_k(2p_iq_i)]/k$, but the average heterozygosity for the entire population (ignoring subpopulations) is $H_{avg}^T = 2p_{avg}q_{avg}$.

So when subpopulations are pooled into one larger population, the heterozygosity in the total population is less than the average among the subpopulations by a factor of $2\sigma^2$. Likewise the homozygosity for the genotypes AA (p^2) and aa (q^2) increase each by a factor of σ^2, the variance in allele frequency among the subpopulations. Similarly, it can be shown that the average homozygosity within the subgroups increases by a factor of σ^2 compared to the homozygosity for the group as a whole, for each homozygote:

$$XX_{avg(S)}: p_{T_{avg}}^2 + \sigma^2$$

$$YY_{avg}(S): q_{T_{avg}}^2 + \sigma^2$$

where again, the p_{avg} and q_{avg} on the right-hand side of the equation represent the values in the total population. Adding $H_{avg} + XX_{avg} + YY_{avg} = 1$ for the subdivided populations, the variance term cancels out ($H_{avg} - 2\sigma^2 + p_{avg}^2 + \sigma^2 + q_{avg}^2 + \sigma^2$) and we are left with the value of 1. It is the change in heterozygosity and homozygosity within the subpopulations that comprises the signature of population stratification, much like it is the excess of homozygotes and excess of heterozygotes in an extended pedigree that comprises the signature of inbreeding.

Equation (1-18) can be rewritten as

$$H_{avg}^T = 2p_{avg}q_{avg}(1 - \sigma^2/p_{avg}q_{avg}) \qquad (1-20)$$

and the term

$$\sigma^2/p_{avg}q_{avg}$$

is the factor by which heterozygosity is decreased among subpopulations under genetic drift, which is referred to as the Wahlund's variance of gene frequencies, after the Swedish statistician Sten Gosta William Wahlund (1901–1976), who was the first to describe the relationship (Wahlund 1928). This term is not as unintuitive as it might seem from brief inspection. The term $p_{avg}q_{avg}$ [or $p_{avg}(1 - p_{avg})$] represents the maximum variance that subpopulations could exhibit due to full differentiation; that is, the variance we would expect after their alleles have reached fixation or elimination in each of the subpopulations. Recall that the variance is the squared average deviation from the mean, and the term σ^2 of course is the observed variance. Thus, Wahlund's variance measures the observed variance relative to the maximum amount of variance we could observe if the populations were fully differentiated with respect to a given locus.

FIXATION INDEX F_{ST} AND POPULATION STRUCTURE

The Wahlund variance of gene frequencies can also be defined as the fixation index or F statistic (F_{ST}) as a measure of genetic distance between hierarchically tiered populations (or subpopulations). When there is population structure due to drift (or any other force), the correlation between two randomly selected alleles in subpopulations is different from that in the total population because as we have seen, there is a tendency toward homozygosity within the subpopulations. The difference we can expect based on this structure is expressed by the Wahlund variance of gene frequencies. However, where the Wahlund variance measures the total variance we observe relative to the greatest amount we could possibly observe, the F_{ST} measures the total variance we observe relative to the total amount of variance that exists, and so is a slightly different measure. Wahlund's variance tells us something about how far along the process of allelic fixation subpopulations are, but F_{ST} tells us how much of the total variation that exists is due to a specific type of variation, such as group-to-group variation or person-to-person variation within groups.

There are two related styles for calculating the F_{ST}, using the delta value or in terms of heterozygosity. The goal of this discussion is to express the F_{ST} in terms of heterozygosity and variance so that we can compare the amount of diversity explained by between-population variation as opposed to within-population variation from a variety of studies for which heterozygote frequencies are available. Considered this way, the F_{ST} is a hierarchical F-statistic that quantifies the influence of inbreeding on extant population structure by comparing the level of heterozygosity expected from random mating at one level of population structure relative to another. If H_S is the average HWE heterozygosity between subpopulations of a metapopulation and H_T is the average HWE heterozygosity in the total (meta) population higher up in the hierarchy, then the F-statistic between the total and subpopulations is defined as:

$$F_{ST} = (H_T - H_S)/H_T \qquad (1\text{-}21)$$

This is the decrease of heterozygosity going from the total population (H_T) to the subpopulations (H_S), normalized by the heterozygosity of the total population ignoring subpopulational divisions for a given hierarchical level of population structure. Our goal in what follows is to illustrate how the variance in gene or allele frequencies is related to the F_{ST}, so that we can express the variance found between groups as a function of the total variance and make some sense out of what the F_{ST} values for various populations actually tell us about population structure.

We have just seen from Wahlund's variance that

$$H_{avg}^{S} = H_{avg}^{T} - 2\sigma^2 \qquad (1\text{-}22)$$

where the variance is that between the subgroups. We can rearrange this slightly to give:

$$H_T - H_S = 2\sigma_{BG}^2 \qquad (1\text{-}23)$$

where we are now denoting that the variance is Between the (sub)Groups with the subscript BG, and H_T and H_S is the new more convenient notation for the average heterozygosity in the total population and among the subpopulations, respectively. We have just expressed the variance in allele frequencies between groups as σ_{BG}^2. The decrease in heterozygosity going from the total population to subpopulations $H_T - H_S$ expressed in (1-22) can be equated to the same from (1-21) to show the relationship between the between-group variance and the F_{ST}.

$$F_{ST} = (H_T - H_S)/H_T$$

So, substituting (1-22) into (1-21) we have

$$F_{ST} = 2\sigma^2_{BG}/H_T \qquad\qquad (1\text{-}24)$$

and you will recognize the term on the right-hand side of the equation as the Wahlund variance in gene frequencies. Given all of these relationships

$$
\begin{aligned}
F_{ST} &= 2\sigma^2_{BG}/H_T \\
&= 2\sigma^2_{BG}/2p_{avg}q_{avg} \\
&= \sigma^2_{BG}/p_{avg}q_{avg}\,(\text{Wahlund's variance}) \\
&= (H_T - H_S)/H_T
\end{aligned}
$$

we can see that the larger the variance in gene or allele frequencies between subpopulations, the greater the F_{ST} among them, the sharper the difference in heterozygosity between them and the larger the drop in heterozygosity experienced going from the metapopulation to the subpopulation. We can also see why Wright called this the fixation index, because in this case, as with the Wahlund variance, it shows us how far along the process of allelic fixation a metapopulation under genetic drift is. Note that we can take the hierarchical comparison down to the level of the individual, where the correlation of genes within individuals relative to subpopulations is the inbreeding statistic f (for more information one can turn to a number of good population genetics textbooks such as Balding et al. 2001 or Cavalli-Sforza & Bodmer 1999).

LACK OF HUMAN DIVERSITY RELATIVE TO OTHER SPECIES

The geographical distribution of human phenotype diversity pales in comparison to that for other species of animals such as great apes or fruit flies (Jobling et al. 2003). It happens to be the case that human populations show less differentiation than populations of other species of animal. In other words, human populations, which originated only within the past 50,000 years or so, have not progressed very far along the route of allelic fixation as measures of the fixation index F_{ST} in Table 1-4 show.

As we can see from this table, the F_{ST} for each species is quite different, as we might expect since the forces of genetic drift, bottlenecks, admixture, founder events, and the age of each species is different. If we were to look at the length of the phylogenetic branches for each type of organism in Table 1-4, we would expect higher F_{ST}'s among species with longer unbifurcating branches. Human branches, which originated merely 50,000 years ago, simply have not had enough time to have drifted to levels of differentiation that we see for other species. The F_{ST} for human populations are generally below 0.07 to 0.08, yet

Organism	Number of populations	Number of loci	H_T	H_S	F_{ST}
Human (major races)	3	35	0.13	0.121	0.069
Human, Yanomama (Indian villages)	37	15	0.039	0.036	0.077
House Mouse (Mus musculus)	4	40	0.097	0.086	0.113
Jumping rodent (Dipodomys ordii)	9	18	0.037	0.012	0.676
Fruit fly (Drosophila equinoxialis)	5	27	0.201	0.179	0.109
Horseshoe crab (Limulus)	4	25	0.066	0.061	0.076
Lycopod plant (Lycopodium lucidulum)	4	13	0.071	0.051	0.282

Table 1-4

F_{ST} across various species.

Source: Principles of Population Genetics by D. Hartl and A. Clark; and Protein electrophoretic data from Nei (1973).

those for the lycopod plant, the jumping rodent, and even the house mouse are notably higher. Table 1-4 shows that there is more evidence for the existence of races of house mice than humans, yet most human beings would have trouble differentiating a house mouse from Sweden from one from Mexico. Physical traits can sometimes belie the existence of races, but using genetic marker data we see that the lack of physical differences can belie their existence.

What then are the generally accepted values for race? There is no objective answer to this question, but what we can say is that there is less evidence for the existence of human races than for other species. This is not to say that the F_{ST} for humans does not qualify our population as having races. The question of whether an F_{ST} of 0.07 qualifies our species as having bona fide subpopulational races may produce different answers among different people, but it certainly means that defining and appreciating them, if they exist at all, would be more difficult than for other species.

Part of this problem, as we will discuss throughout the text, is the subjective nature with which we divide our own species into populations, usually along geographical and social lines that may or may not have as much genetic meaning as we think. Part of the problem, too, is that our population is very complex and populations not only have more or less recent common origins, but have mixed with one another for much of our recent recorded history. Such mixing would tend to diminish F_{ST} values. Though part of the lack of human diversity is due to recent admixture and technologically related changes in human demography, most of it is due to the fact that we are a relatively young population, having reached population sizes greater than 10 million only 10 KYA. One of the consequences of relatively recent population size expansions is relatively low genetic diversity.

If we were able to go back in time and measure fixation indices among human parental populations that lived in different regions of the world 10,000 years ago or more, we might obtain higher values. We know that with humans, like other organisms, we can represent the population in terms of molecular distance in a phylogenetic tree, so there is plenty of evidence that there is some genetic differentiation among human subpopulations. To perform research on molecular photofitting, we must develop methods to recognize and reproducibly measure these subpopulations, which is the topic for the next chapter.

CHAPTER 2

ANCESTRY AND ADMIXTURE

WHAT ARE ANCESTRY AND ADMIXTURE?

There are two ways of thinking about genetic measures of ancestry and admixture that we use, and since they are closely related it helps to discuss them explicitly. First, there is *genealogical ancestry*, or the *expected level of ancestry*, which depends on the population(s) from whence one's direct ancestors are derived. These terms are interchangeable because the genealogical ancestry is an expected level of ancestry determined by genealogical or pedigree-based history for the individual. The genealogical ancestry of a particular person is the average of their parents and is therefore an expected variable. For example, someone whose father is 50% Indigenous American (IA) and 50% West European (WE) and whose mother is 100% European has a genealogical or expected level of ancestry of 25% IA and 75% WE. Due to genetic recombination and independent assortment, however, if these parents had 10 children, they would not all have exactly 25% IA ancestry as measured with genetic markers, even a very large panel of highly informative Ancestry Informative Markers (AIMs). In other words, they would have slightly different *measured genomic ancestries*. There is variation around the actual level of ancestry based on genetic markers, which is why we can think about genealogical ancestry as the statistical expectation. For example, a person with a single American Indian great-great-great grandparent might not have inherited any Native American chromosome segments, so their genetic ancestry would be 0% Native American. However, this does not change the fact that they have an American Indian great-great-great grandparent. Their genealogical ancestry is fixed by history, immune to the laws of independent assortment, and is 1/32 American Indian.

What we do measure with genetic markers is a value we might call the *genomic ancestry* or the *realized level of genetic ancestry*, or the ancestry we can read in a practical sense. At first glance we might be suspicious that there is not a perfect correspondence between the genomic and genealogical ancestry. However in the real world this is not usually the case. Genomic ancestry is the average level of admixture calculated throughout the genome using ancestry informative markers, and its measure depends on how good the AIMs are, how many are used, and how evenly they are spread throughout the genome. Considering the preceding example, and depending on the AIMs used, each of the 10 children in the family will have genomic ancestry estimates that vary, being both higher and lower than the expected level of ancestry of 25%. As the number and coverage of AIMs increase, the variance in genomic ancestry estimates due to measurement error determined with the different sets of AIMs gets closer and closer to zero. If there were an infinite number of independently assorting AIMs the numerical values would converge. However, biologically there are a finite number of independently segregating sites, and at each generation the genetic ancestry changes from the genealogical or expected ancestry, and of course, the genomic ancestry would change from the expected ancestry as well. On top of this, genealogical ancestry is not always evenly distributed among chromosomes of parents and though the expected level of ancestry among offspring of these parents is expected to be the same, the actual genetic ancestry may not be. Even if we used an infinite number of markers and our genomic ancestry estimates were perfect we would still see some substantial variation in genomic ancestry estimates among offspring.

Which of these phenomena we really want to measure is thus a concern for how we generate and interpret the results. If the question is genealogical—for example, "Was grandma really half Cherokee like she always said?"—then the question relates to genealogical ancestry, and we could get a better answer to the question by typing all of her children and grandchildren (if some of the parents were missing) as well as the relevant persons marrying into the family, like grandpa if he's still alive. In this case we would not rely on her *genomic* ancestry, or even her *genetic* ancestry, but would be using the genomic ancestry of multiple individuals in her family to reconstruct the most likely genealogical ancestry. Alternatively, we might be more interested in the particular pheno-type/ancestry relationship for one person at a time. In this case we do not care about the genealogical ancestry; rather, it is the genetic ancestry and our best estimate of the genetic ancestry (the genomic ancestry) that we are concerned with, because these estimates are more directly related to phenotypes. For example, police may have a sample from a crime scene and question, "What is the skin color of the person who left this sample?" Here, provided the genome could be exhaustively screened with AIMs, is a question of genomic ancestry where no additional information is gleaned from typing relatives. Usually it's clear which of these questions are at issue if one considers genea-logical ancestry and genomic ancestry as explicit alternatives.

These ideas of genealogical ancestry and genomic ancestry are analogous to the different ways in which race is presently understood in various cultures. For example, in the United States, where the cultural standard is that of hypodescent or the one-drop rule, race is understood more in terms of genealogical ancestry. You are like your family even if you look different than they do. Persons in the United States who are at the intersection of populations because of how they look and chose to "change" race or "pass for white" always have had to leave their families behind as these ties would out them in this system of inherited race. Alternatively, in a number of Latin American countries, race is more about phenotype and thus more analogous to genomic ancestry. In Puerto Rico, for example, race is much more about the phenotype of the individual, and it is not uncommon to have children of two or three races from the same biological parents. Thus, uses and interpretations of these two types of ancestry vary, and neither should be seen as inherently more valid than the other. In this book we are concerned primarily with the estimation of genomic ancestry measures of genetic ancestry, like biogeographical ancestry, as these are what is being measured directly from the genetic markers and better reflect phenotypic variation and as such are what is of forensic interest. We also are interested, however, in the genealogical applications of these tests as they relate both to the recreational and educational facilities of these tests, and have a section in Chapters 5 and 6 regarding inferences of genealogical ancestry (in a geopolitical sense) from genomic ancestry estimates.

THE NEED FOR MOLECULAR TESTS FOR ANCESTRY

Variation in genetic and therefore genomic ancestry levels among persons in a population, even of the same genealogical or expected level of ancestry, contributes to population structure or admixture stratification. The accurate measurement of population structure within individuals is paramount because it enables us to adjust for a major source of confounding and map disease and/or drug response genes, infer phenotypes from genotype data and design better research projects and clinical studies. Most frequently, scientists measure ancestry as a categorical variable using questionnaires or third-party assessments. In deciding to which race a subject belongs, sample collectors will oftentimes use the subject's notion, or physical appearance in terms of certain anthropometric traits (those traits whose character varies as a function of ancestry such as skin shade, hair texture, facial feature, and the like). Of course subjective data has the benefit of being derived from a source that is exceptionally intelligent and capable of handling multiple variables at once (a sort of real life "biological" neural network), but the downside is that it will be influenced by the past-experiences and beliefs of each person, which of course differ from person to person and culture to culture. For most research purposes, subjective measures of racial or population affiliation leave a great deal to be desired, primarily because of the variation among observers (even in self-assessments) and the categorical nature of the classes (African-American, European-American, Asian, etc.). As currently measured with biographical questionnaires or other survey instruments, little knowledge of population genetic structure other than the obvious is obtained, and only basic connections between population structure and phenotype could be apparent and/or controlled for.

In addition to the subjective nature of subjective classification, there are problems associated with the very definition of what is being classified. Human perceptions of race usually are based on much more than just biology, such as religious, socio-cultural, socio-economic, geopolitical and/or linguistic factors. Because so much more than biology usually is involved in a human assessment of group membership, these assessments tend to obscure and indeed obfuscate the biological variation at question. The self-reporting of race is not as trivial an exercise as the self-reporting of sex, and many people do not know what their race is, or are of sufficient admixture that they have trouble classifying themselves into a single group. Although social factors linked to our perceptions of race often are correlated with biology, nonbiological notions of "race" are not likely to be as tightly linked to human biology as a molecular assessment of some particular phenotype would be. Of course, it is the biological component of race that is of the most interest for forensics professionals, who desire to paint physical portraits from crime scene DNA, or clinical researchers who desire to design better controlled clinical trials.

Biogeographical Ancestry (BGA) is the term we use to express the heritable component of race, and its measure from the DNA is made possible by the evolutionary forces that have shaped our genomes over the past 100,000 years (mutation, genetic drift, natural selection, and admixture). As such, genomic ancestry measures like BGA are strictly biological (i.e., not cultural, religious, economic, or linguistic) and will thus be more directly related to biological factors we may sometimes want to predict or measure. Before we discuss how population affiliation is estimated, we will highlight the problem with using subjective and categorical race classifications for forensics, academic and clinical scientists.

BENEFIT TO CLINICIANS

Clinicians are naturally interested in measuring population structure because the response to many drugs varies from population to population. Interpopulation variation in drug response is so frequently observed that clinical trials usually involve at some level an assessment of race. Most drugs are disposed of through the action of *xenobiotic metabolism genes*. Frudakis et al. (2003) suggested that variation in xenobiotic metabolism genes is more strongly a function of ancestry than for many other gene types, which makes intuitive sense since these genes are involved in dietary metabolism and differences in floral and faunal environments has undoubtedly molded the genes encoding metabolic enzymes for environmental compatibility; a sort of ecological coevolution. For example, we know from decades of observations that BGA is an important component in the variability of drug response (Burroughs et al. 2002; Kalow 1992). The most likely explanation is that regional selective pressures have affected allele frequencies for compatibility with alkaloids, tannins (self-defense chemicals), and other xenobiotics found in their indigenous diets. Most drugs are derived from these chemicals, and so it is no coincidence that the family of enzymes that allow us to detoxify drugs are found at different frequencies across the world.

Although xenobiotic metabolism genes are well known, we cannot say that all of them are known or that the disposition of any one drug does not involve other genes that also may be distributed as a function of phylogeography. Thus, in certain cases, measures of BGA provide useful proxies for measures of drug response proclivities. In other applications, measures of BGA provide a knowledge of population structure inherent to a given study design necessary to conclude that phylogeography has nothing whatsoever to do with the response to a drug. BGA can present a confounding or covariate influence on the trait any given study designs to measure, and if our goal is to prove that a gene sequence is associated with a trait, rather than a covariate of genetic ancestry, we must condition our analysis on BGA measures in our sample. This scenario is not unique to drug response as sequences for other parts of the genome that have nothing to do with drug response have also drifted apart or been subject to selection over time. For example, populations that have lived for millennia at high altitude are significantly better adapted to the hypoxia than other populations. Given our state of knowledge of both the physiology of hypoxic response and population genetics, the most likely explanation for these differences between low and high altitude populations is genetic adaptation at particular genes important in hypoxic response, not the stratification of populations into races per se.

Given that ancestry is expected to have a significant impact on drug response, it would seem that standardization, objectivity, repeatability, and verifiability is of paramount importance for the collection of race or ethnic data in the clinic, because the impact of the assessment could have such an immediate and profound affect on human health. For reasons already mentioned, these are not qualifications we can usually apply to subjective categorical notions of human identity because they are usually not falsifiable. Not only is consistency a significant problem with the self-reporting or third-party estimation of race for clinical trials, but by using "nonanthropologic designations that describe the sociocultural construct of our society'' in the study design process, we invite an unhappy and unscientific situation that can affect poor predictive power and false positive and false negative results. For example, a woman of mainly West African genomic ancestry, raised in Puerto Rico after moving to the United States, may describe herself as Latina or Hispanic and though she socio-culturally, politically, and geographically identifies with Hispanics, her xenobiotic metabolism and drug target polymorphisms would likely be found more frequently among West African populations.

Where a person was raised and lives and what cultural or sociological customs they observe may indeed have an impact on how they respond to a drug and on their proclivity to develop a disease, so nonbiological metrics can and should continue to be assessed, but the growing body of evidence suggests that genomic ancestry or BGA also has a much stronger impact for many drugs, and this, too, needs to be measured in a scientifically accurate and reproducible manner. Notable at this point and to be expanded on is the fact that BGA may well best serve as an intermediate tool and not the definitive means by which to estimate a person's risk to a particular drug–gene interaction. Quite possibly such an interaction may go unnoticed if the study design is carried out without measuring BGA, and although estimates of BGA can help observe and explore the interaction, admixture mapping will be required to identify the specific gene that is variable across populations, ultimately underlying the correlation between BGA and drug response.

Bi-dil, a drug developed for treating congestive heart failure by a company called NitroMed, is a good example (Franciosa et al. 2002). The response rates for this drug initially failed to impress in an eclectic patient population, but based on a more impressive response rate in African-American subset, the study was redesigned by NitroMed to focus on this group alone. The second study, like the first, used subjective racial classification, and the results from the redesigned study were so convincing the FDA stopped the trial early for ethical reasons. Evidently the ancestry component to response was so strong for this drug that even the looseness of subjective racial categorizations proved informative enough to measure it, yet a more objective measure of genomic ancestry likely would have improved the trial results and also may have helped side-step some of the ethical concerns over ''race-based medicine'' that resulted from this FDA approval. As we will show in upcoming chapters, the average African American has 20% West European ancestry—might the trial outcome have been even stronger if the results were conditioned on the proportion of individual West African ancestry rather than ill-defined racial groups? The response for many other drugs may exhibit a similar between-population component of variation, and although our reliance on subjective categorical descriptions of population affiliations usually obfuscate real biological relationships, they usually do not do so completely because the two measures are correlated.

Clearly, then, objective, quantitative, and generally more scientific measures of ancestry or BGA are needed for a variety of professional applications. The question becomes how to develop these methods, which we address throughout the remainder of this text.

BENEFIT TO BIOMEDICAL RESEARCHERS

For a biomedical research scientist, a quantifiably accurate assessment of BGA may enable the more efficient discovery of connections between gene sequence variants and heritable diseases. For the common case/control study design, where the genetic (or other) characteristics of a group of cases (i.e., affected persons) are compared with a group controls (i.e., healthy persons), ancestral differences in study subjects creates population structure that can confound a genetic screen if the trait of interest also varies as function of ancestry (for biological or nonbiological reasons) in the group sampled. Thus, it is usually important to control for population structure lest investigators identify markers that correlate with trait value simply due to admixture stratification rather than markers genetically proximal to phenotypically important loci (Burroughs et al. 2002; Halder & Shriver 2003; Rao & Chakraborty 1974; Risch et al. 2001; Wang et al. 2003). Many common diseases exhibit locus and/or allelic heterogeneity, sometimes as function of BGA, and it is possible that insufficient attention to population structure during the study design step has produced some of the so-called false positive results implicated in the rash of irreproducible Common Disease/Common Variant results obtained to date (Terwilliger & Goring 2000).

To adjust for population structure in a population-based mapping study there are several alternative tests (Cockerham & Weir 1973; Excoffier et al. 1992; Long 1986; Weir & Cockerham 1969, 1984). These methods can be grouped in two main categories, which have been coined genomic control (GC) methods (Devlin & Roeder 1999) and structured association (SA) methods (Hoggart et al. 2003; Pritchard & Donelly 2001). Both methods require a panel of unlinked markers to estimate and correct for the effects of population structure. Which markers are used in these panels can have dramatic affects on their ability to detect and adjust for stratification (Hoggart et al. 2003; Pfaff, Kittles & Shriver 2002). In addition, there are special forms of the SA methods that are designed specifically to harness the statistical information in admixture for a process of gene mapping called Admixture Mapping (AM) (Hoggart et al. 2004; Patterson et al. 2004). AM tests for association between allelic ancestry and trait values, rather than between genotype and trait values, and therefore helps minimize the confounding influence of allelic heterogeneity within populations as well.

BENEFIT TO FORENSIC SCIENTISTS

A forensics scientist has the need to know ancestry for two main purposes, (1) to infer physical characteristics to aid in an active investigation and (2) to assist with the likelihood calculations part of the identity testing procedure once a suspect is in hand. For a forensic scientist, obviously it is not usually possible to ask for race or ancestry, since the individual under consideration often is represented only by a biological specimen such as hair, blood, or semen. Clearly there is a forensics need for a repeatable, testable (i.e., falsifiable), and quantifiable molecular anthropological approach to estimate BGA. Once BGA is accurately estimated, some of the characteristics that the forensic scientist investigates about a suspect can be learned through both direct and indirect methods of inference, as we will describe throughout this book.

Although estimating the individual ancestry components with BGA analysis is straightforward methodologically, the communication of the results to police and their further dissemination of the information are more nuanced. When a result comes back from the lab that a suspect is 85% West African/15% Indigenous American (as in the Derek Todd Lee case, to be discussed later), how can the results be understood? We might report that, ''In the United States, this person would very likely consider himself to be black and be so considered by others.'' This is a start, and it may be enough to turn a case around, as in the Louisiana serial killer investigation, but there is potentially much more information in this sort of data, and it is a better means of communicating the results. One approach is to compile a large database of facial photographs indexed by the BGA estimates of the persons featured in the photographs (see Chapter 8). When an evidentiary BGA estimate is made, investigators can be supplied with selections of photographs of persons who are within the range of the BGA estimate generated for the evidentiary material. A review of these photographs might be a means to help the officers develop a better idea of the gestalt of physical features they should expect to find in the suspect or victim who left the material. These photographs will also impress on the officers the levels of variation within any ancestry range and that based on the BGA estimates alone, they may not be able to jump to conclusions regarding many racial stereotypes.

It is because there is so much variation within any particular ancestry interval that the next phase of research, admixture mapping to identify the genes that determine physical differences between populations, is important. There already has been a significant amount of progress in the discovery of genes determining skin pigmentation variation between West African and West European parental populations using African-American admixture mapping (Bonilla et al. 2005; Lamason et al. 2005; Norton et al. 2005; Shriver et al. 2003). Given the high heritability of skin pigmentation and the limited environmental contributions, it is reasonable to expect that one could define quite precisely the expected skin color of a person. It is also likely that admixture mapping could be used to determine the genetic basis of other traits that are different between parental populations like hair color, hair form, and particular facial features using admixture mapping.

Many phenotypes are shared among populations and are not the type generally considered to be racial. Stature, for example, and many aspects of facial features, ear size and shape, or the relative lengths of the digits of the hands are all variable in overlapping ways across human populations. Our premise is that it will eventually be possible to predict some of these traits to a level that is useful in forensic investigations. We also argue that even for traits that are not dramatically different between populations, BGA estimates will often be important components of models used in their prediction. Most of these traits are polygenic, being defined through the combined effects of variability at several genes. And although some of the alleles determining the trait are equal in frequency across populations, it is unlikely that all will be, so the BGA estimate will be important both to help adjust for population stratification as a means to facilitate valid genotype/phenotype tests as well as a way to bring overall genetic background into the analysis.

ANCESTRY INFORMATIVE MARKERS

Though humans are on average 99.9% identical at the level of our DNA, it is that 0.1% that imparts our biological individuality. It is well documented that the majority (85–95%) of the genetic variation among human individuals corresponding to that 0.1% is interindividual and only a relatively small proportion (5–15%) is due to population differences (e.g., Akey et al. 2002; Cavalli-Sforza et al. 1994; Deka et al. 1995; Nei 1987; Rosenberg et al. 2002; Shriver et al. 2004, 2005). It also has been widely observed that most populations share alleles and that those alleles that are most frequent in one population are also frequent in others. There are very few classical (blood group, serum protein, and immunological) or DNA genetic markers that are either population-specific or have large frequency differentials among geographically and ethnically defined populations (Akey et al. 2002; Cavalli-Sforza et al. 1994; Dean et al. 1994; Roychodhury & Nei 1988). Despite this apparent lack of unique genetic markers, there are marked physical and physiological differences among human populations that presumably reflect genetic adaptation to unique ecological conditions, random genetic drift, and sexual selection. In contemporary populations, these differences are evident in morphological differences between ethnic groups as well as in differences in drug response or susceptibility and resistance to disease. On a basic level, to measure human population structure in terms of biogeographical ancestry (BGA), or the heritable component of race, we first need to identify and characterize those elusive markers with ancestry information content (AIMs), even though they constitute merely 0.01% or less of the genome.

The idea that there are genetic markers present in one population but not others was first presented more than 30 years ago (Neel 1974; Reed 1973). Neel (1974) referred to these markers as *private*, and used them to estimate mutation rates. Reed (1973) used the term *ideal* (in reference to their utility in individual ancestry estimation) to describe hypothetical genetic marker loci at which different alleles are fixed in different populations. Chakraborty et al. (1991) called those variants that are found in only one population *unique alleles*, and showed how allele frequencies could be inverted to provide a likelihood estimate of population, or BGA affiliation (which we will describe shortly). The most useful unique alleles for the inference of BGA are those that have the most uniqueness, or the largest differences in allele frequency among populations (Chakraborty et al. 1992; Reed 1973; Stephens et al. 1994), which others at first designated as Population-Specific Alleles (PSAs; Parra et al. 1998; Shriver et al. 1997), but which are referred to now by the more correct and descriptive term, Ancestry Informative Markers (AIMs; Frudakis et al. 2003a, 2003b; Frudakis 2005; Shriver et al. 2003).

THE ALLELES OF MOST LOCI USEFUL FOR ANCESTRY ESTIMATION ARE CONTINUOUSLY DISTRIBUTED

When working to describe the migrational histories of populations using genetic data, scientists typically eschew those markers known to have been affected by natural selection since such sequences can be unevenly distributed as a function of the local environment and not just demographic history. Rather, it is the neutral alleles that are sought after—not those under the influence of natural selection, but those alleles whose distribution is a function of the process of genetic drift. It is these alleles that arise and spread by the random process of binomial sampling, and are neutral in terms of their fitness values and for which frequency differences between populations exist as a strict function of time since separation and migration rate. Such markers are useful for phylogenetic reconstructions via the molecular clock hypothesis because their distribution is simply a function of statistical sampling and time since most recent common ancestor and the rate of exchange of migrants.

A basic review of this hypothesis can be found in Jobling, Hurles, and Tyler-Smith (2004). This hypothesis states that the rate of evolution and genetic sequence divergence is approximately constant over all evolutionary lineages, which implies that most mutations and polymorphisms are of neither positive nor negative selective value, and due to this lack of functional, biological pressure, the allele frequency distributions among populations take a continuous rather than a discrete form. In reality, even these variants are subject to forces that make it difficult to assign phylogenetic affiliations. The distribution of even neutral alleles across subpopulations derived from a common source is a function of the rate of nucleotide substitution in the relevant region of the genome (which we usually assume to be constant across population groups, but need not be), the architecture of the genealogical tree (who begat whom), and the effective population size of each subpopulation (which influences the allele frequency in each). As many before have pointed out, the regularity of change caused by the random process of genetic drift across populations is in diametric contrast to that caused by the nonuniform change of morphological evolution. Though many traits that have been selected for are correlated with neutral markers as a function of migration history, their rate of change is not necessarily directly informative of phylogeographic history.

SELECTION VERSUS DRIFT IN PRIORITIZING GENETIC MARKERS

We want to ascribe population affiliations in a manner irrespective of present-day regional or geographic differences. The most useful loci for incorporating a population in a phylogenetic reconstruction of the human species, and therefore the most useful for assignment to that population, would be those for which genetic drift has plied its chronological trade in terms of partially fixing or eliminating certain alleles in some populations, not loci that are subject to regional selective pressures even though the distribution for these latter alleles likely would be more discrete and statistically informative. The reason is that loci subject to natural selection could be polyphyletic and uninformative for ancestry. Polyphyletic traits are those that span across multiple, distantly related clades. They are present in evolutionary lineages that derive from many ancestors and do not share a common ancestor to the exclusion of other lineages. The distribution of such a trait among lineages and populations is often a function of some underlying regional selective pressure, rather than the phylogenetic distance we desire if patterns of peopling are our primary concern. Therefore, if a marker is a reporter of geographical selective pressures potentially shared by many distantly related populations, rather than ancestry, its use could perturb our ability to properly infer ancestral origin.

An example of such a polymorphism is the Hemoglobin S (Hb^*S) allele (Haldane 1949). This allele dramatically reduces fitness when present in the homozygous state by predisposing to sickle cell anemia, but in the heterozygous state confers selective advantage by conferring resistance to infection by the *Plasmodium falciparum* form of malaria. Because malaria is endemic to sub-Sahara Africa, sub-Saharan Africans are characterized by a high frequency of the Hb^*S allele, whereas most non-African populations are not. However, the geographical distribution of *Plasmodium falciparum* is not restricted to sub-Sahara Africa. Rather, it is common to many equatorial regions of the world including the Middle East, South Asia, Southeast Asia, and Indonesia. Indeed, the presence of the red blood cell disorders in the human population is spread across these same areas, and the frequency of the Hb^*S allele is likewise distributed, to greater and lesser extents, across populations inhabiting these regions. Choosing the Hb^*S allele as an AIM for the inference of ancestry would be a mistake, because of the fact that this allele could have arisen in multiple populations, multiple times over human history, due to regional, equatorial selective pressures. In other words, the marker would be expected to be identical by state in these populations, rather than identical by descent from a relatively recent common ancestor. If of all these populations, only sub-Saharan Africans are represented in our parental sample due to choice of the population model used by which to calibrate the performance of an admixture test (i.e., calibrated with respect to West African, West European, Indigenous American, and East Asian parental populations, not South Asian Indian, Indonesian, or Australian), the presence of this allele in some non-African equatorial populations would improperly suggest admixture or recent shared ancestry with West Africans.

Thus, as a general rule, markers whose allele frequencies have been molded by the process of natural selection may require more complex models to interpret their patterns of geographical variation. This said, however, it is important to recognize that BGA estimates and many other genetic summary statistics are calculated by averaging across many loci, and so any particular locus that has been under selection will have only a limited effect on the final estimate. Indeed, it may be the case that many of the best AIMs are near loci that have been under selection since the separation of populations, like the Duffy gene null allele (FY^*0). This allele confers resistance to *Plasmodium vivax* malaria and is restricted to African populations, nearly fixed in sub-Saharan Africa but completely absent everywhere else. Due to this disparity, this marker shows among the highest F_{ST} values of any marker known (0.78). The Duffy gene is actually a transmembrane receptor (called DARC) that has served as a blood group antigen for classical geneticists for many years. The FY^*0 polymorphism is a T to C substitution in the GATA transcription factor box located 46 bp upstream of the start site of DARC transcription, and it precludes the initiation of transcription in red blood cells and thus, the expression of the DARC gene product on the red blood cell surface (only, not other tissues such as the brain).

Alleles at polymorphic loci are generally unlike Duffy, being neither fixed nor eliminated across human populations, but varying in frequency. As we have seen in our discussion of genetic drift, most polymorphisms are relegated to this state of suspended animation for long periods of time, depending on the number of founders of a population and the allele frequency in these founders. Some are subject to the forces of natural selection, but most are neutral in terms of their fitness values and their frequencies are a function of the binomial sampling process of genetic drift (see Chapter 1)—which, given the short time since the divergence of human populations, usually means their alleles are all still present in the population and not yet fixed or eliminated (see Figure 2-1).

It is also worth noting that the Hb^*S allele may also be a rather unusual genetic adaptation in that its effect is to create an adhesive form of hemoglobin, which can alter the internal environment of the red blood cells making it inhospitable to the malarial parasite. Perhaps the selective pressures related to infectious disease are potentially more specific than other types of selection pressure, meaning that there are alternate genes or gene variants that can be created to respond to particular selective pressures. On looking further it is clear that based on the haplotype backgrounds, the Hb^*S allele in fact has arisen multiple times in sub-Saharan Africa, and indeed the haplotype markers can be used to explore the levels of contribution from varying parts of Africa to New World populations. Finally, another reason for using neutral markers is that there is always the chance that there might still be natural selection acting on the markers, which could be important especially when populations are migrating across ecological clines and could skew estimates of admixture. Again, averaging across a large number of markers mediates this risk.

ASSIGNMENT OF GROUP MEMBERSHIP OR BINNING

Representing genomic ancestry in terms of group memberships is unsatisfying because it presumes that ancestry is a categorically rather than a continuously

Figure 2-1

There is a wide range in the level of differentiation across the genome. F_{ST} distributions for 313 X-linked and 11,078 autosomal SNPs computed for 12 populations. The F_{ST} is a measure of the allele frequency differential among human populations, with higher values characterized by greater differentiation. Where some markers are characterized with high F_{ST} values and some with low F_{ST} values, the vast majority have intermediate values, indicating that their allele frequencies are neither fixed nor eliminated and at least until recently, were in the slow process of drifting apart in various human subpopulations. From Shriver et al., 2005. Human Genomics, Vol. 2, No. 2, p. 81–89. Reprinted with permission from Henry Stewart Publications.

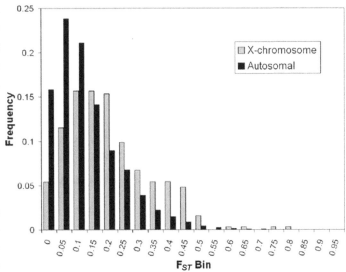

distributed variable. Indeed, if one wants to make an inference on physical appearance, likelihood to respond to a drug, or to use an individual's genetic sequences in a screen for disease loci, human beings are no more properly described as belonging to a particular single group based on overt characteristics or predominant sequences than cars are described as belonging to a certain typological group by their paint colors. Knowing that a car is yellow tells one nothing about how fast it can travel, about its fuel consumption rate, or whether a particular part may need more frequent maintenance than that part on the average car. In fact, in many regions of the globe with recent population expansions (and perhaps in other regions as well), human beings are better described by their position along a multidimensional continuum of ancestry admixture. Human genomic variation is described much better as clinal than discrete.

Though earlier work in the field of human anthropology based on binning has helped establish the foundation upon which better, more meaningful methods could be built, in and of themselves the data they provide is of minimal usefulness for genetics researcher attempting to map disease loci, or a forensics analyst attempting to make inferences on physical appearance. As an example, into what group would an individual of 50% African and 50% European ancestry be binned, and are individuals with even minimal African ancestry to be binned into an African-American group à la the "one drop rule" as is common social practice in the United States? If the binning is being performed as a means by which to make an inference on an element of physical appearance, such as skin shade, the inclusion of individuals of low African admixture on European backgrounds with others showing higher African ancestry is inefficient, because the variability in skin shade in this "African" group would be expected to be almost as great as the worldwide variability. As we will show later, there is a good correlation between the extent of European admixture in African Americans, and objectively measured melanin index and knowing the genomic ancestry of an individual is far more informative for this purpose than knowing whether there is any or even a majority affiliation with the African group. As a another example, because of the fact that many individuals are likely to be of differing degrees of mixed ancestry, a genetics researcher who uses discrete measures of ancestry in an attempt to control for population structure differences between cases and controls would likely do no more to normalize their sample sets for deviations from expected heterozygosity than if they used hair color as a sorting variable.

Nonetheless, various probabilistic methods have been described by various authors (including those for this text) for using interpopulation allele frequency differences to infer the ancestral origin of a DNA specimen in terms of one categorically defined population. We will spend the first part of this chapter describing how binning is accomplished, and then toss it aside for the rest of the book in favor of more informative and meaningful methods for measuring genomic ancestry on a continuous scale. We think it is important to realize the historical development of these concepts and also to be clear on the reasons that these models are inferior to allowing for multiple parental populations.

Most of this work to date has focused on binning individuals into ancestral groups based on the posterior probability of majority affiliation. Some research has been presented using Combined DNA Index System (CODIS) Short Tandem Repeat (STR) loci for this purpose, which, generally speaking, do not fit the criteria of good AIMs, but there have also been papers describing the use of SNP, STR (Shriver et al. 1997) and Alu repeat sequences (Bamshad et al. 2003) that meet the criteria of good AIMs for binning. A somewhat surprising recent paper (Tang et al. 2005) used several hundred STRs to again recapitulate dichotomous race categories.

STR-BASED BINNING AND CODIS LOCI

STRs are loci that often are characterized by dramatic allelic heterogeneity—sometimes 30 or more alleles per locus can be identified and measured in populations. Since these types of sequences usually are found outside of genes, and are assumed to be neutral with respect to natural selection, they are characterized by lower effective population sizes, especially subject to the forces of genetic drift, and therefore often used for a study of the apportionment of human diversity. The CODIS STR loci are a set of tandem repeat array markers also known as microsatellites, which are used by forensic scientists as a sort of identification bar code for matching samples with individuals. Though very useful for this purpose and though their use has transformed the foundation of modern forensic DNA testing as discussed in Chapter 1, they offer little power to distinguish between populations because they were selected from the genome based on intragroup, not intergroup variability. This strategy was chosen for obvious reasons—markers for unambiguous human identification should provide the same discrimination power for everyone lest the results take on different meanings for different subjects of different ancestral backgrounds.

An important property of unlinked markers is that their information can be combined by multiplying across the loci. This is known in forensics as the product rule and can be applied only when allele frequencies are similar in all populations. Nevertheless, several papers have been written concerning the use of CODIS STR loci as AIMs (such as Kashyap et al. 2004; Rajkumar et al. 2004; Sun et al. 2003; and others). When all 13 CODIS STRs are used, success rates of 90% or better are reported, meaning the major ancestry affiliation is correctly identified at least 90% of the time (i.e., a Chinese individual would be assigned membership in the East Asian group 90% of the time). Of course, this unfortunately means that 10% of the inferences are incorrect and 1 out of every 10 Chinese subjects would be categorized with these markers as belonging to some other group. Clearly this error rate makes the reliable inference of physical characteristics from DNA very difficult, but until recently, investigators desperate for information often had nothing else but the CODIS profile with which to work. Generally, there is enough ancestry information content inherent to these markers that population-specific reference databases must be established for proper statistical calculation of the probability of identity, but not enough information content for binning—at least binning with an accuracy high enough to be used as a reliable forensics tool. In summary, although a bin can often be selected, one cannot assign a bin for all individuals and the confidence that the correct bin is selected is low.

Clearly methods based on AIMs would be expected to be more powerful, given that AIMs are selected from the genome based on their ancestry information content. As we will see, this is the case and far fewer AIMs are required to achieve a given accuracy than non-AIMs (i.e., average markers). For example, in terms of correctly specifying the majority ancestry affiliation, STR and SNP loci of modest ancestry informativeness have been able to claim accuracy in inferring majority affiliation in the upper 95 to 99% range, which is a good improvement over the reliability provided by the non-AIM CODIS loci (Frudakis et al. 2003; Shriver et al. 1997). The use of true AIMs (those having large allele frequency differences across populations) described later in this book (mainly, Chapter 5) provides even higher levels of accuracy when we restrict our evaluation to individuals of clearly polarized or equitably admixed ancestral affiliations.

NON-CODIS MARKERS

A number of population assignment or binning studies have been conducted in recent years, including Lewontin (1972; 17 classical markers), Bowcock et al. (1994; 30 autosomal microsatellites), Barbujani et al. (1997; 30 microsatellites), Jorde et al. (2000; 60 microsatellites), Wilson et al. (2001; 39 microsatellites), Romauldi et al. (2002; 8–21 Alu insertions), Bamshad et al. (2003; 100 Alus and 60 STRs), and Tang et al. (2005; 400 STRs). Frudakis et al. (2003) used SNP AIMs (5–30) found in human pigmentation and xenobiotic metabolism genes to obtain accuracies in the upper ninetieth percentile (see Table 2-1; Box 2-A), and Shriver et al. (1997) described a battery of STR AIMs capable of resolving human samples into the expected geographical groups also with an accuracy rate in the upper ninetieth percentile. Overall, these studies indicated that if a large enough collection of markers were used, or smaller samples with higher ancestry information content, then samples taken from specific geographical regions clustered with other samples taken from the same regions.

Jobling et al. (2004) does an excellent job reviewing the ancestry group assignment literature up until now, describing several of the papers just referenced in more detail. A major defect with these methods of ancestry inference is illustrated by these works; most experienced trouble (in terms of lower success rates) for several populations including South Asian Indians, East Africans, and New Guineans. This is because binning samples into single racial groups sacrifices important information about population structure when samples from populations that are potentially admixed or of otherwise complex origins are used. We asked earlier: Into which group should a 50/50 mix be assigned? For individuals having extensive admixture, such a method would seem to produce a wrong answer as many times as a right one. Is there even a right answer to a wrong question? As we will discuss in detail, the inference of ancestry is best accomplished in terms of proportional affiliation, which ameliorates the conceptual problems associated with binning or typological ancestry.

Later works used larger batteries of loci, and got better results although the problem of assigning group membership for admixed samples is omnipresent. In their 2003 paper, Mike Bamshad and colleagues assayed 100 Alu insertion

SNP Loci One

Individual's genotype = AA

	A allele count	T allele count	No. of alleles	Freq. A	Freq. T	LD
Africa	1	1	2	0.500	0.500	1.012
Asia	22	20	42	0.524	0.476	0.918
Basque	2	0	2	0.999	0.001	0.000
Caucasian	10	28	38	0.263	0.737	2.198
Eastern Europe	5	7	12	0.417	0.583	1.378
Mediterranean	3	5	8	0.375	0.625	1.581
Middle Eastern	4	2	6	0.667	0.333	0.450
North Africa	1	1	2	0.500	0.500	1.012
Sub-Continental	7	5	12	0.583	0.417	0.703
Total	55	69	124	0.444	0.556	1.254
				Q1Q2 =	0.247	

Table 2-1

Genotype by population for illustration of discriminant techniques.

and 60 STRs polymorphisms (not selected based on their ancestry information content) in a heterogeneous collection of approximately 565 individuals. They found that to make correct assignment to the continent of origin (Africa, Asia, or Europe) with a mean accuracy of at least 90% required a minimum of 60 Alu markers or STRs and reached 99 to 100% when $>/=100$ loci were used. These authors also highlighted the problem inherent to binning when they showed a less accurate rate of assignment (87%) to the appropriate genetic cluster for a historically admixed sample (or of complex origin) from southern India (the complex ancestry for this group, whether due to complex origins or population admixture, is discussed later in Chapter 6). Nonetheless these results set a minimum for the number of markers that must be tested to make strong inferences about detecting population structure, at least structure considered in terms of Old World population affiliations. Bamshad et al. (2003) noted in their paper that, whereas some proxies correspond crudely, if at all, to population structure, the heuristic value of others was much higher.

The most recent of these categorical clustering papers, Tang et al. (2005), uses a series of 326 STR markers to categorize persons self-identifying as either African-American, European-American, Hispanic, and East Asian into four groups. Using largely the same panel of STR markers and the same software program (STRUCTURE) as the Tang et al. (2005) work, Rosenberg et al. (2002) have performed one of the most comprehensive analyses of human population structure to date. Since this paper and program can, when used to its full potential (not as in Tang et al. 2005), report ancestry in terms of proportional affiliations, rather than by binning, we save discussion of the program for later. Rosenberg studied a sample of 1056 individuals using 377 autosomal STRs, and though 93 to 95% of the total variation observed was between-individual, the remaining variation was adequate to partition the sample into subgroups that strongly corresponded to continent or geographical region of origin (see Figure 3-1).

Notably, this study revealed substantial admixture, or fractional affiliation for several populations that are not thought to be recently admixed including Central Asians, North Africans, Native Americans, and South Asians.

Box 2-A

SNP-based Binning

Frudakis et al. (2003) used a parametric, multivariate linear classification technique (Fisher 1936) with AIMs to bin samples into ancestry groups. With this method, populations are represented by vectors, where the values are average allele frequencies for each of the loci considered. For each locus, they gave a score of 1 if the individual is homozygous for the first allele, score of ½ if the individual is heterozygous, and score of 0 if the individual is homozygous for the minor allele (last allele). For each population, a pooled within-population variance-covariance matrix was computed from

$$S = \sum_{i=1}^{p} \sum_{j=1}^{N_i} (Y_{ij} - \mu_i)(Y_{ij} - \mu_i)' / \Sigma(N_i - 1)$$

(2-1)

where Y_{ij} is the vector of scores for the j'th individual in the i'th population and μ_i and N_i are the vector of means and sample size for the i'th population. By scoring one allele only, they avoided a linear dependence problem, which could lead to matrix singularity (rendering the calculations impossible). The generalized distance of the ij'th individual from the mean of the k'th population was computed from

$$D_{ij,k}^2 = (Y_{ij} - \mu_k)'S^{-1}(Y_{ij} - \mu_k) \text{ for } k \neq i$$

(2-2)

The vector Y_{ij} is used to calculate μ_k, the mean of its own population. To avoid circularity caused by this, Smouse et al. (1977) used a correction when comparing an individual with the mean of its own population:

$$D_{ij,i}^2 = (N_i/(N_i - 1))^2 (Y_{ij} - \mu_i)'S^{-1}(Y_{ij} - \mu_i)$$

(2-3)

With this approach, the ij'th individual is allocated to that population for which (2-2) or (2-3) is minimum. The result of applying (2-2) and (2-3) is an inclusion or exclusion probability matrix for the various populations.

Both linear and quadratic methods can be algebraically simplified for dealing with SNP data. Kurczynski (1970) provided the analytical solution for the inverse of the variance-covariance matrix, and Chakraborty (1992) described the computational equations for using n alleles per loci (when we score n-1 alleles per loci). Here we derived the analytical solution to the linear discriminant function for bi-allelic loci. The i'th individual's discriminant function can be calculated in the following way.

Case 1, if the individual is homozygous for the major allele:

$$D_{ij} = P_{j,2}^2 / (Q_1 Q_2)$$

(2-4)

Case 2, if the individual is heterozygous:

$$D_{ij} = (1/2 - P_{1,j})^2 / (Q_1 Q_2)$$

(2-5)

Case 3, if the individual is homozygous for minor allele:

$$D_{ij} = P_{1,j}^2 / (Q_1 Q_2)$$

(2-6)

where, $Q1$ and $Q2$ are the global allele frequencies (average allele frequencies over all populations for major and minor alleles), and $P_{1,j}$ and $P_{2,j}$ are the major and minor allele frequencies in the j'th population. D_{ij} is the discriminant value of the i'th individual in the j'th population. For L loci, they repeated calculations (2-4) through (2-6), add the sum for each population, and compared the sums for these populations assigning the i'th individual to the j'th population for which D_{ij} is smallest. It is not easy to conceptualize how this method works. Here is the easiest way to explain it: one is computing a penalty for a person that has the major allele rather than the minor allele (if they have the major allele), and this penalty is greatest for incorrect populations. If they have the minor allele, we compute a penalty for having this allele rather than the major allele, and again this penalty is greatest for incorrect populations. For example, look at Table 2-1, which shows an Asian individual's genotype for one locus.

Compare the linear discriminant (LD) score obtained for assignment into the Asia vs. Caucasian groups. Since the individual has the A allele in homozygous form, and this is the global minor allele, we use equation (2-6), and we employ the frequency of the global major allele, $P_{1,j} = P(G)_j$ since the individual is homozygous for the global minor allele (A). Between Asian and Caucasian, the LD is lowest for the Asia group and the genotype AA is more likely to have been obtained from this Asia group because the frequency of the A allele is greater than in the Caucasian group. If the subject were of genotype TT, the major allele, we would use the frequency of the minor allele (A) in the calculations. Among all of the groups, the lowest score is not obtained for the Asian group, but no one locus is expected to be able to result in accurate classifications on its own, and of course we would sum the scores over many loci. Using 56, 30, and 15 AIMs to bin 208 samples, Frudakis et al. (2003) obtained average probabilities of correct classifications in the mid to higher 90% range with higher values obtained using more markers (see Table 2-2).
(Parts taken from Frudakis et al. 2003)

BIOGEOGRAPHICAL ANCESTRY ADMIXTURE AS A TOOL FOR FORENSICS AND PHYSICAL PROFILING

Although binning may effectively resolve majority ancestral origin in most cases, an unacceptable number (5–10%) of classifications are ambiguous (Brenner 1998), and as discussed earlier, the quality of this type of data is in question. Aside from sampling errors caused by rare alleles, and the fact that STRs were not selected from the genome for their ability to resolve population affiliation (i.e., STR allele frequency differentials are not optimally informative for this purpose), the major reason for this is likely to be due to admixture, which is clearly a factor of the genetic variation for many human populations (Cavalli-Sforza & Bodmer 1999; Parra et al. 1998; Rosenberg et al. 2002). For a given problem, whether using self-reported information or DNA marker testing, and whether it is a pharmacogenomics, disease screening, or forensic problem, classifying a patient in a single group sacrifices the subtle but not insignificant information related to within-population structure and substructure.

	African (AA)	Asian (AI)	European(CA)
	Probability*	Probability*	Probability*
56 markers			
AA (n=90)	0.9778	0	0.0222
AI (n=30)	0	1	0
CA (n=88)	0.0114	0	0. 9886
30-best markers			
AA (n=90)	0.9556	0	0.0440
AI (n=30)	0.0333	0.9667	0.0000
CA (n=88)	0.0227	0	0.9773
15-best markers			
AA (n=90)	0.9111	0	0.0889
AI (n=30)	0	0.9667	0.0333
CA (n=88)	0.0227	0	0.9773

Table 2-2

Linear classification probabilities using 56, 30, or 15 markers.

*The lower of the uncorrected and corrected probabilities are shown for classification into the proper group, and the higher of the uncorrected and corrected probabilities are shown for classification into improper groups (17).

CHAPTER 3

BIOGEOGRAPHICAL ANCESTRY ADMIXTURE ESTIMATION— THEORETICAL CONSIDERATIONS

ESTIMATING BY ANTHROPOMETRIC TRAIT VALUE

There are a variety of methods for estimating admixture, some using pheno-types and others using genetic data. Some methods are designed for di-hybrid populations; others are designed for more general hybrids where more than two parental populations are possible. We shall spend most of our time in this chapter describing molecular genetic methods. Conceptually, morphological traits are useful for estimating admixture, but in practice, inferring ancestry this way is problematic. There is significant intragroup variation in anthropo-metric trait value—an example being dark skin. Just because a person has very dark skin does not mean they have sub-Saharan African ancestry, as evidenced by several areas of the world where there are non-African native populations with very dark skin (South Asian Indians, Southeast Asian Negritos, and Island Melaneasians). For other traits that are less easily appreciated, such as nose shape for example, the correlation with ancestry is even worse, making the practice completely untenable.

So why not measure lots of traits? Combining certain traits into a meta-analysis (such as one might do who says, ''that person just looks like they are partly Native American'') is usually haphazard and arbitrary. Further, we don't understand the genetic basis for most anthropometric traits, or how each one relates to ancestry (yet), and it so happens that most are correlated with one another, which further confounds our ability to estimate ancestry. The fact that the genetic basis of the traits are unknown is clearly a problem, since how we compound the information is very different if the traits are correlated because they have a common functional basis (epistasis) as opposed to them being correlated as the result of population substructure. Nonetheless, it is not uncommon for people to estimate ancestry by eye. If a person is of relatively polarized ancestry (not very admixed), then of course guessing ancestry is usually trivial for many observers. The problem, as we will see, is that admixture seems to be the rule rather than the exception, and the more admixed an individual, the more difficult most people find it to be to estimate admixture levels and type. If we took 50 individuals of roughly 50/50 West African/ European admixture and asked 100 people to guess the sub-Saharan African, European, Native American, and East Asian ancestry of each of the 50 people, we would probably find the number of Native American admixture guesses to be high. Admittedly, we are speculating here to illustrate a point, but these indeed would be interesting experiments to carry out, since it is likely that visible population differences are sometimes the result of sexual selection.

Of course, to estimate admixture this way, you need to have a person in front of you, and indeed a major reason we are interested in reading ancestry from the DNA is to gain an appreciation for what a person might look like without having access to the person (such as through crime scene evidence). If we have access to the person corresponding to a crime scene sample, and therefore we know what he or she looks like, then there is no forensic need to measure admixture unless it is to qualify the person for identity testing (that is, using the admixture profile to select which STR allele frequency database to use). Other types of scientists, such as epidemiologists and geneticists, can benefit greatly from objectively quantified BGA for patients they can see, touch, and interview, because as we will see later, lower levels of statistically significant admixture is not always manifested in physical appearance and is not always reported on patient questionnaires, but may still have some potential bearing on the likelihood of complex trait expression. This is called cryptic population structure, and it is especially relevant for the design of clinical trials and genetic and epidemiological studies aimed at identifying disease genes. As we will see, it may also be useful for making indirect inferences on physicality and for use as a covariate in models for predicting phenotype.

The idea that offspring trait value approximates the midpoint of the parents has formed the basis of using phenotypes as indicators of individual admixture levels. This is true when the trait is controlled by additive influences, without dominance or epistasis, and when the trait value is very different between the source populations. Blangero (1986) proposed a multivariate likelihood method that used additive genetic covariance matrix for each population, the mean phenotype trait value vector and a known effective population size for source populations (Chakraborty 1986). The idea was to predict admixture from trait value in a hybrid knowing the trait values in the source populations and the amount of time that has passed since the hybrid population was created from these two source populations. But as Chakraborty (1986) and many others before him have pointed out, phenotypes are subject to selective pressures and since they operate on a geographical rather than lineal framework (an example is skin Melanin content), and since the anthropometric traits often are correlated with one another and therefore effectively under common selective pressures, these selective pressures can produce variations that are not monophyletic and therefore uninformative as to ancestry. When multiple traits are considered the method is expected to be more accurate. Chakraborty (1986) reviewed these methods and suggested that although they are not suitable for the determination of admixture on their own, they are useful in corroborating other types of results, such as linguistic, archaeological, and what we are interested in here, genetic.

ADMIXTURE AND GENE FLOW ESTIMATED FROM SINGLE LOCI

Bernstein was the first to describe (in 1931) how genetic data could be used to estimate the fraction of genes contributed by two populations to a hybrid population. The probability of an allele of frequency P_i in population i being found in a hybrid population (P_h) is dependent on the fraction of population i's genes found in that hybrid population. If this fraction is m, then the probability is:

$$P_h = \sum_{I=1}^{k} m_i P_i \qquad (3\text{-}1)$$

over all possible populations k. For a hybrid population derived from two parental sources P_1 and P_2 in proportions m_1 and m_2, respectively, the frequency of the allele is thus

$$P_h = m_1 P_1 + m_2 P_2$$

But $m_1 + m_2 = 1$ by definition so that

$$\begin{aligned} P_h &= m_1 P_1 + (1 - m_1) P_2 \\ &= m_1 P_1 + P_2 - m_1 P_2 \\ &= m_1 (P_1 - P_2) + P_2 \\ P_h - P_2 &= m_1 (P_1 - P_2) \end{aligned} \qquad (3\text{-}2)$$

So

$$m_1 = (P_h - P_2)/(P_1 - P_2) \qquad (3\text{-}3)$$

(see Bernstein 1931). The validity of Equations (3-2) and (3-3) depends on the assumption that the exact ethnic composition of the parental samples is known, gene frequencies in modern day populations have been subject only to admixture and that there has been no recent genetic drift or selection operating at the loci. To overcome uncertainty imposed by these assumptions, which could contribute to sampling errors, Cavalli-Sforza and Bodmer (original edition, 1971, but also discussed in the current edition, 1999) suggested weighting the estimate of m obtained from each locus by the inverse of its error variance, as a means by which to counter sampling variance within the hybrid population. We will discuss later other methods for how the uncertainty caused by these assumptions can be minimized.

LEAST SQUARES APPROACH

Over multiple markers, an average m can be determined. There is only one correct m value for each population or individual, and so in a two-population model the relationship between $(P_h - P_2)$ and $(P_1 - P_2)$ should be linear. However, estimates for m will naturally vary from locus to locus. If we were to regress $(P_h - P_2)$ values on $(P_1 - P_2)$ values in a Cartesian plot, the slope of the line would yield m, and in choosing the line that best fits the data, we would choose that line for which the sum of squares difference of the estimates is least. This is essentially the approach used for calculating the m_R estimator (Roberts & Hiorns 1962). The linear algebra required to explain this calculation is beyond the scope of this text, but an eccentricity associated with linear algebraic derivations is that they cannot be computed if a matrix is what is known as *singular*, which unfortunately is more likely to occur as more loci are used for the analysis. In addition, the approach sometimes produces negative values for m, and both occasions require special and somewhat arbitrary methods to overcome. Further, because only populations have allele frequencies, this approach is suitable only for use with populations and cannot be used for within-individual admixture. We could encode genotypes with pseudo-frequencies—for example assume that CC is a frequency of 1, CT of 0.5, and TT of 0.0—and use this method as if an individual is considered a very small population, but variation is likely to be extreme, making the determination of the best line difficult.

CALCULATING GENE FLOW IN POPULATIONS BASED ON THE PROBABILITY OF GENE IDENTITY

To estimate the genetic contribution of populations to a hybrid, Chakraborty (1975) derived these equations using an alternative method that relies on the similarity between populations rather than the genetic distance between them. The rationale for this switch in approach is that although with increasing gene flow genetic distance should decrease, the relationship between genetic distance and gene flow is not simple nor linear. He estimated the proportion of admixture in a hybrid population, m, based on the probability of gene identity (Nei 1971). The logic of his approach was based on the notion that admixture causes the same change in frequency for one locus as it does for another in the genome, and that provided there is no selection favoring or disfavoring alleles for certain loci, estimates of admixture in hybrid populations can be expressed in terms of the probability of selecting genes identical by state from parental and hybrid populations. The same logic is applied when calculating an inbreeding coefficient.

If x_{ni} is the frequency of a gene (allele) i in a population n, then the probability of identity in state for a given gene in population 1 is $J_{11} = \Sigma_i x_{1i}^2$, in population 2 it is $J_{22} = \Sigma_i x_{2i}^2$, and when drawn from both populations it is $J_{12} = \Sigma_i x_{1i} x_{2i}$. Generally, if x_{ikl} is the frequency of the kth allele of locus l in population i and x_{jkl} is the frequency of the kth allele of locus l in population J, then the gene identity between the two populations averaged over L loci is:

$$J_{ij} = \sum_{i=1}^{L} \sum_{k=1}^{r} [x_{ikl} \, x_{jkl}]/L \qquad (3\text{-}4)$$

where $r = 1 - L$ (Chakraborty 1975). The gene identity between a hybrid population and population 1 is the probability that of two genes drawn from the hybrid population (a genotype for a locus of course is comprised of two copies), at least one of them comes from population 1. In a two-population model, there are two ways a genotype can be composed of a gene (or allele) that came from population 1: both originated in population 1 (J_{11}), or only one did (J_{12}). Assuming hybrid population h derives m_1 of its genes from population 1 and $(1 - m_1)$ of its genes from population 2, the probability of the first event is m_1 and that of the second is $(1 - m_1)$. Thus, the gene identity between the hybrid population and population 1 is:

$$J_{1h} = m_1 J_{11} + (1 - m_1) J_{12} \qquad (3\text{-}5)$$

where the J_{12} and J_{22} are calculated in (3-4). Likewise, the probability that of two genes of a genotype drawn from the hybrid population at least one came from population 2 is the gene identity between the hybrid population and population 2:

$$J_{2h} = m_1 J_{12} + (1 - m_1) J_{22} \qquad (3\text{-}6)$$

where the J_{12} and J_{22} are calculated in (3-4). This is fundamentally the same as the expression (3-2), which can be seen by substituting for J_{12} and J_{22}, and then multiplying both sides by x_{ikl} and summing over k and l.

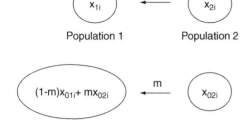

Population 1   Population 2

Hybrid population   Population 2

Figure 3-1

Modeling admixture. Population 2 contributes a certain fraction of genes (m) to population 1 (top) resulting in a change in allele frequency in the subsequent population (bottom). The terms within the circles represent allele frequencies for each population and the arrow indicates the direction of gene flow.

Let us illustrate with an example. Assume that there is a unidirectional flow of m of population 2's genes to population 1, each infinitely sized generation. In the first generation, before flow has occurred, we have two population frequencies for gene i, x_{1i} and x_{2i} (see Figure 3-1, top).

In the next generation, population 1 has changed. The frequency x_{1i} is reduced by a factor $(1-m)$ since m new genes from population 2 have introgressed, and we have added mx_{2i} to population 1 due to the transfer. Since the flow is unidirectional, the frequency of the gene in population 2 remains the same, and after transfer we have two populations of allele frequencies,

$$(1 - M)x_{1i}^0 + Mx_{2i}^0 \text{ and } x_{2i}^0.$$

where x_{1i}^0 and x_{2i}^0 represent the frequency in the preceding generation (see Table 3-1, bottom). Thus, based on (3-2), the probability of drawing the same gene from each population in a two-population model becomes the product of the frequency in the hybrid population and the frequency in population 2:

$$J_{jh2} = \sum_{I=1}^{L} [(1-m)x_{1i}^0 + mx_{2i}^0]x_{2i} = \sum_{I=1}^{L} [(1-m)x_{1i}^0 x_{2i}^0 + m(x_{i2}^0)^2]$$

$$= (1-m) \sum_{I=1}^{L} x_{1i}^0 x_{2i}^0 + m\Sigma_{I=1}^{L}(x_{i2}^0)^2$$

$$= (1-m)J_{12}^0 + mJ_{22}^0 \qquad (3\text{-}7)$$

which rearranges to:

$$m = (J_{h2} - j_{12}^0)/(j_{22}^0 - j_{12}^0) \qquad (3\text{-}8)$$

Here, m represents an admixture estimate obtained from the arithmetic mean of j_{11}, j_{22}, and j_{12}, respectively, over many loci as called for in (3-1). For a single locus:

$$j_{h2} = (1-m)j_{12}^0 + mj_{22}^0. \qquad (3\text{-}9)$$

and

$$m = (j_{h2} - j_{12}^0)/(j_{22}^0 - j_{12}^0) \qquad (3\text{-}10)$$

where m is the admixture estimate obtained using any given locus (reviewed in Chakraborty 1986).

The previous assumption of an infinitely sized population is not realistic for real populations, which are of finite size. Chakraborty (1975) showed that for finite populations (3-8) becomes:

$$(1 - m)^t = [(1 - J_{12}^t) - (1 - 1/2N_2)^{t+1}(1 - J_{22}^0)]/[(1 - J_{12}^0) \\ - (1 - 1/2N_2)(1 - J_{22}^0)] \tag{3-11}$$

over successive generations (superscripts o, t, and t+1), when N_2, the size of the second population, is large. When N_2 is infinity, Equation (3-11) reduces to (3-8), where the accumulated proportion of admixture over the generations m is defined as $m = 1 - (1 - m)^t$. Nei (1978) proposed a bias correction of (3-1) when a small number of genes are used from each population, but according to Chakraborty (1986) the effect of sampling bias is small when sample sizes are reasonable (such as greater than 100). This method was extended using a least-squares approach (Chakraborty 1985) to alleviate inadequacies in stability and application to more than two populations. In populations, estimates of m using gene identity are more stable generally speaking than estimates based on allele frequencies, particularly when large numbers of loci are used (Chakraborty 1985). This is because sampling and drift error are smoothed out by calculating the J_{ij} values. Chakraborty expanded this method to accommodate more than two groups, but since this method involves linear algebra we will refer the interested reader to the paper by Chakraborty, 1985 and a review of the paper Chakraborty, 1986.

LIKELIHOOD

Estimating individual admixture proportions using the Bernstein formula in (3-1) and (3-2) is most easily accomplished with likelihood functions. The easiest way to explain how the likelihood function works to a novice is through analogy. Consider a hypothetical town with four clothing stores. Each store specializes in clothes of a different color—one sells mainly blue clothes, another mainly red clothes, and the others yellow and green. Each of the stores is not exclusive to a color—the red store also sells yellow, green, and blue items—but assume that 95% of the items in the red store are red and so on for the other stores. Assume each person in the town shops exclusively among these four stores. Walking the streets of this town, we can count the number of differently colored clothing items worn by each person and form a probability statement to express their most likely shopping habits—which stores they spend the most time in, and the proportion of time spent in each store. The larger the number of clothing items we consider, the more accurate is our probability statement. This is analogous to the relationship between the number of AIMs and the quality of the ancestry inference with individual ancestry estimations. If the stores are 98% exclusive rather than 95%, our probability statements will also be better. This is analogous to the allele frequency differentials we employ with admixture estimation, and why it is that alleles with higher F_{ST} levels are better than ones of lower F_{ST}.

Though this method would work well most of the time, we may have trouble distinguishing yellow from green clothes at a distance and so we might expect some degree of error caused by the closeness of these colors. This is analogous to the relative difficulty we have in distinguishing between ancestry affiliations for which our AIM panels have less information content—typically among groups of lower genetic distance, such as East Asians and Native Americans. If our assumption that each person shops only among the four stores is wrong—say there is a fifth store—again we would introduce error in our estimation of time spent in each store. This is analogous to the dependence of individual admixture estimates on the quality of our assumptions, namely that there was no genetic drift between the time the parental populations existed and today, or in how admixture took place (for example, did it occur in spurts, or as a continuous process of gene flow?).

Notwithstanding these sources of error, which we will describe in more detail later, the primary question is how exactly we form the hypothesis on how much time each person spends shopping in each store. To do this, we need merely invert the frequencies of the clothing items and incorporate the number present on each person. A small number of red items and the rest yellow present on a person may mean something different if red items are almost never found in these stores than if red items are found frequently in the yellow, blue, and green stores. In the former case, it is relatively more likely that the person does not spend all of his or her time in the yellow store, especially if there are also a couple of green and blue items as well on the person's dress. In the second case, it would be (relatively) more likely that the person spent virtually all of his or her time in the yellow store.

For two populations, the equation for estimating admixture

$$q_n = Mq_c + (1 - M)q_a \qquad (3\text{-}12)$$

is often written for locus g as:

$$p_{Ak} = m_1 p_{1k} + m_2 p_{2k} \qquad (3\text{-}13)$$

where p_{Ak} refers to the allele frequency of the kth allele of locus g in the admixed population, given the admixture level and allele frequency for unadmixed groups 1 and 2 rather than a and c as in (3-12). Equation (3-13) can be rewritten as

$$p_{Ak} = p_{2k} + m_1\delta_k \qquad (3\text{-}14)$$

where $\delta_k = p_{1k} - p_{2k}$. Sampling genotypes from the admixed population (or individual), we can measure the likelihood of m_1 and hence $(1 - m_1)$ using the likelihood function:

$$\ln L_g \alpha \Sigma_k n_{gk} \ln (p_{gAk}) = \Sigma_k n_{gk} \ln (p_{g2k} + m_1\delta_k) \qquad (3\text{-}15)$$

over all alleles k for locus g, if we assume codominance, random mating, and unlinked alleles. Here n_{gk} is the count of allele k in the admixed population, which can be expressed over all loci g as

$$n_{gk} = 2n_{gkk} + \Sigma_k n_{gkl}$$

for all k not equal to 1 ($2n_{gkk}$ is the number of homozygous k alleles, and $\Sigma_k n_{gkl}$ is the number of heterozygous genotypes for this locus with the k allele). Equation (3-15) essentially provides higher (less negative) scores for m_1 values that best fit the observed n_{gk} and p_{g2k} values, and a lower score (more negative) when the m_1 value does not fit the observed n_{gk} and p_{g2k} values. To best understand (3-15) it may be necessary for you to write down several examples of n_{gk}, p_{g2k}, and m_1 values, noticing how the $\ln L_g$ changes as these variables change. For example, consider an individual as a population of size 1. If for a given allele of a given locus, we have a $\delta = 0.5$, $m_1 = 0.1$ and two alleles ($n_{gk} = 2$), we would obtain a value of -0.32 if p_{g2k} is 0.8 and a value of -2.78 if p_{g2k} is 0.2. In this case, the m_1 value of 0.1 fits better with the higher p_{g2k} because the individual has two copies of allele k, and is therefore more like group 2 than group 1.

For a given δ summing (3-15) over all alleles of all loci g, we obtain the $L = \Sigma_g \ln L_g$ which is the likelihood of the m_1 given the data p_{g2}, p_{g1}, and δ for each locus g. We can perform this calculation (3-15) for various m_1 (and thus, m_2) values, selecting that with the greatest L. Alternatively, we can take the first derivative of (3-15) with respect to m_1, and solve for zero to compute the most likely m_1 (and m_2); differentiating (3-15) with respect to m_1 gives

$$\Phi \ln L/\Phi m_1 = \Sigma_g \Sigma_k n_{gk} (\delta_{gk})/(p_{gAk}) \qquad (3\text{-}16)$$

where $p_{gAk} = p_{g2k} + m_1\delta_{gk}$ (Pfaff et al. 2004). The analysis for individuals is performed assuming a sample size of 1 (Chakraborty et al. 1986). For three groups, we have three admixture proportions for each individual and each population.

$$p_{gAk} = m_1 p_{g1k} + m_2 p_{g2k} + m_3 p_{g3k}$$
$$= p_{g3k} + m_1\delta_{g1k} + m_2\delta_{g2k}$$

where $\delta_{g2k} = p_{g2k} - p_{g3k}$ and $\delta_{g1k} = p_{g1k} - p_{g3k}$. The log-likelihood function is the same as shown in (3-15):

$$\ln L = \Sigma_k n_{gk} \ln (p_{gAk}) = \Sigma_k n_{gk} \ln (p_{g3k} + m_1\delta_{g1k} + m_2\delta_{g2k}) \qquad (3\text{-}17)$$

which is summed over all loci g so it is actually

$$\ln L = \Sigma_g \Sigma_k n_{gk} \ln (p_{g3k} + m_1\delta_{g1k} + m_2\delta_{g2k}) \qquad (3\text{-}18)$$

This time we take the partial derivatives with respect to both m_1 and m_2, one at a time, set each to zero, and solve the equations simultaneously to identify the most likely m_1 and m_2, and $[m_3 = 1 - (m_1 + m_2)]$. For j groups, we simply extend (3-18), incorporating j admixture proportions for each individual and each population.

$$p_{gAk} = m_1 p_{g1k} + m_2 p_{g2k} + m_3 p_{g3k} \ldots + m_j p_{gjk}$$
$$= p_{gjk} + m_1\delta_{g1k} + m_2\delta_{g2k} \ldots + m_{(j-1)}\delta_{g(j-1)k}$$

where $\delta_{g1k} - p_{g1k} - p_{gjk}$, $\delta_{g2k} = p_{g2k} - p_{gjk}$ and $\delta_{g(j-1)k} = p_{g(j-1)k} - p_{gjk}$. The log-likelihood function is the same as shown in (3-15) and (3-18):

$$\ln L = \Sigma_k n_{gk} \ln (p_{gAk}) = \Sigma_k n_{gk} \ln (p_{gjk} + m_1\delta_{g1k} + m_2\delta_{g2k} \ldots + m_{(j-1)}\delta_{g(j-1)k})$$
$$(3\text{-}19)$$

which is summed over all loci g so it is actually

$$\ln L = \Sigma_g \Sigma_k n_{gk} \ln (p_{gjk} + m_1\delta_{g1k} + m_2\delta_{g2k} \ldots + m_{(j-1)}\delta_{g(j-1)k}) \qquad (3\text{-}20)$$

Again, we take the partial derivatives with respect to each m_j, one at a time, set each to zero, and solve the equations simultaneously to identify the most likely $m_1, m_2, \ldots, m_j$, where the sum over all $m_j = 1$ and $m_j = 1 - (m_1 + m_2 + \ldots + m_{j-1})$. Alternatively, a more computationally simple iterative numerical approach can be used to determine the m of maximum L, where one simply tries a large number of values of m or even every possible value. Also, one could use the log rather than the ln.

OTHER METHODS APPROPRIATE FOR POPULATIONS OF SAMPLES

We have discussed three different methods of calculating admixture proportions for populations – least squares, gene identity, and likelihood approaches. All of the currently described methods, except that of Bertorelle and Excoffier (1998), are basically the same (Chakraborty 1986) in that they are based on similar assumptions and methods. Each relies on the expectation that allele frequencies in hybrid populations are linear combinations of those in the parental populations that begat them (Chakraborty 1985), and each relies on Equation (3-1), which we refer to as the Bernstein equation, regardless of how it arrives to or uses it. Since this equation requires that parental allele frequencies be accurately specified, and because these frequencies can only be estimated, each approach is subject to sampling error, error caused by genetic drift between the admixture events and the sampling events (which of course vary from problem to problem). Some methods (Roberts & Hiorns 1965) ignore these nuisances, whereas others, as we have described, attempt to minimize their influence (Chakraborty 1986).

In terms of the error inherent to the Bernstein approach, Glass and Li (1953) took into account sampling error, but not error caused by genetic drift (Wang 2003). Long (1991) used least-squares methods that took into account both types of error but only in the hybrid population, and Thompson (1973) considered both types of error in both hybrid and parental populations. Recent advances have allowed for a consideration of most possible sources of error (McKeigue 2000; Wang 2003), which we shall discuss in more detail shortly. Long and Smouse (1983) developed a maximum likelihood-based method that took into account sampling error in both hybrid and parental populations and Chikhi et al. (2001) used a coalescence likelihood-based method that considers both types of error, allowing joint estimates of drift and admixture.

Bertorelle and Excoffier 1998 also use estimates of mean coalescence times of pairs of haplotypes drawn from the populations. Pairs of haplotypes are drawn from the admixed population and their coalescent times, based on genetic distance between them, is compared with that from pairs drawn one from the admixed population and one from the parental population (reviewed in Jobling et al. 2004). The greater the diversity between two haplotypes of a pair, the more ancient the coalescent time between haplotypes drawn from the hybrid population and one of its parentals. The mean age of haplotype divergence can be related to an estimate of admixture called m_Y since genetic distance between two groups is inversely related to the extent of admixture between them. An advantage of this particular method is that it is the only one to account for genetic differentiation within parental groups prior to admixture (Wang 2004).

ADMIXTURE IN INDIVIDUAL SAMPLES

Populations have allele frequencies but individuals do not, so how might we determine admixture percentages within individuals using multiple markers? The admixture of a population is nothing more than the average of individual admixture results for all its members, and calculating the admixture of an individual is the same as calculating the admixture for a population of size one (Chakraborty 1986). A large number of individuals can be used with a small number of alleles to achieve a high quality estimate much like a large number of alleles can be used for a small number of individuals (i.e., 1). Everything we have discussed to this point applies equally for the calculation of admixture in individuals and in populations. However there is one important difference in the style of how we determine admixture on a population versus individual scale—for individual estimates we use genotype frequencies rather than allele frequencies. We will spend time here to describe why this is the case, and provide a brief history of individual admixture estimation.

In the approaches so far summarized, we have used allele rather than genotype frequencies. Approaches suitable for estimating within-individual admixture use the genotypes of individuals (sometimes, unfortunately, called their phenotypes). MacLean and Workman (1973a, 1973b) and Hanis et al. (1986) derived the individual admixture estimation formulae using genotypes as a direct manifestation of the Hardy-Weinberg Theory:

$$Pr\ (A1A1) = [mp_1 + (1-m)p_2]^2 \qquad (3\text{-}21)$$

$$Pr\ (A1A2) = 2[mp_1 + (1-m)p_2][mq_1 + (1-m)q_2] \qquad (3\text{-}22)$$

$$Pr\ (A2A2) = [mq_1 + (1-m)q_2]^2 \qquad (3\text{-}23)$$

where p_1 is the frequency of allele A1 in population 1 and q_2 is the frequency of allele A2 in population 2, and so on, and where Pr (A1A2) is the probability of an A1/A2 heterozygote (reviewed in Chakraborty 1986). The most likely value of m is that which maximizes the probability the individual has the genotype they actually carry (i.e., maximizes the expression 3-21 to 3-23 corresponding to the individual's genotype). The log of the Pr values in (3-21 to 3-23) are the corresponding log likelihoods of admixture value m given p_1 and p_2.

If we were binning samples to groups, we could determine the likelihood of belonging to only one population, population by population, and choose that population for which the likelihood is greatest. Since we are interested in determining the admixture M, we are instead interested in finding the value of m with the highest likelihood given the parameters p_1 and p_2. When

$p_1 \gg p_2$, Pr(A1A1) is high with a high m (fraction of genes derived from p_1) and if p1 $\ll$ p2, Pr(A1A1) is highest with a low m. If the major ancestry from population 1 is m, then the level of admixture with population 2 is $1 - m$. For multiple loci, we can add or multiply the log likelihood values corresponding to the genotype for each to produce a log likelihood for the entire multilocus genotype. As can be appreciated from (3-21 to 3-23), the assumption of Hardy-Weinberg equilibrium is implicit and requires the use of the allele frequencies to calculate the genotype frequency.

Clearly, many loci would be needed to provide estimates of m that are reasonably accurate, and a theoretical treatment of the number of loci required is discussed later in this chapter. Here we will condense Chakraborty's (1986) discussion on the very simple reason why more loci are better than one, which you can extend in your own mind to compare marker sets of larger sizes. For an A1A1 individual, the value of m obtained for a single locus can take on only three values—0%, 50%, or 100% affiliation with population 1. Either the individual received 0, 1, or both alleles from population 1, but the maximum value of m will be 1 if p1 > p2, and 0 if p1 < p2, even if one allele was derived from one population and the other from the other. If the individual is a hybrid of populations 1 and 2, neither of these two answers will be correct. Adding another locus and multiplying Pr values corresponding to the genotypes of the individual, we will find that the maximum m could obtain at values other than 0 or 1, depending on the level of admixture and the allele frequencies in each population. Indeed, the more loci we add, and the greater the δ in allele frequency between populations, the faster we approach the true value of m. This is because the information content in the likelihood function increases with an increasing number of loci. The method we use throughout this text is the computationally economical log likelihood function based on this Hanis et al. (1986) method (see Box 3-A). Essentially, the log likelihood of every possible ancestry percentage is determined and the combination with the highest likelihood is selected.

Box 3-A

A Detailed Description of an MLE Algorithm

The following is an outline of the logic we use to obtain the Maximum Likelihood Estimate (MLE) with SNP AIMs, where we assume a simple two-population model of admixture. Put most simply, the algorithm used relies on the AIM allele frequencies for the parental samples, where the frequencies for each group form a sort of a "template" against which a particular sample can be compared. In lay terms, the parental allele frequencies are inverted to calculate likelihoods of proportional fitting individual genotypes to templates of varying admixture levels.

1. Obtain a list of the estimated allele frequencies for each AIM in each of the reference populations.

2. Compute the probability of an expected SNP genotype in a population of hypothetical samples with the same admixture proportions as that of an unknown sample under the assumption of the Hardy-Weinberg Law.

   2.1 **Expected Allele Frequencies** For example, if "A" and "a" are the two SNP alleles, and p(A,1), p(a,1), and p(A,2), p(a,2) are the corresponding frequencies in the two reference populations (populations 1 and 2), then the *Expected Frequencies* for the alleles A and a in a hypothetical population of samples with the same admixture proportions as that of an unknown sample are

$$P^H(A) = \lambda_1 \cdot p(A,1) + \lambda_2 \cdot p(A,2); \ \lambda_1 \geq 0, \ \lambda_2 \geq 0; \ \lambda_1 + \lambda_2 = 1 \qquad (3\text{-}24)$$

$$P^H(a) = \lambda_1 \cdot p(a,1) + \lambda_2 \cdot p(a,2); \ \lambda_1 \geq 0, \ \lambda_2 \geq 0; \ \lambda_1 + \lambda_2 = 1 \qquad (3\text{-}25)$$

   The mixing proportions λ_1 and λ_2 are treated as unknown parameters.

   2.2 **Expected Genotype Frequencies** for the SNPs in a hypothetical population of samples with the same admixture proportions as that of an unknown sample are:

$$p^H(Aa) = 2^* P^H(A)^* P^H(a) \qquad (3\text{-}26)$$

$$p^H(AA) = P^H(A)^* P^H(A) \qquad (3\text{-}27)$$

$$p^H(aa) = P^H(a)^* P^H(a) \qquad (3\text{-}28)$$

   2.3 **Compute Likelihood** The likelihood of the parameters for a given sample is obtained by taking the log of the product of the probabilities for each observed genotype in the new observation under the assumption of the Hardy-Weinberg Law. If the genotype of a SNP in the sample of unknown admixture is heterogeneous genotype (Aa), then

$$L^H(Aa) = \log(2^* P^H(A)^* P^H(a)); \qquad (3\text{-}29)$$

   If the genotype of a SNP in the sample of unknown admixture is the homogeneous genotype (AA), then

$$L^H(AA) = \log(P^H(A)^* P^H(A)); \qquad (3\text{-}30)$$

   If the genotype of a SNP in the sample of unknown admixture is the homogeneous genotype (aa), then

$$L^H(aa) = \log(P^H(a)^* P^H(a)); \qquad (3\text{-}31)$$

   For multiple loci, we compute LIKELIHOOD VALUE by adding all Observed genotypes of all SNPS. The likelihood of a multilocus genotype of X loci in the hypothetical population H, given an assumed λ_1 and λ_2 is:

L^H(Multilocus Genotype) $= L^H$(Observed genotype for SNP1)

$+ L^H$(Observed genotypes for SNP2)$\ldots L^H$(Observed genotypes for SNP X);

2.4. Since the analytic solution of the likelihood equation does not have a closed form, we have to numerically scan the function over the **parametric space**, so we repeat step 2 for different unknown parameters λ_1 and λ_2.

3. Estimate unknown parameters. The unknown parameters are estimated by those values that maximize the likelihood function, and they represent the most likely mixing proportions.

This procedure is amenable for straightforward generalization (binning) using multiallelic markers, as well as for proportional classification into several groups. It can be applied to both SNP and STR data, both when some of the genotypes were missing as well as when using inferred values for the missing genotypes. *Paraphrased from original text by Visu Ponnuswamy and Siva Ginjupalli.*

USING THE HANIS METHOD ON POPULATION MODELS k > 2

For even a simple four-group worldwide population model (i.e., western sub-Saharan African, European, East Asian, and Native American) the computational load of the log likelihood algorithm is expensive if we test all possible combinations of m_1, m_2, m_3, and m_4. In practice, the large number of possible proportions make the four-dimensional calculations cumbersome, often taking days to complete using a single Pentium IV processor. We have found that simplifying this model for the purposes of computation is convenient. Consider that:

- Individuals of four-dimensional BGA are thought to be relatively rare.
- Calculating three-dimensional BGA proportions is more convenient in a graphical and computational sense than four-model calculations, which take an inordinate amount of computer time and the results from which cannot be collapsed as conveniently onto two-dimensions for visual appreciation.
- Increasing the number of ancestral groups linearly expands the computational resources needed exponentially. By analogy, the number of markers necessary for obtaining good MLE estimates increases rapidly as the number of groups increases. This expectation is derived from observations made from simulations, not a theoretical treatment that would be exceedingly complex. We shall soon see that the relationship between the number of markers used and the standard deviation of an MLE estimate is nonlinear, and it is not counterintuitive that holding the number of markers constant and increasing the number of ancestral groups also would result in a nonlinear effect on the standard deviation of an MLE estimate.

With 71 markers, assuming a three-group admixture model for each sample provides MLEs that are convenient to calculate, taking only an hour or so with a single Pentium IV processor. It so happens that assuming that the ancestry for a sample is no more complex than a mix of three of the four groups also provides better results (lower root-mean-square-error of estimates m) than assuming the sample could be of four-way admixture. This tends to be true when one is using a relatively low number of loci, such as the number required to provide information content for obtaining estimates within 10 percentage points of error, perhaps 50 to 100 high-quality AIMs.

A modified hierarchical version of the MLE algorithm is more convenient computationally. With such a version, we can perform the calculations for four different three-population models rather than a single four-population model. As we will shortly see, if we plot the data from such a simplified calculation scheme on a tetrahedron, and visualize its folding in three-dimensional space to form a three-dimensional pyramid, we can at least visually appreciate the four-dimensional likelihood space inside the tetrahedron—it is the space uniting the likelihood space on the surface of the tetrahedron. Thus, using this scheme, we can conveniently calculate and present the values assuming a three-way population model in such a way that we do not completely lose the four-dimensional part of the space that is part of the answer given that we are using a four-dimensional admixture model. We do lose some information this way, since the four-dimensional likelihood space is not formally quantified; only visualized indirectly and reported percentages will be those of the most likely three-population answer rather than the correct four-population answer. Throughout this text we shall specify whether we have used a three-group or four-group computational model calculating BGA admixture.

Further refinements in the algorithm can be applied to make calculations more computationally efficient. For example, we could first measure the likelihood that the multilocus genotype was derived from an individual of homogeneous affiliation with each of the four groups. Using such a three-way model of admixture, those affiliations with the top three values among the four groups could be selected for the three-way calculation. Proportional affiliation is then determined among these three by measuring the likelihood of all possible m_1, m_2, and m_3 proportions. Scanning all possible m values for this three-way model, and only this three-way model rather than all possible three-way models, we save time that would have been spent looking at m values for less likely three-way models (for which m values would also be less likely). For example, if the top three affiliations for a multilocus genotype were for European, Native American, and East Asian, the likelihood of 100% European, 0% Native American, 0% East Asian is calculated, then the likelihood of 99% European, 1% Native American, 0% East Asian is calculated, and so on until all possible European, Native American, and East Asian proportions are considered. The likelihood of maximum value is selected as the Maximum Likelihood Estimate (MLE). This method ignores all the m values for European, Native American, and sub-Saharan African combinations, or any other combination involving the sub-Saharan African group. Of course, simplification comes at some expense since some individuals may indeed be truly affiliated with all four groups, but if these individuals are rare as they are thought to be, the benefit in terms of convenience could outweigh the penalty, particularly if the laboratory performing the analysis does not have access to large numbers of processors.

When using the four-model calculation, the calculations are performed in the same exact manner, but the likelihood of all possible four-way proportions is determined. These calculations are best performed using clusters of parallel processors. Whether using the three-way or four-way model, when plotting a single MLE on a triangle plot or any other graphical representation, it is important to represent likelihoods that are within a specified confidence interval of the MLE. Throughout this text, we plot the space within which the likelihoods are up to two-fold, five-fold and ten-fold less likely than the MLE, (we could calculate any other X-fold interval and delimit it too), and these proportions are delimited with rings or bars. As an example of the implementation of this algorithm with four groups, this chapter's Appendix shows the use of three AIMs with four populations to calculate the MLE for a sample (ANC30000) of unknown admixture.

It usually is not possible to skip the software development step and simply count alleles in a genotype to reasonably estimate admixture by eye. For each possible combination of ancestry percentages (99% African, 1% European, 0% Native American, 0% Asian; then 98% African, 2% European, 0% Native American, and 0% Asian; then 97%, 3%, 0%, 0%; and so on) the probability distributions of the various possible multilocus genotypes overlap considerably. For this reason, for many genotypes, it is impossible to count Asian markers for example to estimate the right Asian percentage. You might be able to do this for predicting the primary ancestry of individuals of extreme ancestry (as we normally do with linear discriminate method, which works similarly to your eye), but certainly not to estimate admixture. Box 3-B explains why.

Box 3-B

The number of loci favored in one group versus another in a given individual's genotype does not necessarily relate linearly to the likelihood that the individual belongs to that group versus the other. Although linear discriminant procedures treat the problem much like your eyeball would, in reality, it is not possible to count alleles and calculate scores based on fractions of the genotype comprised by certain alleles with certain bias for various groups. The likelihood function addresses this problem. Assume for five haploid markers that Europeans had frequencies for alleles X and Y of 80% X and 20% Y, and Japanese were close, but different: 60% X and 40% Y. An individual with the XXXYY genotype should then be European, right? (since Xs are found more frequently in Europeans). Wrong, and here is why:

> We really are looking only at the likelihood ratio of Bayesian formula. The prior probability of sampling Europeans and Asians and the probability of finding the XXXYY genotype in the total sample are the same for both, and cancel out when making comparisons so we just focus on the p(XXXYY|A) part of the Bayesian equation shown here for the Asian group:

$$P(A|XXXYY) = P(XXXYY|A)P(A)/P(XXXYY)$$

However, a person with XXXYY could be determined to more likely be a full Asian than a full European even though the X allele is more frequently found in Europeans and they have 3 Xs. Why?

$$p(Asian|XXXYY) \text{ would equal } (.6^3{}^*.4^2) = 0.035$$

$$p(European|XXXYY) \text{ would equal } (.8^3{}^*.2^2) = 0.020$$

If we plot the probability of the genotype against the number of Xs in a genotype for *both* Europeans and Asians we would see two offset asymmetrical distributions that seem to be trying to be Gaussian. (You could make the same plots for the probability of a sample being more European or Asian against the number of Xs in the sample.) You would see that the probability of Asian is greater for certain values of X and the probability of European is greater for other values. Consider a genotype of 50 loci, where the frequency of each is 80% X in Europeans and 60% X in Asians, and draw the x-axis from 0 to 50, representing the number of Xs in the genotype. Consider the y-axis to be the probability of ancestry given the number of Xs in the genotype and we plot for both types of ancestry, European and East Asian. With p(X) = 0.8 in Europeans, the mean of the distribution would be at about 40, tapering off gradually to zero at zero Xs in the genotype. For Asians, where p(x) = 0.6, the mean of the distribution would be at about 30, tapering off a bit more steeply down to 0 at 0 x's. At X = 35% (of the multilocus haplotype being X), it so happens that the p(Asian) > p(European) and in fact it is greater for all values of X below this. Above 35%, the p(European) exceeds that of p(Asian). It is not the case that a person with XXXYY is more likely to be European than Asian simply because they have more Xs than Ys, and Xs are found more commonly in Europeans than Asians. In other words, an XXXYY genotype is more like the genotype one would find for the average Asian than European (because 60% of the genotypes are X and the frequency of the X genotype for each locus is 60% in the Asian population), even though the frequency of the X allele is greater in Europeans.

Wang (2003) described the collection of admixture estimation methods as divided among the moment estimators, which use summary statistics on the distribution of estimates about their mean to form generalizations (such as the least squares methods) and the likelihood methods. The advantage of the moment estimators is the ease and convenience of calculation, whereas likelihood methods require substantially more compute time. The disadvantage of the moment estimators is that they are not able to account for genetic drift between parental and sampled populations. According to Wang (2003), the likelihood method is more powerful when the population model is correctly specified but until recently, due to the computational demand of this algorithm, likelihood methods have not been extensively tested under various assumptions like the others.

PARAMETER UNCERTAINTY

The methods described so far have been used principally for the determination of m in populations assuming highly simplified models of population structure. Namely, it is assumed that there was no genetic drift, that no confounding mutations have occurred between the time the historical parental populations existed and modern-day parental samples were obtained, that admixture occurred in sporadic pulses, parental groups were homogeneous populations (no genetic differentiation within parental groups exists), and that allele frequencies are known without sampling error. If any of these assumptions are not correct we introduce error into the estimates. Fortunately, new and relatively advanced methods (namely Bayesian methods) have been developed for accommodating some uncertainty with respect to these assumptions, which require the introduction of more complex admixture models. The determination of the values of M for a more complex population model requires the simultaneous consideration of multiple variables.

At first blush, it may seem that the main sources of error in estimating M are due to

1. Allele frequency estimation errors caused by sampling effect—did we use a large enough parental sample?
2. Error caused by imprecision in the ascription of sample identities—are our parental samples really unadmixed? And are they good representatives of the true parental population—a population that no longer exists?
3. Errors caused by a small number of markers used.

Returning to our analogy of the clothing stores presented at the beginning of the likelihood section, errors 1 and 2 are analogous to errors caused by not perfectly characterizing the frequencies of items of various colors in the stores. For example, did we survey the entire store, or take a sample from only one section of the store? Error 3 is analogous to counting too few clothing items on a person. We will cover the way these three types of errors can be minimized through study-design and planning in the next chapter. For the present discussion, we can reason that higher quality results are obtained when a larger number and/or higher quality of markers is used, but beyond a certain number the increases in precision become too small to be economically justified. In other words, errors caused by a reasonable number of markers used is easily quantifiable and controllable—we know how much error to expect, and we can increase or reduce this error by reducing or increasing the number of markers, respectively. The number we use depends on the nature of the problem (are we attempting to resolve closely related subcontinental populations or continental populations?), and the precision we want to achieve, though there is no right or wrong answer to the question of how precise the estimates should be.

Error associated with allele frequency estimates and sample identities are not as easy to control or quantify, and they often present themselves together to scientists practicing admixture analysis. We could mis-estimate allele frequencies due to having selected too few individuals who are good representatives of the parental group or a large enough group of poor representatives, or a combination of the two. We will soon describe how to calculate the number of samples needed to estimate allele frequencies within certain standard errors, but sometimes it is difficult to obtain suitable numbers of samples for certain populations, which can sometimes run into the several hundreds depending on the allele frequency. In addition, it is always difficult to ensure that the representatives we have chosen are good ones. Aside from the obvious of using huge sample sizes of unquestionable ancestry (an ancestry that can never be guaranteed with certainty), what can we do to minimize the contribution of allele frequency estimation error?

Error in allele frequency estimation caused by sampling effects is relatively easy to handle as long as we are using a fairly large number of loci. We already have touched briefly on the regression methods of estimating admixture, which allow for some accommodation of allele frequency misspecification. Using the m_R estimator, for example, we regress $(q_n - q_a)$ values on $(q_c - q_a)$ values for each locus, and the slope of the line that best fits these estimates is the estimate of M. We will notice a variance of the estimates, and some of this variance will be due to allele frequency errors for certain loci; though the line chosen is the one for which the sum of squares from this line is least, there are still some estimates relatively far from the line and others much closer. The estimates of m_1 in the by now familiar equation

$$P_h = m_1 P_1 + (1 - m_1) P_2$$

depends on the validity of the assumption that the parental allele frequencies are accurately specified in terms of sampling error, and also that the parental populations are perfectly representative of the antecedent populations that did the admixing. Because genetic drift and selection affect loci independently, the quality of the latter assumption can be examined by the similarity in estimates of m_1 obtained from multiple unlinked loci for the same two populations (Blumberg & Hesser 1971; Hertzog & Johnston 1968; Saldanha 1957; Workman 1968; Workman et al. 1963; reviewed in Chakraborty 1986). However, this method is unsatisfactory if there has been admixture in one or more parental populations, since all loci would be affected equally, on average, and simultaneously across the genome and there would exist *bias* in their allele frequency estimates. As for our assumption that the allele frequencies are estimated without error (i.e., no sampling effects), we can again look at the similarity in estimates of m_1 across loci.

If there is considerable variance, this might suggest that the frequencies for some loci were estimated in error more than others. In this case, we can weight the contribution of each locus to the estimate M in a manner that accounts for the precision of each estimate. One way to do this is to use the inverse of the variance as a weighting factor for the contribution of each locus to the estimate of M. The variance of estimated m_1 is obtained from formula (1) as:

$$\sigma^2(m_1) = \left[\sigma^2(p_h) + m^2\sigma^2(p_1) + (1-m)^2\sigma^2(p_2)\right]/(p_1 - p_2)^2 \qquad (3\text{-}32)$$

where the variances are considered as sampling variances (Chakraborty 1986; Kendall & Stewart 1958). Cavalli-Sforza and Bodmer (1971, 1999) suggested that the admixture estimates from multiple loci be weighted by the inverse of their error variance for each locus, summing the result to obtain the weighted m_1. If m_{1i} represents the admixture estimate obtained from the i-th locus, and σ_i^2 the error variance of that estimate obtained from (3-32), for biallelic loci the weighted average m_1 is

$$m_{1avg} = \sum_{i=1}^{L} (m_{1i}/\sigma_i^2) \bigg/ \sum_{i-1}^{L} (1/\sigma_i^2) \qquad (3\text{-}33)$$

There are more sophisticated methods based on more complex statistical ideas that can be used to overcome the various sources of uncertainty such as this. Before we can understand these methods, however, it is necessary to review what admixture is, and what type of admixture model captures most or all of the relevant sources of error. Doing so we can appreciate that there are in fact more than merely three sources of error that we have just described. Indeed, researchers who have studied the error have broken the variance in

M estimates into its various components and concluded that genetic drift has a far larger impact on M estimation error than sampling effects normally do (Wang 2003). Fortunately, there are ways for computationally dealing with all these sources of error so that accurate and precise M values can be estimated.

The ancestry-specific allele frequency of a locus is the probability of that allele given the ancestral state of the locus for a specific gamete. The "for a specific gamete" part of this definition is used because in humans, chromosomes and loci are found in pairs, where one was provided by the mother (who could be of one ancestral background) and one was provided by the father (who could be of another ancestral background). Unlike for a diploid locus, the ancestry of a gamete is by definition unadmixed at each locus. As we have seen, the ancestry-specific allele frequencies are of fundamental importance for estimating admixture. In a laboratory using a model organism (such as mouse) to map genes underlying disease expression, inbred strains are mated in an experimental cross. Here too, the estimation of ancestry admixture is important, but because the mouse strains are controllable (we can keep different strains separate, and inbreed them so that allele frequency differentials for select markers equals 1), the characteristics of the parental populations are easily established and so the allele frequency estimates are known with precision and admixture can be estimated precisely. In natural human populations, which have complicated pasts, there are several conceptual problems with estimating admixture that require advanced statistical treatment to overcome:

1. We cannot know the allele frequencies of the parental groups with precision, since parental groups are not available (they lived in some cases many tens of thousands of years ago).
2. Parental groups cannot be defined precisely; we cannot go back in time and delineate a set of human beings that comprise a specific group. This is a separate problem from not being able to go back in time to genotype parental samples, and calculate their allele frequencies.
3. We cannot know *a priori* the demography and history of modern-day populations from which we *can* sample to estimate parental allele frequencies. Is one ethnic group a better representative than another of a common antecedent population we are calling the parental population?

These problems add to the relatively simple causes of uncertainty that we have already described:

- Allele frequency estimation errors caused by sampling effects
- Error caused by imprecision in the ascription of sample identities (i.e., are we using self-reported ethnicity or does an anthropologist make the decision?)
- Errors caused by a small number of markers used

Uncertainty with respect to these variables creates imprecision in our estimates of ancestry-specific allele frequencies, and hence in our estimates of admixture. This section discusses methods by which we can handle this uncertainty so that we can minimize this imprecision.

Consider the model of admixture presented in Figure 3-2, which was taken from Wang (2003). P_o is the most recent common ancestor to parental groups 1 and 2, of effective population sizes n_1 and n_2, and of allele frequency w. P_1 and P_2 are parental populations, of allele frequency x_1 and x_2, who admix in proportions p_1 and $1 - p_1$ to form the admixed population P_h of allele frequency x_h. The actual populations sampled are P_{1c} (c for current) and P_{2c}, of allele frequencies y_1 and y_2, respectively, and the modern-day admixed population is P_{hc} of allele frequency y_{hc}. These three modern-day populations are characterized by effective population sizes N_1, N_2, and N_h, respectively.

With the admixture methods so far presented, we are using y_1 and y_2 from P_{1c} and P_{2c} to estimate p_1 and $1 - p_1$, but as can be seen in this model of admixture there are other variables that need to be considered. Wang (2003) described a likelihood method for estimating admixture that estimates all of these parameters simultaneously so that a more precise measure of p can be had. His method was applied to populations rather than individuals, and he compared this method to others commonly used, which illustrates the value of considering admixture in more sophisticated terms, as we shall discuss next.

For the likelihood methods we have described so far, we sample from a modern-day population (such as P_{1c}) assumed to be perfectly representative of a parental population (such as P_1) to establish the parental allele frequencies. However, there is usually significant genetic distance between populations P_h and P_{hc}, P_1 and P_{1c}, and P_2 and P_{2c}. This genetic distance could be caused by one or more of the following:

- Genetic drift, creating different allele frequencies between populations due to genetic isolation
- Founder effects, which can alter the representation of alleles in subpopulations relative to their founding populations
- Admixture between subpopulations, which introduce new alleles into the population through gene flow

This genetic distance cannot be known with certainty; in the best-case scenarios with extensive knowledge of population demography and history, it could only be estimated and the quality of our estimates is a function of the quality of our knowledge. This begets a chicken-egg conundrum. How can we measure

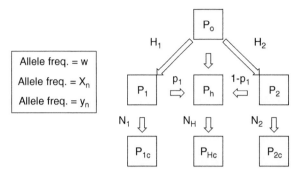

Allele freq. = w
Allele freq. = X_n
Allele freq. = y_n

Figure 3-2

Model of human admixture extending from modern times (P1c, P2c), the parental populations (P1, P2), and the most recent common ancestor to these parental populations (Po). Hybrid populations are shown at both time periods as Phc and Ph, respectively. Effective population sizes for each population are shown next to the arrows and genetic contribution from the parental populations to the hybrid population is indicated with arrows (pI and 1 − pI) as described in the text. Measured or inferred allele frequencies are shown in boxes to the right of the diagram.
Data from J. Wang, "Maximum-likelihood estimation of admixture proportions from genetic data," Genetics, Jun 2003, Vol. 164, No.2, p. 747–765.

admixture for populations characterized by allele frequencies, the calculations of which require a knowledge of admixture? To simplify the problem, until now, we have assumed no admixture, no drift, and no misrepresentation due to founder effects and as a result, there is always a certain amount of uncertainty in estimation of p. Note that this uncertainty influences our estimates of admixture in a manner that is difficult to quantify. The simulations we shall describe in Chapter 5 for calculating bias, root-mean-square estimates (RMSE), and confidence intervals address the imprecision we experience due to the continuous nature of the allele distributions (i.e., the markers are not fixed in one population vs. the others) but do not necessarily address bias caused by the uncertainty in the extent to which P_{hc} represents P_h.

We know from tests on populations of real human beings that by carefully selecting parental samples, based on a knowledge of their history and demographics, we can do a reasonable job of estimating parental allele frequencies as indicated by the coherence of results observed in pedigrees of known structure, concordance of results against expectations for individuals with carefully documented genealogies, and the correlation of certain anthropometric trait values with ancestry admixture (all to be discussed in the coming chapters). By using more sophisticated statistical techniques capable of accommodating this uncertainty, however, we can potentially produce even more accurate estimates of p, and this section discusses some of these techniques.

Before we describe these techniques, let us take a look at the problem with an example. Let P_1 represent Native Americans and P_2 represent Europeans at the time of their admixture to form group P_h Mexican Americans. We obtain modern-day Native American samples from genetically isolated Maya populations in Southern Mexico, modern-day European samples from Mediterranean Europe, and Mexican American samples from admixed populations in urban areas of the United States. In this example, the admixture event took place approximately 400 years ago up until the present, after European expansion to North America. We assume that modern-day Europeans and modern-day Maya are perfectly representative of the European and Native American populations that admixed at this time, and we use them to estimate allele frequencies for P_1, P_2, and P_h so that we can calculate p_1 and $1 - p_1$ (the admixture proportions contributed by each parental population).

However, our assumptions of perfect representation may not be true; P_1 need not equal P_{1c} and P_2 need not equal P_{2c}. For example, if there has been genetic drift between Maya and other Native Americans over the past 400 years (remember, the Maya were selected due to their genetic isolation, which gives us assurance of minimal precontact admixture), or there has been gene-flow into the Maya populations since the time between the admixing event and modern-day samples (they really have not been genetically isolated), their allele frequencies may not be precisely the same as those of Native Americans 400 years ago. If the Maya were a recently founded group from a small segment of the Native American population that existed 400 years ago (which they are not), this too could render our assumptions invalid. Not accounting for the difference in allele frequency between P_{1c} and P_1, and P_{2c} and P_2 runs the risk of introducing bias and error into the estimation of p_1 and $1 - p_1$.

For our other group, Europeans, we know that Mediterranean Europeans are modern-day descendents of Europeans but here we have assumed that Mediterranean Europeans (P_{1c}) are perfectly representative of all the European diaspora who trace their ancestry to the Europeans that lived and admixed with Native Americans 400 years ago, and more to the point, the Europeans that lived and admixed 400 years ago themselves. Indeed, Mediterranean Europeans (depending on the type, such as Greek, Italian, etc.) show different Y-haplogroup ratios than other European populations, possibly due to founder effects, admixture, genetic drift, and patrilocality, or combinations of these processes. If we take Mediterranean Europeans as representative of European parentals, we run the risk that these processes have altered their frequencies in ways not shared by other Europeans, such as Italians, Northern Europeans, or Middle Eastern populations, and therefore are not good representatives of the European parental populations who lived long ago in whom we are interested.

It is possible, even likely, that populations from different European clades harbor slightly different allele frequencies than their European stock. If there has been admixture between Mediterranean Europeans and Africans for example, not shared by other European populations, however minor or ancient, the problem could be even worse and allele frequencies could have been pushed even farther from those of the Europeans who actually admixed with Native Americans 400 years ago. Clearly, the choice of parental samples, and the qualification of their suitability as representatives is crucial for estimating admixture and there is always uncertainty in using allele frequencies from a specific subpopulation as representative of a population that existed some time in the past.

Drift should not influence allele frequency error in the same direction for each locus, since by definition drift is a function of binomial sampling, an inherently random event. Wijsman (1984) proposed a technique for handling this problem in populations, while simultaneously minimizing the sampling effect on the estimates of m. Her technique involves calculating admixture estimates with the normal effects of drift and sampling error, weighting each population by the inverse of the variation from a linear version of formula (3-1) to obtain expected allele frequencies, and then recalculating m by minimizing the squared distance between the estimates of allele frequencies and the expected frequencies that we discussed earlier in this chapter. This method requires an assumption of effective population size Ne, which poses a problem since the uncertainty eliminated in using the method is replaced by uncertainty of the calculated Ne, which is unrealistically assumed to be constant over time. It would be better to estimate all the uncertain parameters simultaneously.

We could approach this problem by attempting to estimate the genetic distance between P_1 and P_{1c}, P_2 and P_{2c} and P_h and P_{hc}, and estimating the parameters that created this distance. The genetic drift can be estimated with knowledge of the mutation rate μ, the effective population size, and the time in terms of the number of generations that have passed between P_1 and P_{1c}, and P_2 and P_{2c}. Assuming we know the mutation rate μ, no back migration and ignoring stochastic processes, the allele frequency at time t for a population is a function of the starting allele frequency, μ and time drift has been taking place (t) using:

$$p_t = p_o \varepsilon^{-\mu t} \qquad (3\text{-}34)$$

The time drift has been taking place, t, can be expressed as a function of the effective population size of the population at time t and the number of generations that have passed (ε, which is expressed as the number of generations separating the modern-day and parental populations):

$$t = \varepsilon/2n_j, \text{ where in our example } j = 1, 2, \text{ and h.} \qquad (3\text{-}35)$$

This allows us to simultaneously estimate the drift that has taken place due to founder effects and time since founding. However, though it may seem simple to discuss in theory, calculating genetic distance for populations that no longer exist is practically difficult—our estimates are only estimates. Using a Bayesian framework, however, we can ensure these estimates are the most likely estimates given the observed data.

The genetic drift, founder effects, and admixture that characterize a population chosen to be representative of a parental population is clearly important for refining the precision of admixture estimation. The other important factor to consider is the actual choice of the P_{1c} and P_{2c} samples themselves. If our choice of P_{nc} populations is a poor one in that the P_{nc} population is admixed or otherwise affiliated with another group to even a small degree, P_{nc} may not be very representative of P_n.

We can estimate the extent to which admixture characterizes our chosen parental representatives (P_{1c}, P_{2c}) using methods not requiring prior knowledge to avoid the chicken-egg conundrum. Pritchard et al.'s (2000) STRUCTURE program and McKeigue's ADMIXMAP programs (more on both later) are two prominent packages capable of individual admixture estimation without a prior knowledge of who the parental populations are and who the hybrid population samples are. Others include GENELAND (Guillot et al., 2005), BAPS/BAPS2 (Corander et al. 2003; Corander et al. 2004), PARTITION (Dawson and Belkhir 2001) (Wu et al. 2007 provides a nice review of the strengths and weaknesses of each of these five programs). STRUCTURE assumes admixture formed by k ancestral subpopulations, which are defined empirically based on genetic distance between samples in a study set provided by the markers used in the analysis. Not specifying the source or parental populations *a priori* in this way, the clustering of samples using a variety of k values can be screened and the k that gives the greatest likelihood, and/or the most distinct subpopulation clusters or the most clearly defined mosaic of ancestry affiliations can be selected as best. If adequate AIMs are used, when these methods are run on samples representative for the major continental populations, they tend to identify clusters at low values of k that fall along continental lines, and clusters at higher values of k that fall along ethnic subdivisions defined by other methods such as Principle Components Analysis (Rosenberg et al. 2002).

Using assumption-free methods to define parental samples allows for an objective definition of not only what parental groups are, and how many parental populations should be considered in the population demography model (which depends on the chronology one wishes to estimate ancestry with respect to), but also the actual samples that should be used as representative of these parental groups. If at k = 4 we see a mosaic of affiliation neatly partitioned along continental lines, and some samples show a low level of admixture but others do not, we can choose to use unadmixed samples defined by the assumption-free approach as our parental representatives P_{nc}. This is the strategy used for the BGA admixture test discussed later in this text, and this process can be made even more objective by using different sets of AIMs for defining parental groups and samples than those used for estimating ancestry and admixture.

Another solution to this problem is to sample across as wide a range of population subgroups or *demes* as possible for the phylogenetic branch corresponding to each parental group. We can do this by using the average allele frequencies over a variety of subpopulations that occupy positions on this branch, which was derived from the P_n parental group. Wahlund's variance of allele frequencies tells us that the average frequency across subpopulations is a good estimate for the allele frequency in the population that gave rise to these subpopulations, and so surveying across Europe and the Middle East would seem to be the best method for estimating European allele frequencies. The risk here though is that the group that admixed with Native Americans, in our example, were themselves unique in some way; not run-of-the-mill Europeans as defined by our samples P_{1c}, and so we are still left with some uncertainty that we must find a way to accommodate if we hope to improve the quality of the admixture estimates. One way to compare alternate parental populations is using Bernstein's formula, plotting the ph − p1 on the y-axis and p2 − p1 on the x-axis, alternating the parental populations (p1 and p2). As described already, the slope of the best-fit line is the admixture proportion and the r^2 is then the measure of the goodness of fit to the parental populations. In this example, one could set up a test of whether modern-day Spanish populations or other European populations are better representatives of the parental population(s) that contributed to the Mexican Americans.

Though consideration of genetic distance between modern-day and parental populations is expected to provide more precise admixture estimates, it also increases the number of variables for the problem, which could have a counter-effect of decreasing statistical precision. Indeed, there are a lot of variables to consider, not just this one, and it would seem logical to use methods that are capable of handling uncertainty in numerous variables such as this one at once. This brings us to Bayesian analysis.

BAYESIAN METHODS FOR ACCOMMODATING PARAMETER UNCERTAINTY

We have just seen that even if we can estimate the drift, founder effects, and admixture for our representatives, we are always left with some uncertainty for any given admixture problem because we cannot be certain whether the admixing population itself was an average of the modern-day subpopulations. Not only can we not precisely define the subpopulation relevant for a given admixture event, or the differences between our chosen representatives and this subpopulation, but we also have uncertainty caused by sampling effects, which we have not even discussed yet. The best solution for dealing with all this uncertainty is to implement methods that are capable of accommodating it—those that can work within the framework of continuously updated posterior probability densities.

PARENTAL REPRESENTATIVES

Recall that our goal is to estimate P_{nc} accurately so that we can estimate admixture between parental groups P_n. We would like to do this in a manner that is irrespective of the particular P_{nc} sample used (i.e., Navaho, Maya, Pima, a mixture of two or all three, etc., as representative for Native Americans, etc.). We necessarily have chosen representatives P_{nc}, which are imperfect representatives, and the problem is that we are not certain exactly what it is about them that makes them so.

Box 3-C

Bayesian Analysis

In Bayesian analysis we attempt to estimate parameters that describe the distribution underlying observed data. We begin with a prior distribution, which can be guesses, likelihoods from observed data, or could even be one value (a uniform distribution) when there is no prior information. We then collect data to produce a distribution of observed values, and this distribution is shaped by parameters of unknown value, which are the values we would like to estimate. Picking different parameter values, we can calculate the likelihood of the observed data conditional on the parameter value. Multiplying this by the prior distribution and normalizing to obtain a unit probability over all possible values, we obtain the posterior distribution. The mode of the distribution can be used as the parameter estimate and probability intervals can be calculated, which are analogous to confidence intervals.

McKeigue (1998) and Hoggart et al. (2004) have developed a Bayesian algorithm for estimating admixture p, which can accommodate uncertainty in the estimation of the parental allele frequencies P_n and the frequencies in their representatives P_{nc}. In Bayesian inference, a posterior probability distribution of estimates is generated from multiplying prior probabilities by likelihood values such as those we have discussed so far (see Box 3-A). We derive the Bayesian posterior probability distribution and x_n for p_n (allele frequencies of the historical parental samples P_n that admixed) from distributions of y_n of p_c (allele frequency of the modern-day sample P_{nc}). Given that there is much uncertainty, which we can call missing data, we specify a full probability model where all the observed and missing data are treated as random variables with appropriate prior distributions assigned as possible and necessary, and the posterior distribution of the missing data given the observed data can be generated by Markov chain simulation (Gelman & Rubin 1992; Gelman et al. 1995; McKeigue et al. 2000). No matter how much genetic distance exists between modern-day samples P_{nc} and the parental sample P_n that lived in some past epoch, we can use the modern-day samples to accurately estimate parental allele frequencies within the framework of Bayesian analysis. To accommodate all this uncertainty caused by genetic distance between the representative samples and the actual parentals, we can:

- Use a variable called a dispersion factor (Hoggart et al. 2004; McKeigue 1998; McKeigue et al. 2000), which helps accommodate uncertainty as to the genetic distance between P_{nc} and P_n.
- Continuously update our allele frequency estimates with data from unadmixed and admixed populations, which helps accommodate uncertainty relating to sampling effects, but also when used with the dispersion factor, relates to differences among modern-day representative subpopulations derived from the parental population.

The dispersion factor is a variable that expresses how the allele frequencies conditional on ancestral state differ between the observed unadmixed samples P_{nc} and observed admixed samples P_{hc}. It can also be employed to measure the difference in allele frequencies, conditional on ancestral state, between observed modern-day representative samples P_{nc} and unobserved, unadmixed parentals P_n. The dispersion factor is related to the Wright F_{ST} statistic by the formula:

$$F_{ST} = 1/(1 + \upsilon) \qquad (3\text{-}36)$$

where (υ) is the dispersion parameter. Larger values of υ imply less dispersion of allele frequencies. If there is no difference in the allele frequencies of a modern-day group P_{nc} from P_n, then the dispersion parameter is high and the dispersion factor is zero. When we select P_{nc} as perfectly representative of P_n as we have done in the preceding sections, we are assuming no dispersion by this criteria. In practice, we do not know what the dispersion parameter is for a given admixture model and it must be determined empirically. It is not necessary to know *a priori* what the dispersion factor is; we can specify a uniform prior distribution. Using a wide range of European samples may require a small or no dispersion factor (potentially only a small difference between P_{nc} and P_n), but using others, such as Greeks or an admixed European population like

Hispanics may require a dispersion model be fit, where less weight (proportional to the inverse of the dispersion parameter) is given to the allele frequency data contributed by the population to our estimates of parental frequencies.

Whether or not a dispersion factor is specified *a priori* in the process may depend on knowledge of the demographics for the populations used, which can be derived from other sorts of data such as archaeological, linguistic, or recorded history. With the ADMIXMAP program (McKeigue 1998), it is possible to test the assumption of no dispersion using a diagnostic test (Hoggart et al. 2004). If there is evidence that a no dispersion model is violated (see Hoggart et al. 2004), we can fit an appropriate dispersion model by systematically comparing the distribution of allele frequencies of the P_{nc} population against the distribution of estimates for P_n obtained from using other populations, since the variance between the distributions should be the same across all loci for each group P_n, and should vary as a function of the genetic distance between P_n and P_{nc} (i.e., the amount of drift experienced by the P_{nc} population, or the degree to which P_{nc} is admixed). In practice, the dispersion parameter v is reparameterized from a consideration of the disparity in allele frequency (μ) between admixed and unadmixed, or otherwise different P_{nc} populations and the allele frequencies for each group $\mathbf{y}^j = \{y_1^j, \ldots, y_{a-1}^j\}$ for groups j. We draw μ from a Dirichlet distribution, $Di(\mu)$, where for loci with a alleles, $\mu = \{\mu_1, \ldots, \mu_a\}$, and reparameterize the dispersion parameter v such that $v = \sum_{i=1}^a \mu_i$. The allele frequencies at each locus are distributed as:

$$p(y^{ij}|\mu, v) = Di(y^{ij}|\mu_1^i, \ldots, \mu_{a-1}^i, v) \qquad (3\text{-}37)$$

where again, i indexes loci and j indexes subpopulations representative of the $j = P_n$ population, or $j = P_{nc}$ subpopulations of unadmixed and admixed samples. Thus, within the Bayesian framework, we have a means by which to objectively qualify and quantify how data from a given population should be used to modify our estimates of parental allele frequencies, whether the new data is derived from admixed samples, or unadmixed samples of significant genetic distance from parental populations. We need not assume that the modern-day sample P_{nc} is the same population, or perfectly representative of the P_n population that actually did the admixing to produce P_{hc}.

Continuously updating parental allele frequencies as more and more admixed and unadmixed samples are analyzed for admixture allows us to counter the problem of picking one P_{nc} representative of P_n, and is better than averaging allele frequencies over P_{nc} populations as we do when we invoke Wahlund's variance of allele frequencies because it is more systematic in its consideration of how each population should contribute. If one population is more admixed than another, its contribution to the estimate of P_n is weighted less. In other words, continuous updating the posterior distribution of y_{nc} from P_{nc} helps minimize the confounding influence of genetic distance between P_n and P_{nc}, which is of different magnitude and caused by different phenomena for different populations n. Working within the framework of Bayesian analysis makes continuous updating straightforward. The posterior distribution of parental allele frequencies determined from one sample can be used as a prior distribution for subsequent analyses of other samples. We can even specify that we have no information on parental allele frequencies, in which the reference prior of 0.5 for each population is specified. This, combined with the likelihood of the data from a survey of unadmixed (or admixed) populations forms a posterior distribution that can be used as a prior for the next.

This cycle can be repeated over and over, continuously building posterior distributions of parental allele frequencies with modes that gradually approach the true value for P_n, especially if a variety of sample types are used. An important feature of this process is that it is not necessary to use unadmixed samples for these cycles—admixed samples can be used in which the contribution to the posterior distribution of parental allele frequencies is weighted by admixture, and conditional on ancestral state for each locus. Since the dispersion factor allows for this uncertainty to be accommodated, it is not necessary that we know beforehand whether a sample is or is not admixed, and we can also use either type of sample that has also experienced substantial genetic drift since the parental P_n sample existed, without being able to precisely parameterize (quantify) this drift.

SAMPLING ERROR

McKeigue et al. (2000) described a score test for the misspecification of ancestry-specific allele frequencies. The test is based on the idea that if no marker frequencies are misspecified, given the ancestral state of a locus, and the level of admixture in a population, variability in the contribution of each locus to our notion of p for the population should be minimal. In other words, all of the loci should say the same thing, if their allele frequencies are correctly specified. McKeigue described a Z statistic (standard normal deviate) for each P_n and a summary chi-square statistic with 2 degrees of freedom for the joint null hypothesis that both ancestry-specific allele frequencies for each biallelic locus are correctly specified. According to him, a positive value of the Z statistic indicates a positive gradient of the log-likelihood function at the null value, and that the most likely value of the ancestry-specific frequency is higher than the value specified. A negative value of the Z statistic indicates a negative gradient of the log-likelihood function at the null value, and that the most likely value of the ancestry-specific frequency is lower than the value specified.

So, given that we can identify which alleles are specified correctly and incorrectly, how do we minimize the contribution of misspecified alleles to our admixture estimation results? When allele frequencies have been specified from small numbers of samples, the same basic principles discussed earlier for handling uncertainty caused by drift and/or admixture in samples representative of parentals apply for uncertainty in allele frequencies caused by sampling effects—that is, the misspecification of allele frequencies caused by sampling error easily is accommodated in the same way (within the framework of Bayesian analysis). With Bayesian analysis, we use a posterior distribution of allele frequencies as a basis upon which to estimate admixture, not a specific and static allele frequency estimate that could be subject to sampling error. We have just described that in Bayesian analysis, the posterior distribution of the current study becomes the prior distribution for the next and the true frequency is approached gradually in an asymptotic fashion as more and more samples are sampled, even if those samples are characterized by varying admixture or drift from the parental population. What this means is that we can start with a very small sample of parental representatives, or even no sample of such representatives, and yet obtain a good posterior distribution of parental allele frequencies from an assay of admixed populations.

We described already that we can start in a state of complete ignorance, specifying uninformative priors to begin with (such as 0.5 for each group), and combine this reference prior with frequencies obtained from our first imperfectly representative sample of parentals P_{1c}, to create a posterior distribution of ancestry-specific allele frequencies. If we start with a sample of perfectly representative samples that is too small, and the ancestry specific allele frequencies misspecified due to sampling effects, we can do the same, improving our estimates of parental allele frequencies by adding the contributions to the posterior distribution from admixed or other samples imperfectly representative of parentals—even experimentals or samples part of the project we developed the method for. The initial posterior distribution of allele frequencies from the small sample—whether small, imperfectly representative, experimental, whatever—is what has value, and the more samples the better the distribution. As with the other parameters in our admixture model in Figure 3-2, the misspecification of allele frequencies caused by sampling effects is not debilitating as long as we employ the power of Bayesian analysis.

The important point of all this is that we can take advantage of the power of Bayesian analysis to accommodate the uncertainty built into our estimation of parental allele frequencies caused by accounting for:

■ Whether or how much drift that has taken place for unadmixed modern-day populations

- The amount of admixture that characterizes the chosen modern-day representatives of the parental populations
- Allele frequency misspecification due to sampling effects
- Allele frequency misspecification due to population heterogeneity in modern descendent populations and inadequate number of subpopulations surveyed

It is therefore not necessary that these parameters be known with ultimate precision (which is clearly impossible), and it is not necessary that we invent a time machine and sample from parentals that may have lived thousands of years ago in order to reliably estimate parental allele frequencies and thus estimate admixture in modern-day individuals or populations.

ASSUMPTIONS ABOUT MARKER LINKAGE AND INTENSITY OF ADMIXTURE IN PARENTS

With the likelihood approach discussed in the earlier part of this text, we have assumed that all the ancestors for an individual were of the same admixture proportions and that all the markers in our battery are unlinked. The first of these assumptions is clearly an oversimplification, as each individual would be expected to have a unique distribution of ancestral identities as a function of their specific heritage. Some individuals with admixture have to go back many generations to find unadmixed ancestors. For example, a Russian with European and East Asian ancestors, that may type with 10% East Asian admixture using a four-population model (West African, European, East Asian, Native American), might have to go back 10,000 years ago or more to identify unadmixed ancestors. If history provided that the admixture occurred during a single pulse, this would correspond to a time when East Asian ancestors (or their progenitors migrating both east and west from Central Asia) constituted 10% of the total number of relatives—the remainder would have been Europeans. Others such as Hispanics need to look back only a few generations since their ancestors were unadmixed (Europeans and Native Americans).

The question of how far back one has to look to find unadmixed ancestors is relevant for determining ancestry admixture because it relates to whether or not, and to what extent, markers are in linkage disequilibrium (LD) with one another. Assuming that the markers are unlinked often produces admixture estimates that show subtle differences from those obtained when assuming the markers may be linked. Not having a knowledge of an individual's genealogy before testing them, we cannot predict the extent to which markers will be in LD beforehand, so we must be able to accommodate uncertainty in the linkage between markers if we hope to improve the precision of admixture estimates.

Ancestry can be distributed among gametes in a salt-and-pepper fashion or in large blocks depending on how many generations have passed since the individual's ancestors were completely unadmixed. It is well known that linkage disequilibrium (LD) extends for greater distances in recently admixed populations than in unadmixed populations or populations that underwent admixture long ago. Within the Bayesian framework, we can survey the chromosomes in order to gain an appreciation for how ancestry is distributed; for example, is the ancestry distributed in a salt-and-pepper fashion or in large blocks? Appreciating the distribution allows for an inference of the number of generations since unadmixed ancestors, since LD between loci decay predictably with generational time.

Hoggart et al. (2004) measures the time since unadmixed ancestors as the variable τ, which is called the sum of intensities parameter. When τ is large for observed gametes, the number of generations that have elapsed since unadmixed ancestors is many and when it is small, the number of generations is few. The variable τ is calculated by comparing the distribution pattern of ancestry states along the gametes with distribution patterns created through a process of simulation. There are two alleles for each binary locus and the stochastic variation of ancestry across loci on each gamete is modeled by k Poisson arrival process, where k is the number of ancestral groups. A Poisson arrival process seeks to sample a large number of possible ways it is possible to get to a fixed result, such as a gamete of specific admixture proportions. The process assumes selection of variables that are distributed as a binomial distribution, such as the alleles of a genetic locus. Simulating a large number of arrivals to a specific admixture percentage for a gamete, we can determine how many of these arrivals show intense patterns of ancestry, and how many show disperse patterns. Doing this for a variety of generational distances, we can form generalizations about how the patterns relate with the number of generations since unadmixed ancestors. If our sample shows a pattern of intensity similar to that provided by the three-generation arrivals, but very different from that provided by two-generation or four-generation arrivals, then the mode of the posterior distribution would be closer to three than to two or four, or any other number and we conclude that there most likely have been three generations since unadmixed ancestors. The rate at which ancestry changes from one population to the next as one travels along the gamete is called the ancestry cross-over rate ρ, and τ is related to the admixture proportion from a given population p based on the following formula:

$$\rho = 2p(1 - p)\tau \qquad (3\text{-}38)$$

PRITCHARD'S STRUCTURE PROGRAM

Pritchard et al. (2000) described a Bayesian clustering approach for the inference of individual population admixture. We have multilocus genotypes for individuals X and we want to know from which populations each individual comes Z and the frequency of alleles in each population P. When there is Hardy-Weinberg Equilibrium and independence between loci (linkage equilibrium), the genotypes we observe in samples are random draws from the

$$\Pr(X|Z, P) \tag{3-39}$$

probability distribution, which is the likelihood of a given multilocus genotype given populations of origin and allele frequencies in those populations. We therefore want to learn about Z and P (which are actually matrices over many populations) given the characteristics of these draws, which is given by the posterior distribution

$$\Pr(Z,P|X) \; \alpha \; \Pr(Z)\Pr(P)\Pr(X|Z,P) \tag{3-40}$$

as a direct expression of Bayes' Theorem. Bayes' Theorem states that the conditional probability of a hypothesis (H) given data (E) is equivalent to the ratio of the unconditional probability of the hypothesis and the data to the unconditional probability of the data:

$$P(H|E) = P(H)P(E|H)/P(E) \tag{3-41}$$

and in this case, we have two data, Z and P. We can expand this model to accommodate another vector for admixture estimates among the populations in Z, and we call this vector Q. Though we cannot calculate the values Z, P, and Q exactly, we can estimate them from a sampling process known as Markov Chain Monte Carlo sampling (MCMC). MCMC is simply a method for taking samples from a stationary distribution, which in this case is Pr (Z,P,Q|X). The larger the number of samples taken, the more closely the distribution of those samples matches the distribution within populations from which they are taken, and it is relatively simple to test the samples for fit and to make sure that samples at different positions of the chain of samples are not too closely tied to one another indicating a deficient distribution within any particular population (i.e., are independent from one another). If they are not we can take a larger number of samples and sample sets.

The method is called a model-based clustering method because it assumes a model by which it classifies samples into groups (either discretely or proportionally).

The model used is the notion that each cluster (population) has its own characteristic set of allele frequencies, and that within populations there is complete Hardy Weinberg Equilibrium for each locus and complete linkage equilibrium among allele at alternate loci (i.e., dependence in state among alleles of loci is taken as an indication or a byproduct of admixture or other population structure). Each allele is considered a random sample from the $Pr(X|Z,P)$ distribution, and over all X in the sample the complete $Pr(X|Z,P)$ distribution is specified. The goal of the program is to create population subdivisions that define groups, within which there exists complete HWE and linkage equilibrium, indicating lack of historical breeding barriers within populations but not between them.

Thus, we are looking for the best way to subdivide the sample (Z), of allele frequencies P, and the best way to partition individuals among these subgroups (Q). For a no-admixture model, we start with the prior probability that for each locus j, the probability of correct assignment is the same for all K groups,

$$Pr(z^{(j)} = k) = 1/K, \tag{3-42}$$

and the prior probability of an allele given population allele frequencies and population of origin is

$$Pr(x_j^{(i,a)} = j|Z,P) = p_{zilj} \tag{3-43}$$

where p_{zilj} is the frequency of allele j at locus l in the population of origin i given the subdivision scheme Z. The probability distribution of an allele in group k is modeled as what is called a Dirichlet distribution, where the frequencies of the alleles in each group sum to 1.

The program uses MCMC to sample

$$P(m) \text{ from } Pr(P|X, Z^{(m-1)}) \tag{3-44}$$

$$Z(m) \text{ from } Pr(Z|X, P^{(m)}). \tag{3-45}$$

We can do this sample after sample, updating the distributions each time, and as higher probabilities arc obtained, new information is learned about the best way to cluster the samples. If enough iterations are used, the output of the MCMC is an approximation of the real distribution $Pr(Z,P|X)$, from which we can infer the best P, and Z for the X in question—in other words, the most likely partitioning of the sample, given the data.

Adding Q to the equation, where each element of Q q_k^i, is the proportion of an individual's genome derived from group k, we expand this to assume an admixture model. The admixture proportions are also modeled as a Dirichlet distribution, since proportions within individuals must add to 1, and the program uses analogous functions to (3-44) and (3-45) incorporating Q, and from the estimated distribution Pr(Z,P,Q|X) we can infer the best P, Z, and Q for the X in question.

The main advantages of this method are:

1. When working with updated distributions of data we gain an ability to logically handle uncertainty in parameter values.

2. We need not define prior population affinities for samples like we do for all the other methods discussed so far (except the Bayesian method of ADMIXMAP). STRUCTURE can actually determine the most likely K and partitioning scheme for this value of K. This is discussed in more detail later in this chapter as well as in later chapters.

IN DEFENSE OF A SIMPLE ADMIXTURE MODEL

So we have learned that when making inference on admixture, there exists uncertainty in the genetic distance between sampled populations and parental populations, caused by genetic drift, founder effects, and admixture. There is a risk that the sampled population is not very representative of the parental population for one or all of these reasons. There is uncertainty caused by sampling error, and caused by assumptions about the intensity of admixture in the parental population samples. Using a simple admixture model that ignores all of this uncertainty would certainly produce biased results, it would seem, but it turns out (as we will soon see in Chapter 6) that estimates obtained using a Bayesian framework and a straightforward likelihood function with static allele frequencies are not grossly (but are subtly) different.

Most of the data we will present in this text will be calculated using the latter, where we have chosen to ignore these uncertainties. Though this introduces bias and error in admixture estimates, as we have seen with the data presented so far, at least for a four-group continental admixture model of the human diaspora, this bias and error is not so debilitating that results do not comport with genealogical expectations, are not associated with anthropometric traits, and do not show patterns of admixture suggested by different methods of analysis such as Y-chromosome and mtDNA haplogroups analysis. Differences in performance between BGA admixture tests based on simple admixture models versus more complex models, as described later, tend to be small. One advantage of using a simpler algorithm is they tend to be computationally more efficient and require less compute time. When typing hundreds of samples, this economy could be important.

Another advantage, which is difficult to quantify or even be certain of, is that the simpler algorithm is tasked with estimating fewer variables and so we might expect fewer opportunities to introduce error or bias (even though the methods are designed to minimize error and bias). Indeed, we have noted what we suspect to be a degree of overclustering when using Bayesian programs whereby the variance within certain populations is decreased, perhaps suppressed, over what is observed with maximum likelihood analyses. For example, with both Bayesian and MLE algorithms we produce results in populations that show a clear trending of non-European admixture as one progresses geographically from Northern Europe to Southeast Asia. The trending is reported by both algorithms but dramatically muted with the Bayesian algorithm. The fact that both algorithm types show the same trending supports the notion that it represents a legitimate element of population structure, which is interesting in light of the fact that Bayesian algorithm disagrees with the MLE algorithm in certain populations part of this pattern. These results are presented in more detail in Chapter 6 and we will discuss them further there, but suffice it to say here that in this case, the MLE algorithm appears to be more sensitive in detecting this particular element of population structure than the Bayesian algorithm.

PRACTICAL CONSIDERATIONS FOR BUILDING AN ANCESTRY ADMIXTURE TEST

Based on our discussion so far, it would seem that the Bayesian methods such as ADMIXMAP or STRUCTURE theoretically would outperform those that rely on simple likelihood calculations where prior frequencies are not used. However, if the goal is to build an admixture assay that runs quickly in a Pentium IV environment, or that is conceptually simple, using formulae that have been written about and studied for decades, there will be a need to identify and characterize parental groups. Either way, there is also the need to identify and characterize AIMs. This section describes some of the practical considerations for doing these things; how many AIMs and samples, which AIMs and samples, and what methods to use for characterization. We will see that for the conservative, who may not want to fully adopt new Bayesian methods just yet, there are ways to incorporate some of the power of Bayesian methods in building a more traditional testing method.

SELECTING PARENTAL GROUPS

The development of an admixture test requires an absolute definition of ancestry, which is difficult to obtain without an admixture test. How could an admixture test then be developed? First, let us reframe the problem for a clearer and deeper understanding.

Most admixture tests are based on discrete population models, and are therefore model-based methods—hence our conceptual problem defining absolute ancestry. For example in Chapters 4 through 6 we discuss one such model, where the parental populations are defined and the relationships between the parental samples and their derived populations are specified or parameterized. We must recognize that there is an inherent problem of selecting and characterizing appropriate parental samples for a given population model that needs to be overcome, because estimates of precision, bias, and indeed, the answers themselves produced by a genomic ancestry test are based on an estimate for absolute ancestry—at least in a statistical genetics sense. If absolute (100%) ancestry is arbitrarily defined from the beginning, then information from it reliably extracted, and error around it can be easily measured, but it is imperative that there be some justification for the estimate if the measures of bias are to have any real meaning. To truly define the bias of the method we must define absolute ancestry—but how do we define absolute ancestry?

We could use subjective criteria, such as whether an individual exhibits any European anthropometric trait character such as lighter skin shade, or types with a different method and different markers to be of at least X% affiliated with a parental group, but these involve the selection of arbitrary definitions that are impossible to justify. If there existed a trait that represented a perfect surrogate for ancestry, then as long as our definition does not bias toward one type of European over another (which could occur if we sampled one part of the European continent but not others), the definition would be reasonable for our purposes. Once the definition of absolute ancestry had been established, it would be possible to then define the error of an admixture method relative to this reference point. We could use our surrogate trait to define parental samples, which we could choose to be those samples that are the least admixed, inferring from quantitative evaluation of the surrogate trait. We could rely on self-reported ancestry for assistance, or more usefully an anthropologist's opinion based on nongenetic factors rather than a single trait. We could take great care in identifying individuals of known and relatively certain recent ancestry on the continental level (obviously) but also on an ethnic level. If we were to collect without regard to ethnicity, for example, we might introduce systematic errors in frequency estimation if there are systematic differences in continental ancestry that are a function of recent population histories, such as differences in recently shared ancestry (i.e., ethnicity) or admixture. The problem is no such trait or traits exist, and using opinion based on nongenetic data poses a circular logic problem that would result in a self-fulfilling prophecy.

There are other problems with this approach as well—recall that the purpose of our effort is to obtain as good an estimate of allele frequency as possible, and though some of the parental samples themselves may be of admixed BGA, such admixture is expected to balance out as long as the sampling is not biased. Since there is no such thing as a completely pure population, how can we be sure there is no such bias—that it really balances out? For example, how can we be sure that using European Americans as a parental group doesn't create a bias by the fact that 90% of the European American population is derived from some location in Europe where admixture is of historical significance (such as perhaps Eastern Europe, where perhaps low levels of East Asian admixture is commonly observed). Clearly, using nongenetic data to define our parental groups poses many problems. Instead we must use a passive, objective analysis of phylogenetic relationships and methods that instruct us as to the best models, parental groups, and representative samples—not the other way around. Fortunately, such methods exist.

For a parental group as we have defined it, absolute ancestry is the average ancestry of the original parental group (that would have inhabited the Earth several thousand years ago), from which all modern-day descendents were derived in whole or in part. Characterizing a parental group with perfect precision is impossible without a time machine, but a genetic law helps us make a good estimate. The allele frequencies in this abstract, indeterminate group is by definition the average multilocus genotype of all diaspora that were derived from this group (weighted for proportional contribution), no matter what its size, how old it is, or what they looked like. This follows from the ideas that:

1. The expected value is the sum of all the possible values times their probability (or frequency). Let us distinguish between an average value and expected value, where the average value is an estimate of an expected value. The expected value is that value we would obtain with an infinitely sized sample and the average value is the value we obtain with the sample size at our disposal. If the sample size is large enough the average value is a good estimate of the expected value.

2. Based on the laws of probability and random sampling, the expected allele frequency in a meta-population differentiating under genetic drift is the same as that for the initial population that gave rise to it (p_0).

3. Idea 2 does not change from generation to generation so that at generational time t, the expected allele frequency $E(p_t)$ across all populations in a branch, or populations across different branches at one point in time is:

$$E(p_t) = p_0 \qquad (3\text{-}46)$$

This is another way to express Wahlund's theorem on the variance of allele frequencies, and is proven mathematically in most population genetics textbooks (e.g., see Section 8.3, Cavalli-Sforza 1999). If we sample from across a broad swath of modern-day ancestral descendents, our average allele frequencies then will give us the values we would find if we went back in time to survey the original parental group that begat the branches.

Given that we can estimate the parental allele frequency considering that of its modern-day descendents, we are still left with the problem of how to define which modern-day groups are bona-fide descendents of the parental groups and which are not. For instance, in estimating the European allele frequency, which modern day groups specifically should be considered? If all the modern-day descendents of a parental population are admixed with other parental populations (such as might be the case if we have surveyed from only part of the modern-day diaspora from a parental group), their unweighted average would not be a good estimate of the initial, parental allele frequency and so we must weight contribution from each with respect to their proportional affiliation with the parental group. So we find ourselves in a position where we not only have to decide which samples are modern-day representatives of historical parental populations, but whether and the extent to which they are admixed with other parental sources as well. A perfect arbiter of modern-day ancestry would seem to be needed to do this, but since we want to do these things in order to create such an arbiter, this presents a sort of chicken-egg problem, for which there are a few solutions:

1. The use of assumption-free approaches to defining parental samples
2. The use of assumption-free approaches to define parental sample affiliation and fractional affiliation with the defined parental groups
3. The use of Bayesian methodologies to handle uncertainty in parental sample ancestry
4. The collection from a wide variety of derived populations for each parental group (i.e., not putting all of your eggs in one lineal basket)

Of these, solution 3 could be implemented after parental samples have been collected, during the calculation of the parameters we need to estimate admixture using the mathematically simpler (non-Bayesian) approaches. The first, second, and third of these solutions can be implemented during the collection of parental samples, when we define what it is our test will report, and so we will describe them in more detail here.

The most common assumption-free, objective approaches are the clustering methods implemented by various authors, most notably methods such as principal components analysis or methods based on genetic distances for clustering such as that of Pritchard et al. (2000). Pritchard's STRUCTURE program, which we presented earlier in this chapter, uses a Bayesian methodology to calculate the likelihood of group membership, or fractional group membership, given a set of genotypes for each individual and starting with an unknown set of groups that best characterize the input sample. The program can be used in a couple of different ways, such as assuming various levels of admixture, in which case the most likely estimate of fractional membership is provided for each sample; or no admixture at all, in which case the samples are binned into the group with which they most likely are part. Conceptually, the Bayesian method is similar to those we have mentioned previously, such as ANOVA, where groups of samples sharing more similarity to one another are identified as clusters. We can think of these clusters as groups of samples for which there exists more between-group than within-group variation. Group membership for samples that do not fit with one of the clusters is apportioned among them by determining the most likely admixture given the genotypes and the frequencies of those genotypes in each sample cluster.

According to Rosenberg et al. (2000),

> (STRUCTURE) assumes a model in which there are K populations (where K may
> be unknown), each of which is characterized by a set of allele frequencies at each locus,
> and assigns individuals... (probabilistically) to a population, or jointly to two
> or more populations if their genotypes indicate that they are admixed.

The number of groups K can be predefined based on other sources of information, such as linguistics, paleoarchaelogy, or genetic (using different markers), or determined empirically by calculating the likelihood of the genotype data (X) for various values of K, and then calculating the posterior probability of K given the observed data (K|X) (Pritchard et al. 2000; also see Pritchard et al. 2001; Rosenberg et al. 2002). The population model used in the first version of STRUCTURE was relatively simple; for example, it assumed no linkage between markers, no genetic drift between parental populations that did the admixing and their modern-day representatives. Falush et al. (2003) describe an extension of this method to account for a slightly more complex model—namely, allowing for correlations between linked loci that arise in admixed populations (admixture linkage disequilibrium). In addition to enhancing the accuracy of admixture estimates, this modification has several advantages, allowing (1) detection of admixture events farther back into the past, (2) inference of the population of origin of chromosomal regions, and (3) more accurate estimates of statistical uncertainty when linked loci are used. They also described a new prior model for the allele frequencies within each population, which allows identification of subtle population subdivisions that were not detectable using the previous method.

When we describe the development of admixture tests in Chapter 4, we will detail how we used STRUCTURE to identify individuals within our parental group who do not exhibit homogeneous ancestry (in other words, they are of detectible, objectively defined admixture and therefore not good parental samples). When we run STRUCTURE with a given battery of markers and a given set of samples, we define the best way to subdivide the sample given the genotype data and so by definition, if our sample is a good one, we establish the most appropriate population model for the markers. In that this particular method does not require a definition of structure *a priori* to infer population structure and assign individual samples to groups, its use obviates the need to define who belongs to a parental group beforehand, or even how many parental groups there are. We can simply analyze a set of samples with a set of markers and choose the most appropriate population model for the samples and markers. When used with samples and markers selected using other prior information, and with respect to a predefined model(s), STRUCTURE allows for a qualification of parental samples from which parameters such as parental allele frequencies can be better estimated. From this data, we can select the best representatives of each subgroup within the best model; that is, those that show the least admixture with other subgroups. There are two choices for proceeding:

- A static approach, where we take allele frequencies provided by our analysis as static allele frequencies for a MLE approach. The output of STRUCTURE could be used to obtain the average allele frequency for the parental representatives, which are actually part of a subgroup defined using other types of prior information, from which we have subtracted the individuals that were determined by STRUCTURE to be affiliated with more than one group (i.e., we get rid of the parental samples that appear in retrospect to not be the best choices). Alternatively, we could avoid eliminating any parental samples and use STRUCTURE's allele frequencies, which account for all individuals and their proportional affiliations in arriving at its estimate of parental allele frequency within the Bayesian framework we have discussed.
- A dynamic approach, whereby we run STRUCTURE each time we perform an analysis, gradually increasing our sample with each run and continuously updating the allele frequency estimates determined in the preceding approach. The advantage here is obviously the increased sample size used to determine the allele frequencies, but the disadvantage is the time it takes to perform each run would increase and quickly become unmanageable without special (and expensive) hardware solutions.

Either way, STRUCTURE (and/or other hypothesis-free clustering-based methods) allows us to successfully overcome most of the chicken-egg problem. However, our chicken-egg problem is not yet completely dismissed. Enjoying the use of hypothesis-free programs like STRUCTURE does not relieve us of the responsibility of selecting our candidate parental samples wisely (solution 3 to the chicken-egg problem). The use of samples from a variety of demes representative of a parental group, discussed earlier, would still be desirable. We can, and indeed should, use other sources of information to help us make our decision; some examples given previously include linguistics (which language family is used by the candidate parentals, and which language did the historical parentals most likely speak), archeology, and historical accounts. However, we must be careful to not impose unwarranted presumptions about what parental samples should be. Since there is a chicken-egg problem to overcome, the answer we get (admixture and population affiliation) is influenced by the definitions of what constitutes a parental sample we ascribe.

To a certain extent, we still have some of the problems we described when using subjective definitions. For example, if we choose modern-day Russians as representative of Europeans, whereby European we mean antecedents for modern-day non-African, non-Asian populations, and if Russians had undergone significant genetic drift or admixture since their derivation from Europeans, our admixture estimates would be biased. Indeed this is not a trivial problem, and as we will see later, many ethnic groups exhibit characteristic continental admixture profiles possibly created by subtle differences in regional subpopulation histories. When defining parental samples is difficult, such as the case where there exist no modern-day populations that have remained isolated since the existence of the antecedent parental population, the best option is to sample from a wide variety of demes derived from the parental group in which we are interested. Though the allele frequencies of any one subgroup may be biased, from Wahlund's variance of allele frequencies, we know that the average frequency of them all should be better estimates, and incorporating assumption-free methods such as STRUCURE will only strengthen the quality of the estimates. We can let assumption-free clustering methods actually define these frequencies, but we must take care to provide a well-qualified sample for this endeavor since the quality of performance for even these methods (relative to the hidden truth) is only as high as the input samples are suited for the problem.

NUMBER OF PARENTAL SAMPLES

The number of samples required to obtain accurate allele frequency estimations has been discussed by many authors before (e.g., Cavalli-Sforza & Bodmer 1999) to be dependent on the expected allele frequency of the alleles. A sample N that provides a good allele frequency estimate provides one that is reasonably similar to that from another sample of the same size, or in other words, an estimate of relatively low variance between samples of size N. In a population of allele frequency p_0, the expected variance in allele frequency estimates between samples of N individuals is given by:

$$V(p_1) = \sigma^2_{p1} = p_0 q_0 / 2N \qquad (3\text{-}47)$$

Here p_1 is the new allele frequency estimate for the sample, obtained by drawing N samples from the population from which p_0 was derived. This equation rearranges to

$$N = p_0 q_0 / 2\sigma^2 \qquad (3\text{-}48)$$

where σ is the standard deviation. For a given σ, N varies with $p_0 q_0$, and since we are concerned with alleles of stark frequency differences among our four continental population groups (i.e., AIMs), the expected product of $p_0 q_0$ will be greater in some populations than others, and the sample required to achieve the same quality of allele frequency estimation will therefore be greater. The typical AIM might have a $p_0 = 0.05$ in one population and $p_0 = 0.45$ in another.

At one extreme, an extremely conservative estimate would require that the standard deviation in allele frequency estimates is within 5% of the *lowest* frequency for any of the ancestral groups. If the lowest $p_0 = 0.05$, the $\sigma = 0.0025$ and the variance (σ^2) = 0.000006. Using (3-48), we obtain N = 15,000 for the average population. However, each of the populations is of starkly different allele frequency, so it is better to perform the calculation using different p_0 values. With $p_0 = 0.05$, this corresponds to an N = 3,800 and when p = 0.45, a huge sample of N=19,800 would be required. Obviously, such sample sizes are not practical for most laboratories.

If we were to strive for a σ that is within 5% of the *expected* frequency across all populations, where the expected $p_0 = 0.25$, the σ would be 0.0125 and for the average locus in the average population we would require N = 600 samples. In a population with $p_0 = 0.05$, we would need N = 152 samples and in a population with $p_0 = 0.45$, we require N = 792 samples.

If we wish to obtain a σ in allele frequency estimates that is within 10% of the expected frequency across all populations, the expected p_0 is 0.25 and the σ is 10% of this or 0.025. Using (3-48) we obtain N = 150 for the average population. Again, each of the populations is of starkly different allele frequency, so it is better to perform the calculation using different p_0 values. For a population with $p_0 = 0.05$, we need N = 38 samples and for a population with $p_0 = 0.45$, we would need N = 198 samples.

At the time we developed our admixture test we in fact had access to 70 samples for each group. Using (3-48), if we assume that the average AIM has an average allele frequency across all groups of 0.25, this number provides a σ of 0.036 for the average AIM. We will use simulations—both simulations of allele frequencies in actual samples and simulation of samples with actual allele frequencies—to show later in Chapter 5 that when many markers are used, and the allele frequency differentials are large, the maximum likelihood method is relatively impervious to allele frequency estimation errors of this magnitude (even systematically biased estimates!). Although it may be desirable to use samples of 20,000 to estimate allele frequencies within less than a fraction of a percent, the simulation results we present later suggest that it is not necessary for the reasonably accurate determination of admixture percentages, and that allele frequency estimates within a couple percent of the true values are suitable.

SELECTING AIMs FROM THE GENOME—HOW MANY ARE NEEDED?

Given that our results are still to a point dependent on the quality of input data, and given (from our discussion immediately preceding) that the type of samples are appropriate, how can we ensure that the number of AIMs we use is adequate for a given admixture problem? Reed showed as early as 1973 that, under a variety of assumptions on selection, mutation, genetic drift, and assuming perfect ancestry information content (i.e., completely private alleles, that each allele at each marker locus in a person is individually classifiable, without error, as coming from either population P_1 of P_2) 18 independent AIMs were necessary to properly estimate admixture proportionality within 95% confidence limits of X. Considering populations P_1 and P_2, the degree to which an individual in P_2 is admixed with genes from P_2 is M_1, and its estimate m_1. If an individual has n ideal AIMs, they have 2n AIM genes or alleles. The variance of m_1, σ_1^2 can be expressed as:

$$\sigma_1^2 = m_1(1 - m_1)/2n \qquad (3\text{-}49)$$

which can be rearranged to express the accuracy of the estimate m_1 by its standard deviation or error, σ, as:

$$n = m_1(1 - m_1)/2\sigma^2 \qquad (3\text{-}50)$$

Clearly, the n required for a given σ will vary with m_1, but not linearly. In other words, for higher admixture levels, fewer markers will be required to achieve an answer of a given certainty than for lower admixture levels. The same is true with how n varies with the standard error. If the standard error of the estimate is to halve, a four-fold increase in the number of markers will be required. For an admixture level of 0.1, we obtain

$$n = 0.045/\sigma^2$$

and for an σ (95% confidence interval of about 0.20), we find that the required number of ideal loci is 18, and for this number of loci, an m_t of 0.10 has a 95% confidence interval of about 0.025 to 0.25 (Pearson, 1962; Reed 1973). For an admixture level of 0.20, and an $\sigma = 0.05$, we find that the required number of ideal loci is 32. If we require only 90% confidence for the same admixture level of 0.20, the required number of ideal loci is only 8 (standard error has doubled, and the number of markers decreased by a factor of 4). To reduce the 95% confidence interval to a width of 0.10, corresponding to an $\sigma = 0.025$, and assuming a $m_i = 0.1$ (a conservative estimate for U.S. African Americans), that no fewer than 72 ideal loci were required as a minimum.

Reed used these types of calculations to show that a large number of ideal loci were required in order to estimate proportional group affiliation with higher confidence. However the genetic distance between the West African and European groups he used is greater than that between other continental ancestry groups (such as East Asians and Native Americans, or East Asians and Europeans), and most loci available for such analysis are not ideal (i.e., not completely private). In other words, for most AIMs, populations share common alleles, and the sharing is more troublesome for other types of comparisons than the West African versus European comparison. Given that only a couple of ideal markers for the West African/European distinction existed at this time, his conclusion was that ''accurate individual estimates of admixture, using genetic markers, are not now possible for this or any other human hybrid population.'' Chakraborty pointed out in 1986 that this is more of a problem for variance in estimates among populations than for variance of estimates within them and said ''. . . even if the loci used are not ideal, and hence, not sufficient to classify unambiguously genes of a hybrid individual into one of the ancestral stocks, there is some merit in computing individual admixture proportions, particularly when a large number of loci for which the allele frequencies in the ancestral populations are substantially different are used.''

Reed performed these types of analyses in 1973, about 28 years prior to the publication of the human genome data, and the automated genotyping platforms, DNA chips, and polymorphism databases like the SNP Consortium and HapMap resources derived from this project. At that time, there were relatively few AIMs (with allele frequency differentials > 0.40) known, and it was not necessarily clear then where the AIMs necessary for accurate within-individual inferences of admixture would come from. Since 2000, however, circumstances have evolved rapidly, as much of this text describes. Using DNA chips of SNPs identified from the human genome project, or massive databases of SNPs from the project with other types of genotyping platforms, it has become possible to efficiently screen the genome and collect large numbers of high-quality (delta > 0.40) AIMs. Today, it is possible to genotype these AIMs conveniently in multiplex to obtain quantifiably accurate inferences of ancestry admixture within individuals, which we spend considerable space discussing later in this text.

Admixture methods and selection criteria for markers quickly evolved to handle more than two groups as we have already described, and to handle the fact that most markers are not ideal, though still contain significant ancestry information. We have described how one can calculate the number of ideal markers necessary for achieving admixture estimates of a particular standard deviation, but most AIMs are not ideal—their allele frequencies are distributed unequally but continuously among populations (to differing degrees depending on the quality of the AIM). These things all conspire to demand more sophisticated methods when selecting and using AIMs for the inference of ancestry and soon, we will discuss some of them, including likelihood functions for calculating the number of nonideal markers necessary to achieve admixture (m) estimates within specific standard error ranges and methods that consider information content across multiple loci simultaneously, in multiple populations. Before we do this, we should review the simplest methods of selecting AIMs, based on the F_{ST} or δ value.

F_{ST} VERSUS δ VALUE

We have previously described the use of the F-statistic (F_{ST}) for measuring population differentiation at the level of the gene. One could argue that the F_{ST} is a better metric for selecting AIMs than the δ value. Fishers F_{ST} is simply a weighted form of the (squared) delta value that corrects the population differences in allele frequency with the actual allele frequencies. Consider two biallelic markers with a delta of 0.4 in Table 3-1.

Here, p_1 and p_2 are the frequency of the first allele in the two different populations; we could just as easily use the q allele, which can be determined by subtracting the values in Table 3-1 from 1. Just considering delta values, the two markers are of equal value or information content because $p_2 - p_1 = 0.4$ for both loci.

	Population 1	Population 2
Locus 1	$p_1 = 0.1$	$p_2 = 0.5$
Locus 2	$p_1 = 0.4$	$p_2 = 0.8$

Table 3-1

Two hypothetical loci in two hypothetical populations exhibiting the same δ value.

Recall that the F_{ST} value is equivalent to the difference in heterozygosity between subpopulations and their metapopulation (all subpopulations together as the total or metapopulation), divided by the heterozygosity of the metapopulation:

$$F_{ST} = (H_T - H_S)/H_T \tag{3-51}$$

If the level of heterozygosity in the total population is high but the level of heterozygosity for the average subpopulation is low, this would indicate extreme genetic differentiation among the subpopulations (i.e., alleles have become fixed or eliminated in many of these subpopulations). When we think of population differences in terms of an excess of homozygosity rather than a deficit of heterozygosity, we have seen that this expression of the F_{ST} is equivalent to:

$$F_{ST} = \sigma^2/(p_{avg}q_{avg}) \tag{3-52}$$

where σ^2 is the variance in gene frequencies among subpopulations, and p_{avg} and q_{avg} are the average p and q for the metapopulation. The variance of gene frequencies between the two populations is equivalent to the delta value between them as:

First line should read:

$$\sigma^2 = \frac{1}{n-1} \sum_{1}^{n} (X_i - X_{avg})^2$$

$$\begin{aligned}
\sigma^2 &= (q_1 - q_{avg})^2 + (q_2 - q_{avg})^2 \\
&= [q_1^2 + q_2^2] - \frac{1}{2}(q_1^2) - q_1q_2 - 1/2(q_2^2) \\
&= \frac{1}{2}(q_1)^2 + \frac{1}{2}(q_2)^2 - q_1q_2 \\
&= \frac{1}{2}(q_1^2 - 2q_1q_2 + q_2^2) \\
&= \frac{1}{2}(q_1 - q_2)^2 \\
&= \frac{1}{2}\delta^2
\end{aligned} \tag{3-53}$$

Thus,

$$\begin{aligned}
F_{ST} &= \sigma^2/(p_{avg}q_{avg}) \\
&= 1/2\delta^2/p_{avg}q_{avg} \\
&= \delta^2/H_T
\end{aligned} \tag{3-54}$$

which shows that given a particular δ value in a two-population problem, the lower the level of heterozygosity (the closer p is to 0), the greater the F_{ST}, and the better the marker serves as an AIM.

In our example in Table 3-1, $p_2 - p_1 = 0.4$ for both loci but the F_{ST} for the first is 0.88 $[.4^2/2(0.1)(0.9)]$ and for the second 0.33 $[(.4^2/2(0.6)(0.4)]$. So we see that two markers with the same delta values do not necessarily have the same F_{ST}. The first marker carries more ancestry informative information, and it has alleles of frequencies closer to fixation/elimination (depending on which allele you are talking about) than the other. As a rule of thumb, given a pair of loci with the same delta values across two populations, the locus with alleles closer to fixation or elimination in one of the populations provides more information than those of more intermediate allele frequencies. Using the F_{ST} as a selection criteria would thus seem better than using the δ.

So, is building a test based on selection of AIMs using the δ value rather than the F_{ST} value illegitimate? The answer is no, because the values are intimately related, but because the F_{ST} is a corrected form of the δ value, it is better to use the F_{ST}. In fact it is notable that as the values of F_{ST} and δ increase, they converge; since in an AIMs panel we want the highest levels, there is even less difference between the measures. In addition, the δ value applies only to binary comparisons, whereas the F_{ST} can be applied for application to larger numbers of subgroups (simultaneously). Consider a four-population problem as we have throughout this text. Using the δ value, we would select markers with the greatest allele frequencies between all possible pairs of populations (there are six of them). Using the F_{ST} to select our panel would maximize the variance of marker frequencies across the four populations (simultaneously).

However, perhaps using the F_{ST} across all populations together is confounding the ancestry information to some extent, since the F_{ST} might be high because only a subset of these six pairwise differences are high. In this case it might be preferable to use a method like the Locus-Specific Branch Length (LSBL), in which the total F_{ST} is geometrically decomposed into population-specific elements (Shriver et al. 2004). Provided there are only three populations, the estimation of LSBL is very straightforward as there is only one topology for the tree relating the three populations (see Figure 3-3). Weir et al. (2005) developed an analogous method they call the population-specific F_{ST}, which is identical for three populations and easier to calculate than the LSBL when the number of populations is greater than three since the topology of the tree relating populations need not be estimated.

For some markers, most of the variance may be due to variation between two specific populations; for others it may be more evenly spread out across all possible pairs of populations. Whether we calculate pairwise F_{ST} or δ, and

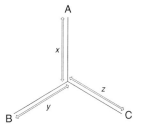

Figure 3-3

Calculating Locus-Specific Branch Length for three populations. To compute LSBL for three populations (A, B, and C), we use the three pairwise delta or F_{ST} levels (d_{AB}, d_{AC}, and d_{BC}). Then using the three equations $[x = (d_{AB} + d_{AC} - d_{BC})/2, y = (d_{AB} + d_{BC} - d_{AC})/2, and z = (d_{AC} + d_{BC} - d_{AB})/2]$ we can express the branch lengths of the phylogenetic tree represented.

assemble our panel of AIMs such that each population pair has a similar summed F_{ST} or δ, or whether we calculate pan-group F_{ST} and juggle the markers such that the value is the same for each population makes little statistical difference in the long run. Our main desire is to balance the discrimination power in all dimensions. One approach to test this balance is to estimate the LSBL (Shriver et al. 2004) or population-specific F_{ST} (Weir et al. 2005) for the marker panel and to ensure that these summed levels are comparable for each of the parental populations under consideration.

In some respects, whether one uses F_{ST} or δ is oftentimes a moot question. The difference in the selection criteria used might be obscured by the variability in the accuracy of allele frequencies in the public databases. Consider Figure 3-4, which for SNPs contributed to the public databases by more than one genome sequencing center, shows the difference in SNP allele frequency obtained for one center versus another. Although the estimates generally agree, there are many examples of extreme difference, and for most SNPs there is some difference. Most selection routines focus on small samples (to cover lots of candidate SNPs), or mine from the publicly available databases. The error built into the public databases is likely to be more significant than defects associated with the use of δ rather than the F_{ST} (or even more complex methods we will discuss later) for selecting AIMs from these databases.

Of course, the goal in building an admixture test is to use so many markers, with so much information content, that estimates approach the true admixture levels as closely as possible within predefined budgetary constraints. At some

Figure 3-4

Comparison of estimated SNP allele frequencies between sites. Several thousand SNPs have been typed by more than one of the six TSC working sites (Celera, Orchid, Washington University, Sanger, Whitehead, and Motorola). For these markers, we show here a plot comparing the allele frequency for one particular population as estimated by one site to the allele frequency for the same population as estimated by an independent site. From the SNP Consortium dataset.

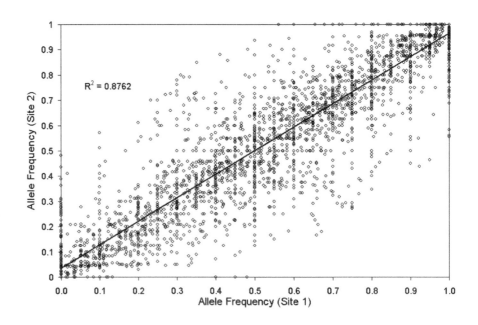

number of markers, the improvement in performance gained by switching SNPs of good power with those of a bit better power will diminish because the greater the markers, the closer the estimate to the true value one will come, regardless of how the markers were selected. Indeed Rosenberg et al. (2002) quite effectively used over 300 markers that were not even selected from the genome based on their Ancestry Information content, and adding another 70 markers to this set that were selected based on their AIM values would be like increasing automobile performance between a 500 horsepower sports car and one with 515 horsepower—would a person sitting in the passenger seat be able to tell the difference, and if so, is it likely to be worth the added expense? Nonetheless, it is difficult to justify not implementing an F_{ST}-based selection criteria when building a test for the inference of individual admixture proportions, because it is just as easy as implementing a δ-based process, but in reality, an admixture test with excellent performance can be constructed using either method.

One case where the subtle differences between F_{ST} and δ methods may be more important is when selecting AIMs for use in gene mapping studies. When conducting an admixture mapping screen, when the ancestral state determined from each marker or small groups of contiguous markers is determined, it is very important to maximize the discrimination power of individual markers. In this case, we need to extract as much information about ancestry from each locus as possible, and it would be preferable to use F_{ST}-based methods since they provide more information about the AIM value of each marker. The problem is that our choice of markers in this case is limited by the availability in certain chromosomal regions, and we may not always have a choice of markers to take. Of course, we could eliminate the selection criteria completely by selecting a much larger number of markers.

Selecting markers irrespective of specific populations is what Rosenberg et al. (2002) did—and this works if one uses enough of them. Theoretically, it doesn't matter whether one uses a small number of powerful markers or a large number of weak markers, and two sets of markers with equivalent summed F_{ST} values will perform the same regardless of how many AIMs constitute the sets. This has been demonstrated with actual data (Rosenberg et al. 2003). However, anecdotal reports have been made of adding ''junk'' markers (with no or little contribution to F_{ST} or δ) to a panel of optimized markers (with high and normalized F_{ST} and δ values across all population pairs), resulting in a degradation of performance, counter to theoretical expectations. These cases may be related to eccentricities associated with the particular software programs used (in this case, STRUCTURE; Frudakis, unpublished observations). Nonetheless, economic factors would seem to promote the development of tests using smaller numbers of highly informative markers rather than larger numbers of less informative markers, and so advancements in the sophistication level in selecting AIMs have been welcomed by the genetics community.

COMPARING THE POWER OF SPECIFIC LOCI FOR SPECIFIC RESOLUTIONS

Reed's estimates assumed that the loci are ideal (i.e., private), and until recently such markers have been exceedingly rare. We can use a simple likelihood ratio to better understand the impact that ancestry information content has on our ability to infer ancestry. Reed, for example, looked at the difference in power imparted to the process of inferring the proper ancestry bin using loci of greater or lesser discrimination power between two groups. For his example, he used European and Africans as the two populations. The A allele of the ABO blood group shows a frequency in Europeans of 0.40 and in Africans of about 0.15. The difference (delta) is thus 0.25, which as one will appreciate further on in this text is a relatively weak AIM. On the other hand, the Fy^a allele of the Duffy blood group system has European and African frequencies of 0.43 and 0.01, respectively, giving a delta of 0.42, which is almost twice that of the A gene. To determine the relative power of the Fy^a allele and the A allele for discriminating between African and European ancestry we calculate the likelihood ratios for each population. We assume that in the entire population $M = 0$; that is, we have two homogeneous populations, one of all gene type 1 and another with no copies of gene type 1:

Duffy Null (p allele is the Fy^a allele):

For the *African group*, $p = 0.43$; therefore, $q = 1 - p = 0.57$. The frequency of Africans who carry the Fy^a (p) allele is therefore

$$f(Afr)_{Fy} = p^2 + 2pq = 0.185 + 0.49 = 0.675$$

For the *European group*, $p = 0.01$, $q = 0.99$, and the frequency of Europeans who carry the Fy^a (p) Allele is

$$f(Eur)_{Fy} = p^2 + 2pq = 0.0199$$

The likelihood that an observed p-allele was derived from an African individual rather than a European individual is therefore:

$$L_{Fy} = f(Afr)_{Fy}/f(Eur)_{Fy} = 0.675/0.0199 = 33.9$$

A Gene (p allele is the A allele)

For the *African group*, $p = 0.40$; therefore, $q = 1 - p = 0.60$. The frequency of Africans who carry the p (A allele) in the African group is therefore

$$f(Afr)_A = p^2 + 2pq = 0.64$$

For the *European group*, $p = 0.15$, $q = 0.85$, and the frequency of A allele-carrying Europeans is

$$f(Eur)_A = p^2 + 2pq = 0.278$$

The likelihood ratio that an observed p-allele was derived from an African individual rather than a European individual is therefore

$$L_A = f(Afr)_A/f(Eur)_A = 0.64/0.278 = 2.30$$

Duffy Null/A Allele comparison

The difference in discrimination power, which we will call the relative information content, $I(\text{group } 1/\text{group } 2)$ is therefore

$$I(Fy/A) = L_{Fy}/L_A = 33.9/2.3 = 15$$

which is to say, the Fy^a allele is 15 times as informative as the ABO group A allele for discriminating African versus European ancestry.

This is the case for individuals of homogeneous African and European ancestry ($m_1 = 1$, $m_2 = 0$ for group 1; $m_1 = 0$, $m_2 = 1$ for group 2 where m_i is the percentage affiliation with population i). Assume Europeans represent group 1 and Africans group 2. If $m_1 = 0.25$ for the African group (AA), then the average African would have 25% European ancestry (similar to the case with U.S. African Americans) and the frequency estimates change for the admixed African group compared to our earlier example where there was no admixture:

Fy^a allele:

$$\begin{aligned} f(AA)_{Fy} &= 0.75(p_{afr}^2 + 2p_{afr}q_{afr}) + 0.25(p_{eur}^2 + 2p_{eur}q_{eur}) \\ &= 0.75(0.185 + 0.49) + 0.25(0.0001 + 0.0198) \\ &= 0.75(0.675) + 0.25(0.0199) = 0.511 \end{aligned}$$

is the frequency of admixed Africans carrying the Fy^a allele, compared to 0.675 when $m_i = 0$, and

$$f(\text{Eur})_{Fy} = p^2 + 2pq = 0.0199$$

is the frequency of Europeans carrying the Fy^a as before. Therefore,

$$L_{Fy} = f(\text{AA})_{Fy}/f(\text{Eur})_{Fy} = 0.511/0.0199 = 25.6$$

A allele:

$$f(\text{AA})_A = 0.75(p_{afr}^2 + 2p_{afr}q_{afr}) + 0.25(p_{Eru}^2 + 2p_{Eur}q_{Eur})$$
$$= 0.75(0.16 + 0.48) + 0.25(0.02 + 0.255)$$
$$= 0.75(.64) + 0.25(0.275) = 0.55$$

compared to 0.640 when $u_i = 0$.

$$f(\text{Eur})_A = p^2 + 2pq = 0.278$$

$$L_A = f(\text{AA})_A/f(\text{Eur})_A = 0.55/.278 = 1.98$$

The relative information content is now

$$I(\text{Fy/A}) = L_{Fy}/L_A = 25.6/1.98 = 12.9,$$

In this case, the Duffy null allele has 13 times as much information as the ABO group A allele for distinguishing admixed African Americans (AA) from Europeans. This is lower than the information for distinguishing between unadmixed Africans and Europeans, and of course, is relevant only if we are binning individuals into ethnic groups of complex composition, rather than determining proportional genomic ancestry. What about in the case of distinguishing between two groups of 50/50 mix? In this case $u_1 = u_2 = 0.5$ for both groups and

Fy allele:

$$f(\text{AA})_{Fy} = 0.50(p_{afr}^2 + 2p_{afr}q_{afr}) + 0.50(p_{Eru}^2 + 2p_{Eur}q_{Eur})$$
$$= 0.50(0.185 + 0.49) + 0.50(0.0001 + 0.02) = 0.347$$

$$f(\text{Eur})_{Fy} = 0.50(p_{afr}^2 + 2p_{afr}q_{afr}) + 0.50(p_{Eru}^2 + 2p_{Eur}q_{Eur})$$
$$= 0.50(0.185 + 0.49) + 0.50(0.0001 + 0.0198) = 0.347$$

$$L_{Fy} = f(\text{AA})_{Fy}/f(\text{Eur})_{Fy} = 0.347/0.347 = 1$$

A allele:

$$f(AA)_A = 0.75(p_{afr}{}^2 + 2p_{afr}q_{afr}) + 0.25(p_{Eru}{}^2 + 2p_{Eur}q_{Eur})$$

$$= 0.50(0.16 + 0.48) + 0.50(0.02 + 0.255) = 0.46$$

$$f(Eur)_A = 0.50(p_{afr}{}^2 + 2p_{afr}q_{afr}) + 0.50(p_{Eru}{}^2 + 2p_{Eur}q_{Eur})$$

$$= 0.50(0.16 + 0.48) + 0.50(0.02 + 0.255) = 0.46$$

$$L_A = f(AA)_A/f(Eur)_A = 0.46/.46 = 1$$

$$I(Fy/A) = L_{Fy}/L_A = 1/1 = 1$$

Thus, both loci have the same information for resolution because the two groups cannot be resolved from one another.

Another way to study the dependence of accurate admixture estimation on allele frequency differential is to compute the variance of the admixture estimate M as a function of the δ. For simplicity, let us consider a diallelic (SNP) system, two populations A and B, and an admixed population where A contributes M genes and B contributes $(1 - M)$ genes. We have seen already that for a given allele of frequency P_A, P_B, in the A population and B population, respectively, the frequency of the allele in the mixed population P_M can be expressed as:

$$P_M = MP_A + (1 - M)P_B = P_B + M(P_A - P_B)$$

and therefore, M can be estimated from Equation (3-3) discussed earlier in this chapter (Bernstein 1931):

$$M = (P_M - P_B)/(P_A - P_B) = (P_M - P_B)/\delta \qquad (3\text{-}55)$$

The determination of the values of M for a more complex population model requires the simultaneous consideration of multiple variables. If we estimate M for each of the loci, we compute the variance of the estimated M values for each of these loci from Equation (3-32) presented earlier:

$$\sigma_M^2 = [1/(P_A - P_B)^2][\sigma_{PM}^2 + M^2\sigma_{PA}^2 + (1 - M)^2\sigma_{PB}^2]$$

where $\sigma_{PM}^2\sigma_{PA}^2\sigma_{PB}^2$ are the variance in allele frequency estimates from the parental admixed A and B populations caused by within-group random sampling error only, and σ_M^2 is the variation in M caused by this allele frequency estimation error (Cavalli-Sforza & Bodmer 1999). From this equation we can clearly see that the variance in M caused by sampling error is inversely related to the square of the allele frequency differential ($P_A - P_B$) among the populations for each locus. The greater the allele frequency differential the lower the variance (and standard deviation) of the estimates. What is important to take from this equation is that the quality of the estimate is dependent on the *square* of the allele frequency differential. Over many loci, an average M can be estimated and the variance of this average M would be dependent on the average allele frequency differential, not linearly but by the square of this difference.

GENOMIC COVERAGE OF AIMs

When selecting AIMs for admixture mapping or even for BGA admixture testing it is useful to assemble as dispersed a collection of markers as possible. At one extreme are the Y and mtDNA tests, which have been instrumental in unraveling population migrations, interactions, expansions, and phylogeographies, but are essentially useless for within-individual ancestry estimation (and hence, for physical profiling or molecular photofitting as well) since they report on only one of the 23 + "chapters" (23 chromosomes + mtDNA) of each individual's genome "book." Autosomal markers of course are inherited (potentially) from all ancestors in the family tree, not just the mother's mother's mother and so on, or father's father's father and so on, and so can report ancestries contributed by all ancestors, not just some of them. Thus, using a small number of AIMs, without good chromosomal coverage, would seem to defeat the purpose of using autosomal markers. One AIM per chromosomal arm would be 46 AIMs, and if using 92 AIMs, each arm should have two AIMs adequately spaced from one another so that they are not in LD.

For the 71 AIM test described later in this text, we gained an average of about three AIMs per chromosome, but since we did not incorporate genomic position as a criteria for selecting AIMs, there is a large variance (some chromosomes have many more and two chromosomes are not covered at all; see Figure 3-5). Augmenting this panel of AIMs to create the 171 AIM BGA test was performed in the same way, gaining better coverage but still leaving chromosome 23 still uncovered (see Figure 3-6). Given an equivalent average or cumulative F_{ST} for each marker among each population pair, the panel would be more informative if the bars for each chromosome in Figure 3-6 were perfectly normalized with respect to chromosome size. Clearly, given that mixed ancestry can be salt-and-peppered throughout the chromosome, the more dense the marker spacing the more precise the estimates of ancestry and the less likely low levels of recent admixture (which might be focused on a small number of chromosomes) would be missed.

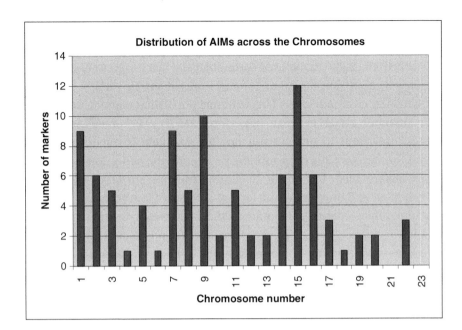

Figure 3-5

Chromosomal distribution of AIMs for the 71 AIM biogeographical ancestry assay described throughout the text. Number of markers is shown on the y-axis for each chromosome, numbered along the x-axis.

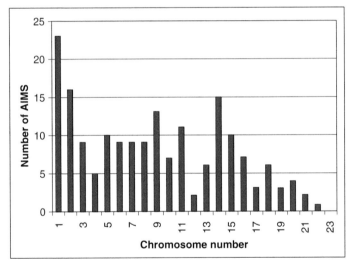

Figure 3-6

Chromosomal distribution of AIMs for the 171 AIM biogeographical ancestry assay described throughout the text. Number of markers is shown on the y-axis for each chromosome, numbered along the x-axis.

MORE ELABORATE METHODS OF SELECTING MARKERS FOR INFORMATION CONTENT

To use the F_{ST} or δ value for cases when using more than two parental groups, one can compute the values for each pairwise group comparison. A panel of markers can then be constructed such as to normalize the F_{ST}, or summed δ value for each type of binary comparison. However, there are better ways to accomplish the task of selecting the optimal panel of markers for a given n-population problem and one of these is the use of Fishers Information. In addition, the methods we have discussed so far are best suited for biallelic markers, but many of the more sophisticated methods can be used for multiallelic markers as well, where some alleles may be quite rare (and therefore measures of δ or F_{ST} subject to considerable estimation error unless very large parental population samples are available).

The utility of a marker as an imperfect AIM ($\delta < 1$), or its "information content," turns out to depend not only on its allele frequency differential between the various pairs of groups, or F_{ST} among the collection of groups, as we have already seen, but on the likelihood function shown as (3-15) earlier in this chapter, and thus the overall level of admixture for the individual and/or population subject of study. Elaborate methods have been designed to maximize the efficiency of selected AIMs given the specific populations they are to be used for, as in an Admixture Mapping study, which tends to focus on recently admixed populations with two main parental groups. Here, the average admixture on the population level can be known, and used along with the δ and allele frequencies to select markers that comprise the most useful and informative panels. Because we are now dealing with three variables (delta, allele frequency, admixture level), information content usually is determined from the likelihood function, which relates all of these variables.

The first derivative of this function is the rate at which the likelihood changes, and the MLE for the level of admixture (m) usually is solved for by setting this equal to zero because it is at the admixture level corresponding to when estimates cease to increase and are about to begin to decrease, that defines the maximum likelihood estimate. The second derivative is the rate at which this rate of change is itself changing. The rate at which the likelihood is changing is an indication of the power or information available for estimating m. Our goal here is to identify markers with allele frequencies and δ values that provide the most information for estimating m, and so we can set the second derivative of the function to zero, and given a known m, solve for these values. If we were screening through large numbers of markers hoping to identify the most informative, we would plug their frequencies and δ values into the log likelihood of Equation (3-15) and all other variables being equal (such as marker coverage), select those that produced the highest likelihood.

SHANNON INFORMATION

One method, the Shannon Information Content, functions in this way. It measures the contribution to the quality of admixture estimates by inserting an allele frequency into the likelihood function. If the population allele frequencies are known, and the level of admixture (m_i) for that population is known, we can determine the Shannon Information Content possessed by the markers for estimating m between two populations by multiplying

- The product of the allele frequency in each population and the level of ancestry in that population, for each allele and each type of ancestry, with
- The logarithm of the same

This is a common metric for measuring information content, and we will refer to equations of this form as fitting the general XlogX information content form. We will run into this form again later. Considering a table with populations as columns and loci as rows, multiplying the product and logarithm for each allele for various ways of grouping the data rows and columns produces:

$$
SIC = -\Sigma(a_{i0} + a_{i1}) \log (a_{i0} + a_{i1}) - \Sigma(a_{0j} + a_{1j}) \log (a_{0j} + a_{1j})
$$
$$
+ \Sigma\Sigma a_{ij} \log (a_{ij}) \tag{3-56}
$$

where for two groups,

$$
a_{00} = (1 - m)(p^A)
$$

$$
a_{01} = mp^B
$$

$$
a_{10} = (1 - m)(1 - p^A)
$$

$$
a_{11} = m(1 - p^B)
$$

and p^A and p^B refer to the frequencies of the allele in two populations A and B. Essentially what this function does is measure whether the information (you can think of this as variability) between groups defined by m (within alleles) and between alleles (within groups defined by m) exceeds the information or variability obtained when not grouping by alleles or groups. When information content is high for partitioning the groups, the variability or information between groups (holding alleles constant) and between alleles (holding groups constant) should exceed the total variability by more than when the information content is low. The use of the XlogX structure for doing this follows from standard information theory, and is a measure of entropy or uncertainty. Here, we measure the degree to which there is shared distinction between two variables; if extreme values of p are associated with extreme values of m, the absolute values for the first two terms of the preceding equation will be more negative, but we take the opposite value with the minus sign ($-$) so the result is that the first two terms $-\Sigma(a_{i0} + a_{i1}) \log (a_{i0} + a_{i1})$ and $-\Sigma(a_{0j} + a_{1j}) \log (a_{0j} + a_{1j})$ become highly positive. The absolute value of the third term is less negative, and so SIC is positive. If less extreme values of p are associated with less extreme values of m, the values for the first two terms of (3-56) will still be more negative than for the last term (i.e., there will be information). The difference just will not be as strong as if the links are between values that are more extreme. As an exercise, you could make up various values of m and p^A, p^B and calculate the terms to visualize this, as we have done in Table 3-2, where the SIC for three scenarios is presented.

Comparing Locus 1 with Locus 2, where the deltas are the same at 0.4, we see that just like with the F_{ST} value, given a common delta, the closer to the state of elimination or fixation in one population, the more information the AIM carries. Comparing Locus 2 with Locus 3, we see that an increased delta value imparts greater information, as expected. When we are building maps of AIMs for purposes that require as complete chromosomal coverage as possible, we can use a more complex form of (3-56) to measure the use amount of information contributed by a new AIM to the collection we had prior to adding this AIM (Smith et al. 2004).

FISCHERIAN INFORMATION CONTENT

Pfaff et al. (2004) also used the second derivative of the likelihood function. They estimated a parameter called Fischer's Information to quantify the information content of markers. The information content of an allele depends on the delta, the frequency in the populations, and the level of admixture. F_{ST} accommodates the first two of these, but not the third. However the method described by Pfaff uses the likelihood function, and so accommodates all three when providing its estimate of information content.

We discussed earlier that by differentiating the likelihood function with respect to admixture m, and solving this differentiated function for zero allows us to determine the most likely level of admixture for an individual or population considering the genotypes of multiple AIMs. Differentiating this first differential creates a second derivative, which is also known as the Fischerian Information function (Pfaff et al. 2004). When the population model involves three populations or more, the Fischerian Information function creates an i X j

Table 3-2

Parameters for three loci for illustrating the SIC metric of ancestry informativeness.

	m	d	p^A	p^B	SIC
Locus 1	0.2	0.4	0.8	0.4	0.027
Locus 2	0.2	0.4	0.5	0.1	0.051
Locus 3	0.2	0.8	0.9	0.1	0.108

matrix where the element in the ith row and ith column is the expected information for the estimate of the parameter m_i, and where the element in the ith row and jth column is the expected shared information for the estimates m_i and m_j. This function is derived from Equation (3-15) as:

$$-E[d^2 \ln L/dm_i dm_j] = \Sigma_g \Sigma_k n_{gk}[\delta_{gik}/p_{gAk}]$$

where

$$p_{gAk} = m_1 p_{g1k} + m_2 p_{g2k} + m_3 p_{g3k} = p_{g3k} + m_1 \delta_{g1k} + m_2 \delta_{g2k}$$

and

$$\Sigma_\varphi m_j = 1$$

The inverse of this matrix, which is calculated across all loci simultaneously, happens to be the expected variance-covariance matrix of estimated admixture proportions. The diagonals of this matrix provides the expected variance in estimates obtained using the battery of AIMs, and the off-diagonals the covariance in estimates for each of the possible population pairs (Pfaff et al. 2004). This matrix is useful for selecting markers because an AIM that is useful for assignment of affiliation to one population is often useful for assignment to other populations as well. Another way to say this is that the information provided by a set of AIMs for one population pair is often not independent of the information provided for others, and the covariance measures the amount of redundancy in the set. Using this matrix function, we can select AIM sets that maximize the values of the diagonals while minimizing the values of the off-diagonals (those that carry the most amount of information with the least amount of redundancy). The Fischer information curve uses the reciprocal of the largest eigenvalue of the information matrix for the maximum likelihood estimation of ancestry coefficients (Gomulkiewicz et al. 1990; Millar 1991). In actually measuring the impact of an AIM or set of AIMs on the precision of admixture estimates, the most direct information desired about the markers as we are screening them is obtained as they are being screened. We can devise algorithms to create new sets of AIMs and test them in this way, one after the other, to identify the optimal set for any given population admixture problem.

INFORMATIVENESS FOR ASSIGNMENT

Rosenberg et al. (2003) described an information-theoretic framework, which also used the concept of entropy described earlier for the Shannon Information

Content method. The basic idea is that knowledge of allele state for an individual gives information about to which group Q the individual should be binned. The entropy or uncertainty Q is greatest with no information about alleles J, and this entropy is reduced due to knowledge of J (the acquisition of information). The difference in information about Q contributed by J is called the mutual information and takes the form

$$I_n(Q;J^L) = H_n(Q) - H_n(Q|J^L)$$

where $H_n(Q)$ is the initial entropy of Q and $H_n(Q|J^l)$ is the conditional entropy of Q given a knowledge of J. It is simply a measure of how much information is gained about the probability of assignment to Q by consideration of an allele J of L loci, as expressed by the reduction in entropy. The calculations are based on the assumption of no admixture (hence the n subscript), which is an appropriate assumption (though not optimal) when considering parental samples as we do when selecting AIMs. Using the XlogX form of expression for data entropy or information,

$$H_n(Q) = -p_i \log(p_i) \qquad (3\text{-}57)$$

and is essentially, in a likelihood framework, equivalent to the expected log likelihood of drawing an allele from an average population, where the allele frequencies are averaged over all groups K. The second term,

$$Hn(Q|J_L) = \sum_{i=1}^{K} (p_{ij}/k) \ \log p_{ij} \qquad (3\text{-}58)$$

where p_{ij} is the frequency of allele j in population i. This term is essentially equivalent in a likelihood framework to the log likelihood of drawing an allele randomly from a defined subset of populations {1, 2, ..., K). According to Rosenberg et al. (2003), these two values $H_n(Q)$ and $Hn(Q|J^L)$ can be added together to obtain the mutual information:

$$I_n(Q;J^L) = \sum_{l=1}^{N} \{-p_i \ \log(p_i) + \sum_{i=1}^{K} (p_{ij}/k) \ \log \ p_{ij}\} \qquad (3\text{-}59)$$

over all loci L, and the information provided by any one locus for assignment into Q is

$$I_n(Q;J=j) = -p_i \ \log(p_i) + \sum_{i=1}^{K} (p_{ij}/k) \ \log \ p_{ij} \qquad (3\text{-}60)$$

By this definition the minimal information provided is encountered when for a fixed p_j the p_{ij} over all subgroups i is the same as p_i, or in other words, there is no difference in the allele frequency within the subgroups compared to the total population. This is very similar to how the F_{ST} works—comparing the frequency of heterozygotes or the variance in subpopulations versus the larger metapopulation. These expressions also obtain from expected log likelihood. According to Rosenberg et al. (2003), the first can be viewed as the log likelihood of drawing the allele from the total population where the frequency is averaged over all K populations, whereas the second is the log likelihood of drawing the allele randomly from the population subsets. If you recall that $\log (X/Y) = \log X - \log Y$, (3-59) and (3-60) can be viewed as the log likelihood of a quotient, where the numerator is the expected log likelihood of the allele being assigned to the subpopulations and the denominator is the expected log likelihood of the allele being assigned to the average population. Equation (3-59) gives the average version of a simpler statistic described by various authors as the Pairwise Kullback-Leibler divergence statistic (Anderson & Thompson 2002; Brenner 1998; Smith et al. 2001). Interested parties can refer to Rosenberg et al. (2003) for details.

Extending (3-59) to diploid genotypes of many independent loci (multilocus genotypes) is accomplished by replacing the frequency (or probability) of the multilocus genotype $\Pi_{l=i}^{L} p_{ij1}^{l} \, p_{ij2}^{l}$ in place of p_{ij}, and the same value summed over all populations K for all possible multilocus genotypes replaced for p_i.

We have previously described that the F_{ST} is related to the δ value through the equation $F = \delta^2/H_T$ (3-54),

$$
\begin{aligned}
F &= \delta^2/H_T \\
&= \delta^2/2p_{avg}(1 - p_{avg}) \\
&= \delta^2/2(p_1 + p_2/2)(1 - (p_1 + p_2/2)) \\
&= \delta^2/((p_1 + p_1)(2 - (p_1 + p_2))) \\
&= \delta^2/\sigma(2 - \sigma)
\end{aligned}
\tag{3-61}
$$

where $\sigma = p1 + p2$. Rosenberg et al. (2003) expressed (3-59) in terms of δ for k = 2 populations, and substituted the value obtained from (3-61) to show the relationship between the $I_n(Q;J^L)$ value and the F_{ST} value for a variety of markers of varying allele frequency sums (σ). The value of a marker in terms of its information content was shown to increase as the σ value approached extremes (very low or very high). As can be appreciated from Table 3-2, for a given F_{ST}, very low or very high values are obtained the more close to fixation or elimination the marker is in one of the populations. Since F_{ST} is directly related to the δ, the same principle was observed expressing $I_n(Q;J^L)$ in terms of δ. As we saw with the Shannon Information Content and the F_{ST}, the closer to fixation or elimination alleles of a locus are, for a given δ, the more ancestry information content. However, the $I_n(Q;J^L)$ varied less across σ values for a fixed F_{ST} than for a fixed δ, indicating that the $I_n(Q;J^L)$ value is more closely related to the F_{ST} than the δ as we would expect since the latter is not corrected for allele frequencies. The advanced consideration of δ and F_{ST} in this paper showed that although it is straightforward to predict F_{ST} from δ, the reverse surprisingly is not true; so even though it is better to use information content statistics, if one were to use the simpler statistics, then the F_{ST} is the preferred statistic.

Rosenberg et al. (2003) also described an Optimal Rate of Correct Assignment (ORCA) method that relies on the assignment of a cost function (such as 0) to the probability of incorrect assignments caused by an allele, otherwise assigning a value of 1. By this method, one attempts to minimize the expected value of loss or maximize the ORCA. This method is expected to perform similarly to the method for calculating $I_n(Q;J^L)$ as described earlier. The informativeness for correct assignment $I_n(Q;J^L)$ and ORCA statistics both assume no admixture, as we normally do when selecting AIMs, but Rosenberg et al. (2003) described an Informativeness for Ancestry Coefficients using an admixture model that represents the assignment of ancestry among the K groups in terms of a K-dimensional vector Q. The derivation of this function is exceedingly complex, and we refer you to Rosenberg et al. (2003) for details, but the performance of the three methods is essentially the same as judged using Spearman Rank Correlation Coefficients on real and bootstrap replicate vectors of AIMs ranked for informativeness. Rosenberg et al. (2003) divided their marker set into subsets including the full set of 377 STRs used in this paper, subsets of the M (where M = various choices of the number of markers) most informative using the I_n criteria and subsets of the M least informative using the I_n criteria. They showed that sets of AIMs selected using the I_n criteria as being the most informative produced admixture estimates far more similar to that obtained with the full set of markers than those with the least I_n information content.

Calculating the number of markers needed to achieve various standard deviation in admixture estimates m, Rosenberg et al. (2003) produced Table 3-3.

Similarity coefficients in admixture estimates for marker subsets with the full set was highest for subsets chosen based on maximizing I_n, reaching 90% with about 80 markers (Rosenberg et al. 2003). In contrast, reaching 90% similarity with subsets of lowest informativeness I_n required about 300 markers. Other indicators of informativeness, such as heterozygosity, were used to select

Standard deviation				
δ	0.2	0.1	0.05	0.01
0.9	4	16	62	1,544
0.8	5	20	79	1,954
0.7	7	26	103	2,552
0.6	9	35	139	3,473
0.5	13	50	200	5,000
0.4	20	79	313	7,813
0.3	35	139	556	13,889
0.2	79	313	1,250	31,250
0.1	313	1,250	5,000	125,000

Table 3-3

Number of markers required to obtain various expected standard deviations in m.
Taken from Rosenberg et al. (2003).

markers, and the subsets selected based on this criteria performed similarly to, but a bit more poorly than, the subset based on maximum I_n (requiring about 100 markers for 90% similarity). Marker subsets chosen randomly performed in between, requiring about 200 to 250 markers to achieve 90% similarity. These results clearly indicate the economic (if not statistical) value of carefully selecting AIMs for the admixture model used.

TYPE OF POLYMORPHISMS

As most others before them concluded, Rosenberg et al. (2003) stated that the markers of highest informativeness for inference of regional ancestry tended to be informative for inference within several geographic regions, indicating that they are relatively old polymorphisms of frequency distributions shaped by genetic drift. But, different types of polymorphisms have different mutation rates, and so the choice of what type of marker to use might best be made based on how deep in the human past one wants to peer. The differential distribution of microsatellite alleles is expected to evolve more quickly given the large number of alleles (many of them relatively uncommon) and the sampling process that underlies genetic drift. Y-chromosome SNP sequences are expected to evolve far more slowly given the lack of recombination and the simple biallelic nature of most SNPs (reduced allelic complexity means rare alleles are less frequent than with microsatellites, which means they start at higher average allele frequencies when populations diverge, taking more generations to become fixed or eliminated). Autosomal SNPs are expected to be somewhere in between.

Microsatellites are loci with varying numbers of repeats, such as an STR (Short Tandem Repeat) locus that might be composed of the basic trinucleotide unit $(GAC)_n$ where n varies from individual to individual. STRs are characterized by many alleles in the worldwide population, sometimes as many as 50, and an STR that is a good AIM shows variation in some specific subset of these alleles as a function of ancestry. SNPs on the other hand are biallelic.

The average (i.e., randomly selected) microsatellite makes a better AIM than the average (i.e., randomly selected) SNP (Rosenberg et al. 2003), a result that is not unexpected given the greater locus diversity for the former. Indeed, the fact that STR alleles can have multiple origins such that they are oftentimes identical in state rather than identical by descent, is a concern for some types of analysis, but does not diminish their greater level of genetic and evolutionary information. However, there are far more SNPs than STRs, and not only can we overcome the identity in state problem with sheer numbers, but the most exceptional SNPs actually make better AIMs than the best microsatellites (Pfaff et al. 2004; Rosenberg et al. 2003). The only practical problem is that these SNPs exceptional in ancestry information content represent a minority of the total number of SNPs, and one has to implement thorough screening protocols to find them. On the other hand, it is a lot harder to estimate allele frequencies for microsatellite markers, since there are so many alleles and usually, some are quite rare. As we have learned so far, accuracy in allele frequency estimates is directly related to the allele frequency, and difficulty in estimating allele frequencies could lead to bias and/or large standard errors in estimating m. Thus, ancestry information content being equal, SNPs usually make better AIMs.

Although Rosenberg (2003) showed that when selected based on information content using I_n (or presumably any other metric, including "quick-and-dirty" F_{ST} methods), the performance of SNP panels was better than for STR panels (presumably because of the availability of exceptional SNPs). He also noted a very interesting difference in the type of information content between STR and SNP markers. As several others before them, Rosenberg et al. (2003) noticed that the microsatellite markers of highest informativeness for inference of between-region ancestry tended to be more informative for inference of within-region ancestry than randomly selected markers, indicating that their AIM character is relatively old (i.e., they were between geographical region AIMs before they became within geographical region AIMs), and that their frequency distributions most likely have been shaped by genetic drift. This same phenomena can be observed with SNP AIMs (Frudakis, unpublished data). Indeed as we will see later, many of the AIMs that are part of the 176 between-continent panel we will be describing in the upcoming chapters as an example of an admixture panel make good within-Europe AIMs as well. Although the allele frequency distributions of SNP and microsatellite loci are subject to forces of different pressure (larger sampling effects for microsatellite loci with the larger numbers of alleles), it is usually not possible to predict which type of polymorphism is best suited for a given population model since probability plays such a large role. For a continental population model (such as West African, European, East Asian, and Native American), one might expect the average microsatellite to show on average greater information content than the average SNPs, but what about the exceptional SNPs?

Since the SNPs have been subject to reduced sampling effects over time and since there are only two alleles, one might expect even the exceptional SNPs to show more residual, relic connections between populations than microsatellites, and data accumulated to date seems to bear this out to a certain extent; the difference between global analyses of admixture using microsatellites and SNPs is described in some detail in Chapter 6. These connections are very useful indicators of population substructure—for example, using basic MLE methods SNP AIM panels reveal that Middle Eastern and other Y-chromosome haplogroup E containing populations show low but reliable levels of Western African affiliation, but other European subpopulations such as Northern Europeans, Russians, and so on, which do not show appreciable levels of haplogroups E, do not (discussed later in Chapter 6). Yet with STR panels no such distinction is observed (Rosenberg et al. 2003).

Although different algorithms were employed for these two studies, the data might suggest that given the same four-population model and similar discrimination power among markers we might seem to gain slightly more information about population structure from a SNP-based test. For some purposes this extra information is useful (such as in forensics, where there are anthropometric trait differences among subgroups), but for others it is not useful and in fact, might be considered to be cumbersome and confusing. So, the answer of what type of marker to choose is best answered by asking what type of answer you want—how old one wants reported ancestry to be, or put another way, how far back along the human evolutionary tree you want to peer.

INTERPRETATION OF ANCESTRY ESTIMATES

In addition to their obvious forensics use, which this book focuses on, individual Biogeographical Ancestry (BGA) or genomic ancestry estimates can and should be very useful for anthropologists and even genealogists for answering certain types of questions. A discussion of the use of BGA data by genealogists in particular allows for the introduction of several important concepts one must have a good grip on in order to responsibly interpret results from an admixture test.

Genealogists collect and interpret information that largely is relevant in a geopolitical context—such as which countries their ancestors are from, what their religions and last names were—rather than in an genetic context, such as whether their ancestors came from one side of an established genetic cline or the other. For this reason a certain amount of discordance between genealogical expectations and actual results is expected (results for genealogists will be discussed in Chapter 5).

Here we focus on the difference between geopolitical and anthropological measures of ancestry. First, we need to understand how ancestry results are obtained. For instance, given the result 90% European, 10% East Asian in a European-American subject, what are the possible explanations for the 10% East Asian result?

There are three main sources for obtaining minority admixture in one's results:

1. Recent exogamous admixture events
2. Ancient admixture coupled with ethnic isolation; that is, affiliations with ethnic groups that are characterized by systematic admixture that took place long ago (perhaps in ancient times)
3. Genetic intermediacy: the primary affiliation is with an isolated group that is genetically intermediate to the four test models (possibly related to the preceding source, but not necessarily)

Each is described in more detail next.

RECENT EXOGAMOUS ADMIXTURE EVENTS

The results of an exogamous event in a recent *genealogical* time frame (arbitrarily we can define this as since 1492 AD, the beginning of the European colonial period). For example, Figure 3-7 shows a hypothetical pedigree that contains a single Chinese great-grandparent and seven European great-

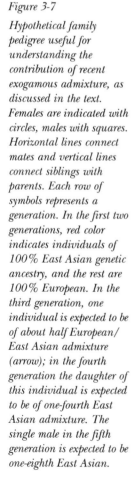

Figure 3-7

Hypothetical family pedigree useful for understanding the contribution of recent exogamous admixture, as discussed in the text. Females are indicated with circles, males with squares. Horizontal lines connect mates and vertical lines connect siblings with parents. Each row of symbols represents a generation. In the first two generations, red color indicates individuals of 100% East Asian genetic ancestry, and the rest are 100% European. In the third generation, one individual is expected to be of about half European/ East Asian admixture (arrow); in the fourth generation the daughter of this individual is expected to be of one-fourth East Asian admixture. The single male in the fifth generation is expected to be one-eighth East Asian.

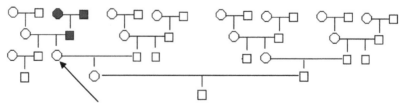

grandparents. In this figure, we show the individuals who are 100% East Asian (Chinese) with red and we are interested in the admixture results for the male (square) at the bottom of the pedigree. A person with a single 100% East Asian great-grandparent and seven 100% European great-grandparents would be expected to have 12.5% (one eighth) East Asian level of genealogical ancestry (as defined at the beginning of this chapter). By the law of genetic assortment, the actual level is expected to fall within a range around 12.5%, with values several percent above and below being possible. The admixture event is fairly recent for the individual in question, and so the block-like character of the DNA would be expected to be relatively high. The grandparent the arrow is pointing to should type as exactly 50%/50% East Asian/European mix since they are completely heterozygous, having one chromosome of European ancestry and one of East Asian ancestry at each locus. The daughter of this person, the subject's mother, is expected to be a 25%/75% East Asian/European mix. In rare circumstances, an admixture test might reliably detect an exogamous event further back than six or seven generations, but usually admixture tests are useful for detecting exogamous events that occurred up to the last four or five generations (i.e., 200 to 250 years back from present time, depending on the average generation length being used and the age of the person in the family being tested).

SOURCES OF ADMIXTURE: ANCIENT ADMIXTURE COUPLED WITH ETHNIC ISOLATION

Admixture could result from ancient sources (in an *anthropological* time frame), which were instrumental in the formation of the population from which the individual comes, and which have been preserved in modern times by endogamous, relatively geographically isolated, close-knit community structure (i.e., ethnicity). For example, we know that modern-day demography is shaped by not only the prehistorical migration process that led to the settlement of the globe, but by admixture between these populations throughout the globe. Given the human population expansions and migrations shown in Figure 3-8, which occurred over many tens of thousands of years, population mixing such as that indicated by arrows in Figure 3-8 could have been instrumental in the formation of some of today's ethnic groups. Specifically, the arrows over western Asia are used to indicate hypothetical and systematic admixture between Europeans and East Asians to form Eastern European ethnic populations, and the possible blending of sub-Saharan African and European populations to form modern-day haplogroup E (YAP+) carrying North African and Middle Eastern ethnic populations (as well as southeastern European populations). We will discuss these hypothetical events, and the evidence for and against them, in Chapter 6, but the point of this section is to illustrate that such admixture could have taken place in pulses long ago, yet still be recognized to this day through application of admixture estimation methods.

Admixture between geographically neighboring groups, after they had developed as distinct groups, usually must be the rule rather than the exception, but just as the news media tends to report bad news because it differs from the norm or expectations, population geneticists tend to focus on the differences between population groups, not their interconnections. Indeed, admixture events are difficult to study. However, there is, from results discussed in this text (in later chapters) and also the work of other population geneticists, budding data for some of these admixture events. For example, preliminary data from application of admixture methods suggests

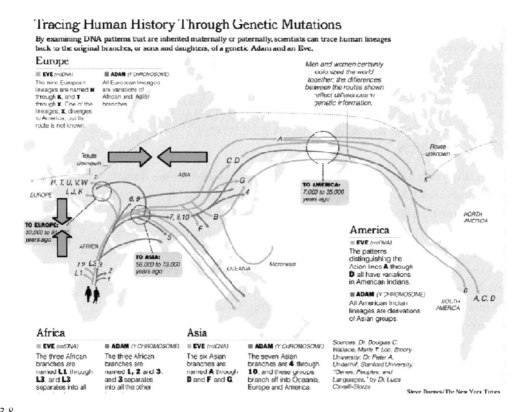

Figure 3-8

Map of human migrations in terms of Y and mtDNA sequences. The history of the Y-chromosome is shown with red lines and the mtDNA genome with blue. Letters indicate haplogroups (see Jobling et al. 2004). Green arrows show gene flow that is known to have or is likely to have occurred in recent history. From Nicholas Wade, "The Human Family Tree," May 2, 2000. Copyright © 2000 by the New York Times Co. Reprinted with permission.

the Mongolian and Hun invasions (or even earlier or possibly later inter-actions) may have contributed to East Asian admixture in Europeans, or the invasions of the African Moors contributed to the sub-Saharan or hap-logroup E admixture in parts of Europe and the Middle East. Currently these hypotheses remain tantalizing mysteries, and we will discuss the evidence for and against them again in Chapter 6, but the point of this section is to illustrate that such admixture could have taken place in pulses long ago, yet still be recognized by a biogeographical ancestry panel if it remained relatively undiluted through ethnic isolation (whether geograph-ical, sexual, or both). When we turn to individual admixture results in ethnic populations later, we will see that East Asian affiliation for certain European ethnicities is not uncommon.

Let us assume for the purposes of the following discussion that Russians exhibit systematic East Asian admixture. Let us also assume that there exists admixture characteristic of East Asians within the Indigenous American population (for which there is actually budding evidence; see Rosenberg et al. 2003; Chapter 6, this text). If so, a person with a fair number of either (1) Native Americans or (2) Russians in their family tree could very well exhibit as much East Asian admixture as an individual with one 100% Chinese great-grandmother and seven other 100% European great-grandparents shown in Figure 3-7! From the small amount of East Asian affiliation, we would not necessarily be able to distinguish between sources (1) and (2). This is shown in Figure 3-9. On the left is a time scale showing the time the most significant migrations occurred. On the right is a very, very large family tree for a single individual, who resides at the bottom apex of the triangle. The tree is so large because it goes back 60,000 years when there existed tens of thousands of ancestors for this person. The time scale for the figure on the left applies to the large family tree on the right as well. The tree is conceptually the same as that shown in Figure 3-7, only much, much larger and without the lines connecting the ancestors (each spot represents an ancestor, but there are so many of them it is not practical to show all the lines connecting the tiny spots). The pink ancestors are Russian, which for the purposes here we will assume are an ethnicity that arose about 18,000 years ago. The red spots represent East Asian ancestors, and we assume here that the average Russian harbors 10% East Asian admixture (no data yet exists to support this—it is an example solely for the purposes of this discussion). The gray spots are the precursors to Russians (whoever they might have been—perhaps continental Europeans 40,000 years ago, and perhaps Eastern Europeans 25,000 years ago).

In this hypothetical example we can see that most of this person's Russian ethnicity comes from the left side of the family tree, which we can denote to be

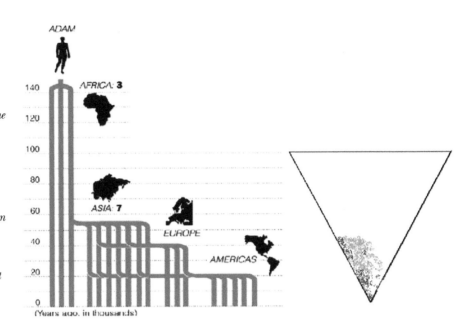

Figure 3-9

Simplified human phylogeography showing the chronology of the spread of populations to various continents. A time scale in hundreds of thousands of years is shown on the left. The inverted triangle on the right is a hypothetical pedigree for a single individual as discussed in the text. Some of the ancestors of this individual are shown as single spots—red for East Asians, grey for Europeans, and pink for the Russian ethnogeographical group—which we are arbitrarily representing here to have become identified as such about 20,000 years ago. From Nicholas Wade, "The Human Family Tree," May 2, 2000. Copyright © 2000 by the New York Times Co. Reprinted with permission.

that part of the tree representative of his father's side of the family. The point of this illustration is that if the average Russian harbors 10% East Asian admixture, and half of the person's family tree is predominantly Russian, the person would be expected to harbor 5% East Asian admixture. East Asian admixture would be significant for this person even though neither their grandmother, grandfather, or any other relative within the past 18,000 years was 100% East Asian. The way to visualize this on the family tree is to count all the red spots and divide them by the total number of spots of any color in the tree for a given slice of time (which would come to about 5%). In other words, relatively homogeneous East Asians represent about 5% of the total number of ancestors for this person, going back to a specific window of time after the formation of the East Asian group. Of course, the family tree for some people involves numerous groups that could be characterized by small degrees of this type of archaic admixture.

Thus, ancestry percentages obtained with the panel of markers we will describe starting in the next chapter (which we call AncestryByDNA) reports genetic affiliations, not necessarily recent genealogical histories the way a genealogist is accustomed to thinking of them. They are anthropology-driven *and* genealogy-driven—a 10% East Asian result for a European population, or person, could have a very simple, genealogical interpretation (i.e., recent ancestors were Chinese) or a more complex, anthropological explanation (i.e., from an ethnic group with a historical connection to East Asians). In the latter case, it may be pointed out that the term East Asian could be part of a nomenclature problem. Both East Asians and Europeans might have been founded by a group that lived in Central Asia for all we know, and so the terms East Asian and European, though picked because they reflect the identity of the modern-day constituents of the parental populations, might be arbitrary and incorrect if we are instead reporting ancestry derived from long-ago populations rather than genetic affiliation with modern-day populations. These points need to be kept in mind when interpreting genome-based admixture tests such as those provided by AncestryByDNA or others (Shriver & Kittles 2004).

SOURCES OF ADMIXTURE: GENETIC INTERMEDIACY

This mechanism is more complex, and we believe it applies only for certain individuals such as South Asian Indians or American Indians—that their primary affiliation is with an isolated group that is genetically intermediate to the four our test is calibrated for. For example, let's say A is East Asians, B are the Maya, Pima, Columbians, and C are the Karitiana and Surul. It is known that group B shows considerable affiliation with groups A and C (East Asian affiliation, our data and Rosenberg et al. 2003), and that group C shares little affiliation with group A. It is also known that group A gave rise to groups B and C. It is unlikely that group A gave rise to group B, who gave rise to group C; rather, currently accepted anthropology states that groups B and C came from group A at different times in different places. In other words, the Maya/Columbians/Pima may be from later waves of migrants across the Bering Strait, and because they were derived from East Asians more recently, they still retain more East Asian sequences than descendents of earlier waves such as the Karitiana and Surul.

Again, we arrive at the nomenclature problem; in this case, a better descriptor of the ancestry common to these groups might be a name for population A, wherever and whenever they lived, such as Central Asian, Prehistoric Asian, or some nongeographical descriptor such as IV or A. Indeed, the use of nongeographical descriptors is the common practice for Y and mtDNA haplogroups identification, precisely for this reason. It is arguable whether this nongeographical convention should apply to SNP-based measures of continental ancestry however, since haplogroups are specific entities whereas SNP allele frequencies are not. We will return again to the nomenclature problem in Chapters 4 and 6.

OBJECTIVE INTERPRETATION

Most genealogists are interested in the type of admixture found in the first source mechanism, where the admixture gives information about geopolitical affiliation of recent ancestors, rather than anthropological information on distant ones. This is because

- There exists little paper data for more distant ancestors compared to recent ones.
- The farther back in time a person goes, the larger number of ancestors exist, making research difficult if not impossible.
- The contribution of single distant ancestors toward a modern person's genetic constitution is less on average per relative than that of more recent ancestors.

Therefore, genealogists tend to seek information such as that which might be produced from the first mechanism—that is, due to recent admixture. For example, if a person is trying to prove or disprove an American Indian rumor or legend in their family, a 10% Native American admixture result would be very useful if it could be assured that the mechanism for this admixture was mechanism 1 not 2. The problem is that it is not possible for us to distinguish between mechanisms 1 and 2 (or even 3 for certain people) simply knowing the results of the BGA test, so our four-group admixture test is not the end to a genealogist's journey, but rather it is another important piece for the genealogist's puzzle that should be considered in light of other evidence.

Genetically, there is no difference in a 10% Native American admixture result from mechanism 1 or 2, except perhaps in the distribution of that ancestry among the chromosomes (recent events showing more clustering of ancestry than older admixture), but to a genealogist the distinction is important. Fortunately for a genealogist, even without a detailed sum-of-intensities analysis to map the ancestry among the chromosomes, the hard work they have gone through can come to bear in a sort of metaanalysis. Indeed, for some genealogists, depending on their family tree, there might exist evidence that strongly suggests whether mechanism 1 or 2, or a combination of the two is the most likely explanation for an admixture result. For example, for a person whose family has paper evidence of a Native American great-grandparent, it would seem logical to assume that a 10% Native American admixture result probably arose due to recent admixture as suggested by the paper genealogy. Alternatively, for a person of genealogically confirmed and homogeneous European ancestry, a 10% East Asian admixture result, assuming it was statistically significant, would probably have arisen due to mechanism 2.

Negative results carry different meaning than positive ones for genealogists. For example, if there is circumstantial, but low quality data suggesting a pure blood West African grandfather, a test result of 100% European would seem to discount that rumor (taking into account the genetic law of independent assortment, of course which would make such a result possible but highly unlikely if the genealogical data in fact was correct, assuming a dense panel of markers was used). However, if a person's family is suspected to have had a Chinese great-grandfather, one cannot *prove* it from a 20% East Asian admixture result, since it is not possible to distinguish mechanism 1 from mechanism 2. Rather, the two data combine to paint a portrait of likelihood—that it is most likely that the person's paper genealogy is correct (depending on the quality of that paper trail of course) and that the suspected Chinese great-grandfather is most likely the source of the 20% East Asian admixture—in other words that it arose from mechanism 1 rather than 2 (though it could also have arisen from a combination of 1 and 2).

In summary, for a genealogist, it is important to bring other information to bear in order to reconstruct the most likely source for an admixture result. In fact, the BGA admixture data serves as an independent clue for someone attempting to reconstruct their family history, and when it is used with genealogical knowledge, the two combine to form evidence that is more powerful than either on their own. This being said, it is important for a genealogist to consider the quality of their paper data when attempting to perform such a metaanalysis. If the paper data is of low quality, the source for a good quality BGA admixture result may be obscured or misconstrued. Fortunately, many genealogists pour endless hours retracing their roots, and for these, it would seem that data quality is as good as is possible to obtain. As with all genealogical research, the preponderance and quality of the evidence should be used to form one's conclusions. Admixture analysis could provide another tool and piece of evidence. But one must always keep in mind the two most common possible sources of minority affiliation: (1) a recent in a genealogical time frame exogamous event, or (2) much older anthropological sources being carried within the population group for many centuries by endogamous customs of that group such as described earlier.

GENETIC MAPPING AND ADMIXTURE

Recall that Linkage Disequilibrium (LD) is a term used to refer to the co-occurrence or dependence between alleles greater than that expected given the propensity of DNA to recombine and shuffle. Thought markers on nonhomologous chromosomes are always distributed at meiosis independently of one another, but the same is not necessarily true for markers on the same chromosome. In fact, neighboring markers on the same chromosome are not always distributed together because there is a mechanism for shuffling the deck, so to speak, with the creation of each new set of gametes. This mechanism normally is referred to as crossing over, caused by the process of recombination. Markers very close to one another, statistically speaking, will have fewer recombination events that take place between them. It is important to remember that LD refers to the statistical connection between markers, not markers and non-DNA entities such as phenotypes, and a typical LD screen attempts to identify markers in linkage disequilibrium with another underlying the determination of a trait called a phenotypically active locus. It is the connection between the marker and the phenotype we measure, even though the dependence between the marker and the phenotype is through the phenotypically active locus.

We call this sought-after dependence (Lm). We call the phenotypically active locus the pseudomarker. The pseudomarker itself may never be observed or measured, and depending on the ascertainment scheme employed for the study and the penetrance of the locus, its character may completely or incompletely correlate with trait value. We call this correlation (Lp). Thus, there are two connections between a measured marker and the trait value. A positive LD result is obtained with an LD scan; what is being measured is an Lm + Lp that is greater than that for the average marker. Measuring the pseudomarker itself will always produce a stronger association than any marker in LD with that pseudomarker.

Since LD (between markers) on average extends for a Kb or so in the human genome, a complete genome scan hoping to achieve full genome coverage would have to incorporate some 2,000,000 markers or so. This is obviously an impediment. Instead most studies aim to survey from several hundred to several thousand markers in order to gain partial LD coverage. Even so, a study of 400 cases and controls, focused on 1,000 SNPs, would require the creation of 400,000 genotypes, which would cost about $200,000 in genotyping reagents using high-throughput methods (i.e., Beckman Ultra-High Throughput genotyping at about $0.40/genotype, as of the date of this writing, not considering labor; Sequenom MassArray at about $0.55/genotype, as of the date of this writing, not considering labor). Various pooling-based strategies have been proposed to offset this cost (Bansal et al. 2002). Though pooling samples introduces allele frequency estimation error, replication of the experiments helps minimize this problem. Of course, other than freeing up resources for including more SNPs in the screen, sample pooling does not really help obtain full genome LD coverage, since coverage is strictly a function of the number of markers used and the extent of LD in the genomes studied, and the difference between the number of markers tested compared to the number required for full genome coverage is often 1,000-fold or more.

LD is neither constant throughout a single genome, nor constant within a genomic region across multiple populations. However, in populations created through a relatively recent process of admixture from two or more others, LD can extend for megabases. The mechanism for creation of this LD is the process of mixing itself, since time is required for chromosomes of different ''types'' to shuffle their sequences through recombination. In recently admixed populations, simply not enough time has passed for this to have taken place. Instead of having ancestral character peppered throughout the genome as we do in an older population created through admixture, in newly admixed populations ancestry tends to exist in large chromosomal blocks. This is most easily visualized through an analogy. Consider a peanut butter and jelly sandwich where jelly is applied to one slice of bread and peanut butter to the other. Opening a sandwich that has recently been prepared one would likely find the jelly stuck mostly to one slice and peanut butter to the other. However, if one let the sandwich sit for a month and then opened it, both slices would likely be a relatively homogeneous mix of peanut butter and jelly.

More formally, we consider the rate of LD decay between two markers as a function of the recombination rate between them and the time in generations. In the absence of selection or other stabilizing effects, the extent of LD between two markers in such a population is an inverse function of the number of generations or time since the admixture event and the recombination rate between them. The rate of LD decay is:

$$\Delta D = (1 - r)^{t}$$

where ΔD is the rate of LD decay, r is the recombination rate that is a function of the chromosomal distance between the markers, and t the number of generations. From this formula, we can see that for closely linked markers, where r might approach 0.00001, it would take 20,000 generations or approximately a couple million years to produce a 10-fold decrease in LD. In contrast, for markers far apart where r = 0.001, then D will decrease by a factor of 10 in 2,000 generations, which corresponds to approximately 60,000 years.

Ancestry as used here is not necessarily restricted to the type of population structure we recognize with our eyes or from census reports, which tends to be on a crude continental level. Rather, ancestry can mean subpopulational (i.e., ethnic) or even finer levels of distinction such as at the clade or even large family groupings. For example, measures of genetic distance and phylogenetic analysis suggest that Aboriginal Australian, Melanesian, and Native American populations are thought to have radiated in a centrifugal fashion from Southeast or Eastern Asian populations from 7,000 to 20,000 years ago (Cavalli-Sforza 1966a, 1966b), and any ethnic population derived from the admixture between East Asians and Melanesians, for example, would be expected to show more than 10-fold LD between markers separated by an $r = 0.001$ or less than in homogeneous East Asians or Melanesians.

Other, finer levels of population structure also exist and are commonly grouped under the term *cryptic population structure*, which is a term used to refer to forms of population structure not easily discernable with the human eye. Clines in gene frequencies are known to exist separating Northern Europeans from Southern, or Mediterranean Europeans. Individuals from populations formed by the blending of different diaspora from a particular ancestral group, such as these two sub-European groups, are expected to show greater LD between markers than within either of the two groups. In nonadmixed populations therefore, LD usually is restricted to regions of DNA quite close to one another. This imposes severe economic impediments on the screening for marker associations with phenotypes. The process of Admixture Mapping (AM) attempts to take advantage of the high level of admixture created in populations of recently admixed ancestry by searching for associations between the ancestry of a marker and phenotype, as opposed to between the genotype of the marker and phenotype. Once such an association has been identified, more traditional mapping approaches usually are applied for finer level screens that serve to hone in on the particular genes underlying the association as well as validate the original admixture mapping result.

APPENDIX

ANCESTRY FREQUENCY TABLE

African frequencies				
SNP1	G	0.8	T	0.2
SNP2	T	0.7	A	0.3
SNP3	C	1.0		

European frequencies				
SNP1	G	0.9	T	0.1
SNP2	T	0.7	A	0.3
SNP3	C	0.8	T	0.2

Native American (NA) frequencies				
SNP1	G	0.6	T	0.4
SNP2	T	0.5	A	0.5
SNP3	C	0.7	T	0.3

East Asian (EA) frequencies				
SNP1	G	0.7	T	0.3
SNP2	T	0.9	A	0.1
SNP3	C	0.9	T	0.1

Example			
SAMPLEID	SNP1	SNP2	SNP3
ANC30000	GT	TT	CT

We first select the three most appropriate populations for the sample; that is, the three groups with which the sample is most likely affiliated assuming the sample is of homogeneous ancestry. We next obtain the mixing proportions by scanning the parameter space and selecting that which has the maximum likelihood value.

Pick three best populations.

African (Assume blind sample 100% African):

SNP1 Alleles: G, T

G allele frequency in African P(G)	$= 0.8$
T allele frequency in African P(T)	$= 0.2$
Expected genotype value for SNP1	$= \log(2*P(G)*P(T))$
	$= \log(2*0.8*0.2)$
	$= -0.4948$

SNP2 Alleles: T, T

T allele frequency in African P(T)	$= 0.7$
Expected genotype value for SNP2	$= \log(P(T)*P(T))$
	$= \log(0.7*0.7)$
	$= -0.3098$

SNP3 Alleles: C, T

C allele frequency in African P(C)	$= 0.9999$
T allele frequency in African P(T)	$= 0.0001$
Expected genotype value for SNP3	$= \log(2*P(C)*P(T))$
	$= \log(2*0.9999*0.0001)$
	$= -3.6990$
Likelihood for African	$= -0.4948 - 0.3098 - 3.6990$
	$= -4.5036$

European (Assume blind sample 100% European):

SNP1 Alleles: G, T

G allele frequency in European	$P(G) = 0.9$
T allele frequency in European	$P(T) = 0.1$
Expected genotype value for SNP1	$= \log(2*P(G)*P(T))$
	$= \log(2*0.9*0.1)$
	$= -0.7447$

European (Assume blind sample 100% European):

SNP2 Alleles: T, T

T allele frequency in European	$P(T) = 0.7$
Expected genotype value for SNP2	$= \log(P(T)^*P(T))$
	$= \log(0.7^*0.7)$
	$= -0.3098$

SNP3 Alleles: C, T

C allele frequency in European	$P(C) = 0.8$
T allele frequency in European	$P(T) = 0.2$
Expected genotype value for SNP3	$= \log(2^*P(C)^*P(T))$
	$= \log(2^*0.8^*0.2)$
	$= -0.4948$
Likelihood for European	$= -0.7447 - 0.3098 - 0.4948$
	$= -1.5493$

Native American (Assume blind sample 100% NA):

SNP1 Alleles: G, T

G allele frequency in NA	$P(G) = 0.6$
T allele frequency in NA	$P(T) = 0.4$
Expected genotype value for SNP1	$= \log(2^*P(G)^*P(T))$
	$= \log(2^*0.6^*0.4)$
	$= -0.3187$

SNP2 Alleles: T, T

T allele frequency in NA	$P(T) = 0.5$
Expected genotype value for SNP2	$= \log(P(T)^*P(T))$
	$= \log(0.5^*0.5)$
	$= -0.6020$

SNP3 Alleles: C, T

C allele frequency in NA	$P(C) = 0.7$
T allele frequency in NA	$P(T) = 0.3$

(continues)

Native American (Assume blind sample 100% NA):	
Expected genotype value for SNP3	$= \log(2^*P(C)^*P(T))$
	$= \log(2^*0.7^*0.3)$
	$= -0.3767$
Likelihood for Native American	$= -0.3187 - 0.6020 - 0.3767$
	$= -1.2974$

East Asian (Assume blind sample 100% ME):	
SNP1 Alleles: G, T	
G allele frequency in ME	$P(G) = 0.7$
T allele frequency in ME	$P(T) = 0.3$
Expected genotype value for SNP1	$= \log(2^*P(G)^*P(T))$
	$= \log(2^*0.7^*0.3)$
	$= -0.3767$
SNP2 Alleles: T, T	
T allele frequency in ME	$P(T) = 0.9$
Expected genotype value for SNP2	$= \log(P(T)^*P(T))$
	$= \log(0.9^*0.9)$
	$= -0.0915$
SNP3 Alleles: C, T	
C allele frequency in ME	$P(C) = 0.9$
T allele frequency in ME	$P(T) = 0.1$
Expected genotype value for SNP3	$= \log(2^*P(C)^*P(T))$
	$= \log(2^*0.9^*0.1)$
	$= -0.7447$
Likelihood for East Asian	$= -0.3767 - 0.0915 - 0.7447$
	$= -1.2129$
Likelihood value for African	$= -4.5036$
Likelihood value for European	$= -1.5493$
Likelihood value for Native American	$= -1.2974$
Likelihood value for East Asian	$= -1.2129$

In this case we drop African and consider the other three for proportions.

Maximum Likelihood Value

Now we begin giving values to the unknown parameters, where I = European, J=Native American, K = East Asian, and I + J + K always equals 1:

---

i = 0;   j = 0;   k=1

{

SNP1 Alleles: G, T.

G allele frequency in European	$P(G,1) = 0.9$
G allele frequency in NA	$P(G,2) = 0.6$
G allele frequency in ME	$P(G,3) = 0.7$
Allele1 Estimated Frequency (A1EF)	$= I^*P(G,1) + J^*P(G,2) + K^*P(G,3)$
	$= 0^*0.9 + 0^*0.6 + 1^*0.7$
	$= 0.7$
T allele frequency in European	$P(T,1) = 0.1$
T allele frequency in NA	$P(T,2) = 0.4$
T allele frequency in ME	$P(T,3) = 0.3$
Allele 2 Estimated Frequency (A2EF)	$= I^*P(T,1) + J^*P(T,2) + K^*P(T,3)$
	$= 0^*0.1 + 0^*0.4 + 1^*0.3$
	$= 0.3$
Expected genotype value for SNP1	$= \log (2^*A1EF^*A2EF)$
	$= \log (2^*0.7^*0.3);$
	$= -0.3767$

SNP2 Alleles: T, T.

T allele frequency in European	$P(T,1) = 0.7$
T allele frequency in NA	$P(T,2) = 0.5$
T allele frequency in ME	$P(T,3) = 0.9$
Allele 1 Estimated Frequency (A1EF)	$= I^*P(T,1) + J^*P(T,2) + K^*P(T,3)$
	$= 0^*0.7 + 0^*0.5 + 1^*0.9$
	$= 0.9$
Allele 2 Estimated Frequency (A2EF)	$= 0^*0.3 + 0^*0.5 + 1^*0.1$
	$= 0.1$

(continues)

Expected genotype value for SNP2 (EGV2) $= \log(A1EF^*A2EF)$

$= \log(0.9^*0.1);$

$= -0.0915$

SNP3 Alleles: C, T.

C allele frequency in European $P(C, 1) = 0.8$

C allele frequency in NA $P(C, 2) = 0.7$

C allele frequency in ME $P(C, 3) = 0.9$

Allele 1 Estimated Frequency (A1EF) $= I^*P(C,1) + J^*P(C,2) + K^*P(C,3)$

$= 0^*0.8 + 0^*0.7 + 1^*0.9$

$= 0.9$

T allele frequency in European $P(T,1) = 0.2$

T allele frequency in NA $P(T,2) = 0.3$

T allele frequency in ME $P(T,3) = 0.1$

Allele 2 Estimated Frequency (A1EF) $= I^*P(T,1) + J^*P(T,2) + K^*P(T,3)$

$= 0^*0.2 + 0^*0.3 + 1^*0.1$

$= 0.1$

Expected genotype value for SNP3 (EGV3) $= \log(2^*A1EF^*A2EF)$

$= \log(2^*0.9^*0.1);$

$= -0.7447$

Likelihood value for unknown parameters $= EGV1+EGV2+EGV3$

$= -0.3767 - 0.0915 - 0.7447$

$= -1.2129$

for European $= 0$; NA $= 0$; Middle East $= 1$; likelihood value is -1.2129

}

Repeat this loop for all possible combinations.

0.0, 0.0, 1.0	−1.2129
1.0, 0.0, 0.0,	
0.0, 1.0, 0.0,	
0.1, 0.0, 0.9,	
0.1, 0.1, 0.8,	
0.1, 0.2, 0.7,	
0.1, 0.3, 0.6,	
0.1, 0.4, 0.5	
and so on.	

Obtain the maximum likelihood value, and the corresponding values are the MLE of ancestry proportions.

CHAPTER 4

BIOGEOGRAPHICAL ANCESTRY ADMIXTURE ESTIMATION— PRACTICALITY AND APPLICATION

All models are wrong; some models are useful.

—*George Box, British Statistician*

We have discussed the conceptual foundation and theory of genomic ancestry estimation, and the technical details to consider when developing a panel of markers for the estimation of admixture (which we will refer to as a *test*). A good way to discuss the practical and applied aspects of developing an admixture panel is to illustrate with examples, and in this section we will describe the development of two such tests for estimating continental ancestry proportions (one with 71 and one with 171 AIMs, the latter of which will also be referred to by two product names, DNAWitness™ 2.5, or Ancestry By DNA 2.5). Details will be presented on our selection of parental samples, and our genetic marker screen for AIMs. Next, we will characterize the performance of this panel, present maximum likelihood estimates of continental BGA admixture proportions in various ethnographically defined human populations, and discuss whether and how data produced from population subgroups might allow us to deconvolute our complex history as a species and make inferences about physical appearance (and biomedical phenotypes) from such DNA-based results.

We will show that when markers are selected carefully from the genome, even a relatively small set provides good estimates of admixture, but as expected, larger numbers of markers provide better estimates. BGA admixture proportions are measurable with significantly improved precision; accuracy and reliability compared to previously described methods based on binning (i.e., Frudakis et al. 2003; Shriver et al. 1997). When using a well-qualified set of markers for continental ancestry estimation, results obtained from parental and hybrid populations generally make sense in terms of geography and history, and the majority BGA affiliations obtained are generally consistent with self-identified categorical groupings. These results show that even a relatively small panel of markers could be helpful for epidemiological study design by helping eliminate or reduce the detrimental effects cryptic or micropopulation substructure in genotype/phenotype association studies or prove useful for forensic scientists wishing to learn about possible physical appearance from DNA. Many of the latter scientists currently use imprecise and sometimes inaccurate means by which to infer race or other metapopulation affiliations.

THE DISTRIBUTION OF HUMAN GENETIC VARIABILITY AND CHOICE OF POPULATION MODEL

Both of the tests we later apply in this text to illustrate basic molecular photofitting principles use a four-group human population model. We are first concerned with identifying the simplest population model with unadmixed parental representatives (from major extant populations), and for which the apportionment of genetic diversity mirrors the apportionment of major anthropometric phenotypes. Of course some researchers interested in certain phenotypes may therefore focus on different population models than others focused on other phenotypes. For most phenotypes we recognize as anthropometric, we fundamentally end up with parental populations that correspond to a time when the extant populations under consideration had become reproductively isolated in continental Africa (West Africans), Europe (Europeans), North/South America (Indigenous Americans), and East Asia (East Asians), and it is precisely as a function of this relative isolation that the use of the ancestry model today is appropriate for forensic applications and the inference of many characteristic phenotypes. The usefulness of the test is presumably due to the lack of intergroup mixture and lower levels of genetic drift within the broad continental regions defined by these well-separated groups, combined with natural and sexual selection on anthropometrical traits (such as level of skin pigmentation, average height, eye shape).

It is important to clearly state from the outset that in what follows, we are not attempting to use a crude continental population model to describe all processes that lead to contemporary patterns of human genetic diversity. The assay based on the four-population model we have chosen to use in the following chapters was developed specifically, and so is most appropriate, for recently admixed individuals and populations from particular parts of the world. This said, it has been very interesting to observe the intermediacy of many populations that have no recent history of admixture. Presumably these results are consistent with the larger emigrational processes, which initially led to the peopling of the planet and subsequent population movements over millennia.

When selecting markers and parental samples, we must have a specific population model in mind, and we must be able to justify that model. The justification for any population model must rely on genetic, and to a certain extent, nongenetic data. Genetic data may include measures of phylogenetic distance or cluster analyses based on the use of various types of genetic markers (SNPs, STRs, etc.). Nongenetic data include linguistic, ethnographic (cultural), archaeological, and historical information.

From a synthesis of many types of data combined (archaeological, linguistic, molecular genetic), we know that the first populations of anatomically modern humans arose in Africa approximately 160 thousand years ago (KYA), and then expanded from somewhere in sub-Saharan Africa into North Africa, the Near East, Central and South Asia, and Australia approximately 40 to 69 KYA. From these populations, others branched to occupy Europe (40 KYA, and then again 10 KYA after the last ice age), East Asia (39 KYA) and the Americas (15–20 KYA). There are many ways to subdivide the world's humans, from the extremes of just one species to 6 billion individuals. Some of our concerns in setting up a pragmatic model are which models best explain most of the variation and which are the simplest (so that we can keep the problem manageable in terms of statistical precision). What we are looking for is the most rational worldwide population model—does the human family tree most clearly divide into two, three, four or more groups?

We have already discussed the work of Rosenberg and collaborators (2002) in Chapter 2, and we return to it here. Recall that Rosenberg used 377 randomly selected microsatellite markers with a worldwide collection of samples (HGDP-CEPH Human Genome Diversity panel) and the clustering program STRUCTURE to address this question. Since the markers in this analysis were selected randomly (without regard to their ancestry information content), the samples an international mixture, and the method a clustering method that does not use *a priori* information about each sample, the results provided definitions of population subgroups that largely follow expectations from phylogenetic distances measured using population allele frequency-based techniques or phylogeographic relationships measured using other nongenetics techniques.

The definitions are meaningful only with respect to the number of groups specified in the model, called the k-value. Rosenberg et al. (2002) showed that with a k = 2 population worldwide model the samples broke out into two groups: group I included the Africans, Europeans, Middle Eastern samples, and group II the Indigenous American and East Asian samples (k = 2 row, Figure 4-1A). However most population samples from East Asia, Central/South Asia, and Oceania showed fractional affiliation with another group, and over a third of samples showed outside-group affinities, so this is not as clear as we

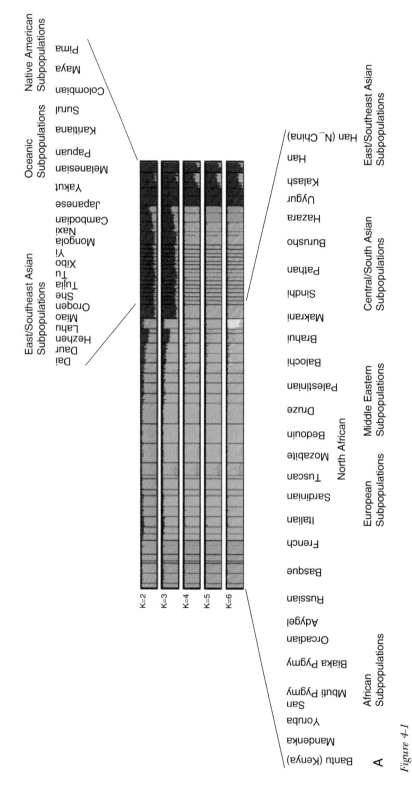

Figure 4-1

A) Stratification of genomic ancestry among various human populations. This figure shows the estimated population structure within individuals, between individuals, and among populations as measured using a collection of 377 autosomal STRs with the program STRUCTURE. Each individual is represented by a thin vertical line, and the populations from which the individual was taken is indicated by the legends above and below the figure. Using STRUCTURE we can specify a population model composed of k groups and express an individual's genomic ancestry admixture as a function of affiliation with corresponding k phylogenetic parental populations (i.e., founder) groups. At k = 1 we see that the populations are broken out largely along continental lines. From Rosenberg et al. (2002). "Gentic Structure of Human Populations." 298: 2381–2384.

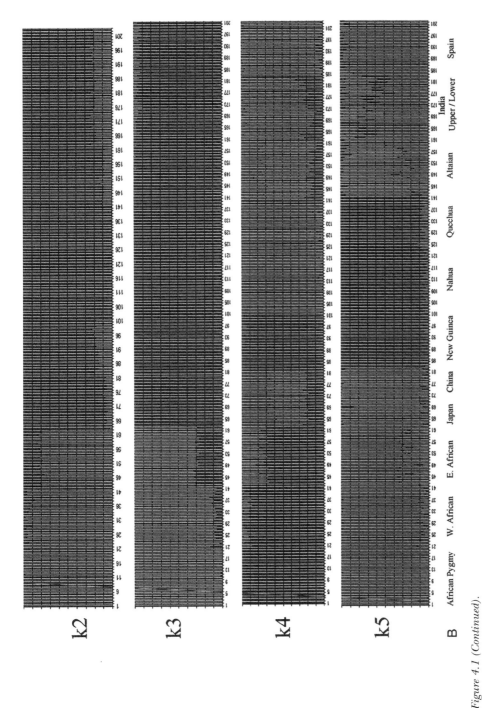

Figure 4.1 (Continued).

B) *Apportionment of global genetic diversity using 500 of the most informative SNPs from the work of Shriver et al., 2005. F_{ST} was used to rank the SNPs with respect to the ability to resolve among the 12 populations in this study. The software program PSMix (Wu et al., 2006) was used, which is faster though not as careful as STRUCTURE. The display of the results was formatted to resemble the output of STRUCTURE, where each individual is represented by a bar and the proportion of ancestry is represented by the amount of color in the bar. Results are shown for four population models (indicated to the left), assuming a model with 2, 3, 4, and 5 populations and the ancestry indicated by a particular color for one is not necessarily the same as for another—for example, "African" ancestry is of blue color at k2, yellow at k3, red at k4 and green at k5. Provided by Jason Eschleman, Trace Genetics, Richmond, CA.*

might have hoped. Given the wide variation in forensically relevant anthropometric trait values within each of these two groups, the k = 2 model is not the model we are looking for in developing a pragmatic ancestry test for the field of forensics, biomedicine, or pharmacology.

Moving to a more complex three-population model, the Europeans and Central/South Asian samples resolved from the Africans to form a Caucasian group, and both the African and Caucasian groups show little intergroup affinity. The East Asian samples all still show substantial fractional affiliation with the Caucasian group (note the mixture of blue European and purple Indigenous American color, k = 3 row, Figure 4-1A). Although the k = 3 model may seem useful since it neatly partitions the most obvious of the forensically relevant anthropometric trait values, the affinity between East Asians and Europeans at least with these markers suggests that this k = 3 population model is not what we are looking for either. However, at k = 4, the incidence of fractional affiliations dropped dramatically and for the first time in the analysis of k-value models, we can see clearly partitioned genetic ancestry groups that comport with specific major population groups. The results were the same as k = 3, except East Asians resolved from the East Asian/Indigenous American combined subgroup found with k = 3, and the overall incidence of outside-group affiliation was much lower (k = 4 row, Figure 4-1A). With this model, populations exhibiting average physical trait values on the extreme ends of the worldwide anthropometric scale (in terms of iris, skin, hair color, and certain facial features, etc.; Europeans, East Asians, West Africans) showed relatively homogenous affiliation with their own metapopulation group.

It is important to emphasize that the STRUCTURE results can indicate multiple group membership for each individual and in each of these models (k = 2, 3, and 4), some groups have persons showing multiple group ancestry. This is what we were looking for in a biogeographical ancestry model for forensics use. Higher k values showed equally clean genetic partitioning but k = 4 was the simplest of the clearly partitioned models, breaking the world population into four groups:

  I. Sub-Saharan Africans, including East Africans, South Africans, and West Africans
 II. Europeans, Middle Eastern, South Asian (the latter of which exhibited fractional affiliation with group III)
III. East Asians, Southeast Asians, and Oceanians
 IV. Indigenous (Native) Americans

For SNPs, effective population sizes are lower than for microsatellites—that is, because they are bi-allelic rather than multi-allelic, it takes longer for SNP alleles to become fixed or eliminated because genetic drift occurs more slowly.

When we look at the global apportionment of diversity with SNP markers we thus would expect to see similar but perhaps slightly different patterns than with microsatellites. Jason Eschleman (Trace Genetics, Richmond, CA) has selected the 500 most informative SNPs (based on Fst) using the Affymetrix 10 K chip data produced by Shriver et al., 2005. With these 500 SNPs, using a likelihood method (PSMix, Wu et al., 2006), as with STRs using a Bayesian method (Structure, Pritchard et al., 2000), we observe a fairly clear pattern, with unadmixed samples existing for each population with each population model (k = 2 through 5, Figure 4-1B). Note subtle differences with Figure 4-1A—notably, the order with which East Asian populations resolve from New Guinean populations from k = 3 to k = 5. With SNPs, where New Guinean populations were included in the model, it appears that a k = 5 model is the simplest that would be forensically useful but without the New Guinea population a k = 4 is the simplest (not shown). Our choice between the two models, then, would depend on our desire to build systems (to be discussed later) that allow for an appreciation of anthropometric phenotype differences between Oceanic and East Asian populations.

We can also turn to Y and mtDNA chromosome haplogroup analysis to paint a portrait of modern human diversity (see Figures 4-2, 4-3) (more on Y-chromosome analysis can be found later in Chapter 4). Due to the fact that patrilocality (i.e., male territorialism) generally exceeds matrilocality,

Figure 4-2

The first assessment of the global distribution and phylogeography of human Y chromosome haplogroups (sequence types). Each haplogroup is indicated by a roman numeral and the same color in the phylogeograpical tree and the global map below. Pie charts reveal the fractional composition of haplogroups among the corresponding populations. From Underhill et al., "The phylogeography of Y chromosome binary haplotypes and the origins of modern human populations" Annals of Human Genetics, 2001, Vol. 65, pp. 43–62. Reprinted with permission of Blackwell Publishing.

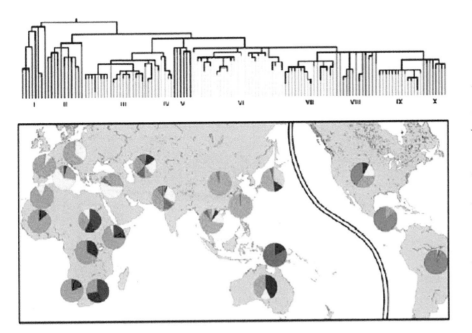

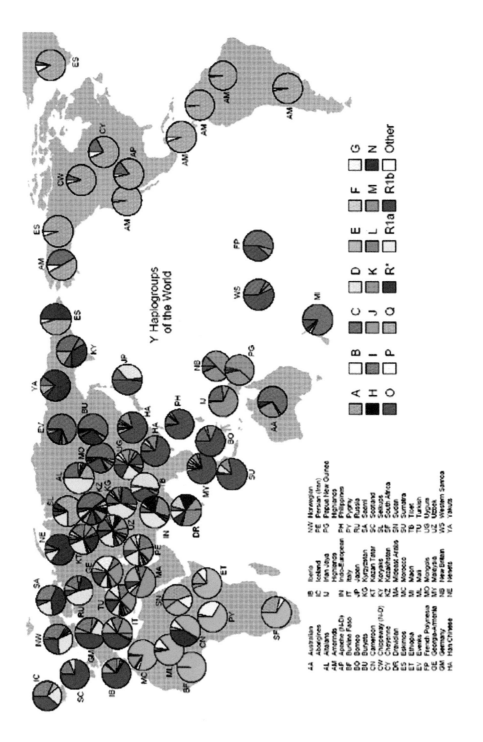

Figure 4-3

A modern depiction of the global distribution of Y chromosome haplogroups. Note that haplogroups are currently referred to by alphabet letters rather than roman numerals, but the difference is largely semantic and the results are essentially the same as that shown in Figure 4-2. Haplogroups are shown with colors in the pie charts for each population. The geographical origin of each population is indicated by the position on the map. The map represents the apportionment of populations prior to the most recent European expansion beginning about 1500 A.D. The identity of the populations is indicated with a two-letter symbol (key is shown at the lower left). Based on published data as of December 2004 and was kindly provided by its creator, J. D. McDonald of the University of Illinois.

populations appear to be more differentiated throughout the globe in terms of Y chromosome than mtDNA sequences. In general, with the Y chromosome data, we can observe a few major partitions—Africa, Europe, Asia (including the Pacific Islands and Australia), and the Americas, with parts of Asia exhibiting extreme complexity and heterogeneity (see Figure 4-2). Breaking the pattern up on a finer level with respect to specific populations, we can distinguish Africa, Europe, Southeast and East Asia, Northern Asia, the Pacific Islands/Australia, and the Americas in terms of characteristic Y-haplogroup proportions (see Figure 4-3). The global pattern of mtDNA haplotypes provides a similar rough grouping of subpopulations into African, European/ Middle Eastern, East Asian, American, and Indigenous Australian. These Y and mtDNA partitions are consistent with those we obtained from the cluster analysis of Rosenberg et al. (2002) (see Figure 4-1), and the similarity in the partitions across these analyses is remarkable given the fact that entirely different markers, chromosomes, and analytical methods are used.

Just as with autosomal admixture analysis we can appreciate diversity on various levels by choosing a k = 2, 3, or 4 population model, with Y and mtDNA analyses we can appreciate diversity of haplotypes using simple and more complex methods of defining the haplotypes. We can use a relatively small number of SNPs to define simple haplogroups or larger numbers of SNPs to define a larger number of subhaplogroups within each. Figure 4-2 shows a relatively simple method of defining haplotypes that captures the bulk of the diversity and establishes rough definitions for the major Y-chromosomal DNA clades. This figure was published as part of the original globewide study of Y chromosome haplotype diversity (Underhill et al. 2001), and assigned multilocus genotypes of the Y chromosome to one of 10 haplogroups.

As more polymorphisms were found, it has become apparent that the complete complexity of the variable Y chromosome can be described using no fewer than 153 different haplogroups, but that the bulk of the Y-chromosome variation throughout the world can be adequately described using a smaller number of core polymorphisms that define about 20 major haplogroups (see Figure 4-3; McDonald (2004) unpublished metaanalysis, reviewed in Jobling et al. 2004). It is through incorporating these haplogroups in a detailed human phylogeny, and consideration of the modern-day geographical distributions for the haplogroups that molecular anthropologists have been able to reconstruct the major human population expansions and migrations over the course of the last 200,000 years.

Let us look in more detail at the wealth of the Y and mtDNA chromosome data that has been accumulated by many authors over many years. Focusing on the Y chromosome fist, we can see that Y chromosome haplotypes A and B represent the ancestral haplotypes found exclusively in Africa (see Figures 4-2, 4-3;

Underhill et al. 2001, reviewed in Jobling et al. 2004) but the predominant African haplotype (E) is not confined to the African continent. All non-African haplotypes and one African haplotype (E) share the derived M168 variant so the founders of the near eastern and Australian populations must have carried this variant. Haplotypes C, G, H, I, J, L, and R are distributed throughout central Asia and Europe, haplotypes C and D are confined largely to East Asia, and Q is almost exclusive to the Americas (see Figures 4-2, 4-3). The derived African haplotype E is found throughout West Africa and in contemporary Middle Eastern and Mediterranean populations. The distributions of these haplotypes and phylogenetic analysis of their sequence states tell us something about their genesis and order of dispersals over the past 50,000 years (see Figure 4-5); namely, that human populations expanded from Africa to the Near East and Australia; from the Near East into Europe, Central Asia, and East Asia; and from Central Asia into the Americas (reviewed nicely by Jobling et al. 2004; Figure 4-5, arrows, where the ovals represent Y chromosome metagroups and the arrows and letters of B represent the order of dispersal of mtDNA sequences; see Figure 4-4).

From an inspection of the phylogenetic analysis of Y haplotype diversity in Figure 4-5, we can appreciate the complexity of the relationship between the derivation status of a sequence and its geographical distribution. For instance, haplogroup E is a derived haplogroup from those containing the M168 and P9 polymorphisms (Jobling et al. 2004), and is found predominantly in Africa (indicated with a red oval), yet most of the other M168 and P9 containing haplogroups are found in central Asia and Europe (yellow, blue, and grey ovals). This would seem to imply that founders carrying M168 left Africa for the Near East, and from there expanded to Central Asia, Europe, and back into Africa, perhaps as part of the Bantu expansion that brought agriculture to vast regions of Southern Africa. However, this interpretation may not be correct, and as reviewed in Jobling et al. (2004), it is actually quite unlikely that such a reverse flow back into Africa occurred. More likely, the original founders for the M168 haplogroups, who gave rise to most non-African populations, them-selves, probably obtained this variant while still in Africa (before spreading). Most likely, M168 carriers originated in a discrete portion of Africa, and their descendents traveled the world founding populations in the middle East, Central Asia, and ultimately, in East Asia, later spreading throughout the rest of Africa. In other words, the pattern of shared haplotype sequences is not necessarily an indication of temporal founding events, but could be reflective of which African populations expanded into certain regions (i.e., the Near East) and which did not.

As another example, many of the M89, m213, and p14 haplogroups (col-lectively called haplogroup F, but containing G, H, I, J, L, M, N, O, Q, and R)

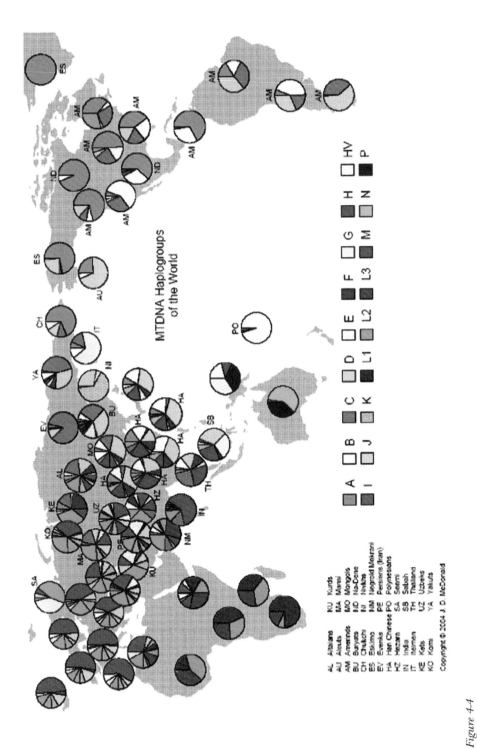

Figure 4-4

A modern depiction of the global distribution of mtDNA chromosome haplogroups. Haplogroups are shown with colors in the pie charts for each population. The geographical origin of each population is indicated by the position on the map. The map represents the apportionment of populations prior to the most recent European expansion beginning about 1500 A.D. The identity of many of the populations is indicated with a two-letter symbol (key is shown at lower left). Based on published data as of December 2004 and was kindly provided by its creator, J. D. McDonald of the University of Illinois.

155

Figure 4-5

*A) Y haplogroup
phylogeny, in terms of
current haplogroup
nomenclature used in
Figure 4-3. The length of
vertical lines is roughly
proportional to time since
origin. The polymorphisms
distinguishing the
geographically significant
polymorphisms discussed
in the text are shown with
their symbols. The color
scheme shows possible
geographical origins for the
haplogroups as shown in
B) suggested by the
distribution among
modern-day peoples. This
scheme assumes a four-
population model B.
Global dispersion of Y-
haplogroups represented as
waves of population
expansion. Haplogroups
are indicated with their
letter as in the top figure,
and colored ovals represent
a model of geographical
origin for each, deduced by
modern-day distributions
such as that shown in
Figure 4-3. Note the
disconnect in the model
between modern-day
distribution of
haplogroups and lineal
origins, which could be
produced by multiple
population expansions at
different times to common
areas. Various other
models are possible,
especially for the European
and far Eastern
haplogroups. Adapted
from figure provided by
D. T. Schulthess,
Gene Tree, Inc.*

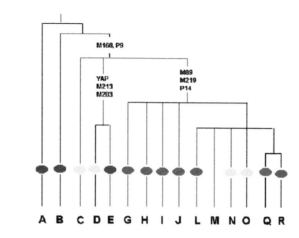

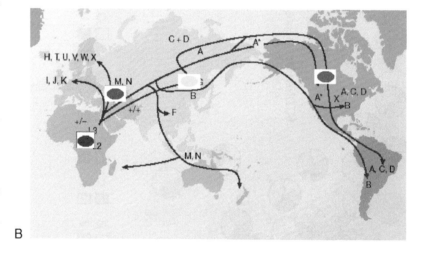

are spread throughout Central, Southern, Eastern, and Northern Asia as well as the Middle East and Europe (see Figure 4-5). From this distribution it is tempting to speculate that populations that acquired the first haplogroup F mutations lived somewhere in the middle of this range. Nested cladistics analysis overlain on geographical maps can be used (Templeton 1998) to estimate geographical origins for elements of diversity by calculating "geographical centers of gravity" for these elements. Such methods have been used to suggest a multiregional theory of modern human origins, and applied here to haplogroup L and its descendents, would seem to suggest that many European, Middle Eastern, South and North Asian populations owe their Y haplogroup character to mutations that arose somewhere in the Near East or Central Asia after the genesis of the most recent non-African common ancestor (MRCA), approximately 50 KYA.

Interpretations such as these from Y chromosome analyses suggest that for practical purposes, the entire human nonrecombining Y chromosome (NRY) phylogeny can be reduced to a few basic groupings of Y haplotypes that explain most of the Y haplogroup diversity of modern-day peoples:

- Proto European/Near Eastern (in the older literature called Caucasian), derived from African founders who carried or acquired G, H, I, J, L, and R Y-haplogroups
- East Asians, derived from founders who carried or acquired D, C, N, and O haplogroups
- West Africans carrying the ancestral A and B haplogroups, and the African populations who carried or acquired haplogroup E (the latter of which eventually, most likely, also founded non-African lineages)
- Indigenous Americans who acquired the Q haplogroup
- Australian derived from founders who carried or acquired haplogroup M and others

The first four of these are represented by colored ovals on the geographical map shown in Figure 4-5, and the phylogeny of the haplogroup classes is indicated by the same colored ovals in Figure 4-6. Note that though the modern-day geographical dispersal of the haplogroups is roughly equivalent in terms of territorial space, the depth and size of the corresponding phylogenetic branches vary substantially. By this reduction, most modern-day European, Middle Eastern, and Western Asian populations, which we refer to here simply as European (but could also be described as proto European/near Eastern), derive from founders that lived in the Near East no sooner than 50 KYA. From these founders also derived East Asians and ultimately Indigenous Americans.

The situation is clearly more complex in time and space than these simple step-wise events, neatly defined, imply. In reality, European populations likely derived from Central Asian/Middle Eastern populations at different times than East Asian and South Asian populations, notably during the Paleolithic approximately 45 KYA and then again during the Neolithic 10 KYA when agricultural technologies spread from the Middle East up into Europe. There may have been two separate waves of population expansion into the Americas from Asia, similar to the known multiple expansions of proto-Europeans into Continental Europe during Paleolithic and Neolithic times. However, the assumption of simplicity, such as that we use when we invoke a single proto European/Near Eastern population, and a single Indigenous American founding population at a single point in time, captures most of the extant continental level genetic sequence variability, which is related to the variables we are most interested in—phenotypes.

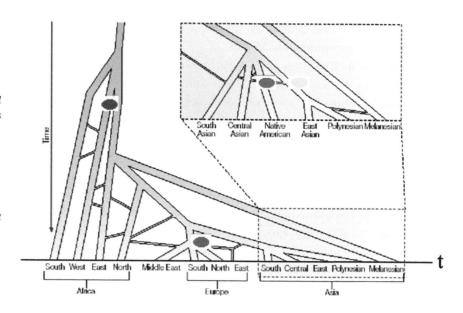

Figure 4-6

Human phylogeny represented with more chronological detail as obtained using the molecular clock theory and methods incorporating this theory for the calculation of time to most recent ancestor. Population descriptors based on modern-day geographical distributions are indicated at the end of each branch at the bottom. Bars connecting branches indicate probable admixture or gene flow events. Colored ovals represent the meta-haplogroups from the model in Figure 4-5 and their application to this diagram is extremely simplistic, and meant to represent that peoples derived from common branches united by a colored oval tend to possess haplogroups represented by that oval in Figure 4-5. Note with reference to Figure 4-5 that a fairly complex branching pattern is required to unite the various lineally unrelated haplogroups by geography suggested by the distribution of modern-day peoples. The placement of the ovals along the Y-axis is a rough guess based on the average of constituent haplogroups for each geographically united meta-group. Reprinted by permission from Macmillan Publishers Ltd: Nature Reviews Genetics, Vol. 5, pp. 611–618, 2004.

Though there were likely multiple founding events (i.e., continuous gene flow), which haplogroups accompanied which founder event is subject to current research and debate. For example, though much work on classical markers such as blood group antigens has suggested the most recent expansion into Europe took place some 10 KYA, this date doesn't tell us which Y haplogroups were contributed during this Neolithic expansion and which already existed due to previous expansions in Paleolithic times. The average proto (European/Near Eastern) ancestor, over this entire time period, was likely an individual who lived in the Middle East or southwestern Asia, since most Y chromosome sequences present in modern-day descendents of proto (Europeans/Near Eastern) populations (notably, Europeans, Middle Eastern, Central Asian) are distributed about these locations in many directions, spread throughout Central Asia, Europe, and the Middle East. This model would explain why it is that Central Asians exhibit the greatest diversity in Y and mtDNA haplogroup composition—various off-shoots from an originally amalgamated population established various distant populations in Europe, South Asia, and so on, with different founder effects causing the differentiation within and between the modern-day populations we see today. Of course, it is equally possible that modern Central Asians are of complex Y and mtDNA identity due to relatively recent admixture, whether the original founders lived in the same area or not, and regardless of where, geographically speaking, drift or founder events took place. Fortunately, for the forensic scientist interested in making inferences about anthropometric traits through an appreciation of individual ancestry, the mechanisms by which diversity has been generated are not so of much importance as the existence of the diversity itself and the fact that this diversity is the principle component of phenotype diversity we observe today.

Most of the rest of this chapter considers one particular admixture assay in detail. This assay was developed and characterized in the laboratory of the author, and is used for many of the molecular photofitting tasks we will describe in later chapters, so it deserves more than just a casual look we afford to most of the other panels discussed herein. In developing our four-population autosomal admixture test based on West Africans, Europeans, East Asians, and Indigenous Americans, we have used a wealth of classical genetics literature, the Rosenberg et al. (2002) analysis, a more recent SNP analysis, and the rich Y-chromosome literature to justify our choice of a four-population model for our purposes. With this model, we essentially slice the human family tree back to a time probably before Europeans as we call them actually lived in Europe, perhaps to the time where their ancestors lived in the Near East and Central Asia as indicated by the blue oval in Figures 4-5 and 4-6. This proto-European population likely contributed to various populations, to varying extents, and the blue oval in Figure 4-5 corresponds to the phylogenetic branchings within the shaded region under the blue oval in Figure 4-7, which shows with grey sectors the modern-day peoples that share elements of each geographically united meta-haplogroup. Given the diversity of these European subpopulations, perhaps a better name for this group would be ''proto Europeans/Near Eastern. We have discussed the difficulty inherent to naming populations that no longer exist in Chapter 2. Grouping populations such as we have focuses only on some of those populations that existed at the time we are looking back to. Since we do not have a time machine, we must infer their character from modern genetic diversity.

An inherent risk to describing the history of populations as molecular anthropologists and human population geneticists do is reifying racial typologies—making abstract population labels more real than they are. We have discussed this before and will continue to discuss later the fallacy of treating ancestry as a dichotomous variable, but aside from the technical problems we experience there are ethical considerations associated with our choice of terms. We emphasize that the labels we use (such as ''European'') in many ways oversimplify the processes by which populations evolve, and describe in sweeping generalities large groups of various peoples who

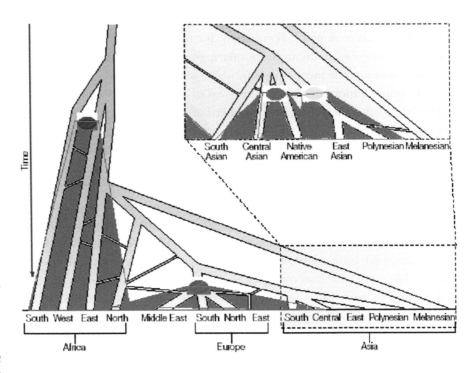

Figure 4-7

Figure 4-6 reproduced with grey sectors indicating the modern-day populations that share elements of geographically united meta-haplogroup identity. For example, modern-day European character in terms of Y meta-haplogroup identity is shared among North African, Middle Eastern, Continental European, and South Asian populations. The shaded regions overlap to a certain extent; North Africans also exhibit character shared by other Africans to the exclusion of non-Africans, possibly due to gene flow with North, South, or East Africa as accommodated by the side branches in this model, or possibly due to an origin of haplogroup E prior to the split of Africans and non-Africans, which would require the blue oval be placed higher in the diagram after the split between East Africans and those that left Africa. Reprinted by permission from Macmillan Publishers Ltd: Nature Reviews Genetics, *Vol. 5, pp. 611–618, 2004.*

more closely approximate metapopulations over both time and space. Such a process is necessary if we are to discuss the past, since the individual parental populations no longer exist and even if they did, they would still have complex relationships among one another. Conversely, however, such a framework could contribute to unacceptably simplistic studies of genetic variation among living populations. For example, the term *Caucasian* often is used as a euphemism for "white," in the biomedical literature and by law enforcement and investigating officers. Sometimes, Caucasian also is used to indicate persons of South Asian or Middle Eastern ancestry along with Europeans. If one is describing a population sample or even an individual in the literature today, Caucasian is thus imprecise, and many scientists feel that with its social connotations, its use has the effect of *racializing* the topic.

In addition to the scientific imprecision of describing population samples with metapopulation terminology, it is also taken by many in the field as a sign of disrespect to refer to a person or sample of people by terms that they themselves would not use. For example, the term *Amerindian* is seen by Indigenous Americans in the United States as an academic term derived from a linguistic category that they never use themselves. That this same term is just fine in many Spanish-speaking countries is further evidence of the complexity of proper language usage.

Another caution is required on the labeling of the gene pool components in the context of existing populations, ethnic and geopolitical groups. Native American, for example is a political designation for the American Indians living within the borders of the contemporary United States. Mexican Indians and *Nativos* from Southern Mexico, for example, are not considered Native American, though both groups share a common recent genetic history. Given these aspects of general usage, the gene pool component that we can express in our models is maybe best referred to as Indigenous American, in an effort to help clarify that the term is not related to the political affiliations of a particular people today, but much more closely an expression of the ancestral genetic groups that occupied the lands of the New World and are today found as substantial components of the gene pools of many different populations throughout North, Central, and South America as well as in migrants from other parts of the world.

Similarly, it is important to be precise regarding the other parental populations that we have used in our model. The West African parental population is a sample of West Africans along with additional ancestral frequency information from African-Americans and so best referred to as West African. The European parental population is estimated from both the European component of large samples of European Americans from Pennsylvania as well as a few other European samples from Northern Europe and so might best be referred to as West European. The East Asian parental allele frequencies are derived from combined samples of Chinese and Japanese and so are best referred to as East Asian. As such, on closer inspection, our parental populations represent the geographical extremes of the regions of these continents. Although this in large effect was done in an effort to best describe the majority of U.S. resident populations, there is an important fortuitous effect of our model—we may be able to describe admixture events and migrational patterns that founded many earlier populations that are not simple admixtures among these existing populations.

In choosing the four-population model with the pragmatically selected (and labeled) parental populations listed earlier, we can perform our analysis with the aim of inferring physical characteristics, since the expression of ostensible, anthropometric traits that are highly variable between groups are labeled this way. From the phylogenetic tree in Figure 4-6, we can see that these four historical parental populations probably lived in different times as well as places. From this figure, it can be seen that the oldest parental metapopulation we are considering is the West African metapopulation, represented by West Africans. The youngest is the Indigenous American metapopulation. As long as the descendents from these parental populations lived in relative isolation in space, as genetic distance among their descendents indicates they did, the date of the various branchings corresponding to an admixture population model need not be contemporaneous nor even necessarily known in order to be useful for the forensic scientist. Nonetheless, for the anthropologist who is concerned with dating the branches, various mathematical methods for appreciating genetic diversity can be used with autosomal AIMs to chronologically characterize the branching patterns (see Jobling et al., 2004).

In terms of the Y-chromosome distribution, in our four-population model, we have sliced the Y-chromosomal DNA phylogeny after the split of Q and R from its precursor P, which occurred about 20 KYA. As shown in Figure 4-7, we have grouped all populations characterized primarily by the light blue haplogroups as European, notwithstanding the difficulty naming this parental population; those by the yellow haplogroups as East Asian; those by the red haplogroups as West African (to include all West Africans, as represented by the West Africans); and those by the grey haplogroups as Indigenous American (notwithstanding that certain Central Asian populations share Y-haplogroup identity with Continental Americans). Though the European group is polyphyletic, the various lineages (South Asian, Middle Eastern, continental European, etc.) combine to form a lineage that is monophyletic with respect to our hierarchical definition. To keep the dimensionality of the test low, we have chosen to work with the k = 4 partitions defined by Rosenberg et al. (2002, Figure 4-1A), and so we omit Pacific Island, East African, and Australian populations as distinct branches.

There is considerable debate whether some of these are admixed populations themselves, and genetic distance measurements do not help us illuminate the situation, unfortunately, since genetic distance is just as easily created by admixture as lineage uniqueness. For example, Y chromosome haplogroups for Pacific Islanders and East Africans are easily recognized as derived haplogroups, not ancestral haplogroups, and this speaks to the fact that they do not represent features with which we would desire to resolve among the deepest phylogenetic divisions such as that indicated for some Oceanic populations in Figures 4-6 and 4-7 (which are trees based strictly on genetic distance). In other words, Y haplogroup sequences characteristic of East Africans are found on European and West African templates, not a special East African template. Y haplogroup sequences characteristic of Pacific Islanders are derived from East Asian and Central Asian templates—that is, templates that appear from modern-day distributions to have originated in Central Asia. The next level of analysis at k = 5 would involve the Papuans and Melanesians as a separate group; East Africans would still cluster with western Africans (Yoruba, Mandenka) (see Figure 4-1A). The fact that at k = 4, Pacific Islanders cluster with East Asians (whether using Rosenberg's panel of markers, or the panel of 176 AIMs we are introducing here) shows that the hierarchy used for a given phylogenetic analysis is arbitrary, and molded by concerns more related to statistical power and general utility for a given purpose than a misguided effort to unambiguously define human "racial" or "genetic" groups. Reporting the ancestry of these three populations in terms of the four main continental groups is not incorrect, only not as resolved as some might prefer (depending on their applications or motives).

Without the use of objective genetic clustering, defining a population model prior to developing an admixture test is somewhat circular in logic, since it is presumably to measure such a parameter for which we are developing the admixture test itself. For example, we might decide that there are two populations based on our own internal biases, such as Siberians and the rest of the world. Using this assumption to select parental samples and hunt for genetic markers would produce a self-fulfilling prophecy; we would obtain markers useful in resolving Siberains from others and such a test would "work" to a certain extent but have little anthropological or anthropometric meaning (which, after all, is what we are after). This said, we explicitly recognize that we have created models that in the immortal words of the statistican Gorge Box, are "all wrong." We hope to demonstrate that at least they are useful.

MARKER SELECTION

At the time we screened the database in 2002, The SNP Consortium (TSC) had contributed data on approximately 27,000 SNPs where frequencies were available on three populations (African-American, European-American, and East Asian). For establishing the parental group allele frequencies, DNAPrint's laboratory genotyped approximately 150 relatively homogeneous descendents of four broadly defined metapopulation groups, which as we just described, existed many tens of thousands of years ago, defined by a coalescence of a simplified human pedigree back to points in time where interbranch molecular distances were at their greatest.

We scanned the genome for a useful panel of biogeographical AIMs using the statistic delta (allele frequency differential) and incorporated the best, namely those with the highest delta levels, that we found with a maximum likelihood algorithm to measure BGA admixture proportions. A larger panel of high-delta SNPs were tested than were ultimately included in the panel. The most common reason markers were left out of the final panel was inefficient genotyping. Some markers were included from one of the only panels of AIMs at the time (summarized in Shriver et al. 2003). We found that not only were a majority of BGA affiliations measurable in a manner consistent with self-held notions of race, but that BGA admixture proportions were also measurable with significantly improved precision, accuracy, and reliability compared to previously described methods (i.e., Frudakis et al. 2003; Shriver et al. 1997). The test we have developed may be helpful during study design by helping to eliminate or reduce the insidious effects that cryptic or micropopulation structure imposes. It may also prove useful for forensic scientists who currently use imprecise and sometimes inaccurate means by which to infer physical appearance from crime scene DNA.

SAMPLE COLLECTION

There exists no ultimate arbiter of BGA affiliation, but this poses an interesting problem—it seems that we must have some prior information on ancestry in order to construct and verify such a test. This is precisely why we turn to assumption-free methods such as Pritchard's STRUCTURE program in defining appropriate population models. Given a preferred (and justifiable) model, we can target selection schemes for parental populations appropriate for the model.

As discussed in Chapter 2, parental group heterogeneity is a potential pitfall; diversity within each subpopulation group needs to be ensured if the allele frequencies are to be precisely defined for our parental groups. In other words, we have to define the diversity within our parental groups and ensure we consider as much of it as we can when establishing parental group allele frequencies. Whether or not we are developing a test for estimating admixture based on predefined population assignments (as we do when using the maximum likelihood algorithm described in Chapter 2) or not (such as using the STRUCTURE program, also described in Chapter 2), collecting without regard to prior notions of ethnicity could result in a foundation possibly lacking diversity and imposing bias. This is because modern-day populations are often heterogeneous even though they derive from a common ancestral population, and sampling from only one particular modern-day subpopulation within a larger one would produce allele frequencies not exactly characteristic for the other subpopulations. In the end, we would produce strange results for these other subpopulations; this would have the same effect as imposing an unreasonable population model with respect to the objectives of the test.

As an example, suppose we have a particular population model, and a particular set of populations, and we want to know admixture with respect to this model (such as West Africans, European, East Asian and Indigenous

American). If our parental samples exhibit systematic admixture with respect to this model (i.e., our European samples were Greeks, and as such not an optimal parental sample for Norwegians), such as might be the case with a continental four-group model if certain ethnic subpopulations systematically exhibit certain types of outside-group affiliations, we could introduce systematic errors in frequency estimation for the parental groups. In this case, our Norwegians might show East Asian or other fractional affiliation as a result of our choice of Greeks as the parental population. Since it is obviously desirable to minimize bias and error, it may be desirable to collect from as wide a variety of subpopulation affiliations (i.e., within continent ethnicities) as possible within each parental group. Sometimes resource limitations make this difficult, but in these cases it is important to recognize the potential limitations of the developed method. Alternatively, one may make a choice to select populations from a particular region within a continent if such a region has historical and/ or genetic evidence for making the model useful. An example for a U.S. admixture panel is using West African as the sub-Saharan representative since most of the enslaved people from Africa were taken from the West African regions during the European colonial periods.

For the development of the admixture assay discussed here, we assumed a four-population model based on the analysis of Figure 4-1A, which includes African, European, East Asian, and Indigenous American, and incorporates Australians, Pacific Islanders, South Asian Indians, East Africans, Pygmy, and Khoisan meta-populations into one or more of these continental population groups (though Figure 4-1B suggests an African, Oceanic, Indigenous American, and European $K = 4$ model, it uses a less sophisticated and probably less reliable algorithm than the Bayesian algorithm of Figure 4-1A). Samples were obtained from corresponding locales around the world, and though the sample sizes are modest, considerable effort was contributed during the collection procedure to ensure samples were affiliated with modern-day descendents of these four populations to as high a degree as possible. This was done through the use of:

- Pedigree questionnaires (requiring all ancestors for at least eight generations on both the paternal and maternal side having been born or are believed to have been born in the relevant region and having described themselves or are believed to have described themselves as members of the relevant ethnic group)
- Anthropologist qualified anthropometric trait evaluation, which is obviously a subjective process, but forms a good double check of sample affiliation integrity when combined with the preceding

The West African samples were obtained in West Africa from people of ostensibly a high degree of West African ancestry, whose parents were reported to be of the same. The European samples were collected from various locales in the United States the same way, and the East Asian samples were collected similarly from Japan, China, and first- or second-generation Asian Americans in the United States. The Indigenous American samples were collected from Nahua and Mixtec *Nativos* inhabiting a remote region of southern Mexico, again, in the same way (Halder et al. 2006). All samples were collected under IRB guidelines for the purposes of genetic studies of human population variation.

As already discussed, we can use model-based methods that do not require prior information about samples, such as STRUCTURE, to test the appropriateness of samples as parental samples. This is particularly important for our European samples since they were collected in the United States rather than in Europe, which might mean some of the samples exhibit some outside-group affiliations. Even in Europe, it is difficult to know which subpopulations might have experienced admixture in the past. We specify the population model we would like to use (i.e., we can break the human population into four groups) and say to STRUC-TURE, "cluster our samples with respect to these four groups," not defining what these groups mean in a geographical, anthropological, or historical sense.

Having collected parental samples based on anthropologist-defined population memberships, we relied on STRUCTURE to identify particular individuals that did not fit the expected categorizations uniformly (i.e., individual samples that were admixed or for other reasons exhibited outside-group affiliation). This ensured that we obtained as good an estimate of allele frequency—as unbiased with respect to our chosen model—as possible. In effect, we used this relatively objective method of defining appropriate parental samples to eliminate suspect samples from our calculations. If we did not use such objective methods, our only defense against bias in parental population composition is to collect from as diverse a subpopulation base within that population as possible—some samples may be admixed with respect to the population model overall, but we would hope that admixture in various directions balanced out. The soundness of such a technique would depend on the level of heterogeneity in the parental population, which is difficult to know without application of an objective method, and on the size of the parental sample.

Pritchard's STRUCTURE method (Pritchard et al. 2000; also see Pritchard et al. 2001 and Rosenberg et al. 2002, presented in Chapter 2) does not require a definition of structure *a priori* to infer population structure and assign individuals to groups. According to Rosenberg et al. (2002),

> It (STRUCTURE) assumes a model in which there are K populations (where K may be unknown), each of which is characterized by a set of allele frequencies at each locus, and assigns individuals ... (probabilistically) to a population, or jointly to two or more populations if their genotypes indicate that they are admixed.

Specifically, we used STRUCTURE to identify individuals who are part of the parental set, but do not exhibit homogenous ancestry and are themselves admixed or otherwise not representative of a good parental sample with respect to our model. Once identified, these individuals were eliminated and the allele frequencies are determined using the remaining individuals—individuals of only homogeneous affiliation to their parental group. It is important to point out that STRUCTURE uses a different but related method to the one we use to determine admixture proportions, and as expected, that the results generally agree with one another (as will be discussed later). The power of STRUCTURE is that it does not require the specification of a population model *a priori*, which is useful for selecting parental samples in such a way as to avoid a circular logic problem.

In Figure 4-8, we show the STRUCTURE results for our parental samples. As for the microsatellite results presented in Figure 4-1A, genetic affiliation among our parental samples largely broke out into continental groups as expected based on the selection of AIMs with power for resolution with respect to a continental model: West African (red), Indigenous American (green), European (blue), East Asian (yellow). Notice one East Asian individual was

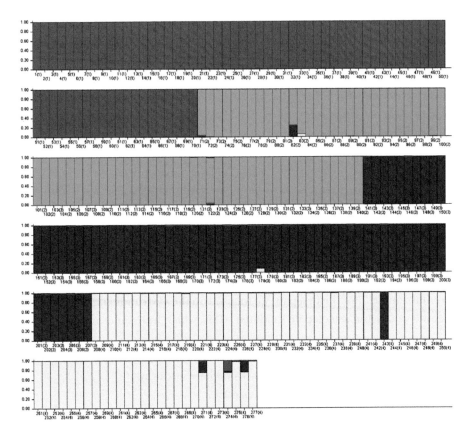

Figure 4-8

STRUCTURE analysis of parental samples selected from the four major continents and genotyped using AIMs selected from the human genome to maximize the information content value with respect to a continental population model. Each sample is represented by a rectangle and the proportion of color in that rectangle is a measure of proportional genetic affiliation with the measured element of population structure. Samples are grouped by parental designation—the first 70 are from the West African parental group (mostly red), the next 70 (mostly green) are Indigenous American samples, the next 67 European (mostly blue), and the last 70 East Asian (mostly yellow). West African samples typed with common affiliation (red), as did European samples (blue), East Asian samples (yellow), and Indigenous American samples (green). Note that several of the East Asian samples exhibited non-East Asian affiliation. Samples such as these were removed prior to calculating parental allele frequencies.

almost entirely European! This particular sample was purchased from a sample supplier who claimed to have collected from recent East Asian immigrants in the United States, but given the European affiliation, it might have been from an admixed East Asian, an East Asian with South Asian admixture (as we will see in a later chapter), or even a person who identifies him- or herself as East Asian for nonbiological reasons. The sample was among those eliminated from the final pool from which East Asian allele frequencies were estimated. We used a cutoff of 2%; samples with more than 2% nonprimary group affiliation were eliminated prior to calculating allele frequencies.

NUMBER OF INDIVIDUALS

At the time we developed our admixture test we had access to 70 individuals for each group. Using Equation (3-48) in Chapter 3, if we assume that the average AIM has an average allele frequency across all groups of 0.25, this number provides a STD of 0.036 for the average AIM. We will use simulations—of both allele frequencies in actual samples and samples with actual allele frequencies—to show later that when many markers are used, and the allele frequency differentials are large, the maximum likelihood method is relatively impervious to allele frequency estimation errors of this magnitude (even systematically biased estimates!). Although it may be desirable to use thousands of individuals to estimate allele frequencies within less than a fraction of a percent, the simulation results we present later in this chapter suggest that it is not necessary for the reasonably accurate determination of admixture percentages, and that allele frequency estimates within a couple percent of the true values are suitable for most purposes.

71 Marker Test

We compiled a list of several thousand candidate AIMs culled from The SNP Consortium (TSC) database (www.snp.cshl.org) for those with a delta level (δ) >0.40 between any two of three available groups used to characterize these SNPs in the database (African Americans, Caucasians—primarily European Americans—and East Asians). After ranking these candidate AIMs by δ level, we genotyped the selected West African, European, East Asian, and Indigenous American parental samples. For the first pass, we focused on the 400 candidate AIMs with the largest δ values and selected 96 validated AIMs, which were determined to be:

- From true SNPs (three clear genotype classes, whereas certain genotyping results such as XX—YY smears are indicative of lack of primer specificity)
- With minor allele frequency greater than 1%
- Of $\delta > 0.40$ for at least one of the group pairs

These AIMs were spread throughout 21 of the 22 autosomal chromosomes (the twenty-third being the X/Y chromosome pair) and the average chromosome was represented by three AIMs (see Figure 4-9). To test the departures from independence in allelic state within and between loci, we used the MLD exact test, which is described in Zaykin et al. (1995). Each AIM had alleles in Hardy-Wienberg equilibrium, and none were found to be in linkage disequilibrium with one another. Of these 73 were selected that fit neatly within multiplex PCR reactions.

As described in Chapter 2, the δ value is an expression of the ancestry information content for alleles of a given marker (Chakraborty & Weiss 1988). For a biallelic marker, the frequency differential (δ) is equal to $p_x - p_y$, which is equal to $q_y - q_x$, where p_x and p_y are the frequencies of one allele in populations X and Y and q_x and q_y are the frequencies of the other. The collection of 73 AIMs used for this work was selected to maximize the cumulative δ value between each of the six possible pairs of the four-dimensional (West African, Indigenous American, European, and East Asian) problem. Table 4-1 shows the delta values for each of the population pairs. Delta levels can be summed over a panel of markers to summarize the overall and relative amounts of information available for particular "axes of ancestry" that can be measured. In these SNPs, we find that there are similar levels of ancestry information (summed δ) for the three African comparisons (29.8, 30.8, and 29.4) for European, East Asian, and Indigenous American, respectively. There is somewhat less information for comparing European to East Asian and

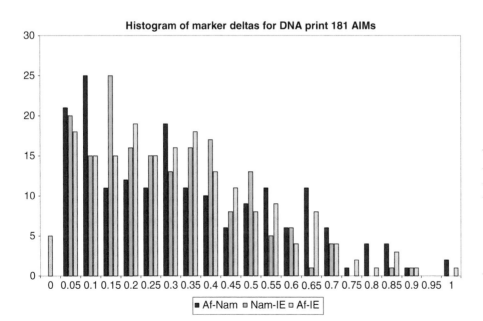

Figure 4-9

Delta values for the 181 AIMs comprising the AncestryByDNA panel described in the text. The number of AIMs is shown on the y-axis and the delta value for three of the four possible binary population comparisons on the x-axis, indicated with colors as specified at the bottom.

Table 4-1

Pairwise δ values for the 71 AIMs used in the BGA test described in the text.

AIM*	AIM code**	Afr-Eur	Afr-EAs	Afr-IAm	Eur-EAs	Eur-IAm	EAs-IAm
rs132560_GC	958	0.336	0.214	0.24	0.12	0.09	0.03
rs913258_GC	960	0.751	0.207	0.64	0.54	0.11	0.44
rs147659_GC	961	0.564	0.549	0.53	0.01	0.04	0.02
rs959929_GC	963	0.286	0.364	0.08	0.08	0.36	0.44
rs9032_GC	964	0.845	0.286	0.36	0.56	0.48	0.08
rs138335_GC	966	0.327	0.464	0.01	0.14	0.33	0.47
rs104132_TC	969	0.321	0.097	0.49	0.42	0.17	0.58
rs716373_TC	970	0.163	0.171	0.06	0.01	0.11	0.11
rs188895_TC	971	0.331	0.7	0.57	0.37	0.24	0.13
rs236919_TC	972	0.497	0.186	0.16	0.31	0.34	0.03
rs67302__TC	973	0.339	0.779	0.62	0.44	0.28	0.16
rs662117_TC	976	0.851	0.252	0.37	0.6	0.48	0.12
rs869337_TC	977	0.372	0.725	0.85	0.35	0.48	0.13
rs725192_TC	978	0.041	0.122	0.18	0.08	0.14	0.06
rs140686_TC	979	0.549	0.268	0.19	0.28	0.36	0.07
rs173537_TC	980	0.606	0.671	0.52	0.07	0.08	0.15
rs984654_TC	993	0.102	0.771	0.61	0.67	0.51	0.16
rs972801_TC	1000	0.623	0.071	0.09	0.55	0.53	0.02
rs101092_TC	1015	0.038	0.1	0.14	0.06	0.1	0.04
rs165099_TC	1022	0.631	0.265	0.06	0.37	0.57	0.2
rs730570_GA	1029	0.674	0.025	0.14	0.65	0.82	0.17
rs729253_GA	1033	0.523	0.714	0.87	0.19	0.35	0.16
rs733563_GA	1034	0.634	0.295	0.6	0.34	0.03	0.31
rs430667_GA	1035	0.172	0.539	0.63	0.37	0.46	0.09
rs3287_GA	1036	0.108	0.449	0.51	0.56	0.62	0.07
rs434504_GA	1040	4.00E−04	0.711	0.17	0.71	0.17	0.54
rs593226_GA	1041	0.65	0.031	0.29	0.68	0.36	0.32
rs100457_GA	1043	0.314	0.535	0.38	0.22	0.06	0.16
rs142620_GA	1044	0.334	0.807	0.52	0.47	0.19	0.29
rs146704_GA	1047	0.343	0.686	0.55	0.34	0.21	0.13
rs251741_GA	1048	0.168	0.537	0.08	0.37	0.08	0.45

Table 4-1
Continued.

AIM[*]	AIM code[**]	Afr-Eur	Afr-EAs	Afr-IAm	Eur-EAs	Eur-IAm	EAs-IAm
rs236336_GA	1049	0.686	0.8	0.96	0.11	0.27	0.16
rs960709_GA	1050	0.606	0.114	0.25	0.49	0.35	0.14
rs110871_GA	1051	0.348	0.586	0.58	0.24	0.23	0.01
rs998599_GA	1053	0.263	0.171	0.26	0.09	0.52	0.43
rs675837_GA	1055	0.534	0	0.27	0.53	0.26	0.27
rs361065_GA	1056	0.361	0.742	0.76	0.38	0.4	0.01
rs716873_GA	1057	0.153	0.022	0.03	0.18	0.18	0.01
rs716840_GA	1058	0.043	0.043	0.04	0	0	0
rs783064_GA	1060	0.392	0.725	0.39	0.33	0	0.33
rs212498_GA	1062	0.729	0.829	0.59	0.1	0.14	0.24
rs741272_GA	1064	0.456	0.193	0.78	0.26	0.32	0.58
rs729531_GA	1066	0.157	0.157	0.16	0	0	0
rs107680_GA	1068	0.043	0.043	0.04	0	0	0
rs115551_GA	1071	0.326	0.013	0.11	0.34	0.44	0.1
rs111110_GA	1073	0.071	0.071	0.07	0	0	0
rs590086_GA	1075	0.704	0.809	0.81	0.1	0.1	0
rs107616_GA	1076	0.464	0.535	0.34	0.07	0.13	0.2
rs361055_GA	1077	0.348	0.728	0.74	0.38	0.39	0.01
rs139927_GA	1081	0.261	0.736	0.61	0.47	0.35	0.13
rs172982_GA	1082	0.343	0.536	0.41	0.19	0.06	0.13
rs133703_GA	1083	0.454	0.821	0.87	0.37	0.42	0.05
rs933199_GA	1084	0.036	0.45	0.17	0.49	0.21	0.28
rs595961_GA	1087	0.74	0.117	0.35	0.62	0.39	0.23
rs2814778_TC	1111	0.985	1	0.99	0.01	0.01	0.01
rs1800498_TC	1113	0.495	0.093	0.15	0.59	0.64	0.06
rs3340_TC	1116	0.133	0.2	0.67	0.07	0.54	0.47
rs3846662_TC	1117	0.548	0.407	0.49	0.14	0.05	0.09
rs3176921_TC	1120	0.567	0.65	0.64	0.08	0.07	0.01
rs2065160_TC	1121	0.368	0.379	0.47	0.75	0.84	0.09
rs285_TC	1122	0.575	0.664	0.59	0.09	0.02	0.07
rs2862_TC	1124	0.177	0.086	0.3	0.26	0.48	0.21

(*continued*)

AIM[*]	AIM code[**]	Afr-Eur	Afr-EAs	Afr-IAm	Eur-EAs	Eur-IAm	EAs-IAm
rs594689_GA	1128	0.416	0.089	0.02	0.33	0.39	0.07
rs2161_GA	1130	0.149	0.1	0.1	0.25	0.25	0
rs3287_GA	1136	0.473	0.634	0.61	0.16	0.14	0.02
rs2242480_GA	1137	0.892	0.667	0.22	0.23	0.68	0.45
rs2695_GA	1138	0.076	0.279	0.69	0.2	0.61	0.41
rs6003_GA	1139	0.612	0.66	0.62	0.05	0.01	0.04
886993_GA	1140	0.435	0.6	0.29	0.17	0.14	0.31
rs3317_GA	1141	0.487	0.564	0.61	0.08	0.12	0.04
rs1800410_GA	1146	0.155	0.571	0.31	0.73	0.46	0.26
	Delta sum	28.6	29.5	28.9	20.9	19.8	12.1
	Aim count	52	47	49	41	40	22

Note: Populations are designated as Afr—West African, Eur—European, EAs—East Asian, IAm—Indigenous American.
[*]Sequences accessible at http://www.ncbi.nlm.nih.gov/entrez/query.fcgi?db=snp.
[**]Internal DNAPrint Genomics, Inc. tracking number. The summed delta levels and number of AIMs with $\delta > 0.20$ for each pairwise comparison is shown at the bottom of the list.

Indigenous American (21.9 and 20.1, respectively) and the least amount of ancestry information in this panel is available between East Asian and Indigenous American (12.8). Details on the performance of this panel, compared to a larger panel, are forthcoming in the following pages (we will sometimes use only 71 of the most reliably genotyped of these 73 AIMs, and when applicable, this number is indicated).

171 Marker Test

The 71-marker test was used as the basis of an Ancestry test at DNAPrint for approximately 18 months under the product name AncestryByDNA 2.0. AIMs for a larger panel of 181 markers were selected in a similar way, a supplementary panel of 110 AIMs were added to the existing panel of 71 markers just described. The distribution of the AIMs along the chromosomes is relatively even (see Figure 4-10). The motivation for increasing the number of markers from 71 to 181 was to increase the precision of non-African–non-African distinctions (such as European–East Asian). Thus, although the goal was to have an additional panel of markers selected based on their European–East Asian, European–Indigenous American, and East Asian–Indigenous American discrimination power, there was no large dataset available on Indigenous

American allele frequencies. As such we had to use an alternative approach in applying the Locus-Specific Branch Length (LSBL) method previously described in Shriver et al. (2004) and in Chapter 2. Briefly, the LSBL approach allows us to combine the information from three pairwise comparisons (i.e., Afr/Eur, Afr/EAs, EAs/Eur) to identify markers that are high-δ between one of them and both of the others. In this way we could focus on the best markers for East Asian/other distinctions and those that are best for European/other distinctions, with the goal of enriching the now expanded TSC database screen for non-African contrast ancestry information. We will distinguish between these two overlapping panels to illustrate differences in performance throughout this text. The allele frequencies for the AIMs that distinguish the 71 marker panel from the 181 marker panel are listed in Table 4-2 (data for 110 AIMs is shown) and a histogram of delta values for the markers in the panel is shown in Figure 4-9.

Only 171 of the 181 markers that genotype the most reliably were selected (and of these, due to vagaries associated with the updating of the publicly available genome reference sequence, precise locations are not known for 5 at the time of this writing, leaving 166 AIMs). The chromosomal distribution of these 166 are plotted in Figure 4-10. We can see that the largest chromosomes (i.e., chromosome 1) tend to have the largest number of AIMs in our panel. Recall that the AIMs were not selected based on knowledge of their genetic locations. From the allele frequency differentials among pairs of the four continental groups, we can calculate the information content for each marker for the distinction between a given pair of populations and we can sum them to obtain the information content for all markers combined. We have described various methods for calculating the information content of AIMs; here we show calculations summed for all six of the continental population pairs (Molokhia et al. 2003; Table 4-3). According to Paul McKeigue at University College Dublin, Ireland, to measure the admixture between any pair of populations with a standard error of less than 10% requires an information content (using this method) of at least 12.5 (personal communication; Molokhia et al. 2003). This is equivalent to 50 markers with 0.4 information content for each. Applying this method to the 171 AIMs we can see that there is adequate information for the measurement of admixture with a standard error of less than 10% for each pair among the four continental ancestries, except East Asian–Indigenous American, where we would expect to encounter a slightly greater than 10% standard error (see Table 4-3). We will address the precision in all of the continental population dimensions later when we discuss the use of simulated genotypes to establish the accuracy and precision of the panel, measured as the bias.

In Table 4-3 we have also included three nontarget population groups—South Asians, Middle East, and Pacific Islanders—and from this table it is clear

Table 4-2

AIMs added to the 71 in Table 4-1 to comprise the 171 AIM panel.

AIM*	AIM code**	Allele	Afr	Eur	IAm	EAs	SAs	MEs	PIs
rs296528_GC	1648	C	0.22	0.36	0.01	0.03	0.19	0.21	0.02
rs953743_GC	1649	C	0.16	0.51	0.16	0.19	0.3	0.38	0.02
rs2010069_GC	1650	C	0.45	0.44	0.81	0.67	0.54	0.51	0.42
rs765338_GC	1651	C	0.94	0.9	0.81	0.45	0.79	0.86	0.6
rs1998055_GC	1654	C	0.92	0.77	0.99	0.92	0.76	0.81	0.94
rs1018919_GC	1658	G	1	0.76	0.75	0.68	0.69	0.8	0.46
rs1015081_TC	1661	C	0.47	0.43	0.42	0.22	0.39	0.48	0.2
rs772436_TC	1662	C	0.79	0.31	0.72	0.81	0.67	0.43	0.52
rs1395580_TC	1663	C	0.39	0.8	0.86	0.51	0.74	0.74	0.3
rs877823_TC	1665	C	0.74	0.76	0.96	0.99	0.86	0.74	0.92
rs1937147_TC	1666	C	0.62	0.52	0.94	0.87	0.65	0.66	0.88
rs770028_TC	1667	C	0.71	0.9	1	0.94	0.87	0.91	1
rs1039630_TC	1668	C	0.52	0.36	0.59	0.67	0.83	0.49	0.74
rs1385851_TC	1669	C	0.33	0.84	1	0.75	0.63	0.84	0.82
rs758767_TC	1670	C	0.49	0.83	0.62	0.74	0.66	0.74	0.92
rs1469344_TC	1671	C	0.9	0.64	0.98	0.95	0.75	0.65	0.95
rs262482_TC	1673	C	0.02	0.31	0.01	0.04	0.17	0.13	0
rs2014519_TC	1675	C	0.35	0.32	0.37	0.65	0.31	0.44	0.42
rs715790_TC	1677	C	0.15	0.42	0.56	0.57	0.36	0.33	0.32
rs1465708_TC	1680	C	0.36	0.3	0.69	0.73	0.43	0.21	0.39
rs1375229_TC	1682	C	0.45	0.02	0	0.08	0.07	0.06	0.1
rs878671_TC	1683	C	0.2	0.39	0.6	0.74	0.54	0.41	0.76
rs1848728_TC	1684	C	0.72	0.61	0.9	0.98	0.74	0.43	0.72
rs715956_TC	1687	C	0.39	0.2	0.34	0.65	0.3	0.12	0.66
rs2045517_TC	1688	C	0.24	0.66	0.51	0.5	0.59	0.73	0.16
rs1368872_TC	1689	C	0.06	0.47	0.13	0.01	0.39	0.5	0.1
rs917502_TC	1691	C	0.41	0.51	0.96	0.87	0.71	0.46	1
rs1005056_TC	1692	C	0.08	0.1	0.45	0.56	0.29	0.05	0.34
rs351782_TC	1693	C	0.69	0.95	0.97	0.81	0.83	0.93	1
rs915056_TC	1696	C	0.06	0.66	0.37	0.46	0.53	0.58	0.33
rs1983128_TC	1698	C	0.94	0.63	0.87	0.96	0.95	0.73	1

Table 4-2
Continued.

AIM[*]	AIM code[**]	Allele	Afr	Eur	IAm	EAs	SAs	MEs	PIs
rs723220_TC	1700	C	0.89	0.65	0.14	0.43	0.56	0.66	0.7
rs1407961_TC	1701	C	0.46	0.91	0.97	0.84	0.87	0.99	0.98
rs174518_TC	1702	C	0.79	0.49	0.37	0.34	0.38	0.66	0.22
rs1546541_TC	1703	C	0.53	0.2	0.56	0.09	0.21	0.13	0.14
rs1296149_TC	1704	C	0.69	0.49	0.37	0.16	0.34	0.6	0.46
rs719909_TC	1705	C	0.86	0.86	0.76	0.56	0.74	0.73	0.48
rs913375_TC	1706	C	0.72	0.43	0.96	0.58	0.69	0.58	0.58
rs721702_TC	1708	C	0.51	0.83	0.81	0.58	0.72	0.74	0.78
rs1426217_TC	1709	C	0.5	0.39	0.57	0.1	0.4	0.36	0.16
rs721825_TC	1710	C	0.12	0.25	0.4	0.84	0.41	0.31	0.88
rs1470144_TC	1712	C	0.25	0.67	0.2	0.77	0.52	0.77	0.9
rs1041656_TC	1713	C	0.49	0.58	0.31	0.24	0.4	0.31	0.16
rs1528037_TC	1714	C	0.8	0.59	0.51	0.19	0.3	0.6	0.06
rs1157223_TC	1716	C	0.31	0.12	0	0.01	0.11	0.07	0
rs1501680_TC	1717	C	0.56	0.76	0.46	0.36	0.67	0.9	0.22
rs763807_GA	1721	A	0.34	0.14	0.3	0.63	0.44	0.21	0.64
rs667508_GA	1726	A	0.33	0.82	0.81	0.86	0.84	0.83	0.96
rs735050_GA	1728	A	0.14	0.76	0.73	0.84	0.92	0.77	1
rs125097_GA	1731	A	0.61	0.46	0.28	0.24	0.33	0.46	0.3
rs1125508_GA	1734	A	0.76	0.31	0.25	0.59	0.17	0.35	0.48
rs725395_GA	1739	A	0.18	0.45	0.83	0.82	0.42	0.43	0.65
rs1221172_GA	1743	A	0.9	0.53	0.14	0.22	0.39	0.49	0.42
rs1431332_GA	1746	A	0.8	0.52	0.98	0.91	0.81	0.68	0.8
rs1447111_GA	1747	A	0.78	0.66	0.84	0.91	0.59	0.63	0.96
rs717836_GA	1748	A	0.24	0.12	0	0.18	0.13	0.23	0.16
rs1039917_GA	1751	A	0.02	0.38	0.51	0.41	0.47	0.35	0.59
rs1125425_GA	1752	A	0.33	0.11	0.36	0.5	0.11	0.15	0.54
rs1437069_GA	1753	A	0.31	0.23	0.51	0.78	0.29	0.31	0.94
rs883055_GA	1757	A	0.01	0.05	0.21	0.62	0.19	0.09	0.63
rs925197_GA	1759	A	0.41	0.09	0.17	0.22	0.37	0.1	0.02
rs409359_GA	1760	A	0.26	0.52	0.55	0.56	0.65	0.44	0.75

(continued)

Table 4-2
Continued.

AIM*	AIM code**	Allele	Afr	Eur	IAm	EAs	SAs	MEs	PIs
rs282496_GA	1761	A	0.46	0.7	0.94	0.76	0.9	0.8	0.96
rs2204307_GA	1762	A	0.01	0.14	0.21	0.44	0.21	0.2	0.58
rs997676_GA	1763	A	0.04	0.12	0.14	0.24	0.26	0.3	0.33
rs53915_GA	1766	A	0.24	0.66	0.31	0.73	0.53	0.49	0.9
rs713503_GT	1767	G	0.34	0.52	0.86	0.9	0.78	0.5	0.8
rs987284_GT	1769	G	0.72	0.57	0.09	0.25	0.53	0.44	0.22
rs1454284_GT	1770	G	0.49	0.38	0.96	0.42	0.2	0.34	0.44
rs275837_GT	1771	G	0.57	0.46	0.19	0.2	0.29	0.31	0.06
rs1040577_GT	1773	G	0.36	0.45	0.25	0.28	0.54	0.57	0.08
rs553950_GT	1776	G	0.99	0.95	0.79	0.46	0.89	0.83	0.33
rs974324_GT	1777	G	0.72	1	0.99	0.93	1	0.99	0.98
rs320075_GT	1778	G	0.68	1	1	1	1	1	1
rs270565_GA	1779	A	0.52	0.9	1	0.87	0.94	0.91	1
rs1446966_GA	1780	A	0.52	0.42	0.46	0.88	0.55	0.44	0.64
rs950848_GA	1782	A	0.11	0.01	0.18	0.28	0	0.01	0.18
rs81481_GA	1784	A	0.75	0.17	0.06	0.29	0.36	0.18	0.42
rs875543_GA	1785	A	0.76	0.83	0.72	0.54	0.81	0.86	0.58
rs1147703_GA	1786	A	0.27	0	0.01	0	0	0	0
rs877783_GA	1787	A	0.59	0.79	0.57	0.47	0.73	0.57	0.24
rs1395579_GA	1790	A	0.1	0.63	0.08	0.1	0.66	0.57	0.18
rs90192_GA	1791	A	0.67	0.85	0.6	0.46	0.5	0.73	0.33
rs697212_GA	1793	A	0.12	0.5	0.22	0.74	0.45	0.44	0.85
rs522287_GA	1799	A	0.32	0.58	0.08	0.01	0.53	0.53	0.02
rs526454_GA	1801	A	0.95	0.93	0.8	0.43	0.59	0.8	0.13
rs892263_GA	1802	A	0.07	0.11	0.08	0.48	0.34	0.21	0.35
rs523200_GT	1804	G	0.94	0.11	0.42	0.37	0.17	0.11	0.38
rs263531_GT	1805	G	0.42	0.54	0.1	0.25	0.3	0.54	0.22
rs559035_GT	1806	G	0.46	0.85	0.53	0.2	0.57	0.73	0.36
rs1030525_GT	1807	G	0.49	0.01	0.47	0.14	0.15	0	0.02
rs1432065_GT	1809	G	0.64	0.78	0.92	0.8	0.71	0.88	0.79

Table 4-2
Continued.

AIM*	AIM code**	Allele	Afr	Eur	IAm	EAs	SAs	MEs	PIs
rs997164_GT	1810	G	0.81	0.58	1	0.98	0.84	0.54	1
rs1533677_GT	1811	G	0.72	0.45	0.7	0.89	0.86	0.55	0.78
rs1034290_GT	1814	G	0.7	0.63	0.26	0.04	0.44	0.71	0.12
rs625994_GC	1816	C	0.73	0.9	0.4	0.57	0.86	0.77	0.82
rs1869380_GC	1817	C	0.16	0.36	0.64	0.81	0.59	0.39	0.96
rs830599_GC	1818	C	0.16	0.31	0.97	0.96	0.6	0.49	0.82
rs920915_GC	1823	C	0.26	0.5	0.66	0.29	0.3	0.53	0.18
rs1113337_GA	1827	A	0.21	0.58	0.42	0.51	0.6	0.57	0.31
rs841338_GA	1828	A	0.54	0.97	0.83	0.81	1	0.94	0.85
rs1407361_GA	1829	A	0.57	0.31	0.66	0.68	0.39	0.23	0.71
rs1414241_GA	1833	A	0.81	0.54	0.82	0.63	0.37	0.4	0.5
rs892457_GA	1837	A	0.59	0.48	0.61	0.92	0.56	0.47	0.81
rs1937025_GA	1840	A	0.2	0.04	0.22	0.51	0.18	0.08	0.71
rs1415680_GA	1843	A	0.77	0.31	0.15	0.04	0.29	0.24	0.06
rs727878_GA	1844	A	0.46	0.28	0.26	0.75	0.51	0.48	0.54
rs959858_GA	1845	A	0.74	0.46	0.24	0.08	0.21	0.4	0.19
rs1155513_GA	1847	A	0.86	0.54	0.98	0.88	0.79	0.66	0.83
rs718268_GA	1848	G	0.24	0.29	0.57	0.56	0.26	0.3	0.75

*Sequences accessible via NCBI dbSNP at http://www.ncbi.nlm.nih.gov/entrez/query .fcgi?db=snp.
**Internal DNAPrint Genomics, Inc. tracking number.

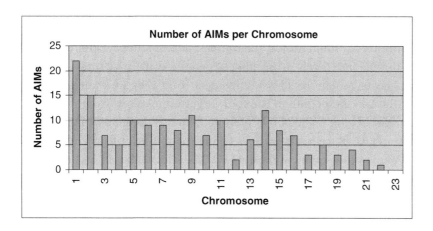

Figure 4-10

Chromosomal distribution of 166 of the most reliably genotyped among the set of 181 AIMs selected from the genome based on their information content with respect to a global four-population model. The number of AIMs is shown on the y-axis and the chromosome on the x-axis.

Table 4-3

Information content for the 171 AIMs described in the text. Numbers in bold indicate information adequate to measure admixture with a standard error <10%.

	West African	Indigenous European	Indigenous American	East Asian	South Asian	Middle Eastern	Pacific
West African	0						
European	27	0					
Indigenous American	31.6	19.7	0				
East Asian	34.9	23.1	11.1	0			
South Asian	24.8	5.9	12.9	13	0		
Middle Eastern	26	2	18.3	20.3	4.7	0	
Pacific	37.2	23.6	14.5	5.4	14.9	20.8	0

that many of these AIMs also have ancestry information for some of the subpopulation pairs. Table 4-2 shows the allele frequencies for these populations as well, and from these the information content was calculated for Table 4-3 (Molokhia et al. 2003); many show good information (>0.25) between some of the pairs of the three nontarget subpopulations, though not nearly as good as in the dimension for which they were selected (continental ancestry). What makes this interesting is that the markers, of course, were selected from the genome based on their information content for continental group, not intracontinental subpopulations (such as Middle Eastern and South Asian populations, which show predominant European ancestry by our definition). Many of the 171 AIMs are of relatively high minor allele frequency in most or all of the subpopulations, and thus it appears they very likely existed as polymorphisms before the ancestral groups separated from one another some 50 KYA. This follows because recent polymorphisms are expected to be less frequent, and more private.

If these AIMs existed as polymorphisms before the populations branched from one another, it would indicate that their allele frequency differentials in our target populations likely arose due to genetic drift or natural selection rather than novel mutation. This would mean that they were already polymorphic before the branching of the four target populations from the ancestral, African population 50,000 or so years ago, and so the genetic drift would be expected to take place in all lineages taking the polymorphisms with them. This would explain why many of these 171 AIMs, selected based on their West African, European, East Asian, and Indigenous American discrimination power harbor adequate information content for certain subgroup discriminations, such as between South Asians and West Africans (24.8), or Pacific

Islanders and Europeans (23.6), even though we did not select them from the genome for this power. The conclusion from these types of observations is that good continental AIMs tend to be better intracontinental AIMs than randomly selected SNPs, which are not expected to carry such information (see Chapter 2).

PRESENTING INDIVIDUAL BIOGEOGRAPHICAL ANCESTRY ADMIXTURE (BGAA) RESULTS

The calculation of BGA estimates has been covered in detail in Chapter 2. We have largely used the Maximum Likelihood Estimation (MLE) approach with a specific attention on reporting the accuracy on the probability space occupied by the multilocus genotype representing any particular person. Since we are working to generate the best profile on one particular person, we felt from the outset that such a profile is not simply the point estimates of BGA components, but instead a clear and precise illustration of the underlying probability space defined by the individual's multilocus genotype in the context of the parental allele frequencies. Next we expand on the alternative methods employed to illustrate the results.

HISTOGRAM WITH CONFIDENCE BAR INTERVALS

As shown in Figure 4-11, one way to figuratively display individual admixture estimates is in the form of a histogram. Bars represent the percentage of affiliation with each parental population estimated, with the two-fold likelihood confidence interval, which corresponds roughly to an 85% confidence interval. The histogram is read in the following manner: a value within the interval could also be the correct answer, but is up to two times less likely to be the correct answer than that indicated by any other value within the bar region. Another way to read the histogram is in saying the probability that the correct answer is found within the confidence range is 85%. Alternatively, we could plot the five-fold (roughly corresponding to a 95% confidence interval) or 10-fold (roughly corresponding to a 98% confidence interval). The advantage of this display is that we are not limited by dimensionality as with some of the other methods described next—we can run the analysis assuming a four-way population model and plot all four percentages on the same plot.

For people of simple or polarized admixture, the bar graph is a useful presentation because it provides a separate, objective view of the possible ancestry percentages for any one particular group, one group at a time. Genealogists seeking to confirm a great-great-grandparent of Indigenous

Figure 4-11

Histogram of BGA admixture proportions for an individual of primarily Native or Indigenous American ancestry. Blue bars indicate the most likely estimate (MLE) and red bars indicate the 85% confidence range as shown in the key below the histogram.

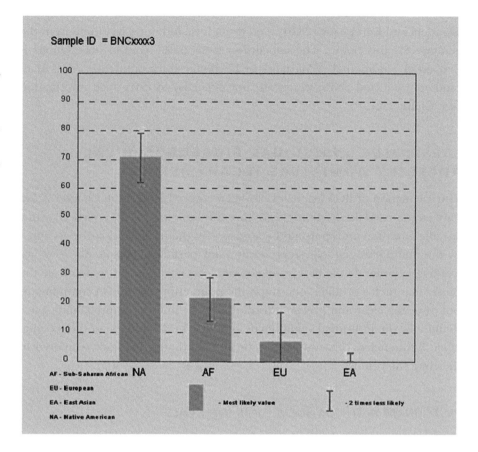

American ancestry who have obtained an MLE showing 100% European find the bar graph useful in understanding the statistical meaning of their results and how it may still be possible (though less likely) that there is a small amount of Indigenous American ancestry if the error bars on the Indigenous American column exceed zero.

The bar graph is made by taking the MLE values from the three-way or four-way MLE algorithms discussed in Chapter 2, where the likelihood of all possible percentage combinations is determined. These are plotted in the 3D (for individuals of polarized ancestry) or 4D triangle plots (for individuals of more complex admixture), to be discussed shortly, but the values are equally useful for the bar plot. The confidence ranges are established for each ancestry group by searching the likelihood calculation of all three-population or four-population admixture combinations and finding the highest and lowest values for each group that fall within the two-fold likelihood range (0.3 Log base 10) of the MLE. For people of more complex admixture (i.e., four-population admixture, such as 30% European, 30% African, 20% Indigenous American, and 20% East Asian, which might be obtained from a person with a Dominican father and a Philippine mother), the MLE is *de facto* computed using a more complex 4-D methodology. Because the triangle plot projects the results in terms of most likely three-population mixture results, and because individuals of four-way admixture are obviously more complex than this, on a group-by-group basis, the bar graph is more informative for some individuals than graphical plots we discuss later.

TRIANGLE PLOT WITH CONFIDENCE RINGS

A different way to plot the data is in the form of a triangle plot, as shown in Figure 4-12. Triangle plots are a common means of displaying 3D data in flat space that is possible only when the sum of the three axes is 1. Within the triangle, we use a grid system to localize the maximum likelihood estimate. As shown in the left-hand side of the figure, the percentage of ancestry can be scaled on a line extending from a base of the triangle (representing 0%) to its opposite vertex (representing 100%). A Maximum Likelihood Estimate (MLE) of 16% Indigenous American ancestry would be plotted 84% of the way along the line bisecting the Indigenous American vertex. Similarly, we can draw the lines corresponding to the other two percentages, in our example, 16% West African and 68% European, and where these three lines intersect is the point within the triangle that corresponds to the position of the MLE—68% European, 16% Indigenous American, 16% West African. Adding scaled percentages in this way for any point inside the triangle gives a total value of 100%, by geometrical definition.

The MLE plotted in the triangle plot is a statistical estimate, and as such should be qualified with confidence intervals. Though the MLE represents the best estimate of the proportions, in reality, the maximum is just one summary of the underlying probability space describing the multilocus genotype. The confidence intervals delimit other point estimates that are also likely. (In order to know a person's genomic ancestry proportions with near 100% confidence, we would have to perform the test for each region of the variable genome, which would make the test prohibitively expensive.)

To illustrate the boundaries of statistical error, we calculate and plot all the estimates that are within two times, five times, and 10 times less likely than the MLE. The first contour (black line) around the MLE delimits the space outside of which the points are more than two times less likely, and the second contour (blue line) around the MLE delimits the space outside of which the estimates are more than five times less likely than the MLE. The third contour (yellow line) delimits the space outside of which the estimates are more than 10 times less likely than the MLE, with the fold decrease in likelihood increasing continuously with increasing distance from the MLE. The yellow circle (10X contour) is also

Figure 4-12

Triangle plot showing a graphical representation of BGA admixture proportions for an individual of primarily European ancestry. On the left, the MLE is shown with a red spot. When this spot is projected perpendicularly onto the three-way grid system shown within the triangle, the position along each axis reveals the percentage of corresponding admixture (arrows). On the right, the MLE is shown with its associated two-fold, five-fold, and 10-fold confidence contours as described in the text.

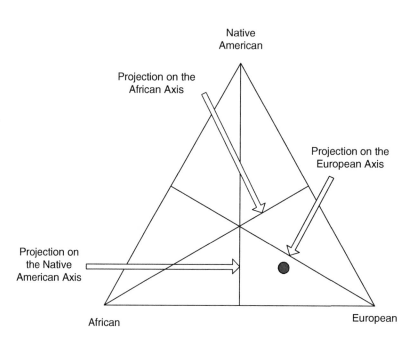

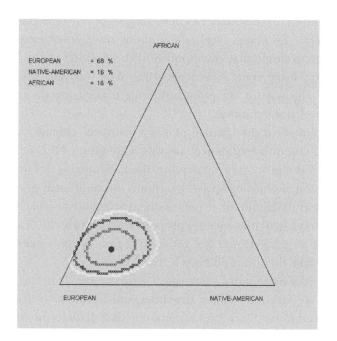

referred to as the one-log interval. Within this first contour, the likelihood gradient proceeds from one-fold (at the MLE) to two-fold (at the contour). Some examples of the size and shape of contour rings around the MLE are shown for three individuals in Figure 4-13 (which are actually compound

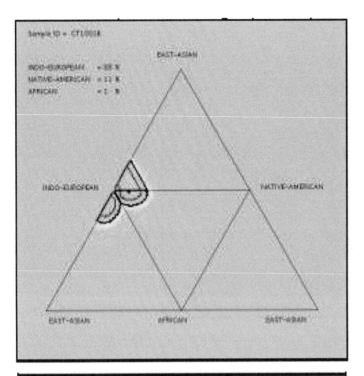

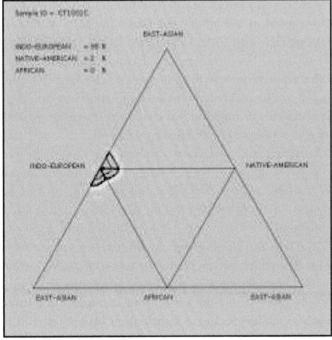

Figure 4-13

Compound triangle (tetrahedron) plots for three individuals showing the diversity of confidence contour shapes that can be encountered. When the compound triangle is folded along the three sides of the internal triangle, such that each of the three vertices in the compound triangle meet behind the plane of the paper, the shape of the contour space is always bilaterally symmetrical. The MLE for the individual at the bottom is actually a four-way MLE at the centroid of the cylinder formed by this folding.

Figure 4-13
Continued.

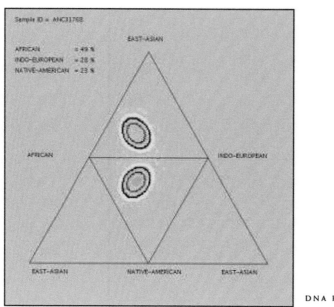

triangle plots called tetrahedron plots as we will describe shortly). The greater the number of AIMs included in the model, the closer the contour lines fit around the MLE point (see Figure 5-13 later in Chapter 5). One could develop a panel that produces estimates with contour lines much closer to the MLE than in Figure 4-13, however this would require us to genotype a much larger collection of AIMs, which of course could quickly render the test unaffordable for large-scale applications. As we will see throughout this chapter, 171 markers provide estimates of confidence level reasonable for most applications.

The advantage of the triangle plot as a type of display is that we can illustrate graphically all of the possible values for a given sample in an easy to comprehend fashion. In addition, we can plot multiple samples in the same triangle, giving us a quick glimpse of the variability within populations and pattern among different populations (see Figure 4-14). The main disadvantage is that we are forced to pick the best three-population model for each sample, since plotting four groups on a two-dimensional plane is not possible. If an individual's ancestry is a function of affiliation with all four of our groups, they either cannot be plotted on the plot or only the result from the best fitting three-group model (three-way admixture estimate with highest likelihood) can be plotted. All of the MLEs shown in Figure 4-14, for instance, either were determined to best fit the West African, Indigenous American, European three-way population model or were forced into this model by taking the highest likelihood score assuming such a model. Of course, since only one of

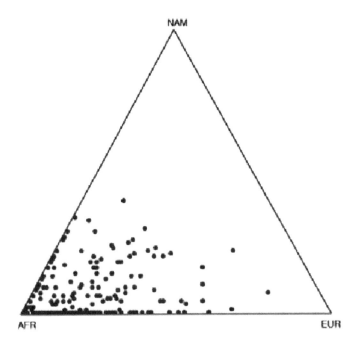

Figure 4-14

A triangle plot allows for an appreciation of within-sample variability. Shown is a plot for a sample of about 150 African Americans. The MLE for each sample is represented by a black dot. Some of the samples show predominantly Western African ancestry, but others show substantial European (EUR) or Indigenous American (NAM) admixture. The use of different colors for different populations would make it relatively easy to visualize significant trending differences.

the four possible three-way models can be shown in any one plot, this increases the number of inherent assumptions, making the model more "wrong" than it has to be and thus, less useful.

PYRAMID PLOT

Figure 4-15 shows an example of an alternative to the triangle plot called a pyramid plot. The pyramid plot essentially combines three triangle plots, but each triangle is no longer equilateral and so the axes are not linearly graded. The advantage of the pyramid plot is that we can show three of the four possible three-way models together, which is important if plotting the MLE of multiple individuals—some of whom may best fit with one three-way model and others with a different three-way model. A single sample can be plotted in only one of the three triangles so we may lose information for samples best described with four-dimensional admixture, but we no longer have to force all samples into a *single* three-way model—the best of three possible three-way models is available to choose from. For single populations, such as Europeans, the pyramid plot works out well because most individuals have at least some European ancestry, which would become the centroid of the pyramid for a European sample plot, and so all the MLEs would plot as a cluster around the centroid. The disadvantage is that it is more difficult to estimate the percentages since the scales are no longer linear, and we are still missing one of the possible four three-way models.

Figure 4-15

A pyramid plot useful for projecting (three-way) BGA admixture results obtained by using a four-population model. Shown are the most likely three-way MLEs plotted for a sample of parental West Africans. In this example, all three triangle plots involving West African (AF) admixture are shown; samples for which the most likely three-way model involves European (IE), East Asian (EA), and West African (AF) admixture are shown in the appropriate (upper) subtriangle with blue spots as indicated in the legend to the lower left. This method sacrifices information available when incorporating a four-population model (and using AIMs with information content for four rather than just three populations). For example, an individual with four-way admixture would be forced into a three-way model and plotted in one of the subtriangles, and a sample lacking AF admixture would not be possible to plot here. Since three-way population models are not as optimal for assessing population structure of most modern populations, this type of plot rarely is used.

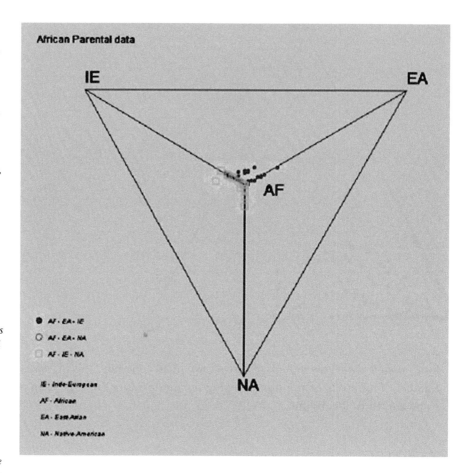

TETRAHEDRON PROJECTION PLOT

The most satisfying graphical presentation of an MLE determined using a four-population model is the tetrahedron plot (see Figure 4-16), because we can see the MLE of all samples plotted using the most suitable three-population model for the sample (whichever it may happen to be). In addition, with the tetrahedron plot we can see the extension of the confidence contours into space corresponding to other three-population models than those corresponding to the MLE, which for many samples would be required for a complete presentation of the likelihood space. The tetrahedron projection holds four smaller equilateral triangles or subtriangles, representing all four possible three-way population models. The percentages for any point in each of these triangles is read as described for the triangle plot.

The centroid of the likelihood space (actually the centroid of the three-dimensional likelihood space, which we will get to shortly) is the MLE, represented with the red dot in Figure 4-16. Since the triangles are equilateral, the scaling along the axes is linear and the percentage admixture corresponding to any point within the

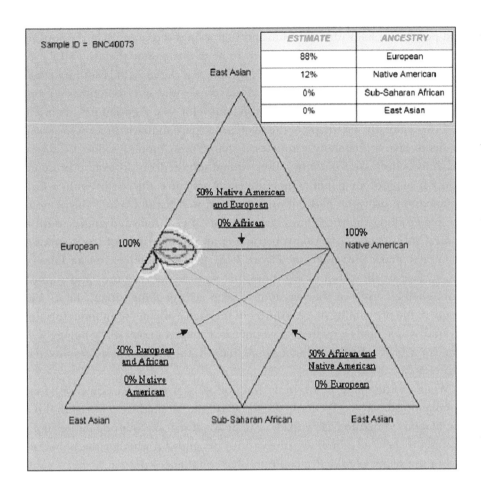

ESTIMATE	ANCESTRY
88%	European
12%	Native American
0%	Sub-Saharan African
0%	East Asian

Figure 4-16

Tetrahedron plot of MLE values for an individual of primarily European admixture. The tetrahedron plot is composed of each of the four possible three-way triangle plots. If the MLE is composed of only three of the four groups, it appears on the surface of the plot in the appropriate subtriangle. As before, the graphical representation of the MLE and its associated confidence contour elements can be extracted through perpendicular projection onto the three axes for the subtriangle within which it appears (colored lines with 50%/50% points indicated with arrows). Since the tetrahedron or compound triangle can be folded along the three sides of the internal triangle, such that each of the three vertices in the tetrahedron meet behind the plane of the paper, we can represent the entire four-dimensional likelihood space with this two-dimensional representation. Four-dimensional MLEs and confidence contours would exist within the volume of this three-dimensional pyramid and so would not be plotted on the surface. In this way, it is possible that the plot for an individual shows confidence contours on the surface but no MLE, as in the bottom tetrahedron plot of Figure 4-13.

subtriangle is determined using the same scaling method described earlier for the (single) triangle plot: we drop a line from each vertex to the base or side to create the scales against which the MLE is projected in order to translate position within the triangle into the percentages of ancestry admixture. By convention we plot the MLE in the center subtriangle of the composite triangle. Any point within the composite triangle represents a mixture of percentage for three groups at a time, but the reason for expressing the data in this new triangle is that each of the four possible groups of three can be presented on a two-dimensional sheet of paper. The MLE is found at the center of the two-fold confidence interval demarked by the black contour.

As with the triangle plot, points within each subtriangle of the composite triangle or tetrahedron that fall within the black line represent BGA proportions that are up to two times less likely to be the true value than the MLE (the greater the distance from the MLE the lower the likelihood). Points within each subtriangle that fall between the black and blue lines represent BGA proportions that are from two to five times less likely to be the true value than the MLE

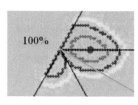

Figure 4-17

Magnification of the MLE and confidence intervals at the European vertex of a tetrahedron or compound triangle plot, showing the detail of the confidence space in two dimensions. Though the MLE is a function of binary admixture, three-way admixture values are found within the two-, five-, and 10-fold likelihood spaces. These create two-dimensional shapes that are continuous from one triangle to the next, but sometimes of strange shape. However, when the tetrahedron plot is folded into three dimensions as described in the text and the legend to Figure 4-16, and shown in Figure 4-18, this volume of the confidence space becomes bilaterally symmetrical.

and points that fall within the blue and yellow lines represent BGA proportions that are from five to 10 times less likely to be the true value than the MLE.

As with the basic triangle plot, the MLE point is the most likely estimate, and the confidence contours are plotted the same way but sometimes assume strange shapes in the tetrahedron projection plot (see Figure 4-17). As briefly described already, the shape of the likelihood space in three dimensions always seems to take a bilaterally symmetrical shape, even though it doesn't appear that this is the case from a two-dimensional tetrahedron projection presentation on a piece of paper. One can easily visualize this point with a little imagination: print the plot out on paper and fold backward into the plane of the paper along the three lines between the outer three subtriangles and the central subtriangle, such that you connect all three vertices of the composite triangle to make a pyramid (see Figure 4-18). The confidence contour lines on the surface of the pyramid are those that you can see, but they also connect through the center of the pyramid to form a three-dimensional shape. The shape of the probability space is different for each estimate, sometimes forming an oval, other times a bilaterally symmetrical shape of complex topology (i.e., you can visualize the three-dimensional shape of the likelihood spaces shown in Figure 4-13).

Whatever the shape, points in the interior of the pyramid are points with four-population admixture. Points on the surface of the pyramid are points with three-population admixture. How likely it is that a three- or four-population point represents the true admixture proportions depends on whether the point falls within the yellow confidence contour, blue confidence contour, or black confidence contour. Whether or not the MLE is visible on a two-dimensional tetrahedron plot depends on whether the point is on the surface (three-population) or interior of the pyramid (four-population). Having the ability to represent four-dimensional problem in three-dimensional space with a two-dimensional diagram, we have the power to visualize the entire likelihood space, not just the part of the likelihood space with the highest likelihood.

The three-dimensional shape of the likelihood space is usually interesting. In fact, some people with four-population mixture would have discontinuous likelihood spaces on the two-dimensional representation of the tetrahedron plot. You will notice that when the triangles are folded and the confidence intervals on the two sides are connected through the center of the resulting solid tetrahedron (a symmetrical four-sided solid), the three-dimensional shape created by the contour lines has a large amount of volume. In fact, for people of two- or 3-population admixture, the ratio of the surface area to volume of the three-dimensional space is relatively high compared to people of four-population admixture, where there is considerably more volume to the shape. Similarly, we have noticed that for persons who have ancestry in

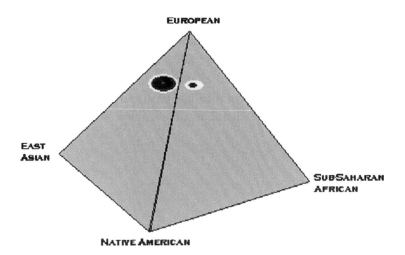

Figure 4-18

Three-dimensional pyramid resulting from the folding of the tetrahedron plot along the three sides of the internal triangle, such that each of the three vertices of the tetrahedron meet behind the plane of the paper. Here, the likelihood volume is clearly weighted against sub-Saharan African admixture but the European individual shown also clearly exhibits some non-European admixture.

populations that are possibly not explained well by the four populations in the model being discussed (e.g., South Asian Indians), the 2, 5, and 10 log intervals define relatively large probability spaces. We speculate that there may indeed be a means of measuring the appropriateness of the model used for the individual in hand from a formal analysis of the shape and size of the probability space defined by one's multilocus genotype.

TABLES WITH CONFIDENCE INTERVALS

The composite triangle and bar graph are tools for the visualization of the data. The results of the MLE are reported in Table 4-4.

In the example in Table 4-4, the best estimate for this person's particular multilocus genotype was determined to be 90% European, 10% Indigenous American, and the MLE is plotted at the position of the triangle plot indicated by these proportions. The bar graph would show solid blue bars at 90% for

SAMPLE-ID = XXXXXXX	
Estimate	**Ancestry**
90%	European
10%	Indigenous American
0%	West African
0%	East Asian

Table 4-4

Tabular presentation of admixture results: straightforward but lacking certain important information.

European and 10% Indigenous American. Although confidence intervals are missing in this particular table, we could easily add a column with the two, five, or 10-fold confidence range. Unlike the table however, the triangle plot and bar graph represent every possible combination of ancestry proportions that reside within the confidence space for an individual plot, and with the table the various combinations of percentages that are possible and less likely are harder to visualize with the mind's eye.

CONCEPTUAL ISSUES

IS THE CHOICE OF GROUP NAME ARBITRARY?

We initially used the term Indo–European to refer to ancestry shared among individuals residing in Europe, the Middle East, and South Asia. From evolutionary analyses, which assume a neutral theory of molecular evolution (Kimura 1968), and certain mutation rates, this ancestry is believed to have been derived from anatomically modern and human (AMH) migrants that left Africa approximately 47 KYA, and colonized the Fertile Crescent in the Middle East. Later (40 KYA) this Fertile Crescent population branched to Europe and (likely) mixed with South Asians, while founding populations in Central Asia approximately 39 KYA. The genetic contribution by these migrants out of Africa to modern-day South Asian, Middle Eastern, and European populations must have been significant because the languages of modern-day South Asian, Middle Eastern, and European populations share a common base, belonging to the Indo–European family of human languages. However, though the use of the term Indo–European may have effectively communicated this shared ancestry among clades derived from this branching out of Africa (i.e., a geographical range that correlates with these descendents, from India—Indo..., to Europe—...European), its use was technically incorrect because of the linguistic connotations it carries. Long before our use of it, the term Indo–European was established as a human language family of a common root. Though languages of this family were and are largely spoken by the diaspora from these original Fertile Crescent migrants, technically speaking, language proclivities cannot be read from the DNA (though they might be inferred, with some degree of quantifiable precision; more on this topic later).

We made the initial choice to use the term Indo–European not because language can be read from the DNA, or because linguistic characteristics correlate perfectly with phylogeny, but to capture the idea that modern day Europeans share a relatively recent common ancestry with other diaspora from these Fertile Crescent-derived populations (indeed, these neighboring populations very likely have had many migrations from the Fertile Crescent over the millennia since the initial founding events).

Consider the Basques, often described (however inappropriately) as descendents from undiluted Paleolithic ancestors and thus, a modern day "Paleolithic relic population." Their use of a language that is classified as non-Indo–European and their genetic uniqueness (Jobling et al. 2004; Mourant et al. 1976) relative to other European populations suggests to many a relatively undiluted Paleolithic ancestry within Europe. This ancestry often is suggested to be indigenous European, and derived from populations who were genetically closer to the modern-day descendents of Paleolithic migrants arriving in Europe 40 KYA than to modern-day East Asians, West Africans, or Indigenous Americans. With an admixture test like the initial AncestryByDNA 2.0 test, calibrated for performance relative to the four main continental groups (West African, Indo–European, Indigenous (Native) American, East Asian, but excluding East Africans, Polynesian, and Melanesian), they would almost certainly type as Indo–European, even though the term Indo–European usually is associated with linguistic groupings, the Basques spoke a language of a different type, and the ancestors of Basques were likely distinct from those of most other modern day Europeans.

Clearly the use of the term Indo–European to refer to evolutionary relationships was, and is, suboptimal. On reflection we also feel that the Indo–European term and underlying concept are more that of a polyphyletic metapopulation, a conglomerate of many populations that are not best explained as having a single common origin to the exclusion of other populations. For example, South Asian Indians have substantial amounts of East Asian ancestry either as the result of their being a persistent progenitor for East Asian populations and/or through admixture with East Asians as is seen today on the borders of the Indian subcontinent with Eastern Asia (Chakraborty 1986). How we know this is explained in more detail later. Likewise in Eastern Europe and Scandinavia, there is clear evidence of Central Asian contribution (Kittles et al. 1998). Obviously, the use of a linguistic term to describe a parental population that could have founded many different populations clearly would be unwise since shared language does not necessarily indicate monophylogeny, and not all diaspora from one group are expected to retain language traditions. On a related note, use of the term Indo–European brings the potential for being understood as a euphemism for the Caucasian racial group.

Given the confusion caused by the use of this term, we switched the terminology to European, since our parental groups representative of this major branch of the human evolutionary bush are, in fact, continental Europeans and American descendents of continental Europeans. The use of the term

European was attractive because when found in South Asian Indians at lower levels, and Middle Easterners at higher levels, it communicated the idea that these groups share a relatively recent common ancestry with other diaspora from the Fertile Crescent migrant farmers who left Africa 47 KYA but are not exclusively the sole contributions to either groups of populations. However, the use of the term European is not a perfect solution, because instead of attempting to express genetic and demographic histories with linguistic group ranges, we are now attempting to do so using geographical terminology. The diaspora of the Fertile Crescent populations did not settle only in Europe, but across Central Asia, Northern Asia, and probably South Asia.

Consider again South Asian Indian populations. We have seen (Figure 4-1A, 4-1B) and will soon see again (in Chapter 6, Figure 6-3) that, using a variety of marker types and algorithms with a k = 4 continental population model, South Asian Indians type with substantial European ancestry or better stated, fractional affiliation. This is not unexpected given what we know about the populations that founded modern day Europeans and South Asians, and their use of languages with a common root. Y chromosome and mtDNA haplogroups show common haplogroups among European, Middle Eastern, and South Asian Indian populations, and nested cladistic analysis suggests a Middle Eastern origin for the related haplotypes among these haplogroups (Jobling et al. 2004). If this is true, then (at least most) South Asian Indians never had ancestors that lived anywhere near Europe and referring to their shared ancestry with modern-day European (and Middle Eastern) populations as European is not technically correct.

Before we consider what the proper terminology should be, if there is a proper terminology, let us consider the markers we choose and the populations we use in choosing them (the parental samples).

Good Ancestry Informative Markers (AIM) arose as distinctive markers (in terms of allele frequency) in populations that lived tens of thousands of years ago, after the origin of the species in Africa. This required genetic drift through reproductive isolation by geography and culture and/or natural/ sexual selection through differential survival and reproduction, and these populations constituted the ancestors for the modern-day populations within which the AIMs are actually measured and applied.

DERIVED AND SOURCE POPULATIONS

Knowing that several populations share common sequence types, directionality of evolution can be determined in theory. To better grasp the problem of choosing geographical or other nongenetic labels for evolutionary clusters, let us review the difference between derived and source populations. For a deeper treatment of the concepts we will discuss here, we recommend Jobling et al. (2004). The difference between derived and source populations is one of chronological order. Genetic distances are calculated from allele frequencies between pairs of subpopulations of a larger metapopulation using statistics like F_{ST} or Nei's distance (D). The date of divergence between populations can be calculated from randomly chosen markers in the human genome using these genetic distances, based on the neutral theory of human evolution and the molecular clock hypothesis. Rooted phylogenetic trees provide a chronological basis for divergence inferred from measures of molecular distance. The sequences normally used to construct such trees are ideally nonrecombining, because relating them within the context of a series of chronological events is made easier due to the lack of shuffling of sequences over time. By way of analogy, it is easier to trace the path of a comet in the galaxy that leaves a clean, neat trail as opposed to a disperse one. A statistic called *rho* (R) frequently is used to date divergence events

$$R = ut \qquad (4\text{-}1)$$

where u is the mutation rate and t the time in generations.

The use of median networks allows the reconstruction of phylogeny considering both the haplotype sequences that are measured as well as those that may be presently extinct. In this way, hypothetical ancestral sequences can be identified, sequences derived from them determined, and the date of divergence among the various derived sequences calculated. The source populations for those we use to select AIMs became distinctive presumably due to their isolation for long time periods, and the older this population, the more likely it gave rise to many subpopulations at different times and in different places. Some of these populations may have mixed with others, and others may have founded new isolated populations that did not mix with but displaced prior residents. Phylogeographic methods attempt to establish the relationship between phylogenetic reconstructions and geography, and through the process of interpolation it is possible to determine a most likely geographical origin for source sequences in the phlyogenetic reconstruction. Nested cladistics analysis (Templeton 2002, 2004) is one method that employs measures of genetic distance, phylogenetic reconstructions, and molecular dating, overlain on geography with statistical analyses, to arrive at a reasonable theory of the spread of human populations and the sequences they carried over the course of evolution.

The distinction between source and derived populations then depends on what sequences one is looking at (are we sure they all abide by the neutral theory of molecular evolution, not being subject to selection?), the statistical methods used, and the assumptions implicit in the implementation of these methods (such as all populations remain isolated after their genesis, with no admixture). Different statistical methods for building phylogenetic trees give subtly different tree topologies; Neighbor-Joining (clustering method; Saitou & Nei 1987), Unweighted Pair Group Method with Arithmetic Mean (UPMGA; Sneath & Snokal 1973), Maximum Parsimony (searching method), Maximum Likelihood (searching method), Minimum Spanning, and Median Networks (both network methods) all use different approaches to phylogenetic reconstruction, and though in most cases they arrive at similar topologies, this is not necessarily the case. In addition, none of these methods accurately account for gene flow (admixture) between branches. As a result, two populations with an ancient or most recent common ancestor that have extensively mixed with one another might produce a lower genetic distance measure than two with a more recent common ancestor that have not. Certainly human evolutionary history has been a function not only of divergence and isolation but migration and gene flow, to different extents for different populations. So then, even though we can formulate a reasonable hypothesis on the chronological order distinguishing between source and derived populations, the complexity of the human genome, modern-day human demography, and our evolutionary past makes it very difficult to accurately reconstruct the true story.

Consider the South Asian Indian populations as an example of the difficulties in deciding whether a population is a source or a derived population, and therefore deciding on the proper terminology to use when discussing ancestry affiliations. Neolithic Middle Eastern farmers spread to blend with (and possibly displace) resident Paleolithic (indigenous) Europeans 10 KYA, and much of the modern-day European gene pool was derived from this Fertile Crescent source population. The same Neolithic Middle Eastern farmers also likely migrated in the opposite direction, not necessarily at the same time, contributing to modern day South Asian Indian populations. On top of this, we are uncertain whether or to what extent there has been mixing between the various diaspora of this Fertile Crescent group. As if more complexity were needed, some of these diaspora very likely mixed with resident populations in the areas to which they migrated to varying degrees.

Relevant for South Asian Indians, it is believed that even earlier migrations of anatomically modern humans (AMH) out of Africa stuck to the Indian Ocean coast 50 KYA. Human migrations or population expansions tend not to be single events, with discrete starting places and destinations, and in this case it is from archaeological evidence that we know the first wave of migration/expansion passed through India and Southeast Asia, eventually through what is today Indonesia and ultimately reaching Australia 50 KYA (reviewed in

Jobling et al. 2004). It is possible that the group we today call South Asian Indians are a blend of descendents that share a relatively recent common ancestry with indigenous Australians as well as migrants from the Fertile Crescent. Indeed, for the former, this may be borne out by the sharing of M* mtDNA and R Y-chromosome haplogroup sequences between modern-day South Asian Indians and Australians. For the latter we point to shared mtDNA (U) and Y-chromosome haplogroups (R, and the P, derived from R). If modern-day South Asian Indians are derived from ancestors associated with the earlier migrations (50 KYA) as well as the expansion of Neolithic farmers no sooner than 10 KYA, the average Indian likely would possess autosomal markers characteristic of each out of Africa branch. So then, we would expect the BGA typing of a South Asian Indian as partially European, by the chosen nomenclature, even though it is likely that none of his ancestors would have lived anywhere close to Europe; and partially West African or even Indigenous American, depending on the route of these migrations, the time they remained isolated, and the degree to which each pool contributed and whether any of these populations later begat populations in the Americas. Indeed, with our 171-marker BGA test, we see that the average South Asian Indian exhibits 58% European, 5.7% West African, 24.9% East Asian, and 12% Indigenous American ancestry (see Chapter 6, Table 6-1).

Looking at the mtDNA, we see that the majority of South Asian Indians possess the M* haplogroup in common with North Africans, Middle Easterners, and East Asians, and U lineages in common with Europeans, Middle Easterners, and North Africans. Looking at the Y chromosome, we see most South Asian Indians harbor R haplogroup sequences in common with Europeans, North Africans, Middle Easterners, and Central Asians, as well as a polyphyletic distribution of E and J haplogroups similarly distributed, but part of different branchings of the Y-chromosome-based phylogenetic reconstruction. So then, are South Asian Indians Europeans, East Asians, Middle Easterners, or a separate group?

The answer to this question is that asking it is the main problem. The question assumes that human populations fall into neat and distinct groups, almost Linnean or typological in character, which we know is not the case given the complexity of human history (multiple foundings, migrations, admixtures, etc.). If indigenous Australian were incorporated as a fifth group for developing and calibrating an admixture test, the average South Asian Indian would possibly show partial affiliation with Australians, even though none of his ancestors lived on the Australian continent. If South Asian Indians were considered as a separate group during development and calibration of an admixture test, most modern-day Europeans would show partial South Asian ancestry, even though virtually none would never have had an ancestor that lived in South Asia.

This nomenclature problem is not unique to South Asian Indians—using an autosomal admixture test, the average Middle Eastern individual shows about 85% European, 10% East Asian, 4% sub-Saharan African, and 3% Indigenous American ancestry (see Chapter 6, Table 6-1) even though none of the ancestors from whom he derives his uniqueness with respect to the rest of the human population likely ever lived in Europe, East Asia, Africa, or the Americas. Clearly then, apart from the fact that they invoke Linnean images, geographically based terms such as European or Australian, or even the other terms we use (West African, East Asian, and Indigenous American) are no more technically correct than those based on linguistic groupings.

Therefore, perhaps it would be better to refer to our ancestral groups as subbranchings of the human evolutionary bush, rather than based on identities associated with modern-day geopolitical terms like Native American or His-panic, or geographical regions like European, East Asian, or linguistic terms like Amerindian or Indo–European. We could use a phylogenetic approach to subdivide our ancestors, and their modern-day descendent groups into clades simply labeled I, II, III, IV, and so on, where each clade describes descendents from the earliest major branchings of humans over the past 100 KYA. This would be similar to how clades are referred to on phylogenetic trees, and technically most correct in terms of achieving what it is we set out to achieve by developing an admixture test—the measure of evolutionary affiliation and population structure. It is in fact the style chosen by others that measure Y and mtDNA sequences for forensics and recreational use, such as Oxford Ancestors Ltd., who use Daughters of Eve rather than racial or continental population groupings.

As with autosomal markers, as we have illustrated with Middle Easterners and South Asian Indians, Y and mtDNA haplogroups are not distributed as a function of neat geography, and groupings based on geography would be fallacies. This is not due solely to the fact that they look back to different points in time than autosomal markers, as autosomal markers show the same types of distributions. Rather it is due to the complexity of human evolution (differing levels of drift as a function of variable sample sizes and demographic histories, multiple migration/expansion events). For example, most every Y and mtDNA haplotype belonging to African haplogroups are also found outside of Africa, the same as classical blood group markers and autosomal AIMs. Virtually every mtDNA haplotype characteristic of Indigenous Americans belongs to hap-logroups also present in East Asia and some even in Central Asia. The drawback of the solution to this problem of changing the terminology to one of roman numerals representing subsets of human clades, and the reason we have not used it, is that most readers, forensic experts, and lay students do not understand terms derived from phylogenetic extrapolation through median network analysis, such as that implicit in the use of terminology such as Daughter of Eve or Son of Adam. Additionally, because of gene flow, human populations are not exactly consistent with a phylogenetic analysis, where there are only divergences and no horizontal connections between populations.

Most lay people, however, do understand colloquial terms used as part of modern-day language, based on (however inaccurate) racial, geographical, and linguistic labels. Naming groups in the most scientifically responsible manner runs the risk of having to educate every beneficiary of this information on the concepts of molecular genetics, phylogenetic inference, population genetics, statistical extrapolation, and paleoanthropology in order for the affiliations with those groups to be understood. We have taken a different, albeit controversial approach of using colloquial though scientifically inaccurate terminology for the express purpose of communicating the interrelationships and connectivity of human populations in terms that most people currently understand or think they understand. We expect that the scientific truth of our results will become plainly apparent and scientifically illustrated to lay users of this technology through the implementation of an empirical method of interpretation, which will be discussed in Chapter 5.

Using BGA admixture results (or ranges) to query databases of digital photographs should illustrate to the forensic scientist that the inference of skin shade purely based on the level of West African ancestry is tenuous. Though virtually all 90% West African subjects would be characterized with high melanin index values, not all individuals with high melanin index values are predominantly or even detectibly West African; South Asian Indians present in the database would harbor very little West African admixture, but very high melanin indices. Further there is a substantial amount of variation in melanin indices for African Americans with particular ancestry levels and as we will see, estimating skin color based on West African ancestry is crude at best. The inference of European physical appearance from a sample typed to be of majority European ancestry is also tenuous (i.e., facial characteristics of South Asian Indians are not quite the same as those of Europeans, although they type as of majority European ancestry). This is not to say that a database-driven method for the inference of physical traits given genomic ancestry admixture with quantifiable precision is not possible, only that the relationship between genomic ancestry and anthropometric features is a difficult problem on its own to understand and implement (as we will see in Chapter 5), and the use of geographical, ethnic, or linguistic terminology in describing elements of genomic ancestry would serve to compound this problem.

A genealogist querying a geopolitical database with BGA admixture results (or ranges) in order to make inference on their ethnicity given that result would quickly learn that a 96% European and 4% West African admixture profile is common for individuals who self-describe as European, Mediterranean, Caucasian, European American, Middle Eastern, and so on. They would realize, without having to study anthropology, statistical extrapolation or molecular genetics in a classroom that dichotomous classifications as European, Middle Eastern do not fit their data, and that they must allow a range of possible scenarios, hopefully understanding the interconnectivity of modern-day populations and the fallacy of dichotomously interpreting ancestry.

PARENTAL GROUPS

Given the imperfect correlation between geography, linguistics, phylogeny, and molecular distance, how can we justify the selection of one population as representative of an entire branch of the human phylogenetic tree? This is a very important question we discussed more fully in Chapter 2, one that has formed the basis for most of the criticism of the test and methods described in this book. The answer is that, optimally, we don't. When searching for AIMs, we strive to identify markers relevant for contrasts across the branches that we are interested in resolving. Given that we cannot sample every human being, or even human beings from most of the branches of the human family bush, we select representatives and allow population genetics theory and statistical methods to provide for more comprehensive parameters. To quote George Box again, "All models are wrong; some models are useful." When considering reporting on personalized genomic histories for forensic, biomedical, and genealogical uses, we are very much interested in this practical aspect of the usefulness of the parental populations and the models implicit in the choice of which parental populations to use.

It is useful to this discussion to recap from Chapter 2. We know from the theory of Wahlund's variance of gene frequencies that getting a good sampling of the diaspora is important for estimating the allele frequency of the population that predated the radiation of this diaspora. Wahlund's principle states that given a metapopulation with many subpopulations, the allele frequency may vary from subpopulation to subpopulation, but that the average allele frequency of these subpopulations corresponds precisely to that for the metapopulation. The assumption is that the metapopulation more closely approximates the source population that lived many thousands of years ago than any one modern (derived) population. To estimate the allele frequency of the founders of modern-day European, Middle Eastern, and (to a lesser extent)

South Asian Indian populations, we are best advised then to sample from across all available subpopulations derived from these founders, which we have done for three of our four groups (European parental sample = European Americans of varying ethnicity; West African parental sample = a variety of West African populations from the Angola, Congo, Nigeria, and Ivory Coast; East Asian parental sample = Japanese and Chinese; etc.). However, even this measure does not provide for accurate parental allele frequencies because it does not account for genetic drift that might have taken place since the founding group for all of these diaspora existed or those subpopulations that were important in the past, but are now just represented by small populations. To account for this parameter, we can use a more complex model of admixture to calculate admixture proportions as described in Chapter 2.

ACCOMMODATING MISSING POPULATIONS AND CONTROLLING BIAS

It is for these reasons we went through the trouble in Chapter 3 of describing the more advanced methods of calculating admixture proportions. With these methods, we don't have to survey across all diaspora daughter populations (which would be a daunting task, and is subject to uncertainty and argument) to do a good job estimating parental allele frequencies for the populations that contributed to contemporary admixed populations. The reason the average frequency of the diaspora from a parental group may not be the same as that for the parental group itself is genetic drift, population bottlenecks, demic infusions (migrations and admixture), and natural selection that has taken place since the parental sample existed (up to 500 year ago for admixed populations like African Americans and Mexican Americans) and when its descendents were sampled (i.e., today). Indeed, the modern-day (derived) populations represent merely a sampling of the parental group's genetic stock.

As described technically in Chapter 3, programs like STRUCTURE and Paul McKeigue's ADMIXMAP program can estimate genetic drift simultaneously with admixture proportions, for individuals as well as populations. We can now return to these methods in light of our nomenclature problem and its underlying foundation, which is based on the difficulty in selecting parental populations. When selecting and characterizing markers of ancestry we are interested in those sequences possessed by parental populations that lived several thousands of years ago. Yet, we cannot sample from these populations directly, so we are forced to sample from modern-day descendents. It is usually very difficult to choose which modern-day populations are most representative of a given parental population because it is infeasible to draw a sample of modern-day descendents with allele frequencies precisely representative of their parental population. Even if it were possible to identify the best modern-day population(s) for a given parental population, the allele frequencies between the modern-day population(s) and the parental populations are expected to vary as a function of genetic drift, mutation, and/or selection since the parental population existed, and if we are estimating admixture, since the admixture event occurred.

To overcome this problem, authors such as McKeigue et al. (1998, 2000) and Hoggart et al. (2004) have developed statistical methods for detecting misspecification of ancestry-specific allele frequencies and respecifying the correct frequencies within the modern-day, often admixed populations available. If you have skipped Chapter 3, we recommend you refer back to this chapter in order to appreciate these methods. The idea is to compare the estimates of admixture in specific populations obtained at single loci against the estimates obtained using a large number (i.e., all available) of loci. Estimates provided by the single locus that vary substantially from the estimate obtained using all loci are likely to be misspecified and within a framework of Bayesian logic, these parental allele frequencies can be corrected.

For example, to model the admixture in a population adequately we must treat the ancestry state of each locus and the distribution of ancestry in the population hierarchically, with the admixture of each parental gamete, the ancestry of the two alleles at each locus in each individual, and any missing marker genotypes as missing data. One approach is to assume that the parental gametes are of equivalent admixture proportions, and that the markers are independent of one another. The former obviously need not be true for any given individual, and the latter rarely is if the individual is of extensive admixture because as we have seen, admixture creates Linkage Disequilibrium (LD) across the genome in unpredictable ways. These variables must then also be treated as missing data, and accommodating all these unknown variables in the classical Maximum Likelihood approach (also called the frequentist approach) we have used throughout this text is difficult.

More advanced methods attempt to estimate the values for these variables simultaneously using a Bayesian approach (MacLean & Workman 1973a, 1973b) where prior distributions are assigned as possible and the posterior probability distribution of missing data is generated using Markov chain simulation (McKeigue et al. 2000). With no prior information, the results from the Bayesian approach approximates that for the likelihood approach with one important distinction; since probability distributions are generated for the missing data we can evaluate these distributions against our expectations from past research (perhaps nongenetics-based research), and we can use the frequentist approach to test whether any of the assumptions should be modified or rejected against a null hypothesis (McKeigue et al. 2000).

McKeigue et al. (2000) used this approach to estimate the admixture proportion of populations of African-American parents (rather, their gametes, one of our missing data) from observations of AIM allele frequencies in African-American individuals, and obtained results that were generally concordant with the proportions reported by others in modern-day African-American populations (Parra et al. 1998; as well as the data herein, see Chapter 5). The inference here is that parental allele frequencies (one important piece of missing data) can similarly be estimated using this approach, even though we cannot go back in time and actually observe their allele frequencies.

However, estimating parental allele frequencies, for parental groups that no longer exist (only their descendent representatives do) is not a trivial mathematical exercise and though they would mathematically represent the "best" estimates, it is difficult to falsify the frequencies even these advanced methods allow us to calculate. It is difficult to know how well the more advanced Bayesian methods accommodate all of these nuisance parameters without a time machine and a detailed and perfectly correct map of human bottlenecks, migrations, expansions, and drift over the past 50,000 years (see Box 4-A).

Fortunately, the brightest in the field are diligently working on these puzzles and it is possible that 20 years from now we will have a much better characterization of human demographic history and parental populations provided by autosomal AIMs. On the level of the population, the gold standard for establishing human migration/expansion patterns and chronologies is the Y and mtDNA chromosome, since the relationships between the haplogroups is more easily inferred in time and space, and since admixture within populations can just as easily be inferred using these sequences. However, we will see in the next chapter that autosomal sequences may provide a "lens" of different power for peering back into human history and as such, represent a platform for filling in gaps and testing Y- and mtDNA-based hypotheses. Until the puzzles surrounding our origins are solved, the choice of nomenclature and parental populations against which to calibrate an admixture assay will remain part "art" and part "science," with a strong emphasis on the latter. However, this situation is not entirely unhappy, for even in art we can find order and reason it is thoughtfully applied.

Box 4-A

In his 2000 paper, McKeigue and colleagues developed a method to test for the misspecification of AIM allele frequencies. A positive slope of the log-likelihood function of parental allele frequency estimates at the null value (true allele frequency in the parental population) gives an indication that the frequency is overestimated, and a negative slope signifies that it is underestimated. A Z statistic and associated p-value are determined for each estimate, and a chi-square p-value is determined with two degrees freedom for the joint null hypothesis that both AIM allele frequencies are correctly specified. In his study, he used 10 AIMs, and determined that the allele frequency for three of these had been misspecified, and that the misspecification was consistent across all eight African-American populations studied. His method is unable to determine for which populations the allele frequency is misspecified. One was underestimated and the other two were overestimated. In this example, it is not the choice of subpopulation chosen to be representative of the parental population that introduced the allele frequency error, but likely the genetic drift that had taken place at these loci between the time the parental proto-West African population existed in western West Africa approximately 15,000 years ago and the time western West Africans founded modern day U.S. African-American populations via migration as part of the slave trade a couple hundred years ago. Alternatively, some of the misspecification could have been due to similar genetic drift in Europeans between these times.

How recently admixture occurred is similar to the question of what the gametes looked like for any given individual. McKeigue et al. (2000) explains that the strength of the statistical association between linked markers that cooccur as a function of ancestry admixture is a measure of how recently the admixture has occurred. This is related to the sum of intensities parameter discussed in Chapter 5 (see Figure 5-33), which is a parameter for measuring how clustered ancestry is on the chromosome (actually the gametes) for an individual. In his sample of African Americans, McKeigue determined that the European admixture present was created between 5 and 9 generations ago, consistent with what we know from recent historical mixing between Europeans and western West Africans in North America. Thus, we can accommodate the uncertainty related to the amount of time that has passed since the preadmixed populations (derived from original parental populations) existed and the time of the admixture event. Work continues to this day toward understanding how these methods perform with older admixture events (for example, is the clustering of ancestry regions among chromosomes we see in Figure 5-33 indicative of recent or old admixture?). Work in the laboratories of the authors suggests that very large numbers of powerful AIMs are necessary to date affiliations with parental populations extending further back than a few generations; the number of markers used to create Figure 5-33 is probably not even close to adequate to address this question.

CHAPTER 5

CHARACTERIZING ADMIXTURE PANELS

PARENTAL SAMPLE PLOTS

With a basic MLE algorithm, we would expect that typing the parental samples would produce admixture estimates similar to those obtained using STRUCTURE as shown in Figure 4-1A. We would expect to see a decrease in variance within parental populations when moving from a 71 AIM to a 181 AIM panel of markers due to an increase in the precision of the point estimates and the increase in ancestry information content. Figures 5-1 and 5-2 show the same exact parental samples typed for four-continent biogeographical ancestry admixture using the basic MLE algorithm and either the 71 marker battery (see Figure 5-1) or the 181 marker battery (see Figure 5-2). For Indigenous American parentals with 71 AIMs, a West African component was relatively low to begin with (the spread of symbols toward the AF vertex is not extreme), but there is relatively high European and East Asian bias relative to the 181 AIM panel (note the tighter clustering in the EU and EA dimensions in Figure 5-2 as opposed to Figure 5-1). Bias for East Asian parentals similarly improved, specifically in the European and Indigenous American dimensions (note relatively tight clustering at the East Asian (EA) vertices in Figure 5-4 as opposed to Figure 5-3).

Figures 5-7 and 5-8 show a similar dramatic reduction in Indigenous American and East Asian bias in European parentals. However, Figures 5-5 and 5-6 show a relatively small improvement moving from 71 to 181 markers for African parentals. This is not unexpected, since the main difference in the two marker panels is the addition of more markers that distinguish among non-African populations; since the 171 marker panel is the 71 marker panel with an additional 100 AIMs selected based on European/East Asian, European/Indigenous America and East Asian/Indigenous American discrimination power. Thus, we would not expect great improvements in African/non-African precision in our parental samples moving from the 71 to the 171 marker panel.

Figure 5-1

Tetrahedron plot of 59 Indigenous (Native) American parental sample MLEs as determined using the panel of 71 AIMs discussed in the text. Symbols are shaped based on within which subtriangle they are found as shown in the key.

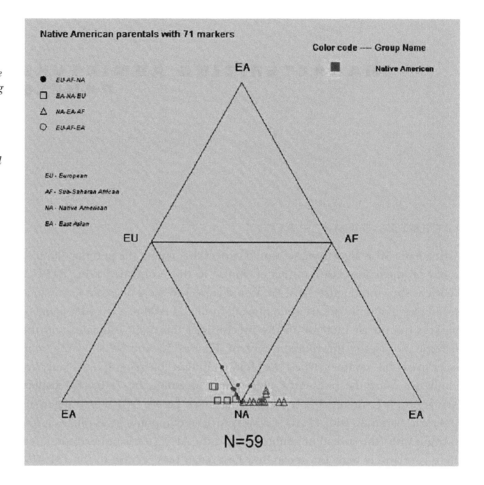

Figure 5-9A shows parental plots for Indigenous Americans, West Africans, and Europeans obtained using a 30-marker test wherein the 30 AIMs are a subset of the 71 and 181 AIM panels just discussed. We can see a reasonable tightness in how close the parental samples plot at their respective vertices, but as expected, not as tight as with the larger marker panels. Figures 5-9B,C, and D show a sample of African Americans, European Americans, and Hispanics, which we will compare for the 71 and 181 marker test results in a later chapter. They are shown here to illustrate how the parental plots should be considered when quantifying results in populations. In this case, the spread of African Americans toward the European vertex is greater than the corresponding spread in West African parentals (compare Figure 5-9B with 5-9A), which indicates European admixture for African Americans (presumably through recent admixture, to be discussed later).

If the spread of West African parentals was similar to what we see with African Americans, we would not be able to conclude bona fide European

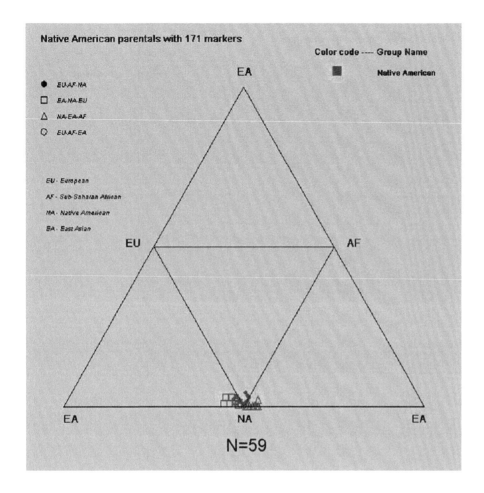

Figure 5-2

Tetrahedron plot of 59 Indigenous (Native) American parental MLEs as determined using the panel of 171 AIMs discussed in the text. Symbols represent individuals from the parental population, and are shaped based on the subtriangle within which they are found, as shown in the key. The spread of samples around the NA axis is much tighter than observed in Figure 5-1, using only 71 AIMs.

ancestry for the African-American sample since the European ancestry for the latter would not be different than the deviations expected from statistical imprecision. Likewise, the Hispanic sample shown in Figure 5-9D is markedly different than either the European or Indigenous American parental samples shown in Figure 5-9A, indicating population substructure not accounted for by statistical imprecision inherent to the process of admixture estimation. As an aside, these results also indicate that as few as 30 well-selected binary AIMs are adequate to estimate the main elements of individual biogeographical ancestry.

MODEL CHOICES AND DIMENSIONALITY

For most of the results and simulations discussed so far we have assumed that the ancestry of each sample is no more complex than a three-way admixture. Although this is computationally convenient, it is clearly suboptimal for

Figure 5-3

Tetrahedron plot of 59 East Asian parental MLEs as determined using the panel of 71 AIMs discussed in the text. Symbols represent individuals from the parental population, and are shaped based on the subtriangle within which they are found, as shown in the key.

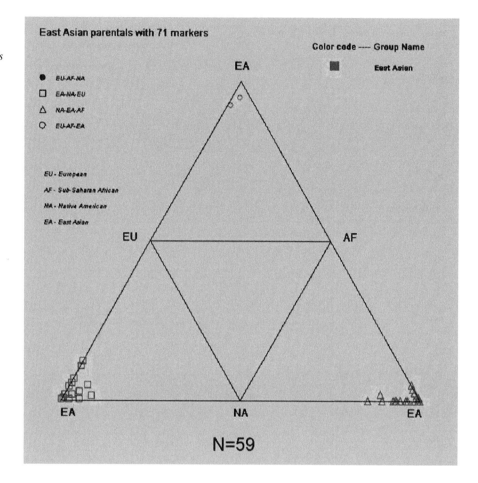

individuals of more complex (i.e., four-way) admixture. Recall from Chapter 3 that there are two ways we used to reduce the computational load required for analysis using a four-population model. We can use the best three-population MLE value assuming the individual is not affiliated with all four populations simultaneously, or in the modified version of the program, we calculate the likelihood score of all possible three-way admixture proportions and compare the highest score for each of the three possible three-way ancestry combinations. If the MLE for the three-group combination with the highest likelihood score is within one log of that for another three-group combination, we discarded the results and reran the algorithm scanning all possible four-way combinations, using the MLE from this scan as the result. If the score for the three-group combination with the highest score was not within one log of the score of another three-group combination, that score was used as the MLE in the first method. We will call this modified method the 3way/4way MLE calculation style.

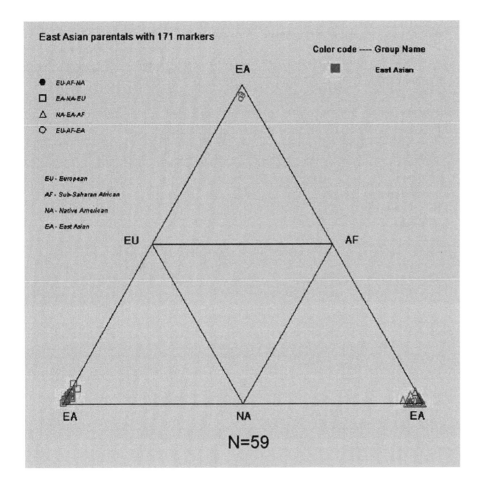

Figure 5-4

Tetrahedron plot of 59 East Asian parental MLEs as determined using the panel of 171 AIMs discussed in the text. Symbols represent individuals from the parental population, and are shaped based on the subtriangle within which they are found, as shown in the key. The spread of samples around the East Asian vertex is much tighter than observed in Figure 5-3, using only 71 AIMs.

All the graphical representations of ancestry we have shown except the histogram are limited in that they require no more complex a population model than three-way (so that the data can be plotted on a two-dimensional surface). What is the penalty we incur by forcing the estimate of ancestry affiliations for a sample into its best three-population model? Using the same AIM panel, MLE determination for a historically admixed group (such as Puerto Ricans) with both three-way and four-way methods would be one way to answer this question. Figure 5-10 shows the European estimates for a group of Puerto Rican samples assuming the most likely three-population model for each (which can be plotted in the triangle plot) versus the European estimate obtained using the four-population continental model (which cannot be plotted on the triangle plot) (Halder et al. 2005). The same basic MLE algorithm was used to obtain the estimates, using the same 171 AIM panel we have been discussing so far.

For the three-population model estimates, we select the most likely three-way mix; if the true estimate of a sample is 50% European, 20% West African,

Figure 5-5

Tetrahedron plot of 58 West African parental MLEs as determined using the panel of 71 AIMs discussed in the text. Symbols represent individuals from the parental population, and are shaped based on the subtriangle within which they are found, as shown in the key.

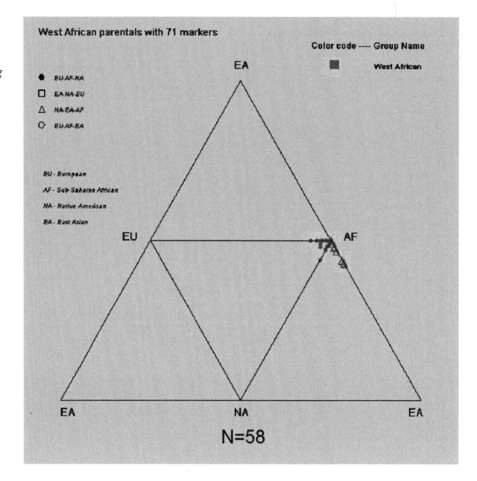

20% East Asian, and 10% Indigenous American, the most likely three-way estimate could be something like 50% European, 20% West African, and 30% East Asian—the Indigenous American affiliation would be lost, and most likely forced into or assimilated with the East Asian category. For the four-population model, the Indigenous American ancestry would not likely be lost—the answer is expressed as the most likely proportions of all four groups and an answer similar to our true values would likely be obtained.

As can be seen in Figure 5-10, the percentage of European obtained when using these two different strategies is similar (correlation coefficient is greater than 95%). The West African ancestry estimates are also very similar between the two methods as shown in Figure 5-11 (correlation coefficient close to 1.0). These results would suggest that using a three-way representation of ancestry rather than a four-way representation, when performing an assay based on a four-population model, does not usually create dramatic admixture estimation error. However, the estimates of Indigenous American ancestry shows

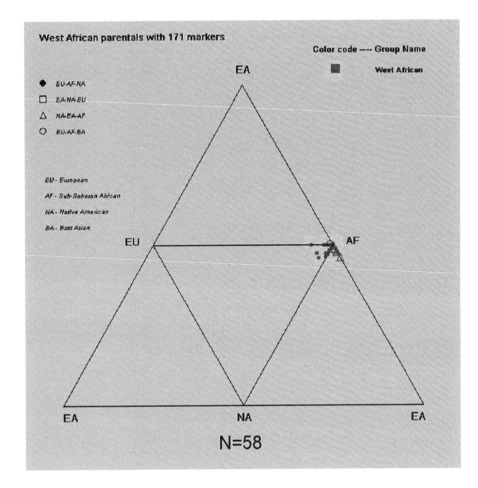

Figure 5-6

Tetrahedron plot of 58 West African parental MLEs as determined using the panel of 171 AIMs discussed in the text. Symbols represent individuals from the parental population, and are shaped based on the subtriangle within which they are found, as shown in the key. The spread of samples around the AF vertex is much tighter than observed in Figure 5-5, using only 71 AIMs.

considerable variation between the two methods (see Figure 5-12), indicating that as we might expect, we in fact do incur a penalty for forcing all the estimates into the three-way model. Evidently from Figure 5-12, the ancestry for almost a third of the samples is better described as a function of four percentages rather than three. It so happens that much of this error is in the East Asian/Indigenous American dimension of the analysis. For example a 10% East Asian and 10% Indigenous American result in an individual with 60% European and 20% West African ancestry would be presented as 60% European, 20% West African, 20% East Asian when using a three-way presentation rather than the four-way presentation. Thus, the penalty we incur appears to be focused in error between affiliation for the two continental populations for which genetic distance is lowest—in this population model Indigenous American/East Asian.

This illustrates that for samples where ancestry is complex, estimating ancestry with the best three-model fit can be treacherous. For others it will be

Figure 5-7

Tetrahedron plot of 70 European parental MLEs as determined using the panel of 71 AIMs discussed in the text. Symbols represent individuals from the parental population, and are shaped based on the subtriangle within which they are found, as shown in the key.

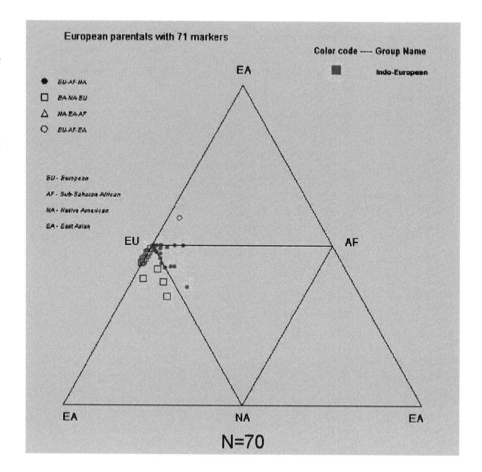

perfectly fine. The trick is to learn to recognize when a simpler three-way MLE calculation scheme is adequate, or else always run the more computationally expensive four-way calculations. As described in Chapter 3, the trick we use with the MLE algorithm is to select the three-way model with the highest likelihood, and if the most likely estimate for the second-best three-way model is within one log of the best three-way model, the four-population MLE calculation method is used and the results presented as a function of four percentages. In other words, we assume that the low level of likelihood difference in three-model likelihoods is an indication that affiliation with all four populations exists for the sample.

SIZE OF CONFIDENCE CONTOURS

Moving from 71 to 171 AIMs, what is the improvement in performance we can expect? Figure 5-13 shows a randomly selected sample of pairs of admixture estimates plotted on the basic triangle plot, whereby the estimate from the

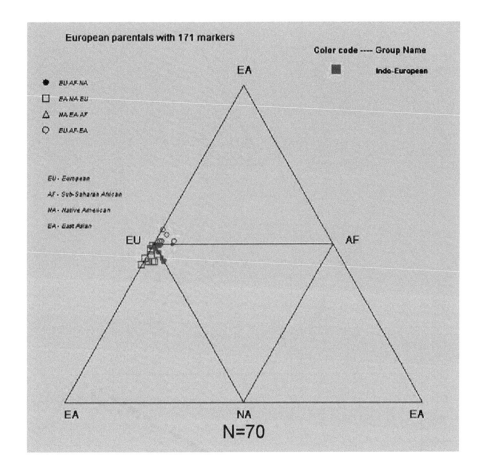

Figure 5-8

Tetrahedron plot of 70 European parental MLEs as determined using the panel of 171 AIMs discussed in the text. Symbols represent individuals from the parental population, and are shaped based on the subtriangle within which they are found, as shown in the key. The spread of samples around the European vertex is much tighter than observed in Figure 5-7, using only 71 AIMs.

71 marker panel is compared to that for the 176 marker panel for the same individual. The error we incur forcing each estimate into a three-way population model (the best fitting three-way estimate—that with the best likelihood estimate of all the possible three-way estimates) should be the same for both presentations, and differences in the plots should reflect differences in the precision of the estimates due to the difference in the number of AIMs. We would expect to see similar (though not exact) MLE positions but smaller confidence contours for the 171 AIM test, and indeed we do. Only H and I show significant differences in the position of the MLE between the two tests; most of the MLE estimates are very similar between the 71 and 171 AIM tests. Overall, the area contained within the confidence contours decreases by about 50% from 71 to 171 AIMs.

Keeping in mind the fact that the 176 AIM test characterized in this text is the 71 AIM test with 100 additional AIMs adept for the non-West African/non-sub-Saharan African distinctions, we would expect moving from 71 to 171 AIM test results, that we would see that the confidence contours are reduced in size

Figure 5-9

Triangle plots for parental and admixed samples obtained using only 30 AIMs. Each spot within the triangles represents the MLD of an individual. West African (AFR), Indigenous American (NAM), European (EUR). A. Parental samples, B. African Americans, C. U.S. Caucasians, and D. Hispanics. Provided by Mark Shriver, the Pennsylvania State University, and produced using data presented in Parra et al. 2004.

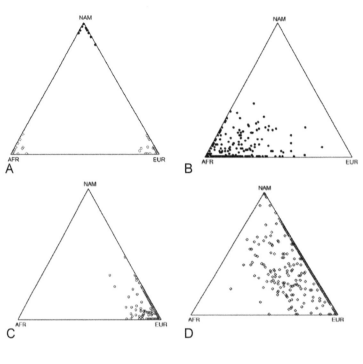

Figure 5-10

Correlation between MLE estimates for a sample of Puerto Ricans, obtained using three-way and four-way MLE algorithms. The same set of 171 AIMs was used for both calculations. Percentage of European (EU) obtained using the four-way algorithm is shown on the y-axis and the percentage of European (EU) obtained using the three-way algorithm is shown on the x-axis.

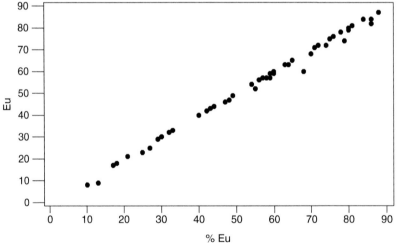

in the East Asian, European, and Indigenous American dimensions but not so much in the West African dimension. Inspection of Figure 5-13H shows an example of such. Here, we see group-specific reduction provided by the directionally enhanced power of the 171 AIM test, whereas the distance from the MLE to the West African 10-fold interval line remains similar, the distance from the MLE to the Indigenous American 10-fold line is reduced by about 50% (i.e., the shape of the likelihood space moves from an oval to a more circular shape). From inspection of larger sets of pairs, this result is not atypical and indeed it is another illustration of the predictability of the admixture estimation process at least with these markers, this analytical method, and this specific four population model.

REPEATABILITY

The quality of admixture estimates depends not only on the power of the marker panel, the appropriateness of the population model, and the statistical considerations we have so far discussed, but also on the DNA extraction step, quality of the genotyping step, and laboratory information management. For example, what impact do varying DNA extraction efficiencies have on the final answers? What impact do genotyping failures (whether caused by variability in sample quantity or quality, or variability in the execution of the genotyping steps like amplifications and washings) have in the final estimates, given that different markers are likely to fail to genotype between different runs? How reliable is the information management process and how often are samples mixed up? When handling hundreds of samples each week during the course of operating a high-throughput genotyping facility, this error cannot be assumed to be necessarily insignificant and must be empirically determined. The best way to

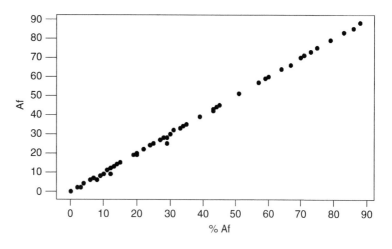

Figure 5-11

Correlation between MLE estimates for a sample of Puerto Ricans, obtained using three-way and four-way MLE algorithms. The same set of 171 AIMs was used for both calculations. Percentage of sub-Saharan African (AF) obtained using the four-way algorithm is shown on the y-axis and the percentage of sub-Saharan African (AF) obtained using the three-way algorithm is shown on the x-axis.

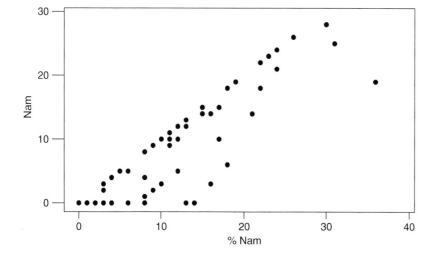

Figure 5-12

Correlation between MLE estimates for a sample of Puerto Ricans, obtained using three-way and four-way MLE algorithms. The same set of 171 AIMs was used for both calculations. Percentage of Indigenous American (NAM) obtained using the four-way algorithm is shown on the y-axis and the percentage of Indigenous American (NAM) obtained using the three-way algorithm is shown on the x-axis.

Figure 5-13

Comparison of the width of the confidence contours between smaller (71 AIM) and larger (171 AIM) marker sets. Markers are those described throughout this chapter. The two analyses/plots for each individual are shown in pairs A through I. Percentages corresponding to the MLE are shown in the upper left-hand corner of each plot.

71 AIM   176 AIM

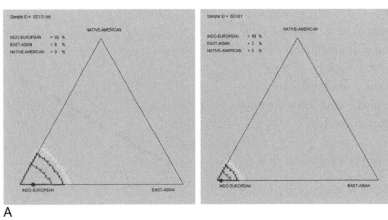

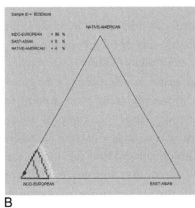
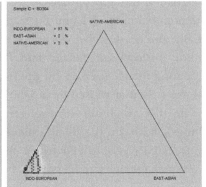

A

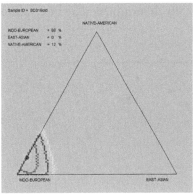
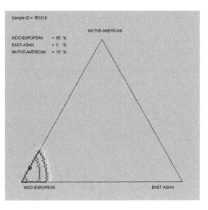

B

C

71 AIM

176 AIM

Figure 5-13
(Continued)

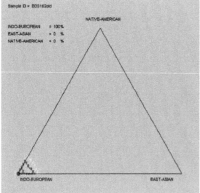

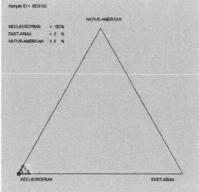

D

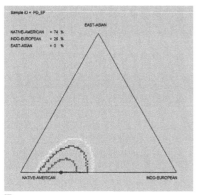
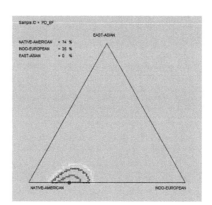

E

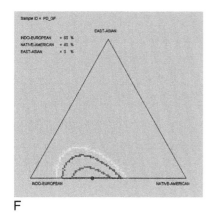
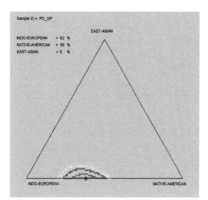

F

Figure 5-13
(Continued)

71 AIM   176 AIM

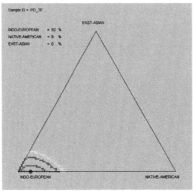
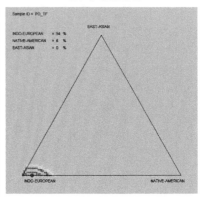

G

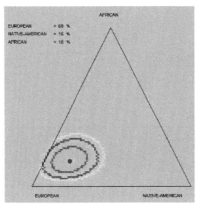
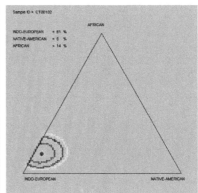

H

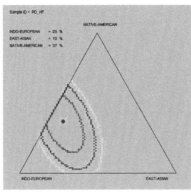
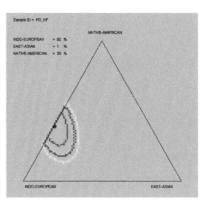

I

determine the variability caused by genotyping error and information management is to genotype the same sample repeatedly (and blindly) in different batch runs (different times, dates, and even personnel).

To measure repeatability in estimates obtained with the 171 AIM panel, we selected five individuals, one of predominant European ancestry, one of East Asian, one of West African, one Hispanic of predominant Indigenous American ancestry, and one Native American person of lower but predominant Indigenous American ancestry. We processed each blindly with the 171 AIM test as part of larger sample runs on 16 different occasions over a three-month period. Variability in ancestry estimates within samples over the 16 tests would be a reflection of variability in the quality of the entire procedure from DNA extraction through the steps required to calculate the computational estimates to the presentation of ancestry proportions. Figure 5-14 shows relatively low within-sample variability from run to run compared to the between-sample difference in average estimates. The admixture results for 16 independent runs are shown in tabular format in Table 5-1 and allow us to quantify the variation within samples from run to run as no more than a couple percent for any population group.

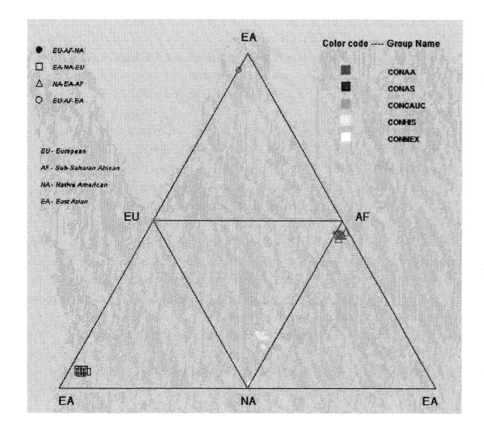

Figure 5-14

Repeatability study of MLE inference. The four-way algorithm was used with the 171 AIMs described in the text. MLEs for five individuals characterized with a relatively polarized genetic ancestry are shown (legend, upper right; CONA—an African-American individual; CONAS—an Asian American; CONCAUC—a Caucasian; CONHIS—a Hispanic; and CONMEX—an American Indian). Data for each of the 16 runs shown was generated from the DNA step onward except CONCAUC, for which DNA was extracted anew for each run. The low variability of MLE percentages from run to run is evident by the clustering of the MLEs for each of the individuals.

Table 5-1

*Admixture values for all
controls from 16 runs.*

Controls	West African	SD	East Asian	SD	European	SD	Indigenous American	SD
African American	91.3	± 0.9	3.6	± 1.4	0		5.1	± 1.4
East Asian	0		83.6	± 2.2	10.2	± 0.6	6.3	± 2.1
European American	1	± 0.4	0		98.9	± 0.4	0.1	± 0.3
Hispanic	22.7	± 1.1	0		7.5	± 1.6	69.8	± 1.8
Mexican	3.75	± 0.4	0		0		96.3	± 0.4

On closer examination of the genotype data that went into creating Figure 5-14 and Table 5-1, we found that most of the variation in scores is due to differences in the markers considered in each batch—markers that did not meet QC and were assigned failed calls, which differ as one would expect from run to run (the software that calculates the maximum likelihood admixture scores ignores failed markers in the calculation). The total number of such failed calls never exceeded 20 markers out of the 171 for these samples and was typically in the 1–10 range, though no errors (miscalling of genotype) were observed for over 31 runs using the 176 markers. The European-American individual was recollected before each run to validate extraction procedures. All other controls are used from stock DNA and have been frozen/thawed multiple times for more than a year.

In one case, the East Asian individual plotted in a triangle distinct from the triangle normally observed for this sample's admixture score (blue sample in upper triangle, Figure 5-14). Again, this alternate plotting originates from slight variations in the MLE calculation dependent on the number of failed loci.

Table 5-2 shows four runs from a similar test of reproducibility for the 71 AIM panel, and Table 5-3 shows the number of genotyping failures encountered in each run—the only type of genotyping error observed was genotyping failure (not incorrect genotype calls). We can see that even though different markers (and numbers of markers) fail in different runs, the admixture estimation process is relatively impervious to these failures (which is expected to be the case if information content of the battery of AIMs is high and spread out over large numbers of markers). Simulations suggest that with the 71 AIM panel, the tolerance of failures with respect to the production of appreciable error appears to be around 10 loci for samples affiliated with at least two ancestry groups. For persons affiliated with a single ancestry group, larger numbers of failed loci are tolerable before error can be recognized, as for

Subject	RUN 1	RUN 2	RUN 3	RUN 4
European American	100% European	100% European	100% European	100% European
African American	93% African	95% African	93% African	94% African
	7% East Asian	5% East Asian	7% East Asian	6% East Asian
Hispanic	77% Indigenous American	77% Indigenous American	74% Indigenous American	74% Indigenous American
	23% European	23% European	26% European	26% European
Indigenous American	100% Indigenous American	100% Indigenous American	100% Indigenous American	100% Indigenous American
Asian	100% East Asian	99% East Asian	100% East Asian	100% East Asian
		1% European		

Table 5-2

Variation in MLE inference in five individuals over four separate runs.

Subject	RUN 1	RUN 2	RUN 3	RUN 4
European American	0	1	0	8
African American	1	1	0	0
Hispanic	0	0	0	1
Indigenous American	0	4	1	5
Asian	0	6	0	0

Table 5-3

Number of failed loci in four independent MLE inferences for a single individual.

binary affiliated persons typed with the 171 AIM panel, which is more robust as we might expect. As we will see in Chapter 6, variability between persons within populations is much greater than the variation within samples, supporting the notion that most of the differences we see between unrelated persons is due to interindividual genetic differences rather than error.

Overall, the data indicates that ancestry estimates obtained using the 71 AIM and 171 AIM panel are highly reproducible from run to run. These control samples had undergone multiple freeze/thaw cycles, and so the results also speak for the robustness of the test relative to common laboratory practices for handling samples. The European control sample part of these tests is extracted from buccal swabs each time. Thus, the reproducibility also speaks for the robustness relative to variation in DNA extractions using this procedure.

SENSITIVITY

To further characterize the 171 AIM panel (DNAWitness 2.5 or Ancestry-ByDNA 2.5 test) we investigated the dependency of genotyping success on the amount of DNA used in the PCR reactions. To determine the minimum amount of DNA required to obtain a reliable genotype call at the majority of loci tested and maintain the biogeographical ancestry (BGA) profile for the sample, sensitivity experiments were performed by titrating three different samples from 2.5 ng/reaction to 0.125 ng/reaction. Samples were extracted either from whole blood or from buccal swabs. In either case QIAgen QIAamp kits were used for DNA extraction. DNA was quantified using the QuantiBlot Human DNA quantification kit from Applied Biosystems. Genotyping was accomplished using a Beckman Ultra-High Throughput platform with their standard single-base primer extension protocol.

The number of failed loci for each sample is plotted in Figure 5-15A and Busing blood and buccal epithelium, respectively, as the starting material. Two

Figure 5-15

Sensitivity of MLE inference with respect to input DNA mass. A. The number of failed loci is shown (y-axis) for replicated genotyping runs of a single individual using various input DNA mass (x-axis). DNA was extracted from blood samples. As expected, the number of failures increase as the starting mass of DNA decreases but not in a linear nor predictable manner. B. Same as A, except data for two individuals is shown and the DNA was extracted from buccal swabs.

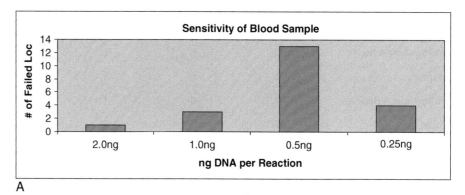

A

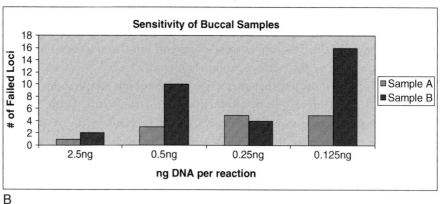

B

separate buccal samples exhibited a consistent genotyping success rate at levels down to 0.25 ng, but at 0.125 ng/reaction of genomic DNA the number of failures differed substantially between the two samples. Over many previous runs using nonlimiting levels of input DNA, the correct genotypes for the samples were known. No incorrect genotypes were observed over the range of DNA concentrations; either the genotype for a locus failed or it gave the correct genotype. No evidence of allele bias was observed for either sample types (that is, GA genotypes did not begin to register as GG or AA genotypes), and the number of failed loci did not strictly follow the trend of decreasing amounts of DNA.

The results of admixture calculation for two individuals of appreciable admixture (and one with no appreciable admixture as a reference) are shown in Table 5-4. From this data we can conclude that the failed loci had little effect on admixture calculations even in cases where admixture is apparent. In all cases, the major ancestry group estimated remained constant, and standard deviations were reasonable. The results in this table also show no appreciable difference in the standard deviation of estimates between DNA extracted from whole blood versus buccal epithelium. In particular for the buccal swab starting material, sample B showed no change in admixture score over the range tested, though sample A showed a small standard deviation.

Sample	Population	Average BGA	SD
Blood663	European	profile 79	2.2
	East Asian	0	0
	West African	3.5	0.6
	Indigenous American	17.5	2.1
Sample A	European	95	2.4
	East Asian	4.3	2.0
	West African	0.8	0.4
	Indigenous American	0	0
Sample B	European	100	0
	East Asian	0	0
	West African	0	0
	Indigenous American	0	0

Table 5-4

Hypothetical admixture results for three individuals.

Indeed sample A's DNA stock was of significantly lower concentration, and this result may illustrate the multilocus genotype-dependent nature of the sensitivity to input DNA levels and number of genotyping failures. This data allows for the conclusion that reliable results can be obtained with rather low levels of input genomic DNA (down to 0.125 ng/reaction) per amplification, amounting to about 2.5 ng total DNA required to achieve 0.125 ng/reaction with current UHT-based assay design (20 multiplex amplifications). However, 1 to 2.5 ng/reaction per amplification would be preferable to minimize the number of failed loci, which would require 20 to 40 ng total DNA with the assay as currently being run as of this writing on a Beckman Ultra High Throughput (UHT) platform.

ANALYSIS OF RESULTS FOR GENEALOGISTS

As detailed in Chapters 1 and 2, and strictly speaking, the measures of Biogeographical Ancestry (BGA) that we have been discussing are measures of genomic ancestry. Genomic ancestry is a summary of ancestry levels obtained directly from genetic markers. We use this term to contrast these genomic or realized ancestry estimates from the genealogical or expected ancestry. Genealogical ancestry is the level of ancestry for a person that is expected based on parental ancestry estimates. For example, a person with both parents being 50% European/50% West African ancestry will have a genealogical ancestry of 50% European/50% West African. However, the realized or genomic ancestry of this person will differ slightly from this expectation of 50% due to the effects of variation caused by the process of marker segregation. Despite these variations for any one individual, the average across many offspring from this family is expected to be 50%. In this light then, we have made explicit comparisons to the expectations of genealogists to the results of the DNAPrint 2.5 test and report on these in this section.

Traditionally, genealogists research family histories using a variety of different tools, including registries of birth, death, marriages, surname databases, state archives, immigration/emigration records, and even web sites and mailing lists. Though much of genealogical research is subjective, and although we cannot judge an admixture test or any other genetic test against subjective metrics, we can say that if all three of the following are true, then the results from the admixture test should be consistent with the average genealogists expectations as well as anthropometrics:

- Biogeographical ancestry determined from the DNA is meaningful in an anthropological sense
- The average genealogist is correct in his or her assumptions of ancestry

■ Genealogical notions of ancestry are correlated with anthropological and not merely geopolitical conception of human physical and cultural variation—whether formulated in terms of race, ethnicity, or heritage

We mean to clarify that it is the average genealogical expectation that has value; though any one genealogist could be wrong about his or her heritage, we expect that the average genealogist is correct in the identification of the major elements of ethnic heritage, and it is against this average that we wish to measure the performance of the test.

We can surmise that the expectations held by some genealogists could themselves be subject to a significant amount of error. Though the individuals have spent considerable effort documenting their ancestry, the estimates of expected ancestry do not usually accommodate chance as required by the law of independent assortment (marker segregation), and this contributes to some error in the expectations. For example, a genealogist who believes his or her great-grandmother was an American Indian of full blood might expect (rightly so) that their expected level of Indigenous American ancestry would be 12.5%. However in reality, the level of genomic ancestry of Indigenous American derivation in each generation is subject to binomial sampling fluctuation so it could be more or less. Second, not all genealogists possess anthropological knowledge, a certain amount of which is necessary for formulating rational expectations based on geopolitical, sociological, or religious affiliations that play a role in determining self-held notions of race. For example, an amateur genealogist might expect 50% Mexican-American ancestry without realizing that Mexican-Americans are an admixed group of largely European and Indigenous American roots, and might be surprised to see a genomic ancestry with respect to a four-population continental model of 50% European. In this case, the genealogists might simply not appreciate that their assessment of ancestral identity is with respect to a more recent phylogeographic history than that for which the admixture test is calibrated.

Lastly, genealogical sources are not as objective a source of data as most scientists are accustomed to, and there is a certain amount of imprecision and anecdotal error inherent to the practice of reconstructing a family tree from government documents and stories. Nonetheless, though any one individual's self-held notions on race and expectations on BGA might not be accurate, we expect that if a sample is large enough, the average expectation would be reasonably precise. For this reason, in the following discussion we focus our attention on the average level of BGA admixture expected and reported, though we also discuss standard deviations for these percentages.

DNAPrint began selling the 71 AIM test under the brand AncestryByDNA 2.0 in late 2002. In 2003, a customer and amateur genealogist established a web site through which customers could report their expected (from traditional lines of research) and observed (from AncestryByDNA 2.0) admixture proportions along with explanations and comments on the nature of their evidence. The following calculations were based on the data taken from this compilation as of Jan. 2004, relevant for the 71 AIM test (AncestryByDNA 2.0). As of this writing, interested readers could view the site at http://www.dnaprintlog.org/.

From a sample of 108 genealogists, we are able to compare the expected and observed ancestry levels and note any interesting trends. As shown in Tables 5.5 and 5.6, the average amateur genealogist expected 6.6% Indigenous American admixture and got 5.1% using the 71-marker test (see Table 5-5). Of all the amateur genealogists, 69% of those expected some Indigenous American ancestry and got a value from the 71 marker test that was within 10 percentage points of the level they expected (see Table 5-6).

All but a few of the genealogists were best described as European Americans; the average levels of European (EUR) affiliation expected and observed were about 86% and 85%, respectively. The average level of total expected admixture was about 14%, and the average level obtained was about 15%. In terms of specific types of admixture, the results were fairly similar from group to group. The average level of detected Indigenous American and West African genomic ancestry are fairly close to the average level expected, but the average level of East Asian BGA detected was more than twice the average level expected by these genealogists.

We next focus on the number of estimates that fall within a 10-percentage point range of the expected values for each ancestry subgroup (see Table 5-6). For example, if someone expected 90% European, 10% Indigenous American, and got 100% European, their result was within 10 percentage points of expected values for both European and Indigenous American, and they were

Table 5-5

*Average obtained and expected results for amateur genealogists (**n = 108**).*

	EUR	EAS	IAM	AFR
Expected	0.858	0.020	0.066	0.056
Observed	0.844	0.055	0.051	0.051

Table 5-6

Frequency of MLE values within 10 percentage points of genealogist expectations using the 71 AIM BGA test (AncestryByDNA 2.0).

	EUR	EAS	IAM	AFR
All samples	0.71 (n = 108)	0.83 (n = 108)	0.81 (n = 108)	0.96 (n = 108)
Samples for individuals expecting admixture only	0.7 (n = 105)	0.7 (n = 10)	0.69 (n = 64)	0.77 (n = 17)

counted as correct for both categories by this 10 percentage point criteria. If they expected 89% European and 11% Indigenous American, they were not correct, and were counted as incorrect by this 10% criteria. We simply counted the number of correct and incorrect results for each group, for all 108 of the samples (all samples, Table 5-6), whether they expected admixture (as one analysis) or not (as a second analysis). As can be seen in Table 5-6, about 70% of the individuals who expected admixture received EUR results within 10 percentage points of their expectations. Over the entire sample of genealogists, including those that did not expect admixture, about 83% received results within 10 percentage points of their expectations averaged over all of the ancestry groups. Although the results show that genealogists usually got results similar to what they expected, this particular table illustrates an important example of the discordance between genealogical expectations and BGA affiliations experienced using the 71 AIM panel.

To determine whether the difference in reported versus expected BGA levels differs among individuals of differing expected levels of admixture, we break the sample into three groups—those expecting homogeneous BGA affiliation, those expecting from 0% to 25% BGA affiliation, and those expecting from 26% to 75% BGA affiliation. Since most of the individuals in the sample of 108 were self-described European Americans, virtually all the individuals expecting homogeneous affiliation were expecting homogeneous EUR affiliation. For these, the average level of European affiliation reported was 92%, 8% lower than the expected level of 100% (see Table 5-7).

The sample sizes for individuals expecting from 0% to 25% affiliation were somewhat better (see Table 5-8). The average East Asian, Indigenous American and African admixture expected (13%, 12%, and 3.6%) was greater than the average level observed (9%, 7%, and 3.2%), even though we shall soon see in simulations that the tendency of the 71 marker test is to overestimate admixture in individuals such as these with relatively polarized BGA. This might suggest that our sample of genealogists, and maybe genealogists in general, are overexuberant in projecting BGA admixture from anecdotal and document-based research (i.e., that the genealogical ancestry tends to be greater

	EUR	EAS	IAM	AFR
Sample	n = 26	n = 1	n = 0	n = 0
Average expected	1	1	N/A	N/A
Average obtained	0.92	0.85	N/A	N/A
Standard Deviation	0.08	0.15	N/A	N/A

Table 5-7

Genealogists expecting homogeneous affiliation (no admixture).

	EUR	EAS	NAM	AFR
Sample	n = 6	n = 9	n = 65	n = 17
Average expected	0.14	0.13	0.12	0.036
Average obtained	0.2	0.09	0.07	0.032
Standard Deviation	0.09	0.09	0.11	0.07

than the genomic ancestry). The average difference between expected and obtained (standard deviation) was relatively high—about as high as the expected levels—which illustrates that a fair number of these individuals received results that were significantly different than expected. We see from Table 5-8, however, that these deviations are similar for each type of affiliation. Thus, though there seems to be a tendency for the genealogical ancestry admixture to exceed the genomic ancestry admixture, the tendency is spread among all types of admixture. This suggests the method of admixture estimation (with these 71 AIMs) is relatively unbiased.

As we will discuss shortly in the simulations section, we indeed know from other sources of information that there should be very little bias toward the 71-marker test incorrectly reporting a certain type of admixture over another, and the results in Table 5-8 support this notion. Given that the test is relatively unbiased from this particular application, the results also suggest that U.S. genealogists of mixed genealogical ancestry are not prone to unreasonably expect a certain type of BGA admixture over another (such as IAM or AFR, for instance, which might logically be predicted given the recent history of U.S. Europeans and Indigenous Americans/Africans).

Looking at individuals expecting greater levels of affiliation, we see that the average individual expecting from 26% to 75% European expected 53% European affiliation but was reported to be of 67% European affiliation (see Table 5-9). The sample size was low for all types of affiliation, but in general, for

	EUR	EAS	NAM	AFR
Sample	n = 15	n = 2	n = 8	n = 7
Average expected	0.53	0.25	0.31	0.69
Average obtained	0.67	0.06	0.04	0.66
Standard Deviation	0.18	0.19	0.27	0.1

this group of genealogists expecting extensive admixture, more European was reported than expected. The reported East Asian and Native American genomic ancestry was significantly less than the expected from the genealogical ancestry, and for two types of admixture, East Asian and Indigenous American, the differences between the two were large (though the sample size for both were low). We will soon see that the explanation for the Indigenous American discrepancy is unlikely to be statistical error in detecting Indigenous American affiliation.

Simulation results with parental and admixed populations show that the estimates are relatively unbiased in nature (to be discussed soon), and no such discordance is observed for Hispanics (Figure 6-9, Chapter 6) or individuals living on federally recognized reservations (see Figure 6-8, Chapter 6). The reasons for this discrepancy between genealogical ancestry expectations and realized genomic ancestry in this and only this specific subset of genealogists are not at all clear, but could be psycho-social in nature. We will see in Chapter 6 that the frequent mismatch between genealogical and genomic estimates of extensive (>10%) Indigenous American admixture is a trend that extends to larger samples of customers expecting mainly European genealogical ancestry. Over 13,000 lay customers had been tested as of this writing, and over 90% of the customer complaints were from Europeans who claimed to have had an American Indian great-grandparent that the test didn't detect (see Box 5-A). From these 13,000 tests, no such phenomena has been observed for any other type of genomic ancestry admixture on any other type of majority ancestry background as even the relatively small samples in Table 5-9 would attest (note that the sample of individuals expecting substantial African admixture was similar in size to that expecting Indigenous American admixture, yet these former individuals realized a relatively small difference between genomic and genealogical ancestry admixture percentages). Overall, the results might suggest that Europeans tend to overreport Indigenous American genealogical ancestry admixture.

Box 5-A

More for Genealogists

Visitors to the www.dnaprintlog.org site also are allowed to read the personal reactions of persons having the test, or enter their own if they wish. The psychology of the customers and their reactions are interesting. Customer reaction to the results was generally favorable, except for the European Americans who expected a low level of Indigenous American admixture, who were generally the most vocal in their displeasure with the results that commonly showed them to have none at the level of the genomic ancestry. Though no complaints were received for missing or overreporting West African or European ancestry, several of the customers complained about the test not reporting their expected (usually but not always dilute) Indigenous American heritage, and a couple were received for the test reporting low levels of East Asian affiliation when none was expected. Many of the genealogists did not understand that what they were getting was an anthropological answer to what they thought previously was only a geopolitical question (see Chapter 2). For example, a primarily West African individual living in Puerto Rico may describe themselves as a Hispanic, and through this connection to others with substantial Indigenous American BGA, may have an expectation of IAM admixture that is rooted in geopolitics rather than anthropology. As another example, many escaped slaves became part of American Indian tribes in the seventeenth through nineteenth centuries, and their ancestors may therefore identify as partially American Indian but this does not mean they necessarily shared genetic affiliation with Indigenous Americans.

Overall, the results show that the average genealogist got a result reasonably close to what they expected. However the standard deviations (average difference from expectations) were relatively high, and so there were many individuals that got significantly different results than they expected. Part of the standard deviation is due to the anecdotal nature of some genealogical expectations; scanning over Mr. Kerchner's site (www.dnaprintlog.org) gives you an idea of the differences in quality of the expectations between individuals—some are better than others. Some of the standard deviation is due to statistical error inherent for the test, as well as other sources of test error we have discussed in previous chapters. One important source of deviation that cannot really be considered an error (i.e., not the fault of the testing method) is in the difference between genealogical ancestry or the expected ancestry based on a person's parents or grandparents vs. the genomic ancestry, which is what we are measuring using genetic markers. We refer you to Chapters 1 and 2 for a more detailed discussion of these types of ancestry. In addition, genotyping error is also involved, which is estimated to be less than 1%.

Another source for the discrepancy between genealogical and genomic ancestry is natural genetic diversity at the level of the ethnic subpopulation. We will see in Chapter 6 that the average results differ substantially between ethnic subpopulations of different recent (and older) histories, almost forming a type of signature for certain subpopulations. Thus, low levels of affiliation can be diagnostic of certain ethnic affiliations rather than recent admixture. Mediterranean, Jewish, and Middle Eastern populations, for example (and as we will discuss more fully next chapter), show on average a few percent Indigenous American and West African affiliation, most likely as a function of shared ancestry with proto-Indigenous Americans in Central Asia and gene flow from North Africa. Virtually none of our Jewish genealogists would have anticipated results emanating from relatively ancient anthropological connections because they are accustomed to thinking in recent, geopolitical terms about ancestry and heritage. So, this type of phenomena might explain the cases where low levels of affiliation are shown but not expected.

What about the cases where low levels were expected but not shown? Another source of discrepancy between genealogical and genomic ancestry is natural individual-to-individual variation. If a certain type of ancestry was recently introduced in an individual's family tree, the ancestry in this individual will be clustered to certain chromosomes and regions of chromosomes. Since we all have two copies of each chromosome, and since it is a function of probability which one of these we inherit from our mothers and fathers to make our own pair, and them from their mothers and fathers, there is a chance that the individual inherited more or less genomic admixture than expected from the genealogical ancestry. Most genealogists are unfamiliar with the law of genetic assortment, and believe that if one parent is 100% European and the second 50% European, then 75% of each child's genes will be European. Indeed, this is what we conceptually refer to as the genealogical or expected ancestry.

Variability about expected levels is not usually built into their expectations on ancestry, but this variability, which is significant even over one or two generations, creates fairly large ranges in expectations for the genomic ancestry. Many of the cases where low levels of ancestry was expected but not observed were obtained for European individuals expecting a low level of Indigenous American heritage. Not only is there a natural genetic variability in the rate of Indigenous American ancestry dilution between and within families through the process of chromosomal assortment, but most genealogists cannot be certain of what the starting level was in the original Native American ancestor; that is, was this person admixed? Most genealogists assume a known American Indian ancestor to have been full blood, but this may not be the case and of course there are usually no records to assist in confirming non-European ancestry. On top of this, there is natural variation in the quality of genealogical research and some genealogists are simply going to be wrong, especially for lower levels.

These are just a few reasons why a genealogist could get results that differ slightly from what they expect, especially if what they expect are lower levels of ancestry (particularly IAM admixture). The multiple sources of error in genealogical research and genomic ancestry estimation indicate why it is best to consider a measure of biogeographical ancestry from the DNA in addition to (rather than instead of) evidence from history, surnames, databases, and opinion. Multiple sources of evidence always allow for a stronger conclusion than any one particular item of evidence, especially for complex problems such as determining the structure of a person's genealogical past. Although we are focusing on the genomic methods in this book, being genomically minded, it would be wrong for us to conclude that the genomic data were essentially superior to other sources of genealogical information. For example, consider a European with an Indigenous American great-grandfather who happened not to inherit any of the Indigenous American genetic material from this great-grandfather. His 100% European, 0% Indigenous American result does not change the fact that he has an Indigenous American great-grandfather and is therefore of mixed heritage. Of course, we cannot measure every chromosomal segment and this person may very well have a low level that a 71 AIM or 176 AIM assay cannot recognize. We firmly believe that a proper genealogical analysis should include several sources of data and that the genetic data, especially when working with minor ancestry components, will rarely provide evidence that a particular hypothesis is necessarily true but generally will provide support that is consistent with other data.

ANALYSIS OF RESULTS FOR NONGENEALOGISTS

To determine whether and to what extent the major proportions obtained with the 71-marker BGA test corroborated with self-reported race, we compared the majority of BGA affiliations with the self-held description of race for each of 1,325 samples (see Table 5-10). This analysis is essentially taking the four-way biogeographical ancestry result obtained with the 71-marker test and binning the sample into the group corresponding to the majority affiliation. Of course

Table 5-10

Comparison of majority biogeographical ancestry determined using 71 AIMs and MLE method with self-reported race.

Self-reported race	European	African	Indigenous American	East Asian
European-American (U.S. white)	1094	2[*]	0	0
African-American (U.S. black)	1[**]	123	1[**]	0
Hispanic	27[***]	0	3	0
Asian American	4[****]	0	1[****]	33
South Asians from India	36	0	0	0

[*]Greatest minor proportion was European.
[**]Greatest minor proportion was West African.
[***]Greatest minor proportion was Indigenous American for each of these 27.
[****]Greatest minor proportion was East Asian.

binning conceals the value of test, and is inappropriate for several reasons we already have discussed, but gives us a different view of the performance of the test relative to how people describe their heritage.

We observed a very strong concordance between the major BGA group determined with the 71 AIM test and the self-reported majority ancestry. Overall, the results comported with expectations; even when a sample was determined to be of majority affiliation with an unexpected group, the results were more or less concordant in that the expected affiliation was the minor affiliation reported by the test rather than being absent altogether. For example, the finding that two African Americans were not of majority West African BGA was unexpected, but in both cases substantial West African BGA was present (just under 50%; asterisks, Table 5-10). For self-described European-Americans (U.S.-born Caucasians) 1094/1096 registered with majority European BGA, the two exceptions being individuals of majority West African and minority European BGA, indicating that these individuals were of substantial admixture. Fourteen percent (156/1096) of the self-described Caucasians registered with less than 7% Indigenous American affiliation, 8% (90/1096) with less than 7% East Asian affiliation, and 6% (64/1096) registered with less than 7% West African affiliation (data not shown). In a sample of self-described African Americans, 123 of 125 showed majority West African BGA as expected, but two individuals showed West African BGA as the secondary affiliation (one on a background of majority European and another on a background of majority Indigenous American), indicating that these two individuals were of substantial West African admixture.

It is clear that consideration of individual ancestry along a multidimensional continuum (i.e., in terms of admixture with respect to a specified parental population model) is better than binning or categorization. Table 5-10 suggests that using the most powerful of AIMs and an admixture method, the concordance between major BGA group and self-identified racial group is 99.2% (the frequency of non-European ancestry as the major ancestry percentage for a European American, non-West African as the major percentage for African Americans, and non-East Asian as the major percentage for East Asians is a combined 0.8%). Recall that Frudakis et al. (2003) and Shriver et al. (1997) calculated correct binning percentages as being in the upper ninetieth percentile, depending on the number of AIMs used (about 98.8% accuracy using 56 AIMs, 96.7% accuracy using 30 AIMs, 95.2% accuracy using 15 AIMs). Table 5-10 would then suggest that about 1% of the error experienced in these studies was due to inadequacies associated with a binning model. A much more significant problem with binning, however, is the information lost by assuming an overly simplistic model that all of a person's ancestors derive from only a single (oftentimes socially defined) parental population.

BLIND CHALLENGE OF CONCORDANCE WITH SELF-ASSESSED RACE

We would expect persons whose type of predominant West African ancestry to appear African American (in the United States) in terms of anthropometric traits that visually distinguish populations. For example, we would expect a darker skin and hair color and darker iris color for the average person who describes him- or herself as African American or black than a person who describes him- or herself as European American, white, or Caucasian. Put another way, we can generally say that an individual typed at, say, 90% West African ancestry should not "look" East Asian or European, even though this might mean, precisely speaking, different things to different people, and there are different (even unscientific) words we often use to describe the appearances or the population groups, which vary in different cultures.

Indeed, concordance with common-sense notions of race is the primary method of evaluation law enforcement officials are likely to use because it makes sense to them and can be communicated effectively with others (although there are very likely striking deficits in terms of what people are actually aware of). DNAPrint has conducted a few hundred blind analyses for law enforcement agencies throughout the United States. For a blind analysis, the laboratory types the DNA sample without access to photographs for the individual, or knowledge of how they or others describe their ancestry. After reporting the results, the photo and geopolitical questionnaire results are supplied and an evaluation of fit with expectations can be made, which, opposed to merely a test that is blind in a computational sense, is truly blind.

To conduct a truly blind test, the San Diego Police Department Crime Lab (SDPD) and the National Center for Forensic Science at the University of Central Florida (UCF) each submitted 10 samples of concealed identity and concealed self-held affiliations to our laboratory. After we performed the BGA test, each group independently evaluated their results and revealed the self-reported population affiliation of the sample to our laboratory along with digital photographs. We found that for this sample, the major percentages determined from the BGA admixture proportions test were not inconsistent with the self-reported population affiliations as Table 5-10 would suggest (see Table 5-11 and Figure 5-16). Figure 5-16 shows the results of one such analysis conducted using the 71 AIM panel for the San Diego Police Department in 2002.

Several of the samples were from individuals affiliated with groups that are considered to be admixed, such as Philippine (SDPD2, SDPD3, SDPD9; Table 5-11), African American or Caribbean (SDPD5, UCF6, UCF8; Table 5-11), Mexican American (Hispanic, SDPD8, SDPD10; Table 5-11), and for each of

ID#		%	Admix ratio	%		%	Self reported race[*]
SDPD1	European	96	East Asian	0	Native American	4	European American
SDPD2	East Asian	53	European	47	Native American	0	Filipino
SDPD3	East Asian	61	Native American	28	European	11	Filipino
Sdpd4	European	100	East Asian	0	Native American	0	European American
Sdpd5	African	69	European	31	Native American	0	Caribbean
Sdpd6	African	67	European	33	East Asian	0	African American
Sdpd7	European	99	East Asian	1	Native American	0	European American
Sdpd8	Native American	57	European	43	East Asian	0	Mexican American
Sdpd9	East Asian	86	Native American	0	European	14	Filipino
Sdpd10	Native American	36	East Asian	28	European	36	Mexican American
Ucf1	European	90	East Asian	0	Native American	10	Ukraine/Italy
Ucf2	East Asian	98	Native American	2	European	0	Chinese
Ucf3	European	88	East Asian	3	Native American	9	Ukraine
Ucf4	European	100	East Asian	0	Native American	0	European American
Ucf5	European	87	East Asian	0	Native American	13	Greek
Ucf6	European	62	Native American	38	East Asian	0	Puerto Rican
Ucf7	African	83	European	17	East Asian	0	African American
Ucf8	European	69	African	31	Native American	0	Jamaican
Ucf9	European	84	East Asian	12	Native American	4	Finland
Ucf10	European	98	East Asian	2	Native American	0	Scotland

[*]Shown is exactly what was reported even if the subject checked Other on the questionnaire and wrote in an ethnic group or geographical location rather than a race.

Table 5-11

Blind challenge of BGA test by the San Diego police department and the center for forensic science at the university of Central Florida.

these, significant admixture was detected. Moreover, the type of admixture detected was reasonable with respect to the anthropological history of the affiliated population, but of course relatively large interindividual variation in genomic ancestry is expected. For example, the Caribbean harbors peoples of West African, Indigenous American, and European descent, but the East Asian influence on this region of the world is not considered to have been great

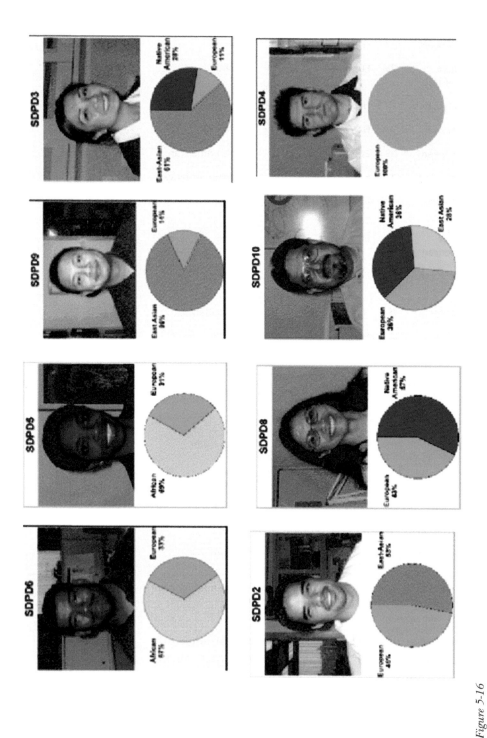

Figure 5-16

A blind inference of Biogeographical Ancestry (BGA) admixture for eight individuals conducted by the San Diego Police Department (SDPD), using the 71 AIM panel described in the text and the three-way admixture algorithm (best three-way result presented even though a four-way population model is used). Genotyping and statistical analyses were performed at the laboratories of DNAPrint genomics in Sarasota, Florida. The BGA estimates for each subject are shown with pie charts. After these results were reported to the SDPD, photographs were supplied for comparison. SDPD 5 described himself as Afro-Caribbean and SDPD 6 described himself as African American; both type as of predominant West African ancestry (and both have substantial European affiliation as well). SDPD9 self-described as Filipino, and was reported by the test to be of predominant East Asian ancestry (as do other Southeast Asians, as we will see in Chapter 6). SDPD3 and SDPD2 also described themselves as Filipinos, and typed similarly. SDPD8 and SDPD10 self-described as Mexican-Americans or Hispanic (see Table 5-11). SDPD4 self-described as Caucasian or European American.

(excepting Trinidad, where many Chinese emigrated). Although, on aggregate, significant West African, European, and Indigenous American BGA was obtained for the three Caribbean samples tested, no East Asian admixture was detected for any of these three samples. In contrast, East Asian influence on the demography of the Philippines is well documented (Jobling et al. 2004), and each of the three Filipinos tested harbored significant East Asian BGA. Note that given the potential discordance between self-held geopolitical notions of race and genomic ancestry, this type of concordance with history need not be found for any one individual, and if not observed, would not necessarily indicate a defective ancestry assessment method.

In general, the results from the 71 AIM test agree well with self-held expectations, as well as from common perceptions derived from anthropometric trait values (which are, admittedly, difficult to put into words, but nonetheless tangible). Blind trials such as these merely allow a potential forensics user of genomic ancestry assessments to evaluate the performance with respect to their own internal metrics, which of course are based partly on commonly held social metrics derived from social constructs and their own internal biases. Though of suboptimal objectivity, not all of these are completely useless for the purposes of constructing a physical portrait from crime scene DNA. Similar results have been obtained using the 171 marker panel, and of approximately 7,000 analyses conducted by DNAPrint as of September 2004, not a single case of a confirmed, extreme error has been observed, where for an individual of relatively polarized self-reported ancestry, their majority affiliation was obviously misdiagnosed based on anthropometric descriptors and self-reported information. For example, if an individual is self-described as three-fourths blood of East Asian ancestry, and exhibited characteristic East Asian anthropometric traits (such as epicanthic eye folds, intermediate skin shade, dark hair and eyes), but was classified as of predominant West or European ancestry, this would represent an example of extreme error. The errors in Table 5-11 would not meet these criteria of extreme error.

CONFIDENCE INTERVAL WARPING

As we already have described, in addition to determining the most likely estimate of admixture (the MLE), we developed our software program to survey the probability space and define that space within which the likelihood of proportional genomic ancestry was two-, five-, and 10-fold less likely to be the correct answer than the MLE. We call these confidence contours, and when single MLEs are plotted in the BGA triangle plot, these contours are plotted as rings around the MLE (see Figures 4-12 and 5-17A). If a mechanic builds a new engine and thinks he understands exactly how the engine he created works, he

Figure 5-17

Robustness of BGA admixture proportion determination imparted by the AIMs used in the analysis. The confidence of the MLE (point) is predictably affected by the elimination of those AIMs from the test that are informative for a particular pairwise BGA group comparison. For both triangle plots, the first contour extending from the MLE defines the triangle plot space within which the likelihood is up to two times lower than that of the MLE, the second contour defines the triangle plot space within which the likelihood is up to five times lower than that of the MLE, and the third contour defines the space within which it is up to 10 times lower. In A, the MLE and its confidence contours obtained by using all 71 AIMs is shown along with the actual percentages in the upper left-hand corner. By eliminating those AIMs from the analysis that are informative for East Asian/Indigenous American distinction, the MLE is relatively unaffected in B, and though the confidence contours along the East Asian/European and Indigenous American/ European sides of the triangle remain undistorted, the confidence contours are distorted along the East Asian/ Indigenous American side.

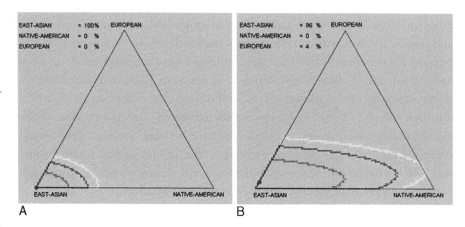

should be able to remove or break a part and predict exactly what will happen to the engine's performance in terms of the chronology and type of failures leading to the final outcome.

One way to test the accuracy of the maximum likelihood algorithm for determining MLE and confidence contours is to observe whether and how these values change when certain of the AIM markers are eliminated from the analysis. To do this, we replaced the genotypes for AIMs of delta (δ) > 0.20 for a given population pair with failure readings and compared the shift in confidence contours obtained with that we expected to see from the change. For example, if for a given sample genotype, we replaced all the genotypes for markers of high delta value for the African/East Asian distinction with failure or no data, leaving only those of good European/sub-Saharan African, European/East Asian, European/Indigenous American, and East Asian/Indigenous American delta or F_{ST}, then we would expect an n-dimensional test to show warped confidence contours in two dimensions and normal contours in the remaining n − 2 dimensions. On a triangle plot, this would look like a scenario where the proportion of the axes covered by the contours was expanded for two of the axes, but not the third. In other words, if the quality of the admixture/affiliation estimates for each group is independent from that for the others, as they should be if the test works properly, then the shape of the confidence contours in a triangle plot should expand only along the dimensions for which power was reduced by the deletion of genotype data.

Figure 5-17A shows the plot of a sample of majority East Asian BGA with its associated confidence intervals obtained using the 71 AIM panel, and Figure 5-17B the change observed upon failing all 24 of the AIMs in this 71 AIM test panel that were of informative (delta > 0.20) for Indigenous American versus East Asian BGA, whereby the MLE and confidence estimates were calculated using the remaining AIMs. From this experiment, we observed the confidence rings to be dramatically skewed in the East Asian and Native American (but not the European) axis—the ovals warp horizontally but not so much vertically along the East Asian toward the direction of the Indigenous American BGA vertex (see Figure 5-17B). Along the NAM axis, virtually the entire axis is covered by the 10-fold confidence contour and the same thing is observed along the EAS axis, but the proportion of the EUR axis remains about the same. This indicates that, as expected, the lack of AIMS with good delta values for East Asian/Indigenous American distinction produced an estimate for which the confidence along the East Asian/Indigenous American component of the European/East Asian/Indigenous American decision was not high but for which confidence in the other dimension (i.e., European/non-European) remained similar. Evidently, because the sample typed as of majority East Asian, and AIMs for the distinction of East Asian/European affiliation were left unmolested, the MLE shift itself was minimal. Most of the uncertainty along the East Asian/Indigenous American axis was apparent in the shift of the contours, not the MLE, presumably since there was little Indigenous American affiliation determined for the sample in the first place. Similar experiments with other samples and AIMs produced similar results (Frudakis, unpublished data).

SAMPLED PEDIGREES

The analysis of admixture estimates in family pedigrees gives a different glimpse into the overall quality of the estimates. The genetic law of independent assortment tells us that although an infinite number of offspring is expected to exhibit ancestry halfway between the two parents (i.e., the genealogical ancestry), we expect a natural variability in the estimates from offspring to offspring depending on what chromosomes they inherited when genomic ancestry is analyzed with AIMs. Substantial admixture observed for one individual of a pedigree not transmitted to offspring would be an example of an unexpected finding and could be a result of estimation error. In contrast, if many members within a pedigree show a level of significant affiliation that is transmitted to offspring within expected ratios, we can conclude that the affiliation is bona fide ancestry and a function of the genetic history of the pedigree. An accurate test would tend to show the latter scenario rather than the former, especially for substantial affiliations of, say, greater than 20% observed for any of its members.

Figure 5-18 shows one such pedigree from real (not simulated) samples, obtained using the basic DNAPrint MLE algorithm with the 171 AIM panel. Paternity and maternity were confirmed for each subject using the ABI Profiler STR system. Individual TF (one of the authors of this book) is a European

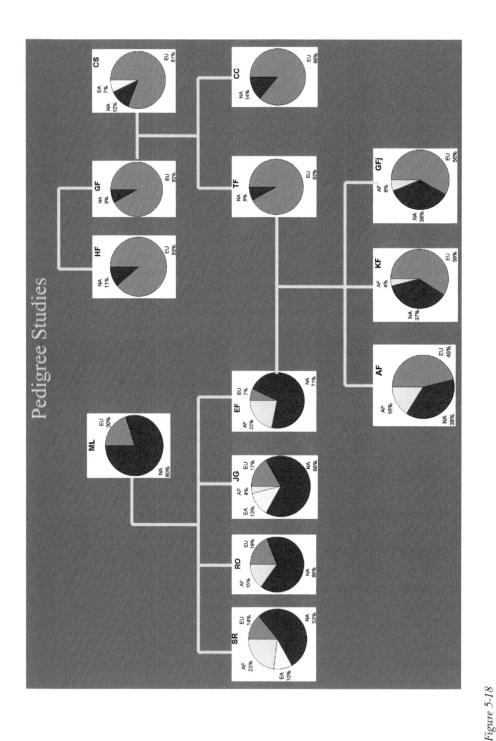

Figure 5-18

BGA admixture proportions are transmitted as expected in family pedigrees. BGA proportions determined using the MLE algorithm with the 171 AIMs and continental population model described in the text. Vertical lines indicate offspring–parent relationships. Rows represent generations and each individual is represented by a two-letter symbol. BGA proportions are indicated numerically next to the relevant regions of the subject's pie chart.

American whose wife EF was born in Mexico and typed of predominant Indigenous American ancestry, with West African and European admixture. Their three children AF, KF, and GFj show affiliations that are roughly halfway between their two parents, TF and EF, but with specific ancestry levels that vary from child to child as we expect. The West African admixture in these three children can be traced back to the father of EF, who was reported by EF and her mother ML to have been part African, with dark colored hair and skin relative to the average Mexican, and kinky hair texture. Indeed, most of EF's siblings show the West African ancestry but the mother does not. Evidently from this pedigree, the father also carried East Asian admixture that was passed to two of ML's offspring but not the other two (presumably via recent East Asian admixture for the father and the law of independent assortment). TF's side of the pedigree shows predominant European ancestry but with a low level of Indigenous American affiliation; the Indigenous American affiliation is present for his sibling, mother, father GF, and father's sister HF, but also for the mother CS. Since it is shown for all the individuals of TF's side of the pedigree, we can conclude that its expression with the 171 AIM test is a function of the genetic character of this particular subpedigree. The low level of East Asian affiliation in CS was not evidently passed to either of her and GF's children (TF and CC).

Approximately 15 pedigrees similar in size to this one have been analyzed with the 171 AIM test using the basic DNAPrint MLE program, with similarly concordant results. Overall these results show that substantial levels of affiliation observed for one individual of a pedigree tend to be shown in levels within expectations by others in that pedigree, supporting the notion that for at least substantial levels of affiliation (to be defined more precisely later), the ancestry estimates are stable in that they report affiliations that are a function of the genetic character of the samples rather than statistical quirks on a sample to sample basis. In other words, pattern such as that shown in Figure 5-18 indicates that the estimates are more likely to be real than for the case when there is no pattern in the pedigree. The level of noise from sample to sample appears to be less than the level of affiliations that are produced by bona fide ancestral relationships between offspring and their parents.

A similar but slightly different result is obtained for some of these same samples using the 71 AIM test (Figure 5-19, where only the European/Indigenous American affiliations are shown). Again, the admixture scores are consistent with expectations based on the genetic law of independent assortment (which would predict a 50% contribution from each parent);

Figure 5-19

Family pedigree including many of the individuals shown in Figure 5-18 and using the same methods, except 71 AIMs were used rather than 171. Indigenous American and European percentages are shown. Overall, we have suffered numerous subtle losses of information with the reduction of markers.

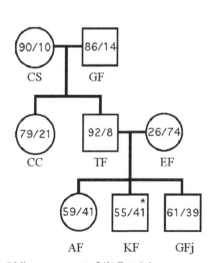

* Minor component of 4% East Asian

the amount of Indigenous American admixture is within predictable patterns at each generation. For example, in the third generation each individual exhibits a score near 40% Indigenous American (expected 41%). Since there are no paternity issues in this pedigree, unreasonably high (such as 75%) or low (such as 10%) results likely would have indicated a high error for this smaller panel of markers. In the second generation, there is more deviation from the expected 12% Indigenous American but within a standard deviation. There are differences from the estimates obtained with 71 and 171 markers, as we would expect since the latter panel is the former panel with 110 additional East Asian/ European, East Asian/Indigenous American, and European/Indigenous American markers. Indigenous American affiliations are lower in the more accurate 171 AIM test and KF revealed minor East Asian admixture in the 71 AIM test, which was not expected, and which disappeared in the 171 marker test shown in Figure 5-18.

SIMULATED PEDIGREES

Although families are relatively expensive to sample, we can simulate thousands of families easily enough in a computer. Figure 5-20 shows one such simulated pedigree. From a large number of such simulated pedigrees, we can determine the distribution of estimates for siblings of simulated individuals,

Figure 5-20

Simulated pedigree—one of thousands created in a computer for the estimation of testing error. Genotypes for four individuals were created in silico by first specifying an ancestry group with which they were 100% affiliated, based on the parental allele frequencies. Siblings were created by random selection of alleles among the parents. A third generation was generated by mating the two siblings to create 24 individuals. The percentage of Indigenous (Native) American admixture present for these third-generation siblings is shown in the histogram at the bottom. Number of individuals is shown on the y-axis and percent of Indigenous American admixture on the x-axis. The average sibling showed 23% Indigenous American admixture, which compares favorably with the 25% we expect based on the ancestry/ admixture of the grandparents.

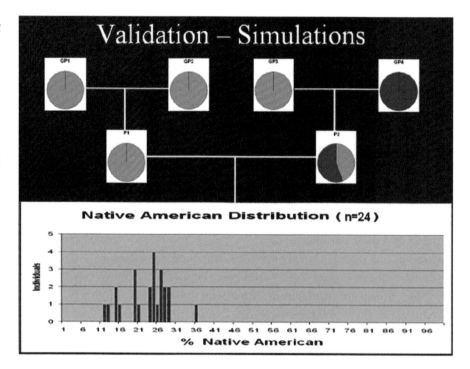

and compared with the distribution of expected values we can calculate testing error for a given panel of markers and methods.

To determine the possible distribution for siblings, simulations were preformed in which family members and offspring are generated using prior ancestry percentages and the known frequencies of alleles in each parental group. In the particular simulation shown in Figure 5-20, the first generation contained three individuals of European descent and one individual of Indigenous American descent, all with scores of 100% affiliation to the appropriate group. We can engineer a variety of different types of matings in the computer, testing all possible combinations of admixture proportions in grandparents, mating them to produce parents, and then mating these to produce offspring. For each generation, mating is accomplished *in silico* by allowing for independent assortment at each marker to generate the second generation. The third generation contains 24 siblings, each generated independently and with respect to proper assortment. We then estimate ancestry proportions for each sibling, and plot the distributions for comparison with expected percentages.

Figure 5-20 shows the distribution of third-generation siblings in terms of their Indigenous American estimates. Based on the grandparents the expected result should have been 25%, and we can see in the chart of Figure 5-20 that the average simulated sibling in fact does measure with about 23.2% Indigenous American ancestry as expected (standard deviation 5.9%). The difference between expected and observed levels is a function of sampling (only 24 siblings are shown) and testing error. Based on the parent's admixture scores, the expected level of Indigenous American ancestry, the genealogical ancestry, in this particular pedigree is 28.5%, which appears close to the mode for the realized genomic ancestry distribution of the offspring. The majority of the offspring range from 24 to 29% Indigenous American with a couple of outliers that do not exceed 20% deviation from expected levels. Considering the expected European levels as well, the majority of the samples cluster near the two expected results, showing good agreement with expectations. We have conducted hundreds of such pedigree simulations, investigating various types of ancestry mixture in various proportions with similar results. These results, which indicate the range of sibling variation in BGA is usually in the mid to upper single-digit percentages, are discussed more fully later in this chapter (simulations of admixed individuals).

COMPARING DIFFERENT ALGORITHMS WITH THE SAME AIM PANEL

How much of the ancestry that we estimate with these methods is a function of the exact method of admixture estimation? If we use a basic MLE program, do

Model	Program	NAM %	EUR %	AFR %	Markers
3-way	MLE	11.4	55.7	32.9	172
3-way	STRUCTURE	6.4	59.8	33.8	172

Figure 5-21

Comparison of BGA admixture results for a Puerto Rican population using the basic MLE algorithm described in the text and the program STRUCTURE (Pritchard et al. 2000, 2001) (right). The same 171 AIMs were used for both. The most likely three-population estimate is plotted for the MLE program and with STRUCTURE, a k = 3 was used. Structure was run using prior population information, with admixture, assuming independent allele frequencies and separate alphas for each population, burn-in of 30 K and with 70 K iterations. The results are fairly similar though the MLE program tends to report more Indigenous (Native) American admixture than STRUCTURE, a result similar to that obtained when compared to other Bayesian methods such as McKeigue's ADMIXMAP (discussed in Chapter 6). Plots provided by Indrani Halder.

we get different answers than when we use a more sophisticated program that samples from a relatively large number of distributions? To answer this, we typed a set of 62 Puerto Rican samples for the 171 AIMs and obtained admixture estimates using a three-way population model (Indigenous American, European, and West African) with two programs: the basic MLE program (discussed in Chapter 3, written in the laboratory of Mark Shriver and optimized/modified at DNAPrint genomics) and Pritchard's STRUCTRE program (Pritchard et al. 2000, 2001). These results are detailed in Halder (2006) and Halder et al. (2005). Table 5-12 shows the results for the average Puerto Rican sample and Figure 5-21 shows the individual estimates in a triangle plot using the basic MLE program (left-hand plot) and Pritchard's Bayesian STRUCTURE program (right-hand plot).

The MLE program estimated higher Native American (Indigenous American) ancestry but the European and West African estimates were within 10% of one another and overall the estimates were reasonably close. The estimates from the two programs was correlated with a Pearsons p value of <0.01.

The basic MLE program produced higher average Indigenous American estimates than STRUCTURE, as is apparent from the greater spread in the NAM dimension of the plot but aside from this the two plots look similar. We will see later a similar tendency for the basic MLE algorithm to estimate higher admixture compared to other Bayesian methods such as McKeigue's ADMIXMAP. However, there were far more similarities between the two results than differences; samples with the greatest Indigenous American, European, and

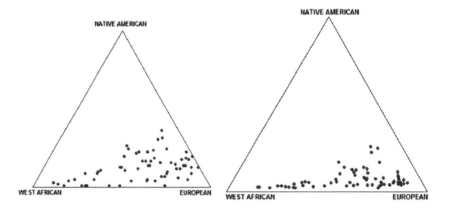

West African percentages with one program were the greatest using the STRUC-TURE Program. Although the average results were similar between the programs, swings of up to 10% were observed between the programs for any individual sample. These results show that estimates obtained are a function of not only the markers used and model, which we have discussed already, but also the particular admixture program used. However, variance in estimates within individuals between programs is far smaller than that between populations or subpopulations using one of them, so as long as a single program is used for the analysis within sample sets, or among various populations, we can be assured that most of the differences we observe are due to the genetic character of the individuals and populations and not due to our choice of programs. Note that these results might suggest a systematic overestimation of admixture for basic likelihood methods compared to Bayesian methods. However, with our basic MLE program, this was observed for individuals of low expected admixture. Hu et al., 2006 showed using PSMix that while their likelihood algorithm probably shows the same thing relative to Bayesian methods when the true admixture level is low, it both over and underestimated admixture when the true admixture was moderate (though the degree of over/underestimation was relatively small). Thus it would not be correct to say that likelihood based methods generally over-estimate admixture compared to Bayesian methods (though each method would clearly need to be studied on its own).

ANALYSIS USING SUBSETS OF MARKERS

Another way to evaluate the performance of the admixture estimation process with the 171 AIMs is to test for correlation in admixture estimates obtained from subsets of the markers. If the ancestry estimates we make are due to real population structure, as opposed to statistical noise, then estimates obtained from sample populations using the markers found on the even-numbered chromosomes, for example, should match those obtained using markers found on the odd-numbered chromosomes. If they do, we can be assured that the population substructure we find from application of an admixture test is bona fide and if not, the reliability of the estimates is considered to be lower and more likely due to statistical error. In other words, correlation between ancestry estimates obtained with the subsets indicates presence of bona fide admixture structure. In addition, we would expect lower precision in the estimates for each subset compared to the entire set of AIMs, since we have cut the marker set about in half, and of course accuracy is a function of the number of markers used. Further, for ancestry components with low levels on average, we expect a greater variation in estimates when cutting the markers in half since affiliation with the group could be focused on a small number of

Figure 5-22

Comparison of MLEs obtained using subset of 171 AIMs present on the even chromosomes versus that obtained using the subset on the odd chromosomes. A correlation between the two is a function of the overall level of admixture and its age, since higher levels and more ancient origins by definition involve a larger number of chromosomes, and for these higher levels of admixture we would expect to see greater correlation (though not exactly the same). The percentage of indicated admixture is shown for each chromosome subset on the y- and x-axes (odd or even chromosomes). As we would expect, since the overall level of Indigenous (Native) American (NAM) admixture is low, we find the greatest variability in estimates for the NAM percentages. Plots provided by Indrani Halder.

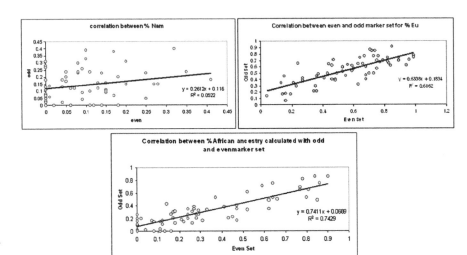

chromosomes, and statistical sampling of relevant markers comes into play. Indeed, an admixture assay that is functioning properly would be expected to show a greater variance of estimates obtained using the two subsets of chromosomes the lower the average level of admixture in the sample.

For these same Puerto Rican samples just discussed, the 171 AIMs were split into two groups, those on the even chromosomes and those on the odd chromosomes, and ancestry was recalculated with each subset. Figure 5-22 shows the estimates obtained for affiliation with a given population using the odd set and the even chromosome set. Correlation coefficients were 0.26 for the Indigenous American ancestry components, 0.74 for the West African ancestry components and 0.64 for the European ancestry components; overall the Pearsons p-values showed a statistically significant correlation for all three types of affiliation (see Table 5-13). As expected, the variance of estimates was greatest for the Indigenous (Native) American (NAM) admixture, which was at the lowest level in the population. This conformance with expectations is one way to suggest that the relatively low levels of Indigenous American ancestry shown in the samples is a significant genetic feature of the Puerto Rican sample, and as shown with the experiments just discussed comparing results

Table 5-13

Correlation and significance of correlation between admixture results obtained using two subsets of 171 AIMs located on the even and odd chromosomes.

Ancestral population	R^2	P
Indigenous American	0.052	0.005016
European	0.616	<0.0001
African	0.743	<0.0001

from the two programs, both programs were adept at detecting this affiliation (though at different levels). Given that there are subtle differences in estimates obtained from sample to sample between the two programs, we can conclude that a laboratory performing admixture analysis is best served by employing a single algorithm consistently rather than shuffling between them. Optimally, this would be the algorithm for which performance could be judged to be superior, based on precision estimates from simulations, performance in parental samples and pedigrees, and concordance with genealogical self-held affiliations.

RESOLVING SAMPLE MIXTURES

For admixture analysis to be useful in a forensics setting, we need to understand how sample mixtures influence the results. We can determine this impact by titrating a known sample of a reference admixture profile with a second one (a contaminant) and measuring the change in admixture estimates obtained against the reference type. Eleven mixtures of two individuals in approximately equal ratio were created for this experiment and were typed using the 71 AIM test. Samples from populations represented in the reference profile were run in conjunction with the mixtures. Scatter plots of the reference population were generated from a Beckman Ultra High Throughput (UHT) instrument platform, and the three genotype classes formed distinct clusters (see Figure 5-23A). When a contaminant is added to the reference sample, the genotype for the mixture represents a hybrid between the reference sample and the contaminant sample and in general, mixed samples genotype in predictable patterns. For example, an XX/YY mixture produces an XY genotype call just like an XY/XY mixture. When an XY/XX mixture is produced, the genotype plots in between the two genotypes of the individuals in the mixture (in

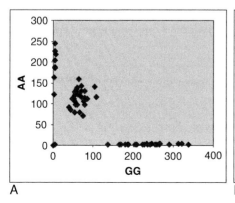

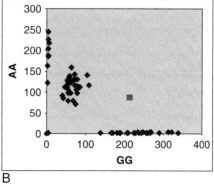

A   B

Figure 5-23

Influence of sample mixture on AIM genotype determination. A reference population was genotyped for one of the AIMs part of the 171 AIM panel discussed in the text, resulting in three genotype classes (AA, GA, GG) as shown in A. Each spot represents an individual. In B, the sample of one of the GA individuals was seeded with an equal mass of sample from a GG individual, resulting in a sample of GA/GG mixture. This mixed sample is shown with the pink spot in B, and genotypes half way between the GA and GG clusters as one would expect (based on the fact that the ratio of G:A alleles is 3:1 in the sample).

between XY and XX), since there are three XX alleles for every one Y allele. An example is shown for one locus in one mixture in Figure 5-23B, where the genotype for the mixed sample falls outside the normal genotype cluster boundaries. In general, a sample mixture would contain many examples of this type of result, making easy the recognition that the sample is mixed (as opposed to there being some other problem with genotyping). Thus, sample mixture is diagnostic from the presence of genotype plots that fall outside the normal genotype boundaries, which should be observed for several of the markers in the panel. The reverse, however, is not necessarily true (that intermediate genotypes always result from sample mixture).

Given a sample mixture, and a reference sample of one of the individuals of that mixture, we can infer the genotype for the second sample (the nonreference sample) and ancestry admixture determined from this inferred genotype. We can then compare the inferred BGA from this second sample against the result we obtain when we genotype and infer BGA for the sample on its own. This was accomplished for 11 1:1 sample mixtures, where two inspectors independently examined the mixtures, and the genotype of the nonreference individual was inferred at each of the 71 loci, and ancestry was then inferred from these inferred genotypes using the DNAPrint MLE algorithm (see Table 5-14). In Table 5-14, we show the results obtained from two separate inspections of the data (different investigators) as well as the known results for the individual.

The results showed that sample mixtures could be successfully resolved when knowing the genotypes of the reference sample in nine of the mix-

Table 5-14

Inferred admixture ratios: mixtures 1 through 11.

MIX 1*							Incorrect calls**
Inspector 1	European	87	East Asian	5	Native American	8	4
Inspector 2	European	91	East Asian	0	Native American	9	2
True Admix Ratio	European	90	East Asian	0	Native American	10	

*Consists of Caucasian (Ukraine/Italy) mixed with African-American reference sample.
**Resolved vs. true admix genotypes from 70 loci.

MIX 2*							Incorrect calls
Inspector 1	East Asian	56	Native American	35	European	9	12
Inspector 2	East Asian	44	Native American	38	European	18	17
True Admix Ratio	East Asian	98	Native American	2	European	0	

*Consists of East Asian (Chinese) mixed with African-American reference sample.

Table 5-14
(Continued)

MIX 3*							Incorrect calls
Inspector 1	European	91	East Asian	0	Native American	9	13
Inspector 2	European	94	East Asian	0	Native American	6	5
True Admix Ratio	European	87	East Asian	0	Native American	13	

*Consists of Caucasian (Greek) mixed with East Asian reference sample.

MIX 4*							Incorrect calls
Inspector 1	European	53	Native American	47	East Asian	0	8
Inspector 2	European	48	Native American	47	East Asian	5	9
True Admix Ratio	European	62	Native American	38	East Asian	0	

*Consists of Hispanic (Puerto Rican) mixed with East Asian reference sample.

MIX 5*							Incorrect calls
Inspector 1	European	40	African	49	Native American	11	13
Inspector 2	African	51	European	39	Native American	10	20
True Admix Ratio	European	62	Native American	38	East Asian	0	

*Consists of Hispanic (Puerto Rican) mixed with African-American reference sample.

MIX 6*							Incorrect calls
Inspector 1	European	69	African	31	Native American	0	0
Inspector 2	European	69	African	31	Native American	0	0
True Admix Ratio	European	69	African	31	Native American	0	

*Consists of Jamaican (one Caucasian parent) mixed with East Asian reference sample.

MIX 7*							Incorrect calls
Inspector 1	European	62	African	38	Native American	0	5
Inspector 2	European	65	African	35	Native American	0	5
True Admix Ratio	European	69	African	31	Native American	0	

*Consists of Jamaican (one Caucasian parent) mixed with African-American reference sample.

(Continued)

Table 5-14
(Continued)

MIX 8[*]						**Incorrect calls**
Inspector 1	European	91	East Asian	9	Native American	0	4
Inspector 2	European	84	East Asian	5	Native American	11	2
True Admix Ratio	European	84	East Asian	12	Native American	4	

[*]Consists of Caucasian (Finland) mixed with East Asian reference sample.

MIX 9[*]						**Incorrect calls**
Inspector 1	European	77	East Asian	6	Native American	17	2
Inspector 2	European	82	East Asian	0	Native American	18	2
True Admix Ratio	European	84	East Asian	12	Native American	4	

[*]Consists of Caucasian (Finland) mixed with African-American reference sample.

MIX 10[*]						**Incorrect calls**
Inspector 1	European	46	East Asian	42	Native American	12	4
Inspector 2	European	41	East Asian	33	Native American	26	4
True Admix Ratio	East Asian	54	European	46	Native American	0	

[*]Consists of East Asian (Filipino) mixed with African-American reference sample.

MIX 11[*]						**Incorrect calls**
Inspector 1	African	78	European	19	East Asian	3	2
Inspector 2	African	75	European	15	East Asian	10	4
True Admix Ratio	African	79	European	17	East Asian	4	

[*]Consists of African-American mixed with European (Caucasian) reference sample.

tures. We observed deviations in two of the mixtures. Mix 2 exhibited a decrease in East Asian components from the true admix ratio but the major affiliation was correctly identified. Mix 5 was a relatively evenly admixed West African/European sample, where the contaminant (the referenced sample) caused a change in the major affiliation of the nonreferenced sample but the ancestry of the sample still was identified correctly as a predominant West African/European mix. In both cases, the inferred genotypes for both mixtures had a large number of incorrect calls contributing to these discrepancies, which probably resulted from unequal mixture and biased genotype calls (we already have described the relationship between genotyping failures and admixture estimates in the repeatability section of this chapter). Overall the results suggest that using a proper genotyping platform, mixtures should be interpreted with caution but this data suggests that mixture resolution can be accomplished with a reasonable degree of accuracy for 1:1 mixtures. This particular data was obtained using a Beckman Ultra-High Throughput genotyping platform and success for others doing likewise would be a function of the semi-quantitative nature of the employed platform—other platforms may or may not possess these same quantitative characteristics.

We can look more deeply at the quantitative nature of the Beckman platform for mixed sample analysis by titrating contaminating samples. The mixtures just discussed were resolved based on the assumption that the two samples were present in roughly a 1:1 ratio and that the assay is semi-quantitative. What happens if the ratio of a mixture deviates from 1:1—is the genotyping assay quantitative enough to enable proper genotype calls? To test the quantitative characteristic of the assay, a titration experiment was performed where a sample of African-American ancestry was mixed with one of a European ancestry. The African-American sample was held constant at 2 ng/reaction and the European sample was titrated down from 2 ng to 0.125 ng/reaction (1:1 to 16:1 ratio). If the assay is quantitative, the genotype plots for each locus should move toward the genotype cluster corresponding to the genotype of the African-American sample the greater the sample-to-contaminant ratio. As the amount of the European sample was decreased, the expected trend was observed. The location of the data point on the scatter plot moved toward the cluster corresponding to the genotype of the reference sample for which quantity was held constant.

Figure 5-24 shows a representative case where CC and TT genotypes are present in the mixture. The 1:1 mixture produces data that plots to the TC cluster (pink symbol). As the quantity of the CC sample is reduced, the genotype of the mixture plots farther along the x-axis toward the XX or TT cluster. The amount of variation in the data points is roughly proportional to the difference in the mixture, indicating that the assay is nicely quantitative. A similar result is seen in Figure 5-25 where TT and TC genotypes are involved. The initial 1:1 data point plots between the XX and XY clusters as expected from the preceding discussion, and subsequent points move toward the TT cluster (x-axis). When the genotypes of the samples are identical at a given marker, all the titration samples plot in the same cluster as expected (data not shown). Admixture estimates followed the same general pattern as shown in Table 5-14 (not shown).

Figure 5-24

Sample mixture titration reveals that the Beckman genotyping platform used for much of the work described herein is fairly quantitative. As the ratio of sample with the TT genotype is increased relative to the CC sample in a sample mixture, the genotype of the mixture plots accordingly. With access to a reference genotype corresponding to the first sample, it is possible to deduce the second genotype and the approximate mixture ratio. From this induced genotype it is possible to infer the MLE of the second (unknown) sample. The level of C signal is shown on the y-axis and the level of T signal on the x-axis. Individuals with the CC genotype plot along the y-axis, those with the TT genotype along the x-axis, and those with the CT genotype in between, where a cluster of samples is seen with appreciable C and T signal.

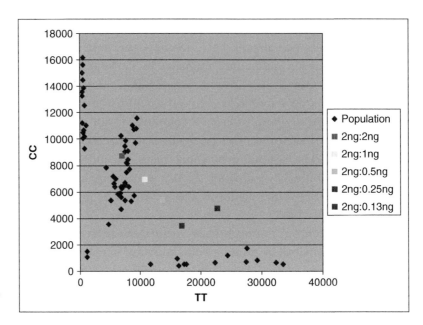

The conclusion from these experiments is that the data generated in the titration experiment is consistent with a semi-quantitative assay. The data shifts in the scatter plots in a manner in the expected direction, roughly proportional to the amount of the mixture; however the relationship is not maintained at larger ratios. This suggests that a mixture in the 1:1 and 2:1 range should be resolvable with the aid of a pure reference sample, which could be useful in a rape case where the victim is available. Mixtures of greater ratios may or may not give accurate admix results, but are at least still identifiable as mixtures. In these cases, results from prior STR testing may aid in the inference of genotypes and admixture estimates (by helping to establish the sample mixture ratio).

SAMPLE QUANTITY

To find the lowest possible amount of DNA needed to determine the correct major MLE proportion, we performed additional titrations with the 71 AIM panel. Three blood samples from individuals of ancestry proportions shown in Table 5-15 (663, 664, 669) were extracted and prepared to reach final concentration of 2 ng, 1 ng, 0.5 ng, 0.25 ng, and 0.125 ng of DNA in each PCR reaction. A fourth sample (CONCAUCA) was a single buccal swab extracted and diluted 1:2, 1:4, 1:8, 1:16, 1:32. Ancestry proportions obtained from the titrated samples were then compared to the result from the 2 ng samples.

As we expected, the number of failed loci generally increased as the samples became more dilute (see Figure 5-26), but these failures did not significantly affect the determination of the major ancestry proportions, as can be seen in

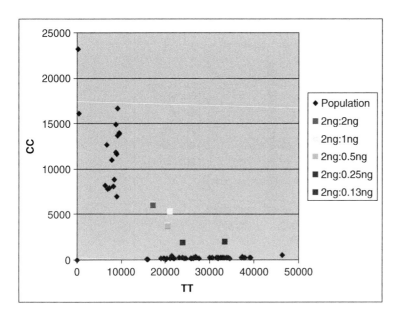

Figure 5-25

Additional sample titration data, holding one sample with the CC genotype constant and increasing the amount of sample with the TT genotype. The level of C signal is shown on the y-axis and the level of T signal on the x-axis. Individuals with the CC genotype plot along the y-axis, those with the TT genotype along the x-axis, and those with the CT genotype in between, where a cluster of samples is seen with appreciable C and T signal. Again, the genotype of the mixture plots at a higher T/C ratio with increasing amounts of sample with the TT genotype relative to the sample with the CC genotype in the mixture.

Figure 5-27. For the 11 multiplex PCR reactions a total of 1.375 ng was used at the lowest concentration without a significant change in major ancestry percentage. These results illustrate that when a suitably sized panel of adequately powerful markers are used, the admixture estimation process is relatively impervious to moderate numbers of failed genotypes. We might have expected this comparing the population and individual results obtained from 30 AIM, 71 AIM, and 181 AIM panels, which are grossly similar.

NONHUMAN DNA

Contamination of nonhuman DNA in samples is a concern given the nature of crime scene evidence, and for an admixture test to be useful in a forensics context, it must be demonstrated that nonhuman contamination of a sample does not influence the quality of the results. To determine if nonhuman DNA

Sample ID	Major	Minor
663	80% European	20% Indigenous American
664	93% European	7% Indigenous American
669	79% African	17% European/4% East Asian
ConCA	100% European	0%

Table 5-15

Admixture of three samples used in dilution series.

Figure 5-26

The influence of input sample mass on the number of genotyping failures across the 171 AIMs described in the text. The number of genotyping failures (y-axis) is shown for four samples (x-axis) at five input DNA masses. The number of failures in this experiment increased for most of the samples with lower input DNA mass, as expected. However, as can be seen in Figure 5-28 and as described in the text, the increase had little impact on the final MLE values inferred for each sample.

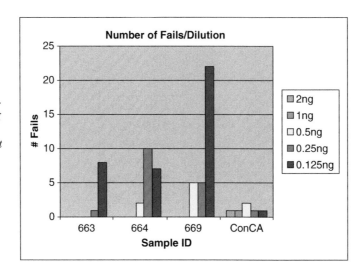

perturbs the calling of genotype or the inference of BGA MLE, mouse (*mus musculis*) genomic DNA was amplified in the 11 multiplex reactions and genotyped exactly as for human DNA. The mouse DNA failed at 68 out of 70 loci we tested (from the 71 AIM panel), indicating that the nonhuman DNA tested did not produce callable genotypes at enough loci to produce an ancestry admixture profile. The 171 markers panel (which we will hereafter also refer to as DNAWitness[TM]) was also tested for cross species specificity with nonhuman DNA. DNA from the higher primates at a concentration of 2.5 ng/reaction was typed using standard procedures for the 171 markers.

As shown in Table 5-16, each sample had significant failed loci that would prevent it from being classified or passing QC review as long as reasonable

Figure 5-27

Major BGA affiliation from the MLE determined for five samples at varying levels of input DNA mass as shown in Figure 5-26. Though the number of genotyping failures increases with lower DNA mass, the percentage of affiliation with the primary parental population remained fairly stable.

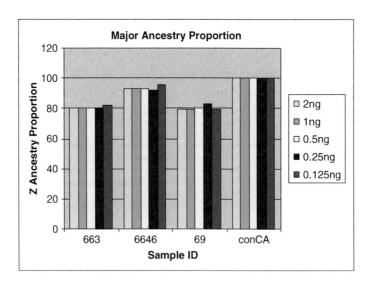

Sample	Number of failed loci
Chimpanzee	157 out of 171
Gorilla	151 out of 171
Orangutan	137 out of 171

Table 5-16

Performance of the 171 AIM panel in nonhuman species.

success rates are required for passing the QC review step. With degraded human samples, there may be a temptation to accept an unusually high failure rate in order to make inference of the MLE possible, and though a 30 AIM panel can be used to infer MLEs with reasonably similar results as 171 AIMs, it must be kept in mind that this panel of 30 AIMs was balanced for information content. A 171 AIM reaction set for which 141 AIMs failed genotyping would leave 30 AIMs for inferring the MLE, but this collection would not be balanced and so such an inference would be hazardous. As Table 5-16 shows, a sample producing such a high failure rate with the 171 AIM panel may not even be one of human DNA, and so should not pass QC review. We typically fail samples exhibiting a greater than 20% genotyping failure rate (i.e., more than 34 failures).

PERFORMANCE WITH ALTERED PARENTAL ALLELE FREQUENCIES

Given that the tests we have described rely on correctly specified parental population allele frequencies, we next investigated the extent to which the admixture proportions produced by the test are potentially influenced by parental allele sampling bias. To do this, we selected a pair of populations, selected the AIMs relevant for resolving affiliation between these two populations (we selected those with the highest delta values for this pair of groups), and then we adjusted the allele frequency for each of these AIMs in one of the groups in a direction toward that of the other group by 20%. This resulted in a diminution of the delta value for each AIM by an amount simulating a 20% error in the estimation of allele frequency in parental samples for the group subject to the changes. This type of bias is roughly what we might expect if we sampled from a hybrid or pseudo-hybrid population such as Greeks or Middle Eastern populations (who tend to exhibit minor non-European affiliations on average, as we will see in Chapter 6).

We normally run our MLE program using static parental allele frequency files. There are six possible pairs for the four BGA groups that are part of the 71 AIM admixture test, and by focusing the adjustments in allele frequency in only one pair of groups at a time, there are 12 modified parental allele frequency files and thus 12 possible degraded versions of the test (the frequency of each AIM informative for resolution of a given population pair changed in this file, changing frequencies for both groups in separate experiments). Using each of these 12 modified parental allele frequency files for each of these 12 degraded tests, we selected 31 samples and inferred BGA affiliation in exactly the same manner as for the original 71 AIM test and then compared the results with those obtained with the unaltered 71 AIM test. If the results from the 71 AIM test were highly sensitive to parental allele frequency error caused by this simulated sampling bias (which might normally be expected to be only on the order of a few percent), the 20% allele frequency changes introduced for the AIMs should result in admixture proportions substantially different from those of the unaltered test.

In fact, the results were similar for each of 12 tests. An example is shown in Table 5-17, which lists the results obtained comparing the degraded test for

Table 5-17

Little influence in accuracy caused by simulated systematic allele frequency estimation bias. MLE obtained using the degraded AIM panel for which allele frequencies were intentionally biased are shown beneath the MLE obtained using the unaltered AIM panel (71 AIMs, best three-way MLE presented).

Sample	Group 1	%	Group 2	%	Group 3	Difference %	%
DNAP1	African	94	European	0	East Asian	6	
DNAP1	African	94	European	0	East Asian	6	0
DNAP2	East Asian	100	Native American	0	European	0	
DNAP2	East Asian	100	Native American	0	European	0	0
DNAP3	European	100	East Asian	0	Native American	0	
DNAP3	European	100	Native American	0	East Asian	0	0
DNAP4	Native American	78	East Asian	0	European	22	
DNAP4	Native American	74	East Asian	0	European	26	4
DNAP5	Native American	100	East Asian	0	European	0	
DNAP5	Native American	100	East Asian	0	European	0	0
DNAP6	European	58	East Asian	0	Native American	42	
DNAP6	European	58	Native American	42	East Asian	0	0
DNAP7	European	52	East Asian	9	Native American	39	
DNAP7	European	54	Native American	40	East Asian	6	3

Table 5-17
(Continued)

Sample	Group 1	%	Group 2	%	Group 3	Difference %	%
DNAP8	European	60	East Asian	0	Native American	40	
DNAP8	European	60	Native American	40	East Asian	0	0
DNAP9	European	92	East Asian	0	Native American	8	
DNAP9	European	92	East Asian	0	Native American	8	0
DNAP10	European	79	East Asian	21	Native American	0	
DNAP10	European	79	East Asian	21	Native American	0	0
DNAP11	European	100	East Asian	0	Native American	0	
DNAP11	European	100	East Asian	0	Native American	0	0
DNAP12	European	100	East Asian	0	Native American	0	
DNAP12	European	100	East Asian	0	Native American	0	0
DNAP13	European	88	East Asian	12	Native American	0	
DNAP13	European	90	East Asian	10	Native American	0	2
DNAP14	East Asian	35	European	38	Native American	27	
DNAP14	East Asian	28	European	45	Native American	27	7
DNAP15	African	73	European	23	East Asian	4	
DNAP15	African	73	European	18	Native American	9	5
DNAP16	East Asian	30	European	34	Native American	36	
DNAP16	East Asian	22	European	40	Native American	38	8
DNAP17	European	79	East Asian	0	Native American	21	
DNAP17	European	79	East Asian	0	Native American	21	0
DNAP18	European	83	East Asian	0	Native American	17	
DNAP18	European	83	East Asian	0	Native American	17	0
DNAP19	European	93	East Asian	6	Native American	1	
DNAP19	European	94	East Asian	5	Native American	1	1
DNAP20	European	81	East Asian	0	Native American	19	
DNAP20	European	81	East Asian	0	Native American	19	0
DNAP21	European	99	East Asian	1	Native American	0	
DNAP21	European	99	East Asian	1	Native American	0	0
DNAP22	European	83	East Asian	0	Native American	17	

(Continued)

Table 5-17
(Continued)

Sample	Group 1	%	Group 2	%	Group 3	Difference %	%
DNAP22	European	83	East Asian	0	Native American	17	0
DNAP23	European	75	East Asian	25	Native American	0	
DNAP23	European	80	East Asian	20	Native American	0	5
DNAP24	European	43	Native American	57	East Asian	0	
DNAP24	European	43	Native American	57	East Asian	0	0
DNAP25	European	86	East Asian	14	Native American	0	
DNAP25	European	89	East Asian	11	Native American	0	3
DNAP26	European	72	East Asian	28	Native American	0	
DNAP26	European	77	East Asian	23	Native American	0	5
DNAP27	European	48	Native American	52	East Asian	0	
DNAP27	European	48	Native American	52	East Asian	0	0
DNAP28	European	87	East Asian	13	Native American	0	
DNAP28	European	90	East Asian	10	Native American	0	3
DNAP29	European	74	East Asian	26	Native American	0	
DNAP29	European	79	East Asian	21	Native American	0	5
DNAP30	European	85	East Asian	15	Native American	0	
DNAP30	European	88	East Asian	12	Native American	0	3
DNAP32	European	48	African	52	East Asian	0	
DNAP32	European	48	African	52	East Asian	0	0

European/East Asian distinction where the allele frequency was changed for the European/East Asian markers away from the European frequency toward the East Asian frequency. Though the average change in allele frequencies resulted in a 20% diminution of delta value between these groups, the average change in admixture proportions was only 1.4% with a standard deviation of the change only 2.4%. The number of AIMs and cumulative δ value is the lowest for the Indigenous American/East Asian population pair, but both degraded tests for the NA/EA pair performed within a couple percent of the unmodified 71 AIM test (2.3% average change, 2% standard deviation, not shown).

We performed the same analysis for the 171 AIM battery. Of the 171 SNPs, 60 or so are particularly informative between European and Indigenous American ancestry and each is informative to a varying extent (measured with the delta value). Adjusting the allele frequencies for just these markers, such that the information content of the panel is diminished by 20% in the European/Indigenous American dimension, resulted in a change in admixture proportions for all groups in actual samples of less than 1% on average. As with the 71-marker test, we have performed this simulation on all possible pairs of groups, with similar results, showing the test is relatively impermeable to systematic allele frequency estimation errors such as those that may be generated in the parental sample selection process. This robust performance is likely the result of using so many markers of such high information content. Of course these combined results do not speak to unbiased frequency estimation errors, which we might simulate with unsystematic changes, such as haphazard bias where one locus is biased in one direction and another in a different direction, but although these changes might be expected to cancel out (making some markers equally stronger on average than others made weaker), the systematic bias approach might be considered a fairly rigorous strategy for testing the influence of allele frequency misspecification on the precision of ancestry estimates.

CORRELATION WITH ANTHROPOMETRIC PHENOTYPES

SKIN COLOR

We can explore the validity of BGA admixture tests by comparing the expression of phenotypes for which there is a strong expectation of an underlying relationship with genetic ancestry. It would be highly improbable for randomly obtained genetic results to strongly reinforce phylogenetic-phenotypic relationships established through appreciation of the linguistic, paleoarchaeological, and historical record. Chief among the traits that could be used to validate the performance of a BGA admixture test are quantitative and/or multifactorial traits such as skin pigmentation. Skin pigmentation is ideal for such a test since it shows a fairly continuous distribution among a variety of parental and admixed populations of diverse origins (see Chapter 9 for a more detailed treatment of pigmentation).

In Parra et al. (2004), Mark Shriver's laboratory recently has employed a small subset of the 171 AIMs that are the subject of this chapter to summarize the relationship between pigmentation and ancestry in five geographically and culturally defined population samples of mixed ancestry. This work will be described later in Chapter 8 when we discuss the basis and methods for indirectly inferring certain anthropometric traits through knowledge of genomic ancestry. Here, we provide basic evidence for the association between anthropometric trait of skin color and estimates of admixture provided by the entire 171 AIM battery we have been discussing up until now.

The study included a sample of 64 Puerto Rican women living in New York. Constitutive skin pigmentation was measured on the upper inner side of both arms on each subject, using a DermaSpectrometer (Cortex Technology, Hasund, Denmark). The melanin content was reported as the Melanin Index (M), which will be described more fully in Chapter 9. Figure 5-28 shows the correlation plots for the M index in relation to the three genomic ancestry components, and the correlations are clear from a visual inspection of the data. The analysis-based 171 markers show a highly significant effect on pigmentation of Indigenous American ancestry (rho = −0.424, p < 0.001; see Table 5-18 and Figure 5-28A), West-African ancestry (rho = 0.602, p < 0.00001; see Table 5-18 and Figure 5-28B), and European ancestry (rho = −0.578, p < 0/0001; see Table 5-18 and Figure 5-28C). Overall, it is clear from the correlations between skin pigmentation and genomic ancestry that measures of genomic BGA must have some anthropological basis and therefore by extrapolation, that is it possible to measure BGA with a certain degree of precision necessary to demonstrate this. As we will see in Chapter 8, this is the case when even relatively small numbers of AIMs have been used to calculate biogeographical ancestry. We will more precisely quantify elements of this precision later in this chapter.

IRIS COLOR

Iris color is distributed globally largely along phylogenetic lines, with the common occurrence of blue and green irises almost exclusive to the European

Figure 5-28

Correlation of skin melanin index (M) and BGA using the 171 AIM panel described in the text. The relationship between M and Indigenous American (A), West African (B), and European (C) BGA proportions in a sample of Puerto Rican women is shown. M index is shown on the y-axis and the percentage of the indicated BGA is shown on the x-axis. Plots provided by Indrani Halder.

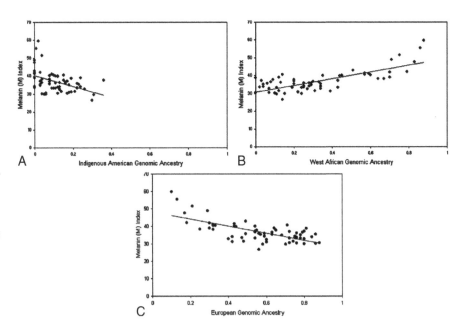

	IAM	EU	WAF
Spearman's rho	−0.424	−0.578	0.602
P Value	0.001	<0.0001	<0.0001
Pearson's r	−0.444	−0.687	0.745
P Value	<0.0001	<0.0001	<0.0001
r^2	0.197	0.472	0.555

Table 5-18

Correlation between M index and genomic ancestry components computed using the AncestryByDNA Panel of 171 AIMs in a sample of Puerto Rican women.

Notes: Genomic ancestry components are coded as Indigenous American (IAM), European (EU), and West African (WAF).

diaspora. If the 175 AIM continental admixture test reports population affiliations that are anthropologically relevant, we therefore would expect to see a strong association between darker iris colors such as hazels and browns and non-European ancestry admixture. Indeed, on a larger scale comparing individuals of predominant West African, East Asian, Indigenous American, and European ancestry, this is easy to show and seems somewhat trivial and similar to our earlier description for this type of association between skin melanin index and West African genomic ancestry in Puerto Ricans (or African Americans) of varying European ancestry.

More revealing than analyses on individuals of polarized ancestry, however, is the association between lower levels of non-European admixture and darker iris colors/skin shades within samples of European Americans. Since there is less of an environmental component to the expression of melanin in the iris colors versus the skin, the most difficult challenge to the anthropological meaningfulness of the four-group continental admixture test would seem to be a test for association between non-European ancestry admixture and darker iris colors. Even more difficult a test still would be to perform this test in a sample of Europeans or African Americans, where the results would be dependent on the relationship of phenotype with lower levels of admixture (as opposed to a gross relationship with primary ancestry). Although with an admixture test of a small number of AIMs, even a few gene sequences might be expected to reveal the associations contrasting individuals of different predominant ancestry, it would likely lack the precision and accuracy to reveal more subtle associations as a function of minor levels of admixture within Europeans. We would expect with a precision of from five to ten percentage points in admixture levels across all ancestry groups, roughly suggested by the pedigree analyses and more formally indicated with simulated parental populations (which we will discuss in more detail later in this chapter), we would expect the 175 AIM panel to do a better job of this, and indeed it does.

Iris color scores were calculated for 860 European American subjects for which four-population (continental) genomic ancestry was determined using standard MLE method and the 171 AIM battery we have been discussing up until now. These samples were collected from police detectives, technicians, and scientists who attended various forensics conferences between 2003 and 2005, and are among the samples we present data for in Chapter 9. Also described in this chapter, the quantification of iris colors is made using digital photography, *in silico* colorimetry, and a value called the iris color score, or iris melanin index. We picked arbitrary threshold percentages, and with contingency tables we divided the sample into two groups about that percentage, testing for nonrandom distribution of iris colors among the two groups.

The results show that lower European ancestry and higher non-European admixture levels are associated with darker iris colors in this sample (see Table 5-19; significant p-values are shaded). What's more, the p-values generally decrease as the percentage of non-European admixture increases for a given type of admixture. This is a pattern we would expect to see from a bona fide association. Blue iris colors were obviously also correlated with the lack of non-European admixture (with higher European percentages, not shown). Logistic regression analyses show similar associations (see Chapter 8, Figure 8-21). These are strong indications that darker iris colors are indeed associated with non-European ancestry, indicating that as expected, iris color is to a certain extent a function of ancestry, but more importantly for our purposes in this chapter, that the genetic affiliations reported by the 171 AIM BGA test described in this text are anthropologically meaningful.

SIMULATIONS

Given the data so far presented, the question naturally arises as to how we can objectively assure the accuracy of minor admixture proportions such as those reported in a test of genomic ancestry, and how such accuracy can be measured given that there is no ultimate arbiter of BGA in existence (certainly, genealogical information would not be considered such an arbiter, given its subjective nature). To measure the precision with which an ancestry admixture test performs, we typically rely on summary statistics from analysis of large samples of simulated genotypes. For example, we can create a sample of 10,000 parental Europeans (10,000 multilocus diploid genotypes), based on the parental allele frequencies, estimate ancestry admixture for each of them, and determine the frequency with which non-European affiliation is detected. This would give us an indication of the error inherent to the process, given that the alleles in question are continuously not discretely distributed.

Row	Group	Threshold	Associated with	Exact-P (two-tailed)	Total European sample (N)	Sample meeting threshold (n)
1	EUR	<99%	Dark (color score < 2.1)	0.01	860	707
2	EUR	<90%	Dark (color score < 2.1)	<0.0001	860	321
3	EUR	<80%	Dark (color score < 2.1)	<0.0001	860	138
4	EUR	<70%	Dark (color score < 2.1)	<0.0001	860	98
5	EUR	<60%	Dark (color score < 2.1)	<0.0001	860	81
6	EUR	<50%	Dark (color score < 2.1)	<0.0001	860	63
7	AFR	>2%	Dark (color score < 2.1)	0.330	860	252
8	AFR	>5%	Dark (color score < 2.1)	<0.0001	860	196
9	AFR	>10%	Dark (color score < 2.1)	<0.0001	860	97
10	AFR	>15%	Dark (color score < 2.1)	<0.0001	860	71
11	AFR	>20%	Dark (color score < 2.1)	<0.0001	860	58
12	EAS	>2%	Dark (color score < 2.1)	0.133	860	214
13	EAS	>5%	Dark (color score < 2.1)	0.005	860	163
14	EAS	>10%	Dark (color score < 2.1)	0.043	860	68
15	EAS	>15%	Dark (color score < 2.1)	0.003	860	34
16	EAS	>20%	Dark (color score < 2.1)	<0.0001	860	21
17	NAM	>2%	Dark (color score < 2.1)	0.488	860	348
18	NAM	>5%	Dark (color score < 2.1)	0.195	860	237
19	NAM	>10%	Dark (color score < 2.1)	0.025	860	110
20	NAM	>15%	Dark (color score < 2.1)	0.009	860	51
21	NAM	>20%	Dark (color score < 2.1)	0.033	860	28

Table 5-19

Association between various levels of European (EUR), West African (AFR), East Asian (EAS), and Native American (NAM) admixture and iris color using the 171 AIM DNAWitness^{TM} panel. Associations are shown among samples meeting the indicated percentage threshold using a contingency analysis of samples meeting/not meeting the threshold and counts of brown and not-brown or blue and not-blue irides as indicated.

CREATING SIMULATED SAMPLES

When taken together, the genotypes for each of the AIM loci combine to form a multilocus genotype (such as G/T, A/C, A/A ... for locus 1, locus 2, locus 3 ...). If we assume that the loci are unlinked and the mating between individuals of a given population is random, the multilocus genotype frequencies in each population are determined from the product rule:

$$P_g = \prod^{L} 2_{gl}^{h} \, p_{li} \, p_{lj} \qquad (5\text{-}1)$$

where the p_{li} and p_{lj} are the frequencies of allele A_{li} and A_{lj} for the lth locus of L total loci. The h of 2_{gl}^{h} equals 1 if p_{li} does not equal p_{lj}, and it is assumed that $p_{li} < p_{lj}$ to avoid considering the same genotype twice (i.e., the frequency of p_{12} GC and p_{21} CG heterozygotes). The likelihood of selecting n_g copies of the gth genotype in a population sample of size n, conditional on the frequency of that genotype in the sample is:

$$\Pr(\{n_g\}|\{P_g\}) = (n!/\Pi_g n_g!)\Pi_g(P_g)^{n_g} \qquad (5\text{-}2)$$

where $(n = \Sigma_g n_g)$. The $(n!/\Pi_g n_g!)$ term is derived from the binomial coefficient, and represents the total number of ways it is possible to arrange n_g genotypes in a total sample of n genotypes, over all genotypes g. It is easiest to think of Equation (5-2) in terms of a single locus, but it is easily extended to multilocus genotypes since Equation (5-1) covers multiple loci and multiple alleles for each locus.

To generate simulated populations based on known genotype frequencies, we choose a total population size, and then pick a genotype, using Equation (5-2) to calculate the likelihood of that genotype. We do this for a very large number of multilocus genotypes and construct a population of multilocus genotypes where the frequency of each is proportional to its likelihood. If N is infinitely sized, all possible multilocus genotypes would be represented at least once, but in practice we make N a manageable number, such as 10,000 or 1,000, and some rare multilocus genotypes may not be represented. Doing so we stipulate panmixia, or complete random mating within the subpopulation, a situation that although highly unlikely in real human populations, does not adversely affect the conclusions we draw in this section. Such a panmictic assumption would be unreasonable were we trying to study methods that relied on the samples being analyzed, like the Bayesian methods or comparing the levels of population structure among different samples.

We simulate populations of multilocus genotypes in this way for each of the four parental populations, using the parental allele frequencies. So doing, by definition, we are simulating populations of 100% ancestry. Since the frequency distribution of the alleles p_{li} and p_{lj} is reflective of real population differences, so too will the distribution of likelihood values and multilocus genotype frequencies among the simulated populations be reflective of the natural distribution in real populations. It is precisely because Equation (5-1) holds that this is so. The loci are independent, and not linked to one another, so the distribution of their various combinations is easily predicted as a function of probability. The representation of a given multilocus genotype in a given simulated population of size N will be similar to that of a real population if N is sufficiently large.

After creating large populations of multilocus genotypes to simulate parental populations, we then randomly draw a large number of genotypes from the simulated populations and determine the Maximum Likelihood Estimate (MLE) of biogeographical ancestry admixture for each.

If using a given battery of AIMs where the MLE calculations were perfect, every simulated sample from a given group would be homogeneously affiliated with just that group. For example, every simulated European sample would score with an MLE of 100% European, 0% Native American, 0% African, and 0% East Asian, and every simulated African would score with an MLE of 100% African, 0% Native American, 0% European, 0% East Asian, and so on. The extent to which the MLE results deviate from absolute homogeneity give an indication of the statistical error inherent to the process of calculating proportional group affiliations in the form of a single set of percentages.

SOURCE OF ERROR MEASURED WITH SIMULATIONS

An individual with the expected European genotype (the average of all of European individuals representing each of the European diaspora) would fall right on the 100% European point if the test were perfect. As we have discussed in Chapter 4, statistical genetics theory instructs that it is not necessary to be able to point to one person and say, that is a pure European, or collect from a population of such individuals in order to define absolute European for the purposes of constructing BGA methods. A good sampling of the diaspora and/or Bayesian methods can be used to estimate absolute ancestry with reasonable accuracy. Statistical genetics theory provides a good estimate of what European is as we choose to define it (see more about the nomenclature problem in Chapters 4 and 6)—that it can be obtained from sampling the European diaspora (as we have defined it) in an unbiased manner. Bayesian methods allow for objective definitions based on genetic distances and we can combine information from genetic clustering with our knowledge of the paleoarchaeological and historical record to establish our definition of European and estimates of parental European allele frequencies. Incorrect or imperfect definitions would contribute error to the test results, but this is not the type of error we measure with simulated genotypes.

The error we measure with simulated genotypes is the error caused by the fact that the AIM alleles are continuously, not discretely distributed among the population groups as we have defined them (see Tables 4-1 and 4-2). An AIM selected because the G allele is present at a frequency of 90% in Africans but only 10% in East Asians would have a very high frequency differential among these two groups (0.80) and thus be of high ancestry information content. However, some East Asians would obviously have the G allele, not because they are of African admixture but because the G allele is not "private to" or only present in Africans. If many AIMs are considered, no matter what the frequency differential, and no matter how many AIMs are used, certain chance East Asian individuals are expected to have enough African alleles by pure chance that their MLE will include some nonzero level of African ancestry in error. Given the number of African alleles required to nudge the MLE away from 100% East Asian (which would obviously depend on the AIMs exhibiting the African allele), the frequency of individuals with this number of African alleles in the East Asian population is calculated using Equation (5-1), and the likelihood of running into such a person within a sample of North East Asian individuals is calculated using Equation (5-2).

The question to answer when developing an admixture test is how many AIMs, and of what quality, are required to minimize the incidence of individuals with, and the magnitude of, artificial or erroneous admixture. How frequently they are encountered for each group, given a specific battery of AIMs, is best determined through simulations because it is difficult to collect parental samples but easy to simulate them once you have learned their allele frequencies. Clearly, the magnitude of MLE error is dependent on the particular battery of AIMs used, which of course is dependent on the population model employed as well as the genetic distance between subpopulations. If quality is defined as discrimination power among the various group pairs, a larger number of low-quality AIMs might provide the same error rate as a smaller collection of higher-quality AIMs. In other words, the likelihood of finding simulated East Asian individuals with large numbers of African alleles decreases as the number of loci increases, but it also decreases as the average or cumulative AIM allele frequency differential between East Asians and Africans increases. Obviously we would like to increase both, but such an increase brings economic consequences and at some point it becomes no longer cost-effective or justified to continue building the marker panel (such as might be the case if a 0.01% improvement in bias can be had by increasing the cost per test from $300 to $600).

RELATIONSHIP BETWEEN ERROR IN POPULATIONS AND WITHIN INDIVIDUALS

The percentage of MLE error determined from simulations is intimately related to the average breadth of the confidence contours in the graphical (triangle plot) representation of individual admixture proportions. They are the same thing, except one is a reflection of uncertainty within an individual genotype, and the other a reflection of uncertainty in a population of individual genotypes.

For an individual's multilocus genotype, we produce many likelihood estimates of genomic ancestry admixture, the highest of which are sometimes not very different from one another even though they involve different percentages and sometimes even different groups. Though we plot the most likely estimate (MLE) as a single answer, in reality, there are usually many others of not much lower likelihood and the best answer is a range of answers for which the likelihoods are all similarly high. This is why in the triangle plot graphical representation of individual admixture proportions, we plot all of those estimates with likelihood values within two-, five-, and 10-fold the value of the MLE. We could eliminate the MLE all together, and define the answer as the range of values corresponding to the five-fold contour (95% confidence interval). Either way, the correct answer is expected to be within the 10-fold confidence intervals about 99% of the time.

In a population of individuals, we have a much larger amount of information to consider. If there are 1 million likelihood values computed for an individual multilocus genotype, there would be 1 billion values calculated for a population of 1,000 multilocus genotypes. By considering only the MLE for each sample, keeping the amount of data per sample down, we can spend our computational resources on a larger number of samples. When we use a single value for large number of simulated samples (their MLEs) rather than a range of values (their 5X confidence spaces), there is a very real chance that the MLE is not exactly correct, or in error for any one sample, but by and large we are using the best estimate for each sample. When we use a range of values for a smaller number of simulated samples we are reporting many estimates we know are not correct just as we do when we report a range of likely values for an individual, within which the correct one almost certainly exists. In a triangle plot, the spread of values between individuals and populations is similar in concept but seemingly not the same; by definition for any particular sample, the likelihood space is larger when plotting the range for each sample rather than just the MLEs.

Figure 5-29 shows the MLE of five multilocus genotypes plotted in a triangle plot A as well as the five-fold confidence contours for the same samples in B. Because the confidence contours cover more space than any single estimate

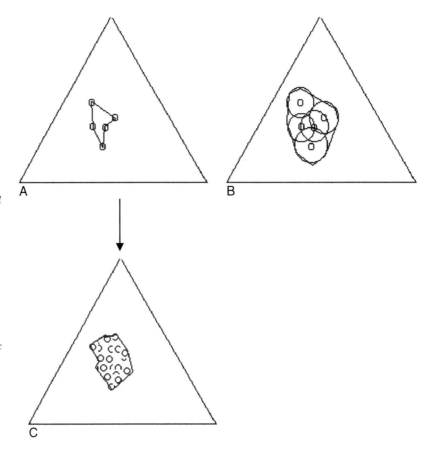

Figure 5-29

Similarity between likelihood space for individual MLE estimates with confidence contours and population of individual MLE estimates without confidence contours. MLE for five individuals is shown without (A) and with (B) confidence contours. In (C) we display an expected distribution of MLE estimates for a population of individuals without their confidence contours. Since the likelihood space for individuals and populations is fundamentally the same, based on the product rule and the population specific allele frequencies, the distribution for the population is expected to be similar to that for the five individuals with confidence contours (sampling effects notwithstanding for the latter).

within them, the area covered by the estimates is larger in B than in A. It would seem that if we were discussing a population for which these five were representative, we should be discussing the region of the triangle plot shown in Figure 5-29B rather than 5-29A. It may seem that the spreads have different meanings because when plotting the MLE for a large number of samples, we are plotting the highest quality result for many samples, but when we plot the range of values for a small number of individuals we are considering many lower quality results and one high quality result per individual.

However, this is only an illusion. The most likely admixture proportions for a multilocus genotype are functions of its sequence only, and the probability of a sequence occurring in a population is a function of the product rule and the population allele frequencies. The likelihood (or uncertainty) of ancestry given a single multilocus genotype is not fundamentally different than the likelihood (or uncertainty) that in a population of individuals of that ancestry there will exist an individual with that multilocus genotype. Both are a function of the ancestry-specific allele frequencies and the product rule for the genotype in question. In Figure 5-29 we can plot more than five MLEs for the same computational expense incurred in plotting the ranges of five genotypes in B. When we add the MLEs of more multilocus genotypes from A to C (arrow Figure 5-29), we see in this hypothetical example that the area occupied by the estimates grows and becomes similar in dimension to that in B. This is the scenario we would expect to see in a population of people with a given ancestry admixture—a plot of 1,000 individual MLEs should overlap closely with the plot of the 10-fold likelihood spaces for 100 individuals.

It is often easier for lay people to understand the concept of estimates for multiple individuals than the concept of multiple estimates within individuals. However it is important to understand that the range of percentages on our list of values within 10-fold likelihood of the MLE, or the breadth of the space of the 10-fold confidence interval in the triangle plot, is proportional to the MLE percentage error obtained from simulations obtained from larger numbers of samples. The primary cause of uncertainty is one and the same—continuously distributed allele frequencies among the parental ancestry groups.

We will consider two types of error:

- The bias, which is the average deviation of estimates from the true value (100% in our simulated parental populations). We can speak of the bias in any one direction (such as toward East Asian) or in all directions, which is the total bias. It can be calculated as the standard deviation (root-mean-square difference) with respect to the true value, or simply the average difference from the true value.
- The root-mean-square-error (RMSE), which is the standard deviation of estimates about their means (rather than the true value, which could be different). As with the preceding, this could be calculated as the average difference from the mean as well.

As long as the panel was properly constructed (i.e., normalized with respect to marker type) and an algorithm based on the likelihood function is used, then a low bias is expected to be accompanied by a low RMSE. Such panels with high bias will tend to have high RMSE, and vice versa. However, it is possible that an improperly constructed panel of AIMs, or a flaw in the algorithm, could produce an RMSE of essentially zero but produce all wrong answers because they are biased toward an incorrect ancestry type. The reverse is true too—the bias could be zero but the RMSE very high. A good panel of markers would have a low bias and RMSE, and when evaluating a panel of markers it is important to measure both.

The statistical error in MLE inference could be its bias or RMSE, as long as the worst one is reported. For example, if the RMSE is significantly lower than the bias, one could use the bias as an approximation of statistical error. More conservatively, a value combining the two could be used, calculated by summing the squared deviation of estimates from the true value, and the squared deviation of estimates from their means, taking the average of these two values and then taking its square root. We present this sum as the Maximum Range of Error later in this chapter.

In addition, one must account for other sources of error than statistical, and the bias and RMSE are not the same thing as total error in MLE inference the battery exhibits. There is also genotyping error, differences in bias as a function of the absolute percentages of ancestry (which we will address later), error contributed by differences in allele frequencies for modern day descendents of populations that admixed and the actual populations that admixed, and so on. Though these errors cannot be measured directly, there are ways to minimize their impact, which we have already discussed in Chapters 3 and 4. For this section, we are focused on statistical error caused by the continuous distribution of allele frequencies.

PRECISION OF THE 71 AIM PANEL FROM SIMULATIONS

It is instructive to observe the differences in expected and unexpected affiliation averages for each simulated population on a group-by-group basis and to discuss in detail what these estimates mean. The total bias of the test is defined as the sum of the average unexpected admixture percentages for samples of a particular simulated population (such as the sum of European, East Asian, and Native American admixture in simulated Africans). Using the 71 best AIMs from our screen we simulated 5,000 samples of each of the four groups (European, sub-Saharan African, Native American, and East Asian), determined the MLE for each simulated sample, and tabulated the average percentage of each type of admixture the simulated samples exhibited (see Table 5-20). The average outside-group or unexpected admixture percentages for the simulated samples was reasonably low; we measured a total bias of 4.7%. Put

Table 5-20

Average admixture results for simulated samples using 71 AIMs.

	AFR	EUR	EAS	NAM	Total	Bias
Africans	98.07	0.96	0.1	0.87	100	1.93
Europeans	0.08	95.63	2.25	2.04	100	4.37
East Asians	0.03	2.45	92.98	4.54	100	7.02
Indigenous Americans	0.01	1.83	3.63	94.53	100	5.47
						Avg. 4.70

another way, using this particular set of 71 AIMs (Ancestry ByDNA 2.0), the average simulated sample exhibited 4.7% admixture when it shouldn't have if the estimation process was perfect (see Table 5-20). Repeated analysis on specific samples showed the average genotyping error to contribute less than another 1%, making a total average MLE error of approximately 6% + X, where X is the amount of error contributed by our inability to mechanically define how admixture occurred (as discussed in Chapter 3).

With the 71 AIMs of AncestryByDNA 2.0, we find that the MLE error is lowest for simulated Western sub-Saharan Africans (1.9%), followed by simulated Europeans (4.4%), Indigenous Americans (5.5%), and simulated East Asians (7.2%) (see Table 5-20). This follows from the greater molecular distance between Africans and non-Africans than between non-African groups, revealed by Y-chromosome and mtDNA analysis (see Figure 4-6; reviewed in Jobling et al. 2004). The average bias for simulated Indigenous Americans and simulated East Asians was significantly greater than for simulated West Africans or Europeans, because of the elevated bias component in the East Asian/ Native American dimension. In simulated East Asians, most of the bias was in the Indigenous American direction, and in simulated Indigenous Americans most of the bias was in the East Asian direction, signifying a greater admixture estimation error along the Indigenous American/East Asian ancestry axis. This is concordant with expectations derived from human anthropological genomic analyses, which show a lower genetic distance between East Asians and Indigenous Americans than between any other pair of these four continental population groups (see Figure 4-6; reviewed in Jobling et al. 2004; also see Cavalli-Sforza & Bodmer 1999). Based on these studies, one would expect that the statistical error in discerning East Asian/Indigenous American fractional affiliation should be greater than for any other pair of groups, and with 71 AIMs it was. This was also one of the primary impetuses for our expanding the AncestryByDNA assay from the initial 2.0 version with 71 AIMs to the 2.5 version with 171 AIMs.

With this panel of AIMs, the bias for simulated Africans was highest in the European direction, again concordant with expectations from phylogenetic analyses and possibly the result of the recently shared component of ancestry illuminated specifically by the Y chromosome E haplogroup (which whether it is due to admixture or shared recent ancestry is not discernable from standard phylogenetic analyses). However, in simulated Europeans the bias was highest for the Indigenous (Native) American group, not the African group as we might expect from the preceding observation. This is almost certainly due to the lower information content for the panel of AIMs with respect to the European-Indigenous American distinction, relative to the European/West African distinction. It may also result to a certain extent from other unintended sources, such as the use of European/American samples as parentals that exhibit a systematic Indigenous American (but not East Asian or West African) admixture or other affiliation (in other words, an asymmetry in structural composition among the parental groups as we have defined them). We will touch on the observation of systematic Indigenous American affiliation for some European populations later when describing the global distribution of genomic ancestry results. For our purposes here, such a European parental population would tend to further obscure the distinction between Europeans and Indigenous Americans (since our European sample would not really be a good European sample, and instead is more like the Indigenous American sample than a normal European sample. This would reduce the effective information content of the markers for the distinction between true Europeans (not systematically affiliated with Indigenous Americans) and Indigenous Americans.

TRENDS IN BIAS FROM THE 71 AIM PANEL

The genes that participate in determining complex, multifactorial physical appearance traits (such as iris color, skin shade, hair texture, etc.) are numerous, and spread throughout the genome. Therefore, from a perspective of a forensics test, one might expect that significant MLE admixture levels, say greater than a few percent, might be informative for certain physical characteristics. We already have shown that certain anthropometric trait values are associated with certain threshold levels of genomic ancestry, and in this regard it would be useful to know the frequency with which simulated individuals from a population exhibit levels of admixture that exceed these thresholds. For example, if low levels of West African admixture in Europeans are associated with darker than average skin melanin indices, we need to define the sensitivity and bias for this type of admixture in this type of individual before we can understand the limits of our inferential capabilities. As we will soon see, admixture levels on the order of 30 to 40% are required for most traits to take on character of the relevant ancestral group such that a person might identify the admixture by eye, but the level depends on the trait. For skin color, strictly speaking (as opposed to African phenotypes in general), it appears that 50% or greater West African admixture is required for one to be reasonably certain the skin color of the individual will be darker than the average unadmixed European (see Figure 5-28B).

As we have seen, even a 71-AIM test can be 95% accurate down to an MLE admixture level of a few percent for most types of admixture on most types of backgrounds. This means that autosomal admixture tests are more than accurate enough to provide information on physical appearance for most traits that differ in value between ancestral groups (with confidence thresholds different for each trait). Whether information for enough of them is available for a forensic scientist to generalize on overall appearance would be determined using an empirical method of inference as discussed later in Chapter 8. Our task here is to define the frequency of simulated samples with certain admixture characteristics so that we can define the limits of sensitivity for the various types of minor affiliations (n = 3) on the various types of backgrounds (n = 4).

First, to assess the statistical accuracy with which the majority group affiliation is properly inferred in the simulated samples, we tally the frequency of simulated samples exhibiting at least 50% affiliation with their own group. Using 71 AIMs of 5,000 simulated Europeans, West Africans, East Asians, and Native Americans (20,000 total samples), all but two of them were determined to be of expected majority ancestry. The two that were not were a simulated East Asian with about 50% Indigenous American bias and a simulated Indigenous American with about 50% East Asian bias. Thus, in terms of correct assignment of majority ancestry, we see that the 71 marker battery is nearly perfect—that all but two (99.99%) of the 20,000 simulated samples were inferred to be greater than or equal to 50% affiliated with their own group.

The samples are simulated to be 100% affiliated with each group, so how many of them show >95% affiliation with their group? 87% of simulated West Africans, 70% of simulated Europeans, 64% of simulated Indigenous Americans, and 57% of simulated East Asians type with greater than 95% affiliation with their own group (see Figures 5-30 and 5-31). This shows that although the average bias is 4.7% for all types of admixture on all types of backgrounds, using this particular 71 marker test in simulated two-way admixed samples, we cannot be 95% certain of a 5% admixture result for any of them; a large number of samples show greater than 5% bias (even though the average sample does not).

A 20% level of bias is quite rare for all of the types of admixture with the 71 AIM test (note the very low values at 20% percentage admixture in Figures 5-30A, B, and 5-31A, B). Of course, all the simulated samples showed greater than 20% affiliation with their own group as indicated with the dark blue line in each of these figures. The average incidence of 15% or greater error in simulated samples is relatively infrequent for most but not all types of admixture. For example, about 10% of the simulated Indigenous Americans show at least 15% East Asian bias (see Figure 5-31C, light blue line), but virtually none of the simulated West Africans show bias this high for any of the three groups (see Figure 5-30A). When we consider 5% admixture error we see that 16% of simulated West Africans, 24% of simulated Indigenous Americans, 36% of simulated Europeans, and 42% of simulated East Asians exhibit greater than

Figure 5-30

BGA results for simulated multilocus genotypes of 100% Western (sub-Saharan) African (A and B) and 100% European ancestry (C and D). The fraction of 5000 samples (y-axis) exhibiting the indicated level of affiliation (x-axis) with the indicated element of continental ancestry is shown for simulated genotypes of 71 AIMs (A and C) or 171 AIMs as described in the text. Virtually all of the samples show >50% affiliation with the expected group, and most show >95% affiliation. Levels of outside-group affiliation (i.e., European, East Asian, and Native American for simulated sub-Saharan Africans) allow for calculation of bias inherent to the panel of markers caused by the continuous nature of the AIM allele frequency distributions among populations.

Figure 5-31

BGA results for simulated multilocus genotypes of 100% East Asian (A and B) and 100% Indigenous (Native) American ancestry (C and D), as in Figure 5-31. The fraction of 5000 samples (y-axis) exhibiting the indicated level of affiliation (x-axis) with the indicated element of continental ancestry is shown for simulated genotypes of 71 AIMs (A and C) or 171 AIMs as described in the text.

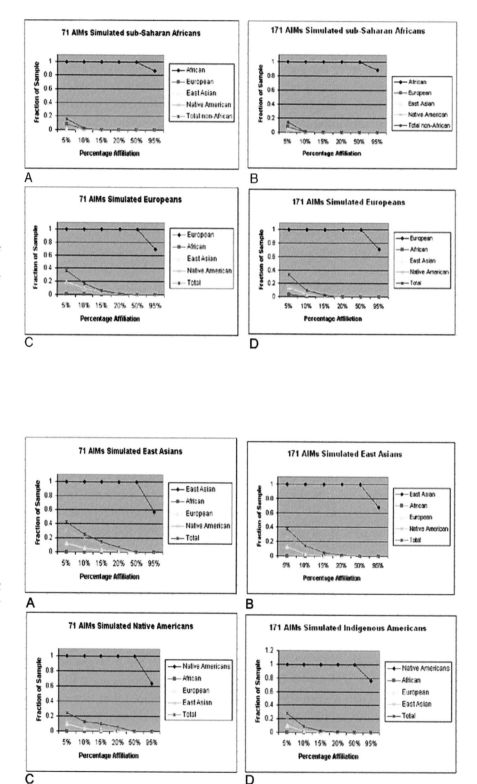

or equal to 5% admixture with at least one other group. Clearly, with the 71 AIM panel, admixture percentages of 5% are not reliable at the 95% level of confidence for most types of admixture.

With the 71 AIM panel, the average frequency of simulated samples exhibiting 5% or greater partial affiliation among all three groups combined as a single admixed group is higher; a whopping 30% on average (purple lines in Figure 5-30A, B, and Figure 5-31A, B). Clearly, we cannot expect the 71 marker test to reliably distinguish individual admixture levels below 5%, yet over 70% of the simulated samples do not show even this low level of admixture error.

95% CONFIDENCE THRESHOLD FOR 71 AIM PANEL

The results, taken together, illustrate that the 71 marker battery is nearly perfect in terms of determining major group affiliation, but 5% admixture levels are essentially meaningless noise with this test because 30% of simulated samples showed 5% or greater admixture with another group. An average 14% of the simulated samples showed 10% or greater admixture with another group, indicating that levels of 10% admixture are significant at the 0.15 level but not at the 0.10 or 0.05 level. The reason that 5% to 10% admixture levels are not significant at the 0.05 level appears to be mainly a problem in the discrimination power between Europeans, East Asians, and Native Americans (simulated samples from these groups most frequently exhibit fractional affiliations among these groups), and in fact it was these results that led us to rescreen the genome again and augment the collection of 71 markers with 100 new markers showing good European, Native American, East Asian allele frequency differentials (results for this 171 AIM panel will be discussed more fully shortly).

From extrapolation of these types of analyses, we can determine the level of affiliation required to conclude with X% certainty that the affiliation is the result of bona fide structure, as opposed to statistical noise (bias). Table 5-21

Table 5-21A

Percentage of affiliation that needs to be exceeded for a sample in order to conclude with 95% certainty that the result is due to bona fide affiliation with that population, rather than statistical noise (71 AIM panel). The best three-population MLE value was used for each simulated sample.

% Required to achieve p < 0.05				
71 AIMs	**AFR**	**EUR**	**EAS**	**NAM**
Africans		8%	<2%	8%
Europeans	<2%		13%	13%
East Asians	<2%	12%		18%
Native American	<2%	9%	18%	

Table 5-21B

Percentage of simulated samples showing greater than 95% affiliation with the expected group (71 AIM panel). The best three-population MLE value was used for each simulated sample.

% Required to achieve p < 0.05				
71 AIMs	**AFR**	**EUR**	**EAS**	**NAM**
Africans	>92%			
Europeans		>83 %		
East Asians			>76 %	
Native American				>79 %

shows the affiliation percentage thresholds for the simulated samples of polarized, binary affiliation we have been discussing so far, using an $X = 95$. For example, 95% of simulated parental samples showed levels of ancestry less than or greater than the percentage shown in each cell (as the case may be). Rows represent simulated samples and columns represent the affiliation type, with cell values therefore representing the percentage of affiliation a sample must exhibit in order to conclude with 95% certainty that the affiliation is bona fide, as opposed to statistical noise (bias). Another way to say this is: values shown for outside group affiliations (artificial admixture) shown in Table 5-21A are those values above which results indicate bona fide affiliation with that group with a $p \leq 0.05$. Values shown for the same groups (primary ancestry) shown in Table 5-21B are those values above which 95% of simulated samples typed and below which it is safe to conclude statistically significant outside-group (artificial) admixture at the $p \leq 0.05$ level.

As we can see from this presentation, with the 71-marker test, the gross aspects of ancestry affiliation is almost always correctly inferred, but depending on the type of admixture, MLE admixture levels less than 12.5% may not be a statistically significant indication of affiliation with that group at the $p = 0.05$ level, even though the level of admixture for a simulated sample is only a few percent on average (see Figures 5-30 and 5-31). What does this mean for a sample with an MLE of 90% European and 10% East Asian admixture obtained with the 171 AIM panel, for example? First, the admixture level for this sample is best considered in terms of a confidence range or space as a range of likely values, rather than the single value represented by the MLE, but if the MLE is used as a simple answer the simulations show what level of reliability we can expect. Not concerning ourselves with the question of which group this individual is from, or what type of admixture they exhibit, overall, the average sample shows about 4.7% admixture (see Table 5-20), so 10% in any given individual would seem to be a significant percentage. However, we see from Figure 5-30C that a little more than 10% of simulated European samples show greater than or equal to 10% East Asian bias or artifactual admixture and that the value of 10% is therefore below the threshold percentage required for 95% confidence that the East Asian affiliation is real for this individual (not due to statistical artifact, also confirmed by reference to Table 5-21A, which shows that a level of 13% or greater is required for 95% certainty). This shows that the 10% admixture is not statistically significant at the $p = 0.05$ level, even though the average simulated sample shows a level of artificial admixture below this level. A 10% admixture result, however, is significant at the 0.10 level meaning that for 9/10 European samples of 10% East Asian admixture, the admixture reading is not due to statistical noise caused by continuous allele frequency distributions. If the reading was 15% East Asian admixture, we can see from Figure 5-30C that only about 5% of simulated European samples exhibit this level of bias so this result would be significant at about the $p = 0.05$ level, assuming that the continuous nature of the allele frequency distributions is the only source of error.

PRECISION OF THE 171 AIM PANEL FROM SIMULATIONS

Though this battery of 71 AIMs provided reasonable resolution power in all dimensions, including the East Asian/Indigenous American dimension, stronger than any other battery of markers yet described, the disproportionate statistical error encountered in the Indigenous American/East Asian dimension of the analysis relative to the other dimensions suggested that an improvement in AIM battery constitution could be made by culling a greater collection of AIMs with high delta values for the East Asian/Indigenous American distinction. We chose to ameliorate bias in non-African/non-African dimension in general, by incorporating a greater number of AIMs. The 71 AIM battery was an optimal list of AIMs that validated with high allele frequency differential among West African, East Asian, Indigenous American, and European parental samples. To assemble this optimal group of 71, we started at the top of the list of several thousand ranked candidate AIMs (by δ value), determined the allele frequencies in parental samples, and validated the δ values in parental samples. Given that it was for the East Asian/Indigenous American distinction that the cumulative delta value was lowest, it is no surprise that the simulations showed the greatest level of MLE admixture error when the majority affiliation was, or admixture percentage refers to, East Asian or Indigenous American.

The previous screen took us through only a few hundred candidate AIMs (400). Now that simulations indicated that more resolution power was needed for East Asian/European/Indigenous American (in particular, the East Asian/Indigenous American) distinctions, we resumed the validation work. We genotyped another several hundred candidate AIMs ranked by East Asian and

Indigenous American Locus-Specific Branch Length (LSBL, Shriver et al. 2004) and selected 105 with the highest validated European/East Asian, European/Indigenous American, and East Asian/Indigenous American information content. Adding these 105 markers to the previous 71, we obtained a new test with 171 markers, for which the cumulative F_{ST} values among all the possible group pairs was relatively normalized. Obviously, the East Asian/Indigenous American cumulative delta value increased the greatest, followed by the European/Indigenous American and European/East Asian deltas.

Five thousand samples were simulated for each of the four genetic groups (20,000 samples total). From Tables 5-21A-B we can see that the average bias using 175 AIMs was 3.3%, an improvement over the 4.7% provided by 71 AIMs. All the simulated individuals showed 50% or greater affiliation with their own group indicating that the error, when applied to individuals of relatively polarized ancestry, is expected to be less than 1/5000 for correctly identifying the major group affiliation (also see Figures 5-30B, D and 5-31B, D). About three-fourths showed greater than or equal to 95% affiliation with their own group. As expected from the fact that the 100 markers added to the previous 71 were selected based on their East Asian/Native American information content, it is no surprise that 95% or greater within-group affiliation for East Asians and Indigenous Americans improved the most (about 10%) (see Figure 5-31A vs. B, and C vs. D, blue line at 95% Percentage Affiliation).

Using the 171-marker battery, we observed relatively low frequencies of simulated samples exhibiting 20% or greater admixture. For example, no simulated West African samples showed this level of non-African admixture, and only a few simulated European samples out of 5000 showed this level of non-European (usually East Asian and Indigenous American) admixture (see Figures 5-30B, D and 5-31B, D). Twenty-two (0.0044) of 5000 simulated Indigenous Americans showed 20% or greater East Asian admixture, but this represented a significant improvement over the simulation data obtained with the 71 marker test (see Table 5-25, where negative values indicate a lower frequency of individuals exhibiting the 20% level of admixture using the 171 marker test). With 171 markers we still see a high average percentage of simulated samples showing 5% or greater admixture (28% on average, see Figures 5-30B, D and 5-31B, D), which indicates that individual results at or below 5% affiliation are still not necessarily illustrative of ancestral affiliation. However, as usual, the error rate was not the same for all of the types of simulated samples, and in fact, 5% West African ancestry in simulated Europeans was encountered in less than 4% of the simulated samples. Similarly, 5% Indigenous American ancestry in simulated Africans was encountered for only 1.4% of the simulated samples, and none of 5000 simulated East Asians or Indigenous Americans were determined to be 5% or greater African admixture. These results imply that for a result such as 94% European, 6% African, the African admixture (i.e., that it is some value greater than zero) is statistically significant at the p = 0.05 level (the range of values delineated by the five-fold confidence contour would be determined to be entirely above the zero line on the triangle plot or bar graph).

MLE THRESHOLDS FOR ASSUMPTION OF BONA FIDE AFFILIATION

Again, we extrapolate from these results to determine the threshold percentage of affiliation (i.e., admixture) with a group that is required for assuring with a p < 0.05 that the affiliation is real. Of course, this threshold will be different for each type of simulated sample. In fact, we can see from Table 5-23A that the threshold value is less than a few percent for most types of samples, and up to 8 or 9% for others with a maximum of 11.5% East Asian admixture in Indigenous Americans and 12.5% for Indigenous American admixture in East Asians. Put another way, and as before, a value of 9 or 10% East Asian admixture on an Indigenous American background (i.e., 90% Native American and 10% East Asian) does not indicate with a p < 0.05 of certainty that the East Asian admixture is real, but a value of 12% East Asian admixture on an Indigenous American background does.

COMPARISON OF 71 AND 171 AIM PANELS

Comparing Figures 5-30A versus B, 5-30C versus D, and 5-31A versus B, 5-31C versus D allows for an appreciation of the reduction in bias we achieved by increasing the number of AIMs from 71 to 171. In general, the levels of bias taper off more rapidly with the larger set of AIMs, for most types of samples with most types of admixture. As expected, the improvement in performance observed between the two panels was in the non-West African/non-West African dimension.

OBSERVED AND EXPECTED BIAS

How do the average bias values compare with those derived from theoretical expectations? As we have already discussed, the variance of admixture estimates M obtained for each AIM allele is a function of the allele frequency differential and the variance of allele frequency estimation in the admixed and parental populations:

$$\sigma_M^2 = 1/(P_A - P_B)^2 [\sigma_{PM}^2 + M^2 \sigma_{PA}^2 + (1 - M)^2 \sigma_{PB}^2] \qquad (5\text{-}3)$$

where P_A and P_B are the allele frequencies and $P_A - P_B$ is the δ value. The equation for computing the variance expected from using multiple loci would be different since the use of more AIMs increases the power of the analysis and terms would have to be added to account for the new AIMs. For a two-way admixture model, we know that the variance is related to the number of AIM loci by the equation:

$$\sigma_1^2 \; \alpha \; 1/2n \qquad\qquad (5\text{-}4)$$

which as discussed earlier, is the variance of any two mutually exclusive variables (such as allele frequencies, or binary admixture percentages within individuals or populations) via binomial sampling theory. For our purposes, consider n as the number of AIMs. Since the variance in admixture estimates is expected to vary with $1/2n$, the variance in admixture estimates using 71 loci is $1/142$ the average variance for each of the 71 loci, with the variance of each locus determined with Equation (5-3) on its own. This makes intuitive sense, since the larger the number of AIMs the more precise the estimate. We have already shown that using 71 AIMs, the standard deviation for estimates of allele frequency, for the average AIM allele, is 0.036 (given an average allele frequency of 0.25). We will assume further that the loci characteristics are the same across AIM loci and that the standard deviation in estimating allele frequencies, and hence the variance of allele frequencies from admixed samples (σ_{PM}^2 in Equation 5-3) to be the same as from unadmixed samples (σ_{PA}^2 and σ_{PB}^2). In our simulations, we create samples homogeneously affiliated with each group, so for Equation (1), M = 1 and (1 − M) = 0 (i.e., there is no admixture). If the average AIM is of an allele frequency differential = 0.40,

$$\sigma_M^2 = 1/0.0162 \; [0.036^2 + 0.036^2] = 0.0162$$

meaning that the average variance produced by each of the 71 AIMs was 0.0162. Dividing this by 142, we obtain 0.00011 and the standard deviation is the square root of this value, or 0.011. Thus, the expected deviation from the mean admixture level using 71 AIMs, each of 0.40 delta and of an average allele frequency estimation error of 0.036, is about 1% (1.1%). Using the 71 AIM test with our simulated samples, the average sample was of no admixture, yet was observed to be on average 4.7% admixture, meaning that the average simulated sample deviated from the true mean by 4.7%, about four times greater than the expected error of 1.1%. The difference is likely explained by the fact that we assumed the standard deviation of allele frequency estimation for each AIM was the same, and that the delta value of all the markers was the same relatively high level of 0.40. In reality, the situation is quite a bit more complex since allele frequency estimation errors depend on the allele frequencies, which are varied, and for any particular pairwise group discrimination, only a fraction of the total number of AIMs is of high delta. Using 175 AIMs, the expected deviation from the mean admixture level, using Equations (5-1) and (5-2) is 0.7% and the realized deviation was 3.3% (see Table 5-22). The observed difference between the realized and expected deviation for both the 71 AIM and 171 AIM batteries is therefore very similar (3.23% for the 171 marker test and 3.59% for the 71 marker test) and higher than but in the same ballpark as those we might expect when we simplify our assumptions and use average δ and allele frequencies.

WHAT DO THE SIMULATIONS TEACH US ABOUT INTERPRETING BGA ADMIXTURE RESULTS?

If the 171 AIM test shows a result of 8% Indigenous American, 92% European for a given sample, how do we interpret the significance of the 8%? From Tables 5-23A-B, we see that this level of Indigenous American admixture in simulated Europeans is not statistically significant at the $p = 0.05$ level. However, if the sample is part of a pedigree, for which admixture results are also available, and the Indigenous American admixture is seen to increase in some or all of the recent ancestors with each generation, and the proportions of

	AFR	EUR	EAS	NAM	Total	Total admixture avg.
Africans	98.21	0.93	0.71	0.15	100	1.79
Europeans	0.4	96.36	1.5	1.74	100	3.64
East Asians	0.08	1.43	95.48	3.01	100	4.52
Native Americans	0	1.16	2.08	96.76	100	3.24
						Avg 3.30

Table 5-22

Average percentages obtained for simulated samples using 171 AIMs.

% Required to achieve $p < 0.05$				
176 AIMs	**AFR**	**EUR**	**EAS**	**NAM**
Africans		<6%	<5%	<1%
Europeans	<4%		<9%	<10%
East Asians	<1%	<8%		<15%
Indigenous American	<1%	<7%	<11%	

Table 5-23A

Percentage of affiliation that needs to be exceeded for a sample in order to conclude with 95% certainty that the result is due to bona fide affiliation with that population, rather than statistical noise (171 AIM panel). The best three-population MLE value was used for each simulated sample.

Table 5-23B

Percentage of simulated samples showing greater than 95% affiliation with the expected group (171 AIM panel). The best three-population MLE value was used for each simulated sample.

% Required to achieve $p < 0.05$				
176 AIMs	**AFR**	**EUR**	**EAS**	**NAM**
Africans	>93%			
Europeans		>87%		
East Asians			>84%	
Indigenous American				>87%

Indigenous American admixture are reasonably close to half-way between that of the parents, this would suggest that the sample is indeed of partial Indigenous American affiliation because:

- Many of the ancestors would presumably be at a level above the threshold in Table 5-23.
- The Indigenous American ancestry would be focused to a greater or lesser degree in certain branches of the pedigree, illustrating a nonrandomness to the result.

Nonetheless, to be conservative, unless there is pedigree data to support a conclusion that a given individual sample is of partial affiliation with a particular group, when the percentage of affiliation for that individual is below the value in Tables 5-23A-B, it must be concluded that the result is not statistically significant. Notable in this respect and more difficult to study systematically using simulations is the confidence intervals. Indeed, the first results that should be consulted in deciding whether a minor ancestry proportion is significant is the shape and spread of the 5X and/or 10X confidence ring for that particular person. We expect that there will be variability across individuals who have the same levels of admixture (e.g., 8% Indigenous American in a European background) with respect to their confidence intervals because the shape of the contours is a function of the multilocus genotype (which is unique for each sample). A more elaborate simulation study is needed to examine the extent of errors in the confidence intervals with respect to stipulated ancestry.

BIAS SYMMETRY

The simulation data shows that the bias is roughly but not perfectly symmetrical (i.e., symmetrical percentages about the diagonal in Table 5-23B). For example, the East Asian bias in Indigenous American parentals is not the same as the Indigenous American bias in East Asian parentals. It would seem that with a perfect test, the markers would be perfectly symmetrical in terms of their information content (i.e., the cumulative delta value for EAS/AFR markers was identical to the EAS/EUR) and the statistics estimated from tables such as Tables 5-23A-B would be symmetrical about its axis (for example, the off-diagonal values in Tables 5-23A, Table 5-21A). There is relative symmetry but not exact symmetry, and the question "why is the test not more symmetrical" is a good one.

Although presented as binary comparisons in the preceding tables (for convenience), the model employed is actually four-dimensional analyses. Recall that the shape of the 4D confidence intervals were bilaterally symmetrical and holding true admixture constant, we expect populations of MLEs to show similar distributions as estimates within the confidence intervals for any one or small collection of individuals (see Figure 5-29). To a certain extent the lack of perfect symmetry is likely related to the simplification of the analysis. However this is not likely the sole explanation. Discrimination power for a given locus in a given pair of populations is related to allele frequency in the populations. With multiple AIMs, the impact of a given allele on the likelihood function is context dependent. For example, by chance the markers uninformative for the East Asian/Native American distinction (such as a European-West African marker) may contribute differently in the determination of a precise Native American/East Asian (East Asian admixture in an Indigenous American, such as 90% IAM/10% EAS) result than when determining a precise East Asian/Native American result (such as 90% EAS/10% IAM) because the allele frequencies for European/West African markers in Indigenous Americans are on average different (perhaps, a net higher frequency) than in the East Asian parental groups. This could produce a subtle, net bias toward certain types of results over their reciprocal results—a bias of the sort we are attempting to measure with the simulations.

As for why the bias is not the same for different types of binary comparisons (i.e., why there is more East Asian bias for Europeans than West African bias in European), this is caused of course by the fact that the discrimination power among pairs of parental populations is not uniform—there is more power for some pairs than others as we can see from Table 4-3 in Chapter 4. This is a byproduct of genetic distance and the difficulty, or ease, of finding markers with certain types of discrimination power. However, it is not merely a matter of effort, and in fact it is virtually impossible to select a panel of markers with uniform power in all pairwise dimension since many good East Asian/Native American AIMs are also good African AIMs due to their age. This creates a situation where we cannot add more East Asian/Indigenous American power without also adding African/Indigenous American or African/East Asian power and so there will always be more African/non-African information content than East Asian/Indigenous American information content, even if we restrict our selection for AIMs based on the latter population comparison.

IMPACT OF MLE ALGORITHM DIMENSIONALITY

For the simulations discussed so far we have assumed that the ancestry of each sample is no more complex than a three-way admixture—that is, we have used the three-way MLE value for each sample. Mostly, this is done for computational convenience. What results do we see for the simulations if we spend the extra energy and time performing four-way MLE analyses, and are they different from what we have been discussing so far? Recall that the three-way/four-way MLE calculation style (see earlier in this chapter, Figures 5-10, 5-11, 5-12; also see Chapter 3) is effectively the same thing as the four-way MLE analysis since individuals of four-way admixture usually show similar MLE values for more than one three-population model. Using the three-way/four-way MLE calculation style to generate 1000 simulated parental samples, as we have just discussed in this chapter, provided average ancestry admixture results that could be compared with expectations and bias could be determined (see Table 5-24).

The average results showed subtle improvement over those obtained using the three-way algorithm. The average parental sample showed more of the expected ancestry and less artificial admixture than with the three-way algorithm, and the average bias decreased from 3.30% (see Table 5-22) to 2.96% (see Table 5-24). When we calculated the threshold percentages as before, we see subtle across-the-board improvement by a couple of percentage points (compare Table 5-25 with Table 5-23A).

Since this algorithm performed the best in simulations we take a more detailed look at the performance. Table 5-26 shows the summary statistics for these parental samples using the 171 AIM test with the three-way/four-way MLE calculation style; we also show the Root-Mean-Square-Error (RMSE, square root of the sum of the variance and squared bias) and the probability that the true level of ancestry resides within the two-fold confidence contours on the triangle plot, which corresponds to an 88% confidence interval.

Table 5-24

Percentage of affiliation that needs to be exceeded for a sample in order to conclude with 95% certainty that the result is due to bona fide affiliation with that population, rather than statistical noise (171 AIM panel). In contrast to the analysis shown in Table 5-22, the three-way/four-way MLE calculation style was used to estimate admixture for the simulated samples.

	AF	NA	EU	AS	Total artificial admixture
AF	98.1%	0.65%	0.79%	0.48%	1.92%
NA	0.34%	97.0%	0.84%	1.9%	3.05%
EU	0.92%	1.36%	96.3%	1.46%	2.82%
EAS	0.52%	2.58%	0.96%	96%	4.04%
					Avg. 2.96%

	AF	NA	EU	AS
AF	>92%	<4%	<5%	<4%
NA	<2%	>88%	<5%	<10%
EU	<5%	<8%	>87%	<8%
EAS	<3%	<13%	<6%	>86%

Table 5-25

171 marker admixture test threshold beyond which the percentage affiliation indicates bona fide admixture for simulated samples of polarized, binary affiliation (with 95% confidence). In contrast to the analysis shown in Table 5-23A, the three-way/four-way MLE calculation style was used to estimate admixture for the simulated samples.

Percentage	Group(s)	Total bias	RMSE	Prob (true estimate within two-fold contour [88% CI] around MLE)
100%	AFR	1.9%	4.3%	85.2%
100%	NAM	3.1%	6.7%	83.2%
100%	EUR	2.8%	7.4%	78.5%
100%	EAS	4.0%	8.2%	81.3%

Table 5-26

Summary of bias and error observed in simulated samples using the three-way/four-way MLE calculation style discussed in the text (171 AIMs).

We have not yet discussed the RMSE in much detail; it is a more conservative indicator of test error. It is more conservative because, as shall be explained here, it overestimates the amount of error a user of an admixture assay is likely to observe in practice. The bias is the deviation of the mean of estimates from the true value used in generating the simulated data. The variance is the spread of data around the mean observed value (not the true value), and represents the square of the standard deviation. What we usually want to appreciate is the average deflection from the true value, since this is the error for the average sample. However any one sample could be greater than the average, and in some cases we may be more concerned with the maximum deflection we could reasonably observe from the true value. Estimating error from consideration of the bias and variance together gives a better indication of this maximum error of the method, notwithstanding we may not always recognize such error in practice depending on the direction the error takes. We call this sum the maximum range of error, or the RMSE and it gives the maximum error caused by the additive influence of bias and imprecision at the same time, since they are not the same thing. For example, if the test on average reports an MLE 2% away from the true MLE on the triangle plot, always toward the center, and if the test has 2% range or spread of estimates around this MLE, the true

MLE and reported MLE could be identical (positive 2% + negative 2% = 0% difference, in which case two errors cancel each other out and we arrive at the correct answer by accident) or 4% (2% + 2% = 4%, where the two errors are reinforcing) or somewhere between. The total statistical error, or maximum deflection we are likely to observe, would be the 4% value and the RMSE would be 4% even if the average estimate is not 4% off from the real value.

The column in Table 5-26 titled True Estimate within two-fold contour (88% CI) shows the probability that the estimate is found within the 88% confidence interval around the MLE. In the simulations we can calculate this because we know that the true value is, by definition, 100%. Using an algorithm that scans the entire likelihood space, we not only determine the MLEs for our simulated samples, but also the ranges of values within two-, five-, or 10-fold the MLE or more. The frequency with which the value (100%) falls within the two-fold confidence range is the probability the true value is within the two-fold confidence contour (the probability shown in Table 5-26). We would expect this value to be roughly 88% from a bivariate Gaussian distribution, and as can be seen from Table 5-26, the calculated values are somewhat close to but a bit lower than this expectation.

SIMULATIONS OF ADMIXED INDIVIDUALS

The bias is particularly informative for simulated samples of extensive admixture. With the simulated samples discussed until now, ancestry was simulated to be 100% and test error would show up as directional bias away from the vertex of the triangle plot toward another vertex or toward the center. Error could be observed in only one direction—away from the vertex (100%). The bias reported here is an apparent bias, that which we can appreciate, since the true value is 100% ancestry and we have imposed limitations on what types of non-100% values to consider. MLE values that plot off scale, with negative percentages for some groups and percentages greater than 100% for others are not considered (more about this later).

For a simulated 100% European sample, there are only three types of error we can observe—artificial levels of West African, Indigenous American, and East Asian affiliation coupled with less than expected European ancestry. With a simulated 50% African/50% East Asian sample, however, there are six types of error we can observe. The true African ancestry could be misreported as East Asian, Indigenous American, or European, and the true East Asian could be misreported as African, Indigenous American, or European.

1. African to East Asian
2. African to European

3. African to Native American

4. East Asian to European

5. East Asian to Native American

6. East Asian to African

The only type of error we cannot observe is reciprocal error—such as African to East Asian error coupled with East Asian to African error at the same time. Therefore, since there are more types of error we could observe, we might expect our measures of bias and RMSE to increase but still be slight underestimates of (but still good estimates of) the real imprecision of the panel.

Indeed, the values are generally higher than for simulated 100% samples (compare Table 5-27 to 5-24). The average RMSE measured for these simulated samples was 9.6% though the average bias was only 2.3%. The two-fold confidence contours for the average simulated 50%/50% sample contains the true value for both estimates 74% of the time, and the true value for at least one of the estimates 92% of the time. The first of these is comparable to the probabilities shown in Table 5-26.

MLE PRECISION FROM THE TRIANGLE PLOTS

The simulations tell us that the admixture estimation process in populations behaves similarly to that within individuals—that is, many individuals plot

		Bias[*]	RMSE[*]	Pr (true value within two-fold contour of MLE [88% CI] for both estimates)[*]	Pr (true value within two-fold contour of MLE [88% CI] for at least one estimate)[*]	Pr (true value within two-fold contour of MLE [88% CI] for neither estimate)[*]
50%/50%	AFR/NAM	2.7%	8.2%	73%	95%	5.1%
50%/50%	AFR/EUR	1.6%	8.4%	77%	91%	9.4%
50%/50%	AFR/EAS	2.9%	7.9%	73%	95%	4.9%
50%/50%	NAM/EUR	2.4%	10.5%	73%	91%	9.3%
50%/50%	NAM/EAS	2.8%	9.7%	70%	92%	8.0%
50%/50%	EUR/EAS	1.2%	13.1%	76%	85%	15%
Average		2.3%	9.6%	74%	92%	8.6%

Table 5-27

Summary statistics from application of the 171-AIM admixture panel discussed in the test with samples simulated to be of 50%/50% admixture.

[*]Only artificial admixture apparent from the simulation results can be considered, for example, EAS, EUR and net (not total) AFR-NAM error in simulated AFR/NAM samples.

about the mean for a population very similar to the distribution of less likely estimates of ancestry about the MLE for a given individual. Both distributions appear to be Gaussian, and having demonstrated this, we can now better define what the confidence contours mean.

CONFIDENCE OF NONZERO AFFILIATION

From a given triangle plot there are a couple of questions that arise naturally. We have just discussed how we can determine from a study of populations whether a minor group affiliation is indicative of real affiliation with that group or merely statistical noise. We will now return to the likelihood space for each individual, which also allows us to make this determination even if there is no research on populations available. Here we will explain in more detail what the confidence contours say about the likelihood that an individual is not affiliated with a parental group represented in the observed MLE.

It is useful to consider the confidence contours when assessing the probability of real affiliation with a group versus the alternative, that a fractional affiliation is the result of statistical noise. It so happens that the contours give simple tests for significance—if the blue contour (five-fold likelihood interval) stands completely clear of the edge or base of the triangle representing 0% along a given axis, or at most just touches the edge, then there is about 95% confidence that the real ancestry for the sample includes a nonzero component of ancestry corresponding to that axis. This is a byproduct of the observed near-Gaussian nature of the error distribution. Similarly, if only the inner contour stands clear of the edge, there is about an 88% confidence of a nonzero affiliation with the group, and if the outer contour stands clear of the edge, this gives about 98% confidence of nonzero affiliation with respect to the group, subject to the definitions of absolute ancestry obtained with the reference samples used.

STANDARD DEVIATION FROM CONFIDENCE INTERVALS

If the average percentage error from simulations of homogeneous individuals is only a few percent, then why is it the confidence rings sometimes span spaces of 20 to 30% in any one direction? Aside from sampling effects, the answer is that the average percentage error in simulated populations should be compared to the standard deviation of estimates in individuals, not the *range* of estimates within five- or 10-fold likelihood of the MLE for the individuals. Recall that in the confidence region, unlike in a population, the estimates are of varying likelihood and not of equivalent value. For example, in the plots, we have arbitrarily chosen to display all values within 10-fold of the MLE, but could have displayed all values within 100-fold, or 1000-fold, which would encompass larger spaces—yet this would not change the bias or RMSE of the test.

To explain this point a bit further, consider that basic statistical theory instructs that the scatter of estimates (precision) about their mean is usually a good indicator of the reliability or accuracy of the estimation process. This usually applies to estimates of equivalent weight or significance. For example, an average hurricane trajectory can be derived from 1000 hurricane trajectory estimates, and if each one of these is equally likely, the variance of the estimates (square of the standard deviation) is a good indicator of the quality of the value represented by this average. If the spread of the 1000 estimates is small, the average of the estimates is more reliable than one derived from 1000 for which the spread is large. Since we have simulated populations of individuals homogeneously (100%) affiliated with one of the four groups, we know that the true value for each estimate should be 0% admixture and that the expected value (mean) of all simulated samples should also be 0% admixture if the test were perfect. Since each estimate considered is an MLE in the simulations, we are working with estimates of equal weight (they are the best possible estimates for each sample, though some are more correct than others). The standard deviation in estimates from the simulations experiments is roughly the average percentage of artificial admixture in populations.

It may seem that in the case of an individual the situation is a bit different than in the case of populations, but as discussed earlier (Figure 5-29) this is really only an illusion (see Box 5-B). Whether for an individual or a population of individuals, the likelihood estimates follow a Gaussian distribution, and since the probability-weighted root-mean-square of the deviation of estimates from the mean of a Gaussian distribution is 0.3999, we in essence expect the standard deviation of those estimates to be equivalent to 0.3999 times the range of those estimates. In the Gaussian distribution of high-likelihood estimates around the most likely estimate, the standard deviation in estimates corresponds to 0.47 (1/2.15) the distance from the MLE to the points on the 10-fold likelihood contour and 0.85 the distance from the MLE to the points along the two-fold likelihood contour.

In a simulated population, we would also expect a Gaussian distribution of estimates about their average (since populations of MLEs are fundamentally the same as a large number of high-likelihood estimates for any one individual as we have discussed). The RMSE of the deviation from the mean value in the simulated populations is fundamentally the same as the standard deviation of within-individual admixture estimations for the average sample within each population. Bias impacts the estimates in an individual the same way it does for a population of individuals, systematically deflecting the observed value from the true value.

Box 5-B

Another way to look at this is to consider that estimates within individuals are not of equal weight, but the same can be said of estimates between individuals of a population. Within individuals, we calculate the standard deviation of the estimates obtained for an individual by correcting for the fact that the estimates are not of equal weight. The estimates out near the 10-fold confidence contour boundary are $x = 10$ times less likely, and those 9/10th of the distance from the MLE to the 10-fold confidence contour are $x = 9$ times less likely, and so on, so we take the weighted average of these estimates. That is, we weight each of the estimates between the MLE and the edge of the 10-fold confidence contour by 1/x and then take the average of these weighted estimates in order to get the standard deviation in likelihood estimates for any given individual. One can see that weighting the estimates 1/10, 2/10, ... 10/10 (for the MLE) and taking the average of all the weighted estimates comes to a value of about 0.4d for any likelihood space within a triangle plot, regardless of its shape or the size of its rings, where d is the average distance between the MLE and the 10-fold confidence contour ring considering the distances for each possible direction. In reality of course, the shape of the distribution is continuous and we must integrate over this shape, but doing so results in a similar number (0.3999 in the average direction).

Since the distribution of likelihood estimates follows a pure bivariate Gaussian distribution, the two-, five-, and 10-fold confidence contours correspond to about 1.17, 1.79, and 2.15 standard deviations from the mean, respectively. The standard deviation in estimates would correspond to 0.47 (1/2.15) the distance from the MLE to the points of the 10-fold likelihood contour. Since the departure from pure Gaussian behavior deteriorates the farther one proceeds from the mean, it's best to work with the inner contour, and the best factor for calculating the standard deviation is 0.85 times the distance to the inner (two-fold likelihood) contour rather than 0.47 the distance to the 10-fold contour. If we were to use these values to compute the standard deviation of estimates within an individual, and compare this to the standard deviation of estimates for a simulated population, the measures should be the same. If we add bias to calculate the RMSE, again, measures between and within individuals should be the same. This is because the probability that a given multilocus genotype "fits" an admixture profile is fundamentally the same as the probability that a given admixture profile "fits" a given multilocus genotype. In a population, the genotypes are a function of probability based on allele frequencies and the likelihood of a given ancestry profile for any one of them is the same as the likelihood of obtaining a given genotype given a particular ancestry profile.

TESTING THE RELATION BETWEEN CONFIDENCE MEASURES IN INDIVIDUALS AND POPULATIONS

We expect the likelihood space on the average triangle plot to be roughly similar in dimensionality or shape to the unsymmetrical nature of the RMSE numbers obtained from the simulations. That is, if the confidence contours in the triangle plots are always more narrow in the African dimension than the

Native American or East Asian dimension (which they usually are), then the RMSE in the simulated populations should be lower for African admixture than for Native American or East Asian admixture. This happens to be the case. In fact, we expect that the standard deviation in estimates obtained from the simulations should be about 0.85 the spread in likelihood estimates within two-fold of the MLE on the average triangle plot (see Box 5-B). It happens that when looking at triangle plots for simulated individuals of no admixture or real individuals of relatively low admixture (the same type of individual as simulated), they seem to be. For example, from our simulations of Europeans, we would expect the distance from the MLE to the Native American and East Asian extremes of the confidence contours would be about twice that in the African dimension. Using the best three-way MLE estimate, we might expect that the distance in the East Asian and Native American dimensions from the MLE would be about three times the distance to the African extreme on the same confidence contour because the African bias in Europeans was 0.4%, but the East Asian and Native American bias was 1.5% and 1.7%, respectively, with the 171 AIM test (see Table 5-22). Using the best four-way MLE estimate, we might expect from simulations that the distance in the East Asian and Native American dimensions would be about two times the distance in the African direction along the same contour—the African bias in simulated 100% Europeans was 0.92% and the Native American and Asian bias was 1.4% and 1.5%, respectively (see Table 5-24).

Inspection of a large number of triangle plots, where the modified three-way/four-way algorithm was used (which is a happy compromise between these two methods, as discussed earlier in this chapter) shows that these expectations tend to be met. For example, compare the distances in Figure 5-32—the shape of the contours is not perfectly circular but ovoid, compressed in the African dimension relative to the Native American dimension (though perhaps not as much as we might expect for this particular sample, keeping in mind that it is only one sample, not an average of samples).

SPACE OUTSIDE THE TRIANGLE PLOT

There is some error outside the triangle plot we have not considered in this treatment, because the 100% level represented at the vertex corresponds to an absolute estimate of ancestry that itself is estimated with a degree of error. Much of this topic was previously discussed in Chapters 3 and 4 but we discuss it again here because it is important to understand what the simulations can and cannot teach us about our admixture panel. Plus, it so happens that comparing results from simulated and real subjects gives us an idea of the magnitude of error caused from parental allele frequency misspecification. We will always

Figure 5-32

Triangle plot for an individual MLE with confidence contours (171 AIMs) reveals that for this individual, as we might expect from the simulation data presented in the text, the uncertainty in the Western (sub-Saharan) African dimension is lower than in the Native American or European dimensions. This can be visualized by measuring the length of a line extending to the black or blue line vertically from the MLE versus one extending 235 degrees toward the European vertex or extending 135 degrees toward the Indigenous (Native) American vertex.

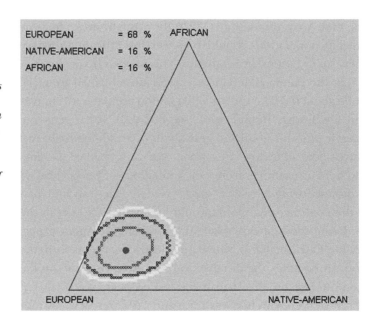

experience error as a result of our inability to identify perfect parental samples. As an extreme example, if we sampled Russians as representatives of European parentals, and if Russians are characterized by an average 10% East Asian affiliation (a hypothetical example), we would observe many individuals from Ireland, Britain, and so on that type of *more* European affiliation than the European parentals (i.e., some value greater than 100% European). Since we are confined by the limit of 100% European ancestry established with our Russian sample we would see the effect of over-clustering. For example, Scottish/Irish and Russians might type with the same European ancestry whereas if the proper parental population was used, Scottish and Irish might type with the same European ancestry but Russians would type with East Asian admixture. Using Russians as the parental group, the plotting is in error in an absolute sense since our goal is to determine MLEs with respect to real parental populations (though not necessarily in a technical sense, since we could plot any samples with respect to any parental populations and get coordinates based on genetic distance that have forensic value). In our example here, the true position for the Scottish and Irish would be outside the triangle whose vertex was established using Russians as the parental population. Since the triangle plot and MLE estimation method consider only affiliation levels between 0% and 100%, we do not account for these off-scale values even though they may be the most likely values (see Box 5-C). In samples exhibiting this type of error, we would expect to see an MLE on a boundary, with crushed confidence contours indicating that most of the likelihood space exists off scale.

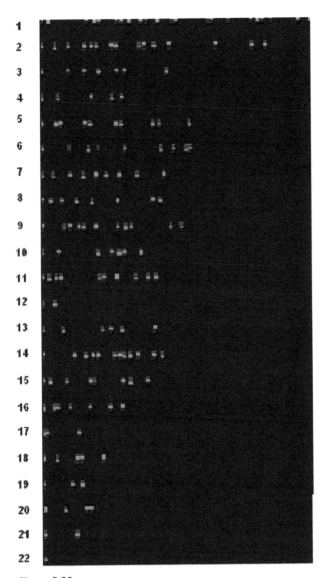

Figure 5-33

Clustering of ancestry admixture on the chromosomes of an individual with European (red), Western African (blue), East Asian (green), and Indigenous (Native) American (brown). The plot was obtained using Paul McKeigue's ADMIXMAP program discussed in this chapter and in Chapter 3 (McKeigue et al. 2000). Chromosomes are indicated with numbers on the left, and the bars represent each of 171 AIM loci discussed throughout this chapter. The relative amount of color in each bar represents the likelihood that the pair of alleles for this locus was derived from the corresponding parental population(s). For example, a bar with 50% red and 50% blue color would indicate an AIM for which one allele was likely derived from the European parental population and the other from the Western African parental population. For this individual, we can see that the presence of the colors is clustered to a certain extent—Western African ancestry seems clustered to chromosome 3, part of chromosome 2, 8, 13, etc. and European ancestry is spread throughout, but notably lacking on chromosomes 19, 20, etc. Though relative comparisons between similar maps of individuals is now possible, the question remaining to be answered with future research is what this clustering tells us about the age of the admixture for this individual in chronological or even generational terms. Work conducted to date by the authors and by McKeigue's group suggest that the density of markers in this analysis will not be adequate to answer these types of questions with any reasonable degree of accuracy. Unpublished work of Timothy Terravainen of DNAPrint Genomics, Inc.

Box 5-C

Illustration of the Relevance of Off-Scale Likelihood Spaces

Consider a two-way admixture test for binary admixture, where the MLE is plotted along a line connecting the two ancestry groups, where the length of the line is a function of genetic distance from some reference point along this line, such as the most recent common ancestor, and information content of the markers. Assume that the plot of an MLE can be known with certainty with respect to the reference point but that the ends of the line are determined based on allele frequency estimates from a small sample. We update this sample by increasing its size, and obtain new allele frequency estimates. With the new estimates, the line becomes larger or smaller in one or both directions as the estimates of absolute ancestry become better. That is, the points defining the two ends of the line moves in or out into space that was originally outside the original line delimited by estimates of absolute ancestry obtained with the previous estimates of allele frequencies. If we saw that the line becomes smaller with better allele frequency estimates then we can conclude that we were previously overreporting admixture with the test. If we saw that the line became larger with better estimates, we could conclude that we were previously underreporting admixture with the test.

Leaving our binary example and returning to the triangle plots assuming all possible three-way admixture models, we could be underreporting or overreporting admixture for individuals with MLEs determined to be at the vertex or along the base of the triangle if our parental allele frequencies were off. On an absolute scaling, the triangle we would be using if the estimates of absolute ancestry were perfect could be larger or smaller than the triangle we are actually using (i.e., we would have been underreporting and overreporting admixture, respectively), or even of different shape where the previous scaling was off. This could also result if the parental sample is representative of the actual parental groups, but we are just using a sample that is too small to obtain accurate allele frequency estimates.

There should be a point of maximum extension into the space outside the triangle because somewhere outside the triangle defined with estimates of allele frequencies, there is a perfect triangle the exact position of which we cannot be certain but beyond which, error estimates have no meaning (because we would have the ability to define absolute European, for instance). MLE simulations cannot measure this type of error because they assume a known absolute ancestry but as discussed in Chapter 2, Bayesian methods are able to handle the uncertainty associated with the specification of parental allele frequencies.

For such samples (real, not simulated), the distribution inside of the triangle would be a skew distribution with a seemingly lower statistical error than we have been discussing until now—the confidence contours would be compressed against an edge of the triangle. For these real samples with skew distributions, the distribution returns to a Gaussian form as we relax the boundary constraints of the triangle plot because we are able to fill in the missing portion of the picture, and the shape of the likelihood space should look more like that for a simulated sample. Likewise, we expect that as MLE values for a real sample plot farther from the base of any particular triangle plot axis, the error becomes more Gaussian in that dimension and the estimates form a plot with more typical contour sizes and shapes like those located closer to the center of the triangle such as those in Figure 5-32.

However, there is a difference between real and simulated data here. For simulated samples, the distribution should always be Gaussian since when creating simulated samples, we are assuming a definition of absolute ancestry and (with the MLE method) perfectly specified allele frequencies. For tests on real samples, we cannot know absolute ancestry and the distribution might be skew because we have used a poor parental sample representative. Therefore, if we were to compare the ''skewness'' of the distributions (or departure from Gaussian) for real samples against that obtained for simulated samples, we may be able to assess not only whether our parental representatives are good ones, and parental allele frequencies correctly specified, but how good/correctly specified they are.

Thus, for real samples, when the contours are compressed against the edge of the triangle plot, one must consider the possibility that the missing portion of these intervals outside of the triangle plot is needed to fully quantify the uncertainty in the MLE. We normally plot only inside the positive likelihood space (inside the triangle), but in the case where there is uncertainty in parental group allele frequency, we could draw the two-, five-, and 10-fold confidence intervals for an MLE constrained at the edge of the plot to extend outside of the plot. In this case one would sometimes see that the centroid of the shape extending beyond the boundaries of the triangle would correspond to an MLE of >100% ancestry for one group and negative ancestry for another. If real samples show a tendency toward these ''scrunched'' confidence contours, relative to the frequency with which they are observed in simulated samples, off-scale likelihood space centroids would suggest that the individual could be more parental than the average parental sample for one of the groups.

In that the sample may be more parental than the parental samples, and given that the purpose of admixture tests is usually to detect nonparental admixture, these results would seem to be not that relevant in our discussion of error (it is a sort of false-negative error, but false-positive error would seem far more insidious). Further, it bears stressing that the relevance of this abstract space outside the triangle plot is directly proportional to the *lack* of certainty about and quality in the reference parental samples, which as discussed is related to the size, subpopulation diversity, and history of the sample. If a good parental sample has been obtained, perhaps as judged using other criteria such as Y chromosome tests, it would seem that significantly less weight should be given to space outside the triangle compared to the space within it.

Aside from this, it is important to point out that many instances of compressed confidence contours would not be related to off-plot MLEs and deficiencies with our parental samples. We expect many cases by chance, and simulations teach us how many cases to expect. Consider that some individuals by chance will have more European alleles than the average European, not necessarily because they are more European. However recall from Chapter 3 that we cannot ascribe ancestry based on simply the number of European or other alleles—it is the likelihood of the multilocus genotype not the allelic count that matters. Nonetheless, the alleles are continuously distributed and each has its own average frequency in each population, and some individuals have statistically unusual genotypes given these frequencies.

Consider an example of the average European population where the average individual is heterozygous for 10 G/T SNPs and the average African homozygous for T/T. Thus the G allele would be the European allele. We would expect to find some Europeans that are homozygous G/G for many of these SNPs, and these would most likely be predicted to be European with more confidence, not predicted to be more European than the average European or more parental than our parental group. The rate with which compressed confidence intervals obtain due to this reason is the same for real and simulated samples.

So, in summary, the simulations don't measure the error contributed by the use of imperfect parental representatives but do allow us to recognize when there is a problem with the parental samples we are using. Also, it is useful to show that the estimates of bias and RMSE are not that much different between simulated 50%/50% mixes (see Table 5-27) and simulated samples of 100% uniform affiliation (see Table 5-26)—as we have shown—though this speaks more to the stability of our mathematical method across various regions of the likelihood space than to anthropological or historical correctness.

COMBINED SOURCES SUGGEST AN AVERAGE ERROR

Whether looking at the simulations for 10,000 individuals or looking at a large number of triangle plots, with either the 71 or the 171 marker test, for individuals with high admixture or none, the spread or precision of admixture estimates amounts to an average of a few to upper single digit percentage points, depending on the majority affiliation and their type of admixture. The 71 AIM test shows greater error than the 171 AIM test both from a sampling of triangle plots and the bias and RMSE from the simulations, and could best be described generally as in the high single digit percentage points, whereas the 171 AIM panel seems to provide an average accuracy of between a few percentage points and high single digit percentage points. To give an exact number we would need to integrate a very large number of results obtained over the entire likelihood space (every type of admixture), but if we were to approximate from a few points as we have done, we would probably be best served to average the results from these few points.

Of course, one loses information when one generalizes by taking the maximum or the minimum distance between the MLE and a point on the standard deviation line (or even the average, for that matter) as the one-and-only valid measure of uncertainty. We have been discussing the standard deviation in likelihood estimates considering an average magnitude of all vectors originating at the MLE and terminating at the confidence contour line. What matters is the distance in the direction of interest. In other words, if you're talking about an Asian admixture, it's the distance in the direction perpendicular to the zero-Asian line (and so on), which could be greater than the distance in the other directions. We also lose information when we use averages from populations to generalize about a specific individual (for example, the individual's genotype might be a fluke, and may show much larger confidence contours than the average sample). For this reason, the confidence contours are the best source of information about accuracy for any given MLE.

CHAPTER 6

APPORTIONMENT OF AUTOSOMAL DIVERSITY WITH CONTINENTAL MARKERS

THE NEED FOR POPULATION DATABASES—WORDS MEAN LESS THAN DATA

Given that we cannot go back in time and observe human populations splitting, fissioning, and migrating great distances, and fusing together again generations or even millennia later, there are some difficulties in modeling the evolutionary genetic past of the species. Who were the parental populations, with whom did they mix, where did they live, how exactly they contributed to today's populations, are all important questions that are difficult to answer definitively. As we have just discussed in Chapter 4, given the depth of our ignorance it's clear that the names we pick for the parental groups are arbitrary in nature. We have used the term European to illustrate this problem—where did their ancestors live, which populations exactly did they give rise to without the aid of admixture from other populations, and which populations were created through admixture with other parental populations? Since we can never know all the answers to these questions without a time machine, the results of an admixture test do not have much literal meaning. Rather, their value is best realized through an empirical consideration.

Take the following result as an example: 50% European, 20% West African, 20% Indigenous American, and 10% East Asian obtained from AncestryByDNA 2.5 (171 AIM test so far described). Given our choice of parental populations, our choice of population model, the type of polymorphisms we survey, various assumptions on the admixture process, and our method of analysis, our definition of European or Indigenous American could turn out to have subtly (or even not so subtly) different meaning from that for a different test or battery of markers. Our definition of European parental group is the monophyletic Paleolithic and Neolithic populations from Europe, the Middle East, and Western and Central Asia that gave rise to various modern-day populations in Europe, the Middle East, Western Asia, Central Asia, and South Asia. With this definition, as you will soon see, South Asian Indians type with significant European ancestry, but the average East Asian (i.e., Chinese) types with none. If in another iteration of such a test, we decided that South Asian Indians represented a parental population group that lived 15,000 years ago, we might see both Europeans and East Asians typing systematically with proportional South Asian admixture.

Although there is somewhat less support for such a model in many genetic studies today, it is really more a mater of for whom the test is being developed. In any case this example shows how sensitive our results are to the model we use, and how small differences in models and assumptions could have less dramatic, but nonetheless significant confounding affects for those attempting to compare results among tests or draw conclusions from absolute definitions of what the group names mean. With our definition, the results would mean one thing in terms of physical appearance, and have different implications for a reconstruction of human migrations and expansions than with another definition. For this reason, we must define what our labels mean exactly. Clearly we can partially do this with words, describing how our markers were selected, where our parental samples were obtained, detailing our evolutionary justifications for making our choices, and so on, but since we do not have a time machine we cannot possibly do a complete job with words alone because even we do not know exactly what they mean. For a forensics scientist or even a genealogist, the best way to describe what they mean to others (and themselves) is through an empirical illustration of what the labels and results mean in modern-day populations. We need to *demonstrate* what they mean. To do this we characterize performance of our models, markers and methods in a world-wide collection of samples so that when the result 50% European, 20% West African, 20% Indigenous American, and 10% East Asian is reported, we do not have to guess what historical and anthropological definitions make this result mean for modern-day individuals—we can observe what they mean in a scientific manner. Without the empirical illustration, the answers could lead to misinterpretation and would be a misrepresentation. Therefore, before we discuss how interpretation of admixture results is best accomplished, how the results can provide information to the anthropologist about human history, or the forensic scientist about what a person might look like, we must build a database of how the test describes a variety of modern individuals.

If we did a good job selecting markers, models and methods, our demonstration of what they mean will produce a body of data that makes reasonable sense in terms of what we know from the paleoarchaeological record, other molecular genetic analyses and linguistic analysis (but even if they don't, the test might still have limited forensics value if used strictly as an empirical tool, just not nearly as much value as a test with a sound anthropological base). In this chapter, we characterize the performance of the 171 marker Ancestry-ByDNA 2.5 panel of markers and methods on a worldwide collection of samples, so that we can learn how to interpret the results in a way that is scientifically, and not just anecdotally, meaningful.

TRENDS ON AN ETHNIC LEVEL: AUTOSOMAL VERSUS SEX CHROMOSOME PATTERN

The use of autosomal AIMs for biogeographical ancestry admixture analysis on the ethnic population level enables an assessment of modern-day demographic structure, which is useful to know if we are going to learn how to make generalizations about physical appearance from ancestry. It is also useful for making inference on population expansions and migrations in our recent demographic past, so it potentially has something to teach us about our history as a species. However, because autosomal loci reside on recombining regions of diploid chromosomes, their alleles provide a different look into the past than other types of markers, such as Y-chromosome and mtDNA markers. Y-chromosome and mtDNA sequences are not useful for physical profiling because they provide information on only a small fraction of an individual's ancestors (patrilineal and matrilineal lines, respectively). However, in populations, the sex chromosome markers are more powerful than autosomal markers in tracing phylogenetic histories and the inference of population level admixture. This is because the effective population size is lower for the sex chromosomes than autosomals and so the forces of genetic drift act more rapidly and decisively. It is also the case that because there is no recombination in the mtDNA or the Y-chromosome (exclusive of the pseudoautosomal regions), these loci are inherited intact and can be studied as phylogenies.

The fact that the mtDNA and the nonrecombining Y (NRY) are merely small parts of the human genome is an impediment for using them to infer individual ancestry, but not the ancestry of populations because they provide information on the ancestry of all the male and female contributors to that population. In a population, we are essentially measuring a very large number of patrilineal and matrilineal ancestries. In contrast, for an individual, there is only one patrilineal and matrilineal pathway, and whether a minor level of admixture present in that individual's pedigree would be present on a sex chromosome is a function of probability. Given the same large number of individuals measured, minor ancestry components are more likely to be reflected in the sex chromosome distribution than the distribution obtained from a well-powered set of autosomal markers, but if the minor ancestry component is real, a well-powered set of autosomal markers should detect it and measures of population constitution from NRY and mtDNA analysis should, theoretically, match that from autosomal marker analysis.

Box 6-A

Sex Chromosomes and Uniparental Markers versus Other Autosomal Chromosomes

The X, Y, and mtDNA have lower effective population sizes than the numbered chromosomes. In the previous chapters we have referred to the effective population size (N_e), which is the number of individuals of an idealized, Wright-Fisher population (no selection, no overlap in generations, equal sex ratios, population of constant size) that would be expected to allow the amount of variation that exists in a real population. Its measure accounts for stochastic nature of gene frequency changes from generation to generation caused by genetic drift, and smaller N_e's reflect populations with more drift and diversity. Using N_e we can compare the amount of genetic drift that has occurred for different populations of known histories. It also allows us to contrast populations that actually exist, which may have been shaped by selection, drift, founder effects, and so on, with those that theoretically would be expected had the only forces acting on the population been genetic drift and mutation, and thereby make inferences on what forces may have shaped modern-day population demographies.

It happens that the effective population size for autosomal chromosomes is different than for haploid chromosomes such as the Y and mtDNA and the semi-haplodiploid X chromosome due to the difference in ploidy (copy number). The average parental contribution to their offspring for autosomal chromosomes is 1 (2 chromosome 1's contributed by two parents). Since only males have Y chromosomes, assuming an equal probability of producing male and female offspring, the average parental contribution to their offspring for the Y chromosome is only 0.25 (one Y chromosome contributed by one parent for male offspring, and no Y chromosomes contributed by either parent for female offspring). Differences in social dynamics related to mate selection also exist between males and females, and males have a higher reproductive variance than females. This is another way to say that although the fecundity for the average female is the same as that for the average male, certain successful males do most of the reproducing while others are left out, but most females are involved to a similar extent.

The higher reproductive variance also impacts the Y chromosome effective size, making it smaller and contributing to more sequence drift and variation between populations. At the same time, the relative lack of reproductive variance among females stabilizes the mtDNA and X chromosome effective population sizes. Like the Y-chromosome, the mtDNA has an N_e that is 25% as large as for autosomes, and the X-chromosome has an N_e that is 75% of the autosomal N_e.

Box 6-B

Uniparental Chromosomes Compared

Y haplogroup frequency clines throughout the world are sharper than those of mtDNA due to patrilocality. A relatively large variety of uniparental sequence variants exist throughout the modern-day globe. Since the NRY and mtDNA chromosomes cannot recombine (they are carried in a haploid state), dispersive genetic forces act relatively quickly to change their sequences as populations diverge from one another. However the distribution of sex chromosome variation is not the same from sex chromosome to sex chromosome. The human sex chromosomes have migrated across the globe in a different manner than the autosomal chromosomes, most likely because males and females have contributed unequally to geographic expansion due to biological and behavioral differences. If new populations are founded by a small number of males and a large number of females, one would expect a stronger founder effect on Y-chromosome sequences than mtDNA sequences. Likewise, if gene flow between populations is due primarily to migration of females between groups, rather than males (i.e., males are more territorial than females), then again we would expect to see stronger frequency clines for Y-chromosome sequences than mtDNA sequences. These stronger clines would make for more disparity in sequence types between geographically distant populations.

In fact, this is the case: mtDNA haplogroups are spread throughout the world far more evenly than Y-chromosome haplogroups. Y-chromosome variation is greater than mtDNA variation, with stronger haplotype allele frequency clines, more disparity in sequence type, and greater genetic distance between population pairs separated by geography. From these observations, we hypothesize that indeed male and female behavior and contribution to population expansions and admixture was unequal, and that in most human societies, mtDNA sequences move relatively freely between populations but Y-chromosome sequences are not only stuck in the mud, so to speak, but possibly repellent to one another. The relative uniqueness of Y-chromosome character from population to population is called *patrilocality*. Many researchers believe that patrilocality is a molecular manifestation of male behavior, specifically territoriality and aggression. Uniqueness of mtDNA sequence character from population to population also exists, and is called *matrilocality*. In fact, the X chromosome (another sex chromosome), at least in females, is diploid, unlike the mtDNA and Y chromosome, and also shows more mobility than autosomal chromosomes. Both males and females have X chromosomes, but since females pass twice as many X chromosomes to their offspring as males, the X chromosome sequence distributions are more a function of female population histories than male.

F-statistics based on autosomal markers tend to be lower than that for Y-chromosomal markers and similar to that for mtDNA markers (reviewed in Jobling et al. 2004). In other words, there is more Y chromosome sequence variation among populations than mtDNA or autosomal sequence variation.

The enhanced variation in Y chromosome sequences is probably due to the smaller effective population size for the sex chromosomes (only one copy per carrying individual), differences in generation times between males and females, and probably most importantly, patrilocality, which has galvanized sharp geographical clines in Y chromosome variation across the globe. Boxes 6-A and 6-B review these basic population genetics concepts.

We have just explained why it is that reconstructed population histories based on Y-chromosome and mtDNA variation should theoretically agree with that based on autosomal variation, but in fact there are some differences between the information provided by the different types of markers. Since the effective population sizes are different for the Y and mtDNA chromosomes, for a given number of markers, we can oftentimes get a clearer picture of population history and structure from Y chromosome analysis than autosomal analysis—as long as we are measuring relatively large samples of individuals. The effective population sizes of uniparental markers lower than autosomal chromosomes, and uniparental chromosomes are carried in the haploid state, where they cannot recombine. As a result their sequences persist unchanged for longer periods of time than autosomal sequences. Thus, we can gain not only a clearer picture of population structure from uniparental analyses, but we also can look back farther in time. Given the lower discrimination power of autosomal markers, on average, autosomal analyses of populations typically involve larger numbers of markers than NRY analyses.

Why would we measure autosomal markers to infer population structure then? In order to measure the relationship between population structure and human phenotype (i.e., drug response trait, a physical characteristic), we need to be able to draw qualitative and quantitative connections, which requires a treatment of individual ancestry rather than in terms of population averages. Using autosomal AIMs, we are focused on measuring structure within individuals rather than populations of individuals. Knowing the ancestral composition of the populations does us no good for physical profiling if we cannot measure the ancestral composition of individuals, and for this sex chromosomes are not useful. The fact that groups of individuals comprise populations and that the autosomal AIMs therefore provide information about populations as a second opinion to the sex chromosome literature is a side benefit. We thus have the opportunity to compare the results we obtain from populations using autosomal and uniparental markers.

If measuring autosomal markers is going to allow us to infer physical characteristics, it must be the case that:

- Autosomal markers must deconvolute population ancestries the same way that sex chromosome markers do, though perhaps with a bit less resolution
- Physical characteristics must correlate with ancestry

The F_{ST} for NRY haplotypes exceeds those from the average autosomal marker, but this is not true for autosomal AIMs, for which the F_{ST} can be maximized for a given battery of markers. We would like to compare the results we obtain from an autosomal admixture test using AIMs and the NRY phylogeny, to be sure they tell us the same story of population structures, admixtures, and recent common ancestries.

WHAT DO CONTINENTAL ANCESTRY AIMs SAY ABOUT ETHNICITY?

Population admixture results obtained from an autosomal test are constrained by a specific population model but those from uniparental analyses are not. Indeed, there are many haplogroups found in various combinations in various populations and the large number of haplogroups makes the analysis multi- (not just four-) dimensional. This would imply that when comparing the results from a Y and autosomal admixture analysis, we might be comparing different chronological snapshots of human phylogeographic history (i.e., rather than partitioning the world population into four, five, or six groups we are looking at perhaps 30 or more). Nonetheless, there should be obvious and logical relationships between the Y chromosome and autosomal analyses if both are correctly reporting elements of the same story. It is thus most informative to compare the results from the two sequence types and interpret them with respect to one another. In addition, where possible, we can incorporate other types of data, such as from mtDNA HVS-1 haplogroup distributions, other nonoverlapping and non-AIM autosomal admixture panels (which we have discussed in Chapter 4), and the paleogeographic record. Before we make these comparisons, we need to discuss the type of information provided by the autosomal admixture test—what types of conclusions are supportable and what types are not.

The first question we ask is what exactly do continental AIMs tell us about within continent *substructure*, rather than just continental *ancestry*, recalling that they were not selected from the genome based on their within-continent or ethnic information content but rather, on their continental information content (i.e., West African, East Asian, European, and Indigenous American)? To answer this we will first turn to the Principle Components Analysis (PCA), and then we will break down the BGA admixture results obtained with continental AIMs in terms of ethnic groups.

Figure 6-1

Principle Coordinates Analysis of continental parental samples— Western African (pink), East Asian (green), European (blue), and Indigenous American (purple)—using genotypes for the 171 AIM panel discussed in the text. Along the x- and y-axes the eigenvalue coordinates for each sample (spot) are shown for the two eigenvectors that explain most of the variation in genotypes among the groups. We observe significantly greater variation between populations than within populations.

From the BGA analysis of the 171 AIM admixture test previously presented in Chapter 5, we saw that the clusters of parental samples were distinct, in that parental samples of one ancestry plot in positions closer to others of the same ancestry than to samples of other ancestry (see Figures 5-2, 5-4, 5-6, 5-8). Using PCA with these same samples and AIM genotypes, we see a strong clustering effect (see Figure 6-1). The lack of overlap between the clusters indicates that the 171 AIMs provide excellent discrimination power, at least in terms of the four-way population model used in their selection. When we look at ethnicity rather than continental ancestry, we see that these 171 AIMs explain less variation, again as we might expect, since in this case the AIMs were not selected from the genome for ethnicity discrimination power and also because there is less genetic distance between these groups than continental groups.

Figure 6-2 shows an example for European ethnicity. There is less discrete clustering, but what is interesting is that the continental markers explain some of the ethnic variation rather than none at all. Greeks and Middle Eastern individuals plot predominantly below the horizontal line, representing the midpoint along the second eigenvector scale, whereas European Americans, Northern Europeans, Italians, and Siberians plot above it. Had we performed this analysis using AIMs selected from the genome based on ethnic rather than continental discrimination power, we would expect much tighter clustering, but the PCA plot using the 171 continental AIMs shows that simply measuring ancestry on a continental level provides information about ancestry on the subcontinental level. Specifically, there appear to exist significant differences within Europe with respect to the distribution of European, West African, East Asian, and Indigenous American alleles. Whether this is due to bona fide admixture or differences in genetic drift between populations is not clear, but as long as physical characteristics are related to the differences (i.e., the subpopulation structure is anthropometrically relevant) and these relationships can be appreciated empirically, the answer is not required for the purposes of physical profiling.

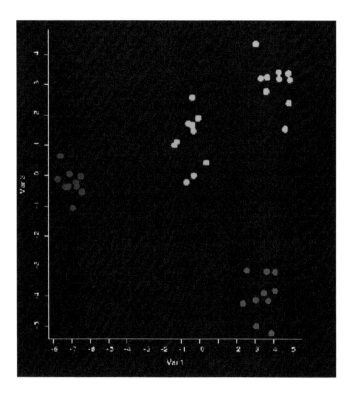

Placing aside the relationships for later, we are focused solely on identifying populations and subpopulations of origin for individuals, and for this goal the continental AIMs are of some (though perhaps not optimal) utility. From an anthropological perspective, rather than an empirical forensics one, it is possible that these differences tell us something about the anthropological origins and/or interactions of various ethnic groups. If we can infer likely subpopulation of origin using continental AIMs, we can then investigate whether ancestry profiles allow us to infer elements of physical appearance that correlate with elements of population structure.

THE SIGNIFICANCE OF FRACTIONAL AFFILIATION RESULTS ON A POPULATION LEVEL

Table 6-1 is discussed more fully later in this chapter, but is introduced here. This table shows that with the 171 AIM continental admixture test, most ethnic groups reveal characteristic admixture results, as might be expected since each likely has its own geopolitical and anthropological history, and since we know from the immediately preceding that alleles for our AIMs are unevenly distributed within subcontinental regions. What is interesting about these fractional affiliations is that they are in terms of continental populations, based on the population model used to build the test. Interpreting distributions of ethnic characteristics defined by events that occurred before and during their genesis is potentially instructive as a means of reconstructing and disentangling their potentially complex origins, and potentially instructive for relating continental BGA admixture profiles to physical appearance.

Of course, as we have described elsewhere in this text, a fractional affiliation (whether for a person, or a collection of people) can be due to many factors of which admixture on a population scale is only one. Other possibilities include intermediate allele frequencies, which are expected to connect incompletely differentiated diaspora from common ancestral stock (which is to a certain extent a function of the population model selected), and statistical artifact (also a function of the population model selected). What we want to know when assessing continental ancestry on a subpopulation level is whether there

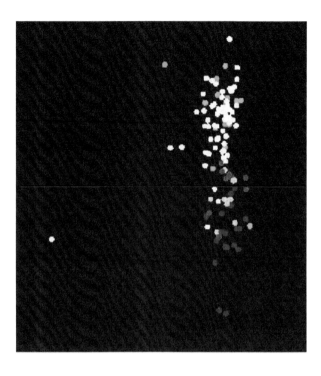

Figure 6-2

Principle Coordinates Analysis of European ethnic samples—Greek (gold), idle Eastern (pink), Northern European (purple), Italian (light blue), Iberian (green), U.S. Caucasian (white)—using genotypes for the 171 AIM panel discussed in the text. Along the x- and y-axes the eigenvalue coordinates for each sample (spot) are shown for the two eigenvectors that explain most of the variation in genotypes among the groups. We observe significantly greater variation between populations than within populations.

Table 6-1

Average continental admixture results (standard deviations) for 25 populations and six metapopulations (171 AIMs, three-way/four-way MLE algorithm).

Ethnic Group (n)	ADMIXMAP				Maximum likelihood (3way/4way algorithm)			
	EUR (s.d.)	SAII AFR (s.d.)	E. Asian (s.d.)	NAM (s.d.)	EUR (s.d.)	SAII AFR (s.d.)	E. Asian (s.d.)	NAM (s.d.)
African American (136)	14.3% (13.3%)	79.6% (14.0%)	2.8% (6.0%)	3.3% (5.1%)	18.2% (15.5%)	77.3% (16.1%)	1.8% (5.3%)	2.8% (3.5%)
North African (7)	77.4% (5.8%)	15.0% (7.3%)	5.6% (5.4%)	2% (3.4%)	85.3% (7.5%)	13.1% (7.3%)	0.6% (0.2%)	1.0% (0.2%)
Puerto Rican (64)	55.0% (20.7%)	32.6% (24.6%)	3.6% (5.4%)	8.8% (8.3%)	63.8% (22.9%)	31.6% (23.8%)	1.1% (1.0%)	3.3% (4.5%)
North Euro subset 1 (10)	97.0% (3.6%)	1.0% (2.1%)	1.9% (3.0%)	0.4% (1.4%)	98.1% (0.6%)	0.8% (0.4%)	0.5% (0.2%)	0.6% (0.2%)
Irish (17)	96.4% (4.3%)	0.7% (2.1%)	1.2% (2.7%)	1.7% (4.1%)	97.8% (0.7%)	0.8% (0.4%)	0.5% (0.2%)	0.9% (0.4%)
Icelandic (12)	93.8% (5.5%)	1.2%*** (2.2%)	0.8% (1.4%)	4.25% (5.0%)	96.6% (1.4%)	2.2% (1.1%)	0.4% (0.2%)	0.7% (0.4%)
Greek (18)	90.4% (4.0%)	4.8% (4.2%)	1.7% (5.3%)	4.7% (4.8%)	95.2% (3.8%)	2.8% (3.9%)	0.5% (0.2%)	1.6% (1.2%)
Iberians (9)	78.8% (21.0%)	6.6% (7.1%)	4.0% (7.6%)	10.7% (16.7%)	85.6% (17.1%)	5.5% (7.4%)	1.5% (1.9%)	7.4% (15.0%)
Basque (10)	93.0% (5.2%)	2.3% (3.6%)	0.8% (2.5%)	3.9% (4.1%)	96.1% (1.6%)	2.6% (1.7%)	0.5% (0.2%)	0.7% (0.4%)
Italian (12)	86.8% (8.9%)	3.2% (4.8%)	2.7% (5.5%)	7.3% (5.9%)	96.5% (1.7%)	1.3% (0.7%)	0.6% (0.2%)	1.7% (1.2%)
Turkish (40)	85.3% (5.4%)	2.3% (3.2%)	7.3% (6.7%)	5.1% (6%)	96.0% (1.6%)	1.4% (1.0%)	0.9% (0.6%)	1.7% (1.8%)
Ashkenazi Jewish (10)	86.8% (5.8%)	4.7% (3.9%)	2.0% (4.9%)	6.6% (3.6%)	94.7% (2.6%)	3.4% (2.2%)	0.6% (0.4%)	1.2% (0.9%)
Middle East vers. I (9)	88.1% (9.7%)	2.8% (5.6%)	4.8% (7.3%)	4.2% (5.1%)	94.9% (5.6%)	2.1% (3.8%)	1.0% (1.2%)	2.0% (1.4%)
Middle East vers. II (11)	82.0% (11.0%)	10.8% (8.9%)	4.5% (7.5%)	2.6% (6.3%)	90.0% (7.0%)	6.9% (6.3%)	1.0% (0.8%)	2.1% (1.9%)
South Asian Indian (56)	58.9% (8.9%)	5.1% (4.7%)	26.9% (10.7%)	9.1% (8.8%)	72.3% (11.2%)	4.5% (4.2%)	14.1% (11.5%)	9.1% (7.2%)
Patel (8)	65.5% (7.7%)	4.0% (6.5%)	25% (10.3%)	5.5% (9.9%)				
Chinese (10)	0.7% (0.9%)	0% (0%)	98.0% (2.4%)	1.3% (2.5%)	3.0% (1.5%)	0.6% (0.3%)	94.9% (1.8%)	1.6% (0.8%)

Population								
Japanese (10)	1.1% (2.6%)	0.4% (1.8%)	95.3% (4.2%)	3.2% (4.0%)	3.1% (1.4%)	0.5% (0.2%)	94.1% (2.1%)	2.3% (1.6%)
Atayal (10)	0.5% (1.6%)	0% (0%)	97.6% (4.2%)	1.9% (4.2%)	2.7% (1.2%)	0.8% (0.3%)	94.5% (1.9%)	2.1% (1.9%)
SouthEast Asian (11)	8.0% (11.1%)	3.6% (7.3%)	82.2% (14.8%)	6.3% (6.7%)	14.2% (9.4%)	2.7% (3.4%)	81.0% (11.5%)	2.2% (1.7%)
Pacific Islander (14)	28.8% (17.6%)	5.6% (5.8%)	57.2% (20.4%)	14.6% (13.0%)				
Aboriginal Australian (8)	63.5% (10.5%)	1.0% (2.4%)	25.2% (8.7%)	10.3% (12.7%)				
American Indian* (223)	41.9% (35.8%)	3.7% (12.4%)	6.7% (8.6%)	47.6% (33.8%)	44.8% (34.5%)	3.0% (10.9%)	3.9% (6.3%)	48.2% (33.2%)
American Indian** (170)	28.6% (27.6%)	2.2% (5.9%)	8.2% (9.2%)	61.1% (27.0%)	33.8% (27.6%)	1.9% (4.5%)	4.8% (6.8%)	59.5% (27.6%)
Mexican (60)	43.2% (19.3%)	5.6% (7.2%)	4.4% (9.3%)	46.8% (18.1%)	50.0% (22.6%)	5.0% (6.3%)	2.1% (2.6%)	42.9% (22.2%)
POPULATION COMBINATIONS								
Combined North Euro/Irish (27)	96.5% (4.0%)	0.81% (2.0%)	1.5% (2.8%)	1.2% (3.8%)	97.9% (0.7%)	0.8% (0.4%)	0.5% (0.2%)	0.8% (0.3%)
Mediterranean (Greek/Italian) (30)	89.0% (6.0%)	4.2% (4.4%)	2.1% (5.4%)	5.7% (5.2%)	95.7% (3.0%)	2.2% (2.6%)	0.5% (0.2%)	1.6% (1.2%)
SE European (Greek/Italian/Turks) (70)	86.9% (5.6%)	3.6% (3.8%)	6.7% (5.6%)	7.8% (5.2%)	95.9% (2.2%)	1.7% (1.7%)	0.8% (0.4%)	1.7% (1.5%)
Middle Eastern (Mid East I + II) (20)	84.8% (10.7%)	7.2% (8.7%)	4.4% (7.7%)	3.7% (5.1%)	92.2% (6.4%)	4.7% (5.2%)	1.0% (1.0%)	2.1% (1.7%)
Ashkenazi Jewish + Middle Eastern (30)	85.4% (8.9%)	6.4% (6.2%)	3.8% (6.6%)	4.4% (50%)	93.0% (5.1%)	4.3% (4.2%)	0.9% (0.8%)	1.8% (1.4%)
European American (207)	90.5% (10.2%)	3.0% (5.8%)	2.8% (4.9%)	3.8% (6.1%)	96.6% (2.2%)	1.6% (1.5%)	0.6% (0.3%)	1.2% (0.9%)

Note: Using 175 AIM battery for continental ancestry admixture. Bold print with dark blue background indicates the primary ancestry.
Light blue background indicates levels of nonprimary admixture that is statistically significant compared to the level in the appropriate simulated parental population using Fisher exact test. Fisher exact test p-values were computed in the following way: percentage m of admixture was multiplied by the sample size (n), and this value combined with the corresponding value for 1-m (non-group admixture) was compared against the m*n and (1 − m)*n for the simulated parental populations. The reason m is multiplied by the sample size is to penalize low admixture estimates contributed by only a small number of samples when the sample size for a group is relatively low.
*Includes individuals from U.S. government-recognized tribes (Sioux, Cheyenne, Cherokee, Arapaho) as well as unrecognized tribes, without regard to blood percentage.
**Includes individuals from U.S. government-recognized tribes only (Sioux, Cheyenne, Cherokee, Arapaho), without regard to blood percentage.

is a systematic or characteristic admixture profile for the population, where characteristic implies that it is statistically different from the relevant parental group (i.e., do Northern and Mediterranean Europeans reveal systematically and significantly different amounts of non-European ancestry?). Such differences will tend to be observed across various studies, using different parental and test samples, different marker sets, and using a variety of algorithms and altorithm types. If this is the case, then we can then hypothesize on how this difference came to be and more importantly for our purposes here, we may be able to empirically use an ancestry profile to infer subpopulation affiliation and thus to infer elements of physical appearance that are different between the subpopulations. We take the average level of admixture for all individuals of a population sample as the characteristic profile for that population, and whether or not it is a meaningful tool for reconstructing certain aspects of population history and physical appearance is best accomplished from empirical observation as we describe later. Here, we will focus on whether the ancestry differences are meaningful. For example, are the ancestry percentages reliable enough indicators of structure to provide information about physical characteristics that correlate with that structure? There are two issues:

■ The statistical certainty with which an element of the measured population structure (admixture or affiliation type) is characteristic of an ethnic group
■ Whether the affiliation type provides information on physical appearance

Not concerning ourselves with the latter for the moment (it is treated later), let us look at the former in more detail. As described in Chapter 5, we can measure precision on the level of the individual by comparing the values with those obtained from simulated parental samples. For any particular individual then, we can compare their values with the threshold percentage required to conclude with confidence that the affiliations are real (and not statistical artifacts)—the threshold percentage levels being unique for each type of admixture, on each type of majority ancestry background, and determined from simulations. We will term this threshold percentage simply the *threshold* from here forward.

As an example of how the threshold can be used to evaluate the usefulness of a particular profile type for understanding population history or making inferences on subpopulation affiliation, consider a hypothetical example: an average 3% West African fractional affiliation for a certain East Asian ethnic group. We know that East Asians, in general, do not show systematic West African admixture, and so our question is how meaningful is the 3%? Based on the simulations, we know that the 3% African would be a significant indication of structure on a population level, since fewer than 5% of simulated

East Asian samples exhibit levels as high as this (actually, any levels up to 2% fractional African affiliation; from Table 5-23A of Chapter 5). In other words, the *average* individual from this ethnic population exhibits a level of African affiliation that is statistically significant because it is observed for fewer than 5% of the simulated homogeneous East Asians parental samples—indeed, the average simulated East Asian shows 0.0008 (0.08%) West African affiliation, which is statistically different than 3% ($p < 0.05$; Table 5-23A, but also see Table 5-32 in Chapter 5). Thus, an average fractional West African level of 3% for this East Asian ethnic subgroup indicates a genetically structural difference between it and the East Asian parental group in general, which by Wahlund's theorem is expected to be the average of all ethnic groups derived from the East Asian parental group.

Although useful in confirming the existence of population structure, the rigid use of threshold values can be misleading in terms of literal meaning,. There are other sources of error than statistical error caused by the continuous nature of the AIM allele frequency distributions; there is a nomenclature problem as we have discussed, and some of our assumptions may not be correct. Low levels of admixture on a subpopulation scale can be difficult to interpret, even if we use advanced statistical methods that are capable of handling uncertainty related to our assumptions of genetic drift, parental allele frequency sampling error, and the like.

On the other hand, what is significant on the level of the population is not necessarily significant on the level of the individual. For example, if we see an 0.5% West African result for 1,000 East Asians of a particular ethnicity, and the result is greater than its variance (indicating that it is seen reliably and relatively often) and not seen for other East Asian ethnicities, then depending on the sample sizes involved, it may be possible to determine using standard statistical techniques (Fisher exact test, for instance, or logistic regression) that the departure from expectations is statistically significant. If it were, it would seem to be a real result and not due to statistical bias even though 0.5% for this type of admixture is below the threshold determined in simulated individuals (see Tables 5-23A and 5-32). In this case, 0.5% West African fractional affiliation for our hypothetical ethnic group may be an indication of admixture or other structure on a *population* level even though 0.5% would not be a significant indication of fractional affiliation on an *individual* level.

Put another way, if one of 1,600 individuals of a 160,000 East Asian subgroup sample type with 8% West African genomic ancestry, we would read a low average of 0.5% admixture level for that subgroup (across all the 160,000 individuals), but in this sample of 160,000 individuals we would have observed 100 samples for which the West African fractional affiliation of 8% was not only above the African threshold (<2%), but well above it. Thus, the fractional affiliation for each of these individuals would be highly significant and their presence in the population would indicate an element of significant population structure. A 0.5% fractional affiliation level could be obtained from a very low level of admixture that is commonly found in the group, or a very high level of admixture found more rarely, or any combination of level and frequency in between.

The lower the standard deviation of affiliation percentages for a given type of admixture relative to the level observed, the more uniform the exhibition of that level of admixture, the more characteristic that level of admixture for the group, and the more likely it becomes that the admixture serves as an effective feature for subgroup discrimination. Of course, the same thing is true the greater the overall level of admixture, and as we will soon see, the greater the overall level the higher the ratio of overall admixture/standard deviation tends to be. Thus, the ratio of these values gives us an indication of the quality or reliability of the fractional affiliation result as a feature of the population subgroup, though this method would not be as reliable a way to measure whether the minor affiliation is significant as proper statistical methods such as those based on contingency tables or better yet, logistic regression.

RECONSTRUCTING HUMAN HISTORIES FROM AUTOSOMAL ADMIXTURE RESULTS

Given the complexities of human migration patterns, the various sources of a fractional affiliation result, and the statistical nature of ancestry admixture determination, making inferences on population histories from autosomal admixture tests, on their own, is tricky and fraught with imprecision. However it would be naive to think that autosomal admixture results on the level of populations do not carry any of this information. We need to

1. Determine what elements of the BGA profiles for populations are significant, on their own, and of these, we then need to
2. Evaluate which are supported by other types of data. We can do a much better job if we have access to other types of information or data which would help support some interpretations and refute others.

We have described how to accomplish (1), and for (2), in many cases, we can turn to the published literature on Y and mtDNA sequence diversity, archaeology, the historical record, and linguistics, and synthesize what they all tell us about human migration events. This information could help support or refute conclusions drawn from our autosomal admixture panel. What we want to confirm is the notion that the fractional result for a population will be useful for the inference of population history, and thus, is reflective of recent population admixture or stratification events rather than statistical noise. As we have alluded to already, although the Y and mtDNA chromosomes are only a fraction of the human genome, and tell only part of the anthropological story for a given individual, the part that is told on the level of the population is generally a more detailed representation of that story because they read history from chromosomes that are inherited from parents as single nonrecombining units of DNA. These qualities mean that their haploid sequences are preserved longer than autosomal haplotypes, and are therefore especially powerful tools for deconvoluting complex ancestries and reconstructing human expansions and admixture, as an alternative or companion to fossil-based records of prehistory.

If the low levels of admixture observed with an autosomal admixture test is a real indication of population structure difference between an ethnic group and the relevant parental group, others looking at Y-chromosome and mtDNA sequences would have been expected to have seen results that are suggestive of the same thing. For example, if the 3% African admixture seen for our hypothetical East Asian ethnic group is a real indication of population structure as the statistical considerations suggest it is, others working with Y-chromosome and/or mtDNA haplotypes would have found a similar frequency of African sequences in East Asia. In evaluating whether previous sex chromosome analyses support autosomal admixture results, we would need to take the average between the Y and mtDNA results, not either one on their own. This is because the assumption that Y and mtDNA chromosomes are inherited in populations over time precisely as autosomal chromosomes is not always correct, as we have already discussed. In many instances, Y-chromosome haplogroup diversity is not equal to mtDNA diversity, due to sex-biased admixture or imbalances in sex ratios for founding populations.

Such inequities undoubtedly result from the physical and biological differences between males and females, the different social roles males and females have played in various populations over human history, and differences in the sexual composition of some population founders caused by statistical sampling in their founding (founder effects). We would also need to make full use of the phylogenetic relationships among the sex chromosomes (both existing and perhaps no longer existing), as well as their geographical distribution, likely origin, and chronology.

SHARED RECENT ANCESTRY VERSUS ADMIXTURE: WHAT DOES FRACTIONAL CONTINENTAL AFFILIATION FOR AN ETHNIC GROUP MEAN?

Imagine a population in Central Asia that might have lived tens of thousands of years ago (we'll call it the founding population). If segments of this population founded others further to the west, south, and east (derived populations), and if the segments were small enough or subsequent genetic drift strong enough after a given period of time, we would expect the genetic character of the derived populations to have become dissimilar to the founding population. Depending on how long the period of time, and exactly how we define dissimilar, the process of becoming dissimilar may or may not have run to completion. If it has not, then each of the derived populations may show some residual affiliation with the founder population, and therefore between each other.

Assume the founder population is ABC, and one derived population takes on character that is mainly B with a little bit of C (which we denote as Bc) and the other C with a little bit of B (Cb). ABC lived many thousands of years ago and is no longer an existing population, but a geneticist reading the two modern-day derived populations would see a pattern that suggested Bc had mixed with Cb when in fact nothing of the sort has occurred. Their shared ancestry affiliations are due to a relatively recent common ancestor (ABC), not due to admixture.

Fortunately we have a mechanism potentially capable of telling the difference between these two mechanisms. Recent admixture creates linkage disequilibrium (LD), whereas with affiliation due to common recent ancestry, the average extent of LD is much lower throughout the genome. Thus, measuring the LD in the genome allows us to date the ancestral affiliation, and to determine whether the population structure observed with the genomic ancestry test was created by historical population admixture as opposed to shared histories, we would simply date the admixture observed in the genome by measuring LD. The date of the admixture may have a critical impact on the usefulness of the admixture for physical profiling purposes. If the fractional affiliation results in neighboring populations correlated with geographical distance (in other words we see a geographical pattern to the result), this result might seem to be an indication that it is real, and it may be, but this geographical sensibility of the result does not mean we can know how ancient or recent the affiliation is without these special methods. Further, a real result may not necessarily leave a pattern, depending on how it originated.

Unfortunately, the methods for dating admixture, which rely on estimating the sum-of-intensities parameter described in Chapter 5 (see Figure 5-33) are still being perfected and good estimates may require that very large collections of markers be typed in each individual. However, even if we do not or cannot measure this parameter accurately, this does not render the information provided by the fractional affiliation result useless. For an anthropologist, it is better to know that certain populations are related to one another even if it is not possible to date the admixture and know the mechanism for its creation, than to not know that a relationship exists. In other words, the relationship may still be meaningful even if it's not possible to establish its origin. In many cases, we can at least delimit a subset of possibilities; we can formulate reasonable guesses, using other types of information such as the Y or mtDNA sequence maps, historical records, and/or the paleoanthropological record. Inference of physical appearance from databases of individuals typed for ancestry would reveal the phenotypic spread associated with our lack of chronological and geographical precision, but would eliminate phenotypes associated with irrelevant parts of the human family bush as we will discuss later in Chapter 8.

Empirical methods for the inference of physical appearance thus do not require that we know which group begat which group, or whether similarity in ancestry profiles for populations is due to one mechanism or the other—only that we are able to reliably and accurately identify the relationship among populations and that there is some correlation with certain phenotypes. Here, we simply ask what it is that the data shows, leaving the explanation for why it shows what it shows for a later time.

We now ask how meaningful continental admixture results are on the level of subpopulations, specifically between them in terms of shared ancestry and admixture. If the connections between populations revealed with a continental autosomal admixture test are statistically meaningful in and of themselves, and supported by other types of genetics and nongenetics analysis (archeology, linguistics, historical accounts, etc.), then the results carry more anthropological validity. If there is more anthropological validity, there is more relevance to the goal of inferring phenotypes; that is, the spread of phenotypes associated with a given ancestry admixture will produce a pattern that empirical methods for inferring physical appearance can operate from. (The empirical method is described in Chapter 8.)

Thus, we turn to what it is that the data for subpopulations (ethnic groups) shows, and attempt to support or refute the significant findings using Y-haplogroup, mtDNA haplogroup, and other autosomal admixture studies on world populations. What we are left with in the end are the BGA characteristics for subpopulations that are most likely to be anthropologically meaningful, and therefore useful for the inference of likely physical appearance.

RETURNING BRIEFLY TO THE NAMING PROBLEM—RELEVANCE FOR INTERPRETING THE APPORTIONMENT OF AUTOSOMAL DIVERSITY

Let us also stress a very important point here at the outset, before we get into the details of geographical pattern in population admixture. Recall that modern populations are the result of both branching events, as two populations emerge from a single one (e.g., when Asians split into Asians and Indigenous Americans) and merging events (e.g., when Europeans and Indigenous Americans merged to form the bulk of current Latin American populations). Depending on the time-depth one looks at and the age of the parental population one desires to make inferences about, one can use different labels or names to denote the ancestry of individuals. For example, at a time depth of, say, 100,000 years, the ancestry of every living human is 100% African. At a time depth of 75,000 years we would likely find three or four different African populations. At a time depth of about 50,000 years we would probably find these African populations along with one non-African population expanding across South Asia (and perhaps elsewhere such as Western and Central Asia) and into Australia. Non-Africans in western and Central Asia contributed to European, Western Eurasian, East Asian, and South Asian populations, and so on, but not necessarily at the same time and to the same magnitude.

A knowledge of the age our markers and methods are capable of reaching back to can be called time-depth, and a knowledge of time-depth is critical for interpreting the results of an admixture test. Returning to our example of the two populations, Bc and Cb that derived from the founder ABC that no longer exists, we would have learned what we know about the allele frequencies in group B and C from sampling within the modern-day Bc and Cb populations. What we call B and C—rather, the specific names we give them—is therefore colored by our modern-day perceptions. If Bc is found on one continent and Cb another, and the name we give to B is strongly associated with that continent such that it connotes certain continent-specific physical traits, politics, religion, and so on, then it may be puzzling to the layperson how the Cb population on one population has the b part from another continent as part of their ancestry.

POPULATION RESULTS

There is a reason we have gone through this example. Returning to Table 6-1, you will note from the MLE algorithm results that we see many geographical and ancestral patterns. For example, Southeastern Europeans, Ashkenazi Jews, Middle Easterners, Southern Europeans, and South Asians have Indigenous American affiliation, and though these levels are lower using a second method, in this case ADMIXMAP, there is still a general trend of increasing Indigenous American affiliation as one proceeds from northwest Europe to southeast Asia (see Figure 6-3 and inset, Figure 6-4). Though the levels were not determined to be statistically significant for most of these populations, as determined through comparison to simulation sample results using a contingency analysis, there is nonetheless a geographical trending to the results, and in South Asia where the results are highest, the levels are indeed statistically significant.

The trending and the observation of significance at the end of the trend are highly suggestive of bona fide population structure. We will discuss this pattern in great detail later in this chapter, but here we want to highlight that if this pattern is indeed indicative of population structure contributed by a common source for each of these populations (say, in Central or Western Asia) then it illustrates another instance of the nomenclature problem. The term Indigenous American is obviously a continent-specific descriptor, and using it to describe part of a Continental European's or South Asian's heritage is semantically awkward. Most people are likely to conclude on first glance that the Indigenous American ancestry in a South Asian is a ridiculous proposition, and that if it existed it must be due to recent admixture. But this is not necessarily the case—it is affiliation, and in this case the affiliation most likely is a by-product of prehistorical population divergences and migrations, the selection of population model with which we measure affiliations among the branches, and the choice of names we have employed.

As we will discuss in detail later, it is highly unlikely that Indigenous Americans back-migrated into the Old World to mix with Asians and Europeans (although this is certainly possible). Rather, if this pattern is legitimate, it would suggest that Indigenous Americans were more likely derived in part from the same ancient populations as South Asians (in this case, those harboring mtDNA haplogroup X- or Y-chromosome haplogroup Q), as well as some European and Middle Eastern populations. These source populations likely lived in Western/Central Asia or the Middle East 20,000 years ago, and gave rise to multiple offshoots. So, to say that mtDNA haplogroup X is Indigenous American or that Greeks score with Indigenous American ancestry is inaccurate as proven by the fact that there is no known back-migration known to have occurred across the Bering Strait is an oversimplification. Rather, the results show *affiliation* with Indigenous Americans, the interpretation of which is somewhat colored by our choice of names, which as we have discussed, is arbitrary (see Chapter 4).

In this case, we choose the term Indigenous American based on the fact that Indigenous Americans were used as the parental samples, but we could just as easily call it Western/Central Asian genetic subgroup I, and call European Western/Central Asian subgroup II, and so on, but this would convey less immediate information to the reader. The point of this section is that while you are analyzing the population results in the pages that follow, you should avoid drawing conclusions based on the names of the elements of population structure we employ and instead base your analysis on the statistical considerations we have discussed, such as whether or not a given affiliation for a population is statistically different than that obtained for simulated samples (see Chapter 5), or whether a lack of geographical trending might suggest an affiliation is due to genetic drift and/or intermediate allele frequencies, for example.

A SAMPLING OF ETHNICITIES USING THE 171 AIM PANEL

Table 6-1 shows the results from application of the 171 AIM genomic ancestry test on various subpopulations, which we will refer to as ethnic groups. Admixture was estimated using the basic MLE algorithm and the ADMIXMAP program, both described in Chapter 3. The latter uses a more complex statistical method to accommodate parameter uncertainty (see Box 6-C). This same data is presented over a global map in Figure 6-3 (MLE method) and Figure 6-4 (ADMIXMAP), and in tetrahedron plots (MLE method for each) to enable visual appreciation of variability and direct comparison between populations (see Figures 6-5 to 6-11).

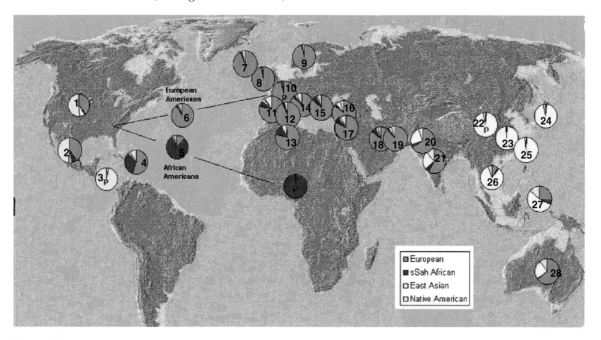

Figure 6-3

Distribution of average BGA proportions in different global populations using the 171 AIM panel and the MLE method (three-way/four-way algorithm). P indicates an ancestral population sample used for all analyses. 1. Indigenous American (includes both recognized and unrecognized tribes); 2. Indigenous American (without individuals from recognized tribes); 3. Indigenous American (ancestral IA); 4. Puerto Ricans; 5. African American; 6. European American; 7. Icelandic; 8. Irish; 9. Northern European; 10. European (ancestral); 11. Iberian; 12. Basque; 13. North African; 14. Italian; 15. Greek; 16. Turkish; 17. Ashkenazi Jews; 18. Middle Eastern (version 1); 19. Middle Eastern (version 2); 20. Indians (Patel); 21. South Asian (South Indians); 22. East Asian (ancestral); 23. Chinese; 24. Japanese; 25. Atayal; 26. South East Asian; 27. Pacific Islander; 28. Australian Aboriginal; 29. Sub-Saharan African (ancestral). Most of this data is derived from Table 6-1.

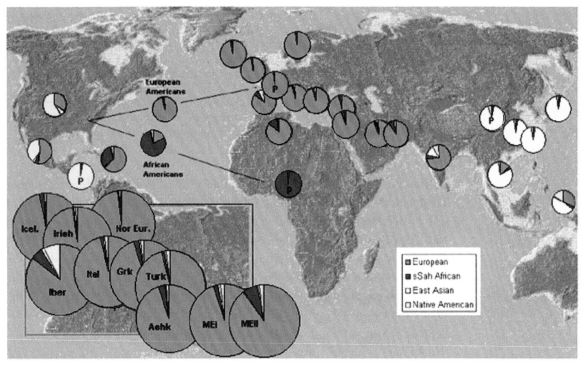

Figure 6-4

Distribution of average BGA proportions in different global populations using the 171 AIM panel and the ADMIXMAP method. Populations are the same as indicated in Figure 6-3, and indicated by geographical origin, though data for some is missing. P indicates an ancestral population sample used for all analyses.

Box 6-C

Why do we present results from two algorithms? The MLE method of individual admixture estimation assumes that the allele frequencies are accurately specified, there is no linkage between markers, each individual's parent was of the same ancestry proportions; and that there has been no genetic drift between populations from which we have sampled and their parental populations. ADMIXMAP on the other hand accommodates uncertainty in some of these variables. A pseudo four-way (MLE method) was used for each individual in Table 6-1 and in Figure 6-3, and a four-way population model was used for each individual of Figure 6-4 (ADMIXMAP). Recall that a true four-dimensional method queries all possible values of West African, European, East Asian, and Indigenous American admixture whereas the pseudo four-way algorithm determines the most likely three-way model. If it is clearly the most likely, it tests all possible three-way combinations with respect to this model and if another is similarly likely it will adopt the four-dimensional approach.

Why did we present the results from both of these different styles of estimating ancestry? Though more sophisticated, ADMIXMAP is a fairly new program, most of the papers describing it from Paul McKeigue's group being published in the past five years, whereas the basic MLE algorithm is relatively simple and has a much longer track record of battle testing. In terms of an analogy, ADMIXMAP is like a new Cadillac automobile, with air conditioning, lots of electronic gadgets and controls, and the basic MLE algorithm is like a model-T, with none of these things. We haven't had the pleasure of using the new Cadillac for very long to fully appreciate its reliability, and though the model-T is not capable of doing some of the things we would like, we keep it in the garage, so to speak, just in case.

Though Figures 6-3 and 6-4 show similar results, the subtle difference in non-European ancestry for continental Europeans raises the interesting question of which algorithm is producing the better result. The results obtained with ADMIXMAP seem to be more consistent with what a layperson might expect, but this does not make this the better or more correct result. If the ADMIXMAP program results are more accurate, the question arises as to why the same basic pattern is observed using the MLE program (increasing Indigenous American, West African, and East Asian affiliation moving from Northwestern Europe to Southeastern Europe and into Southeastern Asia). Indeed, we can refer back to Figure 4-1B, which used a different set of SNP markers and a third MLE type program and see remnants of this same pattern (note the ''Indigenous American''—about 6% and East Asian admixture for the South Asian Indian populations in this figure). Thus, we have multiple sources—different parental samples, markers, algorithms showing a similar pattern and it could be argued thus that this precludes the trivial explanation of statistical noise. Yet the levels are slightly different between Figure 6-3 and 6-4. One might argue that the basic MLE algorithm is more sensitive at detecting the global population structure for Europeans that both algorithms seem to be reporting at different levels. On the other hand, we might expect the more complex ADMIXMAP algorithm, which uses posterior probability distributions, to produce better estimates. In fact, ADMIXMAP also considers admixture in terms of blocks, and it does not assume that each marker is of equivalent information content. The marker spacing and haplotype block structure is considered—some markers are relatively close to one another and found in ancestral blocks and are therefore of redundant information content compared to markers found far apart from one another that report the same ancestral affiliation (irrespective of δ value). Thus, if the structure that the basic MLE algorithm is detecting is real, ADMIXMAP should detect it too but perhaps underemphasize the structural differences among populations (depending on the marker set). Since some of the markers part of the 171 AIM panel are within 0.1cM of one another, ADMIXMAP would seem to be the better choice of algorithms, at least in theory.

In addition, ADMIXMAP is expected to be better at estimating the true parental allele frequency from a survey of modern-day descendents (their branches). One might argue that with the basic MLE program, the parental samples are derived from Northwestern Europe, not Southeastern Europe or the Middle East, so artificial ancestry admixture would most likely obtain for these latter populations due to intermediate allele frequencies with respect to the chosen models. If the Middle Eastern group was the parental group, perhaps the Irish and Northern European would show the most admixture (though, not likely—see our discussion at the end of Chapter 5). The problem with this reasoning is that hypothesis-free methods, which do not require a definition of parental populations *a priori*, show the same results we report here. In other words, with STRUCTURE and these same markers, the Northern Europeans are shown to be of virtually homogeneous European affiliation and the Middle Eastern populations with substantial non-European affiliation.

Although ADMIXMAP is expected to provide better estimates, there are other issues to consider, such as the early development stage of the complex ADMIXMAP software, the possibility of as-yet undiscovered bugs, and so on (see Box 6-C). It is probably best to consider the truth to lie somewhere between the results provided by the basic MLE and ADMIXMAP algorithms, as suggested by some of the other SNP results presented earlier in Chapter 4 (Figure 4-1B) and until more work is done we should consider results from both methods in our discussion.

The results of Table 6-1 and Figures 6-3 and 6-4 (and those using other SNPs, methods and parental samples in Figure 4-1B) seem to be largely influenced by phylogeograpy but also by modern population structure and recent historical events.

We observe several broad characteristics:

1. In terms of the majority affiliation, the results are generally within anthropological and historical expectations, which can most easily be visualized from Figures 6-3 and 6-4; populations united by anthropometric trait character (such as skin color, epicanthic eye folds, hair color variations and textures, etc.) and/or language base show more similarity in ancestry profiles than those that are not. An example of this is the similarity in BGA profiles for North Africans, Northern Europeans, Greeks, Middle Eastern, and South Asian Indians, each of which speak languages with an Indo–European base; each shows majority European ancestry as one might expect from this.

2. In general, the results make geographic sense (not just anthropological or historical sense). There is a directly proportional relationship between the affiliation among ethnic subgroups and geographical distance. For example,

   a) The relatively high levels of East Asian admixture obtained using the MLE algorithm decrease moving from east to west within Asia (yellow color, Figure 6-3).

b) European admixture increases from southeast to northwest within Europe.

c) West African admixture decreases moving from West Africa, into North Africa (relatively high levels of brown color, Figure 6-3) to the Mediterranean and Middle East (intermediate levels of brown color, Figure 6-3) and into Northern Europe (very little brown color, Figure 6-3).

3. The groups that showed statistically significant differences from their corresponding parental populations were predominantly those for which admixture had already been documented by other authors using classical blood-group markers (which we will describe in more detail later), such as African Americans (European admixture, $p < 0.001$ with the MLE method), Puerto Ricans (West African admixture, $p < 0.001$ with the MLE method), Mexicans (European and Indigenous American admixture, $p < 0.001$ with the MLE method), and South Asian Indians ($p < 0.001$). In addition, we detected statistically significant ancestry indicating possible admixture for North Africans (West African affiliation, $p < 0.01$), Turkish (East Asian and Indigenous American affiliation, $p < 0.001$), Middle Eastern (West African affiliation, $p < 0.01$), Pacific Island (European and Indigenous American affiliation, $p < 0.001$), and American Indian (European admixture, $p < 0.001$) populations.

4. For all types of observed admixture with the MLE program, above 5%, the results generally are statistically significant, and the SD was generally lower than the overall level of affiliation (see Table 6-1). Both measures of significance suggest that minor or fractional affiliation results at or above 5% are likely to be reliable features of ethnicity and useful in reconstructing ancient population history and possibly certain physical characteristics.

5. However, although admixture estimates for several other subpopulations were not statistically significant on the population level compared to expectations derived from their corresponding simulated parental samples, some may prove significant with larger samples than we used. This is due to the small sample sizes for many of the ethnic groups—a 4% result in a sample of 100 might be very significant even though a 4% result in a sample of 10 may not be. For example, though none of the continental Europeans such as Irish or Northern Europeans showed significant non-European affiliations, we identified a trend of increasing non-European affiliation as one proceeds from northwest to southeast Europe into the Middle East and South Asia (see Figure 6-3; notice the increasing amount of yellow, blue, and brown wedges in the pie charts as one proceeds from Ireland to India. Also refer to Figure 4-1B which shows some of these patterns in a different sample, with different SNPs and a different program). Toward the extreme of this trend, in the Middle East and South Asia, some of these levels were statistically significant even with the relatively small sample sizes employed (Middle Eastern $n = 11$, South Asians $n = 56$; Table 6-1). Moreover, combining Italians, Greeks, and Turks to form a Southeast Europeans group also showed statistically significant non-European affiliations for the group (West African $p = 0.04$, East Asian $p = 4 \times 10^{-5}$, and Indigenous American $p = 0.009$). Italians, Greeks, and

such independent subpopulations do not show statistically significant non-European affiliations largely due to the sample sizes being too low to perform good statistical measurements. Evidently, combining them into subgroups increased the sample size enough to detect the significance.

6. Thus, certain ethnic groups and/or collections of ethnic groups appear to have a unique mix of admixtures, which could be considered a sort of ancestry signature for that group/subgroup. For example, in addition to the example of Europeans we have just discussed, Southeast Asians show higher levels of non-East Asian affiliation than East Asians, the Chinese and Japanese, who are relatively unadmixed (even though the sample size of 11 was not adequate to demonstrate that these levels were statistically significant). Reminiscent of what was observed with a $k = 4$ or $k = 5$ population model for Central Asians in Figure 4-1B, South Asians appear to be the most genetically complex, with substantial and significant European, Indigenous American, and East Asian ancestry. Substantial affiliation (most likely recent admixture) was observed for American Indian and African-American population samples.

7. Using ADMIXMAP, the results are essentially the same as using the basic MLE program for all the subpopulations, except for Europeans, where the levels of Indigenous American and East Asian ancestry are generally lower than obtained using the MLE program. Even so, the same basic trend of increasing West African, East Asian, and Indigenous American affiliation is seen as one progresses from Ireland and Northern Europe to Southeastern Europe, into the Middle East and South Asia (see Figure 6-4, inset). The lower levels of East Asian and Indigenous American affiliation for Europeans is the primary difference we see when we use ADMIXMAP, and we will return to this briefly.

In the following treatment, we will consider only the statistically significant results, and trends that are supported by statistical significance. We evaluate the average results for a population in terms of its standard deviation but we also consider the results in conjunction with data from Y, mtDNA, and other autosomal tests. We have found that, generally speaking, the higher the [percentage]:[SD] ratio, the easier it is to find support in the literature for the same trends based on Y, mtDNA, and other autosomal admixture methods. These results commonly involve subpopulationwide percentages that exceed the relevant individual thresholds we discussed in Chapter 5, commonly involve samples of individuals with a fair number of individuals that exceed these thresholds (i.e., relatively low variance), and perhaps most importantly, have reasonable paleoanthropological and/or geopolitical explanations. Those less than one usually did not. For this reason, we will arbitrarily refer to fractional affiliations for which this ratio is ≥1 to be of high quality, and those that do not to be of low quality.

INTERPRETATION OF ANCESTRY PROFILES FOR ETHNIC POPULATIONS

In this section we will take a more detailed look at some of the more significant findings we have just presented.

EUROPEAN ADMIXTURE FOR AFRICAN AMERICANS

Both algorithms report African Americans to be of primarily West African ancestry with substantial European admixture. The value (14.3% to 18.2%) was greater than that required for concluding significance at the $p < 0.05$ level for individual admixture ($\geq 6\%$, as defined in homogeneously affiliated West African simulated samples; Table 5-23A) and more than half of the individuals registered with European admixture greater than this threshold level. On the population level, the European admixture level was statistically significant ($p \ll 0.001$), and of high quality based on the ratio of the percentage: standard deviation (it was well above 1). Figure 6-5 shows the individual admixture

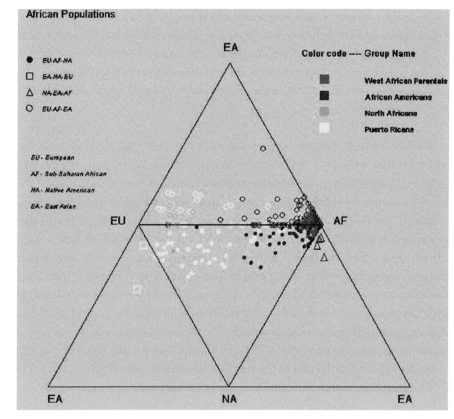

Figure 6-5

Tetrahedron plot of African populations. Samples from each population are color-coded and indicated in the legend, upper right. Symbol shape is determined by the subtriangle in which the MLE falls. Vertices indicative of 100% ancestry for each parental group are labeled with two-letter symbols as indicated to the left. Approximately 70 African Americans (blue) show a greater spread toward the EU axis than parental West Africans (simulated samples), but Puerto Rican and North African populations show even more European admixture. Each population seems to have a unique variance about a characteristic average MLE value.

plots versus 2,000 simulated parental West Africans, and we see a tremendous difference in European ancestry between the two (for both simulated and real parentals the European affiliation was less than 1%. In Figure 6-5, 2,000 samples are shown, and multiple samples plot on top of one another at 100% West African in the plot, so the spread is somewhat deceptive).

Previous studies conducted on the level of the single gene (Cavalli-Sforza & Bodmer 1999) clearly indicate significant European admixture in the U.S. African-American population. As early as 1953, Glass and Li estimated the European M for African Americans to be 0.31 for Baltimore, MD African Americans, which was later revised to 0.216 (Glass 1955), but data from later studies focused on blood group markers showed a variety of M values from 0.07 (Evans and Bullock Counties, GA) (Blumberg et al. 1964) to 0.31 (Glass & Li 1953), with a stark difference in the value of M between Southern and non-Southern African Americans, where the values for the latter were generally greater (see Table 6-2; see Chakraborty 1986 for a review).

The average for these studies, irrespective of region was 16.6%, which is close to the 14.3 to 18.2% we obtained for samples from across the United States. Estimates of m from the gene sequences in Table 6-2 are from loci presumed not to be affected by selection, however all the loci are classical blood group markers, reside within coding regions of gene sequences, and presumably have some functional purpose so even though there was no evidence these sequences were subject to selection at the time they were used, it is possible they were. When loci known to be under the influence of selection are surveyed, very different m values are observed and unless many loci are used, the results might be misleading, as one might expect (Chakraborty 1986; Workman et al. 1963).

INDIGENOUS AMERICAN ADMIXTURE FOR AFRICAN AMERICANS

African Americans typed with significantly more Indigenous American affiliation (3.3% with MLE method) than simulated West African parental samples did ($p < 0.01$). With the MLE algorithm, about one-half of the individuals in Table 6-1 registered with a level of Indigenous American admixture greater than the Indigenous American threshold in West Africans about 1% (see Table 5-23A) and a quarter exhibited levels exceeding 8%. The Indigenous American affiliation result was a low quality result, based on size of the standard deviation relative to the overall level, but in Figure 6-5, there seems to be more spread along the Indigenous American axis for African Americans than West African parentals (especially when considering the difference in sample size—136 African Americans vs. 2,000 simulated parental African samples). This may indicate that the Indigenous American affiliation in the

Table 6-2

Estimates of proportion of Caucasian genes (m) in American blacks by region and locality (modified from Chakraborty 1986).

Region and locality	Allele(s)	m	Reference
Non-Southern			
Baltimore	R^0	30.6	Glass and Li (1953)
Baltimore	R^0	21.6	Glass (1955)
Five areas	R, R^1, Jk^b, T, S	20	Roberts (1955)
Cleveland and Baltimore	Gm^1, Gm^5	31	Steinberg et al. (1960)
Various	R^0, R^1, R^2, r, M, S, Jk^b, k, Fy^b	23.2 – 26.1	Roberts and Hiorns (1962)
Chicago	AK2	13	Bowman et al. (1967)
Washington, DC, Baltimore, and New York City	R^0, R^1, Fy^a	20.0 – 24.0	Workman (1968)
Oakland, CA	Gm^1, $Gm^{1, 2}$, Gm^5	27.3 ± 3.7	Reed (1969)
Oakland, CA	R^0, R^1, R^2, r, A, B, S, Fy^a, P	21.9 ± 2.6	Adams and Ward (1973)
Southern			
Evans and Bullock counties, GA	R^0, R^1	10.4	Workman et al. (1963)
Evans and Bullock counties, GA	R^0, R^1, Fy^a	7.3	Blumberg et al. (1964)
Charleston, SC	R^0, R^1, Fy^a	4.0 – 8.0	Workman (1968)
Charleston, SC	R^0, R^1, R^2, r, A, B, Fy^a, Hb^S	4.0 ± 2.8	Adams and Ward (1973)
James Island, Evans and Bullock counties, GA	R^0, R^1, Fy^a	9.0 – 12.0	Workman (1968)
James Island	R^0, R^1, R^2, r, A, B, Fy^a, Hb^S	15.3 – 14.8	Adams and Ward (1973)
Claxton, GA	R^0, R^1, R^2, r, A, B, S, Fy^a, P, Jk^a, Hp^1, G6PD, Hb^S	12.6 ± 6.9	Adams and Ward (1973)
Sapelo Island, GA	R^0, R^1, R^2, r, A, B, S, Fy^a, P Jk^a, Hp^1, G6PD, Hb^S	8.8 ± 6.9	Adams and Ward (1973)
		Average 16.6%	

African-American population is contributed by low numbers of extensively admixed individuals.

Again ADMIXMAP shows a lower level of Indigenous American than the MLE algorithm (see Table 6-1) and accentuates the point that Indigenous American admixture is not infrequently found for African Americans, but the average level is low and the variance is high. Intermediate allele frequencies are a possible concern with the result since the level is so low, and there is no other data against which to compare this result; African Americans were not addressed in the Y-haplogroup analysis of Underhill et al. (2000), nor the autosomal admixture results of Rosenberg et al. (2001). One might logically suspect that the incidence of Indigenous American admixture would be greater in Afro-Caribbean populations, and the results from Puerto Ricans show just that (Halder et al. 2006); note the spread of yellow symbols toward the NA vertex in Figure 6-5. Thus, substantial Indigenous American ancestry could be indicative of a higher likelihood of Afro-Caribbean versus African-American ethnicity.

WEST AFRICAN AND INDIGENOUS AMERICAN ADMIXTURE FOR PUERTO RICANS

Historically speaking, Puerto Ricans are a blend of European, West African, and Indigenous American admixture; colonial Europeans and escaped and incompletely kidnapped African slaves mixed with each other and Indigenous Americans in the 1700s and 1800s to form what might be termed a modern day Afro-Caribbean ethnic group. With our 171 AIMs and 64 Puerto Rican samples we got very similar results using the basic MLE algorithm or ADMIXMAP; Puerto Ricans type of primarily European ancestry, but with substantial West African admixture (about 30%, Exact p << 0.01; Table 6-1; Figure 6-5). Almost a third (20/64) registered with Indigenous American levels above the Indigenous American threshold for Europeans (12.5%) and both ADMIXMAP and the basic MLE program showed NAM levels that were statistically different from European parentals. Comparing the level of admixture with the standard deviation, the 8.8% Indigenous American admixture MLE result is a high-quality result.

Bonilla et al. (2004) typed Puerto Rican women from New York city with a 30-AIM panel and using a three-way continental population model, obtained similar proportions; 53.3% European, 29.1% West African, 17.6% Indigenous American. Adding the East Asian and Indigenous American percentages together in the MLE results from the 171 AIM panel yields a value that probably corresponds to Indigenous American in the Bonilla study. Our result of 8.8% with the MLE program was higher than the value obtained with ADMIXMAP, which was only 3.3%.

WEST AFRICAN ADMIXTURE IN NORTH AFRICANS

The sample for North Africans was very low, but with both algorithms the results clearly show that North Africans are of primarily European ancestry with substantial West African ancestry as well (see Table 6-1; green vs. blue plots in Figure 6-5). This result is similar to what was observed for East Africans using the most informative of the Affymetrix 10K chip SNPs and the PSMix program (Figure 4-1B). Compared against simulated European parentals, the average level of West African affiliation in our North African sample (about 15%) was statistically significant ($p < 0.01$ with both methods), and the average level was above the individual threshold level derived from parental Europeans (compare the green plots with red plots in Figure 6-5). The standard deviation of African percentage in this small sample was relatively low compared to the average level observed, so the fractional affiliation result is of high quality. Consistent with this observation, each North African tested registered with significant West African ancestry (levels above the African threshold of a few percent), suggesting that North Africans systematically exhibit West African admixture.

Compared with Northern Europeans and Mediterraneans, North Africans as a group show considerably more non-European affiliation, most of this being West African, as do other European ethnic groups in the geographical region (notice that the red symbols for Middle Eastern samples in Figure 6-6 are significantly removed from the European vertex compared to white and green symbols). The average position of North Africans in Figure 6-5 (green symbols) is similar to that of Middle Eastern samples in Figure 6-6, and very different from African Americans or West African parentals, which is not unexpected given similarity in anthropometric traits between North Africans and Middle Eastern individuals (i.e., more similarity in physical appearance—notably skin shade—between Bedouins and Moroccans or Libyans than between Moroccans and Nigerians). The West African affiliation is not unexpected given the geographical proximity of North to West Africa and we will discuss this further, below.

WEST AFRICAN ANCESTRY IN SOUTHEASTERN EUROPE AND THE MIDDLE EAST

We can note from Figures 6-3 and 6-4 that as we proceed from Ireland to India, the level of West African affiliation gradually increases. Though the levels in Southeastern Europe are much greater than in Northern Europe with the MLE program, the sample sizes of these populations in this study do not allow for a demonstration of statistical significance for these levels. However, in the Middle East and South Asia, the West African affiliation reaches levels that are so high that even with low sample sizes we can show statistical significance. The geographical proximity of these regions to Africa may suggest relatively recent population expansions or migrations—note that the further from the African

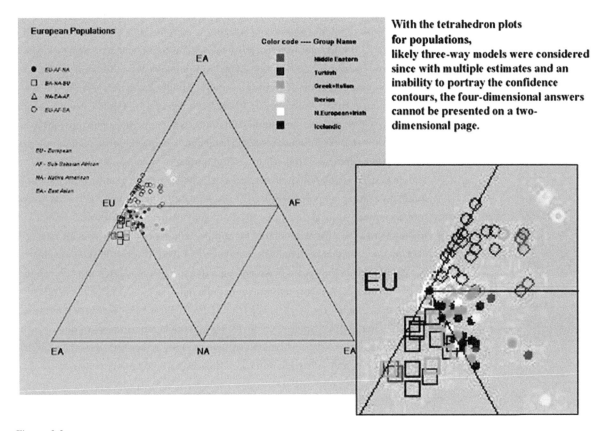

Figure 6-6

Tetrahedron plot of European populations. Samples from each population are color-coded and indicated in the legend, upper right. Symbol shape is determined by the subtriangle in which the MLE falls. Vertices indicative of 100% ancestry for each parental group are labeled with two-letter symbols as indicated to the left. Middle Eastern and Southeastern European (i.e., Turkish) samples show a greater spread from the European (EU) vertex than Northern Europeans. Geographically intermediate populations such as the Greeks and Italians show an intermediate level of spread. Data taken from Table 6-1.

continent one gets the lower the West African affiliation. However, there is no reason to impose a recent timeline on such events.

Other lines of evidence lend support for this finding of low levels of West African ancestry in Southeastern Europeans and Middle Eastern populations.

1. Underhill et al. (2001) showed that the frequency of the YAP+ Y haplogroup commonly referred to as haplogroup E or (III) is relatively high (about 25%) in the Middle East and Mediterranean. This haplogroup E is the major haplogroup found in sub-Saharan Africa (over 75% of all Y chromosomes). Specifically, Europeans contain the E3b subhaplogroup, which was derived from haplogroup E in sub-Saharan Africa and currently is distributed along the North and East of Africa. Thus, whether we refer to the affiliation as African, West African, or North/East African depends on the time

frame we wish to refer to and/or our semantic preference. It appears that the 171 AIM test subject of this chapter may recognize the haplogroup E character as West African.

2. Admixture results using 300 odd autosomal STR markers and a different algorithm from ours show extensive African admixture for Middle Easterners as we have (Bedouins, but not Druze; Rosenberg et al. 2003; see Figure 4-1, Middle Eastern Columns for k = 4). However these same studies did not show West African admixture for Italians or Sardinians (Greeks were not tested). The genetic distance between Mediterraneans and Middle Easterners is relatively low so the finding of West African admixture for Middle Eastern populations in Rosenberg et al. (2001) suggests a similar result might be found for a larger sample of Mediterraneans as we did, probably more frequently as one travels southeast as suggested by the distribution in Figure 6-3. ADMIXMAP showed a trend toward West African admixture for Middle Eastern populations similar to the MLE method. This makes it unlikely that the lack of West African affiliation obtained for Mediterranean Europeans by Rosenberg is due to a genuine difference between the MLE method and Bayesian methods. Rather, the difference is likely due to sampling effects; there might exist heterogeneity within the Mediterranean group—a topic we will return to later in this chapter when discussing subpopulation markers—or difference in markers and/or marker types used (Rosenberg used STRs not selected for their ancestry informativeness, and with a different chronological reach and power to detect).

The significance of our results suggest that the Western African affiliation is a bona fide element of Middle Eastern population structure. Some of this West African affiliation is likely due to recent admixture. Though the enslavement of West Africans by Europeans as part of the Atlantic slave trade gains most of the attention from historians, it is well known that this slave trade was preceded by an equally robust trade across the Mediterranean into the Ancient worlds, including the Egyptian, Greek, Roman, and Arab Empires as well as in other areas (including Iceland). Earlier population expansions of groups such as the Moors could also have contributed to the West African admixture for Mediterranean and Middle Eastern populations. However, it seems unlikely that recent admixture would explain such a substantial distribution of West African character across South Asia and the Middle East, and we probably need to look at the shared common origin hypothesis in a bit more detail.

As mentioned, the distribution of West African affiliation in Europe and the Middle East mirror the distribution of the E Y-chromosome haplogroup (specifically, the E3b haplogroup). E is largely confined to Africa, but is a YAP containing haplogroup that could very well have originated in the Near East or in remote Northeast Africa (Jobling et al. 2004 covers this briefly in Box 9.6, p. 293). It so happens that the E haplogroup branched from other African haplogroups relatively recently compared to other haplogroups, but still many tens of thousands of years ago. The D haplogroup is the sibling of E and is found mainly in East Asia (primarily Japan), yet with the 171 AIM panel discussed in this chapter, we find no West African affiliation for East Asians including the Japanese. Evidently the D haplogroup is more genetically distant from Africa due to extreme genetic drift and/or selection in East Asia and evidently E has not gone through the same pressures, staying relatively close to home in Africa, both in terms of modern-day geographic distribution, and as our results also suggest, in terms of genetic distance. It is undoubtedly the recent origin of E from other African haplogroups that explains why the MLE version of our test, and to a lesser extent, the ADMIXMAP version as well is detecting African affiliation wherever E is found. A recent spread of E through-out Africa (perhaps during the Bantu expansion) could be another factor.

We have no idea what these original YAP carriers looked like, but common sense would dictate that since their descendents currently live in East Asia, Central Asia, Near East, and Europe (populations among which anthropometric traits vary dramatically), and since the source was the same that contributed to most modern-day Africans, they must have taken a decidedly non-European appearance. This is not to say that they must have looked Negroid in appearance. Howells studied 2,800 + skulls, measuring craniometric variation in modern populations and prehistoric East African fossils, and the general conclusions from this work were that prehistoric East Africans (presumably the earliest YAP carriers) were non-African in nature—that is, not Negroid. This is in contrast to skulls from modern-day East Africans. Although East Africa may have been the birthplace of non-African variants who migrated to populate the world, "the ensuing picture for East Africa ... would later have beeen changed through replacement by the expansion of Bantu or other Negroid tribes" (Howells 1995).

Nobody really knows what this means in terms of whose ancestors looked like what, but the West African affiliation detected for continental Europeans and Middle Eastern populations appears to be anthropologically meaningful information that might help define partitions of substructure in Europe. The correlation between AFR (and NAM) affiliation in Europeans with 171 AIM assay and MED and MIDEAS ancestry with EURODNA1.0 panel (to be described shortly) is very high even though the marker sets between the two tests are different and the parental population model and loci are different.

THE CONUNDRUM OF INDIGENOUS AMERICAN AFFILIATION IN ASIA AND EUROPE

The results for the Northern European samples is notable in that they closely resemble the average European parental sample (using either algorithm) and are starkly different from that of other European subpopulations such as

Mediterranean, Ashkenazi Jewish, Middle Eastern, and South Asian populations, which show higher non-European ancestry with the 171 AIM panel. With the MLE algorithm, much of this difference, as can be appreciated visually in Figure 6-6, is in terms of Indigenous American affiliation (see Table 6-1). With the ADMIXMAP algorithm, the Indigenous American affiliation is lower for every European group but we see the same trend toward increasing affiliation with southeastern geography, both within Europe (inset, Figure 6-4) and between Europe and Asia where levels are the highest in South Asia. Thus the basic MLE algorithm and ADMIXMAP both agree that there exists a trend of increasing Indigenous American affiliation from northwestern Europe to South Asia, but they disagree on the overall levels in each subpopulation (and the levels are statistically significant only in South Asians at the extreme of this trend).

It is the trend and the presence of Indigenous American ancestry that is of interest here, not the quantity. Since back-migrations across the Bering Strait are not known to have occurred, the observation of Indigenous American admixture in individuals from the Asian and European continents is at first quite perplexing. Let us consider findings from others that may be relevant to this result:

1. Reidla et al. (2003) showed that the mtDNA haplogroup X2 is found frequently in the Mediterranean and in Indigenous Americans, but infrequently in northern Europe, Central Asia, and Siberia. This haplotype appears to have been derived from Paleolithic Near Eastern populations, branches of which appear to have migrated to North America tens of thousands of years ago but also dispersed throughout Asia. It is possible that these migrations were different migrations/expansions and it is notable that neither East Asians nor Siberians have X2 at appreciable frequencies. The X2 haplogroup seems to connect European and Indigenous American populations through a possible Central Asian population or Near Eastern source (we don't know exactly where they lived but as we will discuss later, evidence points to southwestern Siberia which could be called Central Asian). There are also other mtDNA connections between Asia and America, as can be appreciated in Figure 4-4.

2. Underhill et al. (2001) showed significant frequencies of Y chromosome haplotypes characteristic of Indigenous Americans in Central Asia but not the Mediterranean or Middle East, except Ashkenazi Jews, for whom haplogroup Q is found frequently. Using the old nomenclature, X in general is distributed throughout Central Asia, the Middle East, and South Asia in Underhill et al. (2001) (see Figure 4-2). Using the more modern nomenclature, Q is the predominant Y haplogroup found in the Americas and shared with European populations undoubtedly through its (Central) Asian origin. Thus the Y chromosome also shows a connection between Central Asian and Indigenous American populations, but differs from the mtDNA results in that it does not connect the European populations to the Indigenous Americans (recall that patrilo-

cality usually exceeds matrilocality). Given the mtDNA results of Reidla et al. (2003), it may be that the affiliation with Indigenous Americans was contributed largely by females through sex-biased admixture, or that Central Asian expansion in Asia and Europe involved a substantial bottleneck for males (such as war perhaps?). It should be kept in mind, too, that since the mtDNA often does not show as high a level of differentiation as does the NRY, direct comparisons among these markers are sometimes difficult to make.

3. Rosenberg et al. (2002) did not detect significant Indigenous American admixture for Sardinians, Tuscans, or Italians (see Figure 4-1, k = 4 row), though Greeks were not tested. Thus, the STRUCTURE results are similar to that from ADMIXMAP in that they show very low, seemingly nonexistent levels of Indigenous American, though with ADMIXMAP the low levels observed are part of a trend. Probably more important is the type of marker used by Rosenberg et al. (2002, microsatellites), which is characterized by a lower effective population size and so cannot be considered to be "reaching" back in time as far as the average SNP AIM does.

4. We have already alluded to the results obtained using the most ancestry informative of the Affymetrix 10K SNP chip. Using a completely different sample than that used here for the generation of the data in Figures 6-3 and 6-4, and a different program (PSMix, an MLE program), low levels of "Indigenous American" admixture can be seen for South Asian Indians (about 6%, Figure 4-1B, more for lower caste than upper caste). Though Middle Eastern and Southeastern European populations were not represented in this study, and the bias/RMSE of this panel is not yet known, the presence of the Indigenous American signature in the population at the extreme of the geographical/genetic trend we are discussing suggests that this panel and program would yield the same pattern of increasing Indigenous American affiliation from Southeastern Europe to South Asia.

The combined uniparental marker results, and the geographical trends of the admixture results suggest that the observation of Indigenous American affiliation in Southeast Europe, the Middle East, and South Asia may be a bona fide component of the extant population structure. The questions become what structure, why does it exist, and from whence did it originate?

A POSSIBLE CENTRAL ASIAN SOURCE

Since back-migration across the Bering Strait is not known to have occurred, it seems that the presence of Indigenous American affiliation in Europe and Asia seems to be an excellent example of residual affiliation through relatively recently shared ancestry as opposed to recent admixture. Recall our earlier example of founder population ABC and derived populations Bc and Cb, where uppercase indicates the allele is at a high frequency and lowercase indicates the allele is at low frequency. If Central Asians were ABC, Europeans could have become Bc; South Asians, aC; and Indigenous Americans, Cb; and all three of these populations would be united by a common ancestry, C. However, note that admixture with some populations carrying these relic alleles at varying frequencies could also have occurred to produce the same result. Whether its contribution was through shared ancestry, or admixture has some bearing on when and where they might have lived and is relevant to the nomenclature problem we discussed in Chapter 4, and again later in this chapter, which we could solve by referring to the groups as I, II, III, and IV instead of European, West African, East Asian, and Indigenous American.

Was the Indigenous American parental population really Central Asian, or was it Middle Eastern? We can infer the order of the appearance of the groups using modern phylogenetic methods (Jobling et al. 2004). The use of Principal Components Analysis with classical blood group markers (Cavalli-Sforza et al. 1994) suggest that Northern Eurasians predated the split between Northern Asians and continental Europeans about 47,000 years ago (using a molecular clock theory to date the split), and that Indigenous Americans derived from a split among Northern Eurasians into a Northern, East Asian group and an American group about 15,000 years ago. Y and mtDNA haplotype data generally support this notion though different dates are provided (due primarily to differences in sex-chromosome genetics, reviewed in Jobling et al. 2004).

Given the difference in Y versus mtDNA haplogroup distributions discussed in the earlier points (1) and (2), it would appear that there is potential evidence for greater maternal than paternal Indigenous American contribution to modern-day Mediterranean populations, if we can accept that the term Indigenous American is a misnomer, referring to a genetic distribution that is not optimally aligned with modern-day geographical or political divisions. As we have discussed, Indigenous American Y chromosome haplotypes (haplogroup X or Q) have been detected in Central Asian populations at surprisingly high frequencies (Underhill et al. 2001; see Figures 4-2 and 4-3 though more modern nomenclature is used in the latter). This is a connection between Central Asia and America and the mtDNA chromosome distributions show similar Central Asian connections with Southeast Europe, the Middle East, South Asia, and even Southeast Asia. In fact, a surprisingly complex variety of Y haplogroups are found in Central Asia—V, X, IX, VII, VIII, and VI (where nomenclature is the original used by Underhill in his 2001 paper; Figure 4-2)—and Central Asia is among the most diverse of any population tested by this group (South Asians being equally complex).

The haplogroups shared with modern-day Central Asians differ from group to group. In addition to haplogroup X (also known as haplogroup Q in the modern nomenclature) that connects Indigenous Americans, Central Asians, and South Asians, V is shared among Indigenous Americans, South Asians, Southeastern Asians, and Central Asians. In addition, the VIII haplogroup connects Middle Eastern, South Asian, and Central Asian populations (Underhill et al. 2001; Figure 4-2). VIII is not found at appreciable levels in Indigenous Americans, but is found at high levels with Pacific Island populations. Thus, we have partially overlapping distributions that establish phylogenetic connections between Europeans (as we call them, including Middle Eastern and South Asians) and non-Europeans (such as Indigenous Americans and Oceanic populations) that could have originated ultimately from what could have been Central Asian parental sources. In other words, combining these partially overlapping Y chromosome haplogroup distributions, and comparing the distribution of Indigenous American ancestry in Europe and Asia, it seems that the Indigenous American affiliation observed for Mediterranean, Middle Eastern, South Asian, Southeast Asian, and the Pacific Island populations could be the result of affiliation with an ancient population reminiscent of modern-day Central Asians in their complexity, who themselves are affiliated with Indigenous Americans through a genetic founder event.

We use the overlap in distributions to infer a common source, connecting populations with V, VI, VII, VIII, IX, and X Y-haplogroups together throughout Europe, Asia, and America through a common possible Central Asian source. Of course, with the admixture results and continental AIMs, we expect to read about events contributed by both male and female expansion/migration, not just males. Indeed, we can do the same with mtDNA haplogroups among these populations, including the X, D, B, C, and A haplogroups, but of course, we obtain slightly different patterns of overlap.

Let us look at a couple of examples of how we might relate admixture estimates with Y and mtDNA haplogroup distributions. For South Asian Indians, it is possible that the 9% average Indigenous American affiliation we observe with the MLE algorithm is the result of the presence of V and X—both of which are found in Indigenous Americans, Central Asians, and South Asians, and which add up to about 15% of the total Y-haplogroup diversity in South Asia. IX and VI also are shared among these three groups, but are thought to have been introduced to America through admixture with continental Europeans within the past few hundred years. There are no shared mtDNA haplogroups among these three populations, and so the average of 7.5% (15% shared Y and 0% shared mtDNA) sequence identity among them is reasonably close to the 9% we registered with the autosomal admixture test.

A different explanation would need to be invoked for Middle Eastern populations—the Y haplogroup G, as well as the mtDNA X haplogroup is shared among Middle Eastern, Central Asian, and Indigenous American populations. G represents about 6% of the Middle Eastern Y haplogroup diversity and X represents about 5% of the Middle Eastern mtDNA haplogroup diversity and the average of these two corresponds to about 5.5% of the Middle Eastern diversity that is shared among the three groups. Our estimate from autosomal admixture analysis (with the MLE algorithm) came to about 4%, not too different. Of course, the situation could be a bit more complex—it could be that a substantial amount of IX haplogroups in Indigenous Americans is of ancient Central Asian/Near Eastern origin and founder effects/genetic drift rather than due to recent admixture with Europeans, and if so, this would alter our calculations.

A large-scale analysis of the derivation status of the haplotypes in these populations would help answer the question. For continental Europeans, the level of NAM we see in the southeast of the continent is very small and the mechanism behind the shared Central Asian identity could be more or less complex than that which applies for Middle Eastern and/or South Asian populations. For Europeans, it is harder to explain the low levels of Indigenous American in part due to the low levels, due to the question of whether they are even statistically significant (though due to the trends observed, it seems likely they will be with larger sample sizes), and due to the uncertainty about the extent to which admixture contributed to the extensive R, J, and I Y-haplogroup sharing with American populations. MtDNA X is found at low levels in continental Europe and in the Americas, but again we cannot rule out admixture.

This indirect connection hypothesis through Central Asian-like populations seems very plausible, but there are some problems with this idea we must first overcome. First, the least damning problem is that other workers using autosomal markers and admixture estimation methods have not seen the same Indigenous American character in Europe and Asia (until this work, only one other admixture study had been published—Rosenberg's STRUCTURE analysis in 2001; Figure 4-1A, B). Using 377 microsatellites with STRUCTURE, Rosenberg et al. (2002) did not show significant Indigenous American admixture for Druze Arabs, Palestinians, or Bedouin Middle Easterners, or shared Central Asian derived or other ancestry between Indigenous Americans and Middle Eastern/South Asian populations (though he did show significant West African admixture, like we did with our 171 AIMs). In addition, the ADMIXMAP algorithm on the data we have just presented tended to diminish the Indigenous American affiliation for all these European populations, though not completely (leaving the statistical significance of Indigenous Americans in South Asians intact).

This poses an interesting question then—why does the 171 AIM SNP test with basic MLE algorithm (as well as the Affymetrix chip-derived AIM panel with PSMix in Figure 4-1B) show the connection between Indigenous

Americans and Middle Eastern, Southeast European, South Asian, and Southeast Asian populations (presumably through source populations that resemble modern-day Central Asians) that we also see with Y-chromosome analysis, but methods such as STRUCTURE, using the 377 microsatellite loci of Rosenberg et al. (2001) do not, and ADMIXMAP shows results of significance in between these two? Perhaps ADMIXMAP and STRUCTURE, both relying on more complex Bayesian methods, tend to minimize unexpected findings for some arcane mathematical reason. Perhaps the Bayesian methods are overly reliant on modern-day sequence variability rather than on the qualities of anthropologically qualified parental groups. This seems unlikely since these more complex methods are supposed to be more adept at getting to the true parental allele frequencies.

Perhaps the Bayesian methods are more reliable in reporting the truth, since they do not impose the unreasonable assumption that the information content for each marker is equivalent (see Box 6-C), and the truth is that there just isn't much continental level structure within Europeans. This seems unlikely since the ADMIXMAP program reports the same structure we observe with the MLE program and autosomal markers and from Y and mtDNA analysis, just less of it. Most likely, the explanation is due to differences in the marker types used and their effective population sizes and subtle differences in sensitivity between algorithms. Perhaps the 377 microsatellites of Rosenberg et al. (2001) are reporting less deep divisions in the human family bush because they mutate and drift faster. By this thinking, ADMIXMAP and the basic MLE algorithm with the 171 AIMs described herein both report the lower levels of Indigenous American affiliation as part of a geographical trend because with SNPs we are looking a bit farther back in time. The finding of Indigenous American affiliation in a different sample of South Asians with the Affymetrix chip-derived AIMs (Figure 4-1B) supports this idea.

Indeed, due to their effective population size and allelic complexity, we might expect SNPs to report results somewhere between microsatellites (recent) and Y/mtDNA haplogroups (ancient). Y and mtDNA haplogroup analysis is relatively straightforward, relying on counts rather than on inferences about what generated the counts. Due to maintenance in a haploid state, Y chromosome change is relatively slow, and so we would expect Y haplogroups to show less differentiation than microsatellites, which mutate very quickly due to the type of mutation (stutter) and the fact that they always are found in the diploid state. SNPs are always found in the diploid state, but are not stutter mutations, so we might expect them to evolve more slowly.

Supporting the hypothesis that the 171 AIMs are reporting information more like the Y/mtDNA haplogroups than the microsatellites, we also observed from the 171 SNP AIMs (MLE method) a low level of statistically significant

Indigenous American affiliation for Chinese, Japanese, and Atayal. Underhill showed the same connection between these two populations for Y-haplogroup X, at the same level, but Rosenberg et al. (2002) did not. Comparing our results with the Y and mtDNA distributions, we can rule out the possibility that the autosomal SNPs are reporting deeper ancestry relationships than the Y and mtDNA results. Northern Europeans are characterized by high frequency of Y haplogroups R1b (in the West) and R1a (in the East), but R is not very frequent in the Middle East. Haplogroup R is the phylogenetic sibling of haplogroup Q, which is found in Indigenous Americans, so we might expect that if the 171 AIM admixture test is looking at deep ancestry (contributed primarily by males in the European population), we'd detect more affinity of R-Continental Europeans with Q-Indigenous Americans than between Middle Easterners and Indigenous Americans. In fact, Continental Europeans show less Indigenous American affiliation with our 171 AIM test than Middle Eastern populations, and we encounter similar problems with this hypothesis looking at the mtDNA distribution so it seems untenable.

The reality of the situation is that there are probably recent as well as ancient connections between Europe and America, and that the dates of, and overall balance between male and female contribution of this character differs from population to population. In reality, the answer for the Indigenous American affiliation in European and Asian populations should be found in correlates in Y and mtDNA signatures, but there is nothing to say that the age of the Y-derived NAM character is the same as that for the mtDNA-derived NAM character, or that the precise explanation in one group is the same as that for another. Populations expand and originate at different times from different founders. Although highly informative, the NRY and mtDNA report on only one chromosome and so may have only recorded one aspect of what happened. For example, if during some particular migration, there were no mutations on the NRY or the mutations that did happen were lost to drift, then there would be no defining haplogroup for that migration even though it did happen and it contributed to the composition of the autosomal chromosomes in the Americas.

The reason we have gone through this difficult line of logic is that the nonprimary ancestry obtained for a population, if significant as judged by the existence of supporting pattern, literature, or p values below 0.05, could be useful for making inference on population of origin, which could have relevance for the inference of likely physical appearance. We thus have a need to anthropologically justify the use of these data. For a forensic scientist or a genealogist, the point of an admixture test is not to trace personal phylogenetic history or reconstruct human migration/expansion patterns, but to infer modern-day population of origin—however those populations formed, through whatever myriad interactions and contributions spread out over time and space, which are almost always unknowable with great precision. Indeed, it is the presence of bona fide population structure that interests the molecular photofitter, not necessarily the accurate tracing of specified historical events. Even in this regard, the best methods are empirical in nature (as we will discuss in Chapter 8), and involve the use of multiple different tool types (such as Y, mtDNA, historical evidence, admixture analysis, etc. all combined together) and it is incumbent on any purveyor of admixture testing services to be straightforward about that (Shriver & Kittles 2004).

EAST ASIAN ADMIXTURE IN THE MIDDLE EAST AND SOUTH ASIA

We have discussed the paradox of Indigenous American affiliation for populations of this region, but East Asian affiliation is another similar, though less controversial characteristic. In Figure 6-3 we can also see a trend of increasing East Asian affiliation as one proceeds from Southeast Europe into the Middle East, increasing further in South Asia, and peaking in Southeast Asia. This is the same axis with which we saw the increasing Indigenous American affiliation, but phylogeographically does not seem as difficult to explain. For example, others have seen similarly significant levels of East Asian affiliation for South Asians and by extrapolation these observations may be relevant to the trending we see in Asia:

1. Rosenberg et al. (2002) showed considerable East Asian affiliation in South Asian Indians and other Central Asians (from 10–40%, depending on the ethnicity; see Figure 4-1A, B, Central Asian columns, k = 4).

2. Underhill et al. (2001) showed a significant frequency of East Asian Y-chromosome haplotypes in South Asian Indians and Near Eastern populations (from 5–15% of all Y-haplotypes, depending on the region; X in Figure 4-2, and C in Figure 4-3).

3. In addition, with autosomal markers, there is a rich literature on this particular topic. Chakraborty (1986) reviewed work conducted by 10 different authors that focused on estimating Mongoloid m values for South Asian Indians. Samples from 10 different states were included, and 41 populations were sampled. A range of from 0% to 75% to 95% was observed (Bhasin et al. 1985; Chakraborty 1986; Chakraborty et al. 1986b; Chopra 1970; Ghosh et al. 1977; Gohler 1966; Saha et al. 1974, 1976; Steinberg 1974; Vos et al. 1963; Walter et al. 1980). Most populations showed extensive Mongoloid admixture with lower levels in the south and west than in the north and east. This northeast/southwest distribution would seem to make geographical sense and suggests that Mongoloid migration proceeded from a northeastern or eastern source. Based on measures of genetic distance, East Asian, Central Asian, and Indigenous American populations would be classified as Mongoloid in these works. It bears repeating

however that there are other mechanisms for such affiliation than admixture—South Asians could have acquired East Asian genetic character if they are an offshoot of a Central Asian population that also gave rise to Europeans and East Asians (and not enough time has passed for this connection to have drifted away). Indeed, as we have discussed in the preceding section, Y-chromosome analysis makes it clear that Central Asians are among the most complex populations in the world, suggesting that they are old and perhaps the source of many younger populations.

4. The Affymetrix chip-derived AIMs discussed in Chapter 4 (Figure 4-1B) show extensive ''East Asian'' admixture for South Asians at k = 5 (comparison of these results at k = 4 with ours here is complicated by the presence of genetically distant Oceanic populations in the former analysis).

Notwithstanding, it is possible that the East Asian affiliation in South Asians could represent another example of our nomenclature problem. European affiliation for South Asians is undoubtedly due to the fact that they have partly a Western Eurasian genetic heritage (evidenced by high frequencies of Y-haplogroups R1a/J) and culture (evidenced by the distribution of Indo-European languages). They also have partly a Native South Asian genetic heritage (Indian-specific subclades of mtDNA haplogroup M). It is the Indian-specific subclades of mtDNA haplogroup M that tie South Asians with East Asians (who belong to different, East Asian subclades of M). So, the idea that South Asians are part East Asian may lead some to believe that they're the product of admixture between Caucasoids and Mongoloids, whereas this is not necessarily the case. It could be shared ancestry much like we have invoked to explain the Indigenous American affiliation, which is supported by the presence of different derived variants of haplogroup M (a scenario that is less likely to have arisen due to recent admixture but could have arisen due to very old admixture and subsequent genetic drift). It could be a combination of old and new admixture combined with relatively recent shared ancestry for all we know! In a forensics sense, we are interested mainly in the hybrid character itself, and the range of physical characteristics associated with this hybrid character, not necessarily how the hybrid character came to be. The anthropologist could date the East Asian admixture as statistical methods progress and are validated in the near future.

EXTENSIVE ADMIXTURE OR FRACTIONAL AFFILIATION FOR SOUTHEAST ASIANS AND OCEANIC POPULATIONS

Our sample of 11 Southeast Asians exhibited mainly East Asian ancestry, but with substantially greater non-East Asian admixture than observed in the Chinese or Japanese samples, which were of similar size. Even with these low sample sizes, the difference was significant. This can be visualized in Figure 6-7 by noting the spread of the green symbols (Southeast Asians) away from the East Asian (EA) vertex, in sharp contrast to the black symbols (Chinese). Though below the threshold for European admixture in East Asians (11%), the European affiliation grouping Southeastern Asians (8–14%, depending on the algorithm) seems to be contributed by small numbers of highly mixed individuals (note the green symbols along the EA-EU axis in Figure 6-7). The 4/11 South East Asian individuals that typed with substantial European affiliation had levels greater than the 11% threshold and on the level of the population the European affiliation was statistically significant, signifying that their European ancestry was more than that expected from statistical error caused by continuous allele frequencies.

Progressing to the Pacific Island populations, the levels of European and Indigenous American affiliation observed were substantially greater; even though the sample size was low the levels were so high that they are determined to be statistically significant. This suggests that the incidence of European admixture in Southeast Asia and Oceana is more common than in the Chinese

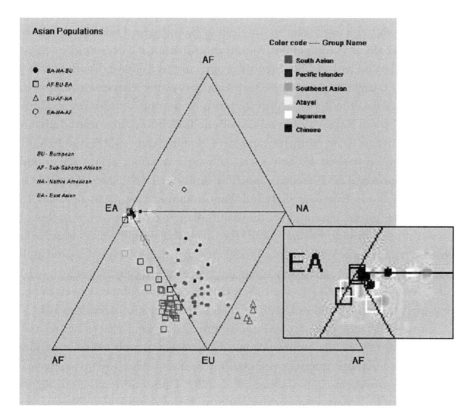

Figure 6-7

Tetrahedron plot of Asian populations. Samples from each population are color-coded and indicated in the legend, upper right. Symbol shape is determined by the subtriangle in which the MLE falls. Vertices indicative of 100% ancestry for each parental group are labeled with two-letter symbols as indicated to the left. Southeastern Asian samples (n = 11) show a greater spread from the East Asian (EA) vertex than Japanese (n − 10) or Chinese (n = 10) samples. Pacific Islander samples show extensive non-East Asian ancestry as do South Asians. Data taken from Table 6-1.

and Japanese, possibly due to ancestry contributed by Central Asian populations (similar to a hypothetical source for South Asian European and Indigenous American ancestry) or recent political events such as nineteenth century colonization by the French. Figure 6-3 puts the European (and Indigenous American) affiliation in a different context—note the spread of European affiliation across the South Asian Indian subcontinent extending into South East Asia and Oceana (even Australia) but not East Asia. Underhill showed in 2001 that about 15% of the Y-chromosome haplogroups in Southeast Asia were commonly found in Europeans (25%–75% of all haplogroups, depending on the population), but are also present in Central Asia at about 15%. Rosenberg et al. (2002) showed considerable European admixture for Cambodians (average about 10%–15%) but not for other East Asians, which agrees with our results and those of Underhill et al. (2001).

The Indigenous American threshold is very high for East Asians due to the AIMs used and the relatively low genetic distance between these two groups, and the level of this type of admixture was well below that threshold. The levels observed with the MLE program were provocative (6.3%, similar to its standard deviation), but with the ADMIXMAP program were relatively insignificant. The low genetic distance between East Asians and Indigenous Americans makes interpretation of this affiliation particularly challenging. We (courageously?) address this difficult topic later in the chapter but will give a brief preview here. As discussed already, Indigenous American affiliation is also found in South Asia, the Middle East, and Southeast Europe, and as we have discussed this is likely due to a single mechanism of shared ancestry rather than recent admixture (i.e., return across the Bering Strait). In terms of autosomal measures of admixture, Rosenberg et al. (2002) showed virtually no Indigenous American admixture for each of 18 East Asian ethnicities and two Oceanic groups (see Figure 4-1A, B), but as we have discussed, his analysis is likely not extending as far back in time as Y and autosomal SNP analyses are. The autosomal SNP study discussed in Figure 4-1B shows extensive shared affiliation among East Asian populations, New Guinea, central Asian and South Asian Indian populations at k = 4, though this was not the best population model for this global sample set (as it included Oceanic populations. With a better model (k = 5), a lower level of shared affiliation among these populations was observed (note the red color, k5, Figure 4-1B).

Underhill et al. (2001) showed the existence of Indigenous American and what might be construed as Central Asian Y-chromosome haplotypes (i.e., V, X) throughout East Asia, from Eastern Eurasia to Japan (see Figure 4-2). Central Asians share the Y-chromosome C haplogroup with Oceanic populations and Indigenous American populations, and also share the Q haplogroup with Indigenous Americans (Jobling et al. 2004; Underhill et al. 2001; Figure 4-3). Though Indigenous Americans share the C haplogroup with Oceanic and

Central Asian populations, there are other indirect connections among these three populations as well, such as the overlap between the distribution of the C and Q haplogroups among these populations. These distributions suggest a relatively recent shared ancestry much in the spirit of our argument explaining Indigenous American affiliation across southern Asia, the Middle East, and Southeast Asia.

MtDNA haplogroup HV* is shared by East Asians, Americans, and Oceanic populations to the exclusion of any others. It thus appears that Oceanic populations and Indigenous American populations may be related through Central Asian populations, but that the diversity of Y chromosome haplogroups is much lower than mtDNA haplogroups, suggesting a relatively low effective population size for founding males compared to females. Indeed, for many populations the effective population size for males is commonly lower than that for females (for example, in Croats and Russians—Barac et al. 2003; Bermishiva et al. 2002; Derenko et al. 2002, 2003; Kittles et al. 1998).

EXTENSIVE ADMIXTURE FOR AMERICAN INDIANS AND HISPANICS

American Indian

Though carefully selected parental Indigenous Americans show little European admixture (see Figure 4-1B, Figure 6-3B light blue pie chart "P"), American Indians showed substantial European/Indigenous American admixture. A dramatic difference in European ancestry was observed for individuals who were part of U.S. government-recognized tribes and living on a reservation (Sioux, Cheyenne, Cherokee, and Arapaho; red and green data points in Figure 6-8) compared to individuals who were part of one particular tribe not recognized by the U.S. government, members of which live in rural U.S. cities (blue and yellow data points, Figure 6-8). Most of the latter individuals claim very dilute Indigenous American ancestry, of uncertain legitimacy. Interestingly, some claim higher blood levels, but do not show substantial levels of Indigenous American admixture, in stark contrast to the U.S. government-recognized tribal members who obtained Indigenous American percentages more or less consistent with the claimed blood level. The remainder of this discussion will focus on the data from the U.S. government-recognized tribes.

The recognized American Indians showed mainly Indigenous American ancestry with substantial European affiliation (exact $p = 1.4 \times 10^{-49}$). The average level of European affiliation was well above the European threshold using both algorithms, and most of the samples tested showed individual levels at or above the threshold (obtained from the simulations of homogeneously affiliated samples; $\geq 7\%$ with the MLE algorithm; Figure 5-31D). The quality of

Figure 6-8

Tetrahedron plot of Indigenous (Native) American populations. Samples from each population are color-coded and indicated in the legend, upper right. Symbol shape is determined by the subtriangle in which the MLE falls. Vertices indicative of 100% ancestry for each parental group are labeled with two-letter symbols as indicated to the left. Samples from U.S.-recognized American Indian tribes are plotted and identified based on the self-reporting of half Amerind blood or greater (red) or less than half Amerind blood (green). In addition, samples from a tribe that is not federally recognized is plotted based on the self-reporting of half Amerind blood or greater (blue) or less than half Amerind blood. Most of these individuals from the latter tribe live in urban areas of the United States. We can see extensive Indigenous American admixture for the samples from the federally recognized tribe but little for the unrecognized tribe. Data taken from Table 6-1.

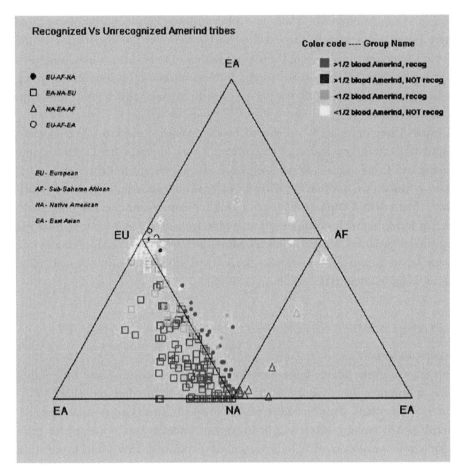

the European percentage result from MLE analysis was high relative to its standard deviation. European admixture for American Indians is not unexpected.

Smith and colleagues (1999) identified the mtDNA haplogroups of 849 individuals that were reported as being full-blood Indigenous American. They found that 99% of the individuals belonged to haplogroups A, B, C, D, or X, but that approximately 1% of the individuals studied belonged to haplogroups found predominantly in Europe or Africa, most likely due to interbreeding between these groups and Indigenous Americans since the time of European Contact (reviewed in Jobling et al. 2004). This 1% is much less than what we showed with American Indians, similar to what we show with Indigenous American parentals and similar to what was shown for other Indigenous American parentals using other SNPs and methods in Chapter 4 (Figure 4-1B). Underhill et al. (2000) showed a substantial number of modern-day North American and Latin American Y-haplogroups were of European origin (about 5%), and although Rosenberg et al. (2002) did not address nonparental Indigenous American populations, they found substantial European affiliation (up to about 15%) for Latin Americans.

For the Rosenberg, Smith, and Underhill work, as well as for the mtDNA work, analyses were performed on Amerind parentals selected for high levels of expected Indigenous American ancestry. The levels we show in Indigenous American parentals here and in Figure 4-1B are equally low (for the former, see the pie chart over Mexico with "P" symbol; Figure 6-3), but the level of European admixture we are showing in American Indian populations obtained in North America is substantially greater than this. Thus, the difference between our Amerind results and those from Indigenous American populations obtained by us and other authors seems to be due to sampling or differences between parental and hybrid populations. Amerinds not selected based on 100% blood expectations would seem to be better described as a hybrid population than a parental population. Indeed, samples from American Indians living on reservations and claiming greater than 50% Native American blood register with consistently greater Indigenous American admixture than those living on reservations and claiming less than 50% Amerind blood (Figure 6-8, red vs. green symbols; note that several incidences of 100% Indigenous American were found, but they plot as a single point in this figure at the Indigenous American vertex).

These samples were the 170 most rigorously qualified Indigenous American samples we could find, using a tribal member living on a reservation as a collector, and even so we clearly see great heterogeneity in Indigenous American/ European admixture proportions (all samples collected from reservations from individuals describing themselves as American Indian). Evidently there is significant heterogeneity in the population and differences in the sampling schemes probably account for the variation in European admixture estimates across studies.

Other work not yet discussed, using autosomal markers (classical blood groups), agree with our results using autosomal AIMs. Chakraborty (1986) reviewed studies conducted in the 1970s, using protein polymorphisms (which might be under selection as described elsewhere in this text) to produce European m values ranging from 3% to 30%, depending on the tribe. Based on this extensive European admixture, he concluded that "... the ancestral Amerindian gene pool is no longer in existence in an unmixed state." Inspection of Figure 6-3 reveals that modern-day Amerinds and Hispanics harbor lower expected primary (Indigenous American) ancestry compared to other continental parental populations. To appreciate this, compare the level of light blue for Amerinds in the pie chart over North America with the level of yellow color for various East Asians in Figure 6-3. A cursory inspection of this figure seems to reinforce Chakraborty's statement and together with the difficulty inherent to locating suitable Indigenous American parental samples (we had to identify the oldest residents in isolated regions of Southern Mexico), paint a somewhat alarming picture; that Indigenous Americans are in the process of disappearing as an unadmixed subpopulation group. The autosomal data produced by others (Rosenberg et al. 2002) shows great European admixture for Indigenous Americans (see Figure 4-1A, B). Whether the admixture was introduced pre- or post-European colonial expansion is the subject of debate, and we cannot rule out precolonial interactions as a source.

In Indigenous Americans, the average level of East Asian admixture was also relatively high, and significantly different from Indigenous American parentals (exact $p < 0.01$ with the MLE algorithm data, Figure 6-8; compare with Figure 5-2 in Chapter 5), but the average level for the population was well below the threshold level for the individual, which is also high due to the relatively recent shared ancestry (and low genetic distance) between these two groups (from Table 5-23A in Chapter 5, it is 11%). This is a good example of the difference in statistical significance on the population and individual level and suggests that although the average Amerind does not harbor enough East Asian admixture to be significant, the Amerind population does. The quality of the East Asian percentage was high relative to its standard deviation. Fifteen percent (26/170) of the individuals showed levels above the 17% East Asian threshold for Indigenous Americans and 8/170 showed levels at or above the 30% level, but the level is nowhere near the level seen for South Asian Indians (compare yellow color for Amerinds and South Asian Indians in Figure 6-3). Rosenberg et al. (2002) showed systematic (occurred in virtually every sample) East Asian affiliation for Colombian, Mayan, and Pima Indigenous Americans, and the level was fairly low at about 10%, not so different from our reading of 7%. As the exact p-value suggests, the average simulated Indigenous American does not show anywhere near this level of East Asian admixture.

Our knowledge of the relationship and low genetic distance between American and East Asian populations makes interpretation of the East Asian affiliation for this group difficult. With autosomal admixture, East Asian/Indigenous American distinctions can be especially tricky to interpret due to possibly small Indigenous American effective population sizes and low intergroup genetic distance. For this reason, the Y and mtDNA data are especially instructive. Underhill et al. (2000) did not show East Asian admixture (in terms of Y haplogroups) for North or Latin American Indigenous American parental samples equivalent to that observed in our admixed samples with autosomal AIMs, but did show a low frequency of East Asian haplotypes in South Americans. Most Indigenous American NRY haplotypes are known to be present in East Asia at very low levels, and we have already discussed the sharing of haplogroup C among East Asian and American populations. The mtDNA, however, sings a slightly different tune. The Indigenous American mtDNA A, B, C, and D haplogroups constitute about 50% of East Asian haplogroup diversity (see Table 6-3 later in this chapter). Combining the mtDNA and Y results, we might expect modern-day Indigenous North Americans to exhibit from about as little as a few percent up to 50% East Asian admixture, but this estimate is complicated by the low genetic distance between the two groups and the possibility that there is no admixture at all, and the sharing of Y and mtDNA haplogroups is also due to a recently shared past. Our estimate of 6 to 7% is much less than the average of the Y and mtDNA results, undoubtedly due to heterogeneity and sampling, but its origin comes with the same limitations and uncertainties attached.

Given the relatively low genetic distance between East Asians and Indigenous Americans, pinpointing a mechanism for this extensive East Asian admixture is exceptionally difficult. One solution may be to date the East Asian character through measurement of LD between East Asian (or all) alleles in Indigenous Americans harboring East Asian admixture. A high LD would indicate a recent event and an LD level equivalent to that for randomly paired markers on the same chromosome would indicate that the East Asian ancestry in Indigenous Americans is more likely due to a residual effect of a relatively recent shared ancestry.

Though most anthropologists support the notion that there was a single migration of East/Central Asians across the Bering Strait into North American about 15,000 years ago, it is possible that the expansion into North America was gradual, or that there was more than one migration into North America. Perhaps the second wave of migrants came from East or Central Asia, and occurred so recently that descendents of this second wave show higher East Asian ancestry (since less time has passed for drift to erase the East Asian character in allele frequencies). Perhaps there was only one migration, and 15,000 years is not enough time to erase all the genetic connections. Lell et al. (2002) and Schurr et al. (2004) suggested that Indigenous American male lineages were derived from two major Siberian migrations. From their paper, ''The first migration originated in southern Middle Siberia with the founding haplotype M45a (10-11-11-10). In Beringia, this gave rise to the predominant Indigenous American lineage, M3 (10-11-11-10), which crossed into the New World. A later migration came from the Lower Amur/Sea of Okhotsk region, bringing haplogroup RPS4Y-T and subhaplogroup M45b, with its associated M173 variant. This migration event contributed to the modern genetic pool of the Na-Dene and Amerinds of North and Central America.''

In contrast, however, Zegura et al. (2004) recently published that the diversity of Indigenous American Y chromosomes in terms of approximately 60 SNPs and a few STRs better supports a single migration event. Indeed Zegura et al.

(2004) narrowed the source of the migration to a region of Western Siberia (i.e., Central Asia).

Our sample of Amerinds also showed West African affiliation levels that approached statistical significance (p = 0.077 average 3%). Some of this information is lost in Figure 6-8 since we are forcing a three-way population model in this figure in order to plot the data in the triangles, but plotting the entire (4D) likelihood space for each individual and folding the tetrahedron projection would show a significant expansion of the likelihood space for the entire population toward the West African vertex. The quality of the West African percentage relative to its standard deviation was high, and this combined with the infrequent occurrence suggests very recent admixture and strong amalgamation within America Indians sampled.

To summarize this section, results obtained by others confirm ours with respect to the European and to a lesser extent the African and East Asian fractional affiliation or admixture for Indigenous Americans. The former two types of affiliations would seem to have come from recent historical events in North America, namely the interaction of Indigenous Americans and both Europeans and West Africans. European colonizers commonly grouped non-Europeans into the same grouping scheme and it is possible that the fractional West African affiliation is due to bona fide admixture precipitated by European-driven social constructions. The East Asian affiliation may be the result of recent shared ancestry or recent admixture and it is difficult to hypothesize which or whether both are at work.

When not part of isolated parental populations, American Indians show primarily Indigenous American ancestry, with substantial European affiliation, less often East Asian affiliation, and even less frequent West African admixture as well.

Mexican

A sample of 60 Mexicans registered with primarily Indigenous American ancestry, as expected, but with almost equivalent levels of European ancestry, which was obviously very different from results obtained in Indigenous American Parentals (exact p = 4×10^{-28}). This basic result was obtained with both algorithms and was not entirely surprising in light of previous admixture analyses of this group by others (described later). The ratio of the percentage affiliation:standard deviation for both Indigenous American and European affiliation was quite high (almost 3), further underscoring the significance of the admixture result as characteristic of this group.

Indeed we know that since European colonization of North, Latin, and South America, Europeans and Indigenous Americans have lived side-by-side for the first time in history, and this combined with the fact that the genetic distance between parental Europeans and Indigenous Americans is relatively great, suggests that the result is due to recent population admixture rather than shared ancestry. There is considerable variability in Indigenous American: European ratios, showing what appears to be a normal distribution about the 50/50 mark (see Figure 6-9). In Figure 6-3 it seems that the constitution of Hispanic Mexicans and American Indians is about the same, but Figures 6-8 and 6-9 belie this hypothesis, indicating that the incidence of extreme Indigenous American or European ancestry is far greater for individuals self-describing as American Indian than Hispanic. In other words, the variability in Indigenous American: European ratios is greater for nonparental American Indians than Hispanic Mexicans.

The average West African affiliation was also high and statistically significant on the population level, but with considerable standard deviation. The 6.5% level obtained with the MLE algorithm is well above the West African threshold for Indigenous Americans (<2%), and in Figure 6-9 we can see a noticeable displacement of the mean estimate toward the West African vertex in the tetrahedron plot. From this figure we can clearly see the high variability

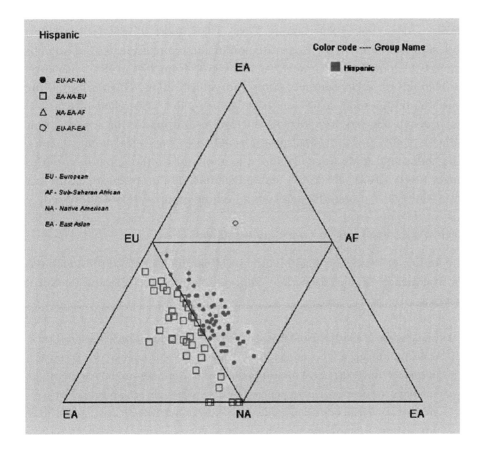

Figure 6-9

Tetrahedron plot of self-reported Hispanic samples collected throughout the United States. Symbol shape is determined by which subtriangle the MLE falls. Vertices indicative of 100% ancestry for each parental group are labeled with two-letter symbols as indicated to the left. The average admixture ratio is about 50% European (EU) and 50% Indigenous (Native) American (NA), but we can clearly see that some self-identified Hispanics are almost entirely European in genomic ancestry whereas others are almost entirely Indigenous American. Substantial sub-Saharan (Western) African admixture is also evident. Data taken from Table 6-1.

in West African affiliation. Indeed, 35/60 (58%) individuals registered with >4% West African affiliation, 17/60 (28%) registered with greater than 10% and 6/60 (10%) registered with >18%. These are not low levels of African affiliation, as the results from the parental populations and simulated populations have shown, and this result indicates convincingly that larger studies on this group will also indicate substantial West African admixture, almost certainly a consequence of political forces that have shaped the migration and socioeconomic structure of African-American and Hispanic populations over the past 300 years. [The wife of the author appears to be one of these admixed Hispanics (EF, Figure 5-18), as well as each of her sisters.] The East Asian level is far below the East Asian threshold for Indigenous Americans, though there are some individuals with substantial East Asian admixture (with levels far above the 95% confidence threshold; see Figure 6-9).

OTHER WORK SHOWING EUROPEAN ADMIXTURE IN HISPANIC POPULATIONS

One of the first papers to describe a comprehensive application of individual admixture estimation methods was that of Hanis et al. (1986) who focused on Mexican Americans of Star County, Texas. Chakraborty (1986a) was another. Both studies confirmed the large variation in admixture estimates indicated by Figure 6-9, reporting Indigenous American levels from 28% to 49%, with levels correlating with socioeconomic status and gentrification. This work used the gene identity method, a two-way population model, and relied on classical blood group markers, very few of which were informative for ancestry. Our estimate of 49%, for individuals not selected based on socioeconomic factors, using autosomal AIMs and a likelihood method, falls at the upper limit of this range. Rosenberg et al. (2001) showed significant European admixture for Latin Americans, though Hispanics as an ethnic group were not addressed.

AFRICAN ADMIXTURE IN MEXICAN HISPANICS

The finding of West African admixture for Mexican Hispanics is presumably as a consequence of the slave-trade importation of West Africans to North America.

1. Underhill et al. (2001) showed that about 10% of the Y-chromosome haplotypes in Latin America are of African origin, a result not dissimilar to our own (see Figure 4-2). In contrast, the frequency of African haplotypes for North American Indigenous American and South American populations was insignificant. This suggests that a primary component of structural difference between American Indians and Mexicans is West African admixture.

2. Rosenberg et al. (2002) did not show significant African admixture for Indigenous American populations, but this group did not test Mexicans or Hispanics specifically, rather they tested relatively homogeneous Mayans, Pima, Colombians, Surui, and Karitiana (see Figure 4-1).

MORE DETAIL ON INDIGENOUS AMERICAN ORIGINS

As for the hypothesis that the Indigenous American affiliation is the result of multiple founder events originating from populations that resemble modern-day Central Asians, let us look in more detail at the origins of Indigenous American and Asian populations, as we know them today. We first describe the extant genetic variation, history of how this variation likely came to be, and then we will illustrate how a residual connection to Central Asia could have been preserved until today. We know from Y chromosome haplogroup analysis that most American Indians have group X/M3 lineages (using the old nomenclature; Santos et al. 1999), the precursor of which is believed to have originated in Siberia, but that they also exhibit lower frequencies of Group V lineages associated with Na Dene speaking populations (Karafet et al. 1999; Ruhlen 1998; Underhill et al. 2003; Figure 4-2). These lineages are believed to have crossed the Bering Strait during the Holocene Period.

Paleoanthropological evidence suggests an expansion of a culture called the Clovis culture into the Americas 11,000 years ago, though South American sites show different archaeological signatures (Fiedel 2000) and it is possible that there were colonizations apart from the crossing(s) of the Strait—perhaps via ocean voyage, or island hopping across the South Pacific. The migration of human beings from Asia across the Bering Strait to the Americas is well documented and intuitive, but at the present time it is difficult to answer the question of whether or not there were multiple or continuous migrations, or whether some of these migrations took place across oceans rather than land or ice masses. Though it is difficult to localize the origin of Y-chromosome types based on modern-day distributions, these distributions can be combined with the paleoanthropological record to suggest a small number of significant migrational events from West to East. Most anthropologists believe there was one such event and it has been recently posited that this event originated in the vicinity of the Altai Mountains in Southwest Siberia (Zegura et al. 2004). Others believe there were two or more, or at least, a continuous flow of genes rather than a sporadic flow (see Jobling et al. 2004 for a review of this debate). The distribution of Y haplogroups cannot exclude a subsequent east-to-west flow (i.e., a return or a reverse migration), but there is no record of such a flow in the paleoanthropological record, which is what makes the presence of Indigenous American genetic signatures in Asia and Eurasia so interesting.

If the answer to the spread of Indigenous American affiliation throughout parts of Europe and Asia is residual relationships due to common recent founders, why is only one Y haplogroup present in each of what we are calling a Central Asian-derived population—a seemingly different haplogroup for each (South Asians with IX, Oceanic with V, American mainly X)? Were genetic drift and/or founder effects so powerful? How could the genetic connections still exist 15,000 years later such that we would read them with an autosomal admixture test? We have discussed the idea of relic distributions and intermediate allele frequencies but we will recast this discussion in hypothetical example, to illustrate the concept specifically for understanding the perplexing issue of how Indigenous American chromosomes could be found in Asia (given that there is no anthropological record to support a reverse gene flow from the Americas back to Asia).

We have already introduced the basic logic relevant here when discussing our hypothetical ABC population earlier in this chapter. Here we will develop this logic a bit more. Let us assume a small, hypothetical group of Central Asians and East Asians contributed to a single migration event across the Bering Strait 15,000 years ago. Assume both groups contributed equally (ratio 50%:50%). If the migrating population was a perfect statistical sample of their original groups, the new established Indigenous American populations would begin their genetic journey in isolation with 50% Central Asian and 50% East Asian character defined by the populations from which they came. If there were four major Y chromosome haplotypes for the former, and another, different four major haplotypes for the latter, the new population would have eight such major haplotypes and the new frequency of each one would be half what it was before. The haplotypes would have been subject to the forces of genetic drift just as they were in their original populations, but perhaps more so after the migration given the relatively low number of founders and the different selection pressures found in the new environment. Perhaps, through genetic drift, three of the original eight haplotypes survived to dominate in frequency. This way, the frequencies of the haplotypes could quickly change and produce a differential representation of common haplotypes between Indigenous Americans and Central/East Asians. For each migrational event (to South Asia, Oceana, etc.) different haplotypes would survive. This would produce the type of result we see from Y-chromosome, mtDNA, and autosomal admixture tests—different frequencies of a common set of Central Asian haplotypes distributed throughout Asia and America. Leaving our hypothetical situation for a moment, most every Indigenous American Y-chromosome haplotype for instance is also found in East or Central Asia (see Table 6-3).

Ripan et al. (2004) called these shared haplotypes basal haplotypes, but reading at a larger number of polymorphisms allows for the discovery of new

Population	A	B	C	D	X	Other
Malays	0	14	0	0	0	86
Vietnamese	0	7	0	0	0	93
Taiwanese	10	20	5	0	0	65
Koreans	8	8	0	22	0	62
Mongolians	5	2	12	19	0	62
Bella Coola	65	11	10	14	0	0
Nuu-Chah-Nulth	45	7	16	26	7	0
TM Chippewa	57	18	18	0	7	0
Pai Yuman	7	67	26	0	0	0
Northern Paiute	0	41	18	41	0	0

Table 6-3

MtDNA Haplogroup frequencies (%) of A, B, C, D, and X in selected Asian and indigenous American groups.

Data from Merriwether and Ferrell (1996); Malhi et al. (2001, 2003, in press).

sequences that are unique to Indigenous Americans. Our test and those from the Y and mtDNA chromosomes are relatively simple tests, which incorporate markers of character that usually have not yet become completely distinct in one group or the other. In other words, if the process of genetic drift is not complete at the time we take our snapshot (which is what we are doing in the present day), chromosome types would have drifted apart in terms of frequency, and even new ones may have appeared (if we know where to measure in the genome in order to read them), but not all the original chromosome types would have disappeared yet and the original and new populations remain connected through these common sequences. Thus, we would have different sequence frequencies in the two groups that came to be through a process that has absolutely nothing to do with population admixture; it is best described as a relic distribution rather than an admixture distribution.

RESOLUTION WITHIN CONTINENTS BASED ON THE FOUR-POPULATION MODEL

Our population model is a four-population model, and though it is one of many possible models, it is especially practical and justified by the results using randomly selected markers and hypothesis free clustering methods (i.e., STRUCTURE). For instance, if an indigenous Australian subgroup were better representative of a fourth genetic subgroup using a four-population model, Rosenberg et al.'s (2002) results at k = 4 would have shown the Oceanic populations breaking out before the Indigenous Americans or East Asian, which they did not. (But were this test developed by the primarily European social establishment of Australia instead of in the United States, wouldn't Oceanians make a better fourth population than Indigenous Americans anyway?) Indeed, to complicate the discussion, we saw in Chapter 4 that Oceanic populations (New Guinea samples) did in fact break out as a cluster before East Asians (Figure 4-1B). This indicates some variation in the order with which signatures contributed by parental groups are recognized by different marker sets, algorithms and polymorphism types and that if Oceanic populations are to be part of a global admixture analysis, the k = 5 model is the best choice. This does not make a k = 4 model incorrect, only most appropriate for global analyses that do not involve Australian and Oceanic populations. Indeed, we have seen that there are significant ethnic and as we will show, anthropometric differences in ancestry reported in terms of these four populations. The issue we address from here is how these differences can be practically exploited to assign ethnic ancestry.

Since the differences in fractional West African, Indigenous American, East Asian, or European affiliation among the ethnic groups is sometimes not great, this is a potentially troublesome endeavor. An individual of 95% European/5% West African ancestry could be an admixed European, a person of Jewish descent, a European of Mediterranean descent, or an individual from the Middle East or Southwestern Asia, with no recent admixture. However, if individuals of these ethnicities are the only ones for which this type of profile commonly is found, we can form a probability statement as to the potential inclusion of our 95% European/5% West African individual into a subgrouping that combines the ethnic groups we have just listed to the exclusion of others (such as Northern European, South Asian Indian, Chinese, etc.).

One way to visualize this probability is through Principle Coordinates Analysis (PCA) shown in Figure 6-2, which is the PCA plot for ethnic status using the 171 continental AIMs. Middle Easterners and Greeks plot below the horizontal line, which represents the midpoint of the first principal component of AIM genotype variation with respect to ethnicity, but Northern Europeans, European Americans, Italians, and Iberians plot above it. Thus, we might imagine plotting an unknown individual on this plot, and if the position of the plot is below the line we might safely apply the classification of Middle Eastern or Greek, not Northern European, Italian, Iberian, or General European.

INTERPRETATION OF CONTINENTAL BGA RESULTS IN LIGHT OF WHAT WE HAVE LEARNED FROM APPLICATION TO ETHNIC POPULATIONS

The results discussed in this chapter showing statistically significant secondary affiliations as a function of ethnicity suggest that we need to rethink our criteria for interpreting genomic ancestry results derived from our simulations in Chapter 5.

Consider an individual showing genomic ancestry levels of 93% European and 7% Indigenous American. On the one hand, a 93% European and 7% Indigenous American BGA profile is characteristic of :

- Unadmixed Europeans (wherein the 7% would represent statistical error)
- Certain admixed Europeans (such as dilute Hispanics and Amerinds)
- Certain ethnic European diaspora (such as Mediterranean, Middle Eastern, Jewish or South Asian Indian, wherein the 7% Indigenous American would come from partial affiliation with populations that share a relatively recent common ancestry with Indigenous Americans, such as Central Asians)
- Certain ethnic Europeans where, due to genetic drift, an isolated subpopulation has moved away from European frequencies toward Indigenous American multilocus genotype frequencies by chance

However on the other hand, the 7% is below the 95% confidence threshold defined by the simulations, which is 10% for Europeans (see Table 5-23A). The likelihood estimates follow a Gaussian distribution, and for a Gaussian distribution, the chance of falling more than 0.5 standard deviations from the mean is more than 6 out of 10. Thus, it would be easy to conclude that the 7% fractional affiliation with Indigenous Americans could easily be due to statistical error and one might prudently conclude that the reading is not significant.

This would be an easy place to leave it, but our discussion of results in subpopulations forces us to reevaluate this interpretation. In particular, the ethnic data suggests that this interpretation may be too simplistic because it does not seem to account for the fact that with this particular panel of markers (71 or 171 marker sets) certain groups show systematic secondary affiliations at this 7% level (from about 5–10%; Table 6-1). At the level of the individual, the readings are not significant indicators of real partial ancestry at the 95% confidence level but at the level of the subpopulation or ethnicity the readings differ from those expected by random statistical noise by a level that is significant. A population of 100 individuals with 93% European and 7% Indigenous American ancestry would indeed be a significant departure from expectations based on the nature of the allele frequency distributions, and so knowing this, we cannot dismiss the 7% in any one individual part of that population as insignificant.

For example, Greeks (one of our diaspora) show systematic fractional Indigenous American affiliation (at levels of about 5%–10%) but not East Asian affiliation (Figure 6-10). Recall that the MLE estimation error in the EAS

Figure 6-10

Tetrahedron plot of various European samples collected throughout the United States and Europe. Samples from each population are color-coded and indicated in the legend, upper right. Symbol shape is determined by the subtriangle in which the MLE falls. Vertices indicative of 100% ancestry for each parental group are labeled with two-letter symbols as indicated to the left. Northern European samples plot with relatively homogenous European (EU) admixture, but Greek, Ashkenazi Jewish, and Middle Eastern samples plot successively with more non-European genomic ancestry. Inset at right is shown for greater detail of the EU vertex. Data taken from Table 6-1.

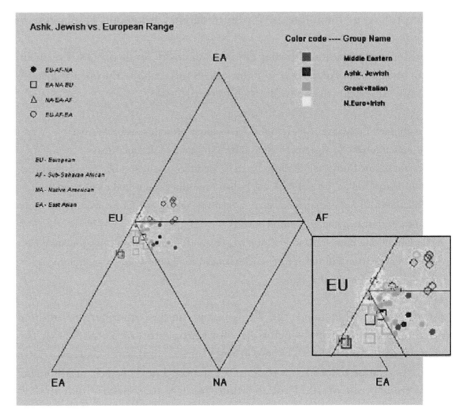

dimension is roughly similar to that in the NAM dimension yet we do not see equal numbers of 5 to 10% EAS as 5 to 10% NAM Greeks—most all are from 5 to 10% NAM. The type of admixture is predictable based on trends we see in neighboring regions. Thus, the Indigenous American admixture for Greeks does not seem to be random noise because random noise would produce low levels of admixture of all types in proportions roughly equal to the magnitude of bias in MLE estimates for each of these types. Table 6-1 shows that many subpopulations show characteristic profiles that differ from parental profiles in a significant way.

The type of result represented by 93% European/7% Indigenous American, on a population level is very difficult to reconcile with what we know about the accuracy and precision of the test on an individual level. This leads to the speculation that the simulations discussed so far may not perfectly report the statistical performance (due to continuous allele frequencies) of the test in the real world. The simulations assume that genotypes are created randomly, but we know that due to genetic drift, sexual selection, natural selection, bottlenecks, and founder effects that the worldwide set of multilocus genotypes is far from a random collection. There is considerable structure and pattern.

We already have discussed these forces as confounding unknowns that introduce error in admixture methods, and for which advanced statistical schemes are required to handle. Though the data in Table 6-1 was created using a straightforward MLE algorithm, using these advanced statistical schemes with the same data, such as McKeigue's ADMIXMAP, produces a very similar table. This suggests that the differences in affiliation thresholds needed to demonstrate statistical significance between simulated individuals and real populations is *not* due to deficiencies with our assumed population model in our MLE method (such as allele frequency misspecification, or a poor choice of parental representatives, etc., as discussed in Chapter 4). Indeed, we used real, observed genotypes from these structured and patterned populations to estimate allele frequencies and define absolute ancestry for our groups using hypothesis-free methods. Simulating *random* genotypes from this real, observed data may not be the best approximation of the real genotypes in the worldwide collection, of course, but this is precisely the type of structure we seek to identify (i.e., that which exists).

APPROPRIATENESS OF A FOUR-POPULATION MODEL

The results presented logically bring up the question of how appropriate many of the ethnic populations tested are for the four-continent population model employed in the analysis. The test described is largely heuristic, and it coalesces the world population into four main anthropological groups that fall largely along continental lines. Though our geographical divisions are consistent with anthropological history, the use of four groups is indeed a simplification of a very complex situation, and though the selection of these groups can be defended from phylogenetic analyses, because we cannot know *a priori* which branches of the family tree of humanity are truly distinct and which are the result of admixture; we cannot be certain that each of the ethnic groups is strictly speaking appropriate for this four-continent population model.

We already have spent considerable space discussing how it is we objectively settled on a four-population model. One could use a test such as that described herein, which coalesces the human population into one group (population model of one group) and the answers provided would show that each sample is 100% affiliated with that group; this answer is not any more wrong than that provided by a four-, five-, or six-population model, just not as useful to use George Box's terms, and maybe uninteresting to most. This said though, it bears repeating anytime one speaks to the media or the lay public that indeed, we are one species with a single recent common origin, and so this single group model may not be as useful as we would first conclude. By selecting the parental groups we did, essentially we have calibrated our test for a particular time scale in human history and the results are a function of this calibration time frame. Choosing other parental samples, representative for other groups, and using different AIMs simply calibrates the test for a different anthropological time scale, and the difference between the answer provided by a test based on four world population groups and a more complex test based on 10 or 25 would be one of anthropological time scale, not accuracy. For instance, using the k4 model of Figure 4-1B may enable the inference of a subset of anthropometric phenotypes possible using the k5 model of Figure 4-1B, but it would be overly simplistic to label one ''wrong'' and the other ''right'' or ''better'' or ''worse'' unless one is referring to application for specific phenotypes in specific populations (such as Australians).

We have discussed the results from others working with STRs that have shown that the human population clusters to some extent along continental lines in a manner that justifies our choice of parental samples and calibration time frame (Rosenberg et al. 2002). Some of the populations we have chosen to work with are themselves complex demographically and anthropologically. For example, phylogenetic trees constructed by Zhivotovsky et al. (2003) have suggested that sub-Saharan hunter–gatherers are as genetically distant to West African farmers as they are to Eurasians, which would suggest our test, which does not incorporate Pygmy, Khoisan, or Eastern Africans as a parental sample, might type Eastern Africans as predominantly European not West African.

Though work on East Africans with our test remains to be done, Rosenberg et al. (2002) showed that much of the extant West African diaspora form a neat cluster when the human pedigree is coalesced to a few (i.e., 3 to 6) groups (see Figure 4-1), and European genetic structure was detected for the farmers using a variety of values of k in STRUCTURE, suggesting that the distinction between Eastern and Western Africans is possibly due to gene flow and admixture rather than divergence from a common West African ancestor. The results with SNPs and PSMix program in Figure 4-1B show the same pattern. If the mechanism is admixture, then our test would be expected to register this admixture accurately. If it is not, the inference of BGA admixture in terms of our four groups for individuals of Eastern African heritage would produce percentages and affiliations that, while genetically correct, may be semantically misinterpreted (not that they are wrong, since we are simply reporting genetic affiliations however those affiliations came to be). Unfortunately, at this time, we cannot say which is the case, or how many other similarly distinct groups not caused by admixture might exist, which is an important reason why a test such as we have described should be used as an empirical, not necessarily a definitive, tool.

This example also speaks to the need for regionally specified genetic models. As we just pointed out, a genomic ancestry test developed for use in the Americas, as the AncestryByDNA 2.0 and 2.5 tests were, using the four populations selected, would not be the best choice for Australia, where one might well substitute Aborigine for Indigenous American or perhaps Southeast Asian for West African. Likewise, in developing such a test for South Africa, the West African component of the 171 AIM (American 2.5) test would likely be informative for the Bantu-speaking derived black population, but the Cape Colored, who are largely Khoisan in extraction, would not be well represented among the four parental populations in the American test. How would they type with the American 2.5 test? Similarly, an Indian test would require different parental populations. Clearly, although the tests described in this book are arguably valid and cosmopolitan in application, they are in no way absolute, and for a truly useful test to be developed, more information, such as the region of the world where the test will be applied, are very important. This is true both for forensic and recreational/educational applications.

DO ALLELE FREQUENCY ESTIMATION ERRORS ACCOUNT FOR THE SECONDARY AFFILIATIONS IN ETHNIC SUBPOPULATIONS?

The simulations show that the interesting secondary affiliation results for the ethnic groups we have just discussed are unlikely due to straightforward statistical error caused by the continuous nature of the allele frequency distributions. What about bias caused by statistical error in parental allele frequency estimation? Recall from Chapter 5 that this could introduce bias in our estimates—bias we cannot discern with simulations. The most powerful evidence that the minor ancestry affiliations we have been discussing so far are bonafide indicators of population structure (however it came to be) is their observation using different SNP panels, software programs and parental and test sample sets. Another line of evidence comes from our study of the statistical behavior of the panel of markers showing these patterns (see Figure 6-11). Since the former evidence is fairly obvious, we will focus on the latter for this section.

Recall from Chapter 5 that when we adjust the true allele frequencies for AIMs relevant for the resolution of affiliation between two groups, such that the parental δ value is decreased significantly for each relevant AIM for a given pair of groups, the overall power of the test described herein is degraded with respect to resolving affiliation proportions between these two groups. However the BGA test produced more or less the same results, both in terms of major affiliation, but more importantly, in terms of minor admixture estimations. Parental sampling bias might cause inaccuracies in parental allele frequency estimation and δ values in this manner, though certainly less than 20% given that our parental sample consisted of about 100 individuals. Further, the error in allele frequency estimation would have to be obtained in the same direction for

Figure 6-11

Another tetrahedron plot of various European samples collected throughout the United States and Europe. Samples from each population are color-coded and indicated in the legend, upper right. Symbol shape is determined by the subtriangle in which the MLE falls. Vertices indicative of 100% ancestry for each parental group are labeled with two-letter symbols as indicated to the left. Continental Europeans (green) plot of similar East Asian and West African admixture as U.S. Caucasians (blue), and both groups show elevated levels relative to simulated European parental samples (red). Data taken from Table 6-1.

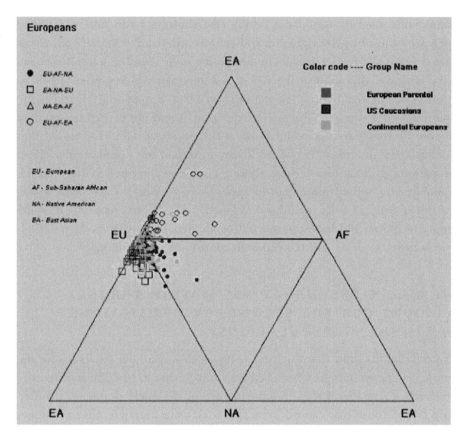

most of the AIMs relevant for resolving affiliation between a pair of groups for a systematic error such as this (parental frequency bias) to exist. Nonetheless, our results from experiments with systematic bias indicate that the BGA test is relatively robust in the face of parental sampling bias and allele frequency estimation.

Might the interesting secondary affiliations observed for the ethnic subpopulations have been produced by allele frequency correlations between groups? Recall from Chapter 5 that when we intentionally diminished the ancestry information content of the battery of markers with respect to a specific pair of ancestry groups, we observed a distortion of confidence contours along only the axis corresponding to that for which genotyping failures were relevant. This shows a healthy disconnect among the interdigitating components (subsets of AIMs) of the test. In other words, if samples are fit to several templates for the determination of proportional fit, the elements of these templates should be independent for the test to be valid (free of bias), and this is in fact what we observed. What this amounts to is that although there may be correlations in allele frequencies between ethnic groups, there is none between parental groups, and the correlation for the former is most likely a function of history rather than statistical error. Nonetheless, Bayesian methods such as STRUCTURE and ADMIXMAP can accommodate correlated allele frequencies and the use of these methods to show the same patterns suggest this is not the explanation.

INDICATIONS OF CRYPTIC POPULATION STRUCTURE

Our findings suggest that BGA admixture is more common than previously believed, and if this result is true it would beg the question of whether finer levels of genomic ancestry components are linked to human biology, such as drug response, disease predisposition, or of interest in this book, physical appearance. Such structure might best be described as *cryptic* structure, which only a molecular test could determine, as opposed to crude population structure, which a questionnaire could accurately measure. Our results aside, we know that there are finer levels of population structure extant than the crude continental levels we measure with questionnaires. As one of many possible examples: significant anecdotal evidence suggests that redheads require 20% larger doses of many common anesthetics, and exhibit a tendency toward hypertension and bleeding while under anesthesia (Liem, personal communication; Cohen 2002). These complex physiological responses would appear to be difficult to explain based on Melanocortin-1 (MC1R) variants that have previously been linked to some of the variation in red hair color (Flanagan et al. 2000; Koppula et al. 1997; Robbins et al. 1993; Smith et al. 1998; Valverde et al. 1995).

There are no doubt specific gene variants responsible for these clinical phenotypes, and if these variants correlate with elements of population sub- or microstructure as they appear to, any study attempting to identify linkages or LD between markers and the relevant phenotypically active loci will be challenged from the outset at the study design step. It is likely that AIMs other than those we describe here are necessary to quantify the elements of population structure relevant for this particular problem, since our AIMs were selected based on their ability to resolve between continental populations, but this example illustrates that precision and objectivity greater than that provided by self-reporting of sociocultural race will ultimately be needed to tease the elements of structure that could interfere with the design and analysis of genetics experiments. The genomic ancestry tests we describe herein represent a first step in this direction. In the next section, we take this approach one step further and look at AIMs for finer levels of population resolution.

CHAPTER 7

APPORTIONMENT OF AUTOSOMAL DIVERSITY WITH SUBCONTINENTAL MARKERS

SUBPOPULATION AIMS AND ETHNIC STRATIFICATION

From the preceding sections, we can appreciate that the continental AIMs have ethnic information content, and that using this information we can form probability statements about gross elements of ethnicity or subpopulation affiliations. Though South Asian Indian, Middle Eastern, Mediterranean, and Northern European individuals reliably show majority European ancestry, each appears to have characteristic levels of West African, East Asian, and even Indigenous American affiliation. For example, with our 171 AIM autosomal test, we showed significant West African admixture for Middle Eastern individuals, and this also was shown by other workers using larger numbers of autosomal, low information content AIMs such as Rosenberg et al. (2002) (about 25%) and Y chromosome haplogroups such as Underhill et al. (2001).

However, not all of the subgroups show such discrete continental admixture characteristics, and the amount of subpopulation information obtained when working with a continental admixture model is really not adequate for ethnic admixture assessment for many European subgroups. Information of a finer nature might be instrumental for further refining the likely physical appearance of a DNA donor. Using the continental AIMs in the preceding sections, in all but the best cases, there is considerable overlap in BGA character among the subpopulation groups. For example, even though a low level of West African affiliation (presumably admixture) on a European background is a common BGA profile for Middle Eastern individuals on a continental level, using this as a feature for identifying Middle Eastern heritage would be subject to significant noise. In Figure 6-10 we can see that a small number of Middle Easterners show profiles more Northern European like (less West African ancestry) than some Northern Europeans. Though this pattern may (and likely does) comport with the expression of ostensible anthropometric phenotypes, we clearly need more ''traction'' if we hope to resolve phenotypes within genetically defined continents and meta-continents such as Europe and Eurasia.

Using a principal components plot such as that shown in Figure 6-2 to estimate Middle Eastern vs. Northern European admixture, or even to bin into either of these two groups, would lead to a large standard error. In other words, using continental level AIMs, there is too much overlap in ancestry profile types to reliably resolve most ethnicities. Though this is not to say that there is no ethnic information provided by continental admixture profiles, or that they are useless for finer levels of physical profiling, but clearly markers of another type will be needed to finely resolve subpopulations to a more reliable degree, which would obviously assist when profiling for physical characteristics and may be essential to adjust for cryptic stratification in large case/control mapping studies.

Until recently, very few autosomal markers have been known with discriminatory power among subpopulations, or within-continental resolution. As we will see, this is not because they do not exist, but rather that they have not heretofore been found. Using continental AIMs (the subject of most investigator interest to date) for the purpose of subpopulation resolution is overly ambitious. With the recent completion of the human genome draft, we have at our fingertips tools for screening through vast numbers of polymorphisms very quickly (though not necessarily inexpensively).

Prior to the beginning of the genome era, for various reasons, most of the work that has been done on the level of the subpopulation has focused on blood group markers and Y-chromosome and mtDNA haplogroups in European populations, though there have been instructive studies in African (Salas et al. 2002), Pacific Island populations (see Su et al. 2000 and Jobling et al. 2004 for a review) and Central/Southern Asian populations as well (Basu et al. 2001; Ramana et al. 2001). Most of these focused on geographically localized subpopulations, defined based on the written record, archaeological data, linguistics, and physical characteristics. Treatments of worldwide levels of structure more detailed than that of the continental level only recently have been achieved, so far only with randomly selected microsatellite loci (Rosenberg et al. 2002), which carry with them certain practical limitations (i.e., they usually are not selected from the genome as AIMs, but randomly selected, and they are characterized by allelic complexity, which imposes the requirement for much larger reference databases).

When identifying and using subpopulation AIMs, we have the same difficulties as we had for continental populations, but they are exaggerated because the genetic distances are lower. The exaggerations of the problems are not just restricted to identifying good AIMs, but deciding on who belongs in our parental populations, and defining which parental populations should be considered. For example, how does one decide who a parental Middle Eastern and Greek is, and who is to say that Greek is a legitimate subgrouping? This is clearly more difficult than deciding who is West African and who is European because of the lower genetic distance, more recent sharing of common ancestors, and lack of phenotypic differentiation, and stands in contrast to our ability to use very clear phylogenetic analyses to define the major continental population groups.

Nonetheless, from archaeology, linguistics, increasingly detailed written records over time, and advances in molecular genetics, we have amassed over the past century vast knowledge concerning the finer topology of our demographic history, and if we can rely on this data to unambiguously define monophyletic subgroups within a continental BGA category, we can take advantage of modern genomics technology to identify and utilize powerful subpopulation AIMs. We can travel the globe to collect samples from various subgroups and screen their genomes for AIMs that are effective in resolving between them just as we did for the continental groups. We would expect our information content to be lower, and standard error to be greater, but we hypothesize that because there are phenotypic differences among these subgroups, and because other information suggests levels of genetic differentiation, that enough genomic information is present in the genome to provide reasonable genomic ancestry estimates.

WITHIN THE EUROPEAN BGA GROUP—A BRIEF HISTORY OF EUROPEANS

The European continental category we have used until now is actually a misnomer, as we have already described. It refers to a proto-European population that probably lived in several places at the same time, most notably throughout regions of the Middle East, otherwise known as the fertile crescent, and possibly parts of Central Asia and Europe. Our parentals for this group were European Americans, who are largely of Northwest European extraction, and yet South Indians type reliably with European ancestry, illustrating that the allele frequencies we have measured and used correspond to those of a population common to these disparate subpopulations. Y-chromosome haplogroups are shared among these subpopulations as we have discussed (see Figures 4-2 and 4-3 of Chapter 4; see haplogroups VI and IX, also see haplogroups J, I, and R* in Figure 4-3, which uses the modern nomenclature).

Indeed, we know from a variety of sources that proto-Europeans diverged from Africa into South Asia and Australia, and into the Middle East some 47 to 50 KYA in different waves and at different times. Middle Eastern populations expanded into continental Europe 40 KYA and established the so-called Paleolithic peoples of Europe, which expanded and mixed with Neolithic farmers after the last glacial maximum, again from the Middle East, about 10 KYA. Neolithic farmers probably also expanded in the opposite direction, toward South Asia, perhaps even Southeast Asia.

The Neolithics were a relatively advanced population that used a different, new (*neo*) tool (*lithic*) type. Paleolithic tools were characterized by small round tools with flakes at one edge, to longer "Mousterian" tools from which large flakes were removed from a rock core, usually with a single long and sharp flake edge (see Jobling et al. 2004 for a review). The people who made these tools were hunter–gatherers, and lived in small, isolated communities with little social structure. In contrast, the Neolithic Period, or New Stone Age, is defined by the use of more sophisticated, polished stone tools and weapons, the manufacture of pottery, and the appearance of relatively large, sedentary settlements. The Neolithic Period arose from the development of agriculture, which produced greater supplies of food than were needed by the producers, enabling specialization of workers and sustenance of larger, denser populations.

Naturally, with greater population density comes population expansion. Indeed, Neolithic culture arose in the fertile crescent of the Middle East, so-called due to the agricultural bounty provided by the region to its innovative inhabitants, and expanded into continental Europe only 10 KYA. This expansion was not a replacement of Paleolithic peoples with Neolithic ones, but a gradual one involving continuous range expansion and gene flow between resident and introgressing populations. To this day, genetic, archaeological, and linguistic signatures remain of this expansion, and from this data we can date the expansion of Neolithic culture from the Middle East to southeastern Europe up until about 7 KYA, into central Europe 6 KYA and eventually into northern Europe only 4,000 years ago. The waves of advancement formed what are observed today as clines of evidence distributed in a southeast to northwest direction. Linguistically, the Indo–European language group dominates continental Europe, the Middle East, and Northern (but not southern) India (see Jobling et al. 2004 for a review of these clines).

In terms of continental ancestry, the genetic diversity across Europe, the Middle East, and South Asia is remarkably similar. Less genetic distance exists among continental European, Middle Eastern and South Asian populations than between them and other populations such as East Asian, West African, or American, and this is detected in the common European ancestry these populations show with continental admixture tests. On a finer level of structure, within the European group, work on classical blood group markers in the 1970s through the 1990s show that the first principal component of genetic variation within continental Europe and the Middle East is along the same southeast to northwest direction, matching the archaeological evidence.

Although mtDNA haplogroups show only weak differentiation as a function of European/Middle Eastern geography, the principal component of variation is also in a southeast to northwest direction. Y-haplogroup sequences show much stronger differentiation in this region of the globe (as with most regions, due to patrilocality), with the principal component again in the southeast to northwest direction (Jobling et al., 2004 contains a number of relevant figures). The pattern is so strong that we can even define specific haplogroups as associated with resident Paleolithic populations and introgressing Neolithic ones; the Gravettian (M170 variant) and Aurignacian (M173 variant) sequences associated with the former, having arisen approximately 22 KYA and 35 to 40 KYA, respectively. Haplogroup J is at high frequency in the south and east, but rare in the west, whereas haplogroup R* is at very high frequency in the west but very low frequency in the east.

Admixture studies that take into account uncertainty with regard to genetic drift and parental character show a sharp northwest to southeast gradient in terms of the Paleolithic and Neolithic contributions to modern-day genetic diversity. This diversity is recognized as having arisen from a combination of postglacial expansions of Paleoliths 18 KYA and Neolithic demic diffusion and admixture, from the Middle East, beginning approximately 10 KYA. The extant difference in mtDNA and Y sequence distributions suggests high patrilocality but low matrilocality in past human demographies; females were evidently much more mobile than males during the past 10 KY, possibly as a result of less territorial and behaviorally isolationist sex-specific character.

In terms of genetic variance, both the unison and diversity among Europeans is strong. In addition to Europe and the Middle East being united with a common language base and common mtDNA sequence types, Y-haplotype and mtDNA haplotype genetic variation, classical autosomal blood group markers, and the archaeological record all sing the same tune. Europeans and Near Eastern populations share a common ancestry to the exclusion of Africans and other non-Africans, but there exists diversity within these groups as well. Admixture commenced during the massive Neolithic population expansion several thousand years ago from the fertile crescent in the Middle East into Europe and this gene flow brought to this region agriculture and a new type of genetic diversity. The trend continues in the opposite direction, where proto-European populations from the Near East admixed with East Asian, South Asian (Dravidian), and Central Asian populations at different times from that characterizing the expansion to Europe. Indeed, the European group as we have defined it is a group with substantial commonality, but also substantial diversity within.

For our purposes, related to physical profiling and anthropological reconstruction, we would optimally want to be able to infer precise subgroup ancestry affiliations. From a forensics perspective, the most useful divisions would be those among which physical characteristics are most distinct. Alas, such subgroups do not always exist, at least not in sharp terms that we would find most useful. Physical characteristics, like eye or hair color, vary in a continuum along geographic axes, which of course, is no accident since genetic diversity also is continuously distributed and gene combinations determine phenotypes. Rather, to develop tools useful for physical profiling and anthropology, we seek to define the most significant clines of genetic, phenotypic, sociocultural, and paleoanthropological variation. If these are concordant with one another, and if certain physical traits correlate with these gradients, as they seem to, we can then justify screening population samples to identify markers informative for ancestry with respect to these clines so that we can learn how to infer physicality from DNA.

HOW DO WE SUBDIVIDE EUROPEANS FOR FORENSICS USE?

Of course as geneticists, we are not of the illusion that European populations can be genetically differentiated among geopolitical lines. Archaeological evidence isn't distributed as a function of modern geopolitical boundaries, and neither are genes. There are almost certainly no French markers, or Italian markers, to the exclusion of any other population residing within Europe. Not only have continental European populations been free to breed across proximal geopolitical lines for most of recorded history, but these lines are ephemeral over time and even the oldest have existed for only a short period when considered on an anthropologically and evolutionary time scale. Rather, the clines that do exist usually are distributed as a function of geography, not with respect to modern-day geopolitics, which often has been facilitated through the imposition of standardized languages and military control through imposed nationalism (though usually accomplished with respect to anthropological history and larger expanses of geography and time).

Indeed, the northwest to southeast clines are the first and dominant principal component of both genetic as well as phenotype diversity, and these clines represent a starting point. Note that this direction is not the direction we might first guess based on phenotype, which would probably be north to south. Even climatologists who might use knowledge of post-glacial patterns of land exposure would likely predict north/south clines as the primary axis of variation. But in fact, the primary clines are northwest to southeast because, presumably, this is how human population expansions and migrations proceeded. There indeed exists variation in the north/south direction, probably due to post-glacial expansions, natural selection related to latitude, and admixture from North Africa, but it is overlain on variation that is patterned in a northwest to southeast direction, contributed by the Neolithic expansions that came from the southeast. On top of that, there exists a component of variation in a west/east direction, most likely due to population expansion and admixture from and to Eastern Europe and West Asia. Autosomal markers (Rosenberg et al. 2002), classical blood group markers (Cavalli-Sforza et al. 1994), Y-haplotypes (Underhill et al. 2001), and archaeological evidence all bring out similar clines. The nature of the genetic clines is relatively complex, but within them can be found an elegant simplicity that agrees perfectly with the archaeological pattern of evidence. The most dominant feature of modern-day genetic variation in Europe is the northwest to southeast component contributed by the Neolithic expansion 10 KYA.

Studies based on the Y-chromosome data have shown that the northwest to southeast pattern of Paleolithic and Neolithic contributions to modern European populations exhibit a clear binary character. Populations north and west of Italy show a high Paleolithic contribution, but south and east of Northern Italy, the sequences are mainly of Neolithic origin (Chiki et al. 2002). Archaeology shows seven to eight major northwest to southeast clines (Jobling et al. 2004), representing time frames of intervals roughly 1,000 years apart. Autosomal markers show seven clines, and Y chromosome variation can be expressed in equally complex terms if we desire to do so. On which level of detail should Europe and the Middle East be divided for the development of genetic ancestry methods?

DEVELOPMENT OF A WITHIN-EUROPEAN AIM PANEL

One way to answer this question is through empirical observation, the method we used. We used sample pools based on geopolitical boundaries as a convenient way to group the samples *a priori*, genotype them for 11,000 SNPs distributed throughout the genome, and then measure the major components of genetic variation, selecting the level of hierarchy for which we have adequate statistical power to resolve (which is determined by application of hypothesis-free clustering methods, such as Pritchard's STRUCTURE program). The eight groups we used would seem to represent most of the Eurasian axes of variation

1. Irish: Obtained from Ireland.
2. Northern European: Obtained from Coriell Institute (precise location indeterminate).
3. Iberian: Obtained from Coriell Institute.
4. Italian: Obtained from Coriell Institute.

5. Greek: Obtained from Coriell Institute.

6. Middle East I: Obtained from Coriell Institute (60% Persian, Arab, Lebanese and 40% Greek).

7. Middle East II: Obtained from Coriell Institute (Arab).

8. South Asian: Obtained from the southern tip of India.

Primary parental samples used to develop the markers we are about to describe (which we will call EURO 1.0) were collected by us in Europe and India, or obtained from a respected sample supplier (Coriell Institute, New Jersey). To the most rigorous extent possible, each sample was anthropologically qualified in that the individual lived in the appropriate geopolitical region, and reported full ethnic affiliation with the parental group back to his grandparents, with no known admixture.

From a genetic screen, candidate AIMs were selected from an Affymetrix HuSNP 10K Mapping Array by ranking SNPs in order of their pairwise F_{ST} values considering four groups of pooled samples (Northern Europeans, combining groups 1 and 2), Mediterranean Europeans (combining 4 and 5), Middle Eastern (combining 6 and 7), and South Asian (9). We also extracted SNPs from the 171 AIM panel that had good information content for certain pairs of these eight groups. For each pair of pooled subpopulations grouped in this way (n = 6 pairs), 150 AIMs were selected based on F_{ST} rankings and the 150 for each pair combined into one list (900 candidates). Those that were redundant to multiple pairs were eliminated to leave a list of 319 candidate AIMs.

To construct a European test we need to define our parental groups objectively so that we can estimate the parental allele frequencies. Geopolitical identity may be a good starting point for screening markers, but may not provide the optimal definitions required for the establishment of parental allele frequencies. We could group the subpopulations using the F_{ST} values as a guide—for example, Northern European and Irish samples should probably be grouped together, as should Middle Eastern subpopulations I and II—though note the Middle Eastern I population is of mixed heritage. Indeed, the fine details on who should be grouped with whom and which parental populations should be used are difficult to tease from F_{ST} values and ethnic descriptors. We instead turn to methods that do not require the *a priori* definition of parental samples or subgroups, such as that used by Pritchard's STRUCTURE program. In essence, we let STRUCTURE cluster the parental samples and define the subgroups for us.

If the geographical pattern of genetic variability upon which the markers were selected effectively resolves populations with respect to this pattern, STRUCTURE should show this and identify individual samples that may be outliers within each subgroup. Allele frequencies for maximum likeli-hood estimation of admixture can then be taken directly from STRUC-TURE's notion of what the proper subgroupings are, or calculated from samples of the same primary affiliation as determined by STRUCTURE. Both methods should produce similar values though the former accommo-dates a larger sample of admixed and unadmixed samples, so should pro-duce a better estimate. Of the various population models tried, the k = 4 model appeared to be the most useful based on two criteria: (1) four groups was the simplest way to maximize the anthropometric trait distinction among European subgroups (primarily pigmentation traits) and (2) using simpler models, the average level of admixture was higher than with a four-way model, indicating that the simpler models did not fit the data as well as the four-way model.

THE EURO 1.0 AIM PANEL FOR A FOUR-POPULATION SUBCONTINENTAL MODEL

The European human population group, as we define it, corresponds to the lineages that contributed predominantly to populations in Europe, the Middle East, and Central and South Asia beginning approximately 40,000 years ago. Most Europeans speak languages derived from the Indo–European family and the systematic distribution of lighter pigmentation traits (skin tone, hair color, iris color) is exclusive to this group. Although it is certainly not true that all Europeans speak an Indo–European language and/or are characterized by light pigmentation, this seems to be as good a population model as any for the subpopulation ancestry-based inference of physical characteristics.

In 2004, DNAPrint's laboratories used the markers we have so far discussed to develop an optimized k = 4 European subpopulation admixture test called EURO 1.0. EURO 1.0 reports European ancestry admixture much like the 171 AIM continental test (ABD 2.5) does, but looks deeper within European lineages for individuals who are predominantly European. As we will see, the choice of k = 4 for a commercial within-European group test was made empirically—that is to say, simulations, analysis of pedigrees, and blind challenges with samples of known recent genealogy revealed that the assumption of a four-group substruc-ture model within Europeans fits the best for the 320 markers (though admix-ture within parental samples clustered as a function of geopolitical identity, and made geographic sense whatever the model). With k = 4, the European group is broken into the following subgroups (Figure 7-1):

- Northern European (NOR)
- Southeastern European (Mediterranean) subgroup (MED)

■ Middle Eastern (MIDEAS)

■ South Asian (SA)

For all the results discussed next, unless otherwise stated, we use the same basic MLE methods and algorithms as presented previously for the continental admixture test ABD 2.5, except of course that the Ancestry Informative Markers (AIMs), loci, and allele frequencies are different. The EURO 1.0 test is comprised of 320 within-European AIMs. Recall that most of the markers were obtained from screening tens of thousands of candidates from human genome database and DNA microchips. A fraction of these 320 are also used in the ABD 2.5 test, discovered based on their high between continent deltas and F_{ST} values; many showed good within-Europe values and we have already touched on the correlation between subpopulation and population information content for relatively old polymorphisms (with respect to the population models).

ESTABLISHING THE OPTIMAL PARENTAL REPRESENTATIVES

We can use STRUCTURE with these 320 European AIMs to determine individual admixture based on a preexisting STRUCTURE model. This requires us to add an unknown sample (for which admixture is to be determined) to an existing sample (comprising the model) each time. With each successive run, our estimates will change slightly, depending on the sample size, assuming we add the unknowns to the model each time or we can specify the reference sample from which allele frequencies are calculated and distinguish them from test samples (using "pop-flags" in STRUCTURE). Alternatively, we can use STRUCTURE to define the best parental samples, calculate static parental allele frequencies from these, and use them with the standard MLE algorithm to calculate the European ancestry estimates. With this set of markers, the difference between the two methods was not great (and shouldn't be unless one is working with very small sample sizes and relatively uninformative markers) and both methods produced a natural coalescence of individuals to genetic subgroups that made sense in geographical and historical terms.

With STRUCTURE, we can either incorporate or not incorporate prior information about population affiliation, and again the results are not very different. Figure 7-1A was obtained without using prior information about population of origin and it takes a similar appearance to Figure 7-1B, where the prior information was used. The EURO 1.0 results that follow are obtained using the prior information.

Looking at the results on the level of the individual, whether using STRUCTURE as in Figure 7-1A, B or basic MLE calculations, variation within

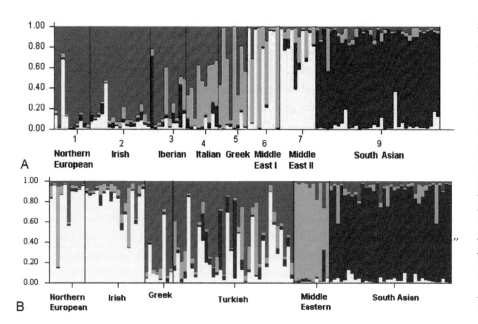

Figure 7-1

A) STRUCTURE plot of European samples using multilocus genotypes for 320 within-Europe AIMs described in the text. Prior information on ethnicity was not considered in this analysis. Bars indicate individuals, and the relative proportion of parental affiliation for four parental groups (k = 4) is shown on the y-axis with red, green, yellow, and blue color. Samples are arranged by self-reported and/or third-party ascribed ethnicity, in a northwestern to southeastern orientation across Europe. We term these groups in the text as, respectively, Northern European (NOR), Southeastern European (MED), Middle Eastern (MIDEAS), and South Asian (SA). B) Similar STRUCTURE plot as in A) except prior population information was used in the Bayesian process and some of the samples are different. Since Middle Eastern I group is mixed Southeastern European and Middle Eastern, this sample was eliminated for this particular analysis. The addition of samples from Turkey allowed greater sample size for "red" ancestry estimates. Overall the results are similar to that in A), and the allele frequencies for each locus within the "yellow," "red," "green," and "blue" groups across all of the admixed and relatively unadmixed samples constitute the parental allele frequencies used later with the MLE program to constitute the "EURO 1.0" assay.

subgroups appears to be the rule, not the exception. This variation is much more pronounced than that observed for continental admixture estimates previously discussed, however the genetic distances between European ethnicities are lower, and the geographic/genetic distance between the samples much lower than between the world's continental populations (due to more recent common ancestors). Partly due to this, and partly due to the complexity of European demography and history, self-held notions of ethnicity are not expected to correspond perfectly with genomic ancestry, and this is in fact what we observed as the following data illustrates.

With STRUCTURE, good separation among the parental samples was obtained. In Figure 7-1A, B the EURO 1.0 results are sorted by geopolitical identity. That is to say that the category along the x-axis (i.e., Irish, Italian) is based on political borders and sociocultural identity rather than genetics or anthropological history. Each colored column is an individual. The percentage of affiliation with each of the four genetically defined subgroups is shown along the y-axis, each genetic group shown with a different color. Figure 7-1B shows the analysis from which parental allele frequencies were taken so we will focus on it but the results between Figure 7-1A and 7-1B are essentially very similar, sample differences notwithstanding. Inspection of this figure shows:

■ Almost all the individuals showing predominant affiliation with the genetically defined NOR subgroup designated by the yellow color (genetic subgroup NOR) are Northern European or Irish.

■ Almost all the individuals showing predominant affiliation with the genetically defined subgroup designated by the red color (genetic subgroup MED) are Southeastern Europeans (Greeks or Turks).

■ Almost all the individuals showing predominant affiliation with the genetically defined subgroup designated by the green color (genetic subgroup MIDEAS) are Middle Eastern.

■ Almost all the individuals showing predominant affiliation with the genetically defined subgroup designated by the blue color (genetic subgroup SA) are South Asians from the Indian subcontinent (i.e., Indian).

For the MLE method, genetically and geographically compatible subgroupings of ethnic samples shown in the x-axis of Figure 7-2A, B, C, were used to group the parental samples for calculating subgroup specific allele frequencies; Irish and Northern European comprised a single Northern or NOR group, Greek and Turkish a single Southeast European or MED group, Middle Eastern II samples as a Middle Eastern or MIDEAS group, and South Asians as a South Asian or SA group.

Again, we see good separation among the parental groups, indicated in bold print in Table 7-1, but we also see trends in the intermediate populations that are also very informative.

Figure 7-2

A) Results from Bauchet et al., 2007 which studied 12 European populations with 11,071 SNPs. Results were obtained using the STRUCTURE program. Bars represent individuals and the color mix of each bar represents the proportions of each of 5 possible ancestry in this analysis. Black lines separate sample sets derived from different regions of Europe as indicated in the legend below. B) Results from DNAPrint's laboratory which studied the same 12 European populations as Bauchet et al., 2007 in A), but using only 1,349 specially selected SNPs from Bauchets set of 11,071 SNPs. Results were obtained using the STRUCTURE program. Bars represent individuals and the color mix of each bar represents the proportions of each of 5 possible ancestries in this analysis. Black lines separate sample sets derived from different regions of Europe as indicated in the legend below. Providing results that are similar to those obtained using all 11,071 SNPs in Bauchet et al., 2007, this marker set was chosen to constitute the EURODNA 2.0 marker panel.

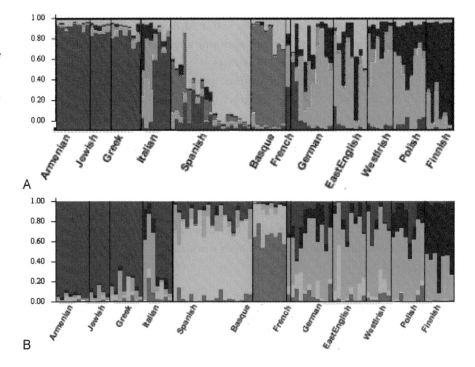

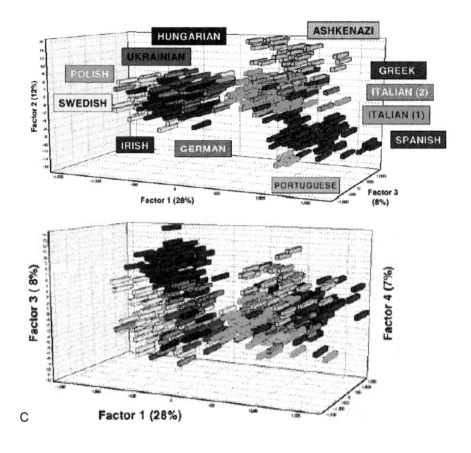

C

Figure 7.2

(Continued)
C) Results obtained by Seldin et al., 2006 for European populations using 749 SNPs chosen for European substructure information. This panel is non-overlapping (different SNPs) with that used in Figure 1, or Figure 2A), B) and uses a factor analysis method for clustering called Genetix (Belkhir et al., 2001). The apportionment of genetic diversity with respect to the first three factors of the analysis are shown in the top histogram and that obtained using factors one, three and four in the bottom histogram. Northern and Continental European populations on the left segregate upon inclusion of factor 4 in the analysis. Figure taken from Seldin et al., 2006.

Sample**	NOR	MED	MIDEAS	SA
Northern Euro	82	5.5	11	1.5
Irish	77.8	7.8	8.5	6
Iberian	48	29.5	7	15.5
Italians	35	46.1	8.9	10
Greek	15	78.1	1.9	5
Turkish	22.9	54.6	8.4	14.1
Middle East I***	3.9	41.7	52.2	2.2
Middle East II	4	7	84	5
South Asian	2.6	4.7	3.7	89
U.S. Caucasians	54.5	22.4	12.6	8.3

*- Levels above 1/4th are shaded in blue.
**- The average sample size for the 10 ethnic populations shown is 41.
***- Sample comprised of 4 Greeks, 1 Lebanese, 3 Arab, and 2 Persians.

*Table 7-1**

Admixture proportions obtained for various European populations using the EURO 1.0 assay (320 European AIMs, 3-way/4-way MLE algorithm).

Let us answer the basic question of what exactly the groups NOR, MED, MIDEAS, and SA mean? Table 7-1 shows the data obtained using the basic MLE method, listing the average admixture percentages for the four genetically defined within-European subgroups. The arrangement NOR, MED, MIDEAS, SA in Figure 7-1A, B as well as in Table 7-1 is in a northwest to southeast orientation, reminiscent of the subgroups defined by clines in synthetic maps of Europe and Western Asia obtained from classical blood group markers and Y-chromosome haplogroups (reviewed by Jobling et al. 2004). Like anthropometric traits such as hair and eye color, and the chronological and geographical distribution of archaeological evidence, the distribution of genetic variation using a four-population European model with EURO 1.0 seems to show the same principal component of variation in a northwest to southeast orientation.

For example, Northern European and Irish have the highest NOR affiliation (Table 7-1 and yellow color, Figure 7-2A, B, C), which suggests that this genetically defined subgroup corresponds to the most northerly subsets of Cavalli-Sforza et al.'s clines (1994; also see Jobling et al. 2004). NOR thus seems to correspond to "Nordic" genetic identity, hence the name "NOR." As we might expect of such, the percentage of NOR affiliation appears to increase gradually from the Middle East to Scandinavia and the United Kingdom, a trend that is most easily visualized in Figure 7-3 (yellow color). Also as we might expect, the percentage of MED affiliation increases gradually, moving from Northern Europe to the Mediterranean Southeast of Europe.

Figure 7-3

Global distribution of European substructure in terms of the parental affiliations shown in Figure 7-1B. Pie charts indicate the relative proportions of Northern European (NOR, yellow), Southeastern European (MED, red), Middle Eastern (MIDEAS, green), and South Asian (SA, blue) ancestry as described in the text and indicated in the legend, bottom right. The location of the pie chart indicates the geographical origin of the populations. (B: Basque; GA: Greek-American). This figure incorporates the data shown in Figure 7-1, as well as data from other similar analyses.

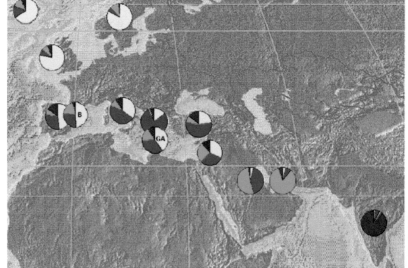

Geopolitical ethnic groups that were not used in developing the EURO 1.0 test, such as Iberians (Spanish, Portuguese) and Italians, type with EURO 1.0 determined ancestry that is intermediate to that of their geographical neighbors, Northern Europeans and Southeastern Mediterranean individuals (notice the yellow and red colors for Iberians in Spain; Italians are about halfway between the levels in Ireland/Scandinavia and Greece/Turkey, see Table 7-1). That Iberians and Italians have levels of NOR and MED ancestry intermediate to that of Northern Europeans and Greeks/Turks would hardly seem to be a coincidence given that they are from geographical regions of Europe between Northern and Southeastern Europe. In addition, the Ashkenazi Jews, not used in the definition of parental samples or calculation of allele frequencies, typed between Middle Eastern and Southeastern European populations in MIDEAS ancestry, and like others from Southeastern Europe and the Middle East in MED ancestry (as a population isolate, we will discuss this group separately later).

Similarly, a Middle Eastern population (MIDEAS I) taken from a different region of the Middle East than the Middle Eastern population used as the parental or reference group shows predominantly Mediterranean and Middle Eastern genetic identity (pie chart over Saudi Arabia in Figure 7-3, though this sample also includes Greeks). These types of results constitute a very important validation of EURO 1.0 because it shows that as we might expect, gene flow or genetic relatedness between subpopulations is directly proportional to geographical distance.

Other very recent autosomal SNP-based studies on European populations show similar patterns. Bauchet et al., 2007 used the Affymetrix 10K chip to partition European populations with the STRUCTURE program. From 9,114 SNPs these authors identified a k = 3 model as most probable (separating Armenian, Jewish, Greeks, and Italians from Spanish and Basque and from Continental, Northern, and Eastern European populations) but nearly as probable was the k = 5 population model which showed these partitions in addition to resolving Continental from Northeastern European ancestry and Spanish from Basque ancestry (Figure 7-2A). The author of this text mined this data to identify the most informative AIMs among these 9,114, using F_{ST} values among populations along the first (northwest—southeast) and the second principal components of variation (east-west). A panel of 1,349 of the SNPs provided essentially the same results as the original set of 9,114 (Figure 7-2B) and this panel was at the time of this writing being prepared as a second commercially available EURO assay (EURO 2.0 or EURODNA 2.0, performance parameters to be discussed later). Another set of authors (Seldin et al., 2006) selected 749 AIMs from an initial set of 5,700 based on European ancestry information content. This group did not calculate admixture percentages, and instead used a factor analysis to cluster samples, which produced three main groups using the 3 top factors—a northern group (Polish, Swedish, Irish, German, Ukranian, Hungarian), a southern group (Ashkenazi, Greek, Italian), and an Iberian group (Spanish and Portuguese) (Figure 7-2C top). The fourth factor added west-east information both in Northern and Southern Europe (Figure 7-2C bottom). Though with this type of analysis it is difficult to discern the primary axes of differentiation, since the first 2 factors separate North from Iberian from Southeastern European this work appears to support the notion we have been discussing up until now that the primary axis of genetic differentiation within Europe is northwest to southeast and the secondary axis is west to east as suggested in a broad sense—throughout Eurasia—in Figure 7-3 (using the 320 EURO 1.0 AIMs).

The same type of geographic sensibility is seen with Y-chromosome haplogroup distributions throughout Europe, and there are many parallels that can be drawn between the autosomal admixture analysis we are discussing here and these Y-haplogroup studies. Studies of European Y-chromosome haplogroup diversity have also revealed a principal northwest to southeast distribution (see Figure 7-4). The distribution of the R1b and I haplogroups, predominantly across northwestern Europe (red and light purple, respectively, Figure 7-4) is reminiscent of the distribution of autosomal NOR genetic ancestry. Note that adding the R1b and I haplogroup percentages in the northwest

Figure 7-4

European distribution of Y-haplogroups. Pie charts indicate the relative frequency of haplogroups, which are identified with colors and indicated in the key at the bottom. The geographical region of the pie chart indicates the origin of the population sample. Much as in Figure 6-14, we see a gradual shift in frequencies along a northwest to southeast axis, but we see a shift along a southwest to northeast axis as well.

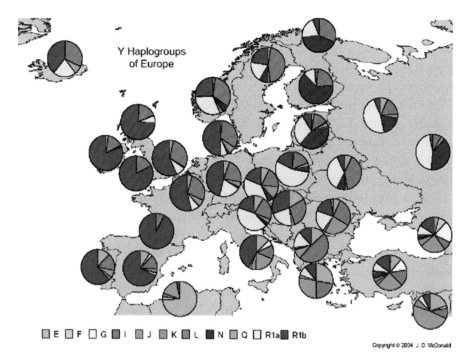

of Europe in Figure 7-4 produces a sector of the pie charts that is about the same size as that of the NOR autosomal genetic identity in Figure 7-3 (yellow wedge). In addition, the frequency of R1b + I in the Mediterranean populations is about the size of the NOR sector from the autosomal analysis, and is also lower in the Icelandic population relative to the United Kingdom and Continental Europe just as the NOR ancestry is. The MED in the autosomal analysis (red color, Figure 7-3) has the same general distribution as Y haplogroup J (green color, Figure 7-4) and the autosomal MIDEAS ancestry from the autosomal analysis (green, Figure 7-3) has the same general distribution as Y-haplogroups E and L (combined) in the Y-haplogroup analysis (turquoise and grey color, respectively, Figure 7-4).

Parallels between the autosomal admixture and Y-haplogroup distributions are what one might expect, and are consistent with yet other types of analyses such as classical blood group antigen studies, or other types of data such as linguistic or archaeological. There is one major difference—the Y-haplogroup analysis, which is based on a 20-haplogroup model, provides more information than the EURO1.0 autosomal admixture panel, which is based on a four-population European model. This explains why the Y-haplogroup analysis shows, in addition to the principal northwest-southeast gradient, an additional southwest to northeast gradient as well (notably, R1a and N). Indeed, as we have seen, increasing the complexity of the population model and inclusion of the appropriate samples allows for the discernment of this secondary axis with autosomal SNP analysis as well (compare Figure 7-2A-C with 7-1A, B).

The concordance between the autosomal and Y-haplogroup results suggests the complexity of autosomal admixture observed for the former is rooted in real population characteristics, rather than statistical noise or sampling effects. In other words, an autosomal test that did not work would probably show results that contradict the patterns observed from Y-haplogroup studies, and may show counter-intuitive results that are geographically confusing for neighboring populations not used in the definition of the population model or calculation of allele frequencies.

That is not to say that there are no unexpected results from a detailed inspection of the autosomal panel we are primarily focused on here. For example, there was substantial South Asian affiliation in the Iberian sample tested (see Figure 7-3), which as we can see in Figure 7-1A (which is the same basic data) can be attributed to a single Iberian individual. Though sampling effects may be involved, the level both in the individual and in the population is greater than that caused by statistical error due to continuous allele frequencies (as we will present data for later), and this type of interesting result suggests that working on such a fine level of population structure, natural variation in ancestry is perhaps to be expected, even in parental samples.

This is not unintuitive—for example, a Roma Gypsy whose ancestors settled on the Iberian peninsula six generations or 500 years ago (and many of the earliest Romani migrants did), may very well call themselves Iberian, Spanish, or Portuguese, and naturally be included in an Iberian parental sample. The Roma are a religiously, culturally, and geopolitically eclectic group, and though their recent ancestry is from the Indian subcontinent, endogamous social structures have maintained this ancestry in a relatively undiluted state (Kalaydjieva et al. 2000). That is not to say that it has remained completely undiluted, and as Figure 7-3 shows for samples not used in the definition of the test, we would expect geographical proximity to eventually beget admixture even in the face of opposing social forces.

BLIND CHALLENGE WITH ETHNICALLY ADMIXED EUROPEAN-AMERICAN SAMPLES

In addition to these samples of *formally* characterized ethnicity, we have typed a variety of samples of *self-proclaimed* ethnic identity. Of course this is not as reliable a test reference as anthropologically qualified samples, but they give us an idea of what to expect in mixed population such as those we find in the United States. Each subject reported the ethnicity of all four grandpar-

Table 7-2

Average EURO 1.0 results (%s) for U.S. test samples of self-described European ancestry. The allele frequencies for the four main elements of European/Eurasian ancestry derived from work shown in Figure 7-1B, and 3way/4way MLE algorithm was used to generate the results.

I	At least half of the grandparents from Ireland, Western Scandinavia, or Great Britain
II	At least half of the grandparents from France, Germany, Denmark, Finland, Holland Poland, Eastern Europe, or Russia
III	At least half of the grandparents from Spain or Italy
IV	At least half of the grandparents from Greece or Turkey
V	A random assortment of U.S. Caucasians regardless of from where their grandparents originated

Group	NOR	MED	MIDE	SA
I	56.4	20.8	16.4	6.4
II	51.2	24.5	15.2	9
III	46.9	33.8	10.6	8.8
IV	45	41	9	5
V	54.5	22.4	12.6	8.3

ents and was binned into one of five arbitrarily defined ethnic groups (I, II, III, IV, V) if at least two of the grandparents came from that group. The average sample size for these groups was 70 and the results are shown in Table 7-2.

Although most of these individuals were of admixed ethnicity, patterns in the results seem to correspond with the major self-held notions of ethnic affiliation. As can be seen from these results, the average individual claiming at least 50% Northern European ancestry (Group I) showed predominant affiliation with the NOR group, as did individuals claiming at least 50% French, German, Danish, Dutch, Eastern European, or Russian ancestry (Group II). In contrast, individuals claiming at least 50% Greek or Turkish ancestry (Group IV) showed significantly greater MED ancestry. As expected, individuals claiming at least 50% Spanish or Italian ancestry (Group III) typed in between Group I and Group IV individuals. Looking at the data in a slightly different way, we tabulate the data in terms of the frequency of observations greater than 25% affiliation with the genetic subgroupings in Table 7-3. The same basic trend is obtained.

All the individuals in this particular sample reporting at least 50% Irish/British/Scandinavian ancestry type of greater than 25% NOR ancestry with EURO 1.0, and all the individuals in this particular sample reporting at least 50% Greek or Turkish ancestry type of greater than 25% MED ancestry with EURO 1.0. Most individuals claiming at least 50% Spanish or Italian ancestry type of both greater than 25% NOR and greater than 25% MED ancestry with EURO 1.0.

As we saw with the parental samples in Table 7-1, the MED genetic group affiliation levels are inversely proportional to the magnitude of distance from the self-reported origin to the Mediterranean Sea—French/Germans show more than British, and Spanish/Italians show more than French/Germans. In general, the MED percentage rises as we proceed from the Northern European group (Group I) to the group for Spain and Italy (III), peaking

Group	NOR	MED	MIDEAS	SA
I	1	0.28	0.17	0.06
II	0.98	0.32	0.17	0
III	0.88	0.75	0.13	0
IV	0.8	1	0	0
V	0.95	0.39	0.14	0.07

Table 7-3

Number of samples with >25% affiliation with each of the four European genetic groups (columns) for the average sample of four types of self-reported ethnicity (rows). Percentages above 50% are shown in blue.

in Greece and Turkey. In contrast, NOR levels increase as we move from American populations of reported Southeast Europe to Northwest Europe heritage. U.S. Caucasians as a consolidated group seem to be a mélange of NOR and MED ancestry, but in unequal proportions as one might expect given the unequal northern versus Mediterranean founder contribution to the Americas.

These results illustrate that most European Americans are of both NOR and MED ancestry (whether or not they call themselves Germans, French, Italians, etc.), and many are of some lower level of MIDEAS ancestry as well (similar to continental Europeans). However, they also show that simply knowing the recent geopolitical origin of a person's heritage is not necessarily a very accurate predictor of anthropological or genetic group affiliation, which is perhaps not surprising given the complex recent political history and interactions between Europe, the Middle East, and South Asia.

POPULATION ISOLATES AND TRANSPLANTS

Icelanders, Basques, and Ashkenazi Jews are relatively isolated with well-documented recent and unique demographic histories. We can take advantage of these characteristics in using samples from these populations to challenge the validity of the markers, methods, and four-group population genetic model so far presented for the EURO 1.0 markers. But we have to keep in mind that because they are genetic isolates, we cannot predict how they will type exactly using the 4-population model. We have information from other studies and sources to guide expectations however (as we will discuss below). For example, with a more complex model, other studies incorporating Basques show that we can expect genetic separation between Basque and Spanish populations with substantial admixture between them (see Figure 7-2) but these same studies mostly cluster Basques with Spanish and secondarily with Continental European samples at k = 4 (Bauchet et al., 2007). Even though Icelanders, Basques, and Ashkenazi Jews have possibly undergone genetic drift from their shared ancestors with other Europeans, their origins are well documented, and unless the drift has been very strong or the EURO 1.0 test especially susceptible to minor allele frequency misspecifications, we should observe that the average member of these isolate groups resemble their geographic neighbors at least to a certain extent. Recall that the principal component of genetic variation for our AIM battery follows a northwest/southeast gradient similar to that observed with classical blood group antigens and Y and mitochondrial chromosome haplotypes. Drawing a line through Europe connecting the origins of these three populations we see a similar northwest/southeast directionality—Iceland in the extreme northwest, Basques from the center of Europe, and the

Ashkenazi from the extreme southeast of Europe, although many Ashkenazi have spent recent generations in Northeastern Europe. If the phylogeographic orientation of the genetic groups NOR, MED, MIDEAS, and SA is anthropologically meaningful, we should be able to predict, to a certain extent (keeping in mind that the groups are isolates), the primary group affiliations for each of these three populations using the 4-population model and EURO 1.0 panel of markers.

ICELANDERS

Iceland is an isolated island of the extreme northwest corner of Europe. Through the use of phylogenetic networks of haplotypes and the frequency-based admixture estimates (m_p), Helgason et al. (2000) used phylogenetic distance to infer Irish Gaelic and Scandinavian origins (mostly from the latter) for Icelanders, which makes geographic sense. Therefore, if our NOR, MED, MIDEAS, SA genetic groupings are phylogeographically relevant, Icelandic samples should type of considerable NOR admixture. We cannot predict how much exactly, since there has been genetic drift since the time Gaels and Scandinavians founded Iceland's population, but we might expect the average level to be similar to that for other Northern Europeans and dissimilar to that for Southeastern Europeans (like Greeks, Turks), Middle Eastern, and South Asian populations. With the EURO 1.0 test, Icelandic samples (n = 12) typed of predominantly NOR ancestry; in a gross sense, the overall proportions were similar to those obtained for Irish and Northern European samples (see Table 7-4, Figure 7-3). The MED ancestry was 27%, significantly higher than the levels observed for Irish and Northern European parentals (which was about 6%), possibly due to the accelerated genetic drift this group must have experienced due to their extreme isolation, possibly rendering the Icelanders not perfectly suited for this particular four-group population model (intermediate allele frequencies for some loci by chance). However, we cannot rule out the possibility of an admixed founder event.

Table 7-4

Average EURO 1.0 results for specific European population isolates. Percentages over 30 are shaded. The allele frequencies for the four main elements of European/Eurasian ancestry derived from work shown in Figure 7-1B, and 3way/4way MLE algorithm was used to generate the results.

	NOR	MED	MIDEAS	SA
Icelandic	65.8	15	12.9	6.3
Basque	49.5	27.5	19.5	3.5
Ashkenazi Jewish	35	29	25	11
Greek American	40.5	33.5	17.5	8.5

BASQUES

Aside from the data we can see in Figure 7-2A, B, C expectations for the Basque population isolate are more difficult to formulate, since the precise origins of this group are not clear. The Basques are an agricultural society indigenous to the Pyrenees of the Central Iberian Peninsula. They speak a non-Indo–European language and are considered therefore by many to represent a Paleolithic relic. However, this may be an oversimplification (Collins 1985; Jobling et al. 2004), and though it is known that they harbor unique genetic characteristics (unusual allele frequencies for some genes) relative to the rest of Europe, the extent to which they have admixed with or shared common recent (within the past 10,000 years) ancestry with Europeans is unknown. Although the evidence that Basques are a genetic isolate are very clear (i.e., retention of non-Indo–European language for longer than most other European populations such as the Galicians or Gascons), it is not so clear that they are any more of a Paleolithic relic than any other European group. Indeed, the Basques Y chromosome is an Iberian specific lineage (Hurles et al. 1999), which one would not expect from a Paleolithic relic population (i.e., they should contain some unique lineages).

The geographic distribution of modern-day Basques was established predominantly north of the Mediterranean coast and existed during the part of European history dominated by the Roman Empire, but their members were known to travel far and wide in the service of this empire (Collins 1985). Given that Neolithic farmers integrated with existing Paleolithic European populations, we can hypothesize that based on all the information available to date, reducing Basque ancestry to a function of NOR, MED, MIDEAS, and SA admixture would probably reveal more affiliation with NOR and less with the MIDEA and SA genetic groups, corresponding to populations that presumably populated Europe before and after the Neolithic expansion. Unfortunately Figure 7-2 doesn't help us much with our predictions for how Basques should type with EURO 1.0, since the population model for EURO 1.0 includes Middle Eastern and South Asian populations, and only two continental European populations whereas the EURO 2.0 data in Figure 7-2A, B, C lacks Middle Eastern/South Asian populations and instead involves five European populations. However the study from which Figure 7-2A came showed that, as we have already mentioned, at k = 4 Basques cluster mainly with other Spanish and to a lesser extent with neighboring continental Europeans (Bauchet et al., 2007). Thus we might predict a similar result with EURO 1.0 and indeed, a small collection of the Basque samples (n = 10) showed an average ancestry proportion that was almost identical to that for other Iberians (about 50% NOR, 30% MED; Table 7-4 vs. Table 7-1, Figure 7-3).

ASHKENAZI JEWS

The Ashkenazim are a relatively endogamous ethnic group that originated during the Bronze Age (between 2000 and 700 B.C) in the Middle East and spread northward throughout Southeastern, Central, Eastern, and Northern Europe. The historical record does not assure us that the Jewish people actually originated in the Middle East as opposed to having been seeded by Europeans from the north, but from Y-chromosome analysis and admixture analysis using 171 continental AIMs discussed previously, it appears that the genetic distance between Ashkenazi Jews and other Middle Eastern populations is relatively low compared to Northern/Central European populations. Thus, we hypothesize that Ashkenazi Jews would show relatively high MIDEA and possibly MED affiliation relative to NOR and SA affiliation (keeping in mind imprecision in the phylogeographic record and that there might exist relatively recent admixture since no group is completely endogamous). Indeed, the STRUCTURE results from Bauchet et al., 2007 showed that whether using relatively simple (k = 2, 3, or 4) or a more complex population model (k = 5, Figure 7-2A) show their Jewish population clustering with Armenians and Greeks so we would predict that EURO 1.0 would show results similar to Greeks and possibly Middle Eastern samples with this assay. Indeed, we see from Figure 7-3 (pie chart over Israel) and Table 7-4 that the average Ashkenazi sample typed more similarly to other Middle Eastern and Southeastern European samples (as Greeks did) than to Northern European or South Asian samples. The NOR (blue Figure 7-3) percentage was greater than expected however, and this may reflect heterogeneity within the global Jewish population—specifically, between Mediterranean and Ashkenazi populations, the latter of which derived more recently in the Middle Ages from continental Europe—specifically, the Rhineland (Germany).

The exercise with the population isolates revealed three informative trends that lend credibility to the anthropological legitimacy of the EURO 1.0 ancestry measurements. Since Icelanders are derived from Scotland and Scandinavia, the Basques are indigenous to the Central Iberian peninsula and the Ashkenazi Jews trace their intricately documented origins to the Middle East, we can define our expectations quite clearly based on a knowledge of the geographical origin for each population. We predicted that the Icelanders would show the most NOR ancestry, and that the Ashkenazi would show the highest MED and MIDEAS ancestry among the three isolates, with Basques falling somewhere in between. Indeed the level of NOR ancestry decreases from Iceland, to Basques to Ashkenazi, and the level of MED and MIDEAS both increased moving from Iceland to Basques to Ashkenazi as we predicted (see Table 7-4).

GREEK AMERICAN

In addition to the samples discussed in Table 7-2 and Table 7-3, we had the privilege to type samples from another group—a collection of self-reported Greek Americans (n = 36). These individuals were all residents of the United States, for whom all four grandparents were born in Greece, Turkey, or Albania, and who claim full Greek heritage. They can be considered as a transplanted population, who would be expected to type with EURO 1.0 as others from southeast Europe do. The average EURO 1.0 proportions were of less MED ancestry than observed previously for the Greek/Turk parental samples (Table 7-4 vs. Table 7-1 and Figure 7-3 GA). This result was somewhat surprising and its significance is not clear.

There was considerable variation in NOR/MED levels reminiscent of the pattern seen for parental Greeks. Recall that Greek/Turk parentals were selected based on high MED ancestry with EURO 1.0 as reported by the hypothesis free STRUCTURE program. Several of the present samples showed MED ancestry in excess of 70% using STRUCTURE or the basic MLE based on the previous Greek/Turk allele frequencies (where parental samples were defined using STRUCTURE), and these samples could conceivably be added to the parental set for the MLE program to augment the parental MED sample size.

To summarize the results from typing samples from populations not used to develop the EURO 1.0 test:

> The more southeasterly the self-reported geopolitical origin, the more similar the average profile is to our original parental samples from the southeast regions of Europe. This constitutes an independent validation of the anthropological relevance of EURO 1.0. Even population isolates and transplants type in a fairly predictable manner given their believed geographical origins.

This would seem to support, or at least not refute, the anthropological relevance of the four-group genetic model of Europe identified and used by the EURO 1.0 panel.

CORRELATIONS WITH ANTHROPOMETRIC TRAITS

If the NOR, MED, MIDEAS, and SA genetic groupings are of legitimate anthropological base, and if there is an association between anthropology and physical appearance (which we know there is from common sense, and studies such as those of Shriver et al. 2003), then there should be a correlation between admixture for certain genetic groupings and the expression of physical traits. When speaking of European populations, two obvious physical traits come to mind—skin melanin content and iris color. Elsewhere in this text we have (Chapter 5) and will again (Chapter 8) discuss the correlation between skin melanin index and West African and European admixture, with higher levels of European admixture correlated with lower melanin index values in admixed African Americans and Puerto Ricans. So, too, is skin melanin index differentially distributed within the European group (mainly along a northwest to southeast gradient).

We have also discussed the association between higher levels of West African and lower levels of European admixture in individuals of predominant European ancestry and darker iris colors (Chapter 5). Here we ask whether the same types of associations can be seen within the European diaspora sorted by genetically defined subpopulation admixture proportions rather than continental ancestry admixture. Since blue iris colors are far more frequently found in Northern European populations relative to Middle Eastern, South Asian populations, then the degree of Northern European admixture should be associated with lighter iris colors; Middle Eastern and South Asian admixture should be associated with darker iris colors. In other words, if the test works, we should see that individuals typing of high NOR ancestry with EURO 1.0 tend to have lighter colored eyes than individuals typing with low NOR ancestry, and individuals typing with high MIDEAS or SA ancestry should show darker eyes on average than those with low levels.

The author has digitally scored iris colors for some of the subjects discussed later in Figure 7-6, and Tables 7-1 and 7-2. As shown in Table 7-5, significant genetic group admixture levels are associated with the expected iris color shades. NOR admixture greater than or equal to 70 to 80% (rows 1–3) is strongly associated with iris color scores greater than 2.2, which are scores corresponding to light colors. Scores of 2.2 or greater correspond almost perfectly to perceived colors lighter than the mean for individuals of majority European ancestry and includes the light hazels, greens, grays, and blues (to be discussed in more detail, Chapter 9). In contrast, SA admixture levels greater than 25% were associated with color scores below 2.2 (Dark), corresponding to color scores below the mean (dark hazels, browns, blacks). MIDEAS admixture was not associated with iris color score, though lighter iris colors are not infrequently found in the Middle East. (Indeed, the genes imparting lighter iris colors may very well have originated in the fertile crescent prior to amplification by genetic drift or sexual selection in Europe starting 45,000 years ago.)

	Genetic groups	Admixture level	Color	Exact-p	Total European sample (N)	Sample meeting threshold (n)
1	NOR (N. Euro + Irish)	>70%	Light (color score > 2.2)	0.003	184	27
2	NOR (N. Euro + Irish)	>75%	Light (color score > 2.2)	0.004	184	21
3	NOR (N. Euro + Irish)	>80%	Light (color score > 2.2)	0.033	184	12
4	SA (South Asian)	>25%	Dark (color score <2.2)	0.060	184	15
5	SA (South Asian)	>30%	Dark (color score <2.2)	0.035	184	8

It turns out that, as we will discuss in detail in a later chapter, iris color can be predicted accurately for European individuals with a direct knowledge of pigmentation gene genotypes. Knowledge of genetic ancestry enables only an indirect inference, which is useful only for individuals of admixed and/or non-European ancestry (see Chapter 8). However, this exercise shows that there is a correlation between genetic groupings with high Northern European ancestry (as EURO 1.0 determines it) and lighter iris colors, as well as with high South Asian ancestry (as EURO 1.0 determines it) and darker iris colors. Note that the latter correlation is between lower levels admixture and iris color in individuals who describe themselves as European-American—not an obvious correlation between very high levels of South Asian admixture and dark iris color, which would be trivial to show and thus not a fair test of the accuracy and sensitivity of the marker panel. The results presented in Table 7-5 suggest that iris color shades are distributed throughout Europe as a function of Northern European ancestry admixture, and that on at least a crude level, the EURO 1.0 panel accurately and precisely measures this ancestry admixture.

For example, if the test did not work (i.e., did not report anthropologically legitimate elements of population substructure), there is no reason why we would see a statistically meaningful association between higher levels of NOR admixture and lighter iris color shades and lower levels of NOR (or higher levels of South Asian admixture) with darker irides in a European-American sample completely unrelated to the reference samples used in developing the test!

TEST ERROR

As with the 171-marker test previously described (AncestryByDNA 2.5), statistical error is caused by a number of phenomena. The most obvious is the imperfect information about ancestry provided by the DNA. Before we present results of experiments used to quantify such error, we will briefly review the source of the error in case you have forgotten or did not read Chapter 5. In scientific terms, the markers we use are distributed continuously among the ancestral groups, not strictly private to any one, and so admixture estimation is determined based on probability. This is largely why the individual MLEs (Maximum Likelihood Estimates of ancestry) are correctly communicated only in terms of their confidence intervals.

Using mathematical simulations, we can quantify the amount of error caused by this imperfection for the average customer. To do this, we use our knowledge of the allele frequencies in each group and the relationships between the alleles to create a large number of multilocus genotypes or simulated individuals. Ancestry admixture is determined for each. If we create 1,000 simulated individuals of 100% Northern European ancestry, and observe that the average sample showed a level of 100% Northern European ancestry, then we would conclude there would be no bias or error in the test caused by continuous allele frequencies. If the average sample showed 90% Northern European ancestry, there would be a 10% error caused by continuous allele frequencies, and any Northern European would be best suited to consider their Northern European result to be accurate to form $\pm 10\%$.

The results from EURO 1.0 simulations are shown in Table 7-6A, B, C. The ancestry for the average simulated sample showed 5.45% cumulative error across all the four European groups. For example, a person with the true values of 55% NOR, 45% MED, 0% MID, and 0% SA and typing with the same error as the average simulated parental sample likely could receive an EURO 1.0 score of 52% NOR, 47% MED, 1% MID, and 0% SA, or 53% NOR, 46% MED, 0%

	NOR	MED	MIDEAS	SA	Bias
NOR	91.35	3.8	2.14	2.71	8.65
MED	2.72	93.72	1.54	2.02	6.28
MIDEAS	0.65	0.41	98.92	0.02	1.08
SA	1.6	2.62	1.64	94.14	5.86
				Average Bias	5.45

Table 7-6

A) Average group-specific and total bias in EURO 1.0 admixture estimates (based on analysis of Figure 7-1B) for simulated parental samples (100% expected ancestry). Numbers are shown in percentages.

(Continued)

Table 7-6

(Continued)

B) Average group-specific and total bias in EURO 2.0 admixture estimates (based on analysis of Figure 7-2B) for simulated parental samples (100% expected ancestry) and 50%/50% admixed individuals.

Average results for simulated samples (n = 100 ea)						
	NEE	**CE**	**IB**	**BA**	**SEE**	**Bias**
100% NEE	0.929	0.022	0.015	0.014	0.021	0.072
100% CE	0.025	0.913	0.015	0.020	0.028	0.087
100% IB	0.013	0.018	0.923	0.022	0.024	0.077
100% BA	0.019	0.027	0.024	0.900	0.031	0.100
100% SEE	0.016	0.022	0.019	0.020	0.924	0.077
					Avg.	0.083
50% NEE / 50% IB	0.456	0.021	0.470	0.020	0.033	0.054
50% NEE / 50% BA	0.473	0.028	0.020	0.452	0.027	0.075
50% CE / 50% NEE	0.476	0.457	0.017	0.020	0.030	0.068
50% CE / 50% IB	0.020	0.453	0.478	0.025	0.024	0.069
50% CE / 50% BA	0.026	0.458	0.026	0.466	0.025	0.077
50% SEE / 50% CE	0.024	0.450	0.024	0.019	0.483	0.067
50% SEE / 50% NEE	0.456	0.028	0.023	0.018	0.474	0.069
50% SEE / 50% IB	0.014	0.026	0.465	0.020	0.475	0.060
50% SEE / 50% BA	0.017	0.030	0.021	0.460	0.471	0.069
50% IB / 50% BA	0.016	0.021	0.477	0.451	0.035	0.072
					Avg.	0.068
					Avg.	0.073

Table 7-6

(Continued)

C) 95% confidence thresholds for concluding bona-fide admixture based on the simulated EURO 2.0 admixture estimates of B).

Admixture required to conclude with 95% confidence admixture is real					
Primary ancestry	**NEE**	**CE**	**IB**	**BA**	**SEE**
NEE		0.061	0.039	0.035	0.041
CE	0.081		0.037	0.073	0.086
IB	0.037	0.05		0.073	0.09
BA	0.054	0.078	0.085		0.076
SEE	0.048	0.073	0.046	0.046	

NEE—Northeastern European, CE—Continental European, IB—Iberian, BA—Basque, SEE—Southeastern European.

MID, and 1% SA or any other percentage combination where the difference between the score and the true values add up across all four groups to be about 5 to 6%. Of course, any one rare individual could show 20% error, or 0% error, distributed across all four groups or even focused in two of them, but the average individual is expected to encounter about 5.45% bias. The bias estimates and 95% confidence limits for the EURODNA2.0 panel we discussed earlier (Figure 7-2B) is also shown in Tables 7-6B and C for comparison with EURO 1.0 in Table 7-6A as well as with the continental assay presented in Chapter 5 (Table 5-23A). Note that while a much larger panel of markers is used for EURO 2.0, the number of populations is greater and there is less genetic distance between them since they are all continental European populations, and so the bias estimates are actually slightly higher for this assay than EURO 1.0.

The simulations just presented are from individuals simulated to be homogeneously affiliated with one group. What do we see when we simulate admixed offspring? Within each population group we simulated four-person pedigrees, allowing for variation in ancestry for each member of the pedigree. Each pedigree comprises two parents and two children. During the simulations, we simulated each pedigree at a time where parent one (P1) was simulated first, followed by parent 2 (P2), using the given population allele frequencies and assuming independence among alleles. Once P1 and P2 had been simulated, the two siblings (S1 and S2) were simulated to have the average BGA of those of the parents. Thus, the design of the simulation relied on correctly specifying the admixture of P1 and P2 individuals, and independent assortment is not a cause of variation from expected levels in offspring. We are measuring only the error inherent to the use of continuously distributed variables when estimating admixture (this time in 50/50 admixed rather than 100% affiliated samples). The general scheme is based on a program written by Mark Shriver's group at Penn State University called SIMSAMPLE. The results from 120 pedigrees using the EURO 1.0 test to infer ancestry of each offspring are shown in Table 7-7.

The notation in the table is read this way: 50NOR–50MED (n = 40) means 40 simulated 50% Northern European (NOR)/50% Mediterranean (MED) individuals. The average 50/50 NOR/MED mix showed 50.47% NOR ancestry and 44.77% MED ancestry, which differed from the expected levels of 50% for each by the level of bias, which equals 4.77%. The average simulated admixed sample, across all types of admixture for the 120 pedigrees showed 6.84% error caused by continuous allele frequencies. Combining the various simulation results discussed we conclude that the error caused by continuous allele frequencies is about 6 to 7%, depending on the type of admixture exhibited, and the type of majority ancestry. For a forensics case, an investigator could match

Table 7-7

Simulated offspring with the EURO 1.0 assay. Numbers are shown in percentages.

	NOR	MED	MIDEAS	SA	Bias
50NOR-50MED (n = 40)	50.47	44.77	2.67	2.10	4.77
75NOR-25MED (n = 10)	79.80	13.60	1.40	5.20	6.60
25NOR-75MED (n = 10)	20.10	75.60	2.00	2.30	4.30
50NOR-50MIDEAS (n = 39)	47.00	5.40	45.58	2.03	7.43
75NOR-25MIDEAS (n = 10)	70.50	2.70	22.40	4.40	7.10
25NOR-75MIDEAS (n = 10)	16.90	2.30	78.20	2.60	4.90
50NOR-50SA (n = 40)	40.15	10.10	3.05	46.70	13.15
75NOR-25SA (n = 10)	70.80	3.90	2.00	23.30	5.90
25NOR-75SA (n = 10)	25.70	0.50	3.40	70.40	3.90
50MED-50MIDEAS (n = 40)	5.25	44.40	47.30	3.05	8.30
75MED-25MIDEAS (n = 10)	1.60	70.70	21.40	6.30	7.90
25MED-75MIDEAS (n = 10)	11.50	11.20	76.10	1.20	12.70
50MED-50SA (n = 40)	2.53	45.00	1.35	51.13	3.88
75MED-25SA (n = 10)	6.60	67.70	3.10	22.60	9.70
25MED-75SA (n = 10)	4.00	23.20	2.60	70.20	6.60
50MIDEAS -50SA (n = 38)	3.42	2.24	48.42	45.92	5.66
75MED-25SA (n = 10)	2.40	7.40	71.90	18.30	9.80
25MED-75SA (n = 10)	0.00	0.60	20.70	78.70	0.60
				Average Bias	6.84

their sample's result to one of the preceding categories that most closely fits the estimate of expected levels of this type of error for that sample. It is for this type of error that the confidence rings in the triangle plot or the intervals of a bar chart or range of values in a table form most closely correspond.

ERROR CONTRIBUTED BY THE NUMBER OF AIMs USED

The 320 AIM EURO 1.0 test is expected to perform better than a test comprised of a smaller number of AIMs. What is the difference in performance we can expect applying a smaller 133 AIM panel with the k = 4 population model compared to the 320 AIM panel (the EURO 1.0 test)? We ran the test assuming the NOR, MED, MIDEAS, and SA population model, using the 133 AIM and 320 AIM panels with the same basic MLE program

	NOR	MED	MIDEAS	SA
Average % difference	3.2	3.3	1.9	1.8
Max RMSE	10.6	10.6	10.6	10.6
Max % difference	15	15	15	15
Min RMSE	0	0	0	0
Min % difference	0	0	0	0
Mode RMSE	0	0	0	0
Mode % difference	0	0	0	0

Table 7-8

Results provided by 133 of the 320 EURO 1.0 AIMs for 114 samples, and difference with the full 320 AIM EURO 1.0 panel.

(3way/4way) for 105 samples. Table 7-9 shows the results for eight randomly selected samples of these and Table 7-8 shows the summary statistics for the comparisons. The average difference in admixture scores between the results provided by the two panels was 3.2%, with similar values for all four genetic subgroups. The largest RMSE observed for the smaller panel was 10.6%, corresponding to a 15% swing in admixture percentages for the average comparison—six of the pairs showed this most extreme difference. As the RMSE indicates, many samples showed substantially lower differences and in fact, the mode was 0% RMSE, 0% difference for all of the subgroups (i.e., most of the pairs showed the same identical results between the two panels of markers).

JUNE0009 of Table 7-9 is representative of a sample showing the largest difference (15% NOR difference), but as with the other 105 that exhibited differences, the difference was distributed more or less equally among the other subgroups and not focused into a single one. For example, the 15% NOR lost by JUNE0009 going from 133 to 320 AIMs did not turn into 15% SA ancestry; it was distributed to MED (10%) and MIDEAS (5%). For only four of the 105 samples did the majority affiliation change, moving from 133 to 320 AIMs, and for each of these four, ancestry was fairly evenly spread among the four genetic subgroups NOR, MED, MIDEAS, and SA. As we expected, there are differences in the results provided by a less powerful panel of markers compared to a more powerful one, but these results indicate that the differences are not extreme and the 133 AIM panel appears to provide reasonable performance relative to the 320 AIM panel. We conclude from this that, although using more AIMs is always better, the error inherent to using a relatively small battery of 320 AIMs is not consequential for determining the major features of European ancestry, and that smaller panels can be used to

Table 7-9

Representative European MLE estimates with a small and large (EURO 1.0) panel of AIMs. Numbers are shown as percentages.

Sample ID	AIMs	NOR	MED	MIDEAS	SA
JUNE0008	133 AIMs	55	40	0	5
JUNE0008	360 AIMs	55	35	0	10
JUNE0009	133 AIMs	40	35	25	0
JUNE0009	360 AIMs	25	45	30	0
JUNE0010	133 AIMs	55	35	10	0
JUNE0010	360 AIMs	50	25	15	10
JUNE0031	133 AIMs	55	40	5	0
JUNE0031	360 AIMs	65	25	10	0
JUNE0032	133 AIMs	35	20	0	45
JUNE0032	360 AIMs	45	10	0	45
JUNE0033	133 AIMs	60	35	0	5
JUNE0033	360 AIMs	60	25	10	5
JUNE0034	133 AIMs	50	50	0	0
JUNE0034	360 AIMs	60	35	5	0
JUNE0040	133 AIMs	50	35	5	10
JUNE0040	360 AIMs	60	25	5	10

obtain similar results. Of course, these conclusions do not necessarily apply to other AIMs; the AIMs we are using herein are exceptionally powerful for (and were selected from the genome based on their) discrimination within the European continental subpopulation.

OTHER ERROR

As discussed already, there are other mechanisms that can cause errors besides continuous allele frequencies, which come from our inability to go back in time and measure precisely how and when admixture occurred in various parts of Europe and what the allele frequencies were for the parental populations that did the admixing. In scientific terms, there may be imperfections in the model we use to estimate admixture. Was the admixture the result of a single pulse at a moment in time, or due to a continuous interaction between parental populations? Are allele frequencies for modern-day descendents of parental populations the same as those for the populations when they did the admixing, or has substantial and directional genetic drift taken place over the years? Scientists debate these issues all the time, and there is no one answer that is guaranteed to be correct for any given admixture problem. We already have discussed how uncertainty in these parameters is handled in a Bayesian context, but unless these methods work perfectly (and we can never be assured that they do), we must consider their influence when interpreting results.

The question of where a 10% MIDEAS result comes from in a given result is similar to the question of where variability in results within populations comes from. We will discuss variability within populations in more detail a bit further on; here we focus on the various sources of error other than that caused by the continuous nature of allele frequency distributions. A given admixture result could come from:

1. **Between-Ethnic Admixture** (not due to statistical error) within the European continental population group, which is likely much more extensive than between-continent admixture. From this we would expect lower F_{ST} values for randomly selected SNPs, relatively low genetic distance between groups, a harder time finding SNPs with good F_{ST} values when screening the genome for good AIMs, and a greater error estimating admixture for ethnic relative to continental admixture.

2. **Incomplete or Imperfect Information Content** for the marker set. This explanation is difficult to accept for EURO 1.0 since the allele frequency differential (δ) and F_{ST} values for our marker set are reasonably high (the average F_{ST} computed by STRUCTURE was 0.10), at least adequate for standard errors in the range of 10 to 20% or so. The cumulative δ value for our 330 markers, among each of the population pairs of our model, is shown in Table 7-10, and the values are exceptionally high for a test as affordable as this (meaning we have excellent power to infer ancestry for such an economical test). Still, there is room for improvement by the inclusion of more AIMs and/or the replacement of existing AIMs with markers having more ancestry information.

3. **Sampling Error in Allele Frequency Estimation**. We assume that the allele frequencies have been properly specified and an adequate parental sample was used. Our average sample size for parentals is on the order of 40 samples. A larger number would ensure greater accuracy, but the results discussed earlier suggest that even so, the accuracy of EURO 1.0 is not unreasonable. Nevertheless, future work with this particular panel of

	Northern European	Greek/Turkish	Middle Eastern	South Asian
Northern European	0	36.7	46.8	44.7
Greek, Turkish		0	47.9	43.5
Middle Eastern			0	49.7
South Asian				0

Table 7-10

Cumulative delta values, 320 AIMs.

markers will undoubtedly focus on increasing our sample size and comparing the results for upgraded AIM panels (i.e., EURO2.0), and if supreme precision is crucial a test can incorporate sample sizes from more parental populations as well. It bears noting, however, that error caused by inadequate parental sample size would most likely manifest itself as a haphazard error (assuming no admixture between parental populations, and no sample size bias resulting from it). However, this is not what we see. For example, the type of unexpected admixture commonly seen for Irish is not Middle Eastern-MIDEAS or South Asian-SA, but Mediterranean (MED), which makes geographical and historical sense.

4. **Defects with the Simple Model of Admixture Assumed**, which is perhaps the biggest concern. When we estimate the admixture we assume that

    a. There has been no genetic drift between modern-day descendents of parental populations and the parental populations themselves.

    b. No linkage exists between markers in admixed individuals.

    c. Offspring are derived from parents of the same ancestry mix.

    d. The markers we use have not been the subject of selection since the time the differentials were created.

    e. The allelic state for each locus among the groups is uncorrelated.

    f. Parental populations represent the movements across Europe that led to contemporary European genomic diversity.

    g. Parental populations are now and have always been homogeneous.

    h. Admixture occurred in a single pulse, rather than through continuous gene flow.

Defects in these assumptions would tend to create imprecision, but could also introduce systematic bias and influence the results on a population scale as they did in the classical gene studies of African/European admixture in the 1960s and 1970s (see Chakraborty 1986 for a review, where blood group antigen loci were used and assumption d. was violated). How might we recognize the symptoms of test error caused by violation of one or more of these assumptions? Unfortunately, it is very difficult to diagnose. It would seem that these types of errors would likely be accompanied by greater within-group variation and more artificial admixture than we have observed but its very difficult to say how much we might expect. For example, Figure 7-3 might make much less sense and the geographic trending we see in this figure might not exist. Further, we might not have observed results for regional population isolates that made geographic sense in terms of the results from neighboring populations.

Another way to say this it that the admixture observed due to this type of error/bias would tend to produce noise that is not necessarily geographically sensible as most of our results seem to be. However, geographic sensibility of results is not a very good test for fit with our assumptions, and it certainly does not prove that there is no error caused by violations in population model assumptions—in fact, we would always expect some error caused by these violations. Indeed, some of the admixture seen in Figure 7-3 could be due to this type of error. The question is how to minimize error caused by making the assumption that the parameter value corresponds to a. through g. We have discussed already how Bayesian algorithms can accommodate uncertainty for various parameters related to the preceding assumptions, so one reasonable means by which we can assess the influence of assuming null values for the parameters in a. through g. is to test whether these programs produce different results than a method that assumes null values. Better yet, we can simulate specific examples and study the impact of each assumption violation in detail. On populations, other authors have shown relatively modest improvements in statistical accuracy when incorporating mathematically complex models that account for uncertainty in the population model, although so far only on a population level (Wang 2004).

STRUCTURE allows for the calculation of admixture values assuming correlated or uncorrelated allele frequencies among the groups (assumption e.). Therefore, to assess whether the assumption of uncorrelated allele frequencies substantially impacts the bias we can compare the results obtained from STRUCTURE, assuming correlated allele frequencies, and a basic MLE program where we assume the frequencies are uncorrelated. STRUCTURE was run on simulated parental samples (see Table 7-12) and simulated admixed samples (see Table 7-14) with correlated allele frequencies and separate alpha values for each population. A burn-in of 50,000 was used with 80,000 iterations, assuming prior populations under the admixture model.

The same samples were run with a basic MLE program (see Tables 7-11 and 7-13). The average bias across the various types of admixture in the various backgrounds was similar for both programs; STRUCTURE had a bit higher bias in simulated parental samples (6.18 vs. MLE program results of 5.47), but a bit lower bias for admixed samples (6.28 vs. MLE program results of 6.87). The similarity of results suggests, at least on a population level with this set of

	NOR	MIDEASED	MIDEAS	SA	Bias
100NOR (n = 50)	91.35	3.80	2.14	2.71	8.65
100MIDEASED (n = 50)	2.72	93.72	1.54	2.02	6.28
100MIDEAS (n = 50)	0.65	0.41	98.92	0.02	1.08
100SA (n = 50)	1.60	2.62	1.64	94.14	5.86
				Average Bias	5.47

Table 7-11

Average group-specific and total bias in EURO1.0 admixture estimates obtained using the basic MLE algorithm (Based on analysis of Figure 7-1B) for simulated parental samples (100% expected ancestry) for comparison with Table 7-12. Numbers are shown as percentages.

Table 7-12

*Average group-specific and
total bias in EURO 1.0
admixture estimates
obtained using the
STRUCTURE algorithm
(based on analysis of
Figure 7-1B) for
simulated parental
samples (100% expected
ancestry) for comparison
with Table 7-11. Numbers
are shown as percentages.*

	NOR	MIDEASED	MIDEAS	SA	Bias
100NOR (n = 50)	92.79	3.08	1.82	2.31	7.21
100MIDEASED (n = 50)	3.18	92.83	1.77	2.23	7.18
100MIDEAS (n = 50)	1.36	1.18	96.67	0.79	3.33
100SA (n = 50)	2.32	2.86	1.82	93.01	7.00
				Average Bias	6.18

markers, very little impact on making assumption e., that allele frequencies are uncorrelated among the parental populations. STRUCTURE also allows for LD between markers, so these results simultaneously test whether assumption b. is crucial to final outcome—evidently not for this test, at least on a population level. On an individual-to-individual level, however, there are differences in the estimates, and so all other things being equal, and lacking evidence to the contrary, it would still be most prudent to assume that the more advanced method in STRUCTURE provides the more reliable estimates.

What about the other assumptions? Assumptions a., f., and g. are particularly troublesome, and fortunately there are programs that allow for uncertainty relating to these assumptions to be handled statistically (Paul McKeigue's ADMIXMAP). Running ADMIXMAP on the same data shows subtle variations from the basic MLE method, but the theme of the comparisons is that the largest contribution of bias error is caused not by the assumptions but by the continuous nature of allele frequencies across the groups, which implies that precision can be purchased by measuring larger sets of markers in larger sets of parental samples. To produce a more perfect test, we can use a thousand or a hundred thousand AIMs, parental samples of hundreds of thousands, using more sophisticated programs that handle parameter uncertainty. Except for using more sophisticated programs, this is a question of balance between expense and accuracy for the test. As programs like STRUCTURE and ADMIXMAP are optimized and validated on model systems, where variables can be strictly controlled and true validation achieved, we can gradually shift from basic MLE methods for EURO 1.0 (and the 171 marker test).

What about the common sense method of assessing whether there is substantial bias introduced due to making assumptions a. through f.? It would seem that one way to answer this question is to compare expected and observed results for people with carefully documented genealogy, but even this is not completely logical. We cannot use genealogy information from the past few generations to evaluate results from an anthropological test that is looking back (potentially) thousands of years. Since most genealogists do not have reliable information going back that far, we simply do not have access to the reference data with which we would need to compare performance against expectations and measure this error, or modify the test to eliminate it. For EURO 1.0 in particular, we are doing what meteorologists are doing when they calculate a hurricane track projection cone. The meteorologist cannot know for certain exactly where the storm will go, but he or she understands how the storms respond to major weather features (like fronts) well enough to form probability statements predicting the storm track. Historically, these predictions are usually quite impressive—they are fairly close to where the storms actually go. The same is true with EURO 1.0—the results suggest that the estimates are fairly close to true values, but almost never exactly correct.

Another way to look at it, when EURO-DNA 1.0 is run for a forensic sample, consider those running and interpreting the test analogous to pilots flying a jumbo jet through low-hanging clouds at dusk. The pilots can see the runway and most of the lights, but not all of them and certainly not the details of the terrain around them. If the pilots and equipment are reasonably competent, there will be a good enough understanding of the surroundings to land the plane safely, but not to draw a detailed map of the area. The confidence rings on the triangle plot, confidence intervals in the bar plot, or range of values in the table of percentages do not communicate this type of error because it is virtually impossible to estimate. Even though we cannot project the precise course a hurricane will take, we can produce a good cone, and we know it is a good one based on analysis of many hurricane tracks after they have been established. In this case, with EURO 1.0, we can never know the exact ''hurricane track'' or true ancestry percentages, but we can evaluate the estimates using various statistical methods on real and simulated data (analysis of results in simulated parentals, real and simulated pedigrees, against a geographical backdrop, against self-held notions, correlations between admixture estimates and anthropometric traits, etc.).

HIERARCHICAL NATURE OF EURO 1.0—PRIOR INFORMATION REQUIRED

While on the subject of error, and having hinted at the idea of population model fit, this is the perfect place to discuss the hierarchical nature of the EURO 1.0 test (which also applies to the EURODNA2.0 panel). EURO 1.0 is a hierarchical test, in that it cannot be used for a sample unless there is prior knowledge about the sample, and incorrect or incomplete prior information could also contribute to test error. Namely, we require that a sample be characterized by mainly European/Eurasian ancestry. The reason for this is that the population model assumed by EURO 1.0 is inappropriate for individuals with low or no European/ Eurasian ancestry. For example, we could force a EURO 1.0 result for a 100% West

African; shoe-horned into the European population model, such an individual would likely type with substantial South Asian affiliation (T. Frudakis, unpublished results). For a genealogist the answer provided would not mean what the customer is likely to think that it means, and for a forensic scientist, the relationship between the ancestry answer provided and phenotype would not necessarily exist because the four-group European population model would clearly be inappropriate for an African person.

It is important to stress that common sense says the EURO 1.0 result in such a case would be wrong, but in reality it is not incorrect *with respect to the population model* since it is based on genetic distance. Nonetheless, it would be very difficult for a lay person to make sense of it, since the ancestry being reported is not apparent from the nomenclature for the ancestral affiliations used.

Looking at this problem in a little more depth, it appears that the type of admixture, or rather, its age and/or source, is relevant. Most ethnic subgroups, such as Greeks, South Asian Indians, Ashkenazi Jews, Middle Easterns, and so on. exhibit some non-European admixture with a continental test such as ABD 2.5 (discussed thoroughly earlier in Chapter 6). We have also discussed the various means by which such fractional affiliation result could be obtained, namely,

- Recent admixture
- More ancient population mixing on a population and ethnic level
- Shared ancestry that has not yet completely dissolved
- Genetic drift and intermediate allele frequencies for groups not entirely appropriate for the population model

Only the first of these can be considered as true admixture, and it is the type represented by the latter three that interferes with EURO 1.0. Persons with non-NOR, MED, MIDEAS, and SA ancestry due to bona fide admixture, such as Indigenous American or African Americans, are not expected to type properly with EURO-DNA 1.0. Their Indigenous American or African genotypes will be misconstrued by the test, which shoehorns the sample into the NOR, MED, MIDEAS, and SA model.

If we require that only samples with more than 50% European ancestry with the 171 marker ABD 2.5 test are suitable for EURO 1.0, what we are trying to do with this restriction is delimit as best we can the eligible sample pool to Europeans, including Middle Eastern and South Asian individuals (who may very well have lower levels of European), while eliminating samples for which there exists recent non-European admixture (where fractional affiliations are due to older population level admixture events, and shared ancestry). Most individuals from the Middle East and South Asia fall into this eligible category based on a 50% European cutoff value. In contrast, a European with higher levels of Indigenous

American, where the Indigenous American was contributed by an American Indian grandmother, would erroneously register with South Asian affiliation with the EURO 1.0 test. If not properly vetted, persons with extensive Indigenous American affiliation not due to South Asian ancestry, such as Hispanics, are also erroneously reported to have South Asian affiliation.

The problem is, a person with a 25% Indigenous American score could derive the score from recent admixture or within-Europe structure and the cut-off value of 50% cannot do the job of eliminating the right samples. How do we distinguish between the various types of fractional affiliation so that we estimate prior ethnicity and know who is suitable for the test? For example, if a person types with 60% European 30% Indigenous American, 10% East Asian with the continental admixture test, then how do we know whether the sample corresponds to a South Asian Indian (for which the EURO 1.0 population model is appropriate) and a Hispanic (for which the EURO 1.0 may not be)? Use of a more complex, and global model may seem to represent the answer but reports have shown that unless the number of markers is increased dramatically, this sacrifices much of the intra-populational power we are interested in (see Bauchet et al., 2007, comparing Figure 1 at high k values with Figure 2 at low k values). There are three better ways:

- Directly through the use of sum of intensities scores to date the admixture from programs such as STRUCTURE or ADMIXMAP. In this case, we could pass the sample on to EURO 1.0 based on a sum of intensities score that indicates a low likelihood of recent admixture (i.e., the Indigenous American and East Asian character is not present in discrete blocks in the DNA).
- Indirectly through clustering with other database entries. In this case, we would ask, is the 60/30/10 profile commonly seen for South Asian Indians and not for Hispanics or vice versa? We can estimate a likelihood value for South Asian versus Hispanic ethnicity, and judge eligibility for EURO 1.0 based on this prior value. Of course this method would not be foolproof.
- Indirectly and categorically through the use of prior information on ethnicity (i.e., self-reported). If we were conducting a clinical trial where this type of information was collected, this method may be useful. Incorporating it would be most appropriate in the context of a different algorithm (i.e., Bayesian) but could be accomplished through a prescreening step, though this would be less precise due to the use of arbitrary cut-off values.

We have discussed the sum of intensities parameter in Chapters 3 and 5 (e.g., Figure 5-33). We could label samples as eligible if their sum of intensities scores are not significantly different from other continental Europeans, and ineligible if they are different, and more similar to that from recently admixed populations. Unfortunately, a very large number of markers is required to compute this score properly (Frudakis et al., unpublished observations; P. McKeigue, personal communication) and so it may not be the most practical solution for application of a small admixture panel.

Alternatively, we can impose the requirement that the continental ABD2.5 results cluster with other nonadmixed Europeans. This can be assessed by screening the result against a database of other results for samples of self- or third-party defined geopolitical identity. If the result calls up a substantial number of admixed peoples, such as Hispanics, American Indian mixes, African Americans, or North Africans, we can eliminate the sample as ineligible for EURO 1.0 evaluation. We could use graphical tools to help us make our decision. In Chapter 6 we showed that South Asian Indians and Hispanics type of apparently admixed ancestry assuming the four-population continental model and the ABD2.5 panel of 171 AIMs. Figure 7-5 shows these samples plotted together. If an unknown sample plotted within the blue data points of this figure, the sample is more likely to be of South Asian ethnicity than Hispanic. If instead, the unknown sample pots among the red samples, the reverse would be true.

Figure 7-5

Tetrahedron plot of Hispanic (red) and South Asian (blue) samples illustrates that clustering an unknown sample among different ethnic groups with similar continental population admixture averages can be used to provide prior information needed to justify application of a European panel of AIMs. If an individual plotted of MLE percentages in the middle of the Hispanic population, it would be more likely that the individual possessed relatively recent non-European admixture and thus is less suitable for the European model assumed with the EURO 1.0 panel of markers described in the text. On the other hand, if the individual's MLE plotted within the blue South Asian samples, we might infer that the sum of intensity parameter for the admixture in this individual was characteristic of more ancient admixture commonly seen among various European populations (little recent non-European admixture), and that the individual was suitable for the European model assumed with EURO 1.0.

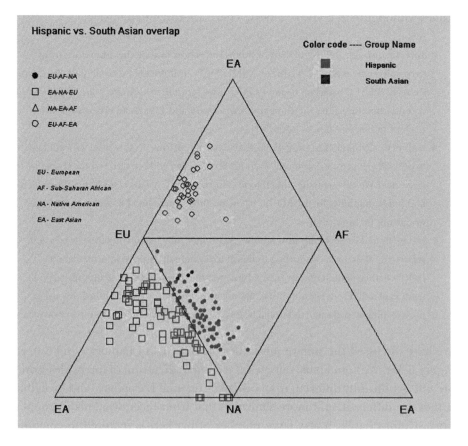

Admixture proportions (n)	NOR	MIDEASED	MIDEAS	SA	Bias
50NOR-50MIDEASED (n = 40)	50.63	44.50	2.88	2.00	4.88
75NOR-25MIDEASED (n = 10)	79.00	13.50	2.00	5.50	7.50
25NOR-75MIDEASED (n = 10)	21.00	75.00	2.00	2.00	4.00
50NOR-50MIDEAS (n = 40)	47.00	5.50	45.63	1.88	7.38
75NOR-25MIDEAS (n = 10)	71.50	2.50	21.50	4.50	7.00
25NOR-75MIDEAS) (n = 10)	17.50	2.00	78.00	2.50	4.50
50NOR-50SA (n = 40)	40.13	10.64	3.08	46.15	13.72
75NOR-25SA (n = 10)	71.00	3.50	2.00	23.50	5.50
25NOR-75SA (n = 10)	26.00	0.50	3.50	70.00	4.00
50MIDEASED-50MIDEAS (n = 40)	5.25	44.5	47.25	3	8.25
75MIDEASED-25MIDEAS (n = 10)	11.5	10.5	77	1	12.5
25MIDEASED-75MIDEAS (n = 10)	2	70.5	21	6.5	8.5
50MIDEASED-50SA (n = 40)	2.38	45.50	1.50	50.63	3.88
75MIDEASED-25SA (n = 10)	7.00	68.00	2.50	22.50	9.50
25MIDEASED-75SA (n = 10)	4.50	22.50	2.50	70.50	7.00
50MIDEAS-50SA (n = 40)	3.13	2.38	48.50	46.00	5.50
75MIDEAS-25SA (n = 10)	2.50	7.00	72.00	18.50	9.50
25MIDEAS-75SA (n = 10)	0.00	0.50	20.50	79.00	0.50
				Average Bias	6.87

Table 7-13

Average group-specific and total bias in EURO 1.0 admixture estimates obtained using the basic MLE algorithm. For comparison with Table 7-14.

Though convenient, the indirect method is a quick-and-dirty method that leaves something to be desired. First, there is overlap in the distributions of ancestry for South Asian and Hispanic samples, and many South Asian samples would be unnecessarily excluded from the eligible classification. Second, cutoff values for a likelihood score to determine eligibility are necessarily arbitrary in nature. What do we do with samples with both South Asian Indian and Hispanic ethnicity, and how much Hispanic ethnicity is acceptable? We know that as the amount of non-European admixture increases, the bias in the results will also increase, but the increase is gradual and establishing a cutoff is arbitrary. The problem is more difficult when one realizes that this is not the only type of affiliation, or the only questionable ethnic groups that this applies to (for example, recent and substantial West African admixture in a Greek would confound EURO 1.0, but not the lower level of West African typically found in Greeks.

The correct way to handle this problem would seem to be to define the regions of each sample's chromosomes that show block-like European ancestry, and then to use only the EURO 1.0 markers within these blocks for the EURO 1.0 analysis. Since the EURO 1.0 test appears to be fairly robust to large reductions in the number of AIMs used (see Table 7-8), this is a reasonable tactic, but as discussed, due to the power needed to perform sum of intensities calculations, we would probably require a much larger panel of markers than we are discussing here. Implementing such an algorithm would allow each sample to be evaluated with EURO 1.0, whether their European scores with a continental test were low or high, whatever their background, though

Table 7-14

Average group-specific and total bias in EURO 1.0 admixture estimates obtained using the STRUCTURE algorithm. For comparison with Table 7-13.

Individuals with NOR and MIDEASED admixture	NOR	MIDEASED	MIDEAS	SA	Bias
50NOR-50MIDEASED (n = 40)	0.52	0.42	0.03	0.02	0.05
75NOR-25MIDEASED (n = 10)	0.87	0.07	0.02	0.04	0.06
25NOR-75MIDEASED (n = 10)	0.16	0.80	0.02	0.02	0.04
50NOR-50MIDEAS (n = 40)	48.47	4.03	45.03	2.50	6.52
75NOR-25MIDEAS (n = 10)	76.18	3.16	18.33	2.35	5.51
25NOR-75MIDEAS) (n = 10)	12.28	3.05	81.62	3.04	6.09
50NOR-50SA (n = 40)	40.62	8.95	2.90	47.53	11.85
75NOR-25SA (n = 10)	78.28	3.59	2.71	15.44	6.30
25NOR-75SA (n = 10)	21.27	3.17	3.84	71.72	7.01
50MIDEASED-50MIDEAS (n = 40)	6.11	43.39	47.77	2.73	8.84
75MIDEASED-25MIDEAS (n = 10)	2.76	77.40	16.12	3.73	6.49
25MIDEASED-75MIDEAS (n = 10)	9.22	8.31	79.88	2.58	11.80
50MIDEASED-50SA (n = 40)	3.16	46.27	1.65	48.92	4.81
75MIDEASED-25SA (n = 10)	6.49	70.47	2.84	20.21	9.33
25MIDEASED-75SA (n = 10)	4.67	16.21	2.86	76.28	7.53
50MIDEAS-50SA (n = 40)	3.48	2.79	44.40	49.32	6.27
75MIDEAS-25SA (n = 10)	2.65	1.91	13.05	82.38	4.56
25MIDEAS-75SA (n = 10)	3.56	6.40	72.31	17.74	9.96
				Average Bias	6.28

the precision of the result would decrease with lower European affiliations due to sampling effects (fewer markers would be used). McKeigue's ADMIXMAP program would be well suited for this endeavor, with some additional code, and in the next edition of this text we hope to be able to discuss some results.

EURO 1.0 PEDIGREES AS AN AID TO INTERPRETING RESULTS

Pedigrees are useful for understanding the strengths and limitations of the European test, and instructive on how best to interpret results. One pedigree for the EURO 1.0 test is shown in Table 7-15.

In the pedigree shown in Table 7-15, none of the four individuals characterized their ethnicity as Middle Eastern. It is clear that there is significant MED and MIDEA ancestry in the children, which we could conclude from the relatively high levels observed, but looking at the results in the entire pedigree it appears that although the MED ancestry in the children was contributed by both parents, the MIDEA ancestry came mainly from the mother. Both mother and father here are real people who described their heritage as continental European, but the MIDEA and SA result may provide a basis for a new genealogical line of investigation, one that is focused in time farther back than most genealogists consider with geopolitical records and surnames. This is true particularly for the mother's side of the family tree (i.e., from where might this substantial and unexpected MIDEA ancestry have come?).

This pedigree was obtained from a real family. Typing actual family members may be difficult and time consuming (mainly, in collecting the samples), so it is much easier to simulate pedigrees. We have simulated over 120 pedigrees, and obtained satisfying results for all but two of them; a sampling of these pedigrees is shown in Table 7-16.

What does an unsatisfying pedigree look like? The pedigree shown in Figure 4-20 (real people, not simulated) is the most discordant pedigree we have observed yet, and it provides an opportunity to discuss how EURO 1.0

	NOR	MED	MIDEA	SA
Mother	44	26	30	0
Father	74	18	8	0
Child 1	70	15	15	0
Child 2	58	19	23	0

Table 7-15
Sample EURO 1.0 percentages for a real pedigree.

Table 7-16

EURO 1.0 percentages for simulated pedigrees.

	NOR	MED	MIDEAS	SA
Gr5-1-P1	96	0	4	0
Gr5-1-P2	0	96	0	4
Gr5-1-S1	35	58	0	7
Gr5-1-S2	39	52	0	9

	NOR	MED	MIDEAS	SA
Gr8-0-P1	40	46	6	8
Gr8-0-P2	0	92	0	8
Gr8-0-S1	12	71	10	7
Gr8-0-S2	18	76	0	6

	NOR	MED	MIDEAS	SA
Gr9-0-P1	97	0	3	0
Gr9-0-P2	0	0	100	0
Gr9-0-S1	41	1	58	0
Gr9-0-S2	40	0	59	1

	NOR	MED	MIDEAS	SA
Gr13-0-P1	69	31	0	0
Gr13-0-P2	0	0	0	100
Gr13-0-S1	19	21	0	60

	NOR	MED	MIDEAS	SA
Gr13-0-S2	32	22	0	46
Gr18-1-P1	0	40	43	17
Gr18-1-P2	0	38	50	12
Gr18-1-S1	0	44	46	10
Gr18-1-S2	0	37	38	25

	NOR	MED	MIDEA	SA
Mother	50	10	15	25
Father	55	40	5	0
Child 1	66	0	34	0
Child 2	35	45	0	20

Table 7-17

Sample EURO 1.0 percentages for a real pedigree.

results should and should not be interpreted. In particular, it illustrates why it is better to draw conclusions about a sample's or individual's anthropological heritage from results of small pedigrees, even if incomplete (such as a child and a mother), rather than individual people (such as in this case, only the mother). In most forensics situations, this is not possible, but nonetheless this concept deserves some attention here because it illustrates how test error can manifest itself.

A mother, father, and two children (STR paternity test positive) are shown in Table 7-17. None of the people had conducted an amateur genealogical study, but as with most people the two parents had a good idea of their predominant European ancestry. Via self-reporting, most (not necessarily all) of the mother's recent ancestors three generations ago were known to have been English and German, and the father reported himself as a little over half Greek (exact percentage unknown).

The father's results were more or less consistent with his expectations (particularly considering that most Greeks, Turks, and Italians show less MED ancestry, a measure of anthropological identity, than they expect from blood, which is derived from geopolitical, sociocultural, and very recent geographical identity). We have discussed the difference between anthropological and geopolitical ancestry as just discussed and in earlier chapters. As expected from self-reported ancestry of "a little more than half Greek," the father shows relatively high MED ancestry with EURO 1.0 compared to the mother, who reported no Mediterranean or Southeastern European ancestry. Not expected from self-reporting in geopolitical terms, however, was the 15% MIDDEA and 25% SA ancestry for the mother. With an error caused by continuous allele frequencies of seven percentage points, it seems unlikely that this result is due to statistical error, and from this result we have good evidence for non-NOR, non-MED ancestry within the mother's family tree—perhaps extending back a few hundred to a few thousand years ago.

Looking at the children, the evidence is stronger. Note that it is possible that levels of non-NOR, non-MED ancestry as high as these (adding to 40%) could be due totally to error, just very unlikely (the average bias is 6–7%, but rare individuals could exhibit 20 or 30% bias, even 40% if the sample is large enough). Child 1 is an example of an individual who obtained results that are somewhat discordant from those of his parents; we would have expected him to have scored somewhere between 5 and 15% MIDDEA, and 0 and 24% SA, but instead he scored 34% MIDDEA and 0% SA. Some of the discrepancy is likely due to the random nature of chromosome inheritance (genetic assortment) but some is undoubtedly due to test error. Child 2 showed less NOR and MIDDEA and more MED than expected, probably for the same reasons.

Why is it that this pedigree so discordant? Recall that the average results are accurate to within 6 to 7% for any individual, meaning some people will exhibit 0% error, some 10% error, and a few 15 to 20% due to continuous allele frequencies. Say the mother has a 15% error in an opposite direction from one of her offspring—her 15% MIDDEA is really 30% and her son's 34% is really 19%. With the father at 5%, the expected results for the son are 5 to 30% and the true value of 19% now make sense.

Of over 126 pedigrees studied, it is unusual for a single child to show such divergence from expectations much less two in the same pedigree like this, our most discordant pedigree, but even in this case we can extract useful information. When we look at the results in the context of the pedigree, it is clear that the mother's non-NOR and non-MED EURO 1.0 results are confirmed in her children (just not in the exact percentages we would expect, given the error associated with the test). The MIDDEA and SA ancestry observed in the children appears clearly to have come from the mother rather than the father. So, concluding that there is significant non-NOR and non-MED ancestry in this pedigree contributed from the mother's ancestors is relatively safe. Note that concluding such from just one of the children, or even from the mother, is less secure. It is in confirming the result and its inheritance in the pedigree as a whole that we obtain the confidence to hypothesize that the mother's and father's anthropological heritage are different.

Given our confidence from the pedigree, we can now ask from where might this MIDEA and SA ancestry have come? It is possible that the mother had a significant number of Western/Central Asian, Middle Eastern and/or South Asian ancestors sometime within the past thousand (or more or less) years. For example, perhaps some of her ancestors were Roma gypsy, or perhaps some were Bedouin Arabs who settled in Europe within the past 1,000 years. It so happens that the mother in this case felt such non-NOR, non-MED ancestry likely came from her mother rather than her father, since much more is known about the latter individual's ancestors, but she could never know enough about either of them to know this for certain. To form hypotheses like these from a single test result, higher levels of MIDEA and/or SA ancestry would be needed, but lower levels are useful for forming such a hypothesis when results are available from multiple individuals within a pedigree because we can see the non-NOR and non-MED ancestry confirmed by the children (even though not in exact percentages we expected). For this reason, to form the most sound hypotheses of anthropological heritage, it is always better to test multiple individuals of a pedigree if possible.

EURO 1.0—INTERPRETATION OF VARIATION WITHIN GROUPS

Variation within groups seems to be the rule rather than the exception. You can appreciate that this applies even to parental groups by studying Figures 7-1 and 7-2 (note the extreme example of South Asian admixture for the single Iberian sample in Figure 7-1). This of course is the whole point of assessing ancestry molecularly and objectively from the DNA, rather than based on phenotype, self-held notions, modern-day geographical origin, or even historical records that are based on ephemeral and relatively recent political boundaries. As a result of the reality that geopolitical and anthropological/genetic heritage do not correspond perfectly, individual results may obtain for some samples that are not expected based on prior knowledge. A sample collected from a person living in Ireland, who describes himself as Irish, could type of a significant level of Middle Eastern ancestry depending on who his ancestors were and where they lived 1,000 years ago. Indeed, with EURO 1.0, one is looking much further back in time (many tens of thousands of years) than one might expect based on relatively recent genealogical records, which are primarily based on geopolitical affiliations.

Most of us are used to thinking about ancestry in terms of recent geopolitical boundaries and events rather than in anthropological terms. We have already discussed this discordance in terms of continental ancestry, but let us take a European example to illustrate the potential discordance between geopolitical and anthropological identity. A blond-haired, blue-eyed person for whom all great grandparents were born in Northern Italy or Greece may very likely be derived from individuals who lived in more northerly climes 10,000 years ago (hence the blonde hair, possibly). Nevertheless, they would most likely describe themselves as Italian, Greek, or Mediterranean even though they may be genetically more close to Nordic peoples of Scandinavia. Because of the age of the anthropological origins we read from the DNA, as opposed to from an archive that may have existed only for the past 1,000 years or so, such a person would not be able to use political records in a library or archive to trace their heritage as far back as with an anthropological test such as EURO 1.0. EURO 1.0 does not care where a person lives, where their great grandparents lived, what language is spoken, or with whom affiliation is felt. Because it is driven by genetic distances, created many thousands of years ago, an anthropological test such as EURO 1.0 looks back to the mixture of one's ancestors that lived in prehistoric Europe, the Middle East, and possibly South Asia.

For the Northern Italian with blond hair and blue eyes we can clearly see the difference in physical characteristics suggesting Nordic heritage, and so a report of predominant NOR ancestry with EURO 1.0 may not be all that surprising, but for many people there are no clues to be derived from physical appearance, and genetic ancestry may not be as easy to understand with the eye. As an objective, independent reporter of anthropological heritage, this is in fact the power of assessing ancestry from DNA rather than from subjective descriptions or even archive records.

What are the *caveat emptors* for such a test in a forensics setting? Given the possible discordance between what the test result means, and what the user may think it means (uncoupling of geopolitics and anthropology), genealogists and forensics scientists may not want to implement a test such as EURO 1.0 if:

- They are uncomfortable with the fact that a test such as this has never before been applied commercially on a large scale, and there is no *de facto* final arbiter of ethnicity, or even another test with which to compare it.
- They cannot accept that there is a difference between anthropological and geopolitical ancestry. In other words, if they feel that *every* person whose grandparents were born in Italy should type with substantial MED ancestry regardless of where their ancestors were derived from 5,000 to 10,000 years ago or longer, then the test results are likely to be misinterpreted.
- It bothers them that we cannot define the precise genetic origins of NOR, MED, MIDDEAS, or SA identity.
- It bothers them that there is no genetic group measured by our test that precisely matches any geopolitical boundaries.
- For a genealogist, any changes in the way they view the heritage of the deeper roots of their family tree would possibly bother them in any way.

These concerns are especially relevant in light of the extant variation observed within ethnic groups. From the detail of Figures 7-1 and 7-2A and B you can appreciate the frequency with which customers may obtain results that do not comport with their geopolitical expectations. Note that there is substantial variation in percentage composition within each of the ethnic populations. South Asian Indians type predominantly of SA genetic affiliation, with little apparent admixture, but Middle Eastern samples show a bit more admixture, and each person is unique. Even more admixture is observed for Mediterraneans and Northern Europeans and it is almost certain that not all of these latter individuals would have expected to see such affiliation. Some Greeks and Italians show more NOR ancestry than MED, and more NOR ancestry than some individuals from Ireland or Scandinavia. Again, these individuals might not have expected to see such results. This *does not* mean they are not Greek or Italian, but it indicates that they are likely to be of different *anthropological* heritage than most Greeks and Italians. In other words, it is likely that on an anthropological time scale their ancestors were relatively new Greeks or Italians. They were more likely relatively recent immigrants to the Mediterranean than a person that scored of high MED ancestry, and unless the movement of the family into the Mediterranean was within the past few generations, genealogical research would probably not have identified the non-MED ancestry. Similarly, but in reverse, some (about 8%) of the Irish/British/Scandinavian samples were characterized by extensive MED admixture although they likely would expect central European or Mediterranean ancestry (note the infrequent incidence of large amounts of red color for certain North European and Irish individuals in Figure 7-1). Many of these people may not have expected such results.

Variation such as this translates into a certain amount of discordance between ethnic expectations and genetic affiliation for some individuals in some groups, a discordance that varies from individual to individual and from ethnic group to ethnic group. When interpreting results it is important to understand the difference between biology/anthropology and sociopolitics/geography, as well as to appreciate the various sources of test error.

The most likely explanation for the bulk of the variability of genetic affiliations within geopolitical groups in Europe is simply that Europe is an admixed continent (on an ethnic scale), of relatively complex interactive history. However, it is impossible to rule out other sources of error we have so far discussed. Inspection of the individual results show most samples are characterized by extensive admixture, that when considered on a population level, is of a not unexpected type, given the complex political and anthropological history of Europe. In other words, geographic pattern in the admixture results suggests sensibility and anthropological meaning. The geographical sensibility of the admixture observed for this data is similar to that observed by Rosenberg et al. (2001) who used STRs with his program STRUCTURE to partition European subpopulations (see Figure 4-1A) as well as that obtained with PSMix on other samples in the work by Eschleman (see Figure 4-1B). In both this work and Rosenberg's study, ethic populations of origin geographically intermediate to Mediterranean and Northern European populations (such as Iberians in our work, and French in Rosenberg's) showed intermediate admixture profiles characteristic of both Mediterranean and Northern European populations (Rosenberg et al. 2001). This would seem hardly to be a coincidence.

However, even if this is true that most of the variation within geopolitically defined ethnic groups is a reflection of the complex interactive history among European populations over the recent past, it does not alleviate the difficulty of explaining to a person who is Greek why they type as significantly Northern European, even if they do have blond hair and blue irises. Indeed, variation in results within the Greeks is particularly interesting; extensive NOR admixture was observed for many of the samples from Greece, Italy, and Turkey; though most of these individuals showed predominant MED affiliation, several showed less than 25% MED affiliation. Rosenberg et al. (2001) showed a similar phenomena with Mediterranean samples; only a fraction of Italians in their analysis showed significant Mediterranean specific genetic affiliations (see low levels of red and tan color of the Europe part of Figure 4-1A, k = 4; Rosenberg et al. 2002). It is doubtful that any of these Greek or Italian individuals would have expected such a heterogeneous result.

We know that there has been extensive gene flow within Europe, and we might expect that in this part of the world there might be an accentuated uncoupling between geopolitical and genetic heritage. We know that there is great anthropometric trait variation within ethnic groups of Europe. The blond, blue-eyed individuals from Northern Italy and Greece for example are characteristically different from the typical brown-eyed, brown-haired Italian or Greek found in the more southern parts of both countries, but such a phenomena is not unique to Greece and Italy—some Swedes and Norwegians have atypical dark complexions, too. From where are their ancestors? Are the Swedes and Norwegians with darker complexions and hair/iris colors the ones showing greater MED, MIDDEAS, and SA ancestry (the latter of which possibly could have come from Roma genetic contributions)? Are Europeans in general better described in genetic terms than geopolitical due to a natural and expected uncoupling between anthropometric trait value and ancestry? These questions are not possible to answer with the analysis available to date, but are the type for which EURO 1.0 was developed.

AN HISTORICAL PERSPECTIVE

Perhaps it is not unexpected that Mediterranean regions of Europe are more genetically diverse than Nordic regions. Both the Byzantine and Roman Empires were known as effective melting pots for peoples throughout the Middle East and Europe, wherein its citizens might have been more united thorough political and sociocultural rather than genetic ties. The Mediterranean represented the center of the civilized world for a long time, and it is plausible that due to this, gene flow from Northern to Mediterranean Europe was greater than in the reverse direction. Between 1900 and 1500 B.C., the Mycenaeans or Achaeans moved southwest from southwestern Russia and invaded/settled wave after wave into modern-day Greece. These civilizations spread to Southern Italy, Libya, Cyrenecia, and the Near East, and they included the fighting groups that attacked the Trojans in Asia Minor 1200 B.C., exploits that were retold centuries later (circa 750 B.C.) in the *Iliad and the Odyssey*.

Later, the classical city-states of Greece emerged, reaching its cosmopolitan peak around the fifth century. Additionally, migration into Europe from the Middle East and most of prehistory Asia is known to have passed through southeastern Europe, hence the northwest to southeast classical blood group marker clines, archaeological record clines, and Y-chromosome clines (all of which are reviewed elegantly by Jobling et al. 2004). This might suggest that gene flow into the Mediterranean has had a relatively strong influence in shaping genetic structure in this part of Europe relative to others. Greece was at one time a Roman province (second century B.C.) until Constantinople fell to the Crusaders in 1204. In 1453, the Turks took Constantinople and made Greece a Turkish province, perhaps providing greater opportunity for East to West migration.

In all, the complexity of Mediterranean history over the past few thousand years may explain why we detect more variation in admixture for people from this part of Europe—some individuals with substantial Middle Eastern genetic affiliation, many others with Northern European affiliation. Of course modern European history is expected to have accentuated the admixture in this part of Europe, but admixture is certainly not unique to Mediterranean Europe. Certainly larger samples and other markers are needed to fully address the myriad possibilities for how or whether the extant Northern European and Middle Eastern admixture in Greece is related to recent historical events. Notwithstanding the mechanism creating this apparent admixture, at least with this panel as well as with Rosenberg's STR panel, it would seem that a fair number of individuals who identify with Greek or Italian ethnicity will show more Northern European and less Mediterranean admixture than they expect.

Some interesting facts about the EURO 1.0 test are:

- NOR scores for non-parental "Caucasians" above 80% are unusual, and when they are obtained, usually are obtained for people with light eye color shades, hair tones, and skin complexions.
- More NOR ancestry is seen in Greeks and Turks than MED in Northern Europeans such as Irish.

- About half of the Greeks tested show less MED ancestry than they expected, but most all Greeks tested so far have shown substantial MED ancestry.
- Middle Eastern subtypes (excluding Turkey) all type with substantial MIDEAS ancestry.
- The amount of South Asian ancestry a South Asian individual exhibits depends on where in South Asia they are from. Anecdotal reports of EURO 1.0 application suggest that high levels of non-South Asian admixture is commonly seen in Northern India, but not Southern India, for instance.
- The average North African has too much non-European ancestry for EURO 1.0 to be used with the accuracy specifications discussed earlier in this chapter.

MORE DETAILED SUBPOPULATION STRATIFICATIONS—k = 7

The same markers used for EURO 1.0 can be used to stratify the European group into more than four groups, but with added complexity comes increased potential for statistical error and bias. In certain circumstances, the added detail may be informative and well worth this penalty. We will now take a look at the results obtained when assuming a significantly more complex population model, where the European group is partitioned in to seven subgroups rather than four. It needs to be stated here at the outset however, that the EURO 1.0 AIMs were not selected using a well-rounded European sample, unlike the EURO 2.0 panel (Figure 7-2), and so we lack the ability to resolve the East-West components of genetic variation with these markers no matter what k level we use for our analysis. What follows is of interest to those in the field who wish to see what happens when we attempt to fit a given panel of markers to an over-ambitious population model. It's not all bad, but not as useful as fitting an appropriate marker panel, of course. Nonetheless there is a certain beauty associated with empirical observation of pattern, whether the marker panel was optimized to report that pattern or not (for example, Rosenberg et al., 2002's microsatellite panel was not selected for optimal information content to their population model either, yet much was learned from that study). As the EURODNA 2.0 panel was brand new, as of the time of this writing, we hope to provide similar results for this more appropriate panel with respect to complex European models in later editions.

Figure 7-6 shows the results from application of the EURO 1.0 panel of markers with a seven-population European model, sorted by geopolitically defined ethnic identity (as before, the ethnic categories along the x-axis are based on political borders rather than genetic information). Each colored column is an individual. The percentage of affiliation with each of the seven genetically defined subgroups is shown along the y-axis, each genetic group

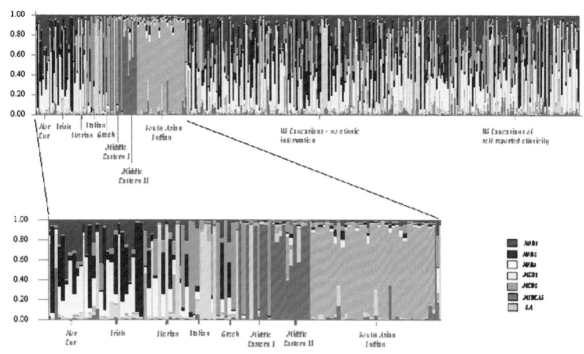

Figure 7-6

STRUCTURE plot obtained from application of 320 European AIMs to Continental European and U.S. populations. Individuals are represented with columns and the proportion of color in each column is representative of the percentage affiliation with each of the seven elements of population structure shown in the key. These elements are named arbitrarily, based on the geographical origin of populations they characterize. The samples are separated into two groups: Parental Europeans (Northern European to South Asian, in geographical order along a northwest to southeast axis across Eurasia) and North American Caucasians. The results for the Parental Europeans are expanded below the main diagram. As one might predict from recent history, the Caucasian samples from the United States show high levels of genetic ancestry shared among Northern European and lower but substantial levels of ancestry shared among Mediterranean populations, but little Middle Eastern or South Asian ancestry.

shown with a different color. The convenience of STRUCTURE for this purpose is that it enables visual presentation of seven-dimensional affiliations within individuals better than could be accomplished with a triangle or tetrahedron plot.

Inspection of the expanded view of the continental European samples in this figure shows:

- Almost all the individuals showing affiliation with the genetically defined subgroup designated by the grey color (Genetic Group 6) are Middle Eastern.
- Almost all the individuals showing affiliation with the genetically defined subgroup designated by the peach color (Genetic Group 7) are South Asian.
- Almost all the individuals showing predominant affiliation with the genetically defined subgroup designated by the green color (Genetic Group 5) are Mediterranean, such as Italians and Greeks.

- Almost all the individuals showing predominant affiliation with the genetically defined subgroup designated by the gold color (Genetic Group 4) are also Mediterranean (Italians and Greeks).
- Almost all the individuals showing predominant affiliation with the genetically defined subgroups designated by the red (Genetic Group 1), blue (Genetic Group 2), and yellow (Genetic Group 3) are Northern European, Irish, and Iberian. Note the lack of resolution between Iberians and Continental Europeans we can achieve with EURO 2.0 (Figure 7-2B).
- U.S. Caucasians show affiliation predominantly with Genetic Groups 1, 2, and 3 (Northern European) and to a lesser extent, 4 and 5 (Mediterranean), but very little affiliation with Genetic Groups 6 and 7 (Middle Eastern, South Asian).

Table 7-18 shows this data in a tabular format, listing the average admixture percentages for the seven genetically defined within-European subgroups detected using STRUCTURE with k = 7 (Table 7-18; admixture percentages greater than 20% are shown in dark blue, and percentages greater than 15% but less than 20% are shown in light blue).

The average F_{ST} values over all loci for each group compared to all others, as determined by Pritchard's STRUCTURE program were:

Mean value of $F_{ST}\_1 = 0.0326$
Mean value of $F_{ST}\_2 = 0.0635$
Mean value of $F_{ST}\_3 = 0.0419$
Mean value of $F_{ST}\_4 = 0.0353$
Mean value of $F_{ST}\_5 = 0.1386$
Mean value of $F_{ST}\_6 = 0.0844$
Mean value of $F_{ST}\_7 = 0.2028$
Average mean value of $F_{ST} = 0.09$

Most of the AIMs are characterized by alleles of high frequency. Using $F = \delta^2/s(2-s)$, where $s = p_{11} + p_{21}$, or the sum of allele frequencies in both groups of the comparison, we obtain average delta values in the range of 0.10 to 0.27 for the marker panel with respect to each group versus total group comparison. The average F_{ST} corresponds to an average $\delta = 0.18$ assuming a frequency sum of the average locus allele of 20%.

WHAT DO THE GROUPS NOR1, NOR2 ... MEAN?

Based on the identity of parental samples showing affiliation with each of the seven subgroups, and with reference to the genetic clines within Europe

Table 7-18

STRUCTURE results of EURO 1.0 admixture analysis applying a k = 7 population model.

		Genetically defined subgroups						
Ethnic sample	sample (n)	Group 1: NOR1	Group 2: NOR2	Group 3: NOR3	Group 4: MED1	Group 5: MED2	Group 6: MIDEAS	Group 7: SA
North Euro	10	0.262	0.261	0.256	0.124	0.026	0.063	0.009
Irish	17	0.187	0.354	0.213	0.172	0.04	0.018	0.016
Iberian	10	0.13	0.151	0.146	0.359	0.109	0.01	0.094
Italian	9	0.097	0.064	0.09	0.312	0.398	0.011	0.029
Greek	8	0.08	0.128	0.062	0.249	0.461	0.009	0.012
Middle East I*	9	0.01	0.008	0.009	0.114	0.311	0.538	0.009
Middle East II	10	0.017	0.015	0.018	0.069	0.024	0.791	0.067
South Asian	35	0.021	0.019	0.021	0.028	0.019	0.032	0.859
Caucasians	190	0.198	0.222	0.183	0.215	0.106	0.035	0.042

*Sample comprised of 4 Greeks, 1 Lebanese, 3 Arabs, and 2 Persians.

presented by others before us (Cavalli-Sforza et al. 1994; also see Jobling et al. 2004), we can ascribe tentative names for each STRUCTURE subgroup:

- Three Northern European subgroups (NOR 1, NOR 2, and NOR 3)
- Two Southern European/Mediterranean subgroups (MED 1, MED 2)
- One Middle Eastern subgroup (MIDEAS)
- One South Asian (SA) subgroup

The STRUCTURE plots shown in Figures 7-1 and 7-6 use the same continental European and Eurasian parental samples. We can therefore conclude that the three NOR subgroups obtained when using a population model of k = 7 are combined into a single NOR group when using the k = 4 model (compare blue, yellow, and red color in Figure 7-6 with the red color of Figure 7-1 for the Northern Europeans, Irish, and Iberians). However, the relationship between the MED1 and MED2 groups using the k = 7 model and the MED group from the k = 4 model is not simple.

EVALUATING THE RESULTS FROM THE k = 7 EUROPEAN MODEL

It may seem natural to expect substantial variation within parental groups, even though we are considering genetic groups that have been mixing with one another for the past several thousand years. In general, the amount of variation within parental samples must decrease with the age of the parental populations. For instance, an African/non-African test probably would show less variation within groups than a Northern European/Central European test, not only because the genetic distance between the latter pair is lower than the former, but because mixing between individuals of the latter pair is more common than between the former pair (due to greater geographical distance between populations, sexual selection, cultural similarity, etc.). In the case of the within-Europe test we have described in this chapter up to this point, we see a fairly rational partition of parental samples along geographical lines that agrees with previous work, but the substantial variation within groups (particularly the Mediterranean groups such as Greeks) was a surprise. We need to evaluate the performance of the panel and to do this we can (as before):

1. Compare the trends in the results against those obtained from other admixture and Y-chromosome studies.
2. Challenge the test with other samples from the same parental groups used to develop the seven-group genetic model.
3. Challenge the test with samples of other groups of known phylogeographical origin that were not used to develop the seven-group genetic model.

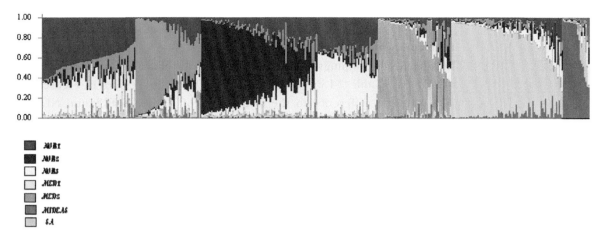

- ■ *NOR1*
- ■ *NOR2*
- ☐ *NOR3*
- ☐ *MED1*
- ▨ *MED2*
- ▨ *MIDEAS*
- ☐ *SA*

Figure 7-7

STRUCTURE plot shown in Figure 7-6, where samples are ordered not by geographical origin, but based on percentage affiliation corresponding to each of the seven elements of population structure discussed in the text and shown in the key with colors. Samples are clustered into groups based on the similarity of proportions.

4. Evaluate the accuracy and anthropological relevancy of the test results through correlation with anthropometric trait values.

5. Evaluate the performance of the test in pedigrees—do results comport with expectations based on independent assortment?

6. Evaluate the precision (root mean square error) and bias of the test using simulations.

COMPARISON WITH PREVIOUS STUDIES BASED ON GENE MARKERS

The distribution of NOR1, NOR2, NOR3, MED1, MED2, and SA affiliation throughout Eurasia, in a northwest to southeast orientation, is reminiscent of the subgroups defined by clines in synthetic maps of Europe and Western Asia obtained from classical blood group markers and Y-chromosome haplogroups (reviewed by Jobling et al. 2004). Like anthropometric traits such as hair and eye color and archaeological maps, the distribution of genetic variation from these older studies shows the same principal component of variation in a northwest to southeast orientation. For example, Northern European and Irish have the highest affiliation with Group 1 (see Table 7-18; red color, Figure 7-8), which suggests that this genetically defined subgroup corresponds to the most northerly of Cavalli-Sforza et al.'s clines (1994; also see Jobling et al. 2004). NOR1 thus seems to correspond to Nordic ethnicity, hence the name NOR1. It is interesting that the percentage of NOR1 affiliation appears to increase gradually from south to north, peaking in Scandinavia (see Figure 7-8, red color).

The percentage of Genetic Group 2 and 3 affiliation also increases from south to north, and both must therefore also correspond to northerly European subgroupings suggested by the genetic clines of Cavalli-Sforza et al. (1994) (see Table 7-18; NOR2-blue and NOR3-yellow colors in Figure 7-8). There is little difference in the distribution of NOR1, NOR2, and NOR3— when you find one you tend to find the others (in populations, not necessarily in individuals). The percentage of NOR1+NOR2+NOR3 is about 75% for all Northern European, British, and Irish samples combined, a bit lower for continental Europeans (French, German), lower still for Iberians (less than 50%), much lower for Mediterraneans (Italians, about 25%; Greeks, the same) and virtually 0% in Middle Eastern and South Asian population samples tested. It is unclear what the anthropological distinction is between NOR2 and NOR3 though both are more commonly found in Northern and Continental Europe (Figure 7-8). Among the samples shown in Figure 7-6, the allocation of NOR1 through SA ancestry was fairly even, though relatively few examples of MIDEAS ancestry were observed due to a small Middle Eastern population sample. This can be appreciated by re-ordering Figure 7-8 based on genetic ancestry rather than geopolitical sample origin (Figure 7-7).

In contrast, Iberians, Italians, and Greeks show substantial affiliation with Group 4: MED1 (tan color, Figure 7-8), which suggests a correspondence with Central European/Northern Mediterranean subgroups of Cavalli-Sforza et al. (1994). Population samples with high affiliation for Group 5: MED2 (lime green color, Figure 7-8) are mainly Italian, Greek, and Middle Eastern, a pattern shifted slightly to the southeast compared to Group 4, suggesting

Figure 7-8

Global distribution of European substructure in terms of the parental affiliations shown in Figure 7-7. Pie charts indicate the relative proportions of affiliation shared primarily among Northern European (NOR1-red, NOR2-blue, NOR3-yellow) populations, those shared primarily among Southeastern European populations (MED1-gold, MED2-green), those shared primarily among Middle Eastern populations (MIDEAS-grey), and that shared primarily among South Asian populations (SA-pink) as described in the text and indicated in the key, bottom right. The location of the pie chart indicates the geographical origin of the populations. This figure incorporates the data shown in Figure 7-7, as well as data from other similar analyses.

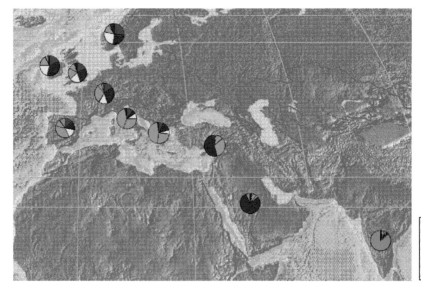

that this group corresponds to a second Mediterranean group corresponding to one or more of the most Southeastern European subgroups of Cavalli-Sforzas synthetic map. Middle Eastern samples show primary affiliation with Group 6 (MIDEAS, dark grey Figure 7-8) and South Asian Indians with Group 7 (SA, peach color, Figure 7-8).

COMPARISON WITH RESULTS FROM OTHER STUDIES

Rosenberg et al. (2002) showed insignificant differences between northwestern European (Orcadian) and Mediterranean populations (Italian, Sardinian) using STRs (see k = 4 bar plot, Figure 4-1; Rosenberg et al. 2002). As we have discussed in Chapter 3, STRs are characterized by lower effective population sizes and may be looking back to a more recent period of time than SNPs or sex-chromosome haplotype markers. Thus, a lack of differentiation among European populations with STR markers may not be entirely unexpected.

The results from Y-chromosome analyses are very similar to the admixture pattern obtained with the 320 AIMs we are discussing (compare Figure 7-4 to Figure 7-8). Semino et al. (2000; also see Jobling et al. 2004) showed a steep northwest-to-southeast gradient in the frequency of six Y chromosome haplogroups. The I (Gravettian) haplogroup is found mostly in Northern Europe. The haplogroups $R^*(xR1a)$ and R1are known as the M173 carrying Aurignacian haplogroups, and they are found more frequently in Northern and Central Europe as well. This distribution corresponds roughly to the NOR1, NOR2, and NOR3 genetic group admixture distribution in Figure 7-8; note however that the mtDNA distribution across Europe is fairly uniform so we would expect autosomal admixture to reveal genetic differentiation across Europe in between the level seen with Y-chromosome and mtDNA analysis. The frequency of NOR2 seems to increase toward the east of central Europe, which is very similar in pattern to that of the R1a Aurignacian haplogroup.

In contrast, the E3b and F^* (x1, J2, G, H, K), G, and J2 haplogroups are found relatively infrequently in northern and central Europe, but commonly in the Mediterranean and northern Middle East. This distribution corresponds roughly to the MED1 and MED2 genetic group admixture distribution in Figure 7-8. The frequency of these Mediterranean haplogroups in Iberians is intermediate to that of Northern Europeans and Mediterraneans, just as the levels of MED2 admixture is intermediate to these two with the 320 AIM admixture test (see Figure 7-8).

Thus, overall, the distribution of ethnic admixture shown in Figures 7-6 and 7-8, and detailed in Table 7-18, is reasonable and within boundaries of expectations given results obtained from mtDNA, Y-chromosome, and STR-based admixture analysis.

BLIND CHALLENGE OF THE k = 7 MODEL RESULTS WITH ETHNIC SAMPLES

In addition to these parental samples of *formally* characterized ethnicity, we included a variety of samples of various *self-proclaimed* ethnic identity admixtures in the STRUCTURE run. After the admixture percentages for the seven groups were determined, we used the self-reported ethnic information to bin the samples into geopolitical population groups; each subject reported the ethnicity of all four grandparents, and if at least two of the grandparents came from one of four arbitrarily defined European metapopulations (I, II, III, IV), the sample was binned to that group. The four groups are shown in Tables 7-19 and 7-20.

Although most of these individuals were of admixed ethnicity, pattern in the results agree with the self-held notions of ethnic affiliation. In particular, the average individual reporting Irish, Scandinavian, and Great Britain heritage (Population I) showed significant affiliation with all three NOR groups, and the first MED group (Group 4, MED1) but the average individual reporting French, German, Danish, Dutch, Finnish, Polish, Eastern European, or Russian heritage (Group II) showed more Group 4, MED1 affiliation. The average individual reporting Spanish, Italian, Balkan, Greek, or Turkish heritage showed substantially more Group 5, MED2 affiliation. This can be seen in

I	At least half of the grandparents from Ireland, Western Scandinavia, or Great Britain
II	At least half of the grandparents from France, Germany, Denmark, Finland, Holland, Poland, Eastern Europe, or Russia
III	At least half of the grandparents from Spain, Italy, Balkans, Greece, or Turkey
IV	Assorted Caucasians collected from the United States

Table 7-19

Admixture among seven European genetic groups (columns) for the average sample of four types of self-reported ethnicity (rows).

Population Sample (n)		Group 1: NOR1	Group 2: NOR2	Group 3: NOR3	Group 4: MED1	Group 5: MED2	Group 6: MIDEAS	Group 7: SA
I	20	0.19	0.25	0.19	0.19	0.12	0.04	0.02
II	36	0.14	0.30	0.15	0.30	0.06	0.04	0.02
III	13	0.19	0.12	0.21	0.16	0.30	0.02	0.01
IV	190	0.20	0.22	0.18	0.22	0.11	0.04	0.04

	n	NOR1	NOR2	NOR3	MED1	MED2	MIDEA	SA
I	20	0.4	0.3	0.3	0.2	0.2	0	0
II	36	0.28	0.42	0.22	0.33	0.08	0	0
III	13	0.47	0.07	0.33	0.13	0.47	0	0
IV	190	0.38	0.25	0.31	0.26	0.12	0	0

Table 7-20

Proportion of samples with >25% affiliation with each of seven European genetic groups (columns) for the average sample of four types of self-reported ethnicity (rows).

terms of the affiliation for the average sample for each group in Table 7-19, or in Table 7-20, which shows the same data in terms of the percentage of samples over one-quarter affiliation with each group. To summarize the results of the blind trial:

> *The more southeasterly the self-reported geopolitical origin, the more similar the average profile is to our original parental samples from the southeast regions of Europe.*

This would seem to support, or at least not refute, the anthropological relevance of the seven-group genetic model. As we saw with the anthropologically defined samples in Table 7-18, the MED1 and MED2 genetic group affiliation levels are inversely proportional to the magnitude of distance from the self-reported origin to the Mediterranean Sea—French/Germans show more than British, and Italians show more than French/Germans. However, simply knowing the recent (on an anthropological time scale) geographical origin of a person is not necessarily a very accurate predictor of genetic group affiliation because as we will discuss later, the political history of Europe is complex, admixture between subpopulations is not uncommon in recent history, and in our original parental groups we observed considerable variation within ethnic groups. This of course is the whole point of assessing ancestry molecularly rather than based on self-held notions or modern-day geographical origin. The average U.S. Caucasian (Group IV; Tables 7-19, 7-20) showed admixture levels more similar to Group I and Group II samples of self-reported Central/Northern European ethnicity, which supports the historical record showing that most of the modern-day U.S. Caucasian population is derived from central and northern Europe, rather than southeastern Europe and Middle East.

In addition to our nine geopolitical groups used to develop the 320 AIM panel, we can test the performance on samples from other countries even though their inhabitants are not isolated or expected to be of homogeneous ancestry. This is not as convincing a challenge, however, since as we have already seen, there is considerable variation within politically defined groups, but we would expect to see consistency with many of the trends seen in Figure 7-8. We genotyped 34 samples obtained in Turkey (which were not anthropologically qualified), and the STRUCTURE results for this group compared to the other blind sample sets can be seen in Table 7-21 (>20% affiliation shaded). Again, most of these samples are mixtures, but the Turkish samples nicely followed most of the geographical trends defined by the parental samples and followed by the Group I through III admixed samples of self-reported affiliation in Tables 7-19 and 7-20. The Turkish samples had substantially less NOR1, NOR2, and NOR3 than Northwestern European mixes (Group I) or Northwestern European parentals (following the trend), and of overall average profile most similar to other Southern and Southeastern Europeans (Group III, Table 7-21), but not as low a level as expected based on those seen for Italians, Greek, and Turkish parentals (Figure 7-8). This could be due to the fact that the samples were from self-reports and current residence in Turkey, and not anthropologically qualified as Turkish.

A similar phenomena was seen for the other self-reported samples of Southeastern European populations in Table 7-21. Regarding the dramatic increase in MED1 + MED2 (or just MED2) affiliation moving from northwest to southeast, the Turkish samples showed relatively high MED2 levels like Italians and Greek parentals, and the Group III test samples (note the blue shading for MED2 in Table 7-21, similar to that seen for Group III mixes). Yet another geographical trend followed was the increase in South Asian admixture from northwest to southeast. The Turkish samples had significantly more South Asian admixture than any of the other blind test samples (groups I–IV) corre-

Group I	At least two grandparents were Northwestern European or from Ireland, Western Scandinavia, or Great Britain
Group II	At least two grandparents were continental European, or from France, Germany, Denmark, Finland, Holland, Poland, Eastern Europe, or Russia,
Group III	At least two grandparents were from southern or southeastern Europe, such as Spain, Italy, Balkans, Greece, or Turkey
Group IV	Assorted U.S. Caucasians

Table 7-21

Addition of EURO 1.0 k = 7 admixture results for Turkish samples to Table 7-20 for comparison.

		Number of samples with >25% affiliation						
Group	n	NOR1	NOR2	NOR3	MED1	MED2	MIDEA	SA
I	20	0.4	0.3	0.3	0.2	0.2	0	0
II	36	0.28	0.42	0.22	0.33	0.08	0	0
III	13	0.47	0.07	0.33	0.13	0.47	0	0
IV	190	0.38	0.25	0.31	0.26	0.12	0	0
Turkish	34	0.38	0.12	0.35	0.18	0.27	0	0.12

sponding to continental Europeans; levels that were similar to those seen in Middle Eastern samples. Because the samples were not anthropologically qualified, the pie chart for these percentages is not placed in Figure 7-8, but this imperfect (though reasonable) blind challenge of the 320 AIM battery seems to confirm the principal components of seven-subgroup admixture variation within Europe.

CORRELATIONS WITH ANTHROPOMETRIC TRAITS

If the NOR and MED genetic groupings are of legitimate anthropological base, and if there is an association between anthropology and any given physical trait, then there should be a correlation between admixture for certain genetic groupings and the expression of that physical trait. When speaking of European populations, two obvious physical traits come to mind: skin melanin content and iris color. We previously have discussed the correlation between

Table 7-22

K = 7 model elements are associated with Iris color

	Genetic groups	Admixture level	Color	Chi-p	Exact-p	Total European sample (N)	Sample Meeting threshold (n)
1	NOR1	≥45%	Light	0.106	0.104	184	15
2	NOR1	≥35%	Light	0.005	0.008	184	35
3	NOR1+NOR2	≥70%	Light	0.09	0.13	184	48
4	NOR1+NOR2	≥45%	Light	0.008	0.011	184	82
5	MIDEA	≥15%	Dark	0.04	0.035	184	9
6	MIDEA+SA	≥35%	Dark	0.021	0.014	184	7
7	MIDEA+SA	≥25%	Dark	0.037	0.034	184	12
8	MIDEA+SA	≥15%	Dark	0.005	0.008	184	25
9	MIDEA+SA	≥10%	Dark	0.002	0.003	184	35
10	SA	≥10%	Dark	0.07	0.06	184	5
11	SA	≥15%	Dark	0.04	0.033	184	9
12	SA	≥25%	Dark	0.042	0.037	184	17
13	MED2+MIDEA+SA	≥50%	Dark	0.019	0.019	184	16
14	MED2+MIDEA+SA	≥45%	Dark	0.003	0.003	184	22

skin melanin index and West African and European admixture—with higher levels of European admixture correlated with lower melanin index values—in African Americans and Puerto Ricans (Chapter 5, discussed more fully in Chapter 8), and so, too, is skin melanin index differentially distributed within the European group (mainly along a northwest to southeast gradient). We also have discussed the association between higher levels of West African and/or lower levels of European admixture in individuals of predominant European ancestry and darker iris colors.

We ask whether the same types of associations can be seen within the European diaspora sorted by subpopulation rather than continental ancestry admixture. Since blue iris colors are far more frequently found in Northern European populations relative to Middle Eastern, South Asian populations, then the degree of Northern European admixture should be associated with lighter iris colors and Middle Eastern and South Asian admixture should be associated with darker iris colors.

Eye color scores (described in Chapter 9) were obtained for some of the subjects shown in Figure 7-6, as well as others. As shown in Table 7-22, significant genetic group admixture levels are associated with certain iris color shades (significant p-values are shaded). NOR1 admixture greater than or equal to 35% (row 2) and NOR1+NOR2 combined admixture greater than 45% (row 4) is strongly associated with iris color scores greater than 2.2 (Light), which corresponds to color scores above the mean for individuals of majority European ancestry, and includes the light hazels, greens, and blues. In contrast, MIDEA and SA admixture levels were individually associated with color scores below 2.2 (Dark), corresponding to color scores below the mean (dark hazels, browns, blacks); these associations were found for combinations of MIDEA+SA at various percentage levels (rows 6–9), and even for groups including MED2 (rows 13, 14), though MED2 admixture on its own was not associated significantly with Dark (or Light) iris color scores (not shown). (Logistic regression shows similar associations, but this contingency table-based analysis should be easier for the lay reader to internalize.)

The p-values are generally significant over a wide range of admixture levels, but trend toward less significance at higher levels of admixture, due in part to lower sample sizes qualifying for these higher levels. It turns out that iris color can be predicted accurately for individuals only with a knowledge of pigmentation gene genotypes, but as with the k = 4 model (EURO 1.0) this exercise shows that, with respect to the seven-population model, there is a correlation between genetic groupings characteristic of Northern Europeans and lighter iris colors, and groupings characteristic of Middle Eastern and South Asians and darker iris colors. Note again that these are correlations between *lower* levels of admixture and iris color in individuals that describe themselves as

Caucasians, not obvious correlations between very high levels of admixture and iris color in a more eclectic sample of polarized ancestry.

There exists a natural variation in iris color shade within the Northern European and Irish geopolitical groups, and though we might expect to see an association between lighter iris colors and predominant NOR1 admixture since Northern European geopolitical affiliations correlate with high NOR1 admixture levels, there is no reason why we would see an association between lower levels of NOR1 admixture and lighter iris color shades in a completely unrelated sample of Caucasians, unless the NOR1 genetic grouping was an anthropologically meaningful designation. In short, the associations between NOR1, MIDEA, and SA genetic group admixture and anthropometric trait value, in this case iris color, reinforce the anthropological validity of the genetic groupings NOR1, NOR2, NOR3, MED1, MED2, MIDEA, and SA. The association with MIDEA group and iris color using the $k = 7$ model stands in contrast to the lack of association between the MIDEAS group and iris color using the $k = 4$ model, and may illustrate the enhancement in information provided by the more complex model.

PEDIGREES

The $k = 4$ model is preferred over the $k = 7$ model based on better performance in pedigree analysis; results of offspring more closely resembled those of the midpoint between both parents, with fewer observed confounding pedigrees (ones where it is highly unlikely an offspring's percentages could obtain from those of its parents). This would seem to be the result of increased statistical error associated with a more complex population model. Nonetheless, European admixture estimates obtained using the $k = 7$ model are not, generally speaking, discordant, just more discordant than obtained using a $k = 4$ model. Given the generally similar performance for $k = 7$ model and $k = 4$ model, we will not present examples of pedigrees here.

SUBSTANTIAL VARIATION IN ADMIXTURE WITHIN ETHNIC GROUPS

The results considered so far for the $k = 7$ grouping scheme are average results for geopolitically defined ethnic populations, and although the 320 AIMs provide good genetic resolution between these groups—resolution that appears to be anthropologically relevant—it so happens that there is substantial variation in percentage composition within each of these ethnic populations. In Figure 7-6 one can appreciate the frequency with which individuals may obtain results that do not comport with their geopolitical expectations.

Note that there is substantial variation in percentage composition within each of the ethnic populations.

If geopolitically oriented expectations are perfect, and if genetic diversity is distributed strictly as a function of geography, there should be little variation in STRUCTURE'S vector $Q = (Q_1, Q_2, \ldots, Q_K)$ within each ethnic population, where Q_i is the most likely ancestry coefficient (percentage) for population i. For some of the ethnic groups this is the case; South Asian Indians type predominantly of Group 7: SA genetic affiliation, with little apparent admixture, but Middle Eastern samples show a bit more admixture and variation in Q values, and Mediterraneans and Northern Europeans even more. Overall, the estimated natural log of the probability of the data provided by Pritchard's STRUCTURE program was −56,810.0, the mean value of the ln likelihood was −56,127.3, and the variance of the ln likelihood was 1,365.4. This variation translates into a certain amount of discordance between ethnic expectations and genetic affiliation for some individuals in some groups, a discordance that varies from individual to individual and from ethnic group to ethnic group. Some Greeks and Italians show more NOR1, NOR2, and NOR3 ancestry than MED1 and MED2. Again, this *does not* mean they are not Greek or Italian, only that it is likely they are of different anthropological heritage than most Greeks and Italians. In other words, on an anthropological time scale their ancestors may have been relatively recent immigrants to the Mediterranean.

Some (about 20%) of the Irish/British/Scandinavian samples were characterized by extensive MED1 admixture although they reported no Central European or Mediterranean ancestry (note the infrequent incidence of gold color in the North Europeans and Irish in Figure 7-6). For example, some (about 20%) of the Irish/British/Scandinavian samples were characterized by extensive MED1 admixture although they reported no Central European or Mediterranean ancestry (note the infrequent incidence of gold color in the North Europeans and Irish in Figure 7-6). Variance within the ethnic groups is likely a byproduct of several things working together.

Looking at the within-European admixture from the perspective of genetic rather than geopolitical definitions, we can also see some interesting features of admixture. Though most individuals of geopolitically defined Northern and Central European origin exhibit substantial affiliation with NOR1, NOR2, and NOR3, when we sort the STRUCTURE run by the seven genetically defined groups, we can see that there exist no individuals with greater than 75% affiliation to any of these three groups (see Figure 7-7). This is another reflection of substantial ethnic admixture, but if the seven genetic groups are anthropologically meaningful as the data so far discussed suggests, this would indicate that whereas some genetic groups have persisted in a relatively unadmixed state over the recent past (such as South Asians), admixture with respect to this model may have been far more extensive within Continental Europe.

ALTERNATIVE STYLES FOR ESTIMATING ETHNIC ADMIXTURE

Recall that the primary parental samples used to develop EURO 1.0 were collected in Europe, the Middle East, and India, and that each sample was assumed to be anthropologically qualified in that the individual lived in the appropriate geopolitical region, reported full ethnic affiliation with the parental group back to his grandparents, with no known admixture. In the European analysis we have been discussing, we have simultaneously analyzed both parental and experimental samples. This has been possible because we have used admixture methods that do not require prior information or definitions (i.e., Pritchard's STRUCTURE program, described in Chapter 3). These methods allow us to see (or not see) an expected coalescence of individuals to genetic subgroups that should make sense in geographical and historical terms. This is an empirical approach, and it is different than the approach used to define parental allele frequencies for the 175 AIM continental admixture test described in Chapter 4.

A very useful feature of STRUCTURE is the ability to define allele frequencies using admixed and parental samples together. Theoretically, this improves the quality of the estimates through increased sample sizes. Including all available samples in a given STRUCTURE run is expected to result in more accurate allele frequency estimates than could be obtained from some of them, regardless of the quality of the prior information (i.e., rigor with which ethnic affiliation was defined) because prior information is not required in these STRUCTURE runs. Rather, the stratifications are strictly empirical in derivation, based on a Bayesian manifestation of the likelihood function, and whether or not a subject knows where his or her ancestors are from, STRUCTURE will empirically fit proportional affiliation with respect to the population model ($k = 7$) based on the subject's multilocus genotypes. Using the ability of STRUCTURE to recalculate parental frequencies for each run, using all available samples, more accurate ethnic admixture results therefore would be expected to obtain the larger input sample size.

When running the panel for forensics purposes it would be desirable to analyze each sample in conjunction with all previously analyzed samples at the same time (adding the unknown sample to the previous set when done). This would produce floating allele frequency estimates, which are expected to drift gradually toward the true frequency value over time. The benefit is that the quality of the estimates is expected to improve, but the drawback from the forensics standpoint is that the quality of estimates produced on one date differ from those in a manner that is difficult to quantify from those on another date. Further, we lack the ability to estimate confidence intervals for the estimates. The justification for applying a floating-allele frequency approach with subcontinental markers but not continental markers is that the genetic distances between subcontinental groups are lower, and it is more dangerous to use self-held notions of ethnicity to define parental groups than for the case of worldwide continental admixture where genetic distances are greater among the groups, and self-held notions of ancestry or visually ascribed affiliations are more accurate and easily verifiable.

The alternative is to use STRUCTURE to establish static allele frequencies from one sample set and to use the likelihood function to determine admixture for unknowns with respect to these statically defined frequencies. The allele frequencies could be updated from time to time as the reference sample increases (a mix of parental and experimental samples) and again the quality of admixture estimates would increase, though this time iteratively rather than continuously over time. The benefit of this latter approach is convenience— less time required for determining each new result, and programs other than STRUCTURE could be used to produce confidence intervals for each estimate. The drawback is that the quality of the most likely estimates increases discontinuously over time, and on average, would be greater using the floating allele frequency method and STRUCTURE.

CHAPTER 8

INDIRECT METHODS FOR PHENOTYPE INFERENCE

ESTIMATES OF GENOMIC ANCESTRY ALLOWS FOR INFERENCE OF CERTAIN PHENOTYPES

The primary reason most forensic scientists want to know about ancestry is to assist them in reconstructing physical appearance; that is, to enable generalizations about overt phenotypes that might help them identify the person, suspect, or victim. Clearly, certain physical characteristics are unevenly distributed among the world's various peoples, such as skin pigmentation levels, hair and eye colors, and even aspects of facial features such as eye shape, lip size, nose shape, and facial height and/or latitude width. Inferring physical trait values from ancestry is an example of an indirect inference, since we rely on measurements of AIMs and not the phenotypically functional loci that underlie the expression of the trait when making such predictions. In this case, we use ancestry as a proxy for the net effect of the character of these loci in individuals, and we do so based on the observation that phenotype character is distributed to a certain but quantifiable extent as a function of ancestry. Making an inference of physical trait value from measurement of the functional loci themselves is an example of direct inference, since we rely on measurements of the (presumably gene) variants that cause trait characters.

We first discuss phenotype variation and then focus on human pigmentation, which provides solid examples of traits for which both the direct and indirect methods are useful. After describing the indirect method in detail, with examples, we will then discuss the direct method in Chapter 9, and then finally present some case studies that serve as examples for how these methods can be used to augment a forensics investigation. As we will show in this chapter, the indirect method can provide some information for traits with strong ancestry components, such as iris and skin pigmentation, but not for traits that show greater intrapopulation variation. We will also show that just because there is a correlation between admixture and phenotype, the range of informative admixture values must be determined empirically and need not cover the entire spectrum. For example, as we will show in this chapter, we can infer skin or iris melanin content of a subject knowing their proportional West African ancestry, without knowing which genes are causing the difference between African and European populations. However, we cannot always use knowledge of BGA to predict precise iris color in cases of low non-European admixture—knowing an individual is 95% European does not allow us to infer their iris color since there is considerable variation in iris colors within the European populations.

In this chapter, we will build on what we have learned in the previous chapters where we described the performance of genomic ancestry tests on ethnic populations, taking this to the next level of practical utility, the indirect method of inferring phenotype. We will describe the implementation of a unique empirical approach to making indirect inferences on anthropometric trait value from ancestry, which relies on the use of databases of digital photographs. This in turn will lead to the direct method of phenotype inference, which, when possible, is the method we really would like to use as objective scientists interested in maximizing precision and phenotype information content provided by genetic sequences.

PHENOTYPE VARIATION AS A FUNCTION OF HUMAN POPULATION HISTORY AND INDIVIDUAL ANCESTRY

The extent of geographical distribution of human phenotype diversity pales in comparison to other species such as great apes (Jobling et al. 2004). This is due to the fact that we are a relatively young species, having reached population sizes greater than 10 million only 10 KYA, with an exponential expansion during the Neolithic Period, which continues unabated to this day. Most of the population explosion experienced during this time was an outcome of the development of agriculture, which enabled the production of more food, a surplus that could support a larger, denser population than the pre-Neolithic carrying capacity of the ecosystem could provide. There were demographic consequences of our ability to sustain populations at higher densities, and social structures diverged from those of predecessor and contemporaneous hunter–gatherers. Some of these include higher densities, better nutrition, longer life span, increased fertility, earlier menarche, and the development of political systems for territorial management (reviewed in Jobling et al. 2004).

Population sedentism emerged, where populations built societies that were specialized to specific geographical regions, gradually displacing (and/or converting) older nomadic, hunter–gatherer lifestyles. Anchors to land brought opportunity to develop societies, and better sociological development accompanied the development of organized trade and facilitated migration between populations. Permanent settlements were associated with specialization and the development of complex, nonportable technologies and social constructs such as bureaucracy, more complex notions of justice, law, and government. Not everyone in the Neolithic culture had to farm and raise their own food; lawyers could focus on law, government bureaucrats on the distribution of resources, soldiers on defense (and offense), teachers on instruction, carpenters on construction, farmers on food cultivation, and so on. The cultural improvements brought about during the Neolithic Period lead to the rapid population growth for our species that continues to this day.

As a consequence of our recent population expansion, human beings exhibit relatively low genetic diversity compared to other species of animals. Human populations are defined as subgroups within which there exists some measure of similarity that is greater than the overall similarity found among subgroups, and for which geographic, cultural, historical, or genetic boundaries allow for the calculation of discrete allele frequencies. In Chapter 1 we discussed the relatively low F_{ST} values characteristic of human subpopulations but pointed out in this and earlier Chapters (2–7) that there is nevertheless considerable population structure inherent to our species, which arose as a byproduct of the reproductive isolation and differential growth of human populations. As described in Chapter 4, autosomal and sex chromosome loci indicate a common African root for the human phylogeny with a time since most recent common ancestor (TMRCA) at 750 (95% confidence range 400–1300) KYA (assuming 20 years/generation; Jobling et al. 2004).

From this root, various human lineages diverged. Some of these lineages remained exclusively African and others distributed throughout the globe. The TMRCA of relatively recent non-African branches ranges from 50 to 200 KYA. Autosomal loci have higher effective population sizes, so we expect a study of autosomal haplotypes to show older TMRCA than a study of sex-linked haplotypes. With these, we again see complete separation of African and non-African lineages with the first two branches leading exclusively to African lineages, and the third containing both African and non-African lineages. The TMRCA for entire phylogeny is 59 (40–140) KYA, and the TMRCA for branch containing both African and non-African lineages is 40 (31–79) KYA. Due to lower effective population size, patrilocality and matrilocality (and possibly other phenomena), the distribution of Y and mtDNA chromosome haplogroups and paragroups is more highly graded than the distribution of autosomal loci, but all of these markers combine to paint a portrait of modern-day human populations characterized by mosaic or patchwork genetic pattern.

Although the human population is relatively undifferentiated compared to other species, we know from our everyday experiences that various human subpopulations exhibit certain distinct, characteristic physical qualities called *anthropometric phenotypes* or *divergent traits*. All the currently proven anthropometric phenotypes are known to be superficial in nature, contributing only toward outward appearance. However, as forensics scientists, it is precisely the superficial phenotypes we are interested in. The mosaic pattern of human genetic variability gives us a framework within which to understand these certain physical differences, and it is for traits found distributed as a function of ancestry, or divergent traits, that we desire to infer values given a measure of individual ancestry. First let us review the mechanisms underlying the distribution of modern-day genetic and anthropometric variability and then we can focus on a few traits of interest.

SOURCES OF PHENOTYPIC VARIATION

During the expansion of the human population, mutation, genetic drift and gene flow, and natural selection provided cooperative and at other times counteracting forces that shaped the development and maintenance of genetic diversity. Unlike most animals, humans inhabit a tremendously broad range of environmental ecosystems. As populations migrated from one environment to another, genetic adaptation and statistical phenomena like genetic drift shaped particular gene frequencies. Darker skin is shared by populations living recently in equatorial regions, such as the South Indians, sub-Saharan Africans, Aboriginal Australians, Melanesians, and so on, presumably because the melanin absorbs potentially harmful ultraviolet radiation before it can reach the nuclei of cells in the epidermis and damage DNA (see Figure 8-1). For populations that expanded to inhabit extreme northerly climes, ultraviolet radiation and other tropical climate-related pressures (including humidity and heat) were ameliorated and selective pressure to maintain darker skin melanin content was relaxed. In fact, it has been hypothesized that upon moving to low-UV environments, humans are subject to a new selective pressure for lighter skin tones so that UV light, which is required to produce the active form of vitamin D, would be sufficient (Grant 2002; Jablonski & Chaplin 2000; Loomis 1967).

In other parts of the world, different but equally extreme environmental forces molded the genetic compatibility of our ancestors in different directions. Not only due to the systematic force of natural selection, natural variation arose as a pure function of stochastic and quasi-stochastic forces such as genetic drift, mutation, recombination, and gene conversion. It is also very likely that sexual selection played a very important role in the determination of patterns of overt genetic variation. The interplay of these forces upon a backdrop of statistical sampling caused by founder effects, migration, isolation, and population bottlenecks undoubtedly established the phenotype diversity that exists across the modern-day branches in the human family tree. Some obvious phenotypic differences among the world's various populations include

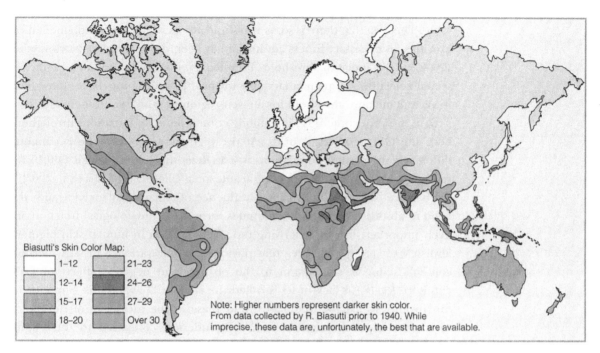

Figure 8-1

Geographical distribution of skin eumelanin content. Average skin melanin content for indigenous populations is indicated with colors, based on the legend (lower left). From Introduction to Physical Anthropology 8th Edition by Jurmain/Nelson/Kilgore/ Trevathan, 2000. Reprinted with permission of Wadsworth, a division of Thomson Learning: www.thomsonrights.com. Fax 800-730-2215.

- **Pigmentation**: skin, iris, and hair color
- **Metabolism**: xenobiotics, lactose tolerance, "thrifty" genotypes
- **Somatic Morphology**: stature, mass, body proportions, hirsuteness
- **Cranial and Facial Morphology**: epicanthic eye folds, other facial features, craniofacial architecture, dental and cranial dimensions, baldness, other skeletal features, and so on

Because sunlight must travel through a thicker atmosphere to reach earth at higher latitudes, and because elements of the atmosphere absorb ultraviolet radiation (UVR), the intensity of UVR is highest near the equator. Due to the harmful effects of UVR, and the equatorial location of the cradle of human evolution, our evolutionary predecessors are generally thought to have had darker skin colors and the variance in skin melanin content among modern-day populations is explained primarily (though not necessarily exclusively) by a differential loss of pigmentation among the human lineages that expanded out of Africa. Recent expansions and migrations could result in a modern population of darker color than required to survive long term at a given latitude, but over time it is expected that the lack of pressure would impact the average trait value. The idea that there is some threshold across which further pigmentation provides no necessary fitness advantage has been put forward by Norton et al. (2006) as the melanin threshold. Presumably, there is a substantial effect of sexual selection on pigmentary phenotypes, and variation both above and below the melanin threshold for a given region may be due to these effects.

Skin melanin content is high among contemporary populations throughout both the Old and New Worlds. At the pigment gene loci, the F_{ST} among differently pigmented populations is 0.6 (Relethford 1992) versus ≤ 0.15 for randomly selected, selectively neutral autosomal alleles (Frudakis et al. 2003b); the balance between the protective influence of melanin balanced against the need to absorb UVR at higher latitudes for normal physiological function are likely important forces underlying the wide variation in human skin pigment levels extant today. The idea that phylogeographic variance in skin melanin content is due to adaptation to the positive and negative affects of UVR exposure levels (or lack of it) is called the vitamin D hypothesis. Support for this hypothesis can be found in the increase in the incidence of rickets for African-American immigrants in North America, compared to their light-skinned European neighbors. Similar phenomena are observed for East Africans living in the Middle East, and South Asian Indians living in the United Kingdom. Jobling et al. (2004) does an excellent job reviewing the vitamin D hypothesis, and other phenomenas which have likely shaped extant melanin distributions such as folic acid hypothesis, immune system suppression, sexual selection, and even nocturnal camouflage.

Skin melanin content is not the only anthropometric trait. Body Mass Index (BMI) is known to vary substantially with ancestry via Bergmann's rule, which states that BMI increases with distance from the equator because northern climates are cold and the ratio of volume to surface area is correlated with the efficiency of heat exchange. High volume relative to surface area allows for more heat retention and less exchange than shapes with high surface area relative to volume (which is why the heat exchangers on engine carburetors and computer hardware are of intricate, tiered, or layered shapes, to maximize surface area). Bergmann's rule seems to explain why populations indigenous to tropical climates (such as Kenyans, Ethiopians, etc.) are tall and thin relative to populations indigenous to the Arctic (such as the Inuit), and is also seen in other animals such as birds and other mammals.

Allen's rule is related to Bergmann's rule, and states that shorter limbs are favored in colder climates because they maximize the volume to surface area ratio, enabling better heat retention. However, as one might expect, there are other forces that have been at work in shaping the morphology of our species. Sexual selection and microclimate adaptation are alternate theories suggested to explain why pygmy African populations do not seem to obey Bergmann's and Allen's rule, and the existence of such populations illustrate why it is dangerous to make broad generalizations without specific data to support them. For example, it would be naïve to claim that a person from India should be thinner and darker than one from Siberia simply because India is closer to the equator. Even if this happens to be true on average, one obviously needs to define populations precisely, using an objective method of measuring them, and refer to specific anthropometric data in order to make any valid inferences relating ancestry and phenotype. As an example, obesity in African-American women has been shown to be positively correlated with West African ancestry (Fernandez et al. 2003), in direct contradiction to Bergmann's rule. The tendency for African-American women to gain weight faster than European-American women may be due to differences in proclivity to store fat as a defense against famine and possibly dependent on differences in resting metabolic rate (Allison et al. 1997). Other populations of equatorial latitude live at high altitudes, where the microclimate is quite cold and the ancestors for some equatorial populations could even have migrated from more northerly climes in recent anthropological time. Variations such as the length of time a population has been *in situ* and microclimate effects all need to be taken into account when developing explanatory models for geographic patterns of phenotypic variation.

High skin melanin content is a polyphyletic trait, meaning the trait is exhibited among populations that correspond to a variety of human phylogenetic lineages (i.e., Papauans, West Africans, aboriginal Australians, South Asian Indians). So even though it may be true that normally pigmented West Africans tend to have skin melanin (M) index values above 35, it is not true that all individuals with a skin melanin index value above 35 are of West African origin or even of partial origin. Traits that owe their origin to the forces of natural selection, such as xenobiotic metabolism traits, hair color, and iris color, are frequently found in many different lineages due to convergent or parallel evolution. We cannot conclude that convergent evolution has occurred simply from similar phenotypes across divergent populations. Convergence at the genetic level cannot be concluded from convergence of phenotype.

An alternate explanation is that gene flow has carried the well-adapted genes from one part of the world to another. To extend this example, it is likely that throughout the 100,000 years since non-Africans left Africa there continued to be some minor level of gene flow between African and non-African populations. Quite possibly then, alleles at high frequency in equatorial Africa that lead to dark skin are some of the same alleles in Southern India that result in dark skin. Likewise, it is thought that the populations from which the Melanesians are descended made their way along the coast of Arabia, India, and Southeast Asia to Melanesia, 40 to 60 KYA. Were this route of migration, these populations may never have strayed into latitudes high enough to promote depigmentation. If this is the case, then what appears to be convergence may actually be the retention of an adaptive phenotype.

One way to test the alternative hypotheses of convergence or retention is to find the genes that determine skin color in the various populations. Are the same genes darkening the skin of South Asians, Melanesians, West Africans, and Australians? Are the same genes lightening the skins of Europeans, East Asians, and the Khoisan-speaking populations of South Africa? In Chapter 9 we will discuss genetic screens for skin pigmentation loci in several populations, and provide answers to some of these questions. In this chapter we pose the question in terms of the ancestry/phenotype relationship, but with molecular analyses, as discussed in Chapter 9, we can go much further and ask if particular functional variants, or tag SNPs indicating particular haplotypes, are similarly distributed across populations with the same pigmentation levels.

Facial features are of particular interest to forensics scientists, since it is primarily through craniofacial morphology that we recognize one another. At first blush, due to obvious complexity, facial prediction would seem to be beyond the scope of molecular photofitting. However, there are clear indications from observations within families that there is a large genetic component to the expression of facial features and that the elements of this component have likely evolved to a tangible extent as a function of phylogeography. So, facial morphology may represent the type of variation that can be appreciated using indirect methods of molecular photofitting (based on a knowledge of genomic ancestry). Indeed, most lay observers using methods other than objective measures of genomic ancestry would agree that the most prominent differences among populations that exist are craniofacial anthropometric features, and if this could be demonstrated formally, and the correlations quantified, we could focus on the gross morphological differences such as skull x, y, z dimensions, interocular distance, chin to brow measurements, and such among the human populations in our attempt to reconstruct what an unknown might look like from their ancestry.

The cephalic index (CI), which is the breadth to length ratio multiplied by 100, is a classic cranial metric. Long narrow skulls are called dolichocephalic, those with CIs over 80 are called brachycephalic, and those between 75 and 80 are called mesocephalic. Other anthropometric cranial features include iris color, hair color and texture, epicanthic eye folding, ear shape, and nose and lip shape—all these traits are thought to vary systematically with phylogeography (reviewed in Jobling et al. 2004). Nose shape in particular shows variation correlated with climate, possibly allowing for air moistening and heating (cold air holds less moisture than warm air). Epicanthic eye folds could help minimize the glare of sunlight off a snow surface. In addition, nonskull bone morphology is highly anthropometric, and the morphology of various bones can be (and actually is) used to infer ancestry. Tooth shapes with shoveling is particularly common in some Indigenous American and East Asian populations, but can also be found in the African Khoisan and European populations to lesser extents.

It is difficult to know whether the apportionment of diversity with respect to these anthropometric traits is a function of adaptive responses to environment, sexual selection, genetic drift, gene flow, or some combination of all of them. Most likely, the answer is different for each trait and even across populations for the same traits. For example, consider that some facial feature, say petite noses, evolved first under selection to protect against frostbite. As these populations migrated out from the colder regions, they mixed with large nosed populations, whose ancestors had been adapted to dry climates but did not really need them anymore, because it is more humid today. Possibly the smaller noses might then become more prevalent as a function of sexual selection. Thus, a trait that initially evolved under the pressure of natural selection could through admixture later in another population evolve under the pressures of sexual selection.

Although there are two main approaches to understanding the genetic basis of phenotypic variation—the direct method from study of genes that mechanistically underlie these phenotypes and the indirect method from correlations and variation in terms of our evolutionary history—both methods target traits that are heritable. Most traits are continuously distributed, and determined by both genetic and environmental factors. Examples of genetic traits include sickle cell anemia, cystic fibrosis, and iris color. Examples of traits that are subject to both genetic and environmental control include skin pigmentation (in exposed areas of the body), BMI, and stature. To study these latter traits, we must tease the environmental sources of variation from the genetic sources of variation. For example, if we are to study skin pigmentation we would need to ensure that spectrophotometric measurements are taken from unexposed areas of the body in our training set and apply our inferences in such a manner as to accommodate environmental variability. Alternatively, we might also be able to predict variability in tanning capacity (which is highly variable) and so make statements as to what extent a person's face may tan under particular environmental conditions (see Wagner et al. 2002).

The best anthropometric traits for molecular photofitting are those that are highly heritable (i.e., the ratio of the genetic to environmental variance V_g/V_e is high), and for which phenotyping can be accomplished in such a way as to be both physiologically meaningful and to minimize environmental variability. The traits need not be discretely distributed—for a quantitative trait we except to be able to identify quantitative trait loci (QTLs) accounting for at least 10 to 15% of total variance in trait expression, given a good study design (Strachan & Read 1999). Whether the trait is discretely or continuously distributed is of little difference for the indirect methods of inferring trait value.

EMPIRICAL OBSERVATION OF ADMIXTURE-BASED CORRELATION ENABLES GENERALIZATION

One would not want to use a literal, anecdotal interpretation of an ancestry profile in order to infer trait value, such as guessing that an individual with 80% European and 20% Indigenous American ancestry should have darker hair, skin, and iris color than a European with 0% Indigenous American ancestry. If we merely assumed that Indigenous American admixture is correlated with higher melanin content because Indigenous American parental populations (or their modern-day representatives) are thought to have had darker pigmentation than European parental populations (or their modern-day representatives), we would be practicing the art of baseless generalization—guessing, without any supporting data. Better to be able to prove that the 20% Indigenous American ancestry is indicative of a higher likelihood of darker skin tone so that we can justify our inference.

To understand whether and how the 20% Indigenous American ancestry is relevant for physical appearance we must first appreciate possible origins for the 20% score. Earlier, we explained three mechanisms for an admixture result. To review, mechanism (1) is recent admixture, what most genealogists are looking for, and the mechanism that usually has the most bearing on the anthropometric value of the result. There are two other possible sources, however, that may make relating the result to physical appearance more difficult. Mechanism (2) is ancient admixture that has remained concentrated in various ethnicities. This is the mechanism that a genealogist would be invoking when making a statement such as ''... research indicates that the Indigenous American affiliation for this person might be from the Sámi or northern Siberian peoples rather than from recent admixture with Native Americans.'' The third mechanism is much less interesting in terms of reconstructing phylogenetic relationships but still useful for molecular photofitting—that the individual comes from a group not strictly, and semantically speaking, appropriate for the population model employed. All three mechanisms might provide information on physical appearance if the expression of physical traits is a function of ancestry, regardless of what type of ancestry, or its mechanism of creation, and for a forensics scientist it is far more important to answer the question of whether this function exists for a particular trait (so that he or she can infer phenotype) than to reconstruct anthropological history. In other words, the forensic scientist might say, ''I don't care why the ancestry allows me to infer a most likely trait value for this trait, only that it be able to do so with quantitatively defined precision, repeatability, objectivity and sensitivity.'' Of course, the fewer mechanisms at work for a given phenotype, set of populations, and a given panel of markers, the more useful it will be to identify an ancestral component to phenotype expression.

An anthropological interpretation of genomic ancestry profile for indirect methods of inferring trait value has some utility but requires a cautious approach. As we discussed in Chapters 5 and 6, the results from carefully constructed genomic ancestry tests (such as AncestryByDNA 2.0 and 2.5) are highly predictable, quantifiable, characteristic, and the results are accurate, on average, to a few percentage points. However, as discussed in Chapter 6, due to the complexity of human migrations and interactions over the past 20,000 years, low levels of apparent admixture can be difficult to interpret. Therefore if lower levels of admixture are to be taken on their own as diagnostic for geographical origin or ethnic affiliation, they should be interpreted with caution (in the case of the 171 AIM panel discussed in Chapter 5, caution should particularly be afforded lower levels of East Asian or Indigenous American admixture in Europeans). Though we may not always have access to other information about a subject/individual, it is far better to interpret lower levels of admixture in conjunction with other prior information when it is available (physical appearance, ethnicity, etc.).

For example, a result of 87% European/13% Indigenous American can be difficult to interpret without other types of information. Using the 171 AIM panel discussed in Chapter 5, virtually every individual tested to have high (>20%) levels of Indigenous American ancestry describes him- or herself as at least fractional Hispanic or American Indian, but lower levels of Indigenous American ancestry (<20%) can also be seen for certain individuals of Mediterranean, Middle Eastern, Jewish, or South Asian Indian ethnicity. We know from Chapter 5 that a reading of 13% Indigenous American with this panel of markers can be statistically significant (depending on the confidence intervals of the person in question), but also from Chapter 6 that certain ethnic groups show systematic levels of Indigenous American admixture at even lower levels, which taken together, are also statistically significant. Thus, an 87% European/13% Indigenous American result does not necessarily on its own provide proof of American Indian ancestry or non-African Hispanic ethnic affiliation versus European ancestry derived from European ancestors. Since Europeans born in Europe take on different average physical features than Hispanics, this lack of discrimination might confound our ability to infer ranges for certain phenotypes. There are two ways to overcome this problem and to different ends:

1. Use other sources of information to assist in an anthropological interpretation of the profile. This is the genealogical approach.
2. Ignore the anthropological interpretation and restrict focus to traits held in common by the polyphyletic (multiple lineage) collection of populations fitting this profile. This is the empirical approach.

For most traits, forensic scientists can use method (2), and we will focus on this method next in this chapter. However, for a genealogist or a forensics scientist desiring to infer trait value for a monophyletic trait, only method (1) will do the job. We simply must be able to infer where geographically the person's recent ancestors have come from. Here we rely on the fact that an 87% European/13% Indigenous American profile can be more meaningful if considered with other information.

For example, if there exists other types of evidence or information, such as Y-chromosome, mtDNA, or paper genealogical evidence suggesting (1) American Indian heritage, and (2) no known Mediterranean, Eastern European, Middle Eastern, or Indo–Pakistani heritage, the result of 87% European/13% Indigenous American would combine with this evidence to paint a likely portrait of fractional Amerind heritage. If for example there is evidence that one of the subject's grandparents was born in Mexico, the result is most likely an indication of Hispanic heritage. If there is no other evidence or information about the subject, their ancestry could be a function of one or any of these ethnicities (non-African Hispanic, American Indian, Mediterranean, Middle Eastern, or Indo–Pakistani, including Indian), and the only way to interpret the results is as potentially diagnostic for any one and all of them. We may lose precision in our quest to infer anthropometric trait value (depending on the trait), but this is much better than making an unjustified or incorrect inference.

Therefore, before inferences about physical appearance or genealogy can be made, one needs to appreciate that interpreting a result involving a low level of secondary or tertiary affiliation needs to be accomplished in a thoughtful manner. One such manner is the empirical database-driven approach that we will describe next. In lieu of this empirical approach, or even as a companion, interpretation is best accomplished in conjunction with other types of information (such as prior information). However, if such information does not exist, interpretation can still be accomplished if we are honest about our lack of information and conservative in our conclusions. The investigator's quest should be considered as similar to a court case, where a preponderance of evidence better decides the case than any single item of evidence. So doing, we apply a convenient form of Bayes' Theorem, which recognizes that certain items of evidence take on different meanings in different evidentiary contexts, and we recognize that lacking prior information, we must interpret the result in a conservative manner, covering all the possibilities.

EMPIRICISM AS A TOOL FOR THE INDIRECT METHOD OF MOLECULAR PHOTOFITTING

From the preceding discussion, one can appreciate that we need to establish a method for objectively interpreting an ancestry admixture result in order to safely use the indirect method of anthropometric trait value inference. Such a method would work no matter why an admixture result obtains (recent admixture, ancient admixture, and isolation or due to affiliation with an intermediate population). In the preceding chapters we described the measurement of continental and intracontinental biogeographical ancestry, and highlighted how certain physical traits (primarily pigmentation traits) are spread throughout our globe as a function of geography and ancestry. This is pertinent to our discussion of molecular photofitting, because it implies that a knowledge of ancestry can impart information about certain aspects of physical appearance, which is what we are after if we are attempting to characterize DNA found at a crime scene. If we want to know how tall a person is, what their skin shade is, or eye color, our first choice would be to use markers that functionally underlie variation in height, skin shade, or eye color, respectively. This is the direct way of making an inference on phenotype, but such genetics tests are only now being developed—we will discuss a few of them in Chapter 9.

However, the direct way of making inferences about phenotype is not the only legitimate way (though it is sometimes politically unpopular to use indirect methods, as discussed in Chapter 11). Genetics tests for the direct inference of most traits of forensic value, such as iris color or skin pigmentation, are cumbersome and expensive to develop. This is because these traits are usually a function of many genes interacting together, each one of which has many variant forms in the population. We speak of this in terms of locus heterogeneity (many genes or loci involved) and allelic heterogeneity (many alleles for a given loci are also involved). The large number of variants and interacting loci make their discovery difficult from a statistical genetics perspective; sample sizes in the thousands, even tens of thousands are necessary, and screens of the entire genome are usually required to find them. Attesting to this, prior to the work described in Chapter 9 for iris color, no panel of markers had been developed for the direct inference of any human phenotype. Genetic screening is expensive and unfortunately (since forensics profiling is not as economically attractive to industry as pharmaceutical development, and since physical traits do not usually relate to human disease directly) many high-powered companies and academic labs simply ignore these most interesting and potentially most instructive genetics puzzles.

Fortunately for the forensics scientist, some phenotypes that impart overt physical characteristics such as pigmentation, hair texture, and the like are precisely those that are known to differ at least to a certain extent as a function of ancestry. In contrast, it is very difficult to prove that ancestral differences exist for other types of traits—such as many diseases, intelligence (however one defines it), or personality—because environmental factors such as diet, socioeconomic status, geographical location, among others play such a large role in their expression (see Chapter 11 for a more detailed discussion). It is primarily because their expression status is only indirectly related to health that the physical traits are among the most evolutionarily pliable in the face of selective pressures, genetic drift, and sexual selection, and this characteristic of the physical phenotypes makes it possible to speak to their values knowing ancestral origin.

As we have discussed, we know that Africans from the sub-Saharan regions are darker in skin shade than Europeans or East Asians, and that East Asians tend to have epicanthic eye folds that most Europeans and West Africans lack, but there is an obvious problem in relating physical appearance to ancestry. If a sample is typed as 100% East Asian, how should the forensics investigator objectively determine what this person is likely to look like? The statement "the individual should look East Asian" is meaningless to a scientist, because it is very difficult to define what the average East Asian looks like. Apart from the fact that we don't know exactly what it means to look East Asian, what about samples that type of East Asian admixture? How much East Asian admixture is required for the individual to look East Asian? If, for example, Europeans with Mediterranean ancestry share physical characteristics in common with East Asians, to the exclusion of other Europeans, how do we identify and quantify these characteristics and how do we communicate how Mediterranean ancestry is related to the expression of these characteristics, how it varies within Mediterraneans, and how Mediterraneans vary with respect to other ethnicities? We know that there are very real differences for some physical traits of our appearance, but communicating these differences with words is very difficult and often ends up as unscientific, no matter how well-intentioned a person speaking them is.

Consider two crime scene results, one of 100% West African ancestry and another of 60% West African and 40% European ancestry admixture. We might assume an age and sex for the individual, and guess that the skin of the former individual would have a melanin index (M) that is:

- Significantly darker than the average 100% European, East Asian, or Indigenous American, the magnitude of difference varying for each comparison
- Somewhat darker than the average African with European admixture

Likewise, we might guess that the melanin index (M) for the second individual is likely to be:

- Somewhat darker than the average 100% European, East Asian, Indigenous American, to different degrees for each comparison
- Somewhat lighter than the average 100% African

But we cannot guess what the precise skin shade should be, cannot prove that the preceding traits are true, and cannot know whether there are any differences between the sexes of persons with this 100% West African profile. We form these expectations, or rather, guesses, based on prior work that shows a correlation between ancestry proportions and melanin index, but if the trait we are interested in has not been characterized by a regression study, and lacking an objective method to infer the answer, we would be forced to form them on subjective opinion. Even in the case of a trait that has been studied with respect to ancestry, what does the ancestry tell us about the most likely boundaries of M values we should expect?

It is probably not unfair to say that the fastest way to make another person question your intelligence (and ethics) is to generalize about physical differences between populations. For reasons we will discuss in Chapter 11, very few studies have been funded for measuring the relationship between ancestry and overt phenotypes, and so there has up to now been little data to rely on for support of such generalizations. Yet as forensics scientists we need the data, and we need to make such generalizations if we wish to paint physical portraits from crime scene DNA. In this regard, the difference between the unsophisticated guess and a scientific inference is data. Thus, our challenge is to devise systematic, yet responsible methods to reconstruct physical appearance from such data. The answer lies in the practice of empiricism.

An empirical result is one originating in or based on observation or experience, or capable of being verified or disproved by observation or experiment. The practice of empiricism relies on observation and experiment due to an underlying tenet that all knowledge originates from experience. We can use empiricism to not only determine what physical characteristics can be inferred from estimates of genomic ancestry, but how to parameterize these characteristics. The primary tools at our disposal for this purpose is the computer database, which can hold a very large collection of physical measurements and digital photographs, as well as computer software for assessing variation in the expression of these measurements as a function of ancestry and for displaying the results in useful ways. Large numbers of individuals can be typed with a continental BGA admixture test, a within-continent BGA admixture test, or any other pan-genome test. Y-chromosome or mtDNA tests might be useful in this situation provided there was clear evidence of either no admixture, but in general, this assumption is not warranted and so the approach is really appropriate only for use with individual genomic ancestry admixture estimates.

For each person in the database, digital photographs, self-reported ethnic affiliations, or other characteristics and/or other anthropometric measurements can be stored alongside genetic information. We can then simply query the database with a genetic result plus or minus a range of expected error from an unknown sample whose physical characteristics we would like to infer (top of Figure 8-2). The return of this query could be a large number of

Figure 8-2

A query (top) and example of a single return (bottom) from a database search using an Individual Genomic Ancestry Admixture estimate. The database queried is composed of a large number of individuals, their IGAA estimates (MLEs and confidence intervals), trait values for a variety of phenotypes, and various demographic data such as country of origin for members of their pedigree and ethnic description of the same. We query the database with the IGAA estimates and their confidence intervals (top), and retrieve all samples with MLEs of values for each parental population that fall within the confidence intervals of the unknown. One such return is shown (bottom). With reference to a large number of similar returns, we can measure summary statistics and learn whether or not any overt physical traits are associated with the admixture profile.

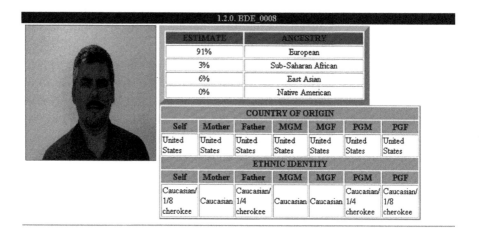

Sampleid : DNAW70078R

Ethnicity	Value	Ranges	
European	91%	83%	98%
Sub-Saharan African	5%	0%	13%
East Asian	4%	0%	11%
Native American	0%	0%	7%

1.2.0. BDE_0008

ESTIMATE	ANCESTRY
91%	European
3%	Sub-Saharan African
6%	East Asian
0%	Native American

COUNTRY OF ORIGIN						
Self	Mother	Father	MGM	MGF	PGM	PGF
United States	United States	United States	United States	United States	United States	United States

ETHNIC IDENTITY						
Self	Mother	Father	MGM	MGF	PGM	PGF
Caucasian/ 1/8 cherokee	Caucasian	Caucasian/ 1/4 cherokee	Caucasian	Caucasian	Caucasian/ 1/4 cherokee	Caucasian/ 1/8 cherokee

photographs or measurements from known subjects that previously had been typed with the same genetic test. An example of a query is provided in Figure 8-2A (top), and an example of the type of return format used is shown in Figure 8-2B (bottom); displayed here is a frontal photograph, the genomic ancestry proportions, country of origin, and ethnic identity extending back two generations for one of the returned individuals and his/her relatives.

The information that is retrieved from the database could be different from this, such as a certain anthropometric measurement, or profile photographs, 3D images, or photographs of irises, and so on. The images or data could be displayed in a sort of collage (multiple independent faces, a fast action anima-tion of faces, or some computer morphed average of selected faces) for the user to appreciate variability, but also displayed in terms of summary statistics when appropriate for the user to quantify that variability objectively. If the sample of individuals is large, then the examples called by the query would provide for a reasonable assessment of the variation of the characteristics within and between ancestry categories. If individuals of 100% East Asian ancestry (as defined using the population model using a particular panel of AIMs) tend to have a greater frequency of epicanthic eye folding than random collections of persons, or other subsets of populations characterized by differ-ent ancestry admixtures (controls, such as those having no East Asian ancestry), then a query of a digital photograph/BGA admixture database with the result

100% East Asian $\pm$ X% should result in a collection of digital photographs of individuals with a higher frequency of epicanthic eye folding than a query of the same database with a control result. If individuals of 100% East Asian ancestry are not characterized by a greater frequency of epicanthic eye folding relative to other ancestry groupings, then the query should produce a set of digital photographs that tell this story as well. We might also expect from our subjective knowledge that the relative proportion of photographs for individuals with epicanthic folding would grow smaller as the percentage of East Asian in the query profile becomes smaller or higher as the percentage of East Asian admixture increases. Whatever the frequency we observe from the database return, we can then invert this frequency using Bayesian logic, or simple likelihood functions to form an empirical statement about the extent and/or likelihood that the individual who donated the unknown sample has epicanthic folds.

In the case of our example, where we have two samples—one with 100% West African ancestry and another with 60% West African and 40% European ancestry—we can query this same database with the objective of inferring likely skin pigmentation level or M index. In the case of the latter profile, if there is more variation in M index within the group of individuals of similar admixture (say, 60% West African and 40% European BGA admixture $\pm$ 10% for either ancestral group) than between this group and other groups (such as 100% West African, or 100% Europeans), then this should be readily apparent from the database output. We should see a similar number of light-skinned individuals in the returned data for the 100% West African, 60% West African/40% European, 100% European groups, and from this we would safely conclude that we can say nothing about skin shade from this result. If it is safe to generalize that individuals of 60% West African and 40% European BGA admixture $\pm$ 10% for either ancestral group will have a higher M index than the average 100% European, but a lower M index than 100% West Africans, then this, too, should be apparent from the output (and in fact, this happens to be the case, as discussed earlier, and later).

Anytime we want to make a generalization about the likelihood of a person exhibiting a particular trait based on their ancestry proportions, we must use the empirical database approach or, more generally, any approach whereby one draws conclusions based on a sufficient collection of data. For example, if an investigator learned that the DNA found at a murder scene was donated by an individual showing 70% European and 30% West African ancestry, he or she would not know precisely what this tells them about skin and hair color unless he or she looked at a large collection of individuals of similar proportions, summarized the statistics relevant to the expression of the trait in this subgroup, and compared these statistics to subpopulations for other types of BGA profiles. The conclusion would not simply be that a person of 70% European and 30% West African ancestry has dark skin, but rather that this person has darker skin than most European Americans, who show on average X% African ancestry (where X is determined through validation of the AIM panel, as in Chapter 6).

The empirical approach answers such questions as: At what percentage of admixture does a person begin to exhibit particular physical characteristics of their minor group? Would a person of 70% European and 30% West African BGA show darker skin than the average European American, or is darker skin not usually observed until the African percentage exceeds 40 or 50%, or even more? How solid is our inference that person A is likely to have darker skin or iris color than person B? How unusual is it for Latinos like Puerto Ricans or Dominicans who have significant amounts of West African ancestry to have blue iris color? The indirect, empirical approach eliminates the need to understand the appropriateness of the population model for the trait value to be inferred, how ancestry profiles should be anthropologically interpreted, whether subpopulation heterogeneity or population histories introduce errors into the admixture estimates, and so on.

If there are problems with any of these questions the entropy of the system increases—a larger error in admixture estimates will be obtained and the database return will simply be less informative than it could have been with better estimates. The error is always toward less information and lower utility (type II error), not incorrect information and false utility (type I error, or false positive results). In other words, if the test is performing poorly, the query result will be a more diverse and less informative sample than we could have if the test were more informative. However, it is possible to obtain more insidious type I error, where one obtains results that seem meaningful but are not. The main culprit for this type of error is a systematically and significantly biased admixture estimation system and small reference databases. However, the use of summary statistics would give one the indication of whether the sample was appropriate or not (sample sizes are integral components of the calculation of any p value). The quality of inferences that can be made are thus dependent on four factors:

- The information content of the ancestry markers and the relevance of the genomic ancestry model being used for a given trait (for example, a two-population model would not be very useful for the inference of skin color)
- The size and breadth of the phenotype and photograph databases
- The level of trait variation across populations included in the genomic ancestry model
- The levels of ancestry structure among the populations included in the genomic ancestry model

For example, this person, with 30% West African and 70% European BGA, has a profile that is consistent with more than one modern-day population. People showing genomic ancestry levels in this range have been observed in population samples of African Americans, Puerto Ricans, Dominicans, and

North Africans, as well as others. As such, it is critically important to have a database with a reasonably broad level of coverage so as to have representatives from all the major groups in the catchment area. Errors could result were there populations that overlapped in ancestral proportions that are not represented in the database and do not overlap sufficiently phenotypically with the populations sharing similar ancestry levels. This last point is critically important and crucial to consider if we want to reduce the risk of type I error when we are making physical measurements or photographic database assessments.

Fortunately, for many phenotypes, most populations that overlap in ancestry levels do also generally overlap in phenotypic assessments. In this example, the average African American with a 30% West African/70% European profile would be relatively light skinned compared to 100% West Africans, as would a Puerto Rican person with the same reading, but lacking representations for Puerto Ricans in the database would not cause type I error in our inference of M index for a Puerto Rican sample. Of course, this may not be the case for all traits and it could be, for example, that the incidence of green eyes is higher in Puerto Ricans with high levels of European admixture than in African Americans. However, we might expect such stark phenotypic differences between genetically similar populations to be exceptional.

The larger problem with a more limited database is that the geopolitical labels could be misleading. Using this same example, were we to have primarily Mexican Americans and African Americans in a database against which the 30% West African/70% European result was screened, the geopolitical query would produce a profile showing likely 100% of persons with this profile are African American or Black, since very few Mexican Americans have as much as 30% West African ancestry. Were we, however, to have had a reasonable representation of Puerto Ricans in the same database, the geopolitical query would have returned a respectable chance of Puerto Rican origin (likely slightly greater than the chances of him being African American; see Figure 6-5 in Chapter 6), and that there is a considerable chance the suspect speaks Spanish and maybe generally refers to him- or herself as Latino/Latina or Hispanic. Absent the Puerto Ricans in the database, we might incorrectly conclude that there is little chance the individual would describe him- or herself as Puerto Rican or even Hispanic. As the database becomes more inclusive, the likelihood that these qualities apply to the individual becomes more accurate.

Assuming we use comprehensive databases upon which to base our inferences, an important point for the indirect approach is that we no longer have to form expectations from our subjective knowledge of the world, and could eliminate such incorrect and unscientific statements such as:

- ''People with West African admixture should show darker skin than people without.''
- ''People with African ancestry have darker skin.''

- "The skin of people with Native American ancestry has brown hair."
- "People with East Asian heritage have narrow skin folding around the eye."
- "Europeans tend to have blue eyes."

Rather we can observe and form generalizations from real data using the empirical method. If the first statement is true, for any given level of admixture, a query of a reasonably sized database and objective measurement of the characteristics for the photographs or descriptors returned should show that it is. Indeed, guessing about phenotype expression as a function of admixture or other ancestral affiliation would seem to be impossible without the use of such an empirical tool, because each phenotype has a different genetic (and environmental) basis and anthropological history, and because the relationship between ancestry and phenotype may differ from population to population.

Indeed, complex genetics traits such as those we are discussing usually are determined by many loci, and for the average trait we can assume that these loci are spread relatively evenly throughout the genome. For some traits, where dominance is at work, only a small fraction of the loci will need to be of Group 1 origin for a person with Group 1/Group 2 admixture to exhibit Group 1 characteristics. However, most polygenic traits are thought to be additive and epistatic, not dominant, and a larger number of the loci will need to be of Group 1 origin for the individual to exhibit Group 1 characteristics. Different loci combinations in different people will produce different results for different traits. One trait may require on average half the functional alleles to be of Group 1 origin, but a different trait may require on average only a quarter of them to be of Group 1 origin before Group 1 character is expressed. In fact, for any trait, the number of loci required depends on the dominance and epistatic influences of the loci, so we cannot even say that a particular trait always requires X% of the underlying loci to be of Group 1 origin for an individual to exhibit Group 1 characteristics—the number depends on which loci you are talking about.

Of course the genetic architecture of most complex human traits remains unknown and so it would not be possible to calculate these values of X for various traits without access to an empirical tool. A short nonexhaustive list of the variables influencing whether a person exhibits a trait value characteristic of one group versus another illustrates the difficulty in forming a simple set of rules for predicting phenotype from ancestry lacking such a tool:

- The locus heterogeneity—how many different loci underlie trait expression
- Allelic heterogeneity—how many different functional alleles exist for each locus
- The degree to which dominance/epistatic components of genetic variance play a role in the expression of the trait

- How evenly across the genome the loci underlying the trait are spread—the alternative to an even distribution being clusters of functional loci
- The admixture percentage and genealogical structure of the person's family—recent versus old admixture
- The level of ancestry stratification in the population to which the person belongs
- Multinomial sampling theory and probability
- Stochastic fluctuations in early development and other sources of environmental variability

For a trait such as iris color, the probability that a 100% West African has blue iris color is essentially zero, but for a 100% European, it could be said to be roughly greater than or equal to 50% (see later in this chapter, and Chapter 5). For a 50% West African/50% European individual, the probability is somewhere in between. If the right genomic regions are European, the person may exhibit blue irises, but if they are West African the iris color could be halfway between blue and brown such as a mosaic hazel/green, or they could exhibit brown irises. For three people of the same European/West African admixture, each will have different genomic distributions of ancestry—a genomic region painted with a large swath of European flavor in one may be scattered with punctate regions of European flavor in the other, or devoid of European flavor altogether in a third, and if this is an important region for the determination of iris color, the individuals might possibly have different iris colors as a result. However at some level of West African affiliation, the percentage of African becomes so high that the probability that these important regions are of European origin or have enough European flavor to override the other West African genes becomes lower and lower, and the incidence of brown irises increases to a point that the frequency of brown irises is qualitatively and significantly different from that of lower levels. It would be the job of the database system and associated software built around it to facilitate estimations of summary statistics or the likelihood of brown irises given different levels of West African ancestry (for example).

Only in cases where ancestry relates directly and predictably to trait value will the database return reveal that a trait value can be inferred from ancestry. In the case of iris color, the probability of brown irises happens to be about 80% for Europeans exhibiting more than 12% West African admixture (see Figure 5-21; to be discussed later). For example, the database return for the query 75% European/25% West African (±10% for either group) would show a collection of Europeans, about 80% of whom have brown iris color. As the level of admixture increases, the probability of brown irises increases until it is very close to one as for parental West Africans. Comparing this return to that for a 90% European/10% West African query (close to 50%), the bias toward brown iris colors for the former would be easily apparent. Summary statistics could be calculated (i.e., is 80% significantly different from the expected level of 50%, given the number of irises returned?), and these summary statistics could be used to determine whether general conclusions can be drawn.

For all traits, variability will always exist within a group of individuals united by admixture profile. Sources of variability include not only genetic complexity as we have discussed, but also admixture estimation error. For some traits there will be more variability than others, and also for some admixture types there will be more variability in phenotype than in other types. The point of the empirical approach and the database system described here is to quantify that variability, so that an investigator knows whether a generalization about physical appearance is warranted from the BGA profile. One can easily envision how this might be valuable for a forensics professional. Imagine a crime has been perpetrated by an Asian American who has more European than East Asian admixture, but epicanthic eye folds. A binned ancestry category from the crime scene DNA provided by the CODIS STR set or Y chromosome SNPs probably would be European, and there would be no hint that the suspect was partially East Asian, and that more importantly, he had certain elements of East Asian character to his physical appearance such as epicanthic eye folds.

In contrast, with an admixture test and associated databases not only is the precise European/East Asian admixture provided but the probability that a person with that mix exhibits epicanthic eye folds would be known. If it is reasonably high, the description used in an investigation and/or disseminated to the public could include this potentially important defining characteristic, which might make the difference in bringing eyewitness testimony "out of the woodwork." Admixture, in this case, provides the tool for obtaining the physical profile, and with respect to the appearance of the eye shape, makes the difference between information versus no-information.

As we have indicated, for this approach to be useful, clearly the database would have to be sizable, representative of a wide variety of populations and admixture ratios, but it need not include everyone on the planet in order to estimate trait values—we need not genotype every person in order to obtain a reasonable estimate of allele frequency or to learn for example that there is a correlation between *APOE* genotype and Alzheimer's disease. The larger the database, the higher the quality of the inference that can be drawn (i.e., the more precise and accurate).

REVERSE FACIAL RECOGNITION USING GENOMIC ANCESTRY ESTIMATES

For the empirical database-driven approach for indirect methods to be useful on a trait-by-trait basis, we must develop methods to objectively quantify trait values. For example, providing description of eye shape in terms of the intercanthal index [$I=(en\text{-}en\text{x}100)/ex\text{-}ex$], or skin color in terms of the M index carries more meaning than using the terms ''epicanthic eye folding'' or ''skin shade,'' respectively. Software tools can be developed to parameterize certain traits well represented in our database. There are a number of anthropometric measures that can be made from standardized photographs as in Figure 8-2 or Figure 8-3. For example, hair texture or form can be measured by the smoothness of the profile of the top of the head. Alternatively, numerical values for phenotypes might be included in the database—in the case of hair texture, we might reference values obtained from microscopic analysis of the cross-section mounts of cut hairs. Summary statistics comparing these values against a control query can easily be generated and a list of phenotypes for which the admixture profile is informative can be created.

Once a reasonable selection of traits have been identified, it is plausible that biometric facial recognition software could be tailored to provide galleries of photos corresponding to the relevant reference sample or synthetic, or create *in silico* average composites of such individuals. This process could be called Reverse Facial Recognition, since we are using very specific information from

Figure 8-3

Example of a three-dimensional photograph for the construction anthropometrics databases. In (A), four different 2D angles of the 3D photograph are shown, but on the computer, moving the cursor rotates the image along the x-, y-, and z-axes. In (B), we use a frontal shot to measure various craniofacial dimensions. Coordinates for defining these dimensions are shown with back spots. Photos were taken using the 3dMD system described in the text. Landmarks are abbreviated as g = glabella, n = nasion, prn = pronasal, ls = labiale superius, li = labiale inferius, pg = pogonion, ex = exocanthion, en = endocanthion, ac = alar curvature, ch = chelion.

A

(Continued)

Figure 8-3
(Continued)

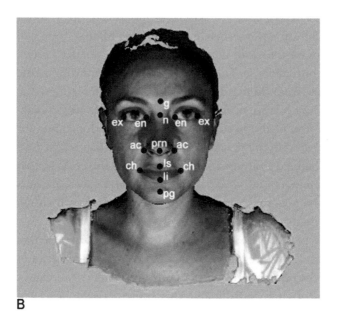

B

individuals (their DNA) to reconstruct crude anthropometric features, which is the opposite of facial recognition software that uses crude anthropometric features for identification. For example, if 100 photos are returned by the database for a given genomic ancestry range, these photos could be distilled by such software to a few expressing various combinations of traits corresponding to the range extremes of the 100—a summary of the 100, so to speak. In this sense, the software tool could represent the forensic artist, and the database could represent the human eyewitness, the result being a computational sketch or series of sketches corresponding to the genomic ancestry profile.

This is very similar to what investigators do when reconstructing a composite sketch from a human eyewitness, except the reconstruction is based on falsifiable, quantitative, and objectively obtained data, which is not the case with a human eyewitness. Mechanically, how might reverse facial recognition work? Given a substantial database, a training set of photographs could be selected from a query return and used to develop a series of formulae (using principal coordinate analysis, discriminate functions, neural networks, genetic algorithms, or other), the inferential value of which could then be applied to suspect photographs. In this way, the selection of suspects can be parameterized, and made more objective.

Obviously this approach is not limited to continuous data such as trait values. We could also distill other types of database returns, such as those involving self-made descriptions of ethnicity, or even geographic birthplace for subjects and/or their relatives within a certain number of generations. If a particular genomic ancestry profile commonly is found among individuals from Trinidad, the East Indies, Puerto Rico, Cuba, and such, a distillation of this information might be had with use of the term Caribbean, and if our encoding for geographical origin in the database was well done we could calculate the likelihood of the unknown's geographical origin. This might be called reverse ethnic description, or geographical description, and would impart different and complementary information to the investigator than reverse facial recognition. Forensic case reports could be drafted using different analyst-specified levels of generalization depending on individual preferences. Alternatively, hierarchical levels of terminology might someday become standardized so that analyses in one laboratory correspond as closely as possible with those of others.

As we have alluded to, the benefit of using reverse facial recognition is that human eyewitnesses often are mistaken, dishonest, or simply inaccurate. Human artists interpreting human testimony are just that—artists. Reverse facial recognition from genomic ancestry is more quantifiable, objective, and unlike most human testimony, falsifiable. We do not know at present how many anthropometric traits will be of utility for the process, however, so it remains to be seen whether the quantity of information would also be superior. It is important to recognize that we should not overlook the innate ability of people to recognize faces and perceive similarities and slight distinctions.

There are a variety of ways in which these abilities might prove useful in the construction and practical application of indirect and direct molecular photo-fitting work. In other words, perhaps we use a person, not a computer to construct the sketches, or perhaps we combine human-derived with computer-derived composites. This option would clearly have its advantages and disadvantages, and one might draw some corollaries with the field of color-metrics where trained observers are used to compare and match colors in especially designed illumination hoods. Presumably, given the high prevalence of color blindness and other variants affecting color vision (Verrelli & Tishkoff 2004), there is variability in who can serve as a good observer. Likewise, there is variation in the ability to recognize faces, the extreme clinical manifestation being face blindness (prosopagnosia). Given there is the extreme version of an inability to recognize faces, one might expect that there is also normal variation in this facility, and indeed personal experience has supported this difference (are both you and your husband/wife exactly equal in your ability to recognize the actors in movies you watch together?).

If there is this sort of variability, it might possibly be related to the ability to recognize similarities among persons such as family relatedness or group affinities. As such, where one works toward using human observers in the context of this type of research, it might be important to first study the degrees of variation among subjects in terms of the ability to recognize faces and the component traits that are affected by genomic ancestry. Similarly, there may be variability in forensic scientists and other investigational personnels' ability to recognize the patterns in the faces that are the results of the database queries. It may also be that there are levels of sensitivity to particular ancestry components, which are learned through particular cultural experiences. This is referred to as the own-race bias or the other-race (Lindsay et al. 1991) effect in social psychology, and indeed we know that facial recognition itself has an ethnic component such that, for example, a European might have a harder time discriminating among Asian faces than an Asian person. Thus, the sensitivity and accuracy of observers would have to be taken into account in planning studies and preparing investigators.

Finally, there also have been sex and age differences in the ability to recognize faces in other-race tests with fewer differences for younger observers and females showing better success (Chance et al. 1982). This area of research in psychology remains active, and one would be well advised to consult the literature thoroughly before planning experiments with observers. Until a good handle on these issues is had, it might be advisable to employ software algorithms so there is at least an element of standardization. There may be other things to consider—here we have tried to indicate only some of the most obvious issues that require appreciation.

ESTIMATING PHENOTYPE FROM 2D DIGITAL PHOTOGRAPHS

Although not ideal for all anthropometric measures, there are a multitude of traits amenable to measurement from digital photographs using software tools. It is also the case that over the past few hundred years we have become well accustomed to viewing faces as still images in two dimensions, and so can observe many of the features that both make us distinctive and show kinship and population affiliations in this media. Measures of distance require a ruler or other reference object of known dimensions be photographed next to the face, in order to ensure proper scaling from photo to photo. For this we routinely have used right angle forensic rulers of the type typically used in evidentiary photography. Some of the traits that can be measured from photographs are:

Hair texture. The smoothness of the surface over the top of the head from the left eye to the right eye is an indication of the curliness or waviness of the hair. For example, for both of the photos in Figure 8-6, a line drawn to trace the contour of the top of the head, not drawing through any hairs that may be sticking up or out, would be wavier and longer when stretched out than a line drawn to trace the contour of the top of the heads in Figure 8-5.

Hair color. Ideally we should measure hair pigmentation using reflectance spectroscopy with measures like the M index and E index, as we will describe later. From digital photographs, a similar value can be obtained using the RGB color scale and luminosity values as we will describe in the next chapter. An important variable in the measurement of color is the illumination, and careful control is required to collect quantitative data from photographs.

Eye color. Color scores using RGB and/or luminosity values are used to quantify hair colors from digital photographs, as we will describe in detail later in the next chapter.

Skin color. As with hair, reflectance spectroscopy is ideal, but with careful control of illumination sources, high-quality digital photos can be generated allowing for quantitative comparisons.

Facial anthropometry. Although ideally one could take direct measurements of the face on the research subject in person, this is not always possible. With careful control of facial positioning some measures of facial anthropometry can be made from 2D frontal and portrait photographs.

a. *Interocular distance* measured in millimeters.

b. *Epicanthal folding* could be expressed as the ratio of the eye height to width, where epicanthal folding would be characterized with relatively low ratios. We could also use the intercanthal index [$I=(en\text{-}en \times 100)/ex\text{-}ex$] (Juberg et al. 1975).

c. *Earlobe connectivity*. The length of contour traced around the bottom portion of the ear to the neck could be used.

d. *Facial width* in millimeters.

e. *Facial height* in millimeters.

f. *Facial hirsuteness* as a binary characteristic (percentage of a given sample with and without facial hair).

g. *Male pattern baldness*. Frontal balding could be measured as a binary characteristic, which is subjective, or with the curvature or length of a line in millimeters, extending from one temporal region to the other or a colorimetric disparity measure comparing the chin or lip vs. the forehead (compare the photos in Figure 8-4).

h. *Chin shape* measured with a horizontal line in millimeters.

i. *Shape of the lips* could be measured in millimeters, comparing the height of the top lip vs. the bottom, or the height of both lips compared to their width.

j. *Nose shape*. Width, depth, and height in millimeters. Standardization and clear definition of where lines are to begin and terminate would be crucial to establish for this measure.

Sampleid : DNAW70078R			
Ethnicity	**Value**	**Ranges**	
European	91%	83%	98%
Sub-Saharan African	5%	0%	13%
East Asian	4%	0%	11%
Native American	0%	0%	7%

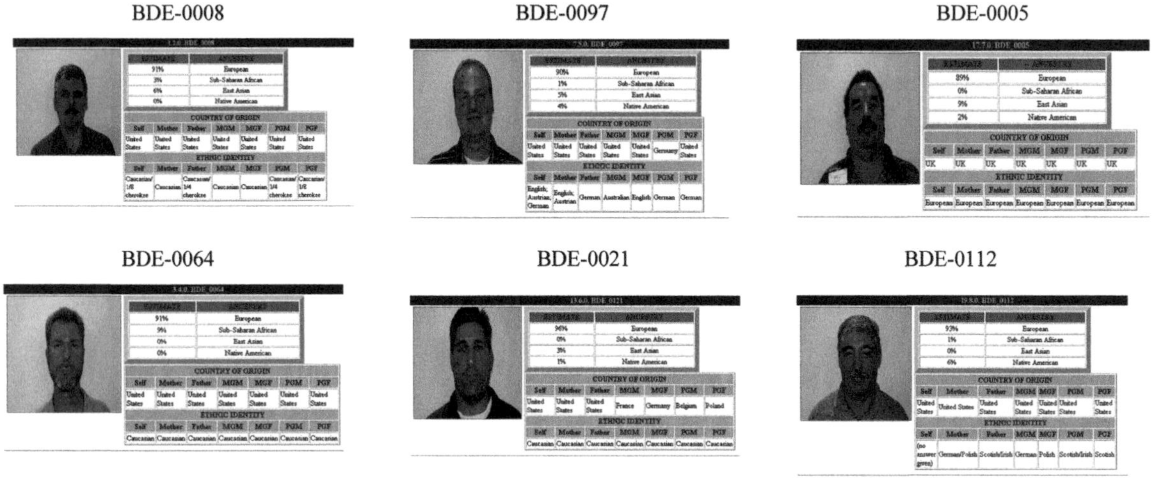

Figure 8-4

Query (top) and six returns (bottom) from a database search using an Individual Genomic Ancestry Admixture estimate for a Caucasian male. As described in the text, the database queried is composed of a large number of individuals, their IGAA estimates (MLEs and confidence intervals), trait values for a variety of phenotypes, and various demographic data such as country of origin for members of their pedigree and ethnic description of the same. The query produces a gallery of individuals with MLEs of values that fall within the confidence intervals of the unknown.

ESTIMATING PHENOTYPES FROM 3D DIGITAL PHOTOGRAPHS

As indicated earlier, facial anthropometrics ideally is carried out on the research subject in person and by an experienced technician. Both the time required and available expertise can be important factors limiting this practice. A useful alternative to direct measures of a subject that is much more informative than 2D photographs is a three-dimensional digital camera system designed by 3dMD (Atlanta, GA) called 3dMDface. The 3dMDface system uses a combination of six digital cameras at fixed angles and a series of randomly generated light patterns projected on the face to record the volume and texture of the surface of the face across 180°. Joan Richtsmeier and collaborators at Penn State University recently have completed a comparison of the 3dMDface system with measurements made directly on the faces of the persons (Aldridge et al. in press). The 3dMD system fares very well in these comparisons showing submillimeter precision in the placement of landmark points. The only measures showing higher levels of error were measures crossing the labial fissure (going between the upper and lower lips), which can be affected by slight changes in the position of the jaw, indicating that subjects should be best instructed to have their upper and lower teeth in contact in order to standardize images in the database.

We (Author and Mark Shriver of Penn State University) have performed some preliminary assessments of facial features using data collected on our own 3dMD system. Figure 8-3 shows an example of a 3dMDface system photograph. The 3D photographs can also be studied or shown with a photorealistic surface, making possible displays with complete information on phenotype, which could be useful in conveying the results of database queries to investigators. Alternatively, we can artificially color the surface, which could be useful both to eliminate the possibility of investigator bias in the placement of the landmarks and to allow for studies of the relative importance of pigmentation and facial features on subjective assessments of ancestry. It is also possible to make small ink spots at the location of the landmarks on the face of the subject, a technique commonly used in 2D anthrophotogramy, before taking the pictures. These ink spots will show up on the photorealistic image facilitating the localization of landmarks. Additionally, since the ink spots will not show up on the false color image, we can use the comparisons between the placement of landmarks on these images to estimate operator landmark placement accuracy.

The main alternative to stereoscopic surface acquisition is laser scanning (e.g., Cyberware, Inc.), which has the inherent risks of laser damage to the eyes and requires on the order of 17 seconds per scan, which is much longer than the fraction of a second required collecting a three-dimensional photograph. The time to acquire an image has many repercussions, including the chance that the subject will move and the chance that the camera will move in a way not intended by the engineers (scanners typically move about the object by design). Additionally, the short time required to acquire a three-dimensional photo means that the subject can be moved into alternate positions, for example into profile for close-up photographs of the ears, and can be asked to make particular facial expressions so that variability in apparent morphology resulting from expressions on the face might be taken into account in the analysis.

There is an extensive literature on the measurement and interpretation of morphological forms. The goal generally is to create measurements and statistics that allow for biologically meaningful summaries. A recent review of the field of morphometrics can be found in a text by Lele and Richtsmeier (2004). One of the methods of analysis that is likely to be very useful in the measurement of faces is known as Euclidian Distance Matrix Analysis (EDMA) and is focused on the identification of landmarks followed by the computation of linear distances among all the combinations of landmarks. These matrices of distances can then be compared across individuals, averaged across population samples (or depending on the research project, species, or treatment groups). One can also test for correlations between grouping variables and particular subsets of the pairwise measures that make up the matrices. These methods have not seen extensive use yet in comparing populations or in other types of genetic analyses although it is clear as to how one might proceed in this respect.

EXAMPLES OF DATABASE QUERIES—GLOBAL CHARACTERISTICS FROM DIGITAL PHOTOGRAPHS

As of this writing, three-dimensional photo databases are just now beginning to be amassed. We return to two-dimensional photographs to show some examples of database returns for actual cases. Figure 8-4 shows a database return from a profile of an actual case sample; the sample was 91% European, 5% West African, and 4% East Asian. We used this percentage with a $\pm$ % range of each ancestry corresponding to the limits of the five-fold confidence interval for each. So the query was:

Query 1:
European: From 83% to 98%
West African: From 0% to 13%
East Asian: From 0% to 11%
Indigenous American: From 0% to 7%

Consider these four ranges as conditions; any individual in the database whose percentages fell within the specified range for all four ancestral groups met all four conditions and was retrieved. Individuals returned from the database had generally light skin colors and a variety of hair and eye colors; six of the returns are shown in Figure 8-4. If we call only females that fall within this specified range of admixture, we see a similar result (see Figure 8-5). The 5% West African is significant at the individual level (the five-fold confidence interval just touches 0% West African ancestry), but evidently from this database return too low to contribute to physical appearance as we see from BDE_0064 (9% West African), BDE_0008 (3% West African), neither of whom to the eye exhibit West African characteristics. The 4% East Asian admixture is not statistically significant at the $p < 0.05$ level (the five-fold confidence interval for the query subject overlaps significantly with 0% West African ancestry) and due to this, as well as the fact that 4% is probably far too low even if it were real, none of the database returns exhibit any East Asian characteristics (to this author anyway). We can see from this collection that our suspect is expected to have what might be characterized as a standard, run-of-the-mill European appearance, with no outright appearance of admixture or ethnicity (loosely speaking). If we don't wish to speak loosely, we could of course measure parameters such as M and E indices and apply summary statistics.

Sampleid : DNAW70078R			
Ethnicity	Value	Ranges	
European	91%	83%	98%
Sub-Saharan African	5%	0%	13%
East Asian	4%	0%	11%
Native American	0%	0%	7%

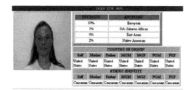

In contrast, using the query:

Query 2:

European: 37–72%

Indigenous American: 0–21%

West African: 14–42%

East Asian: 0–32%

Figure 8-5

Query (top) and nine returns (bottom) from a database search using an Individual Genomic Ancestry Admixture estimate for a Caucasian female, as in Figure 8-4.

the retrieved data shows that an MLE value of from 14–42% West African on a European background does not provide as much useful information about skin shade as we might have predicted based on subjective experience and guess-work. The range of West African ancestry is just too large (using a smaller range based on a different confidence interval value would produce a tighter result in terms of skin color, but with less certainty that the MLE value is correct). Figure 8-6 shows two representative photographs called by this query; one individual with 37% West African and obviously pigmented skin was retrieved but the skin of another with 29% West African was seemingly as pale as people with 0% West African (compare Figure 8-6 with Figure 8-5). In this case,

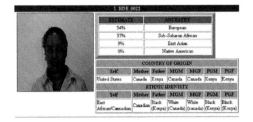

Sampleid : DNAW70076			
Ethinicy	Value	Ranges	
European :	55%	37%	72%
Native American:	0%	0%	21%
Sub-Saharan African:	27%	14%	42%
East Asian:	18%	0%	32%

SAMPLES DOES NOT HAVE IMAGES AND ETHNICITY INFORMATION					
ID	Native	European	Native American	Sub-Saharan African	East Asian
NA17332	Middle East version II	71	0	15	14
NA17336	Middle East version II	67	15	16	2
NA17042	Middle East version I	69	0	16	15
NA17382	North African	66	0	26	8
NA179091	Iberians	55	17	14	14
NA179094	Iberians	59	0	21	20
ANA17184	African Americans	67	1	32	0
NA17709	Mexican American (Hispanic)	48	0	40	12
SA_MS232	South Asian	61	2	14	23
SA_MS432	South Asian	62	15	17	6
SA_MS433	South Asian	60	10	14	16
EC2_1091	European-Americans	67	0	17	16
EC2_1066	European-Americans	45	11	39	5

SUMMARY		
Ethinicity	Count	Total
East African/Cannadian	1	1
Middle Eastern American	1	1
Middle East version II	2	10
Middle East version I	1	9
North African	1	7
Iberians	2	9
African Americans	1	89
Mexican American (Hispanic)	1	94
South Asian	3	34
European-Americans	2	95

Figure 8-6

Database query (top) and return for an individual of substantial European/ sub-Saharan African admixture. Photographs and basic pedigree information is shown for two of the returns (middle). A list of samples from the return is shown with self-reported ethnicity (lower left; only part of the list is shown) and counts for each represented ethnic group are shown in the lower right (Count column). The Total column in the lower right indicates the total number of samples for each category in the database.

the database saves an investigator from improperly assuming that 27% West African admixture means darker skin tone than a 100% European, and more properly puts it in terms of likelihood—the likelihood of darker skin than 100% European is perhaps 50% (from this return of two files), and therefore of minimal inferential value (note that only a couple of returns are shown—obviously these results would be based on larger sample sizes).

Of course, a larger database may provide returns that belie this interpretation from a smaller one, and of course, other traits may be usefully inferred from even this large range of Western African used in the query. For example, note that both individuals have a kinky/curly hair texture. If a large number of samples retrieved from the database had a similar hair texture, the difference in frequency of this character between individuals within 14 to 42% Western African from those with 0% Western African could be statistically significant, in which case we would be justified in making the inference that a person with a level of Western African genomic ancestry measured between 14 and 42% is more likely (and how much more likely) to have a curly hair texture than a person with no measured Western African genomic ancestry.

EXAMPLES OF DATABASE QUERIES—ETHNIC DESCRIPTORS AND GEOPOLITICAL AFFILIATIONS

We have introduced the terms reverse ethnic or geographic description. Querying the database for geopolitical affiliations such as ethnicities or countries of origin provides different information than querying the database for photographs because knowing what a person likely looks like doesn't necessarily tell you anything about his or her own geographical origins (only what parental population his or her ancestors were likely from, which as previously explained, is a different thing). In our examples shown so far we have presented returns that contain a frontal photograph, the continental admixture percentages from 171 AIMs, country of origin and ethnicity for two generations.

Other output styles are possible and useful, including an ethnic summary of the individuals in the database that fall within the specified range. Ethnic descriptors associated with the retrieved reference samples in the database also show a broad range of terms. Consider the terms commonly used by relatively unadmixed individuals of primarily European descent (see Figures 8-7 and 8-8). Figure 8-7 shows the results for the query in Figures 8-4 and 8-5— the ethnic descriptor for 14 individuals retrieved from the database along with their genomic ancestry profiles. The ethnicity and continental admixture percentages for each database hit (individual falling within the range) are listed, hit by hit. Figure 8-8 shows a breakdown of the number and percentage of individuals from a number of retrieved ethnic groups (i.e., terms based on self-descriptions). Figure 8-8 is essentially the same as Figure 8-7, except the returned data is summarized by ethnic group rather than presented individual by individual.

For both presentations, the number of ethnic categories is determined by the number of different groups into which the retrieved individuals placed themselves. In fact, the ethnic groups are taken from the questionnaires verbatim, however incorrectly spelled or redundant, and you will notice different categories for American/European (misspelled by the subject), Caucasian and European. The number of hits within each group is shown in the count column, and the total number of individuals in the database for each group is shown in the total column. The output of the ethnicity (and only the ethnicity) in this way, allows for focus on one measure of likely geopolitical origin.

For the query of Figures 8-4, 8-5, 8-7, and 8-8 (Query 1), the MLE was mainly European, with a low level of West African admixture (possibly East Asian as well), and there were 121 hits in the database. Of these 121, all but seven

Figure 8-7

Database query (top) and return (bottom) for the individual subject of Figures 8-4 and 8-5. The return in this case is not a gallery of digital photographs, or phenotype information, but a list of samples of MLEs falling within the confidence range along with their self-described ethnicity. MLEs for each subject returned are also shown.

Sampleid : DNAW70078R			
Ethnicity	Value	Ranges	
European	91%	83%	98%
Sub-Saharan African	5%	0%	13%
East Asian	4%	0%	11%
Native American	0%	0%	7%

ID	Native	European	Native American	Sub-Saharan African	East Asian
0. 2.0.NA17325	Italian	92	0	6	2
1. 2.0.NA179093	Iberians	92	0	6	2
2. 2.0.EC2_1081	European-Americans	93	0	4	3
3. 3.0.NA17010	Northern European	94	0	3	3
4. 3.0.NA17211	Coriel Caucasian	91	2	6	1
5. 3.0.EC2_1003	European-Americans	94	0	5	1
6. 4.0.NA17371	Greek	91	0	9	0
7. 4.0.NA17372	Greek	95	0	5	0
8. 4.0.NA17007	Northern European	91	3	6	0
9. 4.0.NA17043	Middle East version I	91	4	5	0
10. 4.0.NA17214	Coriel Caucasian	91	2	7	0
11. 4.0.NA17219	Coriel Caucasian	87	0	8	5
12. 4.0.EC2_1017	European-Americans	91	0	9	0
13. 4.0.EC2_1021	European-Americans	91	4	1	4
14. 4.0.EC2_1084	European-Americans	91	3	6	0

self-identified as a category that could be fit as European in derivation. Of the seven, four were Middle Eastern individuals, which is not surprising since we have already seen systematic presence of West African ancestry in this group (see Chapter 6), and two were Hispanic (one from Cuba and the other from Puerto Rico). One retrieved sample reported as an American Indian. Note that only one Hispanic was part of the screened database, and this Hispanic was retrieved as his or her percentages fell within the query range. Also note that the single returned American Indian was one of 38 present in the database. In contrast about half of the European-Americans were returned. Since the database used to make this particular figure is incomplete (i.e., only one Hispanic), we would have to be careful about excluding or even including Hispanic as a term the subject may use to describe himself. This illustrates why it is important to use a comprehensive database.

From this output in Figure 8-8, we can appreciate that the individual corresponding to the sample profile would most likely describe him- or herself as a European of some sort, but that there is a small chance the individual would describe him- or herself as something other than European, such as Middle Eastern or even Hispanic. If we had a better representation in the database used for this particular query, especially for Hispanics, we could invert the

SampleId : DNAW70078R			
Ethnicity	Value	Ranges	
European	91%	83%	98%
Sub-Saharan African	5%	0%	13%
East Asian	4%	0%	11%
Native American	0%	0%	7%

SUMMARY		
Ethinicity	Count	Total
Caucasian/ 1/8 cherokee	1	1
Italian	5	10
Iberians	4	9
European-Americans	52	95
Northern European	6	11
Coriel Caucasian	13	24
Croatian	1	1
Caucasian	10	21
Greek	6	8
Middle East version I	5	9
European	3	6
Irish/Swedish/canadian	1	1
English; Austrian; German	1	1
Middle East version II	4	10
Scotish/Irish/German	1	1
German/English	1	1
Hungarian/Irish	1	1
Dutch/German	1	1
DT-American Indian	1	38
Hispanic	1	1
Hispanic (Puerto Rico)	1	1
NA	1	1
American/Europian	1	1

Figure 8-8

Database query (top) and return (bottom) for the individual subject of Figures 8-4, 8-5, and 8-7. Each individual with MLEs falling within the confidence range is identified computationally, and from this matching subset, a list of ethnic categories corresponding to returned individuals is returned. Each observed ethnic category is returned, along with the counts and the total number of samples in the database corresponding to each group. Ethnic categories returned correspond to the unique subset of self-described ethnic affiliations corresponding to the matching subset.

frequencies to establish a likelihood of ethnic origin, effectively accounting for unequal representation of each group. As described already we could also distill these terms to provide simpler, more general description of the likely ethnicity.

For Query 2, the first few individuals falling within the range are shown in Figure 8-6D (truncated to fit on the page), and the summary counts shown in Figure 8-6E. There were 15 hits in the example database used to create these figures. Five of these 15 were Middle Eastern in ethnicity and a quarter of every Middle Eastern in the database (5/20) were hits for this profile. The fit with African Americans is not good; only one in 89 African Americans were hit by this profile, which might seem surprising given that the range of the West African part of this query was 14% to 42% (recall that most individuals who describe themselves as African Americans have over 50% West African ancestry; see Figure 6-5 in Chapter 6). The number of South Asian hits was also low (only 3/34) and European Americans even lower (2/95). No other groups were retrieved; for example, none of about 30 Hispanics were retrieved, nor any of 95 East Asians and Southeast Asians. Thus, it appears that the profile fits best with individuals reported as Middle Eastern or Iberian, possibly other similar terms for populations nearby, or possibly with individuals reporting of North or East African ethnicity. Most likely, the suspect who left the specimen describes his or her ethnicity as Middle Eastern, North African, or Iberian (possibly also other Southeastern European terms like Italian, Greek, or Turkish, which a larger database for this example might have shown), but there is a small chance the suspects who donated the specimens would describe themselves as African American, Hispanic, South Asian, or European American. Again, we could calculate the likelihood of each conclusion, relative to other conclusions.

VARIATION AND PARAMETERIZATION OF DATABASE OBSERVATIONS

Examples for other types of queries are shown. Photographs for database hits from queries using various West African ranges are shown in Figure 8-9. In this query, we only specified the West African (AFR) percentages and allow the European, Indigenous American, and East Asian percentages to be anything (so we would expect a relatively wide range of anthropometric trait values at low levels of AFR genomic ancestry). The six photographs for hits within the 10% to 33% range show a collection of individuals with darker hair color and iris color, but most would not characterize the average skin tone in this collection as particularly dark. Stepping the percentage up to 33% to 50% we see in our one example photograph a slightly darker skin tone. Stepping up again to between 50% and 75% AFR we see a medium dark skin tone, and with AFR > 75% we see an average relatively dark skin tone. Measuring melanin content from these photographs with computational spectrophotometry tools (such as Adobe Photoshop, using the histogram tool) would allow for a quantification of what is meant by dark, darker, and so on, or better yet, obviate the need for verbal descriptors altogether.

One could calculate the correlation coefficient between melanin index and % AFR ancestry and use the observed relationship to make inference about likely skin melanin content for unknowns, a topic we will treat in more detail later in this chapter. Skin color is not the only anthropometric trait that we can appreciate to differ as a function of West African/European ancestry in this figure. With reference to this figure let us return to the notion of paramaterizing our

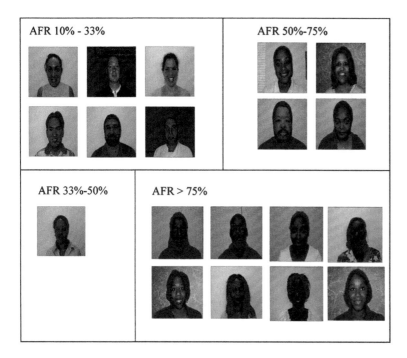

Figure 8-9

Forming hypotheses on overt phenotypes based on the level of sub-Saharan African admixture. A small database was interrogated with four separate queries specifying a range of sub-Saharan African admixture, shown at the top of each of the four boxes. Each individual retrieved from this small database is shown. We note the suggestion that skin melanin content, hair color, and other craniofacial phenotypes may be associated with African admixture from such a display. Such observations form a good basis for the development of hypotheses, which must be tested using regression analyses.

observations. Provided the pictures are taken under controlled illumination, we might quantify the average melanin content in the hair and note that the incidence of lighter index values is greater for the AFR 10% to 33% group than the others. We might also use craniofacial anthropometrics obtained from 3D photographs such as that in Figure 8-3, to identify parameters of statistically significant difference among groups (discretely handling the data) or correlations of high coefficients (handling the data continuously). If the sample size of each group was adequate, and if the trait varies as a function of West African ancestry, we would note such differences in the return. We might note that blue irises are almost never found in any of the groups except the AFR 10% to 33% group and less frequently, the AFR 33% to 50% group. If a sample belonging to an unknown subject is left at a crime scene, and shows a West African affiliation > 75%, then this type of observation (if quantified) would form the basis for our expectation that the unknown subject is very unlikely to have blue irises.

Note however that if the size of the database is small, one would need to be very careful about drawing conclusions—conclusions should be based on the demonstration of statistical significance within samples of good size. For example, we would want to be very careful drawing conclusions on skin pigmentation based on a picture created using 19 individuals as in Figure 8-9 (we would need sample sizes of the magnitude we discuss later in this chapter, or even larger). Likewise, from the small collection of individuals shown in

Figure 8-9, we would not want to conclude that because the only individual with a beard is from the AFR 10% to 33% group, an individual of greater than 33% AFR ancestry is unlikely to have a beard. With a larger sample size, we likely would see this conclusion could not be characterized as statistically significant. However, assume that facial hair was culturally discouraged in the African-American community—we might see with a very large database that facial hair is more likely for individuals with low AFR percentages and that the differences we see are statistically significant.

Is this a legitimate descriptor to have extracted from the database return? It depends on the sample size and summary statistics. There is no reason to accept only anthropometric variables of high heritability and ignore sociocultural variables if they are correlated with admixture, as long as there is sound data to support the correlation. Even though there are no known anthropological or phylogenetic differences between parental Africans and Europeans with regard to the ability to grow facial hair, if one culture shuns the practice and the other does not, would not a crime scene investigator value this information? Recall, the forensics investigator's primary objective is to determine likely/possible physical appearance, not to trace genealogy or personal anthropology. Though it may seem this carries a risk of introducing an unfounded bias into an investigation, forensics investigators are usually mature enough to handle the probabilistic aspects of their cases (i.e., to use Bayes theorem, however informally).

Figure 8-10 shows six hits obtained using the query: "East Asian (EAS) > 80%." One will note that most of these individuals exhibit characteristics commonly ascribed to East Asians, such as epicanthic eye folds and medium skin tone. Also shown in this figure is a single hit obtained with the query "Indigenous American (NAM) > 75%." The similarity in skin tone to the EAS > 80% return is apparent and distinct from the returns for higher AFR percentages as well as European percentages in Figure 8-11, which shows individuals returned from the query "European (EUR) > 90%." We can also see in our relatively small samples that the incidence of lighter hair color is greater for the EUR > 90% query return than the others just discussed, and in fact, though it is difficult to see in the figure, the same is true for lighter iris color. So, even though not every person with European genomic ancestry levels greater than 90% would have lighter colored hair or blue irises, most people that have lighter colored hair and blue irises would seem from these photographic returns to

Figure 8-10

Display of individuals with relatively high East Asian (EAS) or Native American (NAM) admixture, from the same small database as described in Figure 8-9.

EAS > 80%

NAM > 75%

have European genomic ancestry levels greater than 90%.

This may not be as much information as we might like to have, but it is information that may serve as a foundation for the formation of hypotheses, which could be tested in larger samples using regression analyses. To form our hypotheses, we might use computational tools

EUR > 90%

Figure 8-11

Display of individuals with relatively high European (EUR) admixture, from the same small database as described in Figures 8-9 and 8-10.

to measure the width and height of the face, interocular distance, nose width—virtually any variable one wishes to measure—and compare the values obtained between groups of disparate ancestry relevant for a given subject's profile (such as phenotype differences between individuals with EUR > 75% and AFR > 75%, to assess the feasibility of indirectly predicting value for the phenotype in an African American of 50% European and 50% West African ancestry). If the size of the database is good, the differences are statistically significant (whatever the reason why) and the difference between group means is greater than the variation within groups, then we can usually justify the practice of making inference of trait value given the appropriate information about ancestry.

For example if we have a crime scene sample with 50% European and 50% West African ancestry, and we want to know whether we can make an inference on nose width from our database, we could call from the database all individuals matching the 50% European and 50% West African ancestry profile ± X%, where X is determined by the probability space defined by the multilocus AIM genotype. If the sample is large, we then would use a computational tool to measure this variable for each subject, and observe whether there was a correlation in trait value and admixture. If there was, we could make a prediction for an unknown using the regression line. We might alternatively measure the mean and variance of nose widths in the subgroup of similar admixture, and then use the mean and range for this group as the estimate and range of likely values for the unknown individual. The only problem with the second, very straightforward technique is assessing whether the sample size is large enough. With the first approach, we could note whether the correlation coefficient was statistically significant and use the measure of significance as the justification for which phenotypes to infer and which to not infer using a given database with a given unknown.

If more convenient, one could employ a sort of pseudo-regression technique to provide this type of justification—breaking the admixture continuum into bins. For example, assume that the width of the nose is negatively correlated with European ancestry and positively correlated with sub-Saharan African ancestry. To infer nose width for an individual of 50% European and 50% West African ancestry, we could:

- Call from the database all individuals matching the 50% European and 50% West African ancestry profile $\pm$ 10% or so
- Use a computational tool to measure the mean and variance of nose width in this subgroup
- Call other subgroups representing different European/West African ancestry admixtures, such as
    a. 100% AFR
    b. 90% EUR/10% AFR
    c. 80% EUR/20% AFR
    d. 20% EUR/80% AFR
    e. 10% EUR/90% AFR
    f. 100% EUR
- Use the computational tool to measure the mean and variance of nose width for each of these subgroups
- Use the data obtained in b. and d. to
    a. Calculate exact test p-values to observe whether the phenotype varies significantly among these groups
    b. Calculate an F_{ST} to observe whether the phenotype varies significantly among these groups
    c. Plot the relationship between West African or European admixture in EUR/AFR admixed individuals and nose width (similar to Figure 8-19)

If the summary statistics are significant, then we can use the regression line from our subgroups, or the mean and variance for the target group as the estimate and range of likely values for the unknown individual. Even if there is overlap in values between groups, we could possibly identify subsets of the groups a. through f. for which there was a statistically significant difference in mean values (perhaps the extreme bins with polarized ancestry?) and provide guidance for only extreme or polarized samples.

Of course, this entire process could be automated with software, and our laboratories are presently experimenting with such a system, results from which will be covered in later editions of this text.

CAN SOCIAL CONSTRUCT SUCH AS RACE BE INFERRED FROM DNA?

The reflex answer from most scientists to this question would be "of course not," but as we have been alluding to until now in this chapter, the correct answer is something different, more to the tune of "most likely, to a certain extent." Before we go on, we need to make it clear that databases of requisite size to answer this question in a scientific manner have not been constructed yet (by us or anyone else to our knowledge), and as a result we don't actually know yet if it is practical. However, from a theoretical standpoint it is certainly possible.

A common criticism of forensic genetic methods that incorporate measures of population structure are that they cannot be used for the inference of socially determined variables. The concept of race is one such variable, because in addition to having a heritable or biological component for most people, it also incorporates the socioeconomic, religious, cultural, linguistic, and even political (each, to different extents for different people). Of course all these things cannot necessarily be read directly from the DNA. The question is, can they be read indirectly? Given that race is a concept involving so many ambiguous variables, which are defined (unreliably and unpredictably) by human beings, it would seem impossible to learn anything about race from DNA, even if elements of genetic structure correlate with those of social structure. The critics would argue that though there may be a heritable component to race for most people, because race is more than biology, genetics methods simply cannot be useful for its inference.

Our ability to indirectly infer a phenotype is proportional to the simplicity of that phenotype in a genetic sense, as we will discuss more fully in Chapters 9 and 11. At this point, it is difficult to know which human characteristics will be amenable to the indirect method of photofitting and which will not, but we can reasonably argue that for the overt phenotypes of high heritability and low genetic heterogeneity, the method will work, and for characteristics of low heritability and/or high genetic heterogeneity, they will not. The science of molecular photofitting, being an exercise in empiricism or observation, would tell the tale for any given phenotype. By this point in the text, one can appreciate that we have the ability to reliably measure elements of population structure within individuals, and most would agree that since social units of human populations have been subject to relative isolation and unique genetic pressures over the course of recent human evolution, there are correlations between genetic and social elements of population structure. Indeed these correlations may very well be strong enough that we are able to delimit the possible ethnic or racial self-descriptors people use to describe themselves from the international complement. Thus, it is possible that we would be able to infer social elements of population structure knowing genetic elements, at least to a certain level of precision, which for any given connection may be acceptable for some people, but not for others depending on the person and the application desired.

We have just discussed the incorporation of digital photographs in the database of reference subjects, but we also indicated that self-designated descriptors for race could be present for each sample. With the photographs, we screen the database for individuals with similar genetic profiles with respect to a given assay and population model in order to define the range of trait values most likely for any given phenotype. With the self-designated descriptors, we can do the same thing to infer the range of terms a person may likely use when describing his or her race. If we wanted, we could likewise document for each of our reference database samples the race of the relatives two, three, or four generations out, the geographical place of birth, which may be cast in modern-day geopolitical terminology such as Russia or South Korea, or any other variable that is used in determining race, such as primary language spoken, or religion.

Regardless of the variables collected from each reference sample, if they are collected in an error-free manner, and if the correlations exist as we suspect they do, the range of social terms or values corresponding to a given element of genetic structure in the population can be empirically defined. For example, if we query the database with the values 75% ± 10% Western African, 25% ± 10% European ancestry, we would by definition obtain a list of terms that individuals with this type of admixture profile use to describe themselves. If the database is large enough, we would obtain a reasonable representation of the diversity or lack of it corresponding to a given term. In this case, we would likely find that of 100 returned individuals, the following terms would be found to be commonly used:

- African
- African American
- Black

but the terms European, Caucasian, white, East Asian, Oriental, Native American, and so on were rarely represented or not represented at all. This would allow us to state that terms African, African American, or Black (however many terms obtained from the database search) are those the individual likely would use to describe him- or herself, but not likely terms such as European, Caucasian, white, East Asian, Oriental, Native American, and so on. Like science in general, we have used empirical observations to interpret what one thing (genomic ancestry) has to say about the other (race), whatever the mechanism, and no matter how little or how much is actually said, as long as we use a reasonably sized database and can quantify the confidence with which the terms apply to a given genomic ancestry profile, there is nothing invalid about the process. If a given genomic ancestry profile is found in individuals who describe themselves as African American, Black, African, American Indian, Hispanic, and Caribbean, this should be apparent from the database return, and in this case we would be able to infer that the person belonging to a sample is more likely to use these particular race descriptors than others. We might not have the precision we would prefer to have, but we nonetheless have acquired information—information that is quite different than we have considered so far in phenotype.

What are the downsides of such a procedure? Aside from ethical considerations associated with using the indirect method for phenotypes other than the overt (discussed in Chapter 11), there are technical limitations as well. If a genomic ancestry profile is found in three-fourths or more of the human population (in which case, there is probably something wrong with the markers, methods of model used), the database return should indicate this with a very long list of seemingly disparate race descriptors, in which case we probably would not be able to infer much about how the person is likely to refer to themselves. Therefore the downside risk created by using a substandard test is not of erroneous inclusion of racial terms, but the loss of information and resolution among terms. Of course small databases could lead to false positives or the false appearance of good information. It is difficult to estimate how large the database must be for any given question, but as with overt phenotypes, summary statistics could be used to justify an ability to draw a conclusion.

In our experience with relatively small databases, it appears that we might expect good information to obtain. Even ambiguous terms like Hispanic seem to be associated with certain genomic ancestry profiles, which should not be surprising because Hispanics share a discrete (though genetically complex) history. Due to the history and complexity of the term, we might expect that Hispanic would appear for many types of genomic ancestry admixture profile ranges used as queries of a good sized database—perhaps so many that it would be returned using virtually any type of genomic ancestry admixture profile. However, this has not been our experience with modestly sized databases (in the thousands of samples); Hispanic usually is found only in the returns for certain genomic ancestry profiles, usually involving higher levels of European, Native American and low levels of West African admixture (notice the clustering of samples in Figure 6-9, Chapter 6, which is a pyramid plot for a collection of individuals united by the fact that they all self-described as Hispanic). Thus, the presence of Hispanic on a list of terms provided by a database query provides information—maybe not as much as we would like, but information nonetheless.

The entire complement of international racial descriptors is complex, and we simply do not have data for most of them yet, but from this type of result we might expect that for any given genomic ancestry profile, we would be able to use individual genomic ancestry admixture estimates to delimit which of them the individuals would likely use to describe themselves. Note that this empirical process is essentially the same that is used when attempting to determine likely ethnogeographic origin of patrilineal ancestors through Y-chromosome haplogroup analysis (see Figure 4-3, Chapter 4, for instance). Of course, this empirical method is still an exercise in likelihood determination, and there will always be the unique individual with the unique life history, whose self-perception is completely uncoupled with his or her genetic ancestry. However, in science we can rarely provide the exact, 100% correct term for 100% of the subjects, which is why we use summary statistics to express our ideas and communicate to other scientists. As long as the technique is falsifiable and empirical, and as long as we can know how well it works, it is fundamentally no different from any other sort of empirical observation we might make as scientists. For some profiles, more information about race will be available than for others, but for most, querying the database will provide some information where theretofore information was probably lacking. Notwithstanding, some may feel uncomfortable with such an approach, fearing that it may lead to undesirable consequences. This need not be the case however, and we will save discussion of this for Chapter 11.

INDIRECT APPROACH USING FINER POPULATION MODELS

Finer levels than continental population structure might allow for the inference of certain traits that vary within continental population groups. We have described the EURO 1.0 panel as capable of partitioning individual European ancestry into Northern European (NOR), Southeastern European (MED), Middle Eastern (MIDEAS), and South Asian (SA) ancestry. How much can we learn from the empirical database approach about traits that vary within the European group? Recall that our definition of the European part of the population model we have spent most of our discussion on so far includes individuals of Continental European, Middle Eastern, and South Asian descent derived at least in part from a Western Eurasian source that likely existed some 50,000 years ago (50 KYA). Clearly there is great variability in anthropometric traits such as iris color, hair color, and skin shade among these subpopulations so we might expect that application of the EURO 1.0 or similar panels could provide a certain amount of information about these traits for any given individual.

Our discussion on this topic is focused on the EURO 1.0 panel, since data for other panels is scant (for possible reasons, refer to Chapter 11). At the time this book was written, DNAPrint genomics had compiled a database of about 400 unadmixed Europeans for whom EURO 1.0 had been run. We would first need to decide what types of continental genomic ancestry admixture profiles to accept as a hierarchical prerequisite for the sublevel EURO 1.0 analysis. For example, we would not want to subject an individual of 100% Western African ancestry to the test—shoe-horning the individual into European subgroups would be inaccurate. For our purposes here, we define unadmixed Europeans as those with 171 AIM (AncestryByDNA 2.5) results that cluster in PCA analysis with other recently unadmixed Europeans, such as Northern Europeans, Greeks, Italians, Egyptians, South Asians as opposed to African Americans, Hispanics, Asian-European mixes, and so on (see Figure 6-1 in Chapter 6). If we used a reasonably sized circle around the European cluster we might allow for the relatively old secondary affiliations that characterize many European subpopulations (such as low levels of West African affiliation for Middle Eastern individuals) while excluding most individuals of significant non-European admixture.

Clustering can be accomplished using PCA on vectors of BGA percentages against self-reported ethnicity, or by plotting the BGA percentages in a tetra-hedron plot overlain on plots for various ethnicities containing (on average) high European ancestry, such as South Asians, Hispanics, Puerto Ricans, and Northern Europeans. If the plot for the sample falls clearly within the range of observed plots for a reference sample of unadmixed Europeans, but outside the range encompassing 95% of other clusters, the sample could be said to qualify for EURO 1.0 analysis, which considers their history solely a function of European ancestry. For example, compare the distribution for unadmixed Europeans in Figure 6-11, in Chapter 6, versus the distribution for admixed Europeans in Figure 6-9 (Hispanics) or 6-8 (Indigenous Americans in red and green color who belong to federally recognized tribes). Samples that plot in the region of overlap between the plots in Figure 6-11 and Figures 6-8 and 6-9 would be of ambiguous status—not clear if their Indigenous American ancestry is due to recent admixture or ethnic affiliation. For samples that plot near the European vertex like most of those in Figure 6-11, it would seem more likely that any non-European (i.e., Indigenous American) affiliation is of ethnic origin as opposed to being due to recent admixture, and we might more reasonably conclude that they are of unadmixed European genetic ancestry. Other criteria could also be used, and we can parameterize our criteria so that they are not as loose as in the previous discussion, but note that the concept is fundamentally no different from any other application of Bayes theorem we might use in the field of statistical analysis.

Having typed a sample with AncestryByDNA 2.5 and qualifying it for EURO 1.0 analysis, and then typing with the EURO 1.0 panel, we can again use the results to query a database of digital photographs and self-reported ethnicities or countries of origin corresponding to similarly genotyped and phenotyped

Figure 8-12

Figure 8-12

Individuals retrieved from a EURO 1.0 database with South Asian admixture greater than or equal to 25% (SA ≥ 25%, top) or Middle Eastern admixture greater or equal to 30% (MIDEAS ≥ 30%). Percentages of Northern European, Southeastern European, Middle Eastern, and South Asian admixture, respectively, are shown below each photo. Much larger samples would be needed to determine whether there are any overt physical traits that differ between these two subgroups, or between these subgroups and others in Figures 8-13 and 8-14, though we see a suggestion that hair color in the SA > 25% group may differ from that of individuals of high Northern European admixture in Figure 8-14 (namely, the incidence of lighter colored hair).

samples. In one example, calling Europeans with South Asian (SA) percentages greater than 25% from the database, we retrieved five individuals (see Figure 8-12). Though this sample is low, two of the women have medium dark skin tones and all of the individuals have darker hair color and iris color (brown) compared to the result obtained when the subject query was NOR > 75% and SA levels are 10% or below (see Figure 8-14), where several of the individuals have relatively lighter natural (questionnaire confirmed) hair color. Most people would not necessarily be able to identify by eye the low level of South Asian ancestry for any of these individuals in Figure 8-12, but if these darker pigmentation results extend to larger samples, they would illustrate the value of measuring ancestry with molecular genetics (which is anthropologically meaningful) rather than anecdotal or self- or third-party-reported ethnicity (which is not always anthropologically meaningful).

Retrieving individuals from the database that typed of greater than or equal to 30% MIDEAS ancestry we see a collection of disparate phenotypes. Deleting the two individuals with the lowest level of MIDEAS admixture (the two samples with 30%), we have the return for greater than 35% MIDEAS, and several of these individuals have medium skin tones compared to the NOR > 75% query set (Figure 8-12 versus Figure 8-14). Other elements of facial appearance are harder to put into words; computational anthropometric rulers would be needed, and a larger sample of hits to validate whether this result is informative for any other trait. As we gradually increase the MIDEAS percent of the query, we may very well gradually see darker average skin tone and the gradual accentuation of ethnic features we commonly associate with Middle Eastern ancestry, though again these features are difficult to put into words and anthropometric measurements are needed to validate.

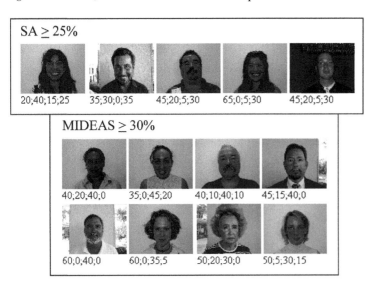

A query of MED greater than or equal to 40% provided 16 hits (see Figure 8-13). Though some might look Mediterranean (i.e., those with darker hair and complexions), there is no standard Mediterranean appearance. Even using subjective notions, it would appear many of the samples do not look any more Southeastern than Northern European, and reminding us of the heterogeneity in Southeastern European group we discussed in Chapter 6 (see Figure 6-14). We might

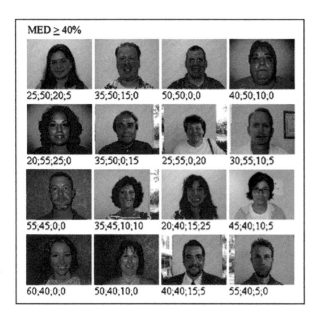

MED ≥ 40%

25;50;20;5   35;50;15;0   50,50,0,0   40,50,10,0

20;55;25;0   35;50;0;15   25,55,0,20   30,55,10,5

55,45,0,0   35,45,10,10   20,40,15,25   45,40,10;5

60,40,0,0   50,40,10,0   40;40;15;5   55;40;5;0

Figure 8-13

Individuals retrieved from a EURO 1.0 database with Southeastern European admixture greater than or equal to 40% (MED ≥ 40%). With such small samples, it is not possible to test for significant phenotype departure from other groups, such as that in Figures 8-13 and 8-15. Note that even if there were any suggestions of disparity between this group and Europeans of lower Southeastern European admixture (see Figure 8-14), putting this into words would be quite difficult.

expect that increasing the threshold percentage for our query would be expected to homogenize or focus the distribution of phenotypes relative to Northern Europeans, if Southeastern European parentals are indeed of different average phenotype than Northern Europeans, though larger databases of photographs than available at the time of this writing would be needed to answer this question. Comparing the high MED group in Figure 8-13 and the low NOR group in Figure 8-15, against the low MED and high NOR group in Figure 8-14, we can see that there is little phenotypically we can say aside from natural hair color seeming lighter for the NOR > 75% group relative to these two (which could be verified with a larger database); other traits such as iris color and skin tone seem to show no difference, though the use of anthropometric rulers in a systematic fashion my reveal some with potential.

These discussions illustrate the importance of database size for making inferences—optimally we would have returns of several hundred subjects to obtain the statistical power necessary to conclude that the difference in a given trait value among different genomic ancestry admixture subpopulations is significant. It also bears noting that samples must be carefully qualified for addition to an anthropometric database. Obviously we want to eliminate (using questionnaires) photographs of individuals wearing colored contact lenses, or dyed hair. Traits subject to environmental influence such as facial skin pigmentation are more difficult to control. Minimally, one would record the date that measures were taken as well as the location (Florida, for example, receiving much more UVR than Pennsylvania). However, for any trait, if our sample is large, we can measure intragroup variability to estimate these environmental

Figure 8-14

Individuals retrieved from a EURO 1.0 database with Northern European admixture greater than or equal to 75% (NOR ≥ 75%). Note that we see a few instances of lighter brown and blond hair color (excluding the elderly or one sample for which it appears hair color has been altered), but the frequency of this phenotype in our small sample of individuals with low NOR admixture in Figure 8-13, and another sample in Figure 8-15 is lower. However, a separate group of individuals with lower NOR percentages in Figure 8-12 (bottom) seems to confound any conclusion about hair color based on NOR percentage alone. Of course, larger database sizes may tease such a relationship but with the sizes shown in these figures, it would not be possible to infer hair color from these levels of sub-European admixture obtained using this particular 320 AIM panel.

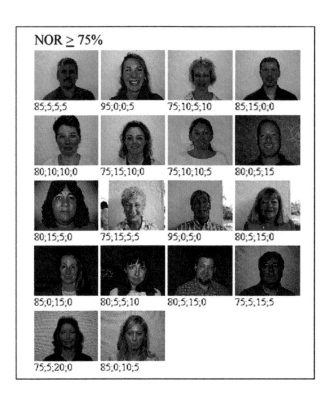

influences, and condition our intergroup comparisons accordingly. As long as the database is reasonably sized, the risk again is that of loss of, rather than the creation of, spurious information.

An important avenue of research is testing the practical ability of observers to take in results from both indirect and direct inference queries and generalize enough to make a correct assessment of likely suspects. The specific question is whether, when provided with a series of photographs (2D or 3D) and some geopolitical or anthropometric summaries, an observer can then select from a lineup those persons that match the group from which their suspect was

Figure 8-15

Individuals retrieved from a EURO 1.0 database with Northern European admixture less than or equal to 30% (NOR ≤ 75%).

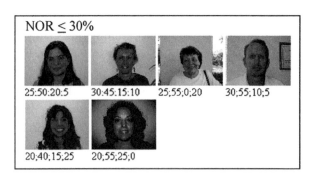

selected. Our discussion throughout this section is focused on specific considerations of single phenotypes in series, one at a time. But this is not how we observe people. We take in many aspects of their appearance simultaneously. It may well be that there is more information in the whole collection of traits than in considering them one at a time (via multivariate analysis, or a complement of separate Bayes analyses).

As discussed in Chapter 1, it is very likely that a number of these traits have evolved in part under the action of sexual selection, and so the human sensorium has coevolved to detect the variability. Even without sexual selection, as most diurnal primates we are highly visual and communicate extensively using facial expression and have the ability to recognize thousands of different people only by their faces (facial recognition). As such it may well be that keen observers can distinguish reflections of the genes controlling variability both within and between populations. We propose that experiments with observers should be included at each stage of molecular photofitting research projects. In some cases the observers are functioning primarily to verify the practicality of the data presentations: for example, that individuals of 75% West African/15% Indigenous American genomic ancestry admixture are distinguishable from those of greater than 50% European admixture. In other cases, the observers are functioning as exploratory probes into the limits of human perceptions regarding superficial cues on identity and relatedness. In either case, properly controlled human interpretation experiments could be designed to test falsifiable hypotheses.

INDIRECT INFERENCE OF SKIN PIGMENTATION

There are a number of traits for which we expect to be able to observe genomic ancestry/phenotype correlations that could be informative in the context of molecular photofitting. Chief among the traits that could be used in this way is skin pigmentation. Skin pigmentation is ideal for such a test since it shows large differences among the primary parental populations for both African-American populations and different Latino groups (viz., Mexican Americans and Puerto Ricans), and as we describe more fully in the next chapter, pigmentation traits are under the control of a relatively small collection of highly penetrant gene variants. The extent of European admixture for many African Americans presented earlier in this book has suggested that proportional West African ancestry might allow us to infer the darkness of skin shade in persons of West African admixture (see Chapter 6, Figure 6-5).

Queries of the ancestry admixture database just discussed also seem to support this notion (see Figure 8-9). For example, in the small sample of Figure 8-9 we can see that if an individual is characterized by greater than 75%

West African admixture, we may be able to infer a darker skin shade than for individuals with less than 33% West African ancestry. However, in Figure 8-9 we can also see clearly, even though we are looking at only a few photographs, that there exists subtle variation in the darkness of the skin within each of the four groups. The phenotypically active loci for most complex traits are distributed throughout the genome. Whether a sufficient quantity of them or the most important ones are of one ancestral origin or another in an admixed individual is a matter of probability. Put another way, whether a person affiliated with group X exhibits traits statistically associated with group X is a function of admixture percentage, probability, and the complexity of the phenotype.

With respect to anthropometric values for multigenic traits, we expect to observe variability among individuals of the same admixture proportions, even though many critics of admixture tests mistakenly point to this type of phenotype variation as evidence of inaccuracy in admixture estimates. We have seen suggestions that a knowledge of genomic ancestry allows for the inference of a *range* of likely anthropometric trait values. For making such inferences, we clearly need to formalize our understanding of the relationship between phenotype and ancestry. The goal of this section of Chapter 8 is to discuss the type of hard data needed to make phenotype inferences using admixture databases. We do this here for skin pigmentation, and then we will discuss human iris color.

Recently published work has summarized the relationship between pigmentation and ancestry in five geographically and culturally defined population samples of mixed ancestry with a wide range of pigmentation and ancestral proportions (Bonilla et al. 2004a; Parra et al. 2004). We present a summary of that work here to illustrate both the potential and some pitfalls associated with the practice of molecular photofitting with indirect methods.

Box 8-A

West African Genomic Ancestry—The Parra et al. Skin Pigmentation Study Methods

Thirty-six AIMs were genotyped either by McSNP (Akey et al. 2001; Ye et al. 2002) or conventional agarose gel electrophoresis. Information regarding primer sequences, polymorphic sites, and population data for a set of AIMs also has been deposited in the dbSNP NCBI database, under the submitter handle PSU-ANTH. Notable is the number of AIMs typed on these samples is much smaller than that used in either the AncestryByDNA 2.0 or 2.5 tests (71 and 171, respectively) we have covered in Chapters 5 and 6, as well as in this chapter up to this point. More information about the AIMs used in this study can be found in (Bonilla et al. 2004a; Parra et al. 1998; Shriver et al. 2003). As one might expect, as we show at the end of this section where the Puerto Rican sample has been assayed for the 171 AIM AncestryByDNA 2.5 test, the number of AIMs used has an important effect on indirect inferences that can be made.

Group admixture levels were calculated with the program ADMIX, written by Jeffrey Long, which implements a weighted least squares method (Long 1991). Individual ancestry estimates were obtained using a maximum likelihood method described in Chapter 3 or alternatively, using the program STRUCTURE 2.0 (Pritchard et al. 2000). The correlation between constitutive skin pigmentation measured in the inner upper arm and individual ancestry estimated with the panel of AIMs was evaluated using the Spearman's rho nonparametric test implemented in the program SPSS (version 10).

The Parra study included a sample of 232 African Americans living in Washington, D.C., a sample of 173 African Caribbeans living in the United Kingdom, a sample of 64 Puerto Rican women living in New York, a sample of 156 individuals from the city of Tlapa, in the state of Guerrero, Mexico, and a sample of 444 Hispanics from San Luis Valley, Colorado. More details about the African-American and African-Caribbean samples can be found in a previous publication (Shriver et al. 2003). Additional information about the Puerto Rican sample from New York, the sample from San Luis Valley, and the Mexican sample from Guerrero have been published, and we refer the interested readers to these manuscripts (Bonilla et al. 2004a, 2004b).

In the African-American, African-Caribbean, Puerto Rican, and Mexican samples, constitutive skin pigmentation was measured on the upper inner side of both arms on each subject, with a DermaSpectrometer (Cortex Technology, Hasund, Denmark), and the melanin content was reported as the Melanin Index (M). In the San Luis Valley sample, constitutive pigmentation was also measured on the upper inner arm with a Photovolt model 575 spectrophotometer (Photovolt Instruments Inc., Minneapolis, MN) and the melanin content was reported as the L^* Index (see Shriver & Parra 2000 for more information about M and L^*).

To estimate the relative West African, European, and Indigenous American ancestry in each individual and in the population samples, Parra of course genotyped at AIMs, but except for the Puerto Rican sample, the AIMs used for this particular study were not the same as those covered in Chapters 5 and 6. For the Puerto Rican sample, this panel of 171 AIMs was used, along with the basic MLE method covered in Chapters 5 and 6. However for the other samples: the number of AIMs characterized was 34 in the African-American and African-Caribbean samples, 36 in the Puerto Rican sample, 24 in the Mexican sample, and 21 in the Hispanic sample from San Luis Valley.

Parra, Bonilla, Shriver, and colleagues recently have studied the relationship between constitutive skin pigmentation and ancestry admixture in five populations. Among European Americans, African Americans from Washington, D.C., and African-Caribbean samples from England, the African-Caribbean sample showed the highest average melanin index (M = 57.8 ± 0.74, Figure 8-16), followed by the African-American sample (M = 53.4 ± 0.63, Figure 8-16), and then the European sample (M about 30) (for methods, see Box 8-A). For comparison, a Mexican sample showed M = 46.1 ± 0.37, and the Puerto Rican sample presented with an average constitutive pigmentation of

Figure 8-16

Constitutive epidermal pigmentation in terms of the melanin index (M) for three separate African populations. Populations are identified with the key, upper right. African Americans (Washington, D.C.) and African Caribbeans (United Kingdom) showed significantly darker skin color than Europeans (Pennsylvania).
From "Skin Pigmentation, Biogeographical Ancestry and Admixture Mapping," by Shriver et al., 2003, Human Genetics, *Vol. 112, No. 4, pp. 387–399. Reprinted with kind permission from Springer Science and Business Media.*

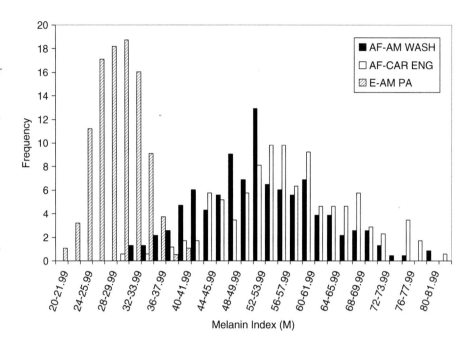

$M = 36.8 \pm 0.75$. The lightness of skin color in Puerto Ricans compared to African-Caribbeans and African Americans is interesting in light of the results we have discussed in Chapter 6, where we showed more pronounced European admixture for a Puerto Rican sample than an African-American sample (compare Figure 6-5 blue versus yellow symbols, Chapter 6). In the Parra study, African-Caribbean and African-American samples showed a much wider range in skin pigmentation than the Mexican and Puerto Rican samples, and clearly distinct in mean values compared to Europeans (see Figure 8-16).

Let us look first at the population averages provided by the lower number of AIMs, where individuals were pooled to their respective populations and estimates made for the population as a whole. The genomic ancestry results for each of the populations (36 AIMs) are shown in Table 8-1 (Bonilla et al. 2004a;

Table 8-1

Average admixture proportions obtained from a small panel of AIMs.

Population	AIMs	% West African	% European	% Indigenous American
African American	34	78.7 ($\pm$1.2)	18.6 ($\pm$1.5)	2.7 ($\pm$1.4)
African Caribbean	34	87.9 ($\pm$1.1)	10.2 ($\pm$1.4)	1.9 ($\pm$1.3)
Puerto Rico	36	29.1 ($\pm$2.3)	53.3 ($\pm$2.8)	17.6 ($\pm$2.4)
Mexican	24	1.3 ($\pm$0.4)	4.2 ($\pm$0.9)	94.5 ($\pm$1.0)
Hispanic	21	3.2 ($\pm$1.5)	62.7 ($\pm$2.1)	34.1 ($\pm$1.5)

Parra et al. (2004) pigmentation study; 36 AIMs.

Parra et al. 2004). We can see that not only are the M values among the samples in this study distinct, but so too are the continental ancestry admixture proportions. The African-American sample in this study revealed about 19% European admixture, not dissimilar to the African-American sample discussed in Chapter 6 that showed about 14% (see Table 6-1, Figure 6-5). The Puerto Rican sample in this study showed only 30% West African admixture and 53% European admixture, similar to the 32% and 55% values obtained for the sample discussed in Chapter 6 (see Table 6-1, Figure 6-5). From analysis on the level of the population, it seems clear that there exists a strong correlation between African admixture and skin melanin index values (M). The next step was to test this hypothesis at the level of the individual.

Parra et al. 2004 estimated the ancestry of each individual using a Maximum Likelihood Estimation (MLE) method. They observed a wide range of individual ancestry values in each of the population samples, reflecting the presence of admixture stratification. The individual admixture values obtained using the program STRUCTURE 2.0 in this study were highly correlated with the MLE values (Spearman's rho values ranged between 0.837 and 0.997, $p < 0.001$).

With individual admixture estimates and M values, it was then possible to search for correlation between admixture and pigmentation among individuals, using regression analysis. Figure 8-17 shows a three-dimensional representation showing the relationship of individual ancestry and constitutive pigmentation values (measured as M-index) for three samples: the African Americans from Washington, the Puerto Ricans from New York, and the Mexicans from Guerrero, Mexico. Individual ancestry is plotted in a triangular format, as described in Chapter 4 (see Figure 4-12), and constitutive pigmentation values are shown in the vertical axis. This plot shows a trend for individuals with greater European and Indigenous American contributions to show lighter skin pigmentation, and shows a clear correlation between M value and West African admixture. The relationship of constitutive skin pigmentation, measured in the upper inner arm, and individual ancestry, estimated with the panel of AIMs, was further explored by this group using bivariate correlation analysis (see Table 8-2).

From the bivariate analysis in this paper a significant positive correlation between melanin index and West African ancestry (see Table 8-2, Figure 8-18) is clear, but so too

Figure 8-17

Three-dimensional triangle plot representing constitutive epidermal pigmentation with respect to European, West African, and Indigenous American admixture from Parra et al. (2004). The melanin content (M) of skin for each sample is shown on the vertical axis. Admixture is represented using the coordinate system described in Chapter 4 (Figure 4-12). Three populations are represented: African Americans (open circles), Puerto Ricans (closed circles), and Mexicans (triangles). From this figure it is clear that M value is correlated with West African admixture in this sample. Reprinted by permission from Macmillan Publishers Ltd: Nature Genetics, Vol. 36, pp. 554–560, 2004.

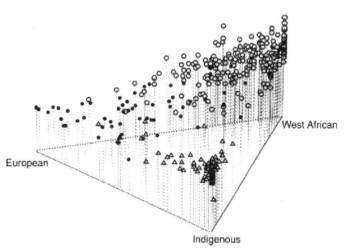

Table 8-2

Relationship of melanin content and individual ancestry in African Americans from Washington, African Caribbeans from Britain, Mexicans from Guerrero, Puerto Ricans from New York, and Hispanics from San Luis Valley.

Sample*	Spearman's rho (95% CI)**	p
Relationship of melanin index (M) and ancestry		
African Americans (W-AF)	R = 0.440 (0.330 to 0.538)	p < 0.001
African Caribbeans (W-AF)	R = 0.375 (0.239 to 0.496)	p < 0.001
Mexicans (I-AM)	R = 0.212 (0.057 to 0.357)	p = 0.008
Puerto Ricans (W-AF)	R = 0.633 (0.457 to 0.761)	p < 0.001
Relationship of skin reflectance (L⁺) and ancestry		
San Luis Valley (I-AM)	R = −0.259 (−0.141 to −0.369)	p < 0.001

*Ancestral proportion axes are indicated by the abbreviations W-AF (West African) and I-AM (Indigenous American).
**Confidence intervals for the correlation estimates are shown in parentheses.
From Parra et al. (2004); 36 AIMs.

Figure 8-18

Relationship between individual ancestry admixture determined using 36 AIMs and constitutive skin pigmentation in a sample of African individuals. Shown is the correlation between constitutive skin pigmentation, measured with the Melanin Index (M), and the proportion of sub-Saharan African ancestry admixture among individuals of three populations. Each symbol represents an individual, and his or her position on the x-axis corresponds to the individual's percent of African admixture. Which population an individual was sampled from is indicated with the key, upper left. We can see a clear trend toward higher M values with greater African admixture. From "Skin Pigmentation, Biogeographical Ancestry and Admixture Mapping," by Shriver et al., 2003, Human Genetics, Vol. 112, No. 4, pp. 387–399. Reprinted with kind permission from Springer Science and Business Media.

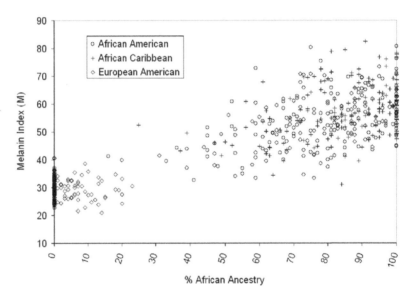

was a weaker correlation with Indigenous American ancestry (see Table 8-2). The correlation was considerably stronger among the Puerto Rican sample (rho = 0.633, p < 0.001), than among the African Americans, African Caribbeans, or Mexicans (rho ranges from 0.212, p = 0.008 in the Mexican sample to 0.440, p < 0.001 in the African American sample). In the Hispanic sample from San Luis Valley, which was measured with the Photovolt reflectometer, there was a significant negative correlation between skin reflectance and Indigenous American ancestry (see Figure 8-19, rho = −0.259, p < 0.001).

However, it is important to note that the strength of the relationship of constitutive pigmentation and ancestry estimated with genetic markers is quite

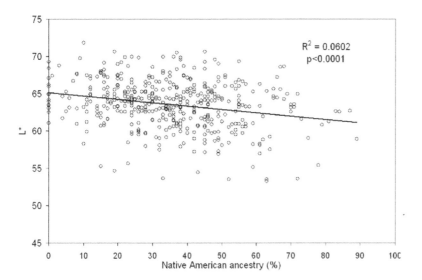

Figure 8-19

Relationship between individual ancestry admixture determined using 36 AIMs and constitutive skin pigmentation in a sample of Mexicans from the San Luis Valley in Southern Colorado. Pigmentation is expressed as a reflectance value L* rather than the melanin index (M) we are accustomed to by this point in the text. The square of the correlation coefficient and its level of significance is indicated in the upper right. The line represents the best fit using linear regression. Halder and Shriver 2003, "Measuring and using admixture to study the genetics of complex diseases." Human Genomics, Vol. 1, No. 1, pp. 52–62. Reprinted with permission from Henry Stewart Publications.

variable in these particular experiments. The correlations between constitutive pigmentation and ancestry estimated with these particular batteries of AIMs can be classified as moderately strong (Puerto Rico, 176 AIMs, rho = 0.633), moderate (African Americans from Washington, rho = 0.440), and weak (African Caribbeans, rho = 0.375; San Luis Valley, rho = -0.259; and Mexico, rho = 0.212). Among the three African populations, the order in association strength is almost the inverse of that obtained when we order the samples based on overall African admixture, and it seems that the power to detect the correlation is weaker the less admixture that exists.

Now we can turn our attention to similar experiments using larger numbers of AIMs—the battery of 171 AIMs discussed in detail in Chapters 4 through 6. A group including the author has recently applied the 171 AIM AncestryByDNA 2.5 panel of markers to the Puerto Rican sample of women from New York City in the Parra et al. (2004) study (Halder 2005; Halder et al. 2006). We obtained a Spearman's rho value of 0.60 (p < 0.0001) for the correlation of M with West African ancestry, a value similar to that which we obtained using the smaller set of 36 AIMs just described (see Table 8-3, Figure 8-20A). The study based on the smaller panel of markers (Bonilla et al. 2004a), though showing significant effects of West African and European ancestry on pigmentation, did not show a significant effect of Indigenous American ancestry on pigmentation level (data not shown). The new analysis based on 171 markers showed the same associations as with the 36 marker panel, but also a highly significant effect of Indigenous American ancestry on pigmentation (rho = -0.424, p < 0.001, Table 8-3, Figure 8-20C).

With the larger panel of AIMs, there is an increase in the accuracy of the genomic ancestry estimates as we have already described in Chapter 5, but here

Table 8-3

Correlation between M index and genomic ancestry components computed using the AncestryByDNA panel of 171 AIMs in a sample of Puerto Rican women.

	IAm	Eu	WAf
Spearman's rho	−0.424	−0.578	0.602
P Value	0.001	< 0.0001	< 0.0001
Pearson's r	−0.444	−0.687	0.745
P Value	< 0.0001	< 0.0001	< 0.0001
r^2	0.197	0.472	0.555

Genomic ancestry components are coded as Indigenous American (IAm), European (Eu), and West African (WAf).

Figure 8-20

Relationship between individual ancestry admixture determined using the 171 AIMs discussed throughout this book, and constitutive skin pigmentation in a sample of Puerto Ricans. Each point within a given plot represents a separate individual. The line represents the best fit using linear regression. Positive correlation between M and West African (A), and negative correlation between European (B), and Indigenous American (C) admixture is apparent. Provided by Indrani Halder, University of Pittsburgh.

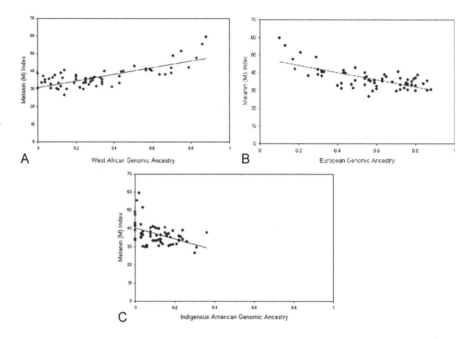

we can see the manifestation of this increased accuracy in terms of our ability to predict a phenotype. How does this increase in precision affect the predictive values of the BGA result for pigmentation level? On examining Figure 8-20A–C, the correlations between skin pigmentation and each of the three ancestry components are immediately evident. Individuals with more West African and Indigenous American ancestry have darker skin and the level of variation at different levels of ancestry is such that there is little overlap in this sample among those who have the highest and the lowest levels of genomic ancestry for these populations.

One way to quantify the dependency of the pigmentation on ancestry is the r^2, or the proportion of the variance in pigmentation explained by the ancestry components, which ranges from 19.7% for the Indigenous American component to 55.5% for the West African component, all three being statistically significant. Since these genomic ancestry components are not independent, we can conservatively say that about 55% of the variance in pigmentation in this sample of Puerto Rican women is explained by individual ancestry.

Before we consider whether these particular results constitute a significant body of evidence by which to infer skin pigmentation based on a prior knowledge of genomic ancestry, let us explore their mechanism and what these correlations might mean.

SOURCES OF THE ANCESTRY-SKIN PIGMENTATION CORRELATION

We observed a significant correlation between constitutive pigmentation and individual ancestry in each of the five admixed samples. The significant correlations between melanin content and individual ancestry estimated can be explained as a result of population structure due to admixture stratification in the samples. Here, we broadly define population structure as the presence of associations between unlinked markers. It is well known that the process of admixture between two populations will create linkage disequilibrium (LD) or nonrandom associations between both linked and unlinked markers, but the linkage disequilibrium between unlinked markers will disappear in just a few generations. LD decays as a function of the recombination rate (θ) between the two markers and the number of generations (n) since the admixture event. This decay in LD can be represented as $D_n = (1 - \theta)^n D_0$, where D_n is the linkage disequilibrium n generations after admixture and D_0 is the initial disequilibrium (Chakraborty & Weiss 1988). For example, after 10 generations of random mating, the LD at unlinked loci will be reduced to 0.1% of the initial level. Thus, we would not expect to find a significant correlation between pigmentation and individual ancestry if the admixture event happened some generations ago and there is random mating in the admixed population. This is another way of saying that with age and random mating, admixture is diluted when there is no new gene flow to maintain it, and we eventually lose our ability to discern phenotype correlations that are a function of this type of ancestry admixture.

In an admixed population, it is possible to envision two factors causing population structure: continuous gene flow and assortative mating. By continuous gene flow, we refer to an admixture model in which there has been an ongoing contribution of one or more parental populations to the admixed population over a period of time extending into the recent past. This creates hidden population structure due to variation in admixture proportions among individuals. Assortative mating can be defined as nonrandom mating according to phenotypic characteristics. If there is assortative mating based on a phenotype (e.g., skin color) or any other factor (e.g., socioeconomic status, education) that is inherited and correlated with ancestry, any population structure originally present in the admixed population will be maintained through the generations.

Although continuous gene flow and assortative mating are very different processes, the end result is similar in that there will be variation of individual ancestry and maintenance of structure from the admixed population. In turn, this will be reflected in significant associations (correlations) between unlinked markers, and potentially, correlations between specific phenotypes and individual ancestry. The results of simulations conducted by Mark Shriver (Pennsylvania State University) and colleagues shows that variation in admixture proportions between individuals has a profound effect on the measured correlations between pigmentation and individual ancestry (Shriver, unpublished observations). A significant correlation is observed in spite of the fact that all the AIMs used to estimate individual ancestry and all the markers determining pigmentation are unlinked.

Given that constitutive pigmentation is relatively independent of environmental factors, significant correlations with individual ancestry are easily explained as the result of genetic factors. It is important to emphasize that for many complex traits and diseases, environmental factors play an important role, and controlling for these factors remains a critical aspect to consider when carrying out this type of analysis (Fernandez et al. 2003). It is possible to envision certain situations in which a positive association between a phenotype and individual ancestry could be misinterpreted as evidence of the presence of genetic factors influencing the phenotype, when in fact the phenotypic differences could be due to environmental factors that covariate with ancestry (Halder & Shriver 2003; Risch et al. 2002). There is no evidence that this is the case for most outwardly appearing traits such as skin melanin content, for which nongenetic mechanisms are more difficult to envision.

It is also important to consider the implications of the wide variation observed in the strength of the correlation between constitutive pigmentation and individual ancestry in these admixed samples. This variability is presumably a reflection of differences in the degree of population structure present in each population and/or the levels of pigmentation differences between the parental populations and the number of genes involved. For example, the strong correlation observed using 171 AIMs in the Puerto Rican sample (see Figure 8-20) seems to indicate that continuous gene flow, assortative mating, or both factors are playing an important role in this population. On the contrary, in the sample from San Luis Valley, the correlation between melanin content and ancestry, although significant, is weak compared to that in the African populations. This is consistent with historical data indicating that the Mexican population is the result of a relatively old admixture event, and independent assortment has greatly decreased the association between unlinked markers created by the admixture process (for more information about the history and admixture dynamics in this population see Bonilla et al. 2004a).

Alternatively, it is also possible that the differences in the extent of the correlation between constitutive pigmentation and ancestry are due in part to admixture histories involving populations with widely different pigmentation levels. The Puerto Rican sample shows significant contributions from three parental groups (Europeans, West Africans, and Indigenous Americans). The African-American and African-Caribbean samples have mainly West African and European contributions, and the samples from Mexico and San Luis Valley show primarily European and Indigenous American ancestry.

The results of this study have implications with respect to the use of pigmentation as a marker of ancestry in admixed populations. Depending on the degree of population structure present in the admixed population, the correlation between constitutive pigmentation and individual ancestry may be strong, weak, or even absent. Recently, Parra et al. (2002) found no correlation of color (described as a multivariate evaluation based on skin pigmentation, hair color and texture, and the shape of nose and lips) and African ancestry (based on 10 AIMs) in an admixed sample from Brazil. Since this negative result was highlighted by many ethicists and journalists as evidence for a lack of correlation between biology and race, it is helpful to understand why this study differed from the Parra et al. (2004) work we have just discussed.

There are a number of important differences in research design and measurement of the phenotype between Parra et al.'s (2002, 2004) study (different investigators), which preclude direct comparison between the two. For example, the former objectively measured pigmentation on a continuous scale using reflectometry, whereas color in the Brazilian paper (Parra et al. 2002) was subjectively assessed by one of several observers and then compressed into three categories (white, intermediate, and black). Parra et al. used only 10 AIMs suitable for only very crude estimates of genomic ancestry, whereas we present herein results using three to 17 times this number. Thus, Parra et al.'s (2002) study simply may have lacked the power necessary to show the associations sought (recall that the correlations between M and admixture improved substantially moving from 30 to 171 AIMs discussed earlier).

These results emphasize the need to be cautious when using negative results from one study to counter positive results in another, especially when dealing with studies of different power. It may also be that there are substantial differences between Brazilian and American/Caribbean populations between the two studies; in contrast to the situation with American and Caribbean populations, it may be that the Indigenous American contribution to average

M value is greater among the Brazilian population and African contribution a minor component.

CAN WE INFER M KNOWING GENOMIC ANCESTRY?

Overall, it is clear from the correlations between skin pigmentation and genomic ancestry admixture we have just presented, that is it possible to measure the latter with precision that is high enough to detect correlation with a trait of disparate character among the parental groups. This is the case when even relatively small numbers of AIMs (e.g., 36) have been used to calculate biogeographical ancestry (though it is really the information content that matters, not the number). Is 55% of the variance explained sufficient to make the indirect method we are discussing useful? This is really the crux of the matter in a consideration of the indirect method and molecular photofitting more generally.

From the plots in Figures 8-17 through 8-20 we can see justification for the notion that the Melanin Index for an individual can be predicted to a certain extent based on their percentage of African admixture. Consider dividing the sample in Figure 8-18 into two groups: those with 25% African admixture and below, and those with 25% African admixture and above. We would be right most of the time if we predicted that an individual belonging to the former of these groups would have an M of 20 to 40, and that an individual belonging to the latter of these groups would have an M of 40 to 85. As can be seen in Figure 8-18, these predictions would not be right 100% of the time, since there is overlap in M value between the two groups, but the overlap is less than 5% of the sample. Though we would be right over 95% of the time, note that there are some individuals for which the inferences would be incorrect. This data was produced using merely 36 AIMs, and at the time of its publication was one of the first direct indications that admixture could potentially serve as a useful tool for the inference of skin color.

Using the 171 AIM panel and data from Puerto Rican women in Figure 8-20, we can see that we might divide the sample into two groups: those with lower M values (say, 38 or lower) and those with higher M values (>38). For samples with less than 50% African admixture, we might infer or predict from this plot an M value of between 25 and 38; for samples with greater than 50% African admixture we would predict an M value between 38 and 60. Again, we would tend to be correct with these methods but there would be some individuals for whom the inferences would be incorrect. Alternatively, using the regression line we might estimate a point value for M and the variance about this line as a measure of confidence, and we would again tend to be correct but with individual anomalies.

For example, if we obtained an admixture estimate of 80% West African and 20% European ancestry for a sample from an unknown donor, we could employ a dataset such as that shown in Figure 8-20, and plot the unknown admixture proportions on the regression line, obtaining an inferred melanin index of about 43 with a range of from 35 to 50 (depending on the confidence intervals associated with the MLE we want to use). If we want to know whether this is significantly darker than the range obtained by plotting a 20% West African and 80% European ancestry result, or a 100% European or 100% West African ancestry result, we could measure various effect statistics for evidence to reject the null hypothesis that the inferred skin melanin content for our unknown sample is not darker than that for the comparison group. The quality of our inference depends not only on the sample size of our database, but on the precision of our admixture estimation method.

For our hypothetical unknown sample of 80% AFR and 20% EUR admixture, the inferred M index range of 35 to 50 from Figure 8-20 is more narrow than that we would obtain from the data in Figure 8-18. The differences could be explained by unequal sample sizes, the fact that Figure 8-20 is of Puerto Ricans (significant Indigenous American admixture) and Figure 8-18 is of African Americans (relatively little Indigenous American admixture), and, most important of all, that 171 AIMs were used for ancestry inference in Figure 8-20 whereas only 36 AIMs were used for the data of Figure 8-18. When making an inference, therefore, it would obviously be wisest to compare apples with apples in terms of population sources and AIM panels. Larger sample sizes are always better than smaller ones, unless one is using data from a large panel to make inference based on an MLE derived from a small panel. In addition, the admixture proportions of a query must be determined with the same method (algorithm) as those of the reference population. As with any scientific endeavor, we need to minimize the number of variables contributing to the system.

At the very least, this work provides a framework of ideas and methods to consider for drawing an inference on skin pigmentation from ancestry. Although we can provide only a range of likely results, our inferences as derived from empirical observations are infinitely more valuable than those based on subjective or anecdotal interpretation (like we would have if we stated that any sample with nonzero African ancestry should correspond to a darker melanin index than one without African ancestry). The anomalies or occasional errors caused by the lack of strict coupling between genetic ancestry and trait value would seem to represent the downside risk associated with inferring trait value indirectly through knowledge merely of ancestry. However even though the method is not foolproof, we can easily appreciate that there is information about M we gain by knowing an individual's genomic ancestry and as long as it is communicated responsibly, this information would be more useful for a forensics investigator as a human-assessed measure of skin color would be, in that the measures are falsifiable. Nonetheless, it is the anomalies we seek to ameliorate as we move from indirect methods to more direct methods in the next chapter.

INFERENCES ON COMPOSITE CHARACTERISTICS

Let us consider the composite of information contained within the indirect method. An important aspect of the database approach that we have defined is that there will be many phenotypes that can be considered together. Thus there may be the ability to compound the information from, for example, skin M index (or observable color in a photograph), hair color, eye color, evident hair texture, and even certain facial features. As we expect that each of these traits will have some level of correlation with ancestry and that (excepting the pigmentation phenotypes) they will be determined by separate sets of genes, we actually have a number of independent physical estimates of the underlying genomic ancestry levels of the person. What we are describing here is really the process by which we as subjective observers recognize physical cues typical of particular populations.

When there is substantial admixture or the parental populations are simply not very divergent, the tendency is to use a holistic assessment of a number of phenotypes. In these cases not all the traits that are on average different between two populations will show up in all members of the populations, but enough of them do so that most persons from one population can be recognized as belonging primarily to one population and not the other, as the case may be. A critical element of the indirect molecular photofitting method discussed in the context of composite features is the communication of results to the investigators. What are the best ways to present these sorts of population summaries to investigation teams? We have discussed some formats earlier in this chapter but others are equally useful; several alternatives are presented in the next section. Though it is possible that composite sketches may be had from admixture database queries, we will have to wait for actual tests with real data sets to know for sure.

WHY NOT USE THE DIRECT METHOD INSTEAD?

What is needed to move forward from the reasonable but imperfect method of indirect inference is the identification of the specific genes underlying pigmentation and the individual scoring of these variants. We describe this superior approach in the next chapter. The problem is the genes and variants underlying most human traits have yet to be identified—indeed, in the next chapter we can discuss in detail only one phenotype for which detailed genetic screens have been fruitful (human iris color). This said, it is reasonable to assume that genomic ancestry estimates will be useful covariates in making estimates of phenotypes like pigmentation even when some of the genes functionally underlying the traits are available. We expect that there will continue to be information in the individual ancestry estimates as long as there are pigmentation genes that have yet to be included in the model for direct inference. It may be for some traits that, due to low penetrance and extreme locus/allelic heterogeneity, there will always be such genes not present in our models. Additionally, given the current evidence for multiple independent origins (phenotypic convergence) for both light and dark pigmentation phenotypes across the world, these genomic ancestry estimates may provide a critical estimate of genetic background giving important context to measurements at pigmentation genes. Indeed what we are discussing, the indirect methods in this chapter as the indirect and in the next chapter the direct methods, will likely need to be combined for effective practical molecular photofitting work.

INDIRECT INFERENCE OF IRIS PIGMENTATION

In Chapter 5 we learned that individual estimates of continental ancestry admixture were associated with iris colors. Levels of West African ancestry greater than or equal to 12%, East Asian ancestry greater than or equal to 16%, and Indigenous American ancestry greater than 25% were associated with darker iris colors on a melanin index scale for iris color less than 2.2, a metric that is explained more fully in the next chapter. We might have expected that iris colors would be associated with ancestry in individuals of polarized (i.e., >80%) ancestry; for example, blue irises would not be expected to be found in West African parentals, or even the average African American, as frequently as in European parentals. However, the illustration of associations with lower levels of ancestry (admixture) in European Americans was a powerful indicator that these estimates of lower levels of ancestry were anthropologically relevant. Let us take a more careful look at this correlation and discuss whether and/or how it may be useful for the inference of human iris color given continental ancestry admixture.

A group led by the author used the 171 AIMs and basic-MLE method discussed throughout this text to determine genomic ancestry admixture for a sample of 695 subjects, almost all of them self-described Caucasians. They then used multiple regression to measure correlation with digitally quantified iris color. Most samples typed of predominantly European admixture (see Figure 8-21A) and European admixture was positively correlated with lighter iris colors (R = 0.28, p < 0.0001; Figure 8-21A). Sub-Saharan African (R = −0.24, p < 0.0001; Figure 8-21B), East Asian (R = −0.1042, p < 0.0059;

Figure 8-21C), and Native American admixture (R = −0.152, p < 0.0001; Figure 8-21D) admixture were correlated with darker iris colors in Caucasians. Since these plots involve Caucasians, most of the samples are clustered at the high-European (EU) end of the plots. Since at high European levels the sample of individuals with iris color scores above the average (C = 2.07) was of similar average genomic admixture as the sample below the average (C = 2.07), we can see that for individuals of low non-European admixture, genomic ancestry estimates are not very useful for inferring iris color.

However, at lower European levels, the situation changes. For example, there are 43 samples with 65% or less European admixture, and all 43 show dark iris color scores at or below C = 2.5 (see Figure 8-21A). There are 31 with 60% or less European admixture, and all 31 show iris color scores at or below 2.25, and 30/31 (97%) show scores 2.0 or below (see Figure 8-21A). Iris color scores at or below 2.0 correspond to those with obvious regions of brown (eumelanin) pigment, and so these results, if extended to larger samples, would indicate that our association between continental genomic admixture using the 171 AIMs and iris color is sufficiently strong to allow for reasonably accurate inferences of iris color based on admixture alone, at least for

Figure 8-21

Relationship between individual ancestry admixture estimates and iris color for a sample of 695 Caucasian subjects. Multiple regression analysis results are shown, and the line represents the best fit to the data. The coefficient of correlation, its confidence interval, and its p-value are shown above each plot. Plots are shown for European (A), African (B), East Asian (C), and Indigenous American (D) admixture. Note that A is a composite of results in plots B to D, and represents the entire data set (all samples are plotted as a function of European and non-European admixture). Many of the samples in B through D are shared.

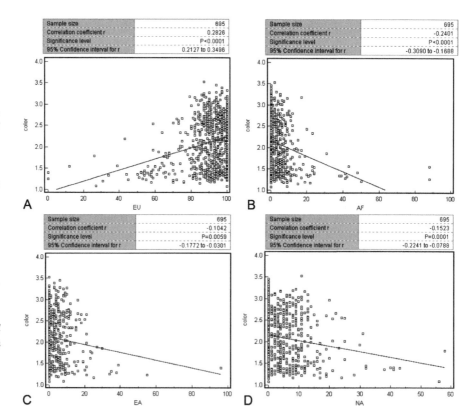

individuals of high non-European admixture. For example, we could use this data to predict that an individual with 60% European/40% sub-Saharan African admixture would have dark (C ≤ 2.0) irides with a high degree of expected accuracy (>95%). As before with the M value, for any particular European/non-European admixture ratios, we could use the regression plot to estimate the expected iris color score value and use the range of scores corresponding to this ratio to estimate the confidence intervals.

The correlation between non-European continental admixture and darker iris color scores with this panel and in this sample appears to be a function of all three elements of non-European admixture: sub-Saharan African, East Asian, and Indigenous American. The correlation between darker irides and sub-Saharan admixture is strongest (R = 0.2401, p < 0.0001; Figure 8-21B). The correlation with East Asian admixture (R = 0.1042, p < 0.0001; Figure 8-21C) and Indigenous (Native) American admixture (R = 0.1523, p < 0.0001; Figure 8-21D) are also significant but not as strong. For an individual of complex admixture we might use the regression plot in Figure 8-21A, but for individuals of European/sub-Saharan African admixture or European/East Asian admixture we would use the plots in Figures 8-21B and 8-21C, respectively. Though we would clearly want to build a larger database, with a larger sample of individuals with low European admixture, the significance of these regression coefficients suggests strongly that this would be a worthwhile endeavor; the p-values associated with these correlation coefficients suggest that the quantitative (or qualitative) inference or iris color given continental admixture (determined with these 171 AIMs) is feasible and would be reasonably accurate.

We can also demonstrate associations between elements of sub-European admixture and iris color—again as we might expect. Specifically, we can use the EURO 1.0 panel discussed in Chapter 7 to demonstrate a correlation between higher levels of South Asian admixture in Europeans and darker iris colors, and between higher levels of Northern European genetic ancestry and lighter iris colors. Table 8-4 shows a contingency table analysis of these results; the two-tailed exact p-values for association between various European subancestry bins were obtained using the EURO 1.0 panel of 320 AIMs and iris melanin index values (C). We can see that the p-values for association with lighter colored irises decrease (increase in significance) as the percentage of Northern European admixture increases. Also, the p-values for association with *darker* colored irises decrease (increase in significance) as the percentage of South Asian admixture increases. The 580 subjects from whom these results were obtained all described themselves as European, and again, we are not discussing merely a trivial observation of significantly different iris colors among different parental populations or individuals of diametrically polarized admixture.

Table 8-4

Association of human iris color with European subancestry. The EURO 1.0 panel of 320 AIMs discussed in the text was used to calculate individual estimates of European subancestry among 580 European individuals. Iris color was graded based on the iris color score discussed in Chapter 9. The sample is broken into 17 bins (rows) based on admixture proportions and p-values for contingency analysis comparing the sample within each bin versus all others not included in the bin was performed. For example, samples with >80% Northern European (NOR) admixture had significantly lighter iris color than samples with <80% NOR admixture (row 4).

	Genetic groups	Admixture level	Color	Exact-p (two tailed)	Total European sample (N)	Sample meeting threshold (n)
1	NOR (N. Euro + Irish)	>50%	Light (color score > 2.1)	0.9301	580	377
2	NOR (N. Euro + Irish)	>60%	Light (color score > 2.1)	0.1977	580	213
3	NOR (N. Euro + Irish)	>70%	Light (color score > 2.1)	0.1181	580	82
4	NOR (N. Euro + Irish)	>80%	Light (color score > 2.1)	0.0003	580	27
5	SE EUR	>5%	Dark (color score <2.1)	0.2621	580	505
6	SE EUR	>10%	Dark (color score <2.1)	0.3018	580	461
7	SE EUR	>15%	Dark (color score <2.1)	0.2084	580	397
8	SE EUR	>20%	Dark (color score <2.1)	0.8003	580	327
9	SE EUR	>25%	Dark (color score <2.1)	1.0000	580	251
10	MIDEAS	>5%	Dark (color score <2.1)	0.986	580	411
11	MIDEAS	>10%	Dark (color score <2.1)	0.3564	580	316
12	MIDEAS	>15%	Dark (color score <2.1)	0.8633	580	221
13	MIDEAS	>20%	Dark (color score <2.1)	0.5574	580	139
14	SA (South Asian)	>10%	Dark (color score <2.1)	0.7893	580	190
15	SA (South Asian)	>15%	Dark (color score <2.1)	0.3949	580	111
16	SA (South Asian)	>20%	Dark (color score <2.1)	0.1495	580	54
17	SA (South Asian)	>25%	Dark (color score <2.1)	0.0203	580	29

Because the frequency of lighter colored irises is significantly greater in individuals of Northern European versus South Asian ancestry, these results validate the anthropological validity of the estimates of lower levels of sub-European genomic ancestry admixture as discussed in Chapter 7. More importantly for our discussion here, they indicate that even though continental ancestry-based predictions of iris color cannot be predicted for individuals of polarized European admixture (see Figure 8-21A), we can expect to predict iris color if their proportion of Northern European or South Asian admixture is high.

Variation in iris colors corresponding to a given EURO 1.0 admixture ratio are not debilitating. Inverting the observed frequencies of light and dark irises in each of numerous discrete ancestry admixture bins, we could form likelihood statements just as we do when using allele frequencies to estimate admixture in the first place. Alternatively, and more correctly, we could treat iris color and BGA as continuous variables as we have done in Figure 8-18, and estimate expected values and confidence intervals for unknown samples using MLE point estimates and the two-fold, five-fold, or 10-fold confidence intervals as queries.

We have thus shown that the indirect method based on an empirical approach is feasible for predicting phenotypes given ancestry admixture for certain phenotypes. With the indirect method we obviously are using ancestry as a surrogate variable for the things we really would prefer to measure directly. Likely we would do a much better job (i.e., more precise and more accurate) if we were able to measure the particular gene variants underlying variability in iris color, which brings us to the next chapter. However it bears stressing that not every anthropometric phenotype will be reducible to a manageable number of predictive, phenotypically active loci, perhaps because they are determined by multiple low-penetrance alleles or simply because we have not yet performed the necessary genome association analyses needed to identify them. The work discussed in this chapter shows that for some of these phenotypes, the indirect method of trait value inference will be a feasible approach, useful to the forensic scientist attempting to learn about physical appearance from DNA.

CHAPTER 9

DIRECT METHOD OF PHENOTYPE INFERENCE

Until now we have considered only indirect methods of inferring anthropometric trait values, using ancestry as a surrogate for the phenotypically active loci that underlie their expression. Since these loci are themselves distributed as a function of ancestry, this is not an unreasonable thing to do, as we have seen, but it is always desirable to use information that directly relates to the problem we are attempting to solve. Unfortunately we do not understand the genetic architecture of most human traits to the point that we could infer trait value from knowledge of DNA sequence and/or environment. The genetic and environmental mechanisms for complex human phenotypes such as behavior/psychiatric conditions, cardiovascular disease, and cancer are exceedingly complex and may involve so many loci, each with such small effects, that it would take sample sizes in the hundreds of thousands or millions in order to detect their associations. However, it happens that the outwardly apparent anthropometric traits—those distributed as a function of ancestry—tend to be *relatively* simple on a genetic level. Presumably this is because social and environmental genetic forces such as founder effects, genetic drift, natural selection, and sexual selection were very powerful in the last 50,000 years of our evolution, acting most rapidly on outwardly visible phenotypes such as skin pigment level, hair color, height, and so on (rather than less visible phenotypes such as cardiovascular function, neural function, etc.).

For example, the expression of the pigment eumelanin in various human tissues is a function of a relatively small number of human pigmentation genes, some of which tend to be connected to expression with high penetrance. As we will soon see, variation in iris colors among Europeans appears to have been facilitated by the generation of rapid and extensive variation in one pigmentation gene in particular. Of course most geneticists recognize that the expression of nonoutwardly visible traits such as disease predisposition or intelligence can never be reduced to single genes or even small collections, which is precisely why we know so little of them at the genetic level. For the forensics professional, it is the outwardly apparent phenotypes that are of interest; it is therefore indeed fortunate that many (though of course not all) of these will be amenable to modern genetics study. That is, for these types of traits, we stand a reasonable chance of detecting associations between polymorphisms and phenotypes because there are fewer underlying loci and the penetrance of the relevant loci is greater.

For our purposes, we want to identify polymorphisms in the most penetrant of the underlying loci for a given anthropometric trait, so that we can form a probability statement on the expression of that trait given the character of the polymorphisms. Once we have identified associations between genes and traits, associated certain polymorphisms in those genes with certain trait values, and validated inferential methods for predicting an individual's trait value given the sequence of polymorphisms in these loci, we gain the ability to infer trait value directly. The value of the indirect method is that it provides some level of phenotypic information when none was previously available. The advantage of the direct method is that it provides higher quality information, but it is not always applicable and far more difficult and expensive to harness for a given trait. The indirect method is always inferior to the direct method if the latter is practicable, because even for anthropometric traits, which are distributed as a function of ancestry, there is usually significant trait variation within each population. With the indirect method we are attempting to project the qualitative and/or quantitative physical traits of an individual through a prism of their individual ancestry rather than reading the sequence of the genes that underlie the trait (presumably because they are unknown). This method is expected to work reasonably well for traits of values distributed as a function of ancestry because the sequence variants underlying expression of the traits are correlated with ancestry (and hence AIM alleles).

The ancestry/trait correlation is also principally dependent on the characteristics of ancestry stratification within and among the populations in the catchment area, and so ideally represented in the database. The data so far presented (and case studies to be discussed) shows that this works quite well if the chosen trait is appropriate for the population model, if our expectations on precision are reasonable and when adequate databases are used. Nevertheless, even when the indirect method is appropriate, the inferences are far from perfect, as evidenced by the fact that they are usually expressed in terms of relatively broad ranges. It is impossible to pinpoint trait value from admixture percentages alone, and we would never achieve the accuracy, sensitivity, and positive/negative predictive power we would expect to achieve if we were measuring the phenotypically active loci directly.

Given the imprecision that is inherent in the indirect method, indirect methods and data are very much an intermediate point in the development of molecular photofitting tests. Indeed, the findings from research and development on indirect tests is not just of use to convey ancestry information or rough physical prediction, but also represents a very important research finding that helps us in our effort to develop more direct methods of gene-based inference. Some important questions can be addressed that direct the next phase of the research: finding the genes. Some of these questions are:

Does trait X vary with ancestry?

Does trait Y show a higher correlation with ancestry than related trait X?

Is the correlation in ancestry for trait X higher in persons belonging to the admixed population with parental populations A and B or those belonging to the admixed population with parental populations A and C?.

These questions are useful for planning admixture mapping methods for honing in on genes that cause phenotypes. Another important question depends on accurate admixture estimation:

Are gene variants we have identified in our case-control study as associated with a phenotype really associated with the phenotype, or are they merely good AIMs themselves?

In fact there are a number of important hypotheses that can be tested to pave the way for gene identification using admixture mapping and related methods for conditioning our genome screening analyses on individual admixture estimates.

This is apparent from the data discussed already in Figures 8-18 and 8-20 of Chapter 8, but one can more visually appreciate the inexact nature of the inferences from Figure 5-9 (different samples than those used to construct Figures 8-18 and 8-20). The small samples in Figure 8-9 suggest that knowing that an individual is greater than 75% West African allows us to infer a darker skin pigmentation than if an individual is less than 33% West African ancestry, but we can clearly see subtle variation in the darkness of the skin within each of the four groups. We wouldn't realize that there was indeed an association between African admixture and M value, and that an admixture mapping approach may be suitable for identifying pigmentation genes, unless we made the necessary measurements as in Figures 8-18 and 8-20.

Aside from its use as an intermediate step toward identifying the genes of interest, as we have seen, this type of data has immediate application in the field of forensics. Although a reasonable forensics investigator might feel that being able to narrow the likely skin shade down to broad intervals described with relative terms such as dark, medium, and such is good enough, and although such knowledge is better than nothing (assuming that the inference was made in a statistically responsible manner and the database was large enough), it is not as good as it could be and it is only part of the way to where we would like to go. With the direct method, we would hope to achieve greater precision when predicting trait values because we would be reading the phenotypically relevant sequences themselves, not proxies, covariates, or surrogates.

If we could identify and measure the genes that specifically underlie variable skin pigmentation within and between populations, we could provide a direct, genetically meaningful inference that would always be more accurate than the indirect inference. However, the direct method of phenotype inference is much more difficult to achieve since complex and expensive genetics screens must be carried out and since there are a number of mathematical and practical impediments to carrying out these screens in a comprehensive manner. In this chapter, we will discuss the state-of-the-art in research aimed at developing more direct methods of inferring human physical trait values for iris color, and progress in the fields of skin pigmentation and hair color genetics.

PIGMENTATION

The most obvious anthropometric trait with which the direct method of inference would seem possible is pigmentation, because pigmentation is easily measured in a physiologically meaningful way, the heritability of pigmentation is relatively high, and because decades of genetic research has shown that although human pigmentation is a complex genetic trait, most overt human pigmentation traits can be explained by a relatively small number of pigmentation genes (Akey et al. 2001; Bito et al. 1997; Box et al. 1997, 2001; Brauer & Chopra 1978; Sturm 1998; Sturm et al. 2001). Before we discuss how accurate inferences can be made for pigmentation traits, we will form a foundation by describing the biology and genetics in some detail. Until now we have considered human pigmentation in broad, lay terms and we have referred to a quantitative measure called the melanin (M) index. In this section, we describe the cell and molecular biology of pigmentation and then describe how pigmentation is measured (phenotyped) and predicted through appreciation of pigment gene sequence variants.

Melanin is the cytochrome pigment whose light absorption and scattering properties impart color to the human pupil, iris, skin, and hair. Melanin is produced in specialized cells called melanocytes that are derived from the melanoblasts, which originate from the neural crest in the second month of embryonic life. Melanoblasts then migrate to the dermis and hair follicles, where they differentiate into melanocytes. In the skin, melanocytes are located on the basement membrane at the epidermal-dermal junction, and in the hair they occur in the bulb above the dermal papilla. In both skin and hair, melanocytes are associated closely with keratinocytes, to which they transfer melanosomes, the cytoplasmic organelles where melanin synthesis occurs (reviewed in Sturm 2002). The number, size, and distribution of melanosomes, as well as the rate of melanosome formation and transfer play important roles in determining skin and hair pigmentation.

The melanocytes have long dendritic processes that extend to and interdigitate with keratinocytes in the skin (Figure 9-1). The melanosomes initially are clustered in the perinuclear space around the melanocyte nucleus but then become distributed along a myosin framework leading to these dendritic processes and eventually, the keratinocytes. Keratinocytes occupy two skin layers, the basal layer and the spinous layer. The keratinocytes absorb these vesicular particles, and upon doing so, complete their differentiation and

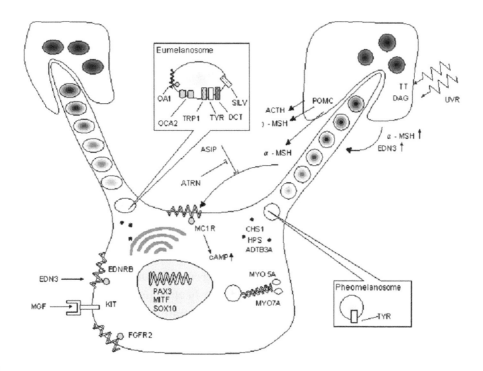

Figure 9-1

Schematic diagram of a melanocyte within which human pigment is synthesized. Gene transcription in the nucleus and translation in the cytoplasm results in the deposition of the pigments eumelanin and pheomelanin into organelles called melanosomes. Two types of melanosomes are shown—one containing mainly eumelanin (left) and another containing the reddish brown pigment pheomelanin (right). In the epidermis of human skin, the melanosomes are transported along dendritic processes and transferred to keratinocytes (upper left and upper right). The various genes involved in these two pigmentation pathways are identified. Receptors and channel proteins are shown on the lipid bilayer of the melanocyte and melanosomes and structural molecules and other signal transduction molecules in the cytoplasm. Messengers that operate between cells are shown outside of the melanocyte, interacting with their preferred receptors. Putative transcription factors are shown inside the nucleus. The helix inside the nucleus represents autosomal DNA and the hemispheric lines above the nucleus represent the endoplasmic reticulum where protein translation occurs. Successive darkening of melanosomes along the dendrite represents organelle maturation. UVR = ultraviolet radiation. Provided by Heather Norton and Mark Shriver, the Pennsylvania State University.

migration to the superficial layers of the skin at the epithelial surface. The production of melanin in the melanosome accumulates as the melanosome completes its migration within the dendrites (shown with successive darkening of the circles along the dendrites in Figure 9-1). Melanin is produced in two forms: pheomelanin, which is the yellow-red pigment and eumelanin, which is the brown pigment. Both types are derived from a common precursor (dopaquinone; see Figure 9-2) and are generally found as polymers of basic repeating units. They are biologically related despite the difference in their spectral properties and structural dissimilarity—both arise from tyrosine and dopa, as well as their conjugate dopaquinone.

Eumelanin is produced in melanosomes called eumelanosomes and pheomelanin in melanosomes called pheomelanosomes. The two varieties of melanosomes have different shapes and internal structures. Eumelanosomes are larger, elliptical, and with a more ordered glycoprotein matrix and fewer cysteine crosslinks (left dendrite in Figure 9-1), whereas pheomelanosomes are smaller, circular, and contain a less ordered glycoprotein matrix with more cysteine crosslinks (right dendrite in Figure 9-1; Nordlund et al. 1989; Sturm 2002). Within the melanosomes, the tyrosinase (*TYR*) gene product catalyzes the rate-limiting hydroxylation of tyrosine to 3,4-dihydroxyphenylanine, or DOPA, and the resulting product is oxidized to DOPAquinone to form the precursor for eumelanin synthesis.

Melanosomes are transported into the keratinocytes in the skin and hair, but in the eye the melanosomes are retained in the melanocytes themselves. In the skin and the hair, it is the light absorbing and scattering properties of these absorbed melanosomes within the keratinocytes that impart colors and shades we are accustomed to viewing in a lighted environment. However, the difference in colors is not due to differences in the number of melanocytes. At the cellular level, variable iris color in healthy humans is the result of the differential deposition of melanin pigment granules within a fixed number of stromal melanocytes in the iris (Imesch et al. 1997). The density of granules appears to reach genetically determined levels by early childhood and usually remains constant throughout later life (see Bito et al. 1997). Thus, variation in pigmentation is caused by differences in the size and distribution of melanocytes but also, the integrity of the pigment molecule itself that is held within.

There are two main protective functions served by melanin—the absorption of ultraviolet radiation (UVR) (particularly the harmful UV-B rays of wavelength 290–315 nm) and the inactivation of free radicals that are formed by UVR and other chemical processes (reviewed in Robbins 1991). UVR causes the photodegradation of nutrients such as flavins, carotenoids, and folate in the skin, and UVR exposure is a well-known risk factor for skin cancers such as melanoma, basal cell carcinoma, and squamous cell carcinoma since mutations to the DNA accompany the painful burns to which some of us are susceptible.

However, UVR exposure is required for vitamin D synthesis, which is important for individuals who don't ingest enough vitamin D through diet (reviewed in Jobling et al. 2004). Dietary vitamin D is found in foods such as egg yolks, liver, and fish oils. The synthesis of vitamin D occurs through UVR exposure to 7-dehydrocholesterol, which is a steroid precursor. Vitamin D itself is a steroid responsible for calcium and phosphate uptake regulation in the small intestine, which is relevant for bone growth and maintenance. People deficient of vitamin D cannot adequately maintain bone mineral density—such as with the childhood disease rickets and osteomalacia. Bone deformities develop, especially in weight-bearing regions in the legs and pelvis, and standing or sitting upright is painful. Aside from the obvious selective disadvantage posed by broken/deformed limbs, deformity of the pelvis could interfere with childbirth, and so the inability to synthesize vitamin D in areas with vitamin D deficient diets could also represent a reproductive disadvantage. Recently there has been a focus in the literature on the effects that insufficient amounts of vitamin D may be having in increasing the risk of several types of cancer including prostate, kidney, and breast. It is presumably due in large part to the balance of physiological needs for vitamin D catabolism and protection from UVR that human skin pigmentation is so unevenly spread throughout the globe (sexual selection could be expected to act secondarily to overall fitness and health; Figure 8-1, Chapter 8).

HISTORY OF PIGMENTATION RESEARCH

In the 1920s, Raper defined the tyrosine-tyrosinase reaction to dopaquinone as a fundamental step in the biosynthesis of eumelanin, and the establishment of the tyrosine pathway to eumelanin was put on firm footing by Mason, who focused on the polymerization of 5,6-dihydroxyindole (DHI) (Mason 1967; for a good review of the history of pigmentation see Prota 2000). The idea that there were pigments other than brown eumelanin can be traced all the way back to the 1870s (Sorby 1878), but the chemical properties of the red pigments were not characterized until the 1940s, by Flesch and Rothman (1945), who observed chemical degradation intermediates from human hair samples.

The pheomelanin biochemical pathway was discovered in the 1960s, using UV spectrophotometry and rudimentary NMR and mass spectrometry of pigments called trichosiderins, which occur in red hairs and hen feathers, and which differ substantially from the darker eumelanin in color (from yellow to reddish brown) and soluability in alkali. Most of the progress in understanding the differences between the melanin and pheomelanin pathway occurred in

Figure 9-2

The anabolism of dark pigment (eumelanin) and red/brown pigment (pheomelanin). Names are given below each intermediate and arrows represent progression along the anabolic pathway. Where gene products catalyze a reaction, the gene symbol is indicated in uppercase bold (TYR, DCT, TYR, TYRP1) along with an arrow indicating the reaction catalyzed. Reprinted from Trends in Genetics, *Vol. 20, by Sturm and Frudakis, "Eye Colour," pp. 227–332. Copyright 2004, with permission from Elsevier.*

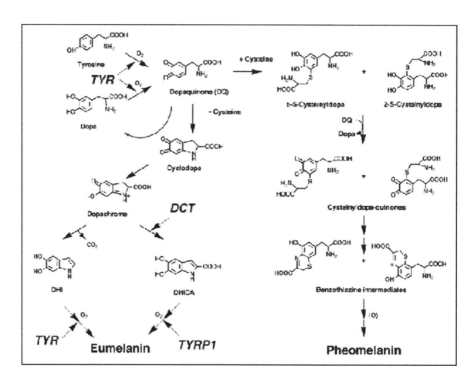

the 1960s, when it was realized by a scientist by the name of Giuseppe Prota that pheomelanins contained relatively large amounts of sulphur, and that cysteine was a fundamental precursor for pheomelanin biosynthesis (Prota & Nicolaus 1967). In the 1970s, anabolic intermediates of pheomelanin, the cysteinyldopas were characterized, and the function of tyrosinase in the synthesis of pheomelanin was established (Ito & Prota 1977). The isolation of the first melanochromes occurred in the 1980s, and the work of Giuseppe Prota was largely responsible for characterizing the anabolic intermediates and biochemical steps part of eumelanin and pheomelanin biosynthesis summarized in Figure 9-2.

THE GENETICS OF HUMAN PIGMENTATION— A COMPLEX PUZZLE

The biochemistry of eumelanin and pheomelanin have been characterized by Prota and his colleagues over the past several decades, but most of what we have learned about the genetics of pigmentation has been derived from molecular genetics studies of rare pigmentation defects in man and model systems such as *Mus musculus* (house mouse) and *Drosophila melanogaster* (common fruit fly). Human pedigree studies in the mid-seventies suggested iris color variation is a function of two loci: a single locus responsible for depigmentation of the iris, not affecting skin or hair, and another pleiotropic gene for reduction of pigment in all tissues (Brues 1975). For example, dissection of the oculocutaneous albinism (OCA) trait in humans has shown that many pigmentation defects are due to lesions in the *TYR* gene, resulting in their designation as tyrosinase (*TYR*) negative OCAs (Oetting & King 1991, 1992, 1993, 1999; see Albinism database, as of this writing located at www.cbc.umn.edu/tad/). Indeed, there are more than two dozen gene products (according to some authors, up to 40 products) thought to be involved in the production, distribution, and metabolism of human melanin. These products function at the level of substrate-availability (tyrosine and DOPA levels), transcription of the gene products, deposition, receptor-ligand interactions involved in the signal transduction pathways, and the migratory behavior of melanosomes (see Figure 9-1).

Since *TYR* is present in both eumelanosomes and phaeomelanosomes, and catalyzes the rate-limiting step of melanin biosynthesis, it is of some interest that the degree to which human irides, skin, and hair are pigmented correlates well with the amplitude of *TYR* message levels (Lindsey et al. 2001). In eumelanosomes, other *TYR*-like proteins are also present including *TYRP1* and *DCT*, which are absent in the phaemelanocyte and thought to be involved in eumelanin but not phaeomelanin production (see Figure 9-1). The *OCA2* gene product is present in both types of melanosomes, and is thought to be required for establishing the proper pH conditions within the lumen of the melanosome (Ancans et al. 2001; Puri et al. 2000). *MC1R* encodes a seven-pass G-protein coupled transmembrane receptor that interacts with the proopiomelanocortin (*POMC*) derived peptide hormones, including α-melanocyte-stimulating hormone (αMSH) and adrenocorticotrophic hormone (ACTH). Upon binding peptide hormone, the MC1R is thought to regulate the pH mediated switching of eumelanogenesis and pheomelanogenesis by altering the levels of cAMP (discussed more later). Keratinocytes that absorb the melanosomes are thought to engage in a feedback communication through the production of these peptide hormones.

The transcription, and ultimately the expression of each of these genes is under the control of the microphthalmia-associated transcription factor (*MITF*), and the transportation of the melanosomes along the dendritic processes are mediated by myosins such as *MYO5A* (myosin 5A) and *AP3D1* (beta-adaptin 3D1).

Though research on pigment mutants has made clear that a small subset of genes is largely responsible for catastrophic pigmentation defects in mice and man (oculocutaneous albinism, or OCA), until recently it was unclear whether or how common single nucleotide polymorphisms (SNPs) in these genes contribute toward (or are linked to) natural variation in eumelanin/pheome-lanin and pigmentation phenotypes. Early genetics research on human pigmentation focused on linkage scanning and candidate gene association methods. With iris color, for instance, a brown-iris locus was localized to an interval containing the *OCA2* and *MYO5A* genes (Eiberg & Mohr 1996), and for hair color, specific polymorphisms in the *MC1R* gene have been shown to be associated with red hair and blue iris color in relatively isolated populations (Flanagan et al. 2000; Koppula et al. 1997; Robbins et al. 1993; Schioth et al. 1999; Smith et al. 1998; Valverde et al. 1995). An *ASIP* polymorphism is reported to be associated with both brown iris and hair color (Kanetsky et al. 2002).

However, the penetrance of each of these alleles appears to be low and in general, they appear to explain only a very small amount of the overall variation in iris colors within the human population (Spritz et al. 1995). Indeed, until recently single gene studies have not provided a sound basis for understanding the complex genetics of any human pigmentation trait. Because most human traits have complex genetic origins, and are qualitatively and quantitatively complex (wherein the wholes are oftentimes greater than the sum of their parts), innovative genomics-based study designs and analytical methods for screening genetic data *in silico* are needed that are respectful of genetic complexity—for example, the multifactorial and/or phase known components of dominance and epistatic genetic variance. The first step however is to define the complement of loci that on a sequence level explain variance in trait value, and of these, those that do so in a marginal or penetrant sense will be the easiest to find.

The abundance of pigmentation gene candidates from studies of mouse and human albino mutants provides a starting point from which we can begin to dissect variation in natural pigmentation phenotypes. However studies focused on these genes until recently had not borne much fruit, highlighting the complexity of the trait and differences in mechanism in different tissues, rather than providing simple, easy-to-understand puzzle pieces that could be generically applied. For example, though *TYR* is the rate-limiting step in melanin production, the complexity of OCA phenotypes has illustrated that *TYR* is not the only gene involved in pigmentation (Lee et al. 1994). Though most TYR-negative OCA patients are completely depigmented, dark iris albino mice (C44H) and their human type IB oculocutaneous counterparts exhibit a lack of pigment in all tissues except for the iris (Schmidt & Beermann 1994). Study of a number of other TYR-positive OCA phenotypes have shown that, in addition to *TYR*, the oculocutaneous 2 (*OCA2*) (Durham-Pierre et al. 1994, 1996; Gardner et al. 1992; Hamabe et al. 1991), tyrosinase-like protein (*TYRP1*) (Abbott et al. 1991; Boissy et al. 1996; Chintamaneni et al. 1991), melanocortin receptor (*MC1R*) (Flanagan et al. 2000; Robbins et al. 1993;

Smith et al. 1998), and adaptin 3B (*AP3B1*) loci (Ooi et al. 1997), as well as other genes (reviewed by Sturm et al. 2001) are necessary for normal human iris pigmentation.

The situation is equally complex for hair and skin pigmentation. For each of these three tissue types across a wide variety of mammals, *TYR* analogues are centrally important, but pigmentation in animals is not simply a Mendelian function of *TYR* or any other single protein product or gene sequence. In fact, study of the transmission genetics for pigmentation traits in man and various model systems suggests that variable pigmentation is a function of multiple, heritable factors whose interactions appear to be quite complex (Akey et al. 2001; Bito et al. 1997; Box et al. 1997, 2001a; Brauer & Chopra 1978; Sturm et al. 2001). For example, unlike human hair color (Sturm et al. 2001), there appears to be only a minor dominance component for mammalian iris color determination (Brauer & Chopra 1978), and there exists minimal correlation between skin, hair, and iris color within or between individuals of a given population. In contrast, between-population comparisons show good concordance; populations with darker average iris color also tend to exhibit darker average skin tones and hair colors.

These observations suggest that the genetic determinants for pigmentation in the various tissues are distinct, and that these determinants have been subject to a common set of systematic and evolutionary forces that have shaped their distribution in the world populations. In *Drosophila*, iris pigmentation defects have been ascribed to mutations in over 85 loci contributing to a variety of cellular processes in melanocytes (Lloyd et al. 1998; Ooi et al. 1997) but mouse studies have suggested that about 14 genes preferentially affect pigmentation in vertebrates (reviewed in Strum 2001), and that disparate regions of the *TYR* and other *OCA* genes are functionally distinct for determining the pigmentation in different tissues.

BIOCHEMICAL METHODS OF QUANTIFYING PIGMENT

One method of quantifying human pigment is a direct biochemical analysis. This is accomplished through an analysis of the degradation products of eumelanin and pheomelanin by High Performance Liquid Chromatography (HPLC) (Jimbow et al. 1991). The method measures the concentration of a chemical called pyrrole-2,3,5-tricarboxylic acid (PTCA), which is a major metabolite of eumelanin resulting from permanganate oxidation. Hydriodic acid hydrolysis of pheomelanins yield amino-hydroxyphenylalanine (AHP) as the major metabolite (Ortonne & Prota 1993). Thus, by measuring the amount of PTCA yielded from a sample in the presence of permanganate, and the level of AHP in the sample in the presence of hydriodic acid, we have a means by which to quantify eumelanin and pheomelanin, respectively. Unfortunately, this technique is rather expensive, slow, and it consumes substantial material so it is not usually practical for large-scale studies. For these, less invasive, quicker, and more convenient methods have been developed.

SPECTROSCOPIC MEASUREMENT OF PIGMENTATION

The levels of skin and hair pigmentation can be objectively measured using a broad class of methods collectively termed *reflectometry* or *reflectance spectroscopy*. Basically, as the name suggests, reflectometry is the controlled illumination of an object and precise measurement of the light that is reflected. There are a number of instruments and methods that have been developed and used for studies of skin and hair using reflectometry. To develop methods for the inference of pigmentation traits, whether through direct or indirect methods, we require the ability to make objective and physiologically meaningful readings of pigment content and spectrophotometric methods generally fit the bill.

Edwards and Duntley were the first to use reflectometry to study skin in 1939. These investigators used an instrument called the Hardy recording spectrophotometer, which used slits and prisms to scan through the visible spectrum (from 400 to 700 nm) and measure how much light is reflected at the different wavelengths. In this way the color of skin and other objects could be represented as a spectral reflectance curve, which is the most complete data on color, other systems being a summary of the spectral reflectance curve. However, the Hardy machine was cumbersome and not well suited to population studies.

In the early 1950s, another reflectometer became available that was much more portable and was used extensively by anthropologists to measure skin pigment level. This instrument was called the E.E.L. (Evans Electroselenium Ltd.) reflectance spectrophotometer and used a series of nine filters to create light of different wavelengths. A profile could thus be generated of the reflectance of skin at a number of different wavelengths. If all nine filters are used the results are equivalent to an abbreviated spectral reflectance curve.

A similar instrument manufactured by Photovolt Corporation also has been used extensively in studies of skin reflectance, particularly studies of New World populations. These early Photovolt reflectance spectrophotometers were similar to the E.E.L. machine in design using six instead of nine filters. One important difference is that the Photovolt instrument had a series of tristimulus filters that allowed for conversions to an international color scale

(CIELab; Hunter Labs, 1996). In the CIELab format, a particular color is designated by three values, L* (white–black), a* (red–green), and b* (yellow–blue). Colors can then be plotted on three-dimensional graphs where white, red, and yellow are positive and black, green, and blue are negative values (see Figure 9-3). When a and b are transformed to the polar coordinate system, the results are the hue angle (h°) and the chromaticity (C), which correspond to perceived color and the intensity of that color, respectively. Although there are several advantages of the CIELab system, including internationally accepted standards of color description and methods for the determination of colors in a fashion analogous to the human eye, this color system had not been used for anthropological or genetic studies of pigmentation until the recently published study of Shriver and Parra (2000). We will see vivid examples of the advantage we gain by using the international color scale when we discuss the measurement of iris colors later in this chapter.

In addition to tristimulus colorimitry, there are two other reflectance spectrophotometric methods that are currently being used in dermatalogic studies of skin pigments. The first method is commonly referred to as narrow-band spectroscopy and is based on work by Dr. Brian Diffey and his colleagues (Diffey et al. 1984; Farr & Diffey 1984). The primary pigments visible in the skin are hemoglobin (oxyhemoglobin and deoxyhemoglobin) contained in red blood cells in the capillaries of the dermis and melanin in the keratinocytes and melanocytes of the epidermis. Diffey and Farr (Diffey et al. 1984) recognized that hemoglobin absorbs throughout the visible spectrum, except in the longer wavelengths, which is why blood is red. Melanin absorbs more evenly across the spectrum. Thus, a reflectance reading at 650 nM is sensitive to the concentration of melanin in a sample and relatively insensitive to the hemoglobin concentration. Alternatively, the level of reflectance at 550 nM is due to absorbance of light by both hemoglobin and melanin. By measuring at narrow regions around these two wavelengths Diffy and Farr computed the melanin index (M) as $100 \times \log(1/(\%$ reflectance at 650 nM$))$ and the erythema index (E) as $100 \times \log[(\%$ reflectance at 550 nM$)/(\%$ reflectance at 650 nM$)]$ (see Figure 9-4). Handheld versions of narrow-band spectrophotometers are available from a number of sources including, Cortex, Inc. located in Hasund, Denmark, who have developed the DermaSpectrometer.

The third type of spectrophotometric method, diffuse reflectance spectroscopy (DRS), is based on having reflectance readings from throughout the visible spectrum. Instruments of this type are commonly called true spectrophotometers and are smaller portable descendents of the Hardy recording spectrophotometer mentioned earlier. Technological developments such as

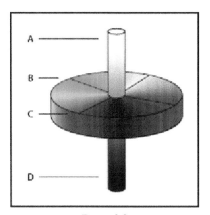

Figure 9-3

CIE color wheel for standardized quantification of colors, hues, and shades. Our goal is to position a pigmented tissue near the wheel where the position indicates the color and luminosity. Progressing around the wheel from green to red, red to blue, and from blue to green color. The 2D projection of the iris on the color wheel can be determined by quantifying the relative level of green, blue, and red relative to the values for the average iris. The position along the third dimension is determined by the darkness, or luminosity, where A is white (+100 luminosity) and D is black (0 luminescence). This representation of color has the largest gamut possible, encompassing all colors in the RGB and CMYK scales.

Figure 9-4

Percent reflectance and apparent absorbance. Three reflectance curves are shown for the same person; constitutive pigmentation (line with no symbols); 24 hours post exposure (line with open circles); and 7 days post exposure (line with filled circles). A. Data displayed as percent reflectance. Raw reflectance data from Microflash is shown. Vertical bars indicate the regions being screened by narrow-band based methods (M index and E index). B. Data displayed as apparent absorbance. Raw data transformed to apparent absorbance as described in the text. Vertical bar indicates the region around 580 nM, where hemoglobin has a maximal absorbance. The methods for calculating AM and AE from the apparent absorbance levels as described in the text are illustrated. From Wagner et al., 2002.

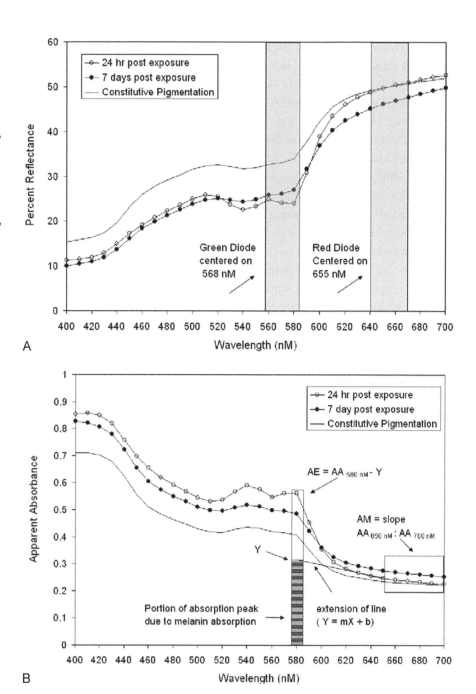

prism-filtered linear photodiode arrays and microprocessors have made possible handheld true spectrophotometers such as the Minolta CM2000 and the Datacolor International Microflash 200D (Lawrenceville, NJ), and more recently, the Ocean Optics 2000 unit (Dunedin, FL). One notable point regarding instrumentation and dermatological methods is the fact that the

DRS instruments are designed to calculate all types of pigmentation measures. CIELab and other standard color systems are usually standard output formats of these instruments. The reflectance levels at narrow bands such as those used by the DermaSpectrometer can be determined. Additionally, the transmission spectrums of the filters used in the E.E.L. and Photovolt systems can be used to integrate across the spectral reflectance curve to generate the predicted reflectance level that would have been read by one of the original E.E.L. or Photovolt instruments. In this way it might be possible to correlate measurements taken today with the hundreds of studies reported using the earlier instruments.

In addition to the larger number of methods that can be implemented using the true spectrophotometer, there are a group of dermatological methods that were developed specifically for the type of spectral reflectance curve data that is available using DRS (reviewed in Kollias & Baqer 1987; Kollias et al. 1991, 1994). Over the past 15 years, Dr. Nikiforos Kollias has been working on applications of true spectrophotometers in the field of dermatology. Dr. Kollias has applied the concept of apparent absorbance to the question of which are the pigments in skin and what their physical properties and concentrations are. For apparent absorbance, one measures an appropriate blank against which to relate the sample, much in the same way in which a spectrophotometer is used in a laboratory to determine the identity and calculate the concentration of a chemical in solution. An appropriate blank for skin might be skin areas affected with vitiligo, the skin of persons with albinism, or unexposed skin for measures of skin response; for hair one might use the white hair of persons with premature graying.

Figure 9-4A illustrates the basis for calculating several of these pigmentation measures. Shown in this figure are the spectral reflectance curves one person at baseline (line with no symbols), 24 hours post UV light exposure (line with open circles), and seven days post UV light exposure (line with filled circles). Relative to the baseline curve, the reflectance curves at both times post exposure show increased absorbance at wavelengths below 650 nm with peak changes at around 540 nm and 580 nm where hemoglobin shows maximal absorbance. Additionally, as the person tans and creates more melanin, there is a clear shift toward greater absorbance (less reflectance) across the spectrum including the wavelengths greater than 650 nm. The shaded bars indicate the regions (maximal emission at 568 nm for the green LED and at 650 for the red LED) where a narrow-band instrument, namely the DermaSpectrometer, measures reflectance in order to differentiate hemoglobin from melanin (see Shriver & Parra 2000; Wagner et al. 2002a).

Shriver and colleagues (Wagner et al. 2002) have investigated several measures of pigmentation level based on the apparent absorbance as illustrated in

Figure 9-4, which shows the same three reflectance curves transformed using the white tile blank to apparent absorbance. As in Figure 9-4A, three time points from one individual are shown: baseline (line with no symbols), 24 hours post UV exposure (line with open circles), and seven days post UV exposure (line with filled circles). The effect of the UV light in causing erythema is clearly apparent at both 24 hours and seven days in increase in the apparent absorbance with peaks at about 540 nm and 580 nm. Dr. Nik Kollias was the first to recognize that the slope of the line at the far right of the visual spectrum was related to the melanin concentration (Kollias et al. 1991, 1994). Additionally, he described a method to allow the dissection of the apparent absorbance curves so that erythema could be measured in the presence of changes in melanin level by extending the slope of the line and then subtracting the difference between the 580 nm peak and the y-axis level of the melanin line at 580 nm. Shriver and colleagues have provided the first thorough test of both of these methods and have compared them to more traditional methods of reflectometry (Wagner et al. 2002).

Because their measurements are interrelatable, and due to differences in the convenience of application, the DermaSpectrometer and Microflash or OceanOptics spectrophotometer have complementary roles in any research project aiming to objectively quantify human skin and hair pigmentation. The DermaSpectrometer is much more field worthy, being lighter, smaller, and more compact and durable. The OceanOptics unit is powered by 120/240V AC, whereas the Microflash requires overnight to recharge the nickel-metal hydride battery, and the DermaSpectrometer can take thousands of measurements on a single 9V transistor battery. Calibration is an important aspect of using reflectometers and is accomplished using black and white standards provided by the manufacturer.

However, though measuring hair and skin pigmentation with the Derma Spectrometer is convenient and straightforward, in the clinic setting, the Microflash or OceanOptics unit is a better choice since these instruments provide more complete data. The optimal pigmentation study would involve using one of these two instruments, and though contacting the OceanOptics lens to living skin is straightforward, hair samples do not always present a flat, homogeneous surface for contact, and we cannot make contact with a human iris (actually, we could, but it would require numbing drops and still be rather uncomfortable for the subject making ascertainment very time consuming). Thus, the measurement of hair and iris pigmentation raise a number of technical issues that must be dealt with and in this chapter, we will describe how these issues have been overcome in the latest research on hair/iris color genetics. We will pay special attention to answer the crucial question many readers may be asking in their heads as they read—''Why not just look at the color and report what it is?''

IRIS COLOR

Only recently has the field of genetics progressed from a quest for gene sequences that cause simple Mendelian diseases (those with recessive or dominant modes of inheritance) to one for factors and sequences that alter the probability that an individual will express a complex trait (those that are polygenic having multiple risk loci). In spite of what we were taught in high school biology, most human traits are complex by this definition and defy attempts at easy dichotomization (e.g., round or wrinkled, yellow or green, tall or short). How we decide to classify an individual with respect to their phenotype depends on the precise question we wish to ask, and why we are asking the question (i.e., what we will do with the information). Whether our goal is to communicate in a standardized manner so that what one forensics investigator says means the same thing as another, or whether we want to map genes with as much statistical power to detect as possible, we must devise a phenotyping method that captures the specific information we want. The most common pitfall with the analysis of complex genetics traits is poor phenotyping and study design (Terwilliger & Goring 2000), and of all the complex traits for which phenotyping is difficult, iris color has to be near the top of the list.

Iris color is a good example of a complex trait. In 1906 the Davenports (operators of the Davenport's Station for Experimental Genetics, at Cold Spring Harbor on Long Island, New York) outlined what is still commonly taught in schools today as a beginner's guide to genetics, that brown eye color is always dominant to blue, with two blue-eyed parents always producing a blue-eyed child, never one with brown. Unfortunately, as with many aspects of the real world, this simplistic model does not convey the complexities of real life, and the fact is that eye color is inherited as a polygenic not a monogenic trait. (Interestingly, the Davenports were racialists, and Cold Spring Harbor at the time was the center for the eugenics movement in the early twentieth century, and so it may be easy to appreciate why they promoted European traits as recessively pure unless contaminated by non-European influences.)

Though we are used to thinking of iris color in discrete terms (brown, hazel, green, or blue, for example), in actuality, iris color is also a quantitative phenotype. Not only is it complex in terms of multifactorialism (horizontal, or locus complexity), meaning many genes contribute to the phenotype so that inheritance is not strictly Mendelian, but also in terms of the number of alleles involved at each locus (vertical, or allelic complexity) and the continuous nature of the phenotype distribution. Indeed, there are not merely two or four iris colors, but an almost infinite variety of color mixtures and patterns ranging from the lightest homogeneous blue to the darkest homogeneous browns.

The physical basis to eye color is determined by the distribution and contents of the melanocyte cell in the uveal tract of the eye (reviewed in Sturm & Frudakis 2004). The iris is composed of several layers, of which the most important for the appearance of eye color are the anterior layer and its underlying stroma (Boissey et al. 1996). In the brown iris there is an abundance of melanin-filled melanosomes in the cells that comprise this anterior layer and stroma, whereas in the blue iris these same layers contain very little melanin. As light traverses these relatively melanin-free layers, the minute protein particles of the iris scatter the short blue wavelengths to the surface. Thus blue versus brown is actually a direct consequence of the structure of the iris in combination with major differences in chemical (melanin) composition. The number of melanocytes does not appear to differ between eye colors (Imish 1997), but the melanin pigment quantity, packaging, and quality does vary, giving a range of eye shades. It is known that most of the pigment found in the iridial epithelium of brown, hazel, and green/brown mixes is eumelanin (Menon et al. 1987; Prota 1998). Green irises contain high levels of pheomelanin but blue irises contain very little of either eumelanin or pheomelanin (Prota 1998). Thus, for the most part, it is the concentration and distribution of eumelanin pigment that determines the major features of iris colors. The common occurrence of lighter iris colors is found almost exclusively in European populations and those showing extensive and recent common ancestry with Europeans (see Table 5-19, Chapter 5; Figure 8-21, Chapter 8) with increasing frequencies in Northern and Western Europe compared to Southeastern (see Table 8-4; Chapter 8). Using biogeographical ancestry it may soon be possible to date the genesis of lighter irises; that is, to distinguish whether lighter iris colors are exclusive to the continental European populations specifically, or also present in unadmixed Middle Eastern or Central Asian populations with whom they share a common ancestry.

IRIS COLOR PHENOTYPING: THE NEED FOR A THOUGHTFUL APPROACH

A few example iris photographs are shown in Figure 9-5. There are light blues (Figure 9-5A) and darker blues (Figure 9-5B), greens with brown rings around pupils (Figure 9-5C) and blue/greens with subtle brown patches and rings (Figure 9-5D), as well as hazels, browns of different intensity (Figure 9-5E), and even blacks. Some blues are more blue than others and some browns are more brown than others. To develop the ability to predict iris color from DNA sequences, we first need a means by which to quantify color, and then we need to screen the genome for markers associated with these classifications. However, as with ancestry, where biological populations are commonly binned

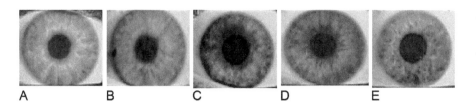

Figure 9-5

Lighter iris colors are found in various patterns and concentrations of melanin. In detail, iris color is far more complex than we realize from a distant appreciation. From a distance the irises shown would appear as light blue (A), blue (B), hazel (C), grey or green/blue (D) and light brown (E).

into metapopulations or races, binning irises into preformed categories poses certain problems.

If we are to bin irises into color groups, then into what category should the iris of Figure 9-5C be put—green, brown, or hazel? From a distance the brown and green seem to combine into a color somewhere between both but up close this effect is not observed and choosing which color is the best is subjective. If the answer is hazel, then how does this capture the difference between individuals with homogeneous distributions of hazel or light brown color versus those like iris Figure 9-5E, where there is a similar level of brown color? Not capturing such a difference might confound a genome study aimed at identifying markers underlying the expression of eumelanin in the iris.

The number of categories would quickly grow if we were to use different categories for different patterns of color rather than just the average color across the iris as observed from a distance. For example, Figure 9-5A may look like a lighter blue than Figure 9-5B from 10 feet away under fluorescent lighting, so we may call it light blue, but upon closer inspection, or under different lighting conditions, we might see that the blue part of the iris is as blue as that of Figure 9-5B, and the difference at a distance was due to the depigmented area around the pupil. We could put iris Figure 9-5C into a green with hazel ring category, but then what exactly constitutes a ring, and what exactly constitutes green? The photos in Figure 9-5 actually were taken from 24 irises shown in an advertisement in a genomics technology magazine. The platform technology being advertised was supposed to provide exquisite quantifications for microarray signals, and the figure of irises, of varying pigmentation pattern and colors, was chosen by the manufacturer to symbolize the nature of the challenge they had overcome. The heading for the figure was ''Quantitate this!'' Evidently this company considered the quantification of iris pigmentation about as challenging a problem as problems get!

Thus, difficulty in phenotyping iris colors is a serious impediment to a successful genetics screen. As has been recently pointed out by genetic historians, genetic screens for complex, multifactorial traits have a history of disappointment and even the most advanced genotyping technology is not going to allow a successful screen if the study is not designed properly. Phenotyping is one of the most overlooked aspects of study design, and a primary reason that genome screens fail to identify good markers that others can independently validate (Terwilliger & Goring 2000). A genetic screen focused on mapping a series of interrelated, or even unrelated traits would require huge sample sizes to ensure representation of variability for each one. If our goal is to map markers for a complex trait such as iris color, so that we can predict iris color from the DNA, we need to do a good job in focusing on a single genetic trait, to define explicitly what we mean by color, and then to define how this relates to the predefined subset of genetic loci we seek to discover. We can then measure color in a manner that imparts the information specifically relevant for the loci we seek to identify.

Considering iris color as a discrete variable is problematic because the color is likely a function of continuously distributed pigment levels, involving more than one pigment and genetic pathway, as well as various anatomical differences. Not only is it more difficult to parameterize iris color as a discrete variable (does this blue/green iris go into the blue group or green group, and does grey come before green on a quantitative scale?), but its unparameterized expression potentially gives too much information about too many unrelated genetic phenomena, making its use in a genetics screen unadvisable since our power to detect associations and linkages is related to the sample size for each trait value class (which, given a constant sample size, decreases for each class as the number of classes increases). With more thought given to the study design, it may be possible to ask a discrete, relatively simple genetic question instead of a question that is almost impossible to answer in genetic terms.

We do not attempt to simplify the trait merely to make it easier to compile adequate samples—our main concern is to parameterize the trait in terms that are more clearly meaningful at a biochemical and/or genetic level than they otherwise would be. Binned iris colors are only partially meaningful at a biochemical or genetic level, not only because of the complexity of the colors and patterns, but because placing an iris into a color bin is an interpretation that often is influenced by nongenetic and nonbiochemical variables such as illumination, surrounding colors of the cornea, color vision of the observer, and personal experience of the observer—all of which are changeable and/or subjective. From a genetics screening perspective, we want to simplify the trait as much as possible without obfuscating the marker relationships we seek. In other words, we want to reduce a complex trait such as iris color to a single or small number of variables that are determined by a discrete group of genes. For iris color, we need to form a very specific genetic hypothesis to explain most of the biochemical differences in color and then test this hypothesis (i.e., can we find associated markers?). Doing so, we reduce the phenotyping problem to one of the quantifications of that variable.

So, given that we would like to reduce the complex variable of iris color to a small number of parameters (perhaps even a single parameter) in order to simplify our study design and maximize our statistical power to detect associated and linked markers, which parameter should it be? The shape of brown, area of blue color, intensity of brown? Eumelanin is the primary pigment that imparts iris color but there may be others at work for green such as the pheomelanins (yellow-red pigments; Prota et al. 2000). Reducing iris color to melanin content (where melanin is meant to signify both eumelanin and pheomelanin), by measuring the luminosity (light reflected) and amount of blue, red, and green light reflected from the iris and combining them in such a way as to indicate the average concentration of melanin is one way to simplify the phenotype (we eliminate tissue ultrastructural and cellular anatomy), and yet retain the continuously distributed nature of the trait. Alternatively, we could not combine the values, but keep them as four separate values and define iris colors by reflectance profile types or use the blue and red reflectance only to define the amount of brownness and blueness. Either of these methods would help us accomplish our goal of making measurements that are relevant for as few loci as possible, and as we have seen, eumelanin and pheomelanin are specific photochemicals, the expression of which can be traced to a discrete genetic pathway with a limited number of players.

Although quantifying melanin content in the iris, regardless of actual color, or the pattern of color or shades, nicely simplifies the problem in a rational way, it requires the simplifying assumption that all shades of iris color are due to differential distributions and amounts of melanin pigment rather than to mixing of pigments (the melanins with those we may not know about yet, for instance), or anatomical differences in iris constitution. Though these are assumptions, they are reasonable if there exists justification in the literature. Given the large variability in the levels of eumelanin and pheomelanin in different colored irises, and the fact that the principal component of iris color variation is most likely absolute melanin levels (Prota et al. 1998; Sturm 2004), this assumption seems reasonable.

MAKING IRIS COLOR MEASUREMENTS

If it were not so difficult to do, binning irises into crude color groups would give good information on melanin content because irises of the brown group have more eumelanin and pheomelanin on average than irises of the hazel or blue group (Prota et al. 1998). Binning would not completely obscure information on the anatomical distribution of the pigment and possibly the layering of fibers in the iris (which as we have described, influences how light is scattered), because irises with larger patches of brown probably would type with more melanin than those with smaller ones. With a limited number of bins, we could avoid overparameterizing the phenotype as we might if we considered all the possible pattern and color types. However, it is very difficult to bin iris colors reliably, as we will soon see, and there are better ways to phenotype for iris color—ways that allow a quantitative measure of color while reducing the complexity of the trait to a smaller number of genetic variables.

One way to objectively reduce iris color to a quantitative variable in terms of melanin content is to biochemically quantify eumelanin and pheomelanin from the tissue as Prota et al. did in 1998. This is not possible to do with living irises since the biochemical process of quantification requires the iris to be destroyed. Given the large numbers of subjects needed to discover and describe the genotype/phenotype relationships, it might seem better to use a spectrophotometric method, but unlike with hair and skin, we cannot contact the iris with a dermaspectrometer or reflectometer, and the spectrophotometric method would have to be a noninvasive method.

One such method employs a digital camera and computer software capable of measuring the type and amount of reflected light. This is the method we have chosen and much of the work discussed will be based on this method of iris phenotyping unless indicated otherwise. The method is straightforward: For each right iris of each subject, a digital photograph is taken under standardized lighting and exposure conditions and imported into a custom designed software application (alternatively, Adobe Photoshop software can be used, with the histogram function). A rectangular box is drawn in each of the four quadrants (top or north, bottom or south, east or right, and west or left) and for each box the average Luminosity, Red, Green, and Blue reflectance values is determined and recorded (see Figure 9-6). A single value for Luminosity, Red, Green, and Blue reflectance can be obtained by taking an average of the four quadrants of the iris—we essentially homogenize the iris *in silico*. In this way green irises have similar amounts of melanin as blue irises with thin brown rings; this *in silico* homogenization allows us to focus only on melanin content, ignoring the pattern with which this melanin is distributed.

The logic of using reflectance values at different wavelengths of light follows the $L^*a^*b^*$ model, which is based on the Comission Internationale d'Eclairage (CIE) in 1931 as an international standard for color. In 1976 this model was refined and named CIE $L^*a^*b^*$. This model originally was developed as a means by which to standardize the interpretation of colors across instruments. Its use here allows for objective quantification of lightness or luminance (higher L^* values), corresponding in this case to a lack of melanin and darkness (lower L^* values), corresponding to an abundance of melanin across the iris (see Figure 9-3), but in a way that accommodates

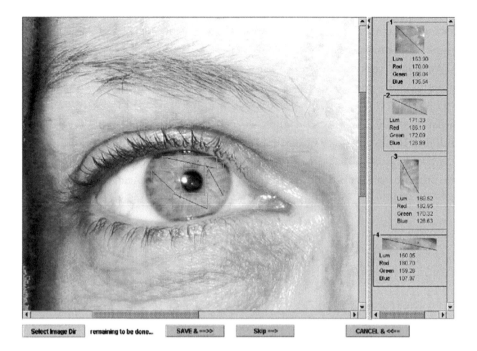

Figure 9-6

Quantification of iris colors using digital photographs and spectrophotometric software. Boxes are drawn in each of four quadrants of the iris, and average pixel luminosity, red reflectance, green reflectance, and blue reflectance determined for each box. Downstream software computes an average among these four boxes to produce four values for each iris that are expected to uniquely represent the melanin content of the iris relatively independent of distribution pattern.

the *amount* of light reflected (the lightness) as well as the *type* of light reflected (the actual color), both of which are likely related to melanin content. Using only the lightness as a surrogate for melanin content would be ill-advised since a dark blue and a light green iris, or a dark green and light brown iris for example, might have similar lightness levels even though they are of totally different melanin content and color, and we would therefore potentially obfuscate the associations our study aims to detect. Using the red/blue/green color scale on its own as the surrogate for melanin content would be equally ill-advised for the same reason. A grey blue iris likely expresses more melanin than a light blue iris, even though the color of the two irises is essentially the same using the color score (only its shade is different).

We could use these four numbers as a profile of iris color descriptors, building a large database of iris profiles and photographs. Calling from the database irises with a certain profile we retrieve an objectively determined bin of irises. For example, in Figure 9-17 later in this chapter, we called all irises that fell within the range of Luminosity 114.61–134.85, Red 129.67–153.69, Green 113.37–135.17, and Blue 71.02–91.14. The database at the time of this query was 596 and 31 irises were retrieved. Some of them were blue irises with large yellow/green rings, others with smaller brown rings. It is likely that both types of irises have similar overall melanin (most likely eumelanin) content—the irises with thin brown rings or patches showing higher concentrations in discrete areas of the iris than those with large yellow/green rings where the eumelanin is more dispersed.

We could reduce the four values into one value to further simplify the problem. One way to do this is to calculate three reflectance ratios: Red/ Green, Red/Blue, and Green/Blue. Summing these ratios gives a crude measure of lightness or darkness we call the color scale, and this can be added to the Luminosity value that represents a different measure of lightness and darkness on a black/white scale (see Figure 9-3). The sum of color scale and luminosity value for the sample comprises what we call the iris melanin index (IMI).

$$\text{Iris Melanin Index (IMI)} = (\text{Luminosity/avg. Luminosity}) + (\text{avg. color scale/Color Scale})$$

How did we choose the ratios red/green, red/blue, and green/blue rather than their inverses? In some respects, our selection of these ratios was similar to the process employed by Reed in 1952 when he was seeking to parameterize red hair color from reflectance curves—he used a feature of the curves that seemed to discriminate between reds and nonreds. Here, we selected these particular ratios because more brown irises showed higher values (compared to the mean value) than hazels, greens, or blues—browns tend to have higher red/green, red/blue, and green/blue reflectance values than hazel irises, which have higher values than green, and so on. Since high levels of both eumelanin and pheomelanin would show a darker color than low levels (they are more colored), the magnitude of each ratio could be considered a rough indicator of the coloration of the iris (not to be confused with its hue), which is precisely what we want to know for a genetic screen of markers influencing melanin content since irises with no color have no melanin and because lighter colors result from less melanin than darker ones. Again, the luminosity is required because irises of the same color could exhibit very different shades or luminosities, such as blue and blue/grey irises. The latter of these two likely has greater melanin content, so it should rank after the former in a ranking from lowest to highest melanin content.

The irises in Figure 9-7 illustrate why it is important to consider luminosity and color together in quantifying iris melanin. This figure shows the same irises shown as in Figure 9-5, but they are ordered by their luminosity (top row), color score (middle row), and combined IMI (bottom row), from lowest to greatest, where the labels A through E are the same for each iris as shown in Figure 9-5. Along the luminosity scale, iris E is the only iris that may classify as homogeneously brown, yet it ranks as second lightest even though it is of brown color, rather than blue or green (Luminosity row of Figure 9-7). This is because this iris is not homogeneously pigmented, but rather a mosaic of brown color with large regions of depigmentation. The high luminosity is coming from the depigmented regions between brown spots (which are white). Going off the color ratio scores would produce a different ranking (Color Score row of Figure 9-7). Here, iris B is the bluest (least brown-like, though not the lightest), and iris E is among

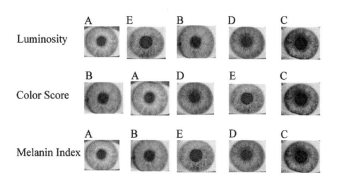

Figure 9-7

Ranking of five irides from lowest to highest luminosity, color score and melanin index. The letter over each iris identifies the iris, not the figure column.

the brownest (most brown-like). Iris C, which is a mix of green and brown, is actually the most brown-like of them all, even though it is a green iris with a brown ring, rather than a homogeneous brown like E; its brown is deeper, with no white spaces in between. This combines with the homogeneous green background, which we assume is due to an even spreading of lower levels of melanin (probably pheomelanin, rather than eumelanin like the brown portion), to make C more brown-like than E. Combining the Luminosity and Color scores, we obtain a sort of a compromise between these two methods of quantifying the absolute melanin content regardless of how it is distributed in pattern or density. The iris with only brown pigment, but dispersed (iris E) ranks in the middle of the melanin index, behind an iris with a deep, dark blue with brown patches (D), and the green iris with the thick brown ring (C) (Melanin Index, Figure 9-7; Table 9-1).

Thus, the IMI assumes that a homogeneous dark blue signifies as much melanin or even more than a disperse brown with substantial areas of depigmentation. Though iris E in Figure 9-7 is the only iris that is homogeneously brown, dubbing it as the iris with the greatest melanin content based merely on its color would likely be a mistake that could confound a genetics study aimed at identifying markers associated with iris pigment expression, and complicate efforts at drawing inferences from DNA. Accounting for the disperse pattern of brown, and the large gaps of depigmentation in this iris, we see that this iris actually seems to express less melanin overall than the deep, dark blue iris (D) and the green iris with a thick and rich brown ring (C). Of course, with this method, we consider all pigments at once, and we are not able to resolve between eumelanin and pheomelanin. More elegant methods of iris phenotyping in the future may help simplify genetic screening even further.

There are, of course, alternatives to the simple summation of luminosity and color for representing iris phenotypes, such as principal components scaling, or the use of vectors for instance, which future work may elaborate on further.

Table 9-1

*Melanin index values
for irides example of
Figure 9-7.*

Photo	Luminosity	Color score	Melanin index
a	1.201216	1.131313	2.332529
b	0.954823	1.230894	2.185717
e	1.050123	0.929774	1.979896
d	0.939949	1.007854	1.947803
c	0.853927	0.812382	1.666308

Though the use of a simple sum of color and luminosity values scaled with respect to the mean as the IMI is quite convenient and easy to visualize, this method is not perfect as averaging values across four quadrants of the iris, of course, conceals the details of the patterns of coloration (see Figure 9-6). Furthermore, great care needs to be taken to ensure that the photographic conditions are the same for each iris because the method is not very robust against fluctuating lighting conditions (Frudakis, unpublished data). A brown iris photo taken under bright lighting conditions could be determined to have a similar melanin index as a blue iris photo taken under dim lighting conditions. Future work may focus on parameterizing the actual pattern of color, which may enable better genetic screening results. This is debatable though, because as we already have mentioned, when screening for markers underlying variation we want to reduce the number of variables, not increase them. In melanin content, perhaps, we actually want to quantify melanin irrespective for how it is laid out in pattern, rather than attempt to measure too much information at once, complicating the parameterization of the trait and reducing the statistical power of our study (by potentially trying to map markers for multiple processes simultaneously).

It bears noting that even the IMI method of iris color homogenization could achieve some consideration of pattern in assessing melanin content since variation in index values within irises is easily calculated. For instance, the four quadrants we have described could be divided further into more equal pie slice segments and the variance calculated across the segments to produce a separate variable. Alternatively, concentric rings of iris could be delimited, averaged, and examined for variability. Given the dark rings generally appear toward the outside of the iris, the statistics could focus explicitly on the difference between the outer most concentric rings and the others.

The methods we have discussed are based on standard digital photography where only the red, green, and blue (RGB) spectral bands are available. There are new methods that recently have been reported that bears mention in the context of measuring pigment levels using photography, called hyperspectral imaging (Stamatas et al. 2004). In RGB imaging, there are basically the three rather broad filters that simulate the color vision pigments of the eye. Hyperspectral imaging works by sampling a much larger number of narrow spectral bands such that the result is analogous to a true spectrophotometer reading for each pixel of the image. Thus, in the same way that a standard digital photo lends itself to colorimeter reading values in the CIE $L^*a^*b^*$ color system, the hyperspectral imaging method lends itself to a spectrophotometric image. Thus, using hyperspectral imaging, one can calculate apparent absorbances and related measure such as the slope of particular portions of the tissue throughout the field of view having implications on both the ability to measure eye color in new ways and for microscopic examination of other tissues such as skin and hair.

POPULATION SURVEYS OF IRIS MELANIN INDEX (IMI) VALUES

Measuring the IMI for a sample of 632 irises (one iris per individual; 622 European Americans and 10 African Americans of European/Western African genomic ancestry admixture) from throughout the United States, we can rank the samples from highest to lowest IMI score, where blue irises have high IMI scores and brown irises have low scores. When we do so, we see a fairly continuous distribution of values, where the ordered values form a continuous line rather than discrete steps with gaps between certain values (see Figure 9-8A). An index of about two marks a rough transition from the greens to hazels and lighter browns.

You will notice a wavy pattern to this distribution, which is a byproduct of the fact that homogeneously blue and brown irises are more commonly found than irises in between blue and brown (the curve flattens out in the blue and brown range), possibly due to population stratification, assortative mating, or the lack of codominance for this quantitative genetic trait. At the low melanin index scores, the values for the darkest samples curve sharply downward; these are the iris colors for the African Americans in this sample, which are on average clearly darker and more colored (brown) than the colors for the darkest Europeans (as we saw in Chapter 5). When we keep the samples in order based on the melanin index values, but plot their color scores (see Figure 9-8B) and luminosity scores (see Figure 9-8C) rather than their IMI values, we can observe the extent to which iris color and luminosity are correlated with the IMI score. As expected, these correlations are good, but since irises of the same color could exhibit very different luminosities, there is variance about the trend line for both plots.

If we traced the position of a single sample in all three plots we would see that if the sample was plotted far from the mean in one direction in the color score plot, it would be found far from the mean in the other direction in the luminosity plot, so the plots are essentially inverses of one another (Figure 9-8B is essentially an upside-down version of Figure 9-8C, and of course, the correlation coefficient between color score and melanin index is the same as that between luminosity and melanin index). Comparing the luminosity score and color score we see a relatively weak but positive correlation between the two (see Figure 9-8D). The variation shows an unusual asymmetric pattern about the mean, with branches shooting upward from the trend line as one moves from right to left in the plot.

What is interesting about this plot are the branches—the seemingly clear bifurcation in the distribution, which illustrates uncoupling between color and luminosity (i.e., knowing the value for one does not necessarily allow for an accurate prediction of the other). Recall that a value of 1 is the average value for both color score and luminosity components of the IMI, since the value in each iris is normalized with respect to the mean value. Many of the samples part of these offshoots are those with high (lighter) colors scores but lower (darker) luminosities, such as grey-blues or darker greens. A grey-blue sample for instance might reflect less light than an average blue iris, giving it a lower luminosity than other blue irises. The uncoupling of the two parameters signifies that each gives different information. If the color of an iris was a function of parameters that influenced luminosity and color in the same way, such that darker irises always have darker luminosity, the variance about the mean would be symmetrical and there would be no need to combine separately measured luminosity and color scores—we could describe the color of an iris with either the color score or the luminosity score on its own.

The distribution of IMI values for a sample of 1,756 subjects is shown in Figure 9-8E, and from this figure we can clearly see the bimodal nature of the distribution suggested in Figure 9-8A. We could bin the samples in this distribution as characterized by lighter (IMI > 2.6), darker (IMI < 1.9), and intermediate (1.9 < IMI < 2.6) values.

RELATION OF IMI TO SELF-DESCRIBED IRIS COLOR

We have taken considerable steps toward the objectification of iris color measurements in terms of an iris melanin index (IMI). The preceding section has shown that iris colors are continuously distributed, not discretely distributed as is the common misperception, and that iris color is best parameterized as a function of two separate variables—color and brightness. If one were to ask a laboratory technician to classify iris colors, this person would be forced to

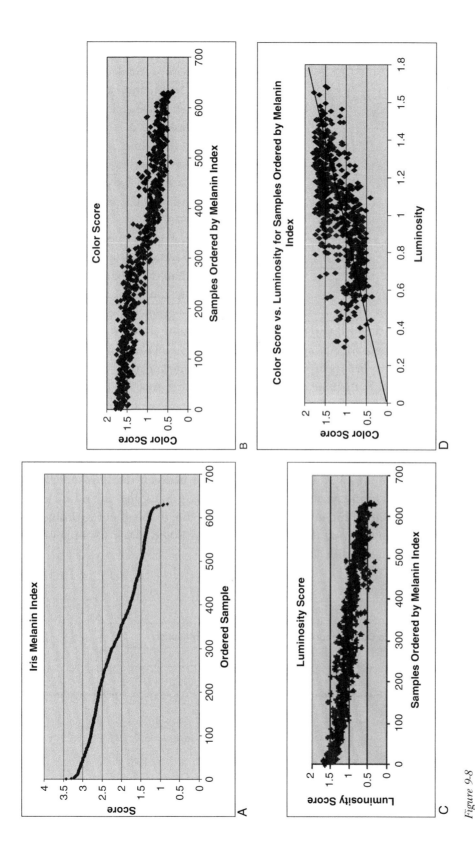

Figure 9-8

Distribution and trending for iris melanin index (IMI) values in a sample of 1,756 U.S. Caucasians. Each spot represents data from a single iris of a single human being. A. The continuity of the distribution can be appreciated when ranked (highest to lowest) IMIs are plotted. Note the lack of steps or breaks in the plot. B. Color score component of the IMI is plotted against the ranked IMI scores (from lowest to highest) illustrating the strong correlation as we expect, but also the discordance for some irides of unusual hue. C. Luminosity score component of the IMI plotted against the ranked IMI illustrating the same as (B). D. Color score versus luminosity for ranked IMI scores. This view of the data shows the most variation; though the correlation coefficient is good, there seems to be two basic color scores but many different luminosity values for each, giving rise to a continuous distribution of colors as judged by the human eye or by the iris melanin index value, which combines both of these variables. (Continued)

525

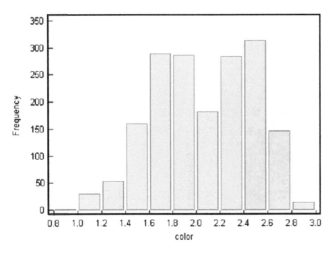

Figure 9-8

(Continued)
E. Distribution of IMI
values is bimodal. The
IMI values are considered
as bins along the x-axis.

Figure 9-9

Relationship between the
iris melanin index value
discussed in the text
(digital color score) and
human assessed colors in
terms of color bins. Data
for 175 samples are
shown. Each symbol
represents an iris from a
unique person. The
collector graded the iris
colors on a continuous
scale from 1 to 6;
1 = blue; 2 = blue/
green, blue with a thin
brown ring; 3 = green,
green/hazel, green with
brown ring around pupil;
4 = hazel, light brown;
5 = brown, medium;
6 = black and dark
brown.

choose one of several boxes, such as blue, green, hazel, and so on. How might these reported classifications relate to our objective measures of the iris melanin index when the same person does the reporting over a series of irises? There is definitely a trend—we see a steady decrease in index with darker categories (see Figure 9-9).

We see that many eyes binned to the green group have a higher (lighter) melanin index than those binned to the blue group. About half the irises placed into the green, green/hazel, green with brown ring categories were of a higher melanin index (actually lighter) than the average iris placed into the blue category. One iris was placed into the black/dark brown box scored as of the same melanin index of several irides placed into the blue category (outlier in category 6 of Figure 9-9). From an inspection of the digital photograph, the iris appears to be a very deep and dark blue/grey/green, and the reporting as dark brown was likely due to the technician using substandard lighting conditions and/or considering the luminosity rather than the color when making their choice for this iris. In general, there appears to be minimal overlap between the blue and brown bins, and it is in the resolution of blue/greens, greens, hazels, and such that there is a problem, and much of this overlap is due to the opposite problem—the technician ignoring the luminosity of the iris and rather focusing on the color when making the classifications.

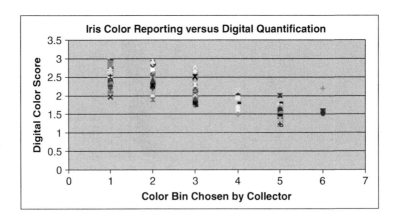

Overall the plot suggests that the technician seemed to have floating criteria for color classification—sometimes the luminosity was used primarily but other times the color was used irrespective of the brightness. This imprecision was observed using a single scientifically trained individual calling the colors—and one can imagine how the precision would deteriorate further if we have a large number of individuals self-reporting their colors. Each individual brings different perceptions and interpretations of color; some perhaps would attempt to accommodate pattern complexity, others not.

HISTORY OF GENETIC RESEARCH ON IRIS COLOR

As a genetic character, iris color has enjoyed significant attention from the human genetics community, at least rare defects in iris color caused by hypopigmentation—diseases such as oculocutaneous albinism (OCA), ocular albinism (OA), and vitiligo. Decades of work on disease phenotypes such as these have given rise to our current understanding of pigmentation pathways, as discussed in earlier in this chapter. However, because of their apparent irrelevance for human disease, normal variation in iris color mostly has been ignored by human geneticists, and until recently it was unclear whether common polymorphic variants for any of these genes were associated with natural distributions of iris colors.

HERITABILITY OF IRIS COLOR

It has been known for many decades that human iris color is largely under genetic rather than environmental control. Larsson in 2003 assessed eye colors in 199 pairs of adult German twins and noted high correlations of 87% and 86% in male and female monozygotic twins. Dizygotic twins showed, in contrast, a low correlation in eye colors (0.26 and 0.43 for male and female, respectively). Zhu et al. (2003) used a research nurse to bin irises into color groups, and showed that the correlation between monozygotic twin eye color assignments was nearly perfect (greater than 99%), though those between dizygotic twins was only about 54%. This showed that the eye color assignments were quite reliable, or at least consistent within their sample using this particular nurse (refer to our earlier discussion on the problems with binning irises into color groups—although this particular study showed consistency in assignment, the assignments do not necessarily relate directly to melanin content). It also showed that the heritability of iris color determination was high, and that there was little influence of a shared environment (specifically, *in utero*) in the determination of iris color, for if there were, correlation between twins derived from two different eggs (of different genetic constitution) likely would have been greater. Measuring the heritability, they obtained a value of 92%, not too different from the value of 87% shown by others previously (Larson et al. 2003), and they speculated that the remaining 8% in unexplained variation in iris colors was due largely to genotyping error, though it is plausible that there is in fact some small role of environmental influence including possibly epigenetic in nature, such as those that cause coat color variegation in model organisms like mice and fruit flies.

EARLY GENETIC SCREENS

The high heritability of iris color makes it a trait suitable for genetic dissection, whereby markers associated or in linkage disequilibrium with underlying variants could be mapped and the relevant genes identified. The first screen for markers associated with natural iris color distributions in fact was not conducted until 1987 and then again in 1996 by Eiberg and Mohr. These authors tested a collection of genomic microsatellite markers located near classical blood group antigens for association with iris colors, using discrete phenotype classes (blue irises vs. others, green/hazel vs. others, and brown vs. others). Their results provided evidence of association of a green/blue eye color locus (GEY/EYCL1, MIM227240) to the Lutheran-Secretor systems on chromosome 19p13.1-q13.11 (Eiberg & Mohr 1987).

Another major locus for brown/blue eye color (BEY2/EYCL3, MIM227220) and brown hair (HCL3, MIM601800) was found on chromosome 15q11-q21 by linkage with DNA markers within this region in families segregating for BEY2 (Eiberg & Mohr 1996), with the OCA2 gene recognized by them as a candidate within this region. Zhu et al. (2004) performed a linkage screen using monozygotic twins, and discovered a chromosome 15 interval containing the OCA2 gene explained no less than three-fourths the variance in natural iris colors among Caucasians (see Box 9-A and Figure 9-10).

Box 9-A

Zhu et al.'s QTL Screen

Zhu et al. (2004) screened microsatellites in a sample of 502 twin families to complete the first complete genome scan in an attempt to map genes responsible for eye color. The approach they used was a linkage screen but it was focused on only a small part of the pedigree structure most linkage screens used to identify patterns in the sharing of alleles between family members of similar and disparate trait values. In some respects, their study was similar to an association mapping study, which is really a linkage study on a large family of indeterminate structure (Terwilliger & Goring 2000). Their most notable result was a peak LOD score of 19.2 on chromosome 15q, which lies directly over the *OCA2* gene (see Figure 9-10), which already had been implicated in brown/blue eye color (Eiberg & Mohr 1996; Frudakis et al. 2003b; Rebbeck et al. 2001). This peak had a long tail toward the telomere, suggesting that other eye color quantitative trait loci (QTL) may lie there, and it is interesting that both the *MYO5A* and *RAB27A* proteins involved in melanosome trafficking are located in this region (Zhu et al. 2004). Zhu and colleagues estimated that 74% of variance of eye color may be due to this single QTL peak and concluded that most variation in eye color is due to the *OCA2* locus (encoding the P melanosomal protein), but that there were likely modifiers at several other loci.

The percentage of variance explained by this interval was consistent with the conclusions of Eibergh and Mohr in 1996 who showed that most brown/nonbrown iris color variation could be attributed to sequences on this chromosome 15 arm. The association studies of Frudakis et al. (2003b) and Rebbeck et al. (2002), and classification modeling by the former, performed on unrelated individuals, clearly implicate the *OCA2* gene under this LOD peak, and the work described in this chapter by Frudakis's laboratory agrees that the *OCA2* gene sequences explain the bulk of iris color variation in the European population. In addition to this 74% of the variance in iris colors due to the QTL at chromosome 15q, an additional 18% was posited by Zhu and colleagues to result from polygenic effects. LOD scores exceeding 2 were shown on 5p and 14q, and LOD scores greater than 1 were seen on 2,3,6,7,8,9,17, and 18. The best LOD scores were seen at 5p, 7q, 8q, 9q, 14q, 15q, 17p, and 18p. Perhaps loci under these peaks explain the variance not accounted for by the classification model described in this chapter, which rely strictly on the *OCA2* gene. Alternatively, it may be that these lower level associations were due to continental ancestry admixture and/or other levels of population structure.

The Zhu et al. (2004) linkage study was impressive in scope, but used rudimentary classification methods, which as we have discussed, provide most of the iris color information one would seek, but not all of it. Rather than consider iris color as a fully continuous trait, they broke eye colors into three main groups: brown, green/hazel, and blue/grey. They employed a single research nurse for the survey of 826 twins, and for many of these they were able to rate color at both 12 and 14 years of age, noting 95.3% concordance in ratings between the two occasions (where discordance was always between adjacent scores). In all, 525 families (419 dizygotic and 83 monozygotic twin sets) were incorporated in the study, with a total number of 1205 individuals (1025 from dizygotic pairs and 180 from monozygotic pairs).

Their screen employed 795 markers and thus achieved 4.8 cM coverage (Kosambi units). This particular screen proved quite effective at revealing linkage between markers flanking the ABO, Rhesus, and MNS blood group antigen loci, which revealed that their screen had good sensitivity, at least for these test loci. By extrapolation, reasonable sensitivity was expected to exist for detecting LD between markers and pigmentation gene loci (whether known or unknown) scattered throughout the genome that presumably produced the variability of iris colors in the twin collection. It is of some interest to note that the data from the Affymetrix GeneChip[R] 10 K array used by DNAPrint (mentioned briefly in this chapter) did not detect chromosome 15 marker associations, indicating relatively sparse LD coverage for this particular system.

There was also another element of the Zhu et al. (2004) analysis that is of interest. They performed their screen using a certain type of linkage approach called quantitative trait loci (QTL) screening. QTL screening usually is performed for traits that exhibit a continuous distribution of values, such as height, skin color, and intelligence, and as we have noted already, eye color is distributed continuously in the population. Quantitative geneticists work in the realm of variance components analysis, where variation in the distribution of a variable such as an LD score is broken into its various components in order to explain the results and understand what they specifically mean. For example, decomposition would illustrate how much of that variance is due to genetic versus other (i.e., environmental) factors. With a QTL screen, the variance in the statistical connection between a marker sequence and trait value can be broken into the variance due to a linked QTL (Q), the variance due to a residual polygenic effect (A), the variance due to shared environment (C), and the variance due to unique environmental variance, including measurement error (E). The variance due to a linked QTL (Q) is the component of primary concern for identifying a marker that is predictive for, and thus, probably near, a gene sequence that underlies the determination of the trait in question.

There are three types of genetic variance Q: those caused by additive sequences in combination, those caused by additive and dominant sequences in combination, and those caused by additive and dominant sequences in combination with environmental factors. The A term represents a measure of how much of the variance Q is genetic but not explained by the specific marker analyzed, such as that due to sets of other markers whose sequences correlate with the marker. The shared environment C term is the component of variance due to nongenetic factors and in this case, the primary nongenetic factor shared between the twins was the maternal, *in utero* environment (whether the mother smoked, drank, or the mother's genotype for maternal effect gene sequences such as those known to influence the maternal establishment of the embryo's dorsal/ventral and anterior posterior axis, etc.). The rest of the variance is caused by unique environmental factors, or factors that applied to one twin but not the other, and includes genotype measurement error that was encountered (whether knowingly or not) for one twin but not the other.

In a twin study such as this, the A and C terms are difficult to resolve in a meaningful way (there is low statistical power to do so), but the QTL effect Q can be estimated by subtracting the variance caused by A, C, and E from that caused by Q, A, C, and E since $(Q + A + C + E) - (A + C + E) = Q$. The same principle applies for simpler models, such as those that omit the C term, or that consider only some Q subterms.

RECENT HISTORY OF ASSOCIATION MAPPING RESULTS

Candidate gene analysis is a hypothesis-driven approach, whereby genes thought to influence variability in trait value are selected based on prior relevant research. For iris color, we rely on research focused on pigmentation diseases or conditions, such as oculocutaneous albinism, or on model systems

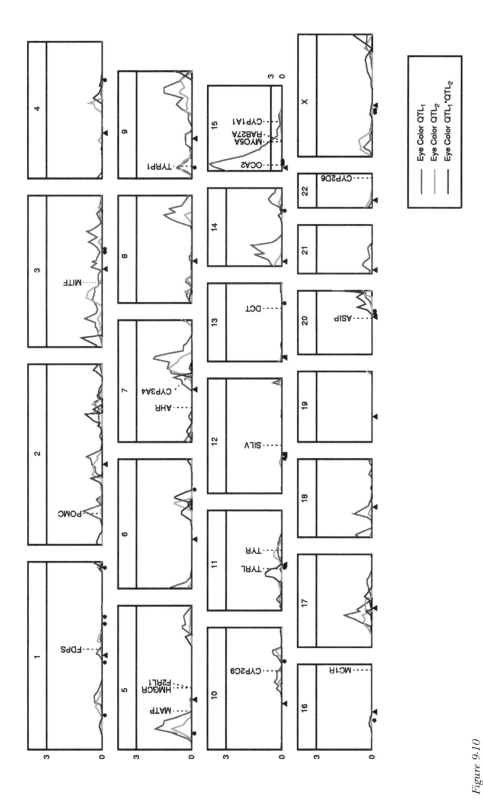

Figure 9-10

Quantitative trait locus analysis of iris color in a set of fraternal twins. LOD scores are shown for the association of genetic markers along each of the 23 chromosomes using a one-locus additive QTL model (red), or a second additive QTL over chromosome 15 (green), and the additive × additive interaction term between the 15p QTL and a second locus (blue). Centromeres are indicated by the black triangles and various human genes are indicated as landmarks (not corresponding to particular linkage markers screened). From Zhu et al., ''A geomne scan for eye color in 502 twin families,'' Twin Research and Human Genetics, 2004, Vol. 7, No. 2, pp. 197–210. Reprinted with permission.

such as mice. Polymorphisms are then screened in these candidate genes for association with the trait. This approach has the advantage of drawing upon prior knowledge to focus the screen toward those regions of the genome for which evidence suggests that focus is warranted, but suffers from its unsystematic nature, for if prior knowledge is incomplete as it often is, even successful association results may explain only a fraction of variable trait expression; the remaining alleles and loci yet to be discovered. Nonetheless, the advantage of using prior knowledge that is extensive and draws from decades of concordant research in model systems and man makes the candidate gene approach a good first step for many genome screening projects.

Using a candidate gene analysis approach in a sample of 629 individuals, the Rebbeck group recently found that two OCA2 coding region variants, R305W and R419Q, were associated with either brown or green/hazel eye colors, respectively (Rebbeck et al. 2003). These same polymorphisms were tested in the twin collection described by Zhu et al. (2004), and each was confirmed as being associated with green/brown but not blue eye color (Duffy et al. 2004). Another locus that has been tested for association for human pigmentation phenotypes by Rebbeck and colleagues is the agouti signaling protein gene (ASIP). The g8818a SNP in the 3′ untranslated region of this gene was genotyped in 746 participants, where the G nucleotide allele was significantly associated with brown eye color (Rebbeck et al. 2002).

These initial candidate gene screens focused on a very small number of candidate loci. The first study to survey a variety of pigmentation candidate genes for polymorphisms associated with binned iris colors was that of Frudakis et al. (2003b). This study focused on pigmentation candidate genes shown in Table 9-2, which were selected based on the fact that rare mutations in each had been associated with hypopigmentation effects in humans (mainly oculocutaneous albinism) as well as for their analogues in mouse.

This study tested for associations using a variety of iris color binning schemes. This is in fact the same group headed by the author that later developed the iris melanin index method described earlier, and it was in part due to knowledge learned during the present screen that motivated this improvement. The candidate gene portion of this study confirmed some associations from the literature and introduced others. The strongest (by far) and most numerous associations were found in the OCA2 gene, though a variety of other associations were found for TYRP1, MATP/AIM, and others that, since they were relatively weak, may have been due to population stratification.

The SNPs, haplotypes, and haplotype combinations (multilocus genotypes) described in the Frudakis et al. (2003b) paper were only part of the story because even though the associations with self-reported iris colors were fairly

Gene	Name	Homology/ model phenotype	References
AP3B1	adaptor-related protein complex 3, beta 1 subunit	mouse pearl human HPS2	Balkema et al. (1983)
ASIP	Agouti signaling protein	mouse Agouti	Kwon et al. (1994) Klebig et al. (1995)
DCT	dopachrome tautomerase	TYR-related protein 2 mouse slaty	Kwon et al. (1993)
MC1R	melanocortin 1 receptor	mouse extension (e)	Robbins et al. (1993)
OCA2	oculocutaneous albinism II	mouse pink-eyed dilution (p)	Lyon et al. (1992)
SILV	silver homologue	mouse silver (si)	Kwon et al. (1991)
TYR	tyrosinase	mouse albino (c), Himalayan	Barton et al. (1988)
TYRP1	tyrosinase related protein 1	mouse brown (b)	Jackson et al. (1988)
MYO5A	myosin VA	mouse dilute (d)	Copeland et al. (1983)
POMC	proopiomelanocortin	mouse Pomc1	Krude et al. (1998)
AIM (MATP or AIM-1)	membrane associated transporter protein	mouse underwhite (uw)	Newton et al. (2001)
AP3D1	adaptor-related protein complex 3, delta 1 subunit	mouse mocha (mh)	none
RAB	RAB27A oncogene	mouse ashen (ash)	Wilson et al. (2000)

Table 9-2

Candidate genes tested for allelic associations with human iris pigmentation.

strong, attempts at using the alleles for blind classification did not work as well as hoped. This group then concluded that the explanation for this failure was one or more of the following:

- Self-reported iris color was not accurate enough for the construction of general classification rules
- Other, informative markers in OCA2 were missing from the model
- Other loci that had not yet been identified were missing from the model

This laboratory did several things to overcome each of these potential limitations. New samples were collected for which the iris melanin index was used, rather than human-interpreted colors. In essence the work of Frudakis et al. (2003b) was repeated from the beginning with these new, better qualified samples. They screened deeper into the pigmentation genes—notably, deeper into OCA2 variation. This gene has an extraordinarily large number of SNPs, and it was known from Frudakis et al. (2003b) and later from Zhu et al. (2004) to explain by far the most variability in iris color shade. In addition, they screened other genes that had been implicated in oculocutaneous albinism and they screened the SNPs part of the Affymetrix 10 K Mapping Array. The most important things this new screen accomplished was objectifying the determination of iris color (using the IMI value) and surveying more of the extant variability in one important gene—OCA2.

About a year after the Frudakis et al. (2003b) study seemed to reduce most of the variation in iris colors to the OCA2 gene, the Zhu et al. (2004) team (led by N. Martin, R. Sturm, D. Duffy, and G. Zhu) performed a more systematic pan-genome linkage scan (for details, see Box 9-A). Their results showed that a whopping 74% of the variance in iris colors was due to a QTL at chromosome 15q. Zhu's results were very clear—LOD scores in excess of 19 were observed for this interval in agreement with Eibergh and Mohr in 1987, who showed that most brown/nonbrown iris color variation could be attributed to sequences on this chromosomal arm. The LOD peaks over *TYRP*, the second most important gene estimated from Frudakis et al. (2003b), paled in comparison and it was clear that of each of the steps that Frudakis's lab took to improve their classification results, the third would probably be among the most important.

OCA2—THE PRIMARY IRIS COLOR GENE

The aim of the second screen by Frudakis et al. (2006) was to completely mine the OCA2 haplotype structure for iris color information, and identify a complement of markers that could be used to predict iris color from DNA. A first impulse would be to rely on hap-tags from the hap-map project. However, LD among SNP alleles comprising major haplotypes from this project is often not complete, and discovery sample sizes employed are low (i.e., Perlegen's discovery set n = 25). This means that although the hap-tags represent most of the major haplotypes, rare (<5%) haplotypes are not reliably detected and much of the finer level of structure remains undiscovered (Clark et al. 2005). For a complex locus that likely has been subject to sexual selection and population bottlenecks, such rare haplotypes may combine to represent a significant fraction of locus variation.

Given our goal of classifying iris color from DNA, we reasoned that we would need to appreciate this finer level of structure and that relying on hap-tag SNP markers for our purposes would be an unwise application of reductionism in light of the complexity of the phenotype, the great length of the gene, and its likely history of selection. Thus, rather than rely on hap-tag SNPs, we used an

Table 9-3

OCA2 SNPs part of the model used for the direct inference of iris color.

SNP ID and haplotype position	DeCode map position	rs Number	Associated allele/ genotype (color)	Self-reported iris colors Fisher's P-value (n)	Reference	Digitally quantified iris colors Fisher's P-value (n)	Associated genotype power ($\alpha = 0.05$)	Associated genotype odds ratio [95% CI]	Sequence
OCA2-A-1	15.46 cM (intron)	rs1874835	T (dark) TT (dark)	N/A	N/A	p = 0.030 (n = 1208) p = 0.087 (n = 604)	0.100	1.5 [0.9–2.4]	CAAACTCCAG AGGAAACAGGG [g/t] AGACGATCGC TTCTGCATTT
OCA2-A-2	15.57 cM (intron)	rs2311470	G (light) GG (light)	N/A	N/A	p < 0.001 (n = 1174) p < 0.001 (n = 587)	0.997	3.0 [2.0–4.7]	TTCATTTAAT TTATTAACTT [g/c] ATGTTTTAA AGCCTCACAA
OCA2-A-3	15.87 cM (intron)	rs1375170	T (dark) TT (dark)	N/A	N/A	p = 0.147 (n = 1212) p < 0.001 (n = 606)	0.950	3.1 [0.8–11.5]	TCTCTCAGCT AAATCCATAC [g/t] CTTGCTGAGG CCATGGGGGA
OCA2-A-4	15.93 cM (intron)	rs12439067	C (dark) CC (dark)	p < 0.01 (n = 851)	Frudakis et al., 2003	p < 0.001 (n = 1214) p < 0.001 (n = 603)	0.738	2.3 [1.2–4.3]	TATCTAACCC TCACTGAGCT [c/t] TGCAGGGGGT ACACAGCCGA
OCA2-A-5	15.95 cM (coding synonymous)	rs1800411	C (dark) CC (dark)	p < 0.01 (n = 851)	Frudakis et al., 2003	p < 0.001 (n = 1214) p < 0.001 (n = 607)	0.989	3.8 [2.0–7.0]	CACATGTTCA TTGGGATTTG [c/t] CTTGTTCTCC TGGTCTGCTT
OCA2-A-6	16.01 cM (intron)	rs10852218	A (dark) AA (dark)	p < 0.01 (n = 851)	Frudakis et al., 2003	p < 0.001 (n = 1212) p = 0.004 (n = 606)	0.998	7.2 [1.63–32.0]	CTTGTCTTCT TCTTTTCCCC [a/g] TAGATGATCT TAGTAGCCAT
OCA2-A-7	16.01 cM (intron)	rs1900758	G (dark) GG (dark)	p < 0.01 (n = 851)	Frudakis et al., 2003	p < 0.001 (n = 1212) p < 0.001 (n = 606)	0.911	3.7 [2.1–6.6]	GTGCATGAAA AGGTGGGGGC [a/g] GTTGAGCCCA CAGCTCACTG

(Continued)

535

Table 9-3
(Continued)

SNP ID and haplotype position	DeCode map position	rs Number	Associated allele/ genotype (color)	Self-reported iris colors Fisher's P-value (n)	Reference	Digitally quantified iris colors Fisher's P-value (n)	Associated genotype power ($\alpha = 0.05$)	Associated genotype odds ratio [95% CI]	Sequence
OCA2-A-8	15.68 cM (intron)	rs1037208	C (dark) CC (dark)	$p < 0.01$ (n = 851)	Frudakis et al., 2003	$p < 0.001$ (n = 1202) $p = 0.001$ (n = 601)	0.998	8.3 [1.9–36.5]	GTGAAATAAT TTCCATGATT [a/c] CTTCCTAAAT ATTGAATATA
OCA2-A-9	16.12 cM (intron)	rs749846	T (dark) TT (dark)	$p < 0.01$ (n = 851)	Frudakis et al., 2003	$p < 0.001$ (n = 1112) $p < 0.001$ (n = 556)	0.998	17.9 [2.4–135.5]	CAGTGTGAAC TGTGTAGGTT [g/t] TGTGTGGTCC CCTGTGCCTG
OCA2-A-10	16.21 cM (intron)	rs895829	C (dark) CC (dark)	N/A	N/A	$p < 0.001$ (n = 1206) $p < 0.001$ (n = 603)	0.972	6.9 [2.1–22.9]	TACTCAGGCA CGTAAGACTC [c/t] GTGAGCAAGT ATAACGCAAG
OCA2-B-1	15.36 cM (intron)	rs1498519	C (dark) CC (dark)	N/A	N/A	$p < 0.001$ (n = 1212) $p = 0.002$ (n = 606)	0.994	1.96 [1.4–2.7]	CTAGTGTCCT CACTAAGCAA [a/c] CCCTCAGGAC ACAGGTTCCA
OCA2-B-2	15.61 cM (intron)	rs1004611	T (dark) TT (dark)	N/A	N/A	$p < 0.001$ (n = 1188) $p = 0.046$ (n = 594)	0.438	1.57 [1.0–2.4]	AATAATGCAA AGTGCATCAA [c/t] GTCAGAAAGG TGCACCAGAG
OCA2-B-3	15.76 cM (intron)	rs3099645	T (light) TT (light)	N/A	N/A	$p = 0.002$ (n = 1146) $p = 0.013$ (n = 573)	0.208	3.7 [1.2–11.3]	ATGAAAAAAA AGTTGTTGTT [g/t] TTTTTTTTT TCCTATTTTT
OCA2-B-4	16.12 cM (intron)	rs3794606	G (dark) GG (dark)	N/A	N/A	$p < 0.001$ (n = 1206) $p < 0.001$ (n = 603)	0.948	8.8 [3.4–22.6]	atgtctgtgt gtgtgcacca [g/t] tgtgaactgt gtaggttgtg

OCA2-B-5	16.12 cM (intron)	rs2305252	G (light) GG (light)	N/A	N/A	p = 0.007 (n = 1210) p = 0.317 (n = 605)	0.995	1.7 [1.2–2.3]	TCCTCCCCG TCTCTTTGC [a/g] TTCAGCGTGA CCCATCTTCC
OCA2-B-6	16.21 cM (intron)	rs895828	C (dark) CC (dark)	N/A	N/A	P < 0.001 (n = 1200) p = 0.142 (n = 600)	1.000	2.0 [1.4–3.0]	AGGTGAGTTG CTGGCCTATC [g/c] CCTTACTCAG GCACGTAAGA
OCA2-C-1	16.01 cM (coding non-synonymous)	rs1800407	A (dark) GA (dark)	N/A	N/A	p = 0.004 (n = 1212) p = 0.002 (n = 607)	0.475	2.0 [1.2–3.2]	CCAGGCATAC CGGCTCTCCC [a/g] GGGACGGGTG TGGGCCATGA
OCA2-C-2	15.49 cM (intron)	rs924314	C (light) CC (light)	N/A	N/A	p < 0.001 (n = 1192) p < 0.001 (n = 596)	0.893	2.4 [1.5–3.7]	GTCAGTGAAA CATCCCAAA [t/c] GGTAAGCTTC AGCATTGTCC
OCA2-C-3	15.50 cM (intron)	rs924312	C (light) CC (light)	N/A	N/A	p < 0.001 (n = 1168) p < 0.001 (n = 584)	0.805	2.3 [1.5–3.6]	TTCTCCCTTT AAGCAAAAGG G/C GGCTCTCTTT TGTTTCCTTT
OCA2-C-4	15.58 cM (intron)	rs2036213	A (dark) AA (dark)	N/A	N/A	p < 0.001 (n = 1212) p = 0.050 (n = 606)	0.422	1.55 [1.0–2.4]	GATGTTATAT GTTTTGAAAG [a/c] AGAAATACTT CAATTAGTAA
OCA2-C-5	16.03 cM (intron)	rs735066	A (dark) AA (dark)	N/A	N/A	p < 0.001 (n = 1182) p < 0.001 (n = 591)	0.955	5.3 [2.0–14.0]	GTGGCCCACA GTGAGTGCTT [a/g] GGACATCTGG ATGCAGCTAT
OCA2-C-6	16.03 cM (coding synonymous)	rs1800404	C (dark) CC (dark)	N/A	N/A	p < 0.001 (n = 1208) p < 0.001 (n = 604)	0.778	3.5 [1.3–9.4]	ATCGTGCACA GAACTCTGGC [a/g] GCCATGCTGG GTTCCCTTGC

alternative approach, aimed not at formally mapping the haplotype structure of the locus but at extracting the iris color information from it in the most efficient manner possible.

Frudakis et al. (2006) examined 147 OCA2 SNPs covering the length of the locus, in a sample of 1,317 subjects broken into two eye color score groups—those with color scores above the mean value of 2.07 (including blue, green, blues and greens with small brown rings/sectors/flecks, etc.) and those below (the darker irides; hazels, blues with thick brown rings, solid browns, etc.). The most strongly associated SNPs (marginally associated, n = 22) are presented in Table 9-3 (OCA2-A-1 through OCA2-C-6). Odds ratios for the genotype—color pair indicated in column 4 of Table 9-3—ranged from 1.5 to 18 and most were significantly above 1.0 (column 8, Table 9-3). Six of these SNPs in Table 9-3 previously were described as associated with iris color (Frudakis et al. 2003) and their p-values were of similar strength as previously reported (compare p-value columns 5 and 7, Table 9-3). Bona fide associations should arise from studies well powered to detect them, and due to the relatively high minor allele frequencies, this particular study was well powered to detect the strong associations in Table 9-3—for most they had a power well over 90% (column 9, Table 9-3).

To investigate whether the alleles of the OCA2 SNPs in Table 9-3 were useful for predicting iris color, the SNPs were arranged into haplotype groups. The groupings indicated by the letters A, B, and C in Table 9-3 were chosen because these SNP combinations produced haplotype alleles that were the most intensely associated with iris colors. An alternative would be to group the SNPs based on physical location, which we will discuss shortly. Using the A, B, and C groupings, they then determined the concordance of iris color among samples sharing the same phased, multilocus OCA2 genotypes across the three groupings (for example OCA2-A, B, C "1/1, 1/1, 1/1" represents a multilocus OCA2 genotype). The concordance was good but not perfect. To improve the concordance among samples with shared multilocus OCA2 genotypes, they augmented the system with additional associated polymorphisms discovered using a hierarchical screening process. Instead of screening for markers marginally (independently) associated with iris color in the entire sample, they focused on associations that seemed to fill in the gaps of information possessed by those in Table 9-4 (see Box 9-B).

Box 9-B

Hierarchical OCA2 Screening

The author and colleagues used a hierarchical SNP screening process for identifying apparent associations within confounded multilocus genotype groups—associations that were context dependent. For example, consider that the haplotypes for each of the three SNP sets in Table 9-3 (OCA2-A, OCA2-B, and OCA2-C) can be numbered 1, 2, 3...n, where n is the total number of haplotypes that exist. Consider the multilocus genotype OCA2-A,B, C "1/1, 1/1, 1/1", where haplotype 1 exists in a homogeneous state for each of the three SNP collections. Assume that of 22 individuals with this genotype, 18 had blue eyes and four had brown. If an OCA2 SNP in the hierarchical screen showed GG genotypes for each of the 18 blue iris samples but GA genotypes for two, three, or four dark iris genotypes then its alleles would appear to explain some or all of the confounders within this particular confounded multilocus OCA2 genotype group. The idea of the screen was that if alleles of a SNP exhibited this type of pattern for many of the confounded multilocus OCA2 genotype groups, whether they were marginally (independently) associated with iris color in the entire sample or not, the SNP likely possessed information necessary for predicting iris color within the context of the OCA2-A/B/C multilocus genotype system (i.e., their alleles were potentially of haplotype context dependent influence but low overall penetrance). Though the associations are merely apparent (due to the complexity of the haplotypes, the sample size of each group is not adequate to demonstrate an association formally), this hierarchical method in the face of high variable/sample ratios is fundamentally similar to that used by algorithms that detect RNA expression signatures from DNA microarray data. Though most of the components of these signatures are not significantly associated on their own, the predictive/diagnostics value of the signatures as a whole have been impressive in cross-validation studies which are described in this chapter.

Frudakis et al. (2006) ranked the 124 OCA2 SNPs based on the number of confounded multilocus OCA2 genotype groups explained and selected the top eight. The top two SNPs (OCA2-D-5, OCA2-D-6, Table 9-4) each explained 33% of the confounded multilocus OCA2 genotype groups on their own, and the alleles for all eight SNPs appear to resolve most of the discrepancies in all but four of the confounded multilocus OCA2 genotype groups (see Table 9-4). To resolve these four, an additional three SNPs were added (OCA2-D-3, OCA2-D-4, OCA2-D-9, Table 9-4). The resulting 11 SNPs were assembled into a fourth haplotype group (OCA2-D) and named by number according to their genetic position along the OCA2 locus (OCA2-D-1 through OCA2-D-11, column 1, Table 9-4). Among these 11 OCA2-D SNPs, the potential power for discriminating among the confounders within OCA2A/B/C groups seemed uncoupled from the association with iris color scores in the entire sample. For example, alleles of OCA2-D-6 appeared to explain some or all of the confounders in no less than 13/40 (33%) of the confounding groups, but were not associated with iris color scores in the overall sample.

Eleven more SNPs were added to the collection (see Table 9-4), giving a total of 33 SNPs spanning the OCA2 locus from 15.05 cM to 16.36 cM along chromosome 15. Most were located within introns, though one (OCA2-C-1) was nonsynonymous, two were synonymous, and three were located in the 3′UTR (column 2, Tables 9-3 and 9-4). Many of the 33 SNPs were in linkage disequilibrium (LD) with one another (see Figure 9-11), but none was completely redundant to any other—that is, at least one recombinant was observed for each pair of SNP alleles.

Figure 9-11 shows the linkage disequilibrium (LD) plot for the SNPs in Tables 9-3 and 9-4, in chromosomal order along the OCA2 locus (bar, top,

Table 9-4

OCA2 SNPs conditionally associated with iris color score.

Haplotype position	DeCode map position	rs Number	Associated genotype (color)	Self-reported iris colors Fisher's P-value (n)	Reference	Digitally quantified iris colors Fisher's P-value (n)	Confounder groups resolved (percent)	Sequence
OCA2-D-1	15.36 cM (intron)	rs4778177	N/A	N/A	N/A	p = insig. (n = 159) p = insig. (n = 159)	8/46 (17%)	GTCTCCTCCC AGCCCAGCAG [c/t] CCTGCTTCCC AGGAGGGCAC
OCA2-D-2	15.39 cM (intron)	rs8023340	N/A	N/A	N/A	p = insig. (n = 159) p = insig. (n = 159)	5/46 (11%)	GTGCGGCCTG GCGACCCCAC [a/g] GTTGGCGGCC TGTTTCTTC
OCA2-D-3	15.55 cM (intron)	rs4778190	N/A	N/A	N/A	p = insig. (n = 159) p = 0.003 (n = 159)	1/40 (3%)	CCTGAGCATT CACCACCCCC [t/c] GCTCCGTCAG AGACTCCTGC
OCA2-D-4	15.76 cM (intron)	rs2871886	T (light) TT (light)	N/A	N/A	p = 0.0075 (n = 159) P = 0.042 (n = 159)	2/40 (5%)	TGGGAAACAA AGGAGTGAGA [t/c] AGGAGCACTC CCTCCCTTAT
OCA2-D-5	15.86 cM (intron)	rs977588	A (light) AA (light)	N/A	N/A	p = 0.096 (n = 159)	13/40 (33%)	TATTTTTCTT ACTACAACAG [c/a] AACATTTTAAA AAGGAAGAT

Marker	Position (region)	rs number	Allele	Genotype			Frequency	p-values	Sequence
OCA2-D-6	15.86 cM (intron)	rs977589	N/A	N/A	N/A	N/A	13/40 (33%)	p = insig. (n = 159) / p = insig. (n = 159)	TGGTGATTCC TTCACCGCTC [t/c] TCTAACACAG ATCACCACAT
OCA2-D-7	15.87 cM (intron)	rs1448490	G (light)	GG (light)	N/A	N/A	7/40 (18%)	p = 0.070 (n = 159) / p = 0.0484 (n = 159)	AATTTCTGGT ACCCTCGGGG[a/c] CTAATAAATG CAGTGATATG
OCA2-D-8	15.88 cM (intron)	rs1545397	N/A	N/A	N/A	N/A	4/40 (10%)	p = insig. (n = 159) / p = insig. (n = 159)	TTGTGAATAT ACTAAAATAC[t/a] CTGAATGATA TAATTTTGCA
OCA2-D-9	16.30 cM (3′ UTR)	rs4778137	G (light)	GG (light)	N/A	N/A	3/40 (8%)	p = 0.092 (n = 159) / p = insig. (n = 159)	ATGTATAAAT TACATACATA [g/c] AACAGAGGCC CAAAAGACAC
OCA2-D-10	16.36 cM (3′ UTR)	rs11855019	T (light)	TT (light)	N/A	N/A	6/40 (15%)	p = 0.016 (n = 159) / p = 0.012 (n = 159)	AATATAACAT ATCAAAATTG[a/g] CAGAACACAG CTAAATCAGT
OCA2-D-11	16.36 cM (3′ UTR)	rs6497268	G (light)	GG (light)	N/A	N/A	10/40 (25%)	p = 0.001 (n = 159) / p < 0.001 (n = 159)	CAGTTTTAA AAAAAACTA[a/c] AATTGCAACT ACCAGCCAAC

Figure 9-11

LD plot for OCA2 SNPs associated with iris colors shown in Tables 9-3 and 9-4. Numbers within boxes indicate the extent of LD between each pair of markers and the color coding represents the extent of LD signified by that number (from red, strongest, to white, weakest). SNPs are identified by name as well as numbered from 1–22. Locations along the OCA2 locus are shown at the top of the diagram. The plot was created using MIT's Haploview program, and the LD blocks selected by this program are indicated with Black diamonds at the top of the plot. However, there are two main areas of intense LD—from SNPs 1–12 and from 13–25.

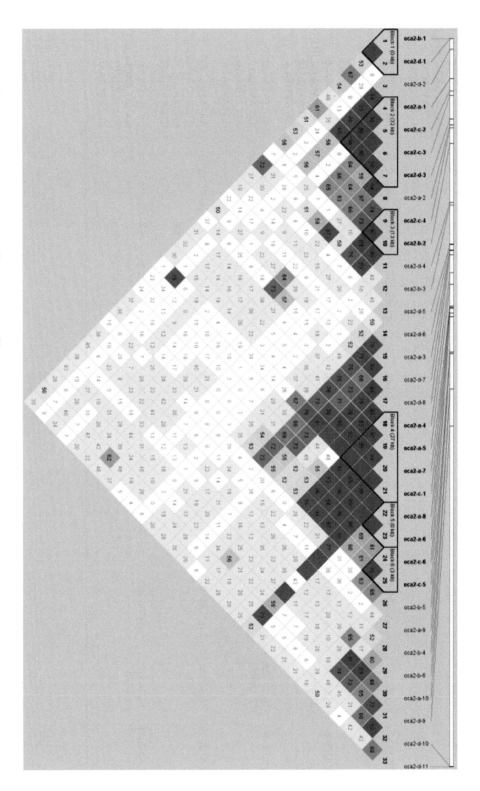

Figure 9-11). Recall that LD is a statistical measure of the allelic association between loci created by departures from random mating (assortative mating, inbreeding, etc., which contribute to population structure) or epistasis among functionally relevant sequences. In lieu of these sorts of systematic genetic forces, LD decays predictably over time.

The program Haploview identified six haplotype blocks, but visual inspection reveals two main regions of extensive LD. The OCA2-A, B, and C SNPs appear to be randomly distributed among the LD regions. Though the information contained about iris color in each of these SNPs was partially overlapping, the fact that none of the SNPs were in complete LD with any other implies that each SNP contributed unique information in our mulitlocus OCA2 genotype classification system (i.e., even the alleles for SNPs in strong LD are expected to appear on unique haplotypes in at least some of the samples). Recall that Frudakis et al. (2006) was attempting to avoid reducing the complexity of OCA2 variation to hap-tags representing most of the variation, reasoning that such a reduction may present an oversimplification for a complex phenotype classification system. In contrast, they aimed to observe all the extant haplotype diversity with respect to iris-color informative SNP alleles and empirically relate these to iris colors through reference to a database of samples—much like we do when attempting to draw indirect inferences from DNA, based on admixture.

AN EMPIRICAL OCA2-BASED CLASSIFIER FOR THE INFERENCE OF IRIS COLOR

If variable melanin content of the iris is a function primarily of OCA2 variation, and if the phase-known alleles of the 33 SNP markers are associated with iris colors and account for most of the variation in this gene relevant to iris color, then this database of phenoypes and multilocus OCA2 genotypes constitutes a rudimentary iris color classification system. That is, the iris color of one sample with a given multilocus OCA2 genotype should tell us the iris color of another sample with the same multilocus OCA2 genotype, and a measure of iris color concordance among samples within shared genotype groups is a measure of the classification accuracy for both model (used to select the SNPs) and validation samples (not used to select the SNPs). Among a separate set of 1,072 samples (602 used in the discovery of the SNPs in Tables 9-3 and 9-4, and 470 validation samples), the concordance of iris colors of matching multilocus OCA2-A/B/C/D genotypes was good (see Figure 9-12). There are examples of discordance (e.g., Figures 9-12B, 9-12I) but for the most part, samples with the same multilocus OCA2 genotype seem to be of similar melanin content. For example, the multilocus genotype (24,48), (1,3), (6,9), (71,151) seems to specify brown color

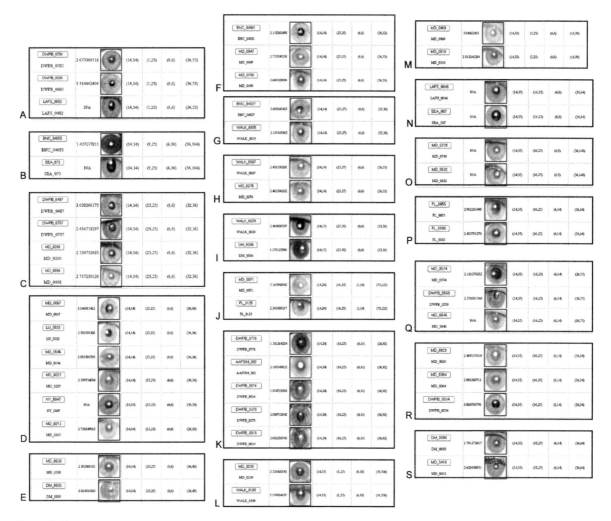

Figure 9-12

Individuals sharing multilocus OCA2 genotypes, defined by the OCA2-A/B/C/D system discussed in the text tend to share gross aspects of ultrastructural iris coloration. Each box represents a unique multilocus OCA2 genotype. The iris melanin index for each iris is shown in the second column of each box and a digital photograph of the iris in the third column. The fourth column from the left is the diploid OCA2-A haplotype combination, the fifth the diploid OCA2-B combination, the sixth the diploid OCA2-C combination, and the seventh the diploid OCA2-D combination, where the SNP constitution of each A, B, C, and D haplotype is shown in Tables 9-3 and 9-4. For the multilocus OCA2 genotype to be the same for each sample, the values in these four boxes must be identical. No information on the phasing of haplotypes among these four regions is discernible in this figure. One will note variability within each multilocus OCA2 genotype set—such as K for instance, where each iris has detectible regions of brown of different size. B represents a discordant pair of irides with the same multilocus OCA2 genotype, A, relatively concordant, and C and D, virtually concordant.

(see Figure 9-12B), whereas the multilocus genotype (14,14), (25,25), (6,6), (36,36) seems to specify blue color (see Figure 9-12D).

However, Figure 9-12 gives merely a visual impression. What we need is a method of quantifiably predicting color and calculating the accuracy of the

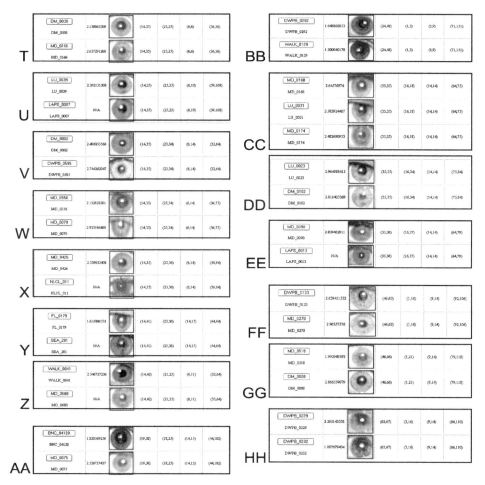

Figure 9-12
(Continued)

predictions. To determine whether the color of an iris A was concordant with that of others {B, C, D, ...} of the same multilocus OCA2 genotype, we can calculate the average color score for {B, C, D, ...} as the point estimate of A, take a range (R) around this estimate, and determine whether the color score of A falls within R. The range of color scores defined by R represents a predicted range of scores for sample A and would more directly allow us to infer melanin content (color being inferred indirectly since irides of the same melanin content likely express slightly different colors and patterns).

The breadth of this range R was determined by defining intervals (I) around each component variable used to calculate R—the Luminance, Red, Green, and Blue reflectance. These intervals usually represent the standard deviation of values obtained from the matching samples {B, C, D, ...}, or if only one matching sample is available, a default value that may be derived from the average I experienced in other matching genotype sets. Using an I = ± 20 units we cover 40 units or about 23% of the observed range for each component variable of the iris melanin index (C) and produce a high and low C defining a range (R) of about ± 0.45 units. For estimates between 1.35 and 2.45 (most irides), this R represents about 45% of the entire iris color score range about the point estimate (refer to Figure 9-8A) and for others, less than 45% (proportionally less the farther along the extremes of the iris melanin index scale).

Using an I = ± 20, R = ± 0.45 about each point estimate for each of the 82 samples, Frudakis et al. (2006) showed that the iris color for 92.9% of the 82 samples fell within the predicted range (see Table 9-5). In contrast, and not inconsistent with expectations, the frequency with which the iris color of randomly selected samples fell within these same ranges was only about 55.8% (see Table 9-5) and the difference in concordance between samples grouped by multilocus OCA2 genotypes and randomly grouped samples was highly significant (p < 0.0001). The rate with which the iris color of model samples fell within the predicted range (92.7%) was similar to that with which the iris color of validation samples fell within the predicted range (92.3%) (see Table 9-5), and both rates were significantly different from that obtained using randomly matched samples (p = 0.0007 and p < 0.0001, respectively).

Using an I = ± 23, R = ± 1.0 about the point estimate for each, they observed the iris color for 96.3% of the 82 samples fell within the predicted range (p < 0.0001), again with a similar rate for model (96.4%, p < 0.0001) and validation samples (96.3%, p = 0.0018, Table 9-5). In contrast, the average frequency with which the iris color of randomly selected samples fell within these same ranges was only 65.9% (see Table 9-5).

Table 9-5

Iris color concordance within multilocus OCA2 genotype groups using two different evaluation criteria.

Comparison	Matched samples (N)	R* = ±1.0 (I** = 22)	R* = ±0.09 (I** = 20)
MODEL OCA match	56	96.40%	92.90%
VALIDAT OCA match	26	96.15%	92.30%
TOTAL OCA match	82	96.30%	92.70%
RANDOM 1	82	67.10%	58.80%
RANDOM 2	82	72.00%	63.40%
RANDOM 3	82	58.50%	45.10%
AVG RANDOM	240	65.90%	55.80%

*Iris color score range.
**Range of iris color score parameters.

We have presented the grouping of samples by multilocus OCA2 genotype in Figure 9-12 so you could appreciate how similar the iris colors are among samples that share genotypes. But for the classification exercise based on these groupings, we need to present the *predictions* in a different way. How do we get a predicted iris color from a match in the database? For example, take Figure 9-12A—if the top sample were a crime scene specimen (for which we had no iris photo), how would we predict the color based on the observation that it fits into a multilocus OCA2 genotype group with the other two below it? We must first determine the average iris color of these two (in terms of the iris melanin index), specify a reasonable range around this average, and ask the database to present a gallery of irides that fall within this range. We actually use the iris melanin index components rather than the index itself and ask the database to return those irides falling within the inferred luminance $\pm$ I, and red $\pm$ I, and green $\pm$ I and blue $\pm$ I reflectance.

The return for this query presents a gallery of digital photographs for irides within the predicted R for the sample, and by inference, of predicted similar overall melanin content. With a database of 1,072 samples, the galleries typically include 100 or more (dependent on the values) iris photographs, but in Figure 9-13 we have presented six photographs (below the line) representative of the range for each of the 82 test samples (iris photograph above the line).

Though the summary statistics in Table 9-5 represent the formal demonstration that the multilocus OCA2 genotypes we have described are predictive for iris color, Figure 9-13 illustrates that the predictions are also visually satisfying (i.e., correct by most reasonable accounts). In visual terms, the results seem similar for validation samples (the 26 boxed photo sets, Figure 9-13) and model samples (unboxed photo sets, Figure 9-13). Those for which the actual color score of the test iris fell outside of the predicted R using an $I = \pm 20, R = \pm 0.45$ (i.e., incorrect) are shown in the lower right of Figure 9-13 (samples 75–80, with boxes or circles around the sample ID). Those for which the actual color score of the test iris fell outside the predicted R using an $I = \pm 3, R = \pm 0.5$ are shown in the lower right with boxes around the sample ID (samples 78–80, Figure 9-13).

Most human observers would probably conclude that many of these latter six predictions were inaccurate as well, though some may consider samples 75 and 76 correct. Though the use of terms such as light or dark are subjective, visually speaking, the predicted ranges tend to be fairly restrictive, either clearly specifying a light (i.e., samples 10–17, etc. Figure 9-13), intermediate (i.e., samples 4, 19, 20, 49, etc., Figure 9-13), or dark color (i.e., samples 3, 7, 8, 9, 47, etc., Figure 9-13). One exception is sample 74, which shows a wide color range including blues and browns, but all of dark hue (like the test sample).

Interestingly, of the 82 multilocus OCA2 genotypes populated with multiple samples, 62 (76%) specified iris color score ranges covering mainly lighter

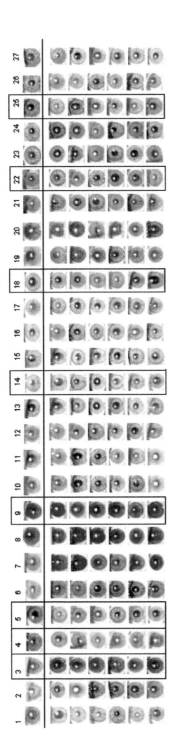

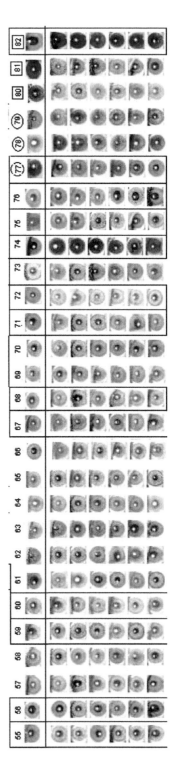

colors and hues. Since the observation of multilocus OCA2 genotype groups is a function of representation in the database, which is a function of haplotype allele frequencies, this observation may suggest that there are a greater number of haplotypes specifying darker colors than lighter colors. If so, this would imply that lighter iris colors are derived and darker iris colors are ancestral, which would fit with our understanding of our recent past, as well as modern day distributions of colors.

In Chapter 8 we discussed that one of the reasons it was important to be able to measure admixture and correlations between admixture and phenotype was so that it is possible to condition a statistical association analysis on individual admixture. Iris color is a perfect application of this technique. Since iris colors are unevenly distributed among world populations we needed to assess whether our associations were merely reflections of (i.e., correlations with) crude, continental population stratification (i.e., false positives). Such correlations would result in an ability to correctly infer iris color as we have just discussed, but for the wrong reasons—because they are good markers of ancestry admixture, which correlates with phenotype, not because they describe variation in the underlying phenotypically active locus (in other words, due to indirect as opposed to direct inferential power).

First we need to determine whether iris color is unevenly distributed as a function of ancestry *in our sample*, which for our purposes here is the only thing that matters (not that iris color is normally distributed unevenly or evenly across the globe). Our sample was almost entirely comprised of Caucasians, and each sample typed with the 171 AIM AncestryByDNA panel (basic MLE algorithm) of predominantly European admixture (see Figure 8-21). Recall from Chapter 8 that in this sample, European admixture was positively correlated

---

Figure 9-13

A visual presentation of predicted and observed iris color for test inferences based on the OCA2-A/B/ C/D haplotypes (see Tables 9-3 and 9-4) and groupings from a database of 1,072 subjects discussed in the text and shown in Figure 9-11. This figure illustrates the general concordance of iris colors among samples with matching multilocus OCA2 genotypes as defined in the text. Inferred Iris Melanin Index parameters are computed for each iris within a multilocus OCA2 genotype group of Figure 9-11 based on others within this same group, as discussed in the text. These parameters were used to query the database for all irides falling within a prespecified range, and of the irides returned, six are shown below the horizontal line. These represent the predictions. The test iris is shown above the line and the predictions for this test iris below the line. Correct inferences are those for which the test iris falls within the computed parameters and appears in this figure to be of similar color or within the range of colors shown below the line. Circles around the sample number indicate that the test iris fell within the range of predicted color scores when determination of the range of the prediction used the $I = \pm 23$, $C = \pm 0.5$ criteria but not the $I = \pm 20$, $C = \pm 0.45$ criteria. Samples with boxes around the sample number indicate that the test iris did not fall within the range of predicted color scores using neither $I = \pm 23$, $C = \pm 0.5$ nor $I = \pm 20$, $C = \pm 0.45$ range criteria and are the most overtly incorrect. All others fell within the predicted range using either of the two criteria.

with lighter iris colors (R = 0.28, p < 0.0001; Figure 8-21A) and sub-Saharan African (R = −0.24, p < 0.0001; Figure 8-21B), East Asian (R = −0.1042, p < 0.0059; Figure 8-21C; also see Table 5-19, Chapter 5) and Native American admixture (R = −0.152, p < 0.0001; Figure 8-21D) were correlated with darker iris colors in Caucasians.

These results raised the possibility that our OCA2 associations were due to ancestry information content—that is, that the OCA2 markers were informative for an element of ancestry correlated with iris colors (i.e., they were merely artifacts of population structure). Thus, it was then incumbent on us to prove that our associations were robust to the admixture in our sample. To do this, we performed multiple regression analysis for each of the OCA2 SNPs in Table 9-3 and 9-4. Genotypes for each of these SNPs were incorporated one at a time into a multiple regression model along with non-European (sub-Saharan African, East Asian, and Indigenous American) admixture as independent variables, to assess the independent association of each toward explaining variation in iris color scores (dependent variable). The sample size for this was n = 695, significantly lower than that used to obtain the p-values in Tables 9-3 and 9-4, but the results were similar.

Recall that up to now we have discussed the association between alleles and elements of population structure as a useful tool for analysis. When demonstrating a functional association between an allele and a phenotype, rather than merely a correlation, one needs to show that the association is independent of such population structure. Multiple regression and Bayesian analysis are useful for this purpose. For the markers discussed in this section, we used multiple regression, which seeks to examine the relationship between a dependent variable (such as phenotype) and multiple independent variables (genotypes, admixture proportions, etc.). With multiple regression we can test the significance of the phenotype–genotype association for all independent variables together in a combined model as well as test the contribution of each independent variable on its own—minus any correlations each variable may have with other associated variables. With this type of analysis, we showed that alleles for most of the OCA2 SNPs (29/33) were associated with iris color scores independently from non-European (or by definition, European) admixture (see Table 9-6). This result indicated that the OCA2 SNP allele associations were not due to BGA admixture and thus population stratification on a continental scale.

Frudakis and colleagues have also constructed the iris color database using haplotypes defined by SNPs part of OCA2 LD clusters, rather than the groupings shown in Tables 9-3 and 9-4, which were those producing the haplotype allele associations with the lowest p-values. Grouping the first 12 SNPs of Figure 9-11 into an OCA2-E set, the next 13 SNPs of Figure 9-11 into an OCA2-F set, and the last eight into an OCA2-G set, we are able to combine SNPs

Table 9.6

Multiple regression of OCA2 genotype values and continental admixture on iris melanin index reveals that most OCA2 genotype associations are independent of admixture.

A

	A-1	A-2	A-3	A-4	A-5	A-6	A-7	A-8	A-9	A-10
OCA2-*	0.015	<0.0001	0.0019	<0.0001	<0.0001	0.0002	<0.0001	0.0005	<0.0001	<0.0001
AF	<0.0001	<0.0001	<0.0001	<0.0001	<0.0001	<0.0001	<0.0001	<0.0001	<0.0001	<0.0001
EA	0.0064	0.0176	0.0037	0.0684	0.0657	0.0044	0.071	0.0045	0.0712	0.1279
NA	0.0004	0.0008	0.0007	0.0016	0.0012	0.0005	0.0014	0.0005	0.0016	0.001
R (adjusted)	0.0876	0.1192	0.0926	0.1172	0.1162	0.098	0.1134	0.0959	0.1276	0.1556

B

	B-1	B-2	B-3	B-4	B-5	B-6
OCA2-*	<0.0001	0.0001	0.3043	<0.0001	0.0001	0.001
AF	<0.0001	<0.0001	<0.0001	<0.0001	<0.0001	<0.0001
EA	0.0135	0.0063	0.0059	0.0476	0.0041	0.0188
NA	0.0003	0.0003	0.0004	0.0004	0.0002	0.0002
R (adjusted)	0.103	0.1003	0.0813	0.1291	0.0993	0.0941

C

	C-1	C-2	C-3	C-4	C-5	C-6
OCA2-*	0.0328	<0.0001	0.0001	0.0001	0.0037	0.0036
AF	<0.0001	<0.0001	<0.0001	<0.0001	<0.0001	<0.0001
EA	0.0027	0.0151	0.0086	0.0058	0.0205	0.0189
NA	0.0003	0.0003	0.0003	0.0003	0.0009	0.0009
R (adjusted)	0.0859	0.1149	0.1012	0.1005	0.091	0.0911

D

	D-1	D-2	D-3	D-4	D-5	D-6	D-7	D-8	D-9	D-10	D-11
OCA2-*	0.0669	<0.0001	<0.0001	0.2947	0.0087	<0.0001	0.0004	0.4883	<0.0001	<0.0001	<0.0001
AF	<0.0001	<0.0001	<0.0001	<0.0001	<0.0001	<0.0001	<0.0001	<0.0001	<0.0001	<0.0001	<0.0001
EA	0.0093	0.0054	0.0119	0.0051	0.015	0.0194	0.054	0.0103	0.0424	0.1194	0.0965
NA	0.0003	0.0002	0.0009	0.0004	0.0003	0.0006	0.0006	0.0004	0.0002	0.0001	0.0005
R (adjusted)	0.0843	0.1095	0.1112	0.0813	0.089	0.104	0.0966	0.0805	0.1121	0.2067	0.2357

*Refers to the genotype id heading each column. Values shown in each cell are p-values for the regression coefficients.

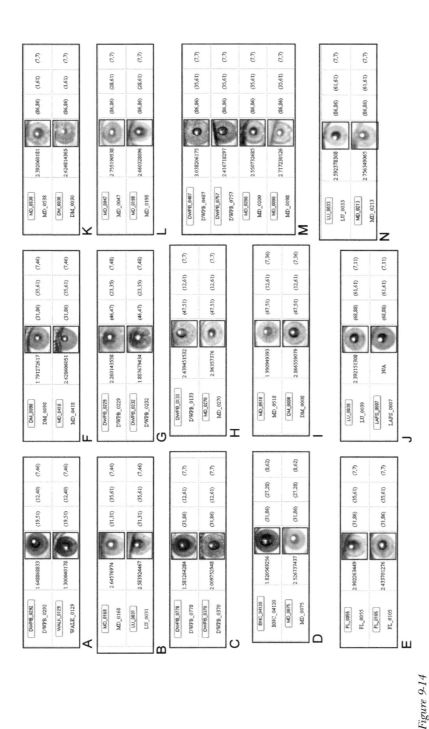

Figure 9-14

Concordance of iris colors among samples of the same multilocus OCA2 genotypes defined by the OCA2-E/F/G SNP grouping scheme discussed in the text, as in Figure 9-12. The OCA2-E/F/G scheme groups SNPs based on their contribution to one of three basic LD regions within the OCA2 gene, as shown in Figure 9-11 (SNPs 1–12, 13–25, and 26–33). Each box represents a unique multilocus OCA2 genotype. The Iris Melanin Index for each iris is shown in the second column of each box and a digital photograph of the iris is shown in the third column. The fourth column from the left is the diploid OCA2-E haplotype combination, the fifth the diploid OCA2-F combination, the sixth the diploid OCA2-G combination. For the multilocus OCA2 genotype to be the same for each sample, the values in these four boxes must be identical. No information on the phasing of haplotypes among these four regions is discernible in this figure. The concordance observed here is good—even among irises of mixed blue/brown color as shown in G—and there are fewer instances of outright discordance compared to the OCA2-A/B/C/D scheme of Figure 9-12, though we observe fewer samples with matching genotypes using this scheme.

in a manner that captures most of the LD among them along the OCA2 locus. Samples sharing multilocus OCA2 genotypes defined in this manner were fewer in number with a database size n = 1,072, but more concordant, indicating that phase inference error plays some role in the cases of discordant observations using the OCA2-A/B/C/D haplotype systems (compare Figure 9-14 with 6-12). Thus, while this OCA2-E/F/G system may represent the best model for predicting iris colors of unknowns, it will require a larger database than that using the OCA2-A/B/C/D system.

THE EMPIRICAL METHOD OF DIRECT PHENOTYPE INFERENCE

Methods of classifier construction based on summary statistics using the aforementioned haplotypes were disappointing. Using neural networks with earlier instances of the database, 80% accuracy predicting iris color shade was achieved, with similar sensitivity and specificity for the two color groups. This compares with about 95% accuracy using the empirical database matching method we have just discussed, though with the statistical methods we can generally provide predictions for most of the test irides, and with the database matching method we can as of the date of this writing make predictions for only 8% or so of new samples.

At this point in the discussion, recall our previous discussion on the lessons learned using the indirect method of trait value inference based on ancestry (see Chapter 8). There, our problem was that of making inferences on various anthropometric trait values knowing the value of a marker (ancestry) that is correlated with the trait. It seems that also, generalizing with summary statistics for the purposes of making classifications and inferences leaves quite a bit to be desired in complex genetics where epistatic interactions are not always predictable from marginal effects. Here, our problem is probably that we are using general summary statistics of SNPs or haplotypes, which do not usually cause traits themselves, to estimate traits caused by multilocus genotypes (i.e., haplotype pairs in different key genes). Generalizing always comes at the expense of lost information, and though this is not usually an impediment when we are attempting to illustrate connections between sequences and traits, as we do in an association scan, it can pose more of a problem when the goal is to harness the complexity of a complex genetics situation involving many alleles that are likely to have different effects in different genetic contexts, as we do when attempting to predict phenotype for unknowns. We have seen that when attempting to match ancestry profiles with anthropometric traits, the genetics of which are likely to be extremely complex, an empirical, database-driven approach performs well and satisfies our needs to accommodate complexity that may be built into our underlying genetics of the system. The same appears true for the direct method of phenotype inference or prediction.

Using an empirical approach with iris color, we consider a multilocus genotype as a single object (which could be composed of SNP genotypes for multiple SNP loci, or haplotypes of SNP alleles for multiple genes). We simply ask the database what samples have the same multilocus genotype as that in question, and whether individuals united by this multilocus genotype are of the same or similar trait value. If so, we can use this trait value as an estimate for our unknown sample. This is not fundamentally different from considering each gene or SNP as an independent variable, but we gain information and reduce assumptions by working at the level of the haplotype. Rather than use a combination of six summary statistics or variables for five different genes or sets of SNPs, for example, we form our generalizations from a single geno-type—a multilocus genotype comprised of the five genes or SNP sets. Of course, when we avoid making generalizations about individual features and form them instead using a multilocus genotype, we suffer in general power because the number of multilocus genotypes is usually large and we must have an adequate number of examples against which to compare, for each type of profile, i.e. a much larger database is required. However, we gain precision and accuracy if the genetics of our system is truly complex, with genes exerting influences that are largely context-dependent, and we obtain the satisfaction that as long as the samples are adequate, we are interpreting the data in the most responsible manner possible given the data available. We minimize the number of assumptions and generalizations, such as those that accompany applications of general principles to specific situations.

We reasoned that given the potential complexity of iris color, where gene variants could have different effects in different contexts, an empirical genotype match method of classification was likely to be better suited for predicting iris colors. This is because the database matching method we have described is expected to be robust to various (and still unknown) components of iris color genetic complexity, such as those one might expect to encounter within large genes. For example, it is possible that haplotypes with a region marginally associ-ated with blue irides and a region marginally associated with brown irides beget green irides in an additive sense, or beget brown irides in a dominant sense. Just because an allele is strongly associated with brown iris color, its penetrance and dominance need not be complete and cause brown in every genotype context.

Dominance exhibited between phase-linked polymorphisms in one haplo-type context may not extend to others, and on top of this, there is dominance in the classical diploid sense to deal with as well, which may not be completely penetrant. Predicting this myriad of potentially complex classification rules using summary statistics or hap-tag SNPs seems challenging for such a complex gene, if not impossible. Yet theoretically each is accommodated by a database-matching system even if they are not yet defined, and no matter how small the database (or large the number of variables).

The empirical genotype match strategy is very simple—either we have seen a multilocus genotype before or we have not. If we have, we can use the characteristics (i.e., in this case the iris colors) of the examples in the database with that genotype to make predictions for the unknown sample. We can require that the group exhibit homogeneity in phenotype value before making an inference, or more practically, we can take the average iris melanin index value and express the quality of the inference using the variance of values within the group. With this method, the more complete the iris color gene set from our genome scans, the better the method would perform. If it were used when only two of say, 10 important genes that underlay a trait are available (because they are the only ones known about), the method would perform poorly, as would any other. If genotypes for all functional variants of all 10 genes underlying the trait variance are available in a large database, and the heritability of the trait was 100%, the method would perform perfectly.

For most genetics traits we will find ourselves somewhere in between, and there will usually be loci of low penetrance (perhaps many such loci) that remain unknown and unrepresented in our database. Nonetheless, as long as the database is large, deficiencies in performance are easily quantifiable, and variability in performance as a function of genotype is also discernible and quantifiable, something that is not necessarily the case with predictive methods based on summary statistics.

For anthropometric traits (and many others) we could capture some of the information present in these loci by drawing upon the indirect method of inference. If some of the important loci are unknown, and if trait value is correlated with admixture, we could incorporate continental genomic ancestry estimates and even European subgroup estimates discussed in Chapters 4 through 7 as variables require a match (see also Chapter 8). These values would then serve, in effect, as proxies for the missing loci if they are distributed unequally among populations of varying iris color. We could use the ancestry levels above those corresponding to the statistical thresholds of association discussed earlier, the levels we discussed before when drawing indirect conclusions— such as ≥12% West African (associated significantly with brown irises), ≥16% East Asian (brown), ≥25% Indigenous American (brown), ≥68% European (blue), ≥75% Northern European (blue), and ≥30% South Asian (brown). We might alternatively encode the admixture variables in a more quantitative sense.

DATABASE SIZE

The simple database method we have described might prove equally useful for analysis of multigene classification systems. Of course, the method is the most basic imaginable, and although it enjoys an empirical power not usually afforded to methods that rely on generalizations based on population averages, it suffers from the requirement of very large sample sizes if one hopes to be able to classify most new samples encountered (particularly for large genes and/or systems involving many genes). As a result of the size of our database (n = 1,072) for the work described earlier, we could classify only about 8% of the iris colors in our sample (those with multilocus OCA2 genotype matches—82/1,072).

If we assume that our results extend to other samples, as our validation sample results suggest they do, then improving on this rate is simply a matter of building the database. Though adding samples is expected to increase the number of observed haplotypes, and multilocus OCA2 genotype groups, by definition with 33 SNPs there are a finite number of both represented in the human population and increases in complexity due to increase in database size is expected to plateau. Indeed, an earlier instance of this database compiled with n = 835 samples revealed 64 OCA2-A, 37 OCA2-B, and 18 OCA2-C haplotypes, which is a similar level of haplotype complexity observed in the sample of 1072 (70 OCA2-A, 38 OCA2-B, and 19 OCA2-C haplotypes), indicating that with 1,072 samples we are already well along the plateau phase. Thus, rather than increase the complexity of the classification system by creating more haplotypes, added samples are expected to increase the number and representation of multi-locus OCA2 genotypes, enabling us to classify a larger fraction of new samples.

No publication has previously described the correct inference of a multi-factorial phenotype from DNA for even a fraction of samples, and in this regard our findings are worthy of dissemination. Nonetheless there are some who believe that any calculation of classification accuracy must involve all samples, not just those one feels comfortable or able to classify. If this were true, the classification of all but the simplest of phenotypes from DNA is likely to wait many years until the cost per genotype comes down and research groups can afford to build databases tens of thousands of samples strong. However, we have ample evidence that focusing technology on subsamples (e.g., in this case, samples of genotypes we've seen before) is not only acceptable in a theoretical sense but highly effective in a practical sense as well. For example, the drug Herceptin[R] is used to treat only a small fraction of breast cancer patients (those with Her2 positive cancers) because it performs well in that subpopulation. Had Herceptin[R] efficacy calculations been mandated in the clinical trial to include all breast cancer patients, the drug would likely not have been approved and papers on its effectiveness summarily rejected. Thus, the case could be made that classification accuracy rates for subpopulations are legitimate, as long as they are labeled as such.

It is likely that we are going to have to come to accept the limitations that database representation imposes on our ability to predict phenotype from DNA. Prior to the human genome era, the prediction of complex phenotypes from genetic measurements had not been possible. The introduction of DNA microarrays broke this barrier and enabled the detection of highly characteristic and predictive RNA expression signatures for various phenotypes. In ovarian cancer alone, predictive gene expression signatures have been identified as features of tumor metastasis (Ramaswamy et al. 2003), malignancy (Ouellet et al. 2005), tumor drug resistance (Helleman et al. 2006), and cancer prognosis (Bild et al. 2006; Meinhold-Heerlein et al. 2005).

However, genotype signatures of multifactorial human traits have been far more elusive, probably because DNA is less directly related to phenotype than RNA. As a result, to my knowledge, no accurate multifactorial DNA-based classification systems had been described prior to this one. Part of the cause for this must be due to the dimensionality of DNA-based problems. In our earlier work, the number of variables produced from phasing the 33 SNP genotypes (1,018 multilocus genotypes) was about the same as the number of samples (1,072), and on the surface, it would seem that any attempts at classification would be challenging. However, this type of challenge has not prevented the successful classification of disease subtypes, drug response, and prognosis using RNA expression signatures. In the case of iris color, if each of the multilocus genotypes we observed unambiguously specified a specific iris color (within a reasonable range), then the empirical database-matching method we described should perform well regardless of the size of the database—the only penalty for a small database is that fewer predictions or tests of the hypothesis could be executed.

VARIATIONS ON THE EMPIRICAL METHOD

If the size of the database is limiting, and we are dissatisfied with the ability to provide inferences or predictions for only a small percentage of new test samples, we can relax our matching criteria. Rather than require an exact match across all OCA2-A, B, C, and D haplotype pairs, we might specify groupings based on partial matches among three of these, or even two. With iris color, this approach is justified by the observation of characteristic similarity in iris colors for samples with partial matches, and seems to work reasonably well (see Table 9-7). Of course, whereas we can expect to make predictions for a larger number of test samples with an incomplete matching criteria, the penalty incurred is expected to be lower overall accuracy. We expect to be able to make more predictions because the number of samples matching partially is expected to exceed the number matching completely. We expect lower accuracy since we are using samples with different pan-OCA2 genotype as reference samples for the predictions and the inferences represented then become educated guesses. This method may work better for OCA2 genotypes with dominant sequences, but worse for those that contribute to iris color strictly in an additive sense.

Results: Using: sample* (n)	Correct	Incorrect	Accuracy	OCA2-A	OCA2-B	OCA2-C	TYRP	ASIP	AIM	Continental ancestry	European sub-ancestry
100	76	20	79%	yes	no	no	no	no	no	no	no
596	132	11	92%	yes	yes	yes	no	no	no	no	no
596	136	7	95%	yes	yes	yes	no	no	no	yes	yes
596	38	1	97%	yes	yes	yes	yes	yes	yes	yes	yes

*Data is derived from an early instance of the database (n = 596) used for the results discussed in the text and figures which employed a database with n = 1,072.

Table 9-7

Prediction accuracy using weighted incomplete OCA2 versus complete OCA2 matching criteria with and without information for other loci or about admixture.

We could blend the exact and partial matching criteria. For example, given a sample we desire to predict the iris color for, we collect the photographs for all the irises with exact genotype matches in the following groups (see Box 9-B for more detail):

■ Hierarchy 4: across all the regions of all the genes
■ Hierarchy 3: across any three haplotype regions of the *OCA2* gene, such as *OCA2-A+ OCA2-B+OCA2-D*
■ Hierarchy 2: for the pairs *OCA2-A + OCA2-B, OCA2-B + OCA2-C* and *OCA2-A + OCA2-C*

If there are five examples in Hierarchy 4, we could take the average of each value from the collection in Hierarchy 1 as the profile estimate for the unknown and $\pm$ the standard deviation as the range. If there are not five, we include those of Hierarchy 3, and if this new collection contains at least five examples, the average of it and Hierarchy 4 is used similarly. If there are still not five examples, the irises from Hierarchy 2 are used—using the average luminosity, red, green, and blue among all irises in the collection meeting Hierarchy 4, 3, and 2 as the estimate and their standard deviation as the range. Whichever Hierarchy applies for a given sample, the profile estimate then is used as a subject in a database query where all photographs for irises that fall within all four ranges are called. We could also attach more weight to the parameters from samples with more complete matches, which we refer to as a weighted incomplete OCA2 matching approach. Using this type of method with an earlier instance of the iris color database, we see the predicted drop-off in accuracy and increase in percentage of samples for which we are able to predict color [(Correct + Incorrect)/Sample, Table 9-7].

CASE REPORTS

When reporting the results it would be desirable to present the raw genotypes, haplotypes, inferred iris melanin index components with their confidence

Actual:

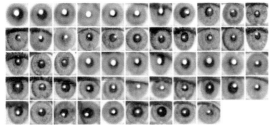

Predicted:

LUMN = 151.25 -- Range (142.25 160.25)
RED = 157.83 -- Range (145.70000000000002 169.96)
GREEN = 152.28 -- Range (143.26 161.3)
BLUE = 127.82 -- Range (110.38999999999999 145.25)

** (3 examples)

Figure 9-15

Example of part of a case report used to communicate the inference or prediction to investigators. Haplotype information is shown at the top, corresponding to the test iris. In this case, we have the photograph for the test iris but in an actual case we would not. Other variables, such as genotype for other genes or admixture could be presented here. In the Predicted section, we can present the inferred components of the iris melanin index score and their confidence intervals, calculated as described in the text. Below this, we can present a gallery of irides from the database that fall within each of the four ranges. We can thus visually communicate the range of colors and patterns corresponding to the inferred level of melanin represented by the numerical predictions above. This iris was predicted to be of a range of colors falling on the light end of the spectrum, and the prediction was correct. This inference was performed using the hierarchical method described in the text, and the level of multilocus OCA2 genotype match from which the inferences were possible is shown to the lower right, where the number of stars represents the number of matching OCA2 haplotypes (** samples matching at two of the four OCA2-A/B/C/D haplotypes were used in calculating the values). The number of samples (examples) meeting this criteria is also shown in the lower right.

intervals, and the photo gallery of samples falling within these inferred ranges. We are essentially saying in such a report, "here are the color value ranges most likely corresponding to your sample, and here are a set of irises that fall within these ranges." Figure 9-15 shows an example of an unknown predicted with the partial OCA2 match method to be of a profile shared among irises that are predominantly blue in color. The unknown (AAFS04_034) was also blue. Figures 9-16 and 9-17 show examples of unknowns (AAFS04_019 and DWPB0748) correctly predicted with the same method to be of intermediate color (a mixture of browns and green/blues in different ratios), and Figure 9-18 shows an example of an unknown (DWPB0490) correctly predicted to be of dark green to brown in color (the lowest hierarchy level upon which the inferences were drawn and the number of samples incorporated in the hierarchical set is shown in the lower right hand corner of each figure, where *** corresponds to Hierarchy 3 and ** to Hierarchy 2, etc.).

Collections of irises such as those in Figures 9-15 to 9-18 could constitute the most practical answers for forensic cases, which could be interpreted two ways:

1. **Ranging**: One could appreciate the range of colors and patterns present in the phenotypic range of the gallery. If more than one type of pattern/color is presented, the proportion of each pattern/color is reflective of the likelihood the sample's iris is characterized by such a pattern/color. For example, if the collection of photographs for a sample is a blend of 90% clear blues and greens, with 8% blues with small brown rings/patches/flecks and 2% light hazels, the iris of the subject is most likely clear blue, but there is a chance there is a small ring/patch or fleck of brown color when looking closely at the iris, and a smaller chance of a more homogeneous light hazel.

Actual:

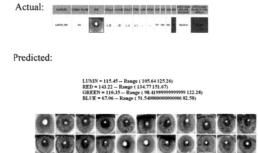

Predicted:

LUMIN = 115.45 – Range (105.64 125.26)
RED = 143.22 – Range (134.77 151.67)
GREEN = 110.35 – Range (98.41999999999999 122.28)
BLUE = 67.06 – Range (51.540000000000006 82.58)

*** (19 examples)

Figure 9-16

Example of part of a case report used to communicate the inference or prediction to investigators as in Figure 9-15. This inference was performed using the hierarchical method described in the text. This iris was predicted to be of a range of colors falling in the middle of the color spectrum, and the prediction was correct.

Figure 9-17

Example of part of a case report used to communicate the inference or prediction to investigators as in Figure 9-15. This inference was performed using the hierarchical method described in the text. This iris was predicted to be of a range of colors falling on in the middle of the color spectrum, and the prediction was correct. Note that the predicted color differs from that of Figure 9-16, although both are of medium color along the spectrum shown in Figure 9-8A.

2. **Blending**: One could stand back and appreciate the entire phenotypic range at once, considering each photograph as a pixel for a very large square or rectangular iris. The blending of colors together creates an average result, the color of which gives a similar estimate for how the sample's iris will appear. For example, if the collection of photographs for a sample is a blend of 90% clear blues and greens, with 8% blues with small brown rings/patches/flecks and 2% hazels, the collection of all these together as small pixels of a large iris would produce a single iris that is almost entirely blue/green with a very small amount of brown coloration that may appear as small spots, a small ring, or a small patch.

Overall, the results from our efforts to predict iris color have taught a few interesting lessons:

- A complete *OCA2* haplotype is needed to infer iris colors, and the improvement in results achieved was due primarily to a more complete survey of *OCA2* gene variation.
- More sensitive and objective quantification of iris color most likely will assist in improving the results.
- Using ancestry correlated with iris colors as an independent variable may help, though it is not yet clear that all the information sought is not inherent to the OCA2 locus on its own. The use of admixture would seem better advised than the use of slippery markers from structure-based associations that are associated with iris color indirectly through correlation, which likely only obliquely measure this ancestry.
- Other genes may be involved, but if so, only as secondary determinants and only for the refinement of *OCA2*-mediated iris color.

None of this is particularly surprising for a multifactorial, quantitative genetics trait. The upside to this approach is that it seems to work well, and continued study will more clearly illuminate just how well. The downside is that because the number of haplotypes for each SNP set is large, huge database sizes will be needed to provide answers for a larger fraction of test samples, tease all the components of variation, and fully characterize the

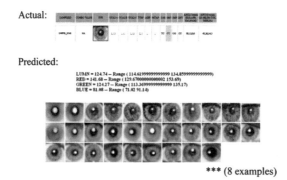

Actual:

Predicted:

LUMIN = 124.74 – Range (114.61999999999999 134.85999999999999)
RED = 141.68 – Range (129.67000000000002 153.69)
GREEN = 124.27 – Range (113.3499999999999 135.17)
BLUE = 81.08 – Range (71.02 91.14)

*** (8 examples)

Actual:

SAMPLED	COMBO VALUE	EYE	OCA2-A	OCA2-B	OCA2-C	TYRP	ASIP	MTAP	2168	1867	1869	2193	ANC2.5 scores (EUR,AFR, EAS,NAM)	ANC3.0 scores (IB,NE,OR-TUK, MED,SA)
DWPB_0490	NA		13.12	.15	L.11	L.1	.	.	CC	GG	GG	TC	91,9,0,0	71,17,12,0

Predicted:

LUMN = 106.13 -- Range (83.99 128.26999999999998)
RED = 137.29 -- Range (113.41999999999999 161.16)
GREEN = 101.02 -- Range (77.16 124.88)
BLUE = 48.54 -- Range (36.3 60.78)

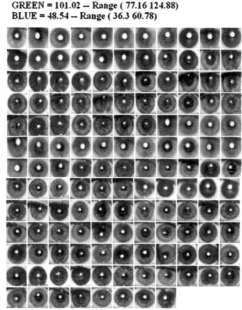

** (3 examples)

Figure 9-18

Example of part of a case report used to communicate the inference or prediction to investigators as in Figure 9-15. This inference was performed using the hierarchical method described in the text. This iris was predicted to be of a range of colors falling in the darker region of the color spectrum, and the prediction was correct. Note that in this case, the predicted iris melanin index value components encompass a variety of colors and patterns that include dark greens, hazels, and browns. It may be that the exact color imparted by this particular multilocus OCA2 genotype is dependent on other loci and/or variables.

method. What about the subtle variation in colors and patterns we see within multilocus OCA2 genotype groups? It is possible that other iris color loci exist. If so, the requisite size of the database for refinement of color prediction would increase—perhaps to tens or hundreds of thousands of samples.

Many epistatic interactions are characterized by relatively low penetrance, and if an epistatic interaction requires 1,000,000 samples in order to appreciate then it will be of such low penetrance and/or explain such a small amount of trait value variability that it will probably not be very useful as a classification variable, and indeed, we would not recognize it as important in our classification scheme using practical database sizes. On the other hand, if the epistasis is highly penetrant and recognizable with a low sample size of, say, 1000 samples, its influence already would be part of the performance parameters we have obtained so far.

HAIR COLOR

ANATOMY OF THE HAIR FOLLICLE

Hair is produced by hair follicles, and the color of the hair follicle is a function of the melanin content of the hair shaft. Cells called keratinocytes are directly involved in the production of the hair structure and coloration in that they secrete the necessary keratins, proteins, and melanosomes into the developing hair follicle. The melanin content of the keratinocytes is imparted through the transfer of melanosomes from neighboring melanocytes. Dopa-positive (i.e., pigmented) melanocytes occupy positions along the wall of the pilary canal and the upper parts of the bulb, within which the keratin and proteins that constitute the hair follicle are secreted. Regardless of hair color, melanocytes in the lower parts of the bulb are generally dopa-negative and depigmented. In mice, the bulb population of melanocytes is part of a perpetuating system. Undifferentiated melanocytes are contributed by the epithelial column to the hair germ (luminal layer of the bulb) at specific stages of the hair bulb life cycle. As bulbs and differentiated melanocytes are lost by hair shedding or hair replacement, those in the remaining tissues dedifferentiate and proliferate to occupy the columnar epithelium from which a new one arises. This process is called the hair cycle.

At least in mice, the synthesis of tyrosinase is under tight regulation during the hair cycle, occurring in a phase called anagen, levels peaking later and coming to completion before telogen, the rest phase of hair growth. The melanocytes themselves undergo ultrastructural modification; the cytoplasm volume increases, dendricity increases, Golgi complex and rough endoplasmic reticulum (RER) develop, and in darker haired mice, the size and number of melanosomes increases. Prior to anagen, the cytoplasm is relatively low in volume, and the Golgi and RER are less well developed. During anagen, the melanocytes transfer melanosomes to follicular epithelial cells lining the lumen of the bulb. The size of the melanosomes is generally two to four times larger than for skin epidermis (Ortonne et al. 1993).

As the hair follicle is constructed, melanosomes are secreted and broken, with melanin granules and other cytoplasmic materials from the melanosomes becoming embedded in the cortex of the keratin matrix. Little melanin is found in the surrounding structures or cell layers. Most of the pigmentation is found in the upper part of the outer root sheath with little found in the lower part of the outer root sheath and upper or lower portions of the inner root sheath. This pattern is distinctive from that found in epidermis, and it has been suggested that these histological differences arise due to differences in signal transduction between keratinocytes and other cells in skin and hair follicles (reviewed in Prota 1993).

That the process of follicle pigmentation is largely distinct from that in skin epidermis can be appreciated from certain phenotypes; bleached patches of skin found in vitiligo patients give rise to normally pigmented hairs, and bleached hair follicles are known to arise from normally pigmented skin. As another example, the common trait of premature (or mature) hair graying does not accompany a corresponding bleaching of the skin. This has given rise to the idea that the pigmentation systems of hair and skin are compartmentalized into an epidermal compartment and a follicular compartment. However from wound healing, where epidermal melanocytes contribute to newly formed hair shafts, and experiments on model organisms dealing with melanocytes injected into dermis but ending up in the shaft, it is known that these two compartments are not completely closed to one another (Ortonne & Prota 1993; Staricco 1963).

PIGMENTS IN DIFFERENT HAIR COLORS

In humans, hair color is distributed as a continuous trait, but several categories are commonly recognized. Each of these groups or classes generally can be characterized from ultramicroscopy by the size and shape of the melanosomes, as well as the deposition of melanin within the intracellular matrix of the melanosomes (Ortonne & Prota 1993). Most red hair is characterized biochemically as pheomelanic, which is another way to say that high levels of pheomelanin are found relative to eumelanin, the latter of which often is not found at all. Since both pheomelanins and eumelanins derive biochemically from a common source, they often are found together in human hair, at different levels depending on the color.

In pheomelanic hair, the melanocytes are characterized by small, round pheomelanosomes within which exists a proteinaceous matrix containing spotty depositions of pheomelanin. Some cases of red hair are associated with melanocytes of mosaic character, with partially differentiated eumelanocytes, which contain both pheomelanosomes and eumelanosomes in the same cell. Blond hair in humans is characterized by follicular melanocytes, which have the same number of melanosomes as black or brown hair, but of smaller size and not as densely melanized. The lack of melanin content is apparent for melanosomes in the soma of the melanocyte as well as in the dendrites. Brown and black hair follicle melanocytes have melanosomes that are large and elliptical, with dense distributions of eumelanin depositions in the intracellular matrix. Light brown hair follicle melanocytes show smaller melanosomes, and less dense depositions of matrix eumelanin. In grey or white hair, it is often the case that melanocyte number is greatly reduced, and those that exist show little or low melanin content. Immunohistochemistry shows that little tyrosinase protein exists.

Humans and some animals possess a yellow/reddish pigment that is different in a chemical and structural sense from the pheomelanins, being of lower molecular weight and chemical structure. In fact, these pigments are now known to be oxidized variants of eumelanins, arising from partial peroxidative cleavage of eumelanin anabolism intermediates (Prota 1992). Prota (1993 review) described that shades of red hair are obtained in some humans through a peroxidase-mediated oxidation of eumelanin. In other words, individuals of varying shades of eumelanic red hair would have corresponding levels or activity of peroxidase or other oxidation capability within the melanosomes, keratinocytes, or both.

One can envision the scenario as follows: when people with brown hair bleach their hair with peroxide, the progression from brown to blond proceeds through various shades of red (anyone who has ever tried to bleach their hair could confirm the importance of carrying out the procedure to completion, lest reddish blonde or brown hair result). Sunlight exposure creates a similar scenario because eumelanin is photosensitive—however in this case free-radicals perform the oxidation rather than peroxidase. A brown-haired surfer who spends much of his or her time on the beach may have hair with blonde tips but between the blond and brown sections of the hair shaft exist ranges of reddish brown and blond. What this means for genetic screens aimed at developing tools for molecular photofitting is that not all red heads are necessarily phaeomelanogenic and that some, capable of the genetic process of peroxidase mediated bleaching, are completely eumelanic. From hair color phenotype alone, we would expect melanocytes for such individuals to have smaller, less densely filled melanosomes like those for individuals with lighter brown colored hair, but the reality is that some individuals exhibit a mixed type of eumelanic and phaeomelanic phenotype with both normally shaped and filled ellipsoid eumelanosomes and smaller less densely filled phaeomelanosomes. This complication has minimal impact on our phenotyping, however, because we are still mainly interested in quantifying the amount of light reflectance in the red and brown wavelengths; we need only be mindful that some genetic heterogeneity is expected for samples with redness to their hue because some redheads will be partially or completely eumelanogenic and have high eumelanin/pheomelanin ratios like brown-haired people.

PHENOTYPING HAIR COLOR: THE CHALLENGE

As with iris color, our primary goal in phenotyping hair for color is to standardize descriptions of colors between investigators, and to facilitate genetic screens aimed at identifying functional markers in the genome. For the latter, we want to simplify the complexity of the phenotype, and enhance the statistical power of our genetic analysis. In the pigmentation section of this chapter we learned that human skin coloration is due primarily to levels of three main pigments, the brown eumelanin, the red/yellow pheomelanin, and red hemoglobin. As with iris color in the preceding section, we would like to target our genetic screen for identifying markers underlying the specific bio-chemical or physiological determinants of variable color, not necessarily to the color itself (although we understand the direct relationship between the two fairly well, unlike for iris color).

Though age and environmental variables such as sunlight exposure and exposure to some drugs are thought to impact the expression of hair color, we can design around, or condition our analyses on the former and exclude cases of the latter with proper ascertainment. The primary determinant of hair color is melanin (eumelanin and pheomelanin) content in the hair shaft, and so we want to design our study so as to detect markers in LD with loci that impart melanin expression in the shaft. The process of binning is inexact as we have seen with iris color, and by reducing the complexity of hair color to subjective categorical descriptions rather than biochemical quantities, we introduce a very significant source of error. For mapping or otherwise using genes that underlie quantitative traits, concise, sensitive, and physiologically meaningful measures are required, and we would like to simplify without loosing too much information. The best way to do this for hair color, given what we have learned from decades of classical genetics work in the area of pigmentation, is to take measurements that represent good proxies or surrogates for eumelanin and pheomelanin expression. After we have found markers predictive of and associated with melanin expression in the hair shaft, we can then figure out how levels of these two pigments relate to hair colors perceived by the eye as a separate problem.

For molecular photofitting, then, our primary goal is to develop a classifica-tion tool for predicting the amount of melanin in the hair shaft. If we assume correctly that the only two types of melanin are eumelanin and pheomelanin, there will be two components of error for this tool:

■ The error associated with predicting melanin content based on marker sequences
■ The error associated with predicting hair color based on melanin content

Note that if our ability to measure melanin content is not perfect then the first error will be more pronounced. If our assumption is incorrect and hair color is a function of (for example) five different pigments, the second error will be extreme. For the classification tool to work, our hypothesis that hair color is primarily a function of melanin content must be true and we must be able to measure melanin content consistently and with precision and accuracy. What is the evidence that hair color is determined primarily by melanin (eumelanin and pheomelanin) levels, as opposed to other pigments or physical properties of the hair shaft?

NATURAL VARIATION IN MELANIN LEVELS AND HAIR COLORS

As we have described, differences in phaeomelanin/eumelanin content are thought to underlie the natural distribution of hair shades such as blond, light brown, red, brown, and black. This belief is based primarily on the strong correlation between color and brown/red pigment levels in the hair follicle and melanosomes/pheomelanosomes of the ductal epithelium, as we have just described. We have discussed the difference in melanosome size, shape, and color among different hair colors, but biochemical studies give us information that is more relevant for understanding the phenotype. Applying the biochemical technique of quantifying eumelanin and pheomelanin to human hair (described earlier in this chapter), it has been discovered that virtually every type of hair contains both eumelanin and pheomelanin (Jimbow et al. 1992; Thody et al. 1991). The relationship between phaeomelanin and eumelanin is linearly inverse in the range from black to dark blonde hair color. That is, darker hair colors are associated with high eumelanin but low phaeomelanin content and lighter hair colors up to dark blond are associated with lower eumelanin and higher phaeomelanin content relative to the average color. Red hair does not fit this pattern, since about a third of subjects are pheomelanogenic and the remaining two-thirds eumelanogenic. Lighter blond hair is characterized by mixed type melanogenesis—both phaeomelan and eumelanin are present. From dark blond to light blond, the ratio of phaeomelanin to eumelanin remains similar but higher overall levels of both pigments are found the darker the hue.

So, most hair colors have both eumelanin and pheomelanin, and it is thus the quantity and ratios of the two that distinguish the various colors. Since both the eumelanin and pheomelanin pathways derive from the same biochemical tyrosine starting point (see Figure 9-2), this would suggest that switching between the pathways is involved in determining hair color. Studies of the activity of the TYR gene product show that tyrosinase is as active, and even more active in blond and red hair follicles than in brown and black hair follicles, the difference presumably being that the specific downstream pathway (eumelanin vs. pheomelanin) consumes the products of the TYR mediated reactions. It bears noting that the general ultrastructural configuration of brown/black hair melanocytes and melanosomes is the same across all human phylogenetic clades and so, differences in hair pigmentation among the world's various peoples must derive from differences in genetic control of the pheomelanin/eumelanin pathway. Thus, we have all of the ingredients we need for a successful genetic study—a phenotype that can be related directly to a relatively small number of biochemical entities, which are encoded by a discrete number of genes part of a small number of genetic pathways (actually two TYR mediated pathways). The problem now is to determine how best to quantify eumelanin/pheomelanin.

One gold standard for quantifying eumelanin and pheomelanin quantities in hair shafts is the direct biochemical method, which we have described earlier in this chapter. Biochemistry, however, is not always the most expedient method; for one it is expensive, requires organic chemistry expertise and equipment, and relatively few laboratories exist that are capable of performing this type of analysis. Spectrophotometry on the other hand is convenient, requiring only a spectrophotometer, and can be performed by any reasonably skilled laboratory technician. Naysmith et al. (2004) showed that spectrophotometric methods are good substitutes for biochemical measures of melanin/pheomelanin ratio—this paper showed a good correlation between the biochemically determined LEPR ($\log_e$ eumelanin/pheomelanin ratio) and the CIE tristimulus a* (red-green, $R^2 = 0.51$), b* (yellow-blue, $R^2 = 0.47$), and L* (luminosity, $R^2 = 0.22$) reflectance values.

Hair reflectance has been measured since the 1950s, when British researchers such as T. Reed took a keen interest in the expression of red hair color (Reed 1952). Though the biochemistry of melanin production enjoyed the attention of certain organic chemists since then (such as Giuseppe Prota and colleagues, from whose work and reviews much our discussion of the biochemistry and ultrastructural components of hair color has been derived), the *genetics* of biosynthesis suffered from neglect after geneticists realized that human pigmentation phenotypes were under complex controls. With the molecular biology revolution and completion of the human genome sequence, the interest in the genetic basis for pigmentation has been revived but in humans, most of the melanin work has focused on the skin, since a variety of diseases are related to the expression of pigmentation in this tissue (i.e., melanoma, squamous cell carcinoma, basal cell carcinoma, vitiligo, etc.). As might be expected, since no human diseases result from hair color anomalies, very little attention has been focused on quantifying pigment levels from hair shafts.

SPECTROPHOTOMETRY OF HAIR COLOR

Good metrics for the use of reflectometers in quantifying eumelanin and pheomelanin have been established from trial and error. With iris color, we are forced to use noninvasive and noncontact means by which to measure melanin content because contact with the iris is unsafe and uncomfortable for donors. In contrast, for hair we have a completely different situation. Reflectometry is a relatively straightforward method for objectively assessing hair color, since a reflectometer can be held in contact with a lock of hair whether or not it is still attached to a person's head. However, if we have decided to use the spectral properties of human hair to indicate the levels of pigment, then how do we standardize the way in which we take our readings? What specific wavelengths should we use, or should we use all the visible wavelengths? Is apparent absorbance with respect to a control hair color a better tool to use than absolute reflectance?

Box 9-C

Measuring Hair Color with a USB2000 Spectrometer

Hair color is better measured in a controlled laboratory environment, where the surface of the plane through which the readings are obtained can be standardized from sample to sample. Reading hair colors by holding a full spectrum spectrophotometer (such as an Ocean Optics USB2000 instrument) to a person's head seems suboptimal for this purpose, so we need to clip hair samples and we need to measure them all in the same way on a laboratory bench. Managing hair samples for spectrophotometry is best done through a plastic bag. Plastic has the advantage that the reflectance properties can be appreciated through it and the orderliness of the hairs is easily maintained. When reading reflectance spectra from hair inside a plastic bag, one has to be sure that every sample is read through the same type of plastic because the plastic mutes the reflectance a bit, and different thicknesses or compositions of plastic will mute the reflectance to differing extents. As long as the plastic is the same type and thickness for each sample, and unwrinkled in each instance, the reflectance will be muted equally for all samples and so the relative values are valid.

Whether reading through plastic or not, it is very important to comb the hair inside the bag. This is because the amount and type of light reflected back to the reflectometer is not merely a function of the chemical composition of the hair, but also the ultrastructural topology or arrangement of the hair shafts. Disorderly hair shafts reflect less light back to the instrument than neatly ordered hair shafts, and it is possible that a disorderly light brown hair sample will show a reflectance spectrum similar to an orderly dark brown one. Combing, or ordering the hairs through the plastic, can help control this confounding variable but cannot eliminate it altogether. Sometimes a device is needed to hold the lock in a fully extended configuration within the bag, but oftentimes two fingers and some effort suffice. The same problem presents itself in the difference between curly and straight hair, and for very kinky hair good quality readings will be difficult. The OceanOptics USB2000 spec reads at an oblique 45 degree angle and at this angle, the reflectance off the surface of the bag is negligible. In contrast, the reflectance off the surface of the bag is not negligible when reading perpendicularly to the sample, so only the 45 degree angle should be used.

The earliest studies of hair color using reflectance spectroscopy demonstrated the utility of the approach and the ability to distinguish red hair from brown and blond (e.g., Gardner and MacAdam 1934; Little & Wolff 1981; Sunderland 1956). With reflectance spectroscopy, we illuminate the hair with controlled amounts of full spectrum light and measure the amount of light at each visible wavelength that is reflected. Blond hairs will reflect light at different wavelengths than brown hairs or red hairs. Figure 9-19 shows the reflectance spectra obtained with the Ocean Optics USB2000 spectrophotometer for a variety of hair colors. Each reflectance spectrum was obtained from a randomly selected sample of a larger 3,000-sample collection (all Europeans, using the broad biogeographical ancestry definition discussed throughout this book), following the protocol discussed in Box 9-C.

All hair samples (regardless of color) show a peak reflectance at 650 nm or so, but the precise position of the peak varies from color to color type. Blond hair samples show the greatest reflectance at 650 nm ranging from the high 1,000s (dirty blonds) to over 3,000. The light browns show similar spectra as reds, with maximum reflectance values ranging from about 800 to 1,500, and the browns and blacks show relatively low reflectance at 650 nm in the range of a couple hundred to the high hundred reflectance units (see Figure 9-19). Blond hair looks lightly pigmented to the eye because it reflects lots of light of various different wavelengths, and therefore it absorbs little light. We can see this in the spectrum of Figure 9-19 by noting that more light at most of the wavelengths is reflected for blondes than for other color types. Brown hair on the other hand absorbs lots of light in a wide range of wavelengths, reflecting little. We can see this in Figure 9-19 by noting that the reflectance units are lower for all wavelengths in brown hair than lighter hair colors.

Figure 9-19 shows considerable intragroup variability, but Figure 9-20 shows an example of an experiment that indicates that a good part of this is due to measurement error. Some of the error within samples is undoubtedly due to

Figure 9-19

Reflectance spectra curves obtained by the author from a variety of hair samples using the Ocean Optics USB2000 spectrophotometer. Each curve is represented by a different colored line, and corresponds to a unique sample, and shows the units of reflectance obtained for each wavelength indicated on the x-axis (scanned at 1 λ intervals). Hair samples of blonde color show the greatest reflectance overall followed by dirty blondes, reds, light browns, and browns as indicated with arrows. Note that red samples fall into two separate profile types, probably due to the fact that there are two forms of red hair (eumelanogenic and phaeomelanogenic). Note also that it is not only the overall reflectance that varies by color, but the position of the peak on the wavelength scale.

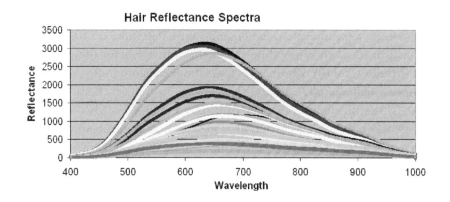

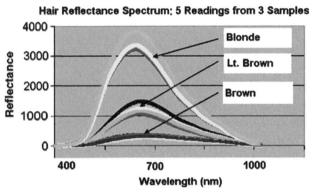

Hair Reflectance Spectrum: 5 Readings from 3 Samples

Figure 9-20

Reflectance spectra curves obtained from three separate hair samples on different occasions. Five readings are shown for three samples. All five readings for the blonde sample show reflectance maxima above 3,000, those for the light brown samples between 1,000 and 1,500, and those for the brown below 500, as indicated with the arrows. Variation within samples represents an estimate of measurement error, such as that which changes in color from tip to root of a lock of hair may cause, or variation in the stat of the plastic bag from position to position on the lock.

quantitative variations in hair color from location to location within a given lock. In Figure 9-20, three hair samples (a blond, light brown, and brown sample) were measured on repeated occasions, and though variation between the samples is significantly greater than variance within samples, the variance within samples is very clear.

M INDICES FOR HAIR

In contrast to iris color where the concentration of eumelanin probably causes most of the differences in apparent color, hair color is a function of the eumelanin/pheomelanin ratio, so we would like to parameterize hair color by calculating two values from the hair specs—a eumelanin index value and a phaeomelanin value (note that this ratio is also relevant to iris colors, and might be used to improve on the system earlier described). Wagner et al. (2002) described how to calculate brown eumelanin content in terms of a melanin index from the reflectance spectrum. Basically, we take the reflectance value at 640, 650, 660, and 670 and use the following equation:

Hair Melanin Index

$$(\text{HMI}) = 100^* \ln \frac{1}{0.01^*((PR_{640} + PR_{650} + 0.5^*PR_{660} + 0.5^*PR_{670})^* \frac{1}{3})}$$

We can also call this the eumelanin index (HMI, or hair ''M'' value), since the readings at these wavelengths are mainly a function of eumelanin content. We could also use the log rather than the ln, and values determined using the ln can be transformed to corresponding values that would obtain using log by taking the ln(value)/ln(10). Figure 9-21 shows the eumelanin index scores for 2690 European hair samples ordered from lowest to highest. Recalling the wavy shape of the distribution of eumelanin index values for iris color (see Figure 9-8A), this one is relatively smooth. One can note an interesting shape suggesting that very blond and very dark are less common in the sample set. The distribution of M values shown in Figure 9-22 shows that the distribution is fairly normal but may be bimodal much like that for iris color (see Figure 9-8). Whether or not this bimodal character is a general quality of the Caucasian

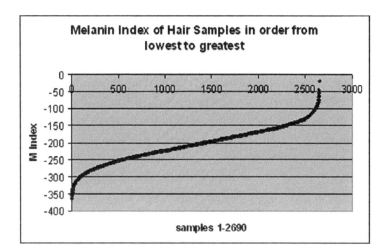

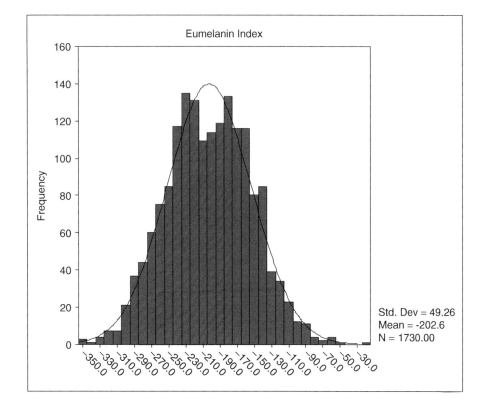

Figure 9-21

Melanin (eumelanin) index (y-axis) of hair samples in order from lowest to greatest (x-axis) in a sample of 2,690 Caucasian fraternal twins. The author generated the spectra for each of the samples using an Ocean Optics USB2000 spectrophotometer. M index for hair was calculated as described in the text. The continuity of the distribution is apparent, and samples with extreme levels of melanin appear to be rarely found. Samples provided by Nick Martin, Queensland Institute of Medical Research, Brisbane, AU.

Figure 9-22

Distribution of the eumelanin index in a sample of 1,730 Caucasian fraternal twins. These are a subset of the samples and data as shown in Figure 9-21. Note that even though the distribution is largely normal in shape, there appears to be an element of bimodality for this particular sample. Figure provided by Sri Niranjan Shekar and Nick Martin, the Queensland Institute of Medical Research, Brisbane, AU.

distribution (as opposed to the distribution within twin collections) remains to be seen. The distribution shows that eumelanin level in the hair is continuously distributed and that most persons of European descent have intermediate levels of eumelanin, with observations of very low or very high levels relatively infrequent.

APPARENT ABSORBANCE

Shriver and colleagues used an apparent absorbance method that was developed for skin (Kollias & Baqer 1987) to study hair. This method uses the average spectral reflectance curves of two persons with precocious graying of the hair (MIM: 139100) as a reference corresponding to the absorbance levels for hair with no pigment.

The apparent absorbance is simply:

$$AA = \log\left(\frac{PR_{580}\,blank}{PR_{580}\,object}\right)$$

where $PR_{580nm}\,blank$ is the percentage reflectance of the precocious grey hair sample. Essentially this function compares the reflectance of the hair sample at 580 nm versus the reference, a completely grey hair sample. This is better than comparing it to a universal white object such as a chip, because the reading from the object is influenced by nonspecific characteristics of hair samples (such as the shape of the hair shaft).

Pheomelanin Index

With the M values, red hair resembles brown and light brown hair; the red hair samples shown in Figure 9-19 indicate little difference between the red and lighter brown hair colors. More subtle measurements are thus needed to resolve the spectral properties of red hair color from other colors. Shown in Figure 9-23 are the apparent absorbance levels for persons with four different

Figure 9-23

Apparent absorbance (AA) levels for persons with four different hair types: black (open circles), brown (lines with no symbols), blond (open squares), and red (filled circles). Note that characteristic curve for red hair samples. Observations from these types of plots gave T. E. Reed the idea that the rate of AA change in the region from $\lambda = 500$ to $\lambda = 600$ was diagnostic for red hair. Provided by Mark Shriver, The Pennsylvania State University, PA.

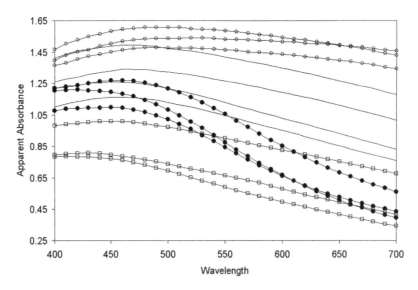

hair types: black (open circles), brown (lines with no symbols), blond (open squares), and red (filled circles). Due to variance within groups and the quantitative nature of hair color, discerning a red profile from a brown one based merely on the shape of the apparent absorbance spectra in Figure 9-19 is very difficult, but the apparent absorbance red hair samples in Figure 9-23 show a steeper slope from 500 nm to 700 nm.

In the reflectance spectra shown in Figure 9-19, this steeper slope is difficult to detect—the transformation of the readings into apparent absorbances against precocious grey hair blanks is needed to see it as we do in Figure 9-23. The problem with this method of resolving red from other hair colors is that not all red hair samples show this slope difference and the sensitivity and specificity of using the apparent absorbance slope in this range to define red hair samples does not appear to be very good. More work would seem to be required therefore, to fully appreciate and quantify the spectrophotometric difference between pheomelanin- and eumelanin-rich samples.

Fifty years ago, T. E. Reed struggled with this problem, and noticed that there was a sharper inflection in the percentage reflectance curve for red hair in the range of 500 nm to 600 nm, with the inflection point tending to reside around 550 nm (see Figure 9-24). Beyond the inflection point, the rate of change for red hair is greater than for other hair colors. Reed described a formula for determining an R value that takes advantage of this observation. The R value uses the ratio of values found above 530 nm to those below to give low values to samples for which the reflectance curve changes its slope rapidly around 530 nm (reds), compared to others like browns and blonds.

$$R = 100^* \left(\frac{y_{530} - 0.243 y_{400}}{y_{650}} \right)$$

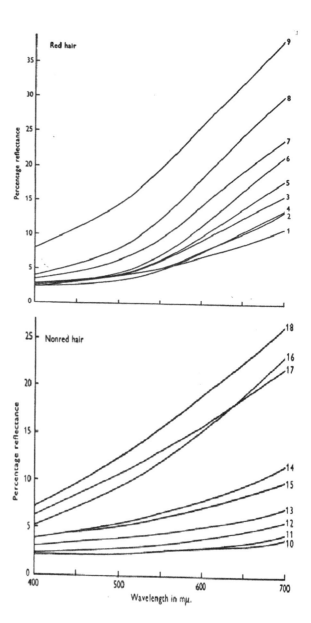

Figure 9-24
Percentage reflectance curves for hair samples studied by T. E. Reed in the 1950s shows a unique inflection for red hair samples in the region of $\lambda = 500$ to 550. Nine red hair samples are shown in the top curves and nine nonred hair specimens in the bottom set of curves. T. Reed, Annals of Eugenics, 1952, Vol. 17, No. 2, pp. 115–139. Reprinted with permission of Blackwell Publishing.

where y_{530} for example is the reflectance value at 530 nm. Essentially what this function does is measure the change in the reflectance moving from 400 to 530 nm—rapid changes give high R values and are characteristic of red colors. The distribution of about 3,000 R values from spectrophotometric readings obtained from an Ocean Optics SB2000 is shown in Figure 9-25; it appears to be normal like the HMI value. It nevertheless seems that we cannot resolve reds versus other colors using this particular value on its own since phaeomelanin is present in various shades of brown hair as well. Expressions of the eumelanin/phaeomelanin index ratio also does not appear to separate reds clearly from others, though blond hair samples show the lowest HMI values (< -225) and browns and reds show intermediate values (most are between -275 and -150), with black hair samples showing high M values (> -150) (see Figure 9-26). What is needed is a different method for expressing the R value relative to the M value. Plotting the hair samples on a two-dimensional plot, with M values on one axis and R values on another, we see the red hair samples are now fairly well resolved from the browns (see Figure 9-27). There is a continuity between the red and brown distributions, but so too is there a continuity in the redness of brown hair or the darkness or brownness of red hair as judged by the eye. Of all the methods discussed so far, this appears to be the best for resolving the hair samples in a manner consistent with visual assessments.

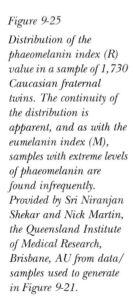

Figure 9-25

Distribution of the phaeomelanin index (R) value in a sample of 1,730 Caucasian fraternal twins. The continuity of the distribution is apparent, and as with the eumelanin index (M), samples with extreme levels of phaeomelanin are found infrequently. Provided by Sri Niranjan Shekar and Nick Martin, the Queensland Institute of Medical Research, Brisbane, AU from data/samples used to generate in Figure 9-21.

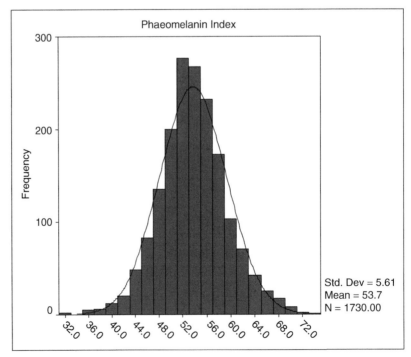

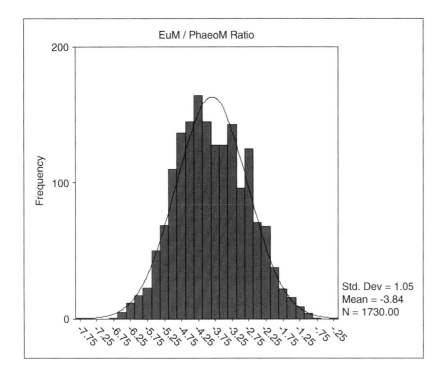

Figure 9-26

Distribution of the eumelanin (M): phaeomelanin index (R) ratio in a sample of 1,730 Caucasian fraternal twins. The continuity of the distribution is apparent, appearing largely normal, but with a suggestion of multimodality we might expect given the shape of the eumelanin distribution in Figure 9-22. As with the eumelanin index (M), samples with extreme levels of phaeomelanin are infrequently found. Darker colored samples comprise much of the middle and right-hand part of the spectrum, and blonde samples the left-hand side. Red samples reside in the middle of the spectrum but cannot be resolved from the brown samples in this diagram. Provided by Sri Niranjan Shekar and Nick Martin, the Queensland Institute of Medical Research, Brisbane, AU from data/samples used to generate in Figure 9-21. From Shriver et al. (2005).

THE RED HAIR COLOR PHENOTYPE

At least some of the complexity of human pigmentation genetics (at least for skin and hair) is contributed by the well-known melanocortin 1 receptor (*MC1R*) gene. Davenport and Davenport hypothesized from scant data as early as 1909 that red hair was caused by one dominant autosomal gene. We now know that this gene is *MC1R*, which is a seven-pass transmembrane G protein coupled receptor, and a key control point in melanogenesis. Loss-of-function mutations at the *MC1R* are associated with a switch from eumelanin to phaeomelanin biosynthesis, resulting in a red or yellow coat color in model animals. Activating mutations, in animals at least, lead to enhanced eumelanin synthesis. Mice carrying a transgene for Agouti (the murine homologue of the human *ASIP* gene) under a keratinocyte specific promoter express a dominant yellow coat color, which has been ascribed to constitutive phaeomelanin expression (Smith et al. 2003). Dominant mutations in the noncoding region of mouse Agouti cause yellow coat color as a result of Agouti chronically antagonizing the binding of alpha-MSH to MC1R.

In man, a number of loss-of-function mutations in the *MC1R* gene have been described but no activating mutations have been found yet. The majority of red-haired humans are compound heterozygotes or homozygotes for up to five

Figure 9-27

Plot of hair melanin index (HMI, or hair M) value versus phaeomelanin index (R) allows for discrimination of red samples from others to a certain extent. Each spot represents a unique sample; red squares = red hair color, brown circles = brown color (various shades), open green circles = blonde hair color, and black triangles = black hair color. Contours represent odds ratio scales for each hair color; the color of the contour indicates the color of hair the ratio corresponds to (same color coding as the symbols). For example, the farther a sample plots to the left on the coordinate system, the higher the likelihood the sample is of red color and the farther the sample plots to the lower right the higher the likelihood the sample is of blonde color. Note that although the resolution of colors is good, there is still overlap. Some of this overlap is undoubtedly explained by difficulty binning colors into four rigidly defined hair color groups. Provided by Sri Niranjan Shekar and Nick Martin, the Queensland Institute of Medical Research, Brisbane, Australia from data/samples used to generate in Figure 9-21.

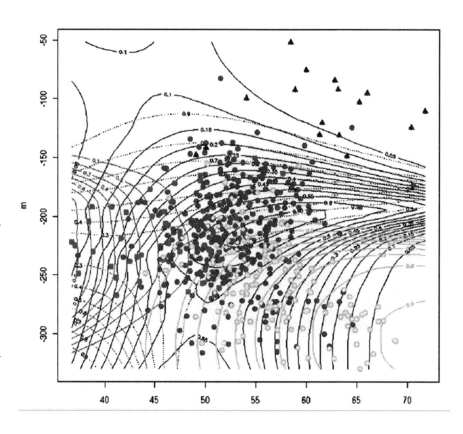

frequently found loss-of-function mutations and they share other common phenotypes such as fair skin and freckling, collectively termed the red hair color (RHC) phenotype. The first association between *MC1R* variants and the red hair color (RHC) phenotype was from a case-controlled study in 1995 (Valverde et al. 1995), but subsequently numerous authors have confirmed and extended these results using sib-pair analyses in twins (Box et al. 1997; Duffy et al. 2004), several large population studies (Bastiaens et al. 2001a, 2001b; Box et al. 2001; Kennedy et al. 2001; Palmer et al. 2000; Smith et al. 1998), and family inheritance studies (Flanagan et al. 2000) (reviewed by Sturm 2002). From these studies, a total of nine relatively common alleles have been identified as associated with the RHC phenotype (see Table 9-8).

Individuals containing two of these polymorphisms (whether in compound heterozygous or homozygous states) are called MC1R homozygotes, and frequently found in populations with European ancestry. Individuals carrying only one of these polymorphisms (whether in the heterozygous or homozygous state) are said to be MC1R heterozygotes, and individuals with none of these polymorphisms (sequence matches the consensus exactly) are considered

Table 9-8

MC1R variants associated with the RHC phenotype (adapted from Sturm 2002).

MC1R variant	Freq (%)	Box et al. (1997) sib-pair	Smith et al. (1998) population	Palmer et al. (2000) population	Flanagan et al. (2000) population family	Box et al. (2001) population	Bastiaens et al. (2001) population	Kennedy et al. (2001) population	Duffy et al. (2004) twins
Val60Leu	12.4	NA	NA	NA	Partial recessive	NA	NA	OR = 2.3 (1.0–5.4)	OR = 6.4 (2.8–14.9)
Asp84Glu	1.1	—	—	—	Recessive with RHC allele	—	p < 0.004	OR = 5.1 (1.4–18.3)	OR = 62.8 (17.6–223.7)
Val92Met	9.7	NA	NA	NT	NA	NA	NA	NA	OR = 5.3 (2.2–12.9)
Arg142His	0.9	—	NT	NT	Recessive	—	p < 0.0001	OR = 49.2 (16.8–145.5)	—
Arg151Cys	11.1	p < 0.001	p = 0.0015	p < 0.01	Recessive	p < 0.0001	p < 0.0001	OR = 20.7 (11.5–37.3)	OR = 118.3 (51.5–271.7)
Arg160Trp	7.1	p > 0.05	p < 0.001	p < 0.001	Recessive	p < 0.0001	p < 0.0001	OR = 12.5 (7.1–21.9)	OR = 50.5 (22.0–115.8)
Arg163Gln	5	—	—	NT	NA	NA	NA	NA	OR = 2.4 (0.5–11.3)
His260Pro	NT	NT	NT	NT	NT	NT	p < 0.008	OR = 9.9 (2.6–37.0)	NT
Asp294His	2.8	p < 0.05	p < 0.005	p < 0.001	Recessive	p < 0.0001	p < 0.0001	OR = 12.4 (3.8–40.5)	OR = 94.1 (33.7–263.1)

NA — not associated, NT = not tested.

— = insufficient sample size for statistical association.

MC1R wild-type. Most of the polymorphisms cause coding changes, and are associated with quantitative diminution of signaling rather than obliteration of gene function. Using the data presented in Table 9-8, Duffy et al. (2004) designated the ASP84Glu, Arg151Cys, Arg160Trp, and Asp294His alleles as the strongest RHC alleles (shaded MC1R variants, Table 9-8) and designated the presence of any of these in a genotype as a single allele, R. The Val60Leu, Val92Met, and Arg163Gln were designated as weak RHC alleles and the presence of any of these in a genotype was designated with the lowercase "r."

We can combine the odds ratios of each allele from Table 9-8 in a linear multiplicative inheritance model in order to predict the number of redheads we should observe from a given sample of R/R, R/r, r/r individuals. Duffy et al. (2004) made these types of predictions and showed that the observed number of RHC phenotypes were similar to the number expected based on these odds ratios (chi, p = 0.02 for goodness of fit; Table 9-9). In this work, and that of Naysmith et al. (2004), there was a clear dosage effect; heterozygotes show intermediate freckling, lightness of skin, redness of hair relative to wild-type and homozygotes, who show normal and lightened pigmentation, respectively. From the Duffy et al. (2004) work, it is clear that the R/R genotype is highly predictive of the RHC phenotype—over 90% of the R/R carriers have red hair. Some did not, and it is possible that these are the eumelanogenic redheads, for whom other mechanisms than melanin/pheomelanin switching are at work (and hence, other genetic loci).

Table 9-9 shows that knowing a person's MC1R status makes it possible to form a probability statement on whether the individual expresses the RHC phenotype. These MC1R polymorphisms do not explain all the extant variability in human hair color, but they seem to be highly penetrant for pheomelano-

Table 9-9

Expectations and observations of RHC phenotype based on MC1R genotype.

MC1R genotype	No.	Expected %, red	Observed No. Red	Observed % red
"+/+"	425	0	0	0
"r/+"	412	0.2	4	1
"R/+"	344	2.5	5	1.5
"r/r"	111	1	1	0.9
"R/r"	203	11.8	22	10.8
"R/R"	73	62.1	49	67.1
Total	1,568		81	

genic red hair in the European populations that have been studied so far. Most of these individuals, whether living in Ireland or Australia, are commonly of Irish ancestry, and it seems likely that the MC1R associations are not merely artifacts of population structure. It is probably not the case that the R and r polymorphisms are associated with RHC merely because they are good Irish AIMs, and because RHC is common among the Irish. We can conclude this based on functional assays showing that these polymorphisms result in altered *MC1R* receptor activity (Frandberg et al. 1998; Schioth et al. 1999), that other AIMs have not been identified as associated with RHC, that eumelanin/pheomelanin switching is mediated by MC1R (Barsh 1996), and that variations in eumelanin/pheomelanin content underlies variability in hair colors (Napolitano et al. 2000; Thody et al. 1991).

These results lead one to suspect that MC1R is to hair color what OCA2 is to iris color; for the latter we could point to one gene as accounting for the majority of the color differences we can see with our eye. But do MC1R variants explain more about hair color than merely the RHC phenotype? Naysmith et al. (2004) showed that MC1R genotype was predictive of the biochemically determined eumelanin/pheomelanin ratio (LEPR score), with homozygous, heterozygous, and wild-type designations accounting for a whopping 67% of the variance in hair colors. This result would suggest that MC1R is involved in the determination of other colors as well. Duffy et al. (2004) described relatively low odds ratios for fair/blonde hair based on their R and r genotypes—the odds ratio (OR) for R was 1.4 (1.2–1.8), and the OR for r was 1.0 (0.8–1.3). Further, Zhu et al. (2004) conducted a pan-genome LD screen for categorical hair colors as well as the spectrophotometrically determined eumelanin/pheomelanin (HMI/R) ratio we have discussed earlier in this chapter (see Figure 9-27), and did not find any significant peaks including over the MC1R locus at chromosome 16q24.3 (see Box 9-D, Figure 9-28).

This negative result was in stark contrast to the result they obtained with iris color, as we have discussed already (see Figure 9-10), which detected a very large LOD peak over the *OCA2* gene, extending a very long distance 5′ and 3′ of this gene. Though there may have been gaps in the coverage of their screen, the discrepancy in results between hair and iris color was very similar to that obtained by Frudakis and colleagues at DNAPrint (unpublished data) from pigmentation candidate gene association screens, where in contrast to the situation for iris color, no good hair color markers could be found in the pigmentation genes. Evidently, hair color could be more complex genetically than iris color (perhaps a larger number of less penetrant alleles). The unfortunate consequence of this is that it seems unlikely that a concise set of highly penetrant markers will be found that explain enough hair color variance for inferring colors (other than pheomelanogenic red) from DNA.

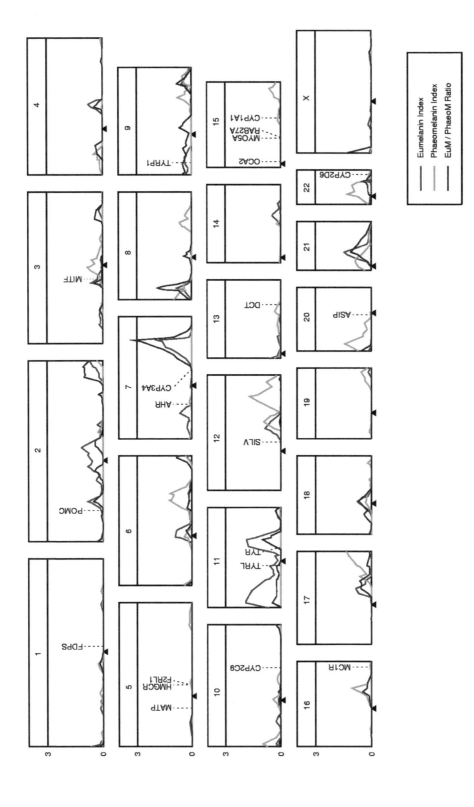

Figure 9-28

Quantitative trait locus (QTL) analysis of iris color in a set of fraternal twins. LOD scores are shown for the association of genetic markers along each of the 23 chromosomes using a one-locus additive QTL model with the eumelanin index (red), the phaeomelanin index (green), or the eumelanin/phaeomelanin ratio (blue). Centromeres are indicated by the black triangles and various human genes are indicated as landmarks (not corresponding to particular linkage markers screened).

580

DIFFICULTY FINDING HAIR COLOR MARKERS: NOT LIKELY A PROBLEM WITH PHENOTYPING

Recall that from our *in silico* spectrophotometry, we were able to distinguish irides with higher phaeomelanin/eumelanin ratios (e.g., greener colors) based on the iris melanin index (e.g., Figure 9-17) much like we could distinguish red hair color from other colors using the HMI versus R values (see Figure 9-27). In addition, association screens using self-reported iris colors (Frudakis et al. 2003b) versus digitally quantified iris colors (Frudakis et al. 2006) resulted in the discovery and confirmation, respectively, of most of the SNPs eventually associated with the continuous iris melanin index (IMI) values. Zhu et al.'s (2004) linkage screen for iris color also operated from nurse-reported but categorical colors, and they had no problem detecting the huge OCA2 LOD peak. Categorizing hair colors rather than properly phenotyping them, both Frudakis and colleagues (unpublished data) and Zhu et al. (2004) had difficulty identifying loci associated with hair color. Using properly phenotyped hair samples (HMI vs. R values, Figure 9-27), Zhu and colleagues reported similar difficulties (unpublished data; Box 9-D). Thus, our choice in method of phenotyping may have power implications, but probably does not explain our inability to find good hair color markers.

Box 9-D

Linkage Scan for Hair Color Markers—Not as Clear Cut as for Iris Color

The HMI/R value was used in a genetic linkage screen conducted by Zhu et al. (2004), though no good peaks were found, in stark contrast to the screen conducted by this group for iris color where several large LOD peaks were observed (one of them very large, Figure 9-10). Zhu et al. (2004) repeated their screen using the actual HMI/R values shown in Figure 9-26 (same samples), and obtained a single peak extended above the statistical significance line (Figure 9-28, chromosome 7). Interestingly, this peak was obtained for the HMI/R value but not using the HMI or R values on their own, and not using nurse-reported hair colors. It is interesting that by using nurse-reported hair colors rather than spectrophotometric HMI/R values, virtually all the peaks present in Figure 9-28 are diminished in height and significance. Nonetheless, the results shown in Figure 9-28 are probably not very useful in a forensic sense. They are disappointing in that it seems we are not able to identify powerful and penetrant markers for hair color, as we have for iris color. However, they are encouraging in that we can see an improvement in the results provided by using a more objective, quantitative phenotyping method.

Why are hair color markers so difficult to find compared to iris color? Possibly this is due to one of the following:

- Different genes in the TYR pathway, perhaps genes that remain to be described, are involved for determining hair color. We know that iris and hair color are genetically uncoupled. Although this idea might explain why the candidate gene approach failed, it would not seem to explain why the linkage approach failed and so it seems unlikely to be the sole answer.

- Environmental or developmental nongenetic influences may be stronger for hair color.
- Some essential epistatic interaction is critical in the determination of hair color such that tests based on single locus linkage or associations fail. Zhu et al.'s (2004) linkage analysis also tested for QTL X QTL interactions without much success, so if this is part of the answer the epistatic interactions must be quite complex.
- Our phenotyping method for hair color may still be too imperfect—perhaps expensive and inconvenient biochemical or microscopic quantification will be required.
- Perhaps hair color is determined by a much larger number of gene variants, each of relatively low penetrance in contrast to iris color, where we are lucky that there is a single major gene variant (*OCA2*).

If MC1R is so important for hair color, what about the hormones that bind it? The pro-opiomelanocortin hormone (POMC) that binds to MC1R is crucial for the differentiation of melanocytes and so it is reasonable to suspect that this gene may also underlie some variation in hair colors. As we have discussed, the pro-opiomelanocortin derived peptides adrenocorticotropic hormone and alpha-melanocyte stimulating hormone are the principal mediators of human skin pigmentation, acting through the melanocortin-1 receptor to influence the decision a melosome makes between the eumelanogenic and pheomelanogenic pathways (see Figures 9-1 and 9-2). The POMC gene thus represents a good candidate gene for additional population surveys. In support of this idea there have been case reports of individuals of red hair and very light skin caused presumably by a POMC mutation (Krude et al. 1998). Nonetheless, a cursory screen of POMC SNPs in the pigmentation gene candidate study of Frudakis et al. (2003b) did not illuminate any highly penetrant variants, and Zhu et al.'s (2004) screen did not show impressive LOD peaks over the POMC locus at chromosome 2p23.3.

PREDICTING HAIR COLOR (NOT JUST THE RHC PHENOTYPE) FROM DNA

Ultimately, what we would like is to identify a few genes with a small number of highly penetrant alleles, and use these—perhaps along with biogeographical ancestry—as variables with the empirical database approach described for iris color. However, at present, we seem to be lacking "the few genes" part of this puzzle, having really only one (MC1R) that accounts for a fraction of the expression of red hair. Perhaps Zhu et al.'s (2004) peak over chromosome 7 indicates another? Even if so, the height of the LOD peak is not impressive, and suggests that this marker will not be penetrant enough to allow for accurate predictions (it would fall far short of the standard set by OCA2 for iris color—note the height of the OCA2 LOD peak in Figure 9-10 compared to the height of the chromosome 7 peak in Figure 9-28).

Maybe more thorough screens with better LD coverage of the genome are needed. Perhaps other approaches entirely, such as admixture mapping, will be required to identify the major brown-blond hair color genes. It may be that they will never be identified because there are simply too many, each with low marginal effects. Given the difficulties screeners have had identifying hair color genes other than MC1R, and until future work identifies additional gene variants linked to hair color variation, it appears that, at least for the time being, hair color will be predicted based only on:

- MC1R variants and the RHC phenotype, where for in each case, only an odds ratio will be possible.
- The indirect method of anthropometric trait value inference covered in Chapter 8 (though this data remains to be generated).

Much of the results needed to implement the indirect method are still unpublished and immature, but there are preliminary findings that the NOR, MID-EAS, and SA genetic subgroups from the within-European panel of markers (EURO 1.0 test) described in Chapter 7 are associated with lighter and darker hair colors, respectively—even admixture in Caucasians seems to be associated (compare the frequency of lighter hair colors between Figure 8-14 vs. 8-12, top). At the time of this writing, it was not known how valid these associations are (due to low database sample sizes), but future editions of this text should include a discussion based on summary statistics from much larger population surveys. Until more of the variation in hair colors can be reduced to variants from a few key genes, the direct prediction of hair colors will continue to be confined to RHC phenotypes based on the MC1R-pheomelanogeic alleles, and the indirect method of inference based on continental and European ancestry. Indeed, this scenario illustrates why the direct method of inference is not always an alternative to the indirect method based on measures of population structure.

SKIN PIGMENTATION

We already have discussed the molecular and cell biology of skin pigmentation. Here we discuss the progress that has been made in identifying the genetic determinants that underlie variable skin pigmentation with an eventual aim toward developing direct gene-based methods for inferring skin color from DNA.

THE GENETICS OF SKIN PIGMENTATION

Pigmentary traits such as lack of tanning response to UVR and the propensity to freckle constitute other components of the red hair color, or RHC, phenotype and have been linked as risk factors for melanoma, basal cell carcinoma (BCC), and squamous cell carcinoma (SCC) (Sturm 2002). So far, MC1R is the only gene that has been associated with variable skin pigmentation in humans. As we have described already, MC1R encodes a seven-pass transmembrane domain G-protein-coupled receptor that is expressed by skin melanocytes. It activates adenylate cyclase to elevate cAMP levels in response to stimulation by the proopiomelanocortin (POMC) derived peptides α-melanocyte stimulating hormone (α-MSH) and adrenocorticotrophic hormone (ACTH) (Thody et al. 1998). Activation of MC1R signaling is responsible for the switching between eumelanogenesis and pheomelanogenesis routes of the melanin synthetic pathway from the limiting substrates Dopa and tyrosine (see Figures 9-1 and 9-2).

Individuals with the MC1R mediated RHC phenotype are compromised in their ability to respond to UV by tanning response. In non-RHC individuals, UV exposure leads to release of αMSH or ACTH from skin keratinocytes, then to an increase in the number and distribution of eumelanosomes and activation of eumelanocytes along the eumelanogenesis pathway, which we recognize with our eye as the tanning response. We can thus expect that there might be genetic associations between RHC alleles at the MC1R locus and fair skin, and indeed, Duffy et al. (2004) showed that such associations, though weak, do in fact exist.

Table 9-10 shows the odds that an individual with each of the nine major MC1R variants will express a fair or pale skin phenotype relative to the average European. We can see from this data that some of the polymorphisms, such as D84E confer a very significant risk for fair or pale skin, whereas others seem to confer more

Table 9-10

Odds of fair skin based on MC1R genotype.

MC1R variant allele	Fair/pale skin OR
V60L	1.7 (1.3–2.1)
D84E	12.5 (4.8–42.8)
V92M	2.3 (1.8–3.1)
R142H	1.8 (0.6–5.7)
R151C	4.4 (3.3–5.7)
I155T	2.1 (1.0–4.3)
R160W	3.2 (2.4–4.4)
R163Q	2.0 (1.4–2.9)
D294H	7.5 (4.4–13.7)
r (V60L, V92M, and R163Q)	1.9 (1.6–2.3)
R (D48E, R151C, R160W, and D294H)	4.2 (3.4–5.2)

modest risk. Carriers of the most significant RHC alleles, D48E, R151C, R160W, and D294H (denoted as R genotypes) are four-fold more likely to express pale and or fair skin than the average European. These results are impressive evidence of a link between MC1R sequences and fair skin, and are probably strong enough to serve as a basis for accurately predicting skin shade from DNA—at least for the small fraction of the population with the RHC phenotype.

It is true that the association is not corrected for ancestry and we are uncertain whether the results are due to population structure correlations or an underlying functional geneotype–phenotype relationship, but the same was true for hair color and given the solid work that has shown a direct relationship between MC1R and pigmentation biochemistry and cell biology, we can probably feel safe concluding that the results indicate that MC1R variants are one piece of the skin pigmentation puzzle (though probably a relatively small piece).

Given these results, we might expect that in Europeans at least, variation in the MC1R gene is a determinant of sun sensitivity and a genetic risk factor for melanoma and nonmelanoma skin cancer. Indeed, recent work suggests that MC1R also shows a clear heterozygote effect on skin type, with up to 30% of the population harboring loss-of-function mutations and an association with skin melanomas, basal cell carcinomas, and squamous cell carcinomas (Sturm 2002). Given the strong ancestry component with which these pigmentation diseases and phenotypes are expressed, MC1R is thought to be particularly informative and tractable for studies of human evolution and migration. In particular, the study of the MC1R may provide insights into the lightening of skin color observed in most European populations. The worldwide pattern of MC1R diversity is compatible with functional constraint operating in Africa, whereas the greater allelic diversity seen in non-African populations is consistent with neutral predictions rather than selection (Harding et al. 2000). Whether this same conclusion will be drawn from studies of other pigment genes will be of great interest for studies of human skin color evolution.

Our primary objective herein is to make inferences on phenotype from DNA, and from the moderate odds ratios and low allele/RHC phenotype frequencies in the European populations just discussed, it is clear that the bulk of skin color variation within Europeans remains to be explained. We need to be able to measure more than just the MC1R genotypes to make inferences and because skin color varies so much with biogeographical ancestry (on a continental level), one approach that could be useful is the admixture mapping approach.

SKIN COLOR MARKERS

Mark Shriver and colleagues at Pennsylvania State University have performed precursor admixture mapping studies aimed at identifying functional loci underlying variability in skin color (Bonilla et al. 2004; Shriver et al. 2003). They focused on both admixed populations of West African and European ancestry (African Americans from Washington, D.C. and African Caribbeans from the United Kingdom) as well as an admixed population with European and Indigenous American ancestry (the Spanish Americans/Hispanics of the San Luis Valley in Southern Colorado).

Admixture mapping is appropriate for traits that exhibit significant phenotypic difference between parental populations. For the work of Shriver et al. (2003), the samples and data shown in Figure 8-18 (African American and African Caribbeans) and Figure 8-19 (Spanish Americans) were used. From Figure 8-18 and Figure 8-20, we saw that West African ancestry explains a significant amount of the variability in skin color among African Americans and Puerto Ricans, respectively, but we also see that there exists other variability not seemingly explained by ancestry. Although only 32 AIMs were used to measure ancestry for Figure 8-18, and we would therefore expect a rather large standard error in genomic ancestry admixture estimates, interindividual variation within genomic ancestry levels are also seen in Figure 8-20, for which 171 AIMs were used. With the larger panel, we see that the correlation does increase, but not to such an extent that ancestry alone will be sufficient to pinpoint the precise melanin index for an individual. Presumably, the data missing from these models and necessary for pinpointing melanin index values from DNA is that contributed by the underlying phenotypically active loci.

Shriver et al. (2003) and Bonilla et al. (2004) used admixture mapping to find some of these loci. The ultimate goal of admixture mapping is to exhaustively screen the genome for loci responsible for phenotypic variation across the parental populations. In many ways this study of admixture and pigmentation was the direct conceptual descendant of the much cited and reanalyzed data of Harrison and Owen (1964).

Shriver et al. (2003) tested for differences in the average pigment levels by genotype for 33 AIMs typed in the aforementioned two populations of West African/European ancestry. The panel of AIMs included three candidate gene markers (OCA2, TYR, and MC1R) and others located in nonpigmentation genes as well as between genes. Individual biogeographical ancestry was estimated two ways: (1) with all the AIMs and (2) with all AIMs except one (each AIM left out in turn, which was then tested for association with phenotype). Analyses to test for relationships of the markers on skin pigmentation were performed three ways: (1) with no consideration for the individual admixture estimates (ANOVA), (2) after conditioning to control for the effect of individual ancestry leaving out the locus under consideration (ANCOVA/IAE minus marker), and (3) using the complete individual ancestry estimate for the conditioning (ANCOVA/IAE).

These results are presented for an African American and Afro Caribbean sample in Table 9-11A and a Spanish American sample in Table 9-11B. In the African sample, a total of 17/33 (53%) of the AIMs show significant differences ($p < 0.05$) in skin M index among the three genotypes, including two of the four candidate gene markers (OCA2 and TYR). When using an alpha level of

Marker[1]	AF-EU DELTA[2]	ANOVA[3]	ANCOVA[4] IAE-MARKER	ANCOVA[5] IAE	Bayesian score test probability[6]
TYR-192	0.444	**0.000**	**0.000**	**0.006**	**0.000**
OCA2	0.631	**0.000**	**0.008**	0.249	**0.030**
MC1R-314	0.350	0.552	0.945	0.731	0.666
MID-575	0.130	0.491	0.257	0.219	0.683
MID-187	0.370	0.086	0.837	0.958	0.743
FY-NULL	0.999	**0.000**	0.012	0.979	0.389
AT3	0.575	**0.030**	0.646	0.737	0.260
F13B	0.641	**0.005**	0.280	0.954	0.348
TSC1102055	0.441	**0.019**	0.056	0.052	0.445
WI-11392	0.444	**0.038**	0.094	0.283	0.343
WI-16857	0.536	0.573	0.868	0.228	0.806
WI-11153	0.652	**0.000**	**0.025**	0.098	0.150
GC	0.697	**0.046**	0.808	0.574	0.720
SGC30610	0.146	**0.017**	0.195	0.202	0.284
SGC30055	0.457	**0.001**	0.223	0.948	0.193
WI-9231	0.017	0.647	0.214	0.157	0.459
LPL	0.479	**0.000**	**0.008**	0.308	0.151
WI-11909	0.075	0.066	**0.058**	0.156	0.709
D11S429	0.429	**0.023**	0.102	0.171	**0.019**
DRD2 Taq D	0.535	0.147	0.311	0.522	
DRD2 Bcl I	0.080	**0.047**	0.089	0.400	**0.025**[7]
APOA1	0.505	**0.032**	0.935	0.673	0.060
GNB3	0.463	0.223	0.771	0.542	0.858
RB1	0.611	0.697	0.696	0.071	0.989

Table 9-11A

Testing for an effect of single-locus genotypes on pigmentation in the combined African-American and African-Caribbean sample. Significant values ($\alpha = 0.05$) indicated in boldface.

Table 9-11A
(Continued)

Marker[1]	AF-EU DELTA[2]	ANOVA[3]	ANCOVA[4] IAE-MARKER	ANCOVA[5] IAE	Bayesian score test probability[6]
WI-14319	0.185	0.322	0.133	0.136	0.086
CYP19	0.045	0.547	0.492	0.565	0.310
PV92	0.073	0.064	0.422	0.559	0.756
WI-14867	0.448	**0.001**	0.116	0.230	0.967
WI-7423	0.476	**0.003**	0.464	0.359	0.966
Sb19.3	0.488	0.285	0.496	0.916	0.490
CKM	0.150	0.828	0.505	0.128	0.524
MID-154	0.444	0.182	0.543	0.963	0.347
MID-93	0.554	0.111	0.361	0.796	0.286

[1]Marker indicates the AIM used in the test. Markers shown in bold and italics are in or near candidate genes for pigmentation (viz., *OCA2, MC1R, TYR*).
[2]Delta is the allele frequency difference between African and European populations.
[3]Analysis of variance significance level where sex is the only covariate.
[4]Significance level for a one-way ANCOVA analysis using individual admixture estimates (M) where the tested locus was excluded as the covariate.
[5]Same as footnote 3 except the M is based on all markers.
[6]Bayesian admixture mapping one-sided probability.
[7]p value for DRD2 TaqD-Bcl haplotypes.

0.05, only 5% of the markers tested are expected to yield significant results by chance alone. The observation of 53% significant results is consistent with the presence of admixture stratification in the population studied, and pigmentation differences between the parental populations (Parra et al. 2001; Pfaff et al. 2001, 2005) as we discussed in the last chapter.

In order to test for functional effects and not just admixture stratification, Shriver et al. (2003) needed to adjust for these effects by conditioning on the individual ancestry estimates (IAE). This test is conceptually analogous to that described by Paul McKeigue in 1998, where he outlined that a test conditioning on parental admixture was equivalent to an interspecific cross (see Box 9-E: McKeigue 1998; McKeigue et al. 2000). Using this Bayesian approach, Shriver et al. (2003) described four loci that show evidence of linkage to skin pigmentation at an alpha level of 0.05 [TYR ($p < 0.0001$), OCA2 ($p = 0.03$), DRD2 ($p = 0.025$), and D11S429 ($p = 0.019$)] (see Table 9-11A). It is notable that D11S429 is located 8 cM telomeric to TYR and DRD2 approximately 15 cM centromeric to TYR. The concordance between the ANOVA results and the Bayesian admixture mapping results at the TYR and OCA2 loci is a strong indication that these associations are real and not artifacts of either analytical method.

Table 9-11B

Testing for an effect of single-locus genotypes on pigmentation in the Spanish-American sample. Significant values ($\alpha = 0.05$) indicated in boldface.

Marker[1]	Delta[2] NA vs EU	ANOVA[3]	ANCOVA[4] IAE minus marker	ANCOVA[5] IAE	Bayesian score test p-value[6]
MID-575	0.546	0.240	0.383	0.248	0.93
TSC1102055	0.744	**0.027**	**0.023**	0.366	0.39
WI-11153	0.628	**0.012**	**0.036**	0.406	0.12
SGC-30610	0.427	0.093	0.133	0.565	0.13
WI-17163	0.521	0.192	0.268	0.967	0.17
WI-4019	0.296	0.875	0.762	0.319	0.93
WI-11909	0.663	**0.026**	**0.020**	0.146	0.13
TYR-192	0.417	**0.006**	**0.012**	0.129	0.07
DRD2Bcl	0.485	0.254	0.517	0.827	0.35
DRD2TaqD	0.582	0.418	0.461	0.550	0.35
D11S429	0.376	0.235	0.511	0.545	0.65
WI-14319	0.494	**0.036**	0.054	0.061	0.30
CYP19E2	0.423	**0.000**	**0.000**	**0.002**	**0.001**
PV92	0.624	0.183	0.276	0.591	0.50
WI-7423	0.402	0.426	0.318	0.309	0.64
CKMM	0.545	0.257	0.569	0.579	0.55

[1]Marker indicates the AIM used in the test. Markers shown in italics are in or near candidate genes for pigmentation (viz., TYR-192 and CYP19E2 near MYO5A and SLC24A5).
[2]Delta is the allele frequency difference between Indigenous American and European parental populations.
[3]Analysis of variance significance level where sex is the only covariate.
[4]Significance level for a one-way ANCOVA analysis using individual admixture estimates (M) where the tested locus was excluded as the covariate.
[5]Same as footnote 3 except the M is based on all 21 markers.
[6]Bayesian admixture mapping one-sided probability.

When the complete individual admixture estimate is used to condition (ANCOVA/IAE), only one locus shows significant average differences among genotypes—the tyrosinase gene (TYR). When a less conservative conditioning approach is taken where the locus under consideration is left out of the individual ancestry estimate (ANCOVA/IAE minus marker), there were five significant results: OCA2, TYR, FY, WI-11153, and LPL. At the very least, these results suggest that variants of TYR (not just the ancestral origin of the variants) underlie some of the variation in skin color we see in African Americans, and by extension, in the human population.

Variants in or around the OCA2, FY, WI-11153, and LPL loci may also be linked to West African/European variability in skin pigmentation. Shiver et al.

(2003) concluded that to augment these results, what is required is a genome-wide screen for admixture linkage signals. Hoggart et al. (2004) details the sample size requirements for given levels of statistical power in such a screen. One approach to such studies is to have 1,500 evenly distributed AIMs in the screening set, and significant linkages could then be followed up by a denser analysis of additional AIMs.

Bonilla et al. (2004) also performed a preliminary admixture mapping screen for admixture linkage to skin pigmentation in a panel of Spanish Americans/ Hispanics from the San Luis Valley (SLV) in Southern Colorado (Bonilla et al. 2004). These subjects were ascertained as part of the long-standing San Luis Valley Diabetes Study directed by Richard Hamman through the University of Colorado Health Science Center (Hamman et al. 1989). Using 22 AIMs, the ancestral proportions of the SLV Hispanic population were estimated as $62.7 \pm 2.1\%$ European, $34.1 \pm 1.9\%$ Indigenous American, and $3.2 \pm 1.5\%$ West African—proportions not dissimilar to other Hispanic populations we discussed in Chapter 6 (Table 6-1, Figures 6-3 and 6-4).

In this study, there was very little association of unlinked markers, and the absence of correlation between calculated initial and observed disequilibrium suggested that there is little genetic structure in this population. However, there was a small but highly significant correlation between individual Indigenous American ancestry and skin pigmentation ($R^2 = 0.06$, $p < 0.0001$), which indicates that in fact there may be a low level of population structure in this group. Additionally, an analogous test based on the correlation of individual ancestry estimates also shows a significant relationship that is most likely due to variation in ancestry levels or population structure. (These results highlight the relevance for assessing population structure at the level of ancestry admixture, rather than strictly with LD measures.)

The results for admixture mapping tests (both the ANOVA-based tests and the Admixture Mapping tests) are presented in Table 9-11B. Six markers appeared to be related to the variation in skin pigmentation with the standard ANOVA method (no adjustment for BGA) and five AIMs were significant with the ANCOVA/IAE, but most of the significant signals disappeared when the Bayesian method was applied. In contrast, two markers, TYR-192 and CYP19-E2, continued to be significant even with the most stringent test. CYP19-E2 was highly significant with both approaches, is located on chromosome 15 about 1 MB from MYO5A, 3 MB from SLC24A5, and 4 MB from RAB27A. These three genes are all reasonable pigmentation candidate genes. MYO5A and RAB27A are both mutated in patients with Griscelli syndrome, an autosomal recessive disorder characterized by pigmentary dilution among other features (Pastural et al. 1997). MYO5A codes for myosin V, a protein that participates in organelle movement throughout the cell (OMIM 160777). The RAB27A protein product has GTPase activity and is likewise known to function in organelle movement. SLC24A5 has recently been shown to be the cause of the Golden phenotype in zebra fish, where it functions in the determination of melanosomal ultrastructure and to be associated with skin pigmentation in humans (Lamason et al. 2005; Sturm 2006, who also conditioned their results on admixture).

Shriver and colleagues subsequently built on these results in determining the patterns of variation across the world at four genes that have each been shown to affect levels of pigmentation in West African/European admixed populations. These genes are discussed later, in the context of admixture mapping (see Box 9-E). In their most recent work, Shriver and colleagues showed that the genes for tyrosinase (TYR), the human P gene (OCA2), the Agouti gene (ASIP), and membrane-associated transporter (MATP) each have measurable effects on pigmentation in African Americans after adjusting for variation in genomic ancestry.

Shriver did two sets of experiments to study how these variants are distributed across the world: (1) comparing levels of F_{ST} and locus-specific branch length (LSBL) at markers in question with an empirical distribution of thousands of SNPs, and (2) typing the variants in question in the Human Genome Diversity Project (HGDP-CEPH) cell line panel of 52 populations from throughout the world. An assumption for the test in the first approach is that most SNPs are evolving via selectively neutral mechanisms and so comprise a demographically appropriate null distribution for tests of natural selection using F_{ST} and LSBL. These methods (Akey et al. 2002; Shriver et al. 2003) are recent extensions of a conceptual framework laid out in 1964 by Luca Cavalli-Sforza. There have been many important developments in this area of genomic screens for signatures of natural selection and we refer the interested reader to an excellent review by Storz (2005). The LSBL approach, though relatively straightforward theoretically, has a major advantage in making very few assumptions regarding the demographic histories of the populations in question.

Shriver and Norton tested these four genes by selecting AIMs located within each gene that had been reported previously in the literature. They then used these AIMs in an admixture mapping screen, testing against the empirical distributions of F_{ST} and LSBL for signatures of genetic adaptation (i.e., looking for markers that show associations over and above that which we would expect based on the correlation between trait value and ancestry) (Norton et al. 2005). Overall the results provide compelling evidence for concordance between tests of genetic adaptation and admixture mapping, which, given some consideration of what is being measured, is expected.

Table 9-12 shows the matrix of pairwise F_{ST} computations and the resulting p-values based on comparison with the empirical data from the Affymetrix

Table 9-12

Pairwise locus-specific F_{ST} levels for four pigmentation candidate genes tested using ancestry informative markers.

	Island Melanesian	East Asian	South Asian	Native American	European
TYR					
East Asian	0.000 (1.000)	—			
South Asian	0.019 (0.566)	0.019 (0.0436)	—		
Native American	0.000 (1.000)	0.000 (1.000)	0.009 (0.606)	—	
European	**0.347 (0.045)**	**0.347 (0.017)**	**0.274 (0.004)**	**0.333 (0.035)**	—
West African	0.000 (1.000)	0.000 (1.000)	0.010 (0.646)	0.000 (1.000)	**0.333 (0.043)**
ASIP					
East Asian	0.037 (0.445)	—			
South Asian	0.061 (0.361)	0.000 (1.000)	—		
Native American	0.269 (0.115)	0.124 (0.147)	0.094 (0.231)	—	
European	0.127 (0.243)	0.024 (0.502)	0.009 (0.436)	0.039 (0.464)	—
West African	0.149 (0.282)	0.323 (0.065)	**0.377 (0.023)**	**0.687 (0.011)**	**0.496 (0.011)**
OCA2					
East Asian	0.037 (0.444)	—			
South Asian	0.011 (0.633)	0.003 (0.580)	—		
Native American	0.042 (0.492)	0.000 (1.000)	0.005 (0.646)	—	
European	0.069 (0.388)	0.001 (0.637)	0.019 (0.345)	0.000 (1.000)	—
West African	0.053 (0.523)	0.167 (0.212)	0.114 (0.235)	0.175 (0.253)	**0.348 (0.039)**
MATP					
East Asian	0.004 (0.634)	—			
South Asian	0.031 (0.490)	0.013 (0.488)	—		
Native American	0.027 (0.580	0.010 (0.562)	0.000 (1.000)	—	
European	**0.747 (0.001)**	**0.717 (0.000)**	**0.624 (0.000)**	**0.636 (0.003)**	—
West African	0.020 (0.545)	0.004 (0.751)	0.000 (1.000)	0.000 (1.000)	**0.654 (0.003)**

Shown pairwise F_{ST} levels are calculated using Weir and Cockerham (1984) unbiased statistic. The p-values are computed as the proportion of SNPs, from the 11,078 autosomal SNPs on the 10 K Mapping Array that have pairwise F_{ST} levels higher than those observed for the marker in question. Markers showing significant results (p-value > 0.05) are shown in bold.
Data from Norton (2005).

(Santa Clara, CA) 10 K Mapping Array (Shriver et al. 2005). Five populations are shown based on both their distribution across the world and relative skin pigmentation levels. It is clear that both MATP and TYR have SNPs and show significant allele frequency differences primarily between European and all other populations. These matrices of pairwise differences can also be diagramed as population trees. Figure 9-29 shows the average tree across all 11,078 autosomal SNPs on the 10 K mapping array constructed using the neighbor-joining tree method (Saito & Nei 1987), and Figure 9-30 shows the Principle Coordinates Plot for various populations using this mapping array. They created similar trees to Figure 9-29 to model the pigment gene polymorphisms among the populations shown in Table 9-12.

Doing this gene by gene, they observed a striking pattern. For MATP and TYR, the Europeans have very long branches separating them from all the other populations. Alternatively, for ASIP and OCA2 the pattern is more clinal, with the most substantial difference in OCA2 being between European and West African with all the other populations in between. A similar situation was obtained with ASIP. These results mirror those we might visually estimate (in the mind's eye) from the F_{ST} data shown in Table 9-12.

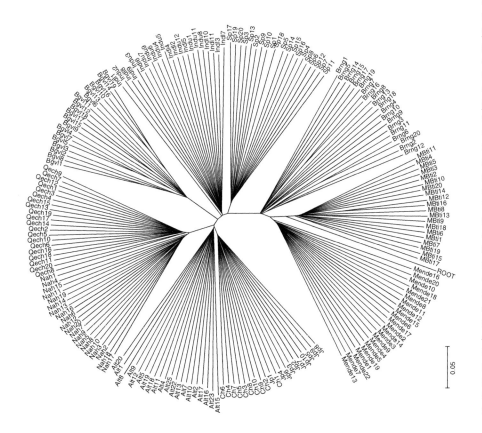

Figure 9-29

Phlyogenetic tree of 203 individuals from 12 populations based on genotypes from 11,555 SNP loci. This data was used for discerning associations with skin pigmentation from ancestry correlations via admixture mapping. The tree was constructed using the neighbor-joining technique and an allele-sharing distance matrix. The genotype of the root was taken from loci of homozygous state among a small collection of primates (chimpanzees, gorillas). Mbti (Mbuti), Brng (Burunge), Sp (Spanish), Indl (South Asian Indian lower caste), Indu (South Asian Indian upper caste), Bgvl (Nasioi), Qech (Quechua), Nah (Nahua), Alt (Altaian), Ch (Chinese), Jp (Japanese).

Figure 9-30

Principal components plots for the same data presented in Figure 9-29, also based on the allele-sharing distance matrix. The first three of the four significant principal component axes are shown for all populations in A as indicated with the key to the upper right. B shows an expanded view of the European and Asian segment of A. From Shriver et al. (2005).

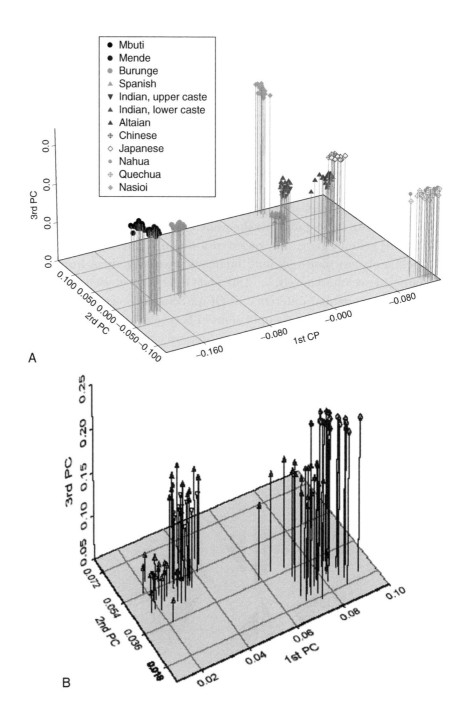

Shriver and colleagues also tested these four pigmentation gene SNPs in the HGDP-CEPH panel of 52 population samples. This panel is most well known for its use by Noah Rosenberg and colleagues in their award winning *Science* paper in 2002 (see Figure 4-1 in Chapter 4). Although there are some clear deficits to the HGDP-CEPH sample collection (e.g., small sample sizes, no South Asian populations, and no admixed populations), they do represent an excellent resource for studies of human genetic variation. Shriver noted differences in the patterns of variability across the globe and used the results as an aid for understanding whether there has been convergence or conservation at the pigmentation loci in question. Specifically, they showed that the clearest evidence for the case of convergence/conservation is for a separate evolutionary history in terms of skin lightening events for European and East Asian populations for both MATP and TYR. There was also some indication for shared allele variation among the West African and Melanesian populations for all four of these genes.

Box 9-E
Bayesian Admixture Mapping

Paul McKeigue at University College, Dublin, performed the admixture mapping for the pigmentation research discussed in this chapter. For Bayesian admixture mapping, he used what he calls a Bayesian full probability model for admixture and marker genotypes (McKeigue et al. 2000). McKeigue's ADMIXMAP program is used not only to estimate individual admixture, as we discussed in Chapter 3, but for testing the dependence (or independence) of allelic associations on ancestry admixture. A model is used incorporating the ancestry of each marker allele, the admixture of each individual, and the distribution of admixture in the population—all treated as missing data. These parameters are modeled as random variables with a hierarchical structure (e.g., accommodating dependence among them we would expect in any data set).

The distribution of individual admixture in the parental generation is assigned a noninformative prior distribution and then the stochastic variation of ancestry between two states (European and African) on each set of chromosomes inherited from a parent is modeled as a Markov process. The posterior probability distribution of the missing data, given the observed marker data, is generated by Markov chain Monte Carlo simulation. With a large sample and noninformative priors, the posterior mode (or mean) and 95% central posterior interval obtained in this Bayesian analysis are asymptotically equivalent to the maximum likelihood estimate and 95% confidence interval, which have served as the basis for most of the admixture results presented so far in this book (Carlin et al. 1995; McKeigue et al. 1998). As discussed in Chapter 3, when estimating individual admixture, the model takes into account the uncertainty with which individual admixture can be estimated using a given number of markers, and estimates the slope of the relationship that we would observe if individual admixture were measured accurately.

McKeigue (1998) has shown that if we condition on parental admixture, a test for association of a trait with ancestry at a marker locus is a specific test for linkage. An assessment (a score test) of the dependence of pigmentation on the proportion of alleles of African (or other) ancestry at each marker locus is constructed by averaging over the posterior distribution of missing data (McKeigue et al. 2000) and essentially we are calculating the most likely values given the data at hand. Score tests for linkage are based on testing for an independent association of pigmentation with number of alleles of European ancestry at each locus, one at a time, in a regression model that includes individual admixture (estimated from marker data).

Included in Tables 9-11A and 9-11B are the one-sided probabilities for the score tests (Bayesian test score p-value, Tables 9-11A and 9-11B). The average proportion of European admixture for the data in Figure 8-18 of Chapter 8, which was the basis for the pigmentation study of Shriver et al. (2003), was estimated at 0.23 (95% posterior interval 0.20–0.23)—a value not dissimilar to that we presented from MLE calculations (see Table 6-1, Figure 6-3) or with ADMIXMAP (see Table 6-1, Figure 6-4) with larger AIM panels in separate African-American samples.

In other work, Norton et al. (2005) showed that the MATP gene similarly shows admixture linkage ($p < 0.0001$) to skin pigmentation as measured by the M index in the African-American and African-Caribbean population samples used in the Shriver et al. (2003) work. In contrast to ASIP, MATP shows a pattern of codominant inheritance whereby the CC homozygotes have the lowest M index (42.6), the GG homozygotes the highest (57.4), and the heterozygotes are intermediate (51.0).

FINAL CONSIDERATIONS FOR THE DIRECT INFERENCE OF SKIN PIGMENTATION

The progress, therefore, for defining markers informative for skin pigmentation is in a similar though more advanced stage as for hair color. We thus have evidence that variation in a number of human pigmentation genes (MC1R, TYR, ASIP, MATP, SLC24A5, and OCA2) is associated with skin pigmentation in various populations, over and above their information for ancestry, and we can thus safely conclude that variants in the genes play some direct role in the determination of skin pigmentation status. We can point to a number of other candidate genes such as POMC as deserving of more research. What is needed to move forward and contrast these alternative candidate genes is a fine scale admixture mapping panel, whereby a denser selection of ancestry-informative markers are typed across whole regions spanning these genes. Presumably, a multipoint analysis will indicate more support for one of the genes compared to the others. These studies should help identify the best markers within these genes for use in predicting M values of unknowns.

Currently however, variants in these genes do not seem to explain enough of the variability in skin pigmentation to serve as stand-alone classification variables. Methods incorporating these sequences (optimally, the best markers within these genes once defined) with individual ancestry estimates may prove the most useful for this purpose in the near term. At this point in time, it's not clear that we have all the puzzle pieces to predict skin color directly as we can iris color. However, results demonstrating whether markers in these genes are adequate to predict skin color with reasonable accuracy simply have not been collected yet. Databases for using the empirical method as with iris color discussed earlier in this chapter simply have not yet been built.

Placing this aside for the moment, the strength of the associations we have discussed for skin pigmentation are not of the same order as those observed for iris pigmentation. Linkage scans for skin pigmentation conducted by Zhu et al. (2004) show weak LOD scores like those observed for hair color, in contrast to the OCA2 score observed for iris color. It seems likely, therefore, that empirical tests of predictive power may not produce results as compelling as those so far obtained for iris color. Thus, it is probably the case that more comprehensive candidate gene and genome screens need to be conducted to identify all the puzzle pieces for predicting skin color. If more powerful markers can be identified, the databases could be built for use with the empirical method of inferring skin pigmentation directly.

Although there is still much progress to be made in developing predictive tests for skin color, it is clear that for the time being the indirect method is capable of providing good guidance. It may be the case that there are numerous, low penetrance alleles at work for skin pigmentation, and so genomic ancestry estimates may represent critical covariates in predicting skin pigmentation levels even for direct methods. The results so far in hand for skin pigmentation suggest considerable genetic heterogeneity. What is becoming evident is that we may need to be able to delineate the "genetic evolutionary architecture" underlying skin color in order to design robust predictive assays, since evolutionary convergence in skin pigmentation among various populations may be more the rule than the exception. In other words, we may be seeing results that suggest populations from different parts of the world that are light-skinned are not light-skinned because they share a common core set of alleles and haplotypes, but because they have separately evolved to be light-skinned. Likewise, dark-skinned populations do not always share the same functional mutations that result in dark skin. These differences might have resulted from separate changes in the pigmentation of populations once they separated and were subject to natural or sexual selection.

Alternatively the notion of phenogenetic drift is worth mention. Phenogenetic drift (Weiss & Fullerton 2000) is a process of change in the genetic determinants of a trait when there is no change in the phenotype. Were there, for instance, constant selective pressure for dark skin in two populations with a recent common ancestry, where each was subject to unique constraints (drift, bottlenecks, etc.), one could expect after time that there would be some difference in the underlying determinants of the dark skin. In other genes than those providing dark skin in the ancestral population, mutations could arise in one of the daughter populations but not the other, and over time these might replace the alleles that initially had made both populations dark-skinned.

Some of these points on the evolutionary genetic architecture are quite academic and of little interest to the forensic scientist in search of a practical test, but they have bearing on whether these tests appear anytime soon. It is not clear yet why skin (and hair) pigmentation might be so much more heterogeneous in genetic root than iris color, though we might guess that since skin and hair coloration have more direct implication on the expression of disease as a function of geography, the evolutionary forces at work may have been somewhat more complex. For all we know at present, there may have been a single eye-color change that spread throughout Western Europe, but many skin (and hair) lightening events at different times and places.

At least, for skin and hair pigmentation, we can detect both indirect associations with pigmentation and map pigmentation genes using admixture mapping. What is left to do is really the simple part conceptually: collect the population samples, measure the phenotypes, and perform more genetic analyses. An exhaustive effort to this end will not be possible with just a few admixed populations (African Americans and Mexican Americans, for example). These two populations will help us identify those genes that are functioning to make skin pigmentation different between West African and European, and European and Indigenous American populations, but this leaves out many important populations. Unlike with iris color, it seems likely that these other populations need to be considered for reaching our eventual goal of a skin pigmentation test. Ideally, we need to sample both admixed (to capture the within population variance) and unadmixed (for capturing the between population variance) populations. We might set our sights on Middle Eastern, South Asian (tribal and Hindi), Central Asian, Eastern European, East Asian, and South and East African populations for these studies.

Until we have a fundamental understanding of the genetic determinants for skin pigmentation we can rely on inferences made with the indirect approach, based on our understanding of the relationship between ancestry and skin pigmentation. Though this is less reliable than the direct method, with accurate individual ancestry estimates, as we have seen, we can make reasonable inferences that should be quite useful to the forensics community and indeed represent the initial and fundamental step in the discovery of the genes functionally underlying these traits.

CHAPTER 10

THE FIRST CASE STUDIES OF MOLECULAR PHOTOFITTING

CASE REPORTS

Methods of inferring phenotypes are relatively new to the forensics community, but we already have first applications of molecular photofitting technology. We give some details for these first applications in this chapter. Most of these applications of molecular photofitting have been accomplished using the indirect method, where we extract phenotype information from a sample based on genetic ancestry admixture. Iris color is the only phenotype that we can predict using the direct method, based on OCA2 haplotypes, but as of this writing the size of the only existing OCA2 database was only adequate to provide information for about 10% of forensic samples (by the time you read this, the size of this database may be much larger). The first cases to employ molecular photofitting techniques occurred from 2002–2003. For each of these cases, investigators sent crime scene samples to the laboratory of DNAPrint Genomics located in Sarasota, Florida, which has built and operates the only existing database for indirect methods of molecular photofitting (as well as for the direct methods for iris pigmentation). It is from these databases that much of the material in Chapters 4 through 9 came and today's databases are based on the AIM panels that appear in Chapter 4. DNAPrint refers to the bundling of the AIM assay with database- and software-based indirect interpretation services as DNAWitness^{TM} 2.5 (the 171 AIM panel discussed in Chapters 4 through 6) and DNAWitness^{TM} 1.0 (the 71 AIM panel discussed in Chapters 4 through 6).

LOUISIANA SERIAL KILLER MULTIAGENCY HOMICIDE TASK FORCE INVESTIGATION

Soon after the development of the 71-AIM version of DNAWitness^{TM} 1.0, DNAPrint Genomics employed the test for one of the highest profile cases at the time (October 2002)—the Louisiana Serial Killer case. Formally, this case was called the Louisiana Multiagency Homicide Task Force Investigation, since it involved not only local agencies but federal as well. Someone was raping and murdering young, light-skinned college co-eds in the state and the case was frequently the subject of national news coverage. The murders began in Baton Rouge in September 2001, when Gina Wilson Green, 41, was found strangled near the campus of Louisiana State University. In May 2002, Charlotte Murray Pace, 22, was found stabbed to death in her home, which was also near the LSU campus. In July 2002, Pam Kinamore, 44, was abducted, her throat was slit, and her body was dumped 30 miles outside Baton Rouge. The body of Trineisha Dene Colomb, 23, was found in a wooded area November 24, about 20 miles from where her abandoned car was discovered. At a news conference on March 17, 2003 in Baton Rouge, Louisiana, police announced that the fifth victim, LSU graduate student Carrie Lynn Yoder, was linked to the south Louisiana serial killer. Ms. Yoder was the last of the victims; her body was found in an area near Whiskey Bay, Louisiana, where the body of an earlier victim attributed to the serial killer was found eight months previously. The cause of death was asphyxiation; an autopsy found Yoder was strangled after being beaten and raped. DNA was found associated with many of these rape/murders and an STR profile obtained. According to a task force member, a national cable news program reported the DNA tests showed ''highly unusual genetic markers'' and that the frequency of the entire profile was expected to be on the order of ''something like 1 in 4 billion.'' Unfortunately there were no matches to this STR profile in the FBI's CODIS national offender database, meaning that the donor had never before been convicted of a sex-related felony, or if he had, was convicted prior to the establishment of CODIS (though many states are attempting to expand the list of felonies, at present, samples are taken only from sex-crime offenders).

By the time of the fifth murder, the serial killer case was receiving national attention and causing considerable discomfort for area women. Police urged women to take measures to protect themselves, including taking a self-defense course being offered by Baton Rouge police. ''We want to stress to the women of Baton Rouge the importance of exercising caution in every aspect of their lives,'' Cpl. Mary Ann Godawa of the police force said. ''It is absolutely neces-sary that we not only watch out for ourselves, but we must also watch out for each other.''

There is no doubt the task force was doing everything possible to bring the case to a timely resolution. FBI Special Agent-in-Charge Kenneth Kaiser con-firmed that two senior profilers in the Behavioral Science Unit back at FBI headquarters were assigned to the case. A story in 2003 reported that ''They have reviewed the case files, they have reviewed the crime scenes, they have looked at all the forensic evidence and they are assisting the task force.'' Others had tried to help as well. On March 18, 2003, the Pittsburgchannel.com reported that a North Carolina investigative psychologist unconnected to the case claimed that the crime scene locations can pinpoint where the Louisiana serial killer may live, and that he hoped detectives take notice of the geographic profile he has constructed that delineated a triangular region around the LSU campus. Dr. Maurice Godwin said his work predicted that the serial killer likely lives closer to the victims than police might suspect.

The task force was obviously focused on the DNA evidence. The task force took two approaches with this evidence. Due to the rarity of the STR profile, standard STR binning practices for the inference of ancestry and indirect inference of anthropometric phenotypes were not satisfactory. Indeed as we have learned so far in this book, the CODIS STR panel is not suitable for the inference of genomic ancestry to a qualitative or quantitative extent necessary to infer anthropometric phenotypes. The other approach involved combing the area for persons of interest to donate DNA in a type of dragnet. Two eyewitnesses had reported seeing a Caucasian male in the general vicinity of one of the crime scenes, and so the task force logically focused their dragnet on white males (see Figure 10-1). This dragnet garnered considerable attention in the media, not all of it positive, but its application indicated that the task force

A. March 2002-March 2003

B. March 2003

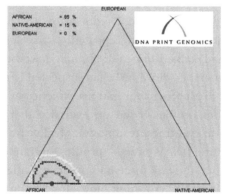

BGA profile of serial killer
obtained from DNA found at one
of the crime scenes.

C. May 2003

The individual linked to the
Louisiana Serial Slayings by
CODIS STR profile.

Figure 10-1

Indirect molecular photofitting was instrumental in the 2003 resolution of the Louisiana Multiagency Homicide Task Force. A. Eyewitness testimony suggested a Caucasian profile, which led to an investigative focus that consumed about a year and was eventually fruitless. DNAPrint Genomics of Sarasota, Florida then applied the DNAWitness 1.0 (71 AIM) panel of markers and reported that the genomic ancestry admixture of the crime scene sample donor was predominantly African, with about 15% Indigenous American admixture (B). C. The individual was arrested one month later, and since, convicted in two of the rape/ murders.

was indeed doing everything it could to resolve the case. Unfortunately, none of the sampled STR profiles matched the crime scene profile. It was approximately a year after this dragnet began, after approximately 1,000 individuals had been tested, and immediately after the introduction of the 71-AIM DNA-Witness™ 1.0 assay to the forensic market in October of 2003 that DNAPrint became involved in the case (Stanley 2006 is a good historical account of the case).

The DNAWitness™ 1.0 assay indicated that the individual who donated the crime scene specimen was most likely 85% sub-Saharan African, 15% Native American (see Figure 10-1b). Confidence intervals around this MLE indicated that the probability of no Native American ancestry was low, and that if the subject had European ancestry it was quite low. Based on this, and the data presented earlier in this text associating skin pigment content with European admixture for African Americans (see Chapter 5), the company advised the task force that the skin shade for the subject was most likely of average to darker than average tone relative to that of the African-American group. The task force had commissioned the analysis of about 30 test samples, and it was based on the results from these samples that ultimately led to their acceptance by the force. A new composite sketch of the serial killer was drafted, which lacked overt European features. The DNAWitness™ result caused a 180-degree inflection in the investigation, and reminded investigators of a previous suspect. Derrick Todd Lee was considered by law enforcement of the area to be bad news for many years prior, with previous arrests for peeping and burglary. One woman repeatedly had suggested to the task force that Lee, who had stalked her in 1999, may have been involved in the murders. Task force members previously had suspected Lee's involvement in the 1998 murder of Randi Mebruer, but had no evidence at the time and had become focused on other profiles based on evidence they were able to accumulate (mostly based on eyewitness testimony). Mr. Lee was included in a short list of suspects that fit the DNAWitness™ profile—and indeed, investigator Mixon later stated that when the task force had publicly changed their focus toward a black man, it became the "crowning blow" for his pursuit of a subpoena to acquire DNA from Mr. Lee.

While Lee's sample awaited testing, a link was made between the serial killings and a rape/attempted murder of a white woman by a black man in July of 2002. This link was not previously made with the serial killings because the perpetrator of these crimes was black, and the serial killer case was focused on a white man (Stanley 2006). A partial STR match was obtained linking the cases. The new composite sketch, previously denoting an individual of ambiguous ancestry, then was refined through testimony from the survivor of the attempted murder and the use of a sketch artist. The new drawings coupled with information about the suspect's car quickly elicited a number of tips implicating Derrick Todd Lee. Lee's sample was prioritized in the DNA lab, and very soon thereafter the STR match was made between him and the samples from several of the rape/murder scenes. Lee had fled the area after his DNA sample was taken, and quickly was tracked to Atlanta using cell phone records. The FBI agent received a call from an informant who reported possibly seeing Lee at an Atlanta tire shop. Three officers went there and found a man fitting Lee's description at the rear of the store, who was arrested and subsequently booked on charges of first-degree murder and aggravated rape. Lee eventually was charged with murder and aggravated rape in two of the cases. In the Charlotte Murray Pace case the charges were upheld by a jury and the sentence was death.

The use of DNAWitness^{TM} in the case has been the subject of numerous news stories since then (ABC's Good Morning America, PrimeTime Thursday, the *New York Times,* and CBS's Evening News). How the task force used the DNAWitness^{TM} information to redirect their investigation and eventually identify Derrick Todd Lee was murky until the publication of Stanley's 2006 book on the case, which incorporated official and unofficial testimony from a variety of people involved. The case had been bogged down for a year prior to the application of DNAWitness^{TM}, focused on white males, but solved two months later after the apprehension of an individual that fit the very different DNA-Witness^{TM} anthropometric profile. Various pieces of the puzzle fell into place after the change in direction, such as the link between the attempted murder case and the serial killings, which resulted in a refined composite sketch. Tipsters who had suspected Lee's involvement based on prior behavior or comments were motivated by the change in profile to report and/or re-report their suspicions. Subpoenas were inspired. Samples were prioritized. The end result was resolution a couple months later. As confirmed unofficially by several members of the task force since, DNAWitness^{TM} was instrumental in redirecting the focus of the investigation. In so doing, this case represents the first where genomic ancestry admixture analysis assisted in the resolution of a criminal investigation.

OPERATION MINSTEAD

DNAWitness^{TM} was used by the United Kingdom's New Scotland Yard and London's Metropolitan Police as part of a 12-year investigation known as Operation Minstead. Operation Minstead was established in 1998 to apprehend a rapist/burglar who had been terrorizing south London women since 1992. By 2004, over 80 attacks had been documented—most all of them on elderly women, between 68 and 93—and the case has become, by far, the biggest serial rape investigation the London Metropolitan Police has conducted in terms of the number of victims, the number of suspects, and the amount of hours devoted.

DNA evidence from the cases indicated a single perpetrator. Once inside, the burglar used a standardized operating procedure; first he disabled the phone and electricity, removed light bulbs from selected rooms, and eventually when the victim returned home, or upon waking a victim at night, he would ask for money. He usually spends several hours in the victim's home. At the time of the first attack in 1992 the perpetrator was already an accomplished burglar.

There were three sources of information about the perpetrator. The first was provided by the unusual chronological pattern with which the attacks occurred.

In December of 2003 New Scotland Yard Detective Superintendent Morgan said, "After the initial attack in October 1992, no incidents were reported until 1997 when the first spate of assaults occurred. After a particularly vicious rape on August 5 1999 the offences ceased again until October 13 2002, nearly ten years to the day after the first attack. We do not know why the attacks stopped during these periods. The perpetrator may have been out of the country or in prison. Again we would appeal for officers to think carefully if these dates correspond to any individual's whereabouts or brings anything to mind."

The second was provided by the victims, who had described the perpetrator as black, in his mid-30s, 5 ft. 11 in., and of athletic build. Due to the fact that all of the crimes occurred at night, in rooms without lighting (bulbs had been removed), there was considerable uncertainty about the perpetrator's race and appearance.

The third was from DNA left at the crime scene. Investigators were not able to match the STR profile within their databases and had even sampled a few thousand individuals from the south London area without finding any matches. In March of 2004, DNAPrint utilized the 171 AIM DNAWitness[TM] assay (see Chapters 4–6) for the case and obtained an MLE of 82% sub-Saharan African, 6% European, 12% Native American, and 0% East Asian. Analysis of the data was carried out with the assistance of Dr. Paul McKeigue of the University College, London. Applying his ADMIXMAP program the sum-of-intensities parameter (which corresponds roughly to the number of generations since unadmixed ancestors) was 3 per 100 centimorgans, with a 95% posterior interval of from 1 to 7.5. Though the density of markers provided by 171 AIMs is too low to estimate the sum of intensities in such a manner as to impart reliable chronological information, the estimates were reliable for gamete to gamete comparisons. The estimate suggested a relatively clustered pattern of admixture and it gave some information about the structure of the perpetrator's family tree. Posterior distributions of admixture proportions on each gamete revealed that one parent had considerably more sub-Saharan African ancestry than the other, indicating that the individual's admixture was contributed through the latter parent. Native American admixture is not common for sub-Saharan Africans in Europe and based on its presence, investigators suspected the admixed parent may have been a recent immigrant to the United Kingdom. Dr. McKeigue then proceeded to survey the demographics of the U.K. Afro-Caribbean community. From the case report:

> The high proportion of Native American ancestry is unusual among African-Caribbean migrants to the UK. The main Commonwealth Caribbean countries in which individuals with a high proportion of Native American ancestry still exist are Guyana, Trinidad, Dominica, Belize and St. Vincent. Our own research has shown that Native American admixture proportions are relatively high in Trinidad (avg. 12%) compared with Jamaica, the main source of Caribbean migrants to the UK. In Trinidad, however, there is a relatively high proportion of European ancestry (Molokhia et al. 2003) for which there is no evidence in this individual.
>
> Admixture between African and Native American, with very little European admixture, is consistent with the demographic history of a group known as the Black Caribs, or Garifuna. This ethnic group, numbering some 200,000, is to be found mainly on the Caribbean coast of Central America with smaller numbers in St. Vincent. The Black Carib population was formed by admixture between escaped African slaves and Carib Native Americans on the island of St. Vincent during the 17th century. In 1797 most were deported to Honduras, wherefrom they spread to Guatemala and Belize, maintaining their language and identity. Since WWII, many of this group have migrated, mainly to the United States.

Interested in determining whether the genomic ancestry results could give any additional geographical information, the Minstead team sent DNAPrint hundreds of carefully annotated Afro-Caribbean samples. Each of these was analyzed with the 171 AIM DNAWitness™ assay. The idea of this project was to determine whether there were any systematic differences in average genomic ancestry admixture composition among the various islands. Place of birth was recorded for each relative extending to grandparents three generations back, and if the island of birth was the same for all four grandparents, the genomic ancestry admixture data for the sample was used in calculating the average for a given island. Figure 10-2 shows the results. It appears that the level of European genomic ancestry admixture increases gradually the closer we get to North America, except that as McKeigue had suggested, the level is exceptionally high in Trinidad & Tobago. It also appears that the level of Indigenous American (NAM) increases gradually as we move in the islands from the west to the

Figure 10-2

Distribution of continental genomic ancestry admixture throughout the Caribbean. Afro-Caribbean samples were collected as part of the Operation Minstead investigation. The goal of the analysis was to empirically determine whether Indigenous (Native) American admixture observed for one of the suspect's parents could be localized to any particular region of the Caribbean. Sample sizes indicated for each pie chart in the table at bottom of figure; island to which the pie chart corresponds is indicated by the boxed names closest to the chart. Though Indigenous American levels are generally low, we can see a trend among the four islands tested toward greater percentages moving from the northwest to the southeast, returning to a lower level in Guyana. In contrast, European admixture seems to increase in the opposite direction, toward North America.

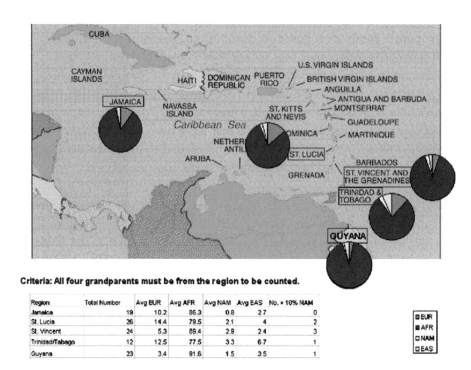

Criteria: All four grandparents must be from the region to be counted.

Region	Total Number	Avg EUR	Avg AFR	Avg NAM	Avg EAS	No. > 10% NAM
Jamaica	19	10.2	86.3	0.8	2.7	0
St. Lucia	26	14.4	79.5	2.1	4	2
St. Vincent	24	5.3	89.4	2.9	2.4	3
Trinidad/Tabago	12	12.5	77.5	3.3	6.7	1
Guyana	23	3.4	91.6	1.5	3.5	1

□ EUR
■ AFR
□ NAM
□ EAS

southeast, toward South America. The rapist had a relatively low level of European and a relatively high level of Indigenous American genomic ancestry, and so we might infer that his grandparents were more likely to have originated from the windward (southeastern) islands than the western or northern Caribbean. However, this would be a tentative conclusion since only a few islands had been sampled from and incomplete data could be misleading (just as low databases are when inferring anthropometric traits indirectly using genomic ancestry admixture).

The Minstead team consulted with the FBI and Interpol about the case and more than 3,000 suspects had been eliminated from the inquiry prior to the involvement. Then, from application of DNAWitness™ to a tiny amount of semen left at the crime scene, not only was the ancestry of the suspect apparent but the type of structure in his recent family tree as well. Three things were now known to be true or likely true about the rapist:

1. His non-African ancestry was contributed mainly by one parent.
2. His ancestry (as well as that of the admixed parent) is likely Afro-Caribbean rather than any of the myriad other African ancestries present in Britain.
3. The parent contributing the non-African ancestry more likely came from the Windward Islands area of the Caribbean than Jamaica, but other western islands cannot be ruled out.

Metropolitan Police Detectives proclaimed that the test gave them their biggest lead yet in the 12-year hunt. "After 12 years we feel this is a major breakthrough and are confident that the new DNA results should lead us to the suspect or even his relatives," Detective Superintendent Simon Morgan of London's Metropolitan Police told reporters.

Soon after these results were provided, in July of 2004, detectives invited 1,000 men to provide DNA samples as they try to identify the rapist. "Officers have narrowed down a list of possible suspects to 1,000 men who meet a geographical and racial profile. They started off with 10,000," said the *London Times* OnLine. Main benefits provided to New Scotland Yard and the London Metropolitan Police included a reduction in the number of samples and possible leads that needed to be investigated, saved investigative time and a guided allocation of investigative resources. As of this writing (January 2007), the case is still unresolved, but only a small fraction of the 1,000 men have agreed to submit samples.

THE BOULDER, COLORADO CHASE CASE

In January 2004, the 171-AIM DNAWitness™ test was also strategically employed in an investigation by the Boulder, Colorado Police Department. Susannah Chase, 23, was brutally beaten and raped in December of 1997. Her body, as well as a DNA specimen, was found several days later but there were no matches in the STR databases. DNAWitness™ indicated that the donor of the sample was predominantly of Native American ancestry. A representative of the Boulder Police Department stated, "DNAPrint reported that the DNA investigators submitted is indicative of someone exhibiting features that are common to Hispanics or Native Americans. This information will assist detectives with prioritizing tips and possibly generate more investigative leads." Again, the test was applied by investigators to infer physical information about the suspect from the crime scene DNA, allowing them to focus precious investigative resources on high value leads related to the crime. Though this case also remains to be solved, since the results point to a member of a segment of the community, there was no need to perform large scale, indiscriminant dragnets in the community, saving time, money, and political capital.

OTHER CASES

In two other cases the DNAWitness™ test proved instrumental. At Mammoth Lakes, California in 2002, investigators discovered a decomposed body that they believed belonged to a Southeast Asian gang member. They followed up this notion through investigation into the local LA gang community. DNAWitness™, however, determined the ancestry was 100% Native American (not East

Asian). Such high levels of Native American are usually found in relatively isolated American Indian reservations and Latin/South American population isolates, such as the Mayan communities—not Southeast Asia (as discussed in Chapter 6)—and so the results were quite informative. The results stimulated the memory of a park ranger who responded to a domestic dispute some time before the body was discovered, because the complainant was a Native American woman. The woman had complained that her companion was physically abusive. The park ranger visited their campsite, interviewed the man, and returned to the station not thinking about the incident again until the DNAWitness™ results were provided. Sketch artists were called to produce images of the man the ranger recalled speaking with at the campsite but as of July 2004 the case remained unsolved.

In 2002, in Concord California, a body of a woman was found on a running trail alongside a water viaduct. Given that the homeless community based along this trail is mainly composed of white males, investigators immediately assumed that the perpetrator was a white male. DNAWitness™ indicated the DNA found at the crime scene was indeed from an individual that was predominantly European, enabling the investigators to proceed along this line with confidence. This case, too, remains unsolved.

We have thus provided some of the specifics of the very first applications of admixture methods in the field of forensics. Mistakes early in an investigation can be the most crucial mistakes of them all, since it is early in the investigation when the trail is the warmest. Honing in on a segment of the population based on eyewitness testimony or preconceived notions, as we have seen, is a perilous exercise. In these cases, investigative effort was properly apportioned based on empirical observations from the crime scene—something that forensics investigators always strive to do. Because time spent tracking down inappropriate leads not only wastes time and money, but opportunity in following up the appropriate ones, and as we saw clearly with some of these cases the ability to discern which leads merit attention and which do not may make the difference between resolving a case or not. These are merely the first cases, and there are likely to be many more in the future as the size of admixture databases grow and indirect methods of phenotype inference gradually proliferate. As well, direct methods of inference are expected to gradually supplant the indirect methods for some phenotypes as the relevant databases grow. Nonetheless, elements of the media and academic community were not particularly supportive of the application of these methods in these cases. Their reactions were in some cases interesting, in others frustrating, but all of them were helpful in defining some of the social barriers that need to be overcome before the use of indirect (and even direct) methods of phenotype inference become widespread. Thus, we arrive at a discussion of the social dynamics, politics, and ethics of molecular photofitting, the subject of the next chapter.

CHAPTER 11

THE POLITICS AND ETHICS OF
GENETIC ANCESTRY TESTING

Measuring genetic components of race and ethnicity, and using them to infer phenotype from DNA comes with obvious political and social overtones. Not surprisingly, work in this field is not easily funded, and as most scientists who work in this area will attest, achieving grant support for work involving *genetic* measures of population structure that correspond to racial or ethnic stratifications is notoriously difficult (much of the work in this book was made possible from private funding sources). The same does not necessarily apply for non-genetic measures of population stratification, as in the social sciences, and one will find that an entirely different set of standards and attitudes are applied to work at the level of DNA. For example, rarely do we test human eyewitness testimony for accuracy, or quantify the general reliability of human eyewitness reports, yet DNA-based methods for the inference of phenotypes are expected to pass grueling validation exercises before the law enforcement and justice system will use them (such as some of those described in this book).

To the geneticist, this is not a problem since scientists generally hold their own work to standards higher than those we apply in our everyday lives. When a genetic technology is new, it is forced to "jump through the hoops" that older, accepted technologies already have, and though individual admixture methods have been around for the past 70 years, until recently they have been based on measures from within well-known gene sequences. As we have discussed, gene sequences are not adequate for estimates of individual admixture, since genes are subject to forces of evolution (such as convergent evolution). In addition, there just have not been that many of these gene sequences available, with good apparent ancestry information. The result of this is that until recently, individual admixture estimation has languished as a soft-science and has not been taken very seriously by geneticists, anthropologists, forensic scientists, and historians. It is only recently, with the development of the human genome project over the past decade that the resources have become available to allow estimation of individual admixture with large panels of true AIMs, and the types of experiments such as those described in this book have become possible (i.e., we can begin jumping the technology through the hoops). Graduating from a soft to a hard science though, is not just a matter of scaling scientific hurdles, but social and political hurdles as well.

As with any novel technology, perhaps none more so than methods for inferring genetic ancestry, there will be an acceptance curve. Resistance that exists within the fields of forensic science, basic academic research, and in legal circles is presently not low. Part of this is misinformation and/or politics, but another part is due to our desire as a society to get things right. In this chapter, we discuss many of the more common criticisms, and provide analysis and some opinion on the matter. We find it useful to refer to previously published articles or manuscripts for defining the most common complaints, and then addressing interesting concepts introduced by some of them one by one— much like a lawyer would dissect elements of an argument in a court of law. We intend this section to be an interesting portrait of various points of view, not didactic or necessarily instructive. There are considerable opinions of the criticisms and rebuttals, and you are invited to come to your own conclusions.

RESISTANCE

Where does the resistance to molecular photofitting and individual admixture estimation come from and why does it exist? Undoubtedly much of it must be a byproduct of the fact that, from our tribal past as hunter–gatherers to recent times, we have been killing one another based on our differences. Distinctions among lines of race, culture, religion, and so on are easy to make and conveni-ent organizational metrics and indeed, even catalysts for an overarching Mal-thusian struggle for limited resources. Differences lead to separation, the remnants of which are evident in the amalgamation of modern-day popula-tions on a global scale (which we discussed in Chapter 6; see Figures 6-1 through 6-4) or even on subregional scales such as within large urban areas of any western nation. Sometimes the conflict is at a low but civilized level, as in democratic cultures where even a constant undercurrent of tension among separated populations is addressed at the ballot box, not on the battlefield. Other times it is overt; even in recent times we have seen holocausts in Eastern Europe, ethnic cleansing in the Balkans, genocide in Rwanda and the Sudan, and wars between sides that largely align with culture/religion.

Many of those who resist the use of the technology focused on in this book, feel that it lends new tools for the continuation of our unhappy past. They wonder (vociferously, sometimes), "how might it have enabled the limpieza de sangre ('cleanliness of blood') doctrines that underpinned the Spanish Inquisition" or "what would Hitler have done with an ancestry test," and "doesn't technology for discriminating elements of human population structure reinforce the problem?'' It is clear from history that Hitler didn't seem to need help from a test for ancestry admixture in order to execute the ''Solution to the Jewish Question'' first expounded in a speech to the Reichstag on January 30, 1939. But had it been available, would this solution have been easier to broach, or its execution more efficient? Perhaps more solutions would have been offered for more questions, throughout Europe and perhaps in other parts of the world.

The answer to these questions is not entirely clear, but there is reason to believe that more thought is needed on the topic than that we might arrive at with our first knee-jerk reaction. As we have presented in this book, treatment of genetic ancestry admixture as a continuous variable teaches us the fallacy of using dichotomous descriptors such as Caucasian or Jewish. This is not to say that Caucasian or Jewish are not legitimate cultural terms, and not to say that there is no population genetic distinction between Caucasians or Jews and others, only that there is often little genetic basis for the ''ethnic'' terms *a priori*. Indeed, we have seen that admixture is the rule rather than the exception, variation within most groups (sometimes with extreme examples) and that generally speaking, there is a disconnect between our social/political notions of identity and genomic measures of our anthropological past. We have seen that the ancestry for most individuals is genetically best described as residing somewhere along a continuum of admixture rather than falling into a prefabricated bin or box like white, black, Caucasian, and the like. Were Adolph Hitler himself to look at his results from some of the assays described in this book, and compare them to the average Jew or gypsy, he may have been very surprised. Quite likely he would have seen little difference—particularly if any of his distant ancestors were derived from Central Asia or the Middle East. This is not to say however that Hitler would have stopped the holocaust based on such an observation. He may then have claimed that what he was after were finer genetic distinctions—more recent in origin. If Y-chromosomal testing was available to Hitler, he might have targeted the Cohanim sequence, which is exclusively found in a small percentage of the Sephardic and Ashkenazi Jewish. Even so, it's very possible he would have had difficulty distinguishing his genetic background from that of the majority of individuals he sent to the concentration camps. Recall from earlier in this text, as we look at the genetics we see that these divisions of people along such amorphous lines often don't have much of a genetic basis, or if they do, there is oftentimes considerable interpopulation variation, making classification of individuals difficult. Hitler may have noted that his estimates were not very different from those of the average Jew or gypsy, and upon study of other people's results he may have been dismayed at the extent of intrapopulation variation and overlap in results between populations. He very likely would have seen little genetic base to support his notion of Aryan, or Caucasian, and if he used the data in an attempt to draw new notions of white supremacy he most likely would have had great difficulty in defining corresponding elements of genetic ancestry that would have accommodated him and all of his collaborators to the exclusion of those he found undesirable. Faced with the realization that in terms of the past few thousand years, his genetic background was no different than the average Jew or Gypsy, perhaps he would have lacked the basis from the outset to purge Europe of, as he put it, the "racially inferior." Alternatively, he may have shifted the emphasis from the hard science of genes and biology to the social realm of politics (I maintain that politics is not a science, since political hypotheses are not unambiguously or fully testable and indeed as discussed later, is the opposite of science since its root is advocacy not objectivity). Lacking an objective, logical basis for his notion—how might history have been different? Perhaps access to knowledge and information in this case would not have made any difference. After all, Hitler was a politician and advocate and unlike scientists, politicians are well known to never allow facts to get in the way of a good argument. The real question here however is whether or not access to this type of information could have exacerbated the Holocaust. For all we know, Hitlers biogeographical ancestry might have been very different from the average Jew or Gypsy, such that the use of genomic technology would have bolstered his claims, exacerbating the holocaust. And, if we can predict iris color, or skin shade, and form a probability statement on ethnic origin given admixture as discussed in this book, can't at least some of the methods described herein be misused?

The answer to this question is yes. However, in my opinion, technology does not commit murder or cause war crimes, people do. Our problems as a species are a function of human behavior, not the tools at our disposal or the environment in which we live. Indeed, our decision as humans to engage in genocide or murder rarely are based on scientific facts or logical applications thereof; more often they are based on the concept of advocacy (for an ethnic group, political group, etc.), and thus they often incorporate the non- or quasi-scientific concepts such as ethnicity, socioeconomics, religion, and so on. We are used to blaming tools for our behavior, rather than the general personality of our species. It is easier to blame tools for our problems, as it prevents us from having to look into the mirror.

In addition, it should be pointed out that virtually every new technology could be banned for fear of misuse. Airplanes could be banned because terrorists might use them as missiles and fly them into buildings. Drugs could be banned out of fear of them being used as poisons for the commission of homicide. Knives and pencils could be banned as potential instruments of murder. Though measures of genomic ancestry admixture could certainly be misused, it could be argued that they are easier to control than knives or pencils, and are completely benign when used responsibly. Indeed, measurements of admixture offer to help improve clinical trial design, help identify important disease genes in the human genome (through admixture mapping) so that new drugs can be developed targeting them, help capture dangerous criminals, and aid us in our understanding of our history of a species (as well as our personal histories).

ARTICLES—INSIGHT INTO PUBLIC REACTION

Let us look at various news stories written on the technology we have discussed in this book, as a means by which to appreciate the attitudes of others in the field of science as well as the lay person. To start with, consider a recent *Toronto Globe and Mail* story published in June of 2005.

MOLECULAR EYEWITNESS: DNA GETS A HUMAN FACE

Toronto Globe and Mail, June 25, 2006, p. A6. Reprinted with permission.

Canadian police have been quietly using a controversial new genetic technology to reveal the racial background and physical appearance of criminals they are hunting, according to the Florida company that sells the test.

Officials with DNAPrint Genomics, a biotech firm in Sarasota that has offered the test since 2002, say four separate forces in Canada—including the RCMP—have used the technology to narrow their search for suspects. This spring, two Canadian investigators made the unusual move of hand-delivering a crime-scene DNA sample to the Florida lab.

Unlike the more familiar forensic test that tries to match DNA found at a crime scene with samples from known suspects, this test is based on a single recovered sample and has the potential to tell police if the offender they are looking for is white, black, Asian, native, or of mixed race. The company then supplies photos of people with similar genetic profiles to help complete the portrait.

The company says the so-called DNAWitness test has been used in 80 criminal investigations by law-enforcement organizations worldwide, including the U.S. Federal Bureau of Investigation, the U.S. Army and Scotland Yard.

"This could be helpful in solving crimes, more helpful than human eyewitnesses," said Anthony Frudakis, the company's chief scientific officer. "Our technology serves as a potential molecular eyewitness. It's objective."

It's also advancing at a dizzying pace. This spring, the company launched a new DNA test that can discern a person's eye colour with 92 percent accuracy. Meanwhile, the prospect of learning other physical—even psychological—traits could soon follow.

But while law enforcers seem to be embracing the new science, it has received a chilly reception from others who compare the technology—a similar version of which has been developed in Britain—to racial profiling in the genomics age.

"You still have to make a leap that what you're getting from the DNA correlates to visual characteristics," said Mildred Cho, associate director of the Stanford Center for Biomedical Ethics in California. "Then in order to round those people up, you have to say to the police department, or the force, 'Go find people who look like this, someone who looks black, or someone who looks half black and half Asian.'"

"This technology really overstates the ability to classify people by race and ethnicity."

For this reason, as well as the risk of tipping off criminals who could try to alter their appearance, Dr. Frudakis said police are loath to discuss their use of the technology—which appears to be the case in Canada.

Officials at DNAPrint, who sign confidentiality agreements with police, say they cannot reveal details of the Canadian cases and investigators they contacted on behalf of the *Globe and Mail* have not responded to requests to discuss the test.

But as far as the company knows (and police do not generally keep them updated), Dr. Frudakis said, the test has contributed to six arrests internationally. The most prominent example comes from Louisiana where detectives used it to catch a serial killer.

Eyewitness accounts of a white man driving a white pickup truck, as well as an FBI psychological profile [author's note: this is incorrect—it was based on two eyewitnesses], had suggested it was a Caucasian man who was raping and killing women in the Baton Rouge area in 2002.

But crime-scene DNA the Florida company tested indicated the offender was 85 percent sub-Saharan African and 15 percent Native American. In short, the test told police they should not be looking for a white man. Two months after the shift in focus, Baton Rouge police arrested Derek Todd Lee, a black man now on death row for the slaying of six women.

Asked whether RCMP hunting a possible serial killer in Edmonton might consider using the technology, spokesman Corporal Wayne Oakes would say only that investigators on the case are "aware of this technology, but it's not one they have had occasion to use."

Sergeant Don Kelly of the Baton Rouge police force said in an interview that one of the detectives involved in their serial killer case has made a presentation to police in Edmonton. But he could not say if the technology was discussed.

In some cases, the test has been helpful in identifying victims of a crime.

Police in Southern California, for example, had been targeting Asian gangs after discovering skeletal remains at Mammoth Lakes Park that bone-structure experts felt belonged to an Asian woman. But the Florida test found the woman was largely Native American, prompting park rangers to recall that a woman who fit that description had complained about her husband's abusive behaviour.

The test, which costs $1,000 (U.S.), scans 176 particular genetic mutations that each offer information about a person's continent of origin. The results then break DNA inheritance down into percentages of four geographic groups: sub-Saharan African, East Asian, European, and Native American.

The company refers to the process as an estimate of "biogeographical ancestry" and from this, investigators can indirectly infer key physical traits—in particular skin, eye and hair colour.

Of the 8,000 DNA samples they have tested by this method in the course of their research and work, 95 percent of people turn out to be of significant mixed heritage, said Zach Gaskin, a technical coordinator of forensics at the company.

Still, Mr. Gaskin said, once a DNA sample suggests that at least 30 percent of a person's heritage belongs to a particular racial group, a person starts "to exhibit features consistent with that population."

But in a paper published in *American Psychologist*, U.S. sociologist Troy Duster and ethicist Pilar Ossorio caution that the test has risks: "Some percentage of people who look white will possess genetic markers indicating that a significant majority of their recent ancestors were African. Some percentage of people who look black will possess genetic markers indicating the majority of their recent ancestors were European.

"Inferring race from genetic ancestry may mislead police rather than illuminating their search for a suspect."

For these reasons, company officials in Florida do not actually interpret test results by trying to describe shades of skin or hair color. Instead, they provide photographs taken from their sample database of 2,500 people who match the genetic mix of suspects.

Stanford's Prof. Cho criticized this technique, however, arguing that even children from the same family can look very different from one another.

Toronto police Detective David Needham of the major sex crimes unit applauded the technology and said it is scheduled to be presented at a conference the force is holding in October.

"If the science is reliable and it can be accepted and established in the courts, it's going to be great," he said.

But he knows firsthand about the controversy it attracts.

A year ago, Det. Needham was hunting two men who had abducted and raped a woman who had seen only one of her attackers. To narrow his search for suspects, Det. Needham asked experts at Ontario's Centre for Forensic Sciences to try to give him a sense of racial background based on semen samples.

"They said, 'We can't do that, that's racial profiling,'" Det. Needham said. "If someone said they had seen a white man or a black man leaving the scene of the crime, we would use that information. So what's the difference?"

Bruce O'Neill, spokesman for the Ministry of Community and Safety and Correctional Services, which oversees the CFS, said the technology at hand has nothing to do with racial profiling.

Tim Caulfield, director of the Health Law and Policy Institute at the University of Alberta, noted that if the technology is indeed a sound tool for determining a suspect's physical appearance, it could turn out to be more reliable than eyewitness accounts.

"With a witness, there may be a whole set of social stereotypes that come out," he said. "If this technology is providing information that is factual, and people don't use it to make unwarranted presumptions, then it could be worthwhile. We need to be careful about how we let politically correct concerns colour our views."

Yet such concerns are bound to grow right along with the power of genetics.

The Florida firm, for example, is now developing 3D technology to read gene types to infer physical traits such as hair texture, skull shapes, or the distance between the eyes. Dr. Frudakis predicted that such technology might allow them within the decade to generate a crude sketch of a suspect from a DNA sample.

If all this sounds more like an episode of CSI: Crime Scene Investigation, the popular television program that plays up the power of forensics, it is just a brief trailer for the plot lines to come.

Profs. Duster and Ossorio note police will eventually be able to discern psychological characteristics from DNA samples and generate behavioral profiles of subjects.

The criticisms in this article deal mainly with the indirect method of molecular photofitting, and stem from a misunderstanding by nongeneticists of how the process does and does not work. The author did a good job of describing the empirical process in a single sentence—not a trivial exercise for a nonscientist. One critic in the article stated: "This technology really overstates the ability to classify people by race and ethnicity."

But this person was probably unaware that the primary purpose of measuring genomic ancestry is not to classify people by race and ethnicity, but rather, through an appreciation of admixture, to learn something about likely physical appearance. Nonetheless, as we have described, if one were querying a database of geopolitical descriptors for people who have been analyzed with the genomic ancestry assay, rather than digital photographs, one could determine the range of races or ethnicities corresponding to a given genomic ancestry admixture profile and thereby infer the most likely values. If this person is right in the insinuation that genomic ancestry estimates provide no or little information about race and ethnicity, then for a given admixture profile, the database return would indicate this. However, typically what we would find is that the genomic ancestry result helps to substantially delimit the field of self-descriptors likely to be used. Though we might not be able to pinpoint whether an individual was likely to refer to themselves as American Indian or Hispanic, we would likely be able to exclude terms such as African or South Asian, as the case may be. Even so, the inference would not be a classification of race and ethnicity, but rather an empirical observation—a collage of self-applied descriptors that people with similar genomic ancestry admixture proportions tend to use (as may be the case). As we have already described, we could look at ancestral identity from a different angle if we queried a database of subject birthplaces.

Consider the ethicist who said: "Some percentage of people who look white will possess genetic markers indicating that a significant majority of their recent ancestors were African. Some percentage of people who look black will possess genetic markers indicating the majority of their recent ancestors were European," and "Inferring race from genetic ancestry may mislead police rather than illuminating their search for a suspect."

This is an example of a criticism borne from a lack of understanding of the concepts discussed in this book, which is not necessarily the fault of the interviewee because at the time, little literature existed on the methods and assays for molecular photofitting. What this ethicist is insinuating is that racial binning is too imprecise to reveal information of adequate resolution for forensics investigations, which is very true, and why one needs to infer genomic ancestry admixture instead. Another insinuation made here, seemingly, is that genomic ancestry admixture estimates are of no use for learning about physical characteristics such as skin melanin content. As a newcomer to the concepts discussed in this text, the interviewee was uninformed that categorical binning is not the current state-of-the-art, and also that genomic ancestry admixture estimates serve as an adequate base for the inference of certain, though not all, overt physical characteristics.

Many forensics professionals are likely to have the same misimpressions, until they read this book or some of the manuscripts we cite. Notwithstanding these mistakes, the ethicist is partially correct nonetheless, since for some cases this insinuation will be true and for others it will not be. However, the database method described in Chapter 8 would tell this tale and we run very little risk of drawing incorrect inferences for such phenotypes.

Next, take the following criticism: ''... criticized this technique (the database-driven method for relating genomic ancestry estimates to phenotype), however, arguing that even children from the same family can look very different from one another.''

What this critic seems to be saying is that (1) children from the same parents have identical admixture proportions, and (2) unless the indirect (or direct) method can pinpoint the exact phenotype value for a given trait, then it is useless. We explained in Chapter 8 that (1) is simply not true (e.g., see Figure 5-18). Since siblings do not share identical genetic ancestry admixture characteristics, even if ancestry were perfectly coupled with phenotype (which it is not), some natural variation in phenotype among siblings would be expected. Aside from this, very few scientific methods for inferring or predicting outcome for a multifactorial or otherwise complex problem can pinpoint the exact value—our favorite example being tropical cyclone forecasts and advisories that generally provide ranges rather than discrete values, yet work well enough that their use helps save lives each year. For some cases, knowledge that a person with 40% sub-Saharan African/60% European ancestry is likely to have an intermediate skin color to a person with 100% sub-Saharan African/100% European genomic ancestry will be very useful (as was the case in Kansas City, Oklahoma, as you recall from Chapter 7, and which you will read about in the next story). For others, such as when the suspect pool is already determined to consist primarily of admixed individuals, this information will not be useful and more resolution would be needed.

This critic's sweeping statement on the forensic use of ancestry information, or lack of it, would seem to suggest that broad, general levels of information are not helpful to a forensic investigation. However, we know, of course, from centuries of prior experience that human eyewitnesses often serve as invaluable in assisting investigators, yet nothing could be quite as imprecise or inaccurate as subjective human testimony. Some eyewitnesses do not realize what impact lighting conditions have on whether a person's skin is dark or light, and others cannot tell the difference between a Native American or an admixed African American from across the street. Some eyewitnesses provide false testimony that is in their best interests. Of course, human eyewitness testimony is used with some peril, and many innocent people have suffered due to faulty eyewitness testimony, but this is because unlike that from DNA, human testimony is not falsifiable, not because the information provided is broad and/or general in scope. Notwithstanding limitations on the utility of eyewitness testimony, if broad conclusions from human eyewitnesses have proven so helpful in the field of forensics up until now, why would superior conclusions from DNA be any less useful? In this day and age of STR typing, the question of culpability would be addressed separately after the investigative process had led to high-quality suspects. Unlike human eyewitness testimony, it is undesirable and unlikely that molecular photofitting, as a presumptive assay, would ever put anyone in jail.

On August 17, 2005 *USA Today* ran a front-page cover story that develops some of these ideas further, addressing some of the criticisms just discussed.

DNA TESTS OFFER CLUES TO SUSPECT'S RACE

USA Today, August 17, 2005, cover story.
Reprinted with permission from *USA Today*, a division of Gannett Co., Inc.

Police seeking the killer of an unidentified girl who was found decapitated in Kansas City, Mo., four years ago kept a secret from the public.

The child, dubbed "Precious Doe" by local residents, appeared to be black. But new DNA tests that can determine a person's heritage indicated she was of mixed ancestry—about 40% white. That meant she almost certainly had a white grandparent.

This year, a tip led police to an Oklahoma woman who had not reported her young daughter's disappearance. When the woman was found to have both a black and a white parent, police moved in. Further DNA tests determined that the woman, Michelle Johnson, was the girl's mother. Johnson and her husband, Harrell Johnson, the victim's stepfather, have been charged in the slaying.

Precious Doe was identified as 3-year-old Erica Michelle Marie Green of Muskogee, Okla. During a trip to Kansas City, prosecutors allege, her stepfather kicked her to death because she wouldn't go to bed on time.

In the past 12 years, police across the USA have identified thousands of suspects by testing DNA profiles in blood, sweat, semen, or skin tissue left at crime scenes, and then comparing them to the profiles of known offenders on file in government databases. But as the Kansas City case showed, advances in DNA testing are allowing investigators to learn more about suspects whose profiles are not in the databases. Tests that can identify a suspect's ancestry are being used not to identify the suspect by name, but rather to give police an idea of what he or she looks like.

DNA ancestry testing "made a huge difference" in the Precious Doe case by helping investigators sort through reports about possible suspects, says Dave Bernard, a Kansas City police detective. "It allowed us to prioritize our tips, to give special attention to tips about mixed-race children, for instance. It was invaluable."

How the test works

DNA is a cellular acid that carries a person's unique genetic code. The company that invented the ancestry test, DNAPrint Genomics of Sarasota, Fla., says that by examining tiny genetic markers on the DNA molecule that tend to be similar in people of certain population groups, it can tell whether a suspect's heritage is European, sub-Saharan African, Southeast Asian, Native American, or a mix of those.

The test works, the company says, because population groups developed different DNA characteristics after splitting off from common African ancestors more than 60,000 years ago.

In 2003, police in Louisiana used ancestry testing to help find the suspect in seven rape/murders. Since then, police in Missouri, Virginia, Colorado, California, and the United Kingdom also have used such tests to develop leads in more than 80 other homicide, rape, and missing-persons cases, according to DNAPrint Genomics and USA Today research.

Using the same genetic principles, DNAPrint Genomics is developing tests aimed at determining a suspect's eye color from a DNA sample. In the United Kingdom, meanwhile, the government's Forensic Science Service has begun examining DNA samples for indications of hair color.

DNAPrint Genomics also sells the test to people who want to trace their roots. The test, which costs $219, has been especially popular among those seeking to determine whether they are descended from Native Americans, lab director Matt Thomas says. DNAPrint Genomics charges police departments $1,000 for each ancestry test, because testing crime scene evidence for DNA can be particularly difficult.

Bernard and many other police detectives hail the ancestry tests as a breakthrough in crime-fighting. But medical ethicists, defense lawyers, and even some police officials are troubled by the push to use DNA tests to identify suspects by what amounts to their race.

Some, such as Terry Melton, president of Mitotyping Technologies of State College, Pa., say the reliability of ancestry testing remains unproved.

William Shields, a biology professor and genetics specialist at State University of New York's College of Environmental Science and Forestry in Syracuse, says that even if the tests are correct, a person's ancestry often is a poor predictor of what he will look like. Human beings, Shields adds, are too scientifically similar to one another to be distinguished by a "layman's term" such as race.

Some defense lawyers say they fear that using ancestry testing to determine suspects' heritage could lead to genetic racial profiling, or promote the idea that certain races are more inclined than others to commit crimes.

"How far are we from having (ancestry tests) used to justify taking DNA from any black man on a street corner, because we think a sub-Saharan African committed the crime?" asks Ingrid Gill, a Chicago lawyer who has lectured on ancestry testing at the American Academy of Forensic Sciences.

George Rhoden, a detective with London's Metropolitan Police and president of the force's Black Police Association, also is a skeptic. He says that in a society in which marriages between people of different ethnic backgrounds are increasingly common, racial designations often are "very broad" and "don't do us coppers much good."

Rhoden points out that suspects with similar genetic ancestry can look significantly different from one another. A person whose profile is 75% sub-Saharan African, for example, may have skin color that is nearly identical to someone whose profile is 35% sub-Saharan African.

"As a detective, I don't care where (a suspect's) grandfather came from," Rhoden says. "I want to know what he looks like."

Mark Shriver, an anthropological geneticist at Penn State University and a consultant to DNAPrint Genomics, acknowledges that "there's a huge sensitivity about race in our society. We are making a strong attempt to be sensitive to the issue." But "that doesn't take away the reality that people often describe each other in terms of race. We're saying: Let DNA be the witness."

Beyond standard DNA tests

Conventional DNA analysis compares 13 relatively large areas on the molecule where the DNA sequence is known to vary greatly among individuals. If two DNA samples match at all 13 positions, statistics maintained by the FBI say it's highly likely they came from the same person.

Ancestry tests, by contrast, examine 176 mutations in which the DNA varies at only one position. Some of the mutations, called single nucleotide polymorphisms (or SNPs), have been found to occur only in certain ancestral groups. Others tend to cluster in one group more than others because of centuries of geographic separation and inter-marriage. Together, Thomas says, SNPs are "highly informative of ancestry."

In 2003, DNAPrint Genomics began to license its test to police agencies. The scientists realized, Thomas says, that knowing a suspect's race or ancestral background "may not be great for, 'Who do we arrest?' but could help police determine, 'Who do we question?'"

The company's test was first used in a criminal investigation in the Baton Rouge area, where a series of at least seven rape/murders had authorities stumped. Witnesses had reported seeing a white man in a white truck near the scene of two of the killings. Police had taken DNA samples from more than 1,200 white men in the area and had not found a match to samples from the crimes.

Then the DNAPrint Genomics ancestry test found that the unknown attacker was mostly of sub-Saharan African ancestry with a smattering of Native American.

That led authorities to focus on Derrick Todd Lee, a black man with convictions for burglary and stalking. Additional testing matched Lee's DNA to samples taken from victims. He has been convicted in two of the slayings.

In 2004, police in Charlottesville, Va., used ancestry testing to confirm the race of a suspect in six unsolved rapes that began in 1997. Police had been criticized for seeking DNA samples from local black men based on victims' descriptions of the assailant. The testing indicated that he indeed was of sub-Saharan African descent.

Ancestry testing also has been used on a female skeleton that was found in the snow near Mammoth Lake, Calif., in May 2003. The slain woman initially was misidentified as southeast Asian, based on witnesses' descriptions of a woman seen in the area. DNAPrint Genomics found she actually was a Native American, a finding confirmed by analyses of her diet and bone composition and further DNA tests.

The ancestry test "turned around the whole investigation," says Paul Dostie, the police detective investigating the case. "We're still looking for the killer, but we know a lot more now."

New technology "scares me"

For all the promise of ancestry testing, there are increasing concerns about how police will use such information.

Defense lawyer Bruce Unangst, who defended Lee in his second murder trial, says the new technology "scares me. It's supposed to be new and foolproof, but that's traditionally what they say about all new" crime-fighting innovations. "By the time we find out there are serious questions ... a whole bunch of innocent people have had their DNA searched."

Last year, London police sought DNA samples from officers of Afro-Caribbean backgrounds to compare them with evidence from nine unsolved rapes. The suspect's accent and the neighborhood in which he operated suggested to police that he was a black man with Caribbean roots.

Working with DNAPrint Genomics, London police hoped to develop a database of DNA characteristics that are particular to Afro-Caribbeans to confirm their suspicions and to help them find suspects in other cases.

Rhoden, as head of the Black Police Association, urged members not to cooperate. "In our view, this promoted racial stereotyping while adding little to the investigation."

Melton, the private lab president from State College, Pa., says inferring a suspect's appearance by examining only 176 ancestry markers is "more than (labs) ought to be doing."

Because scientists have identified thousands of SNPs, Melton says, many more should be tested.

DNAPrint Genomics reviewed about 25,000 DNA markers before choosing the 176 that were "most informative of ancestry," Thomas says. The company now has a test that can tell whether a European's DNA came from a northern or southern European, he says.

For detectives who use its service, the company provides photographs of people whose ancestral profile matches that of the detectives' suspect.

"What does a Northern European, Native American, and Southeast Asian mixture look like? That's a fair question," Thomas says. "We're told the photographs are extremely helpful."

The company's research is continuing. After Afro-Caribbean police in London refused to donate DNA samples, DNAPrint Genomics collected about 150 samples from police on Caribbean islands.

More samples are needed, Thomas says, but the DNA profiles collected so far suggest there are markers that distinguish Afro-Caribbean blacks from others in the sub-Saharan group.

London police, Thomas says, were "on the right track" in their rape investigation.

Rhoden draws a different lesson from the episode. He notes that few of the Caribbean officers who gave DNA samples were willing to have their photographs added to the company's files.

"Even for these guys, who wished to be helpful, that was going too far," Rhoden says. "We should take notice of how nervous it makes such people before we endorse any kind of mass DNA taking from ordinary people."

The comment that ancestry testing "remains unproved" was interesting—new technology is always unproved in its early days and this technology is no different. Eventually, work such as that which appears in this book and in the peer-reviewed literature combine with real case studies to lay the foundation of credibility for the forensics community. For example, this book is the first ever written on the subject of molecular photofitting. So, this is not so much a criticism as much as it is a statement that the technology is new, which is obvious.

Again, we encounter the criticism "even if the tests are correct, a person's ancestry often is a poor predictor of what he will look like. Human beings (the critic) adds, are too scientifically similar to one another to be distinguished by a 'layman's term' such as race."

Ignoring the distinction between race and genetic ancestry, this person seems to be saying that the biological basis for race is too complex to be described in modern day "layman's" terminology. This is indeed correct, and an astute observation, but it is incorrect to imply that methods for inferring genetic ancestry are not useful for inferring elements of physical appearance. Of course, the inference of genetic ancestry admixture produces highly individualized data, which are related in populations and individuals to certain phenotypes in meaningful ways (see Chapters 5 and 8). It is not a description of race, but data that relates to race only indirectly that is useful for molecular photofitting, and then, only through the use of empirical database-driven systems. The first sentence of this criticism suggests that the relationship between genetic ancestry and phenotype is too ambiguous to be useful for forensic science, which of course is a matter of opinion, but not entirely consistent with what we have shown in this text, where we have seen we can make broad statements about trait value given certain levels of admixture for certain phenotypes. Indeed, the utility of human descriptions of physical appearance in terms of race would seem to suffer the same drawbacks, and however imperfect they may be, the standard use of human reports on physical appearance seem to belie this assertion. In other words, if it were not useful it would seem that the practice of questioning human eyewitnesses would not have survived the past couple hundred years as a standard forensic operating procedure.

Broad statements are not as desirable as precise statements, but that does not mean they are not useful. Assume the application of molecular photofitting techniques indicate that, based on genetic ancestry admixture, one is looking for an individual likely to express phenotypes A, B, and C. Just because more than one of the world's populations exhibit A, B, and C does not logically destroy the value of this information. With our information from ancestry, we have delimited the likely possibilities, which is what investigation is all about. In addition, using phenotype and genetic ancestry together delimits them further. For example, the value of knowing that one is looking for an African with dark skin color in a given crime is not diminished by the fact that some *Europeans* (i.e., South Asian Indians) also have dark skin color like Africans. Properly constructed, the profile is based on empirical observation and reports what can and cannot be stated from statistical analysis.

Consider the criticism "... suspects with similar genetic ancestry can look significantly different from one another. A person whose profile is 75% sub-Saharan African, for example, may have skin color that is nearly identical to someone whose profile is 35% sub-Saharan African."

In Chapter 5, we discussed that the relationship between anthropometric trait expression and ancestry is a function of probability and so it is indeed possible (though not necessarily likely) that a person with 75% sub-Saharan African ancestry could have the same skin tone as one of 35% sub-Saharan African. On average, the data discussed in Chapters 5 and 8 suggest that we can gain useful information as long as a good panel of AIMs is used and the database is suitably sized. For example, Figure 8-18 shows that the average individual with 75% sub-Saharan African ancestry will have significantly higher melanin index (darker skin) than one of 35% sub-Saharan African. Figure 8-16 shows a similar phenomenon, using only 30 AIMs, and Figure 8-17 (30 AIMs), Figure 8-18 (30 AIMs), and Figure 8-20 (171 AIMs) show that this phenomena is observed in multiple African derived populations. Figures 8-21 and Table 8-4 show the same for iris color. For those wishing the sample sizes for these latter two figures were larger, recall that measures of statistical significance for the correlation coefficients or contingency analyses are a function of the sample size, whether one feels they are too small or not, and note that demonstrating associations between minor elements of population structure (e.g., admixture) and phenotype is considerably more impressive than between primary ancestry and phenotype (which is fairly obvious).

Granted that these associations do indeed exist, this criticism then boils down to the same argument we have discussed earlier that was made by one of the critics in the last article—that we cannot pinpoint the *exact* level of skin melanin with indirect methods, and so the indirect method of molecular photofitting is not useful. Indeed, the words in the *USA Today* article from the Kansas City Police in the Precious Doe case seem to belie this argument; as this case shows, a knowledge that a person with 40% sub-Saharan African/60% European ancestry is likely to have an intermediate skin color to a person with 100% sub-Saharan African/100% European genomic ancestry was very useful in this particular investigation.

The fear that ancestry testing will lead to racial profiling is a legitimate concern, but not really specific to DNA based inference—it is a concern that should also be (but often isn't) discussed in the context of human eyewitnesses as well. Racial profiling is something that can happen upstream or downstream of the information gathering phase of an investigation. Specifically, it refers to the application of racism to crime fighting in ways that violate human dignity and civil rights. For example, assuming that individuals of heritage X are more likely to be criminals without any specific prior evidence for a given case in general does not justify targeting individuals of heritage X for roadside searches. However, if we know that an individual of heritage X was involved in a crime down the street 30 minutes ago, this information obviously needs to be applied in the investigation if for no other reason, to avoid infringing on the privacy of individuals who could not possibly have been involved (i.e., suspects who are not of heritage X).

Of course, there are other reasons, more important than avoiding inconvenience, such as getting criminals off the street before they hurt other people. Using information about ancestry in criminal investigation is not racial profiling because there is specific data linking phenotype to a case. Indeed the courts generally agree with this in that this type of data often constitutes part of the probable cause needed by judges to issue search warrants. Indeed, it would seem to be common sense that one would use information available to help solve a crime, and of course it is the job of the forensic investigator to do so. If an eyewitness said that a rapist was a very tall person, would using this information be "height profiling" and discriminatory against tall people? If an eyewitness said that a rapist was a male, would that be "sex profiling" and discriminatory against males?

There is the real possibility, as the defense lawyer in the article pointed out, that ancestry testing could promote the idea that certain races are more inclined than others to commit crimes. Indeed we must recognize that the distribution of socioeconomic status with respect to ancestry and race is not random in modern society, and that there are disadvantaged groups of people for whom recent political climates have been hostile (such as African Americans who only a couple hundred years ago had to endure the ultimate in socioeconomic suppression—slavery). Ancestry testing would seem to have little to add to, or have little to subtract from this already obvious scenario, but it would seem important to recognize the scenario for what it is if the etiological forces are to be rectified. Indeed as we have discussed, ancestry admixture data can be misused, but perhaps, too, it may help elucidate social problems that are not readily apparent using ambiguous notions of social belonging. An argument can be made that even if it turns out that the truth hurts, perhaps we should learn this truth so that we can properly deal with it, if possible. Though rates for certain crimes and behaviors may correlate with elements of population stratification, it is most likely that such correlations are the result of social dynamics as opposed to genetic proclivities. However, even if this is not true, as we will discuss in more detail later, there is virtually no chance that admixture analysis could prove a genetic ancestry basis for certain crime behaviors since it is impossible for these phenotypes to separate social and environmental factors from biological.

The criticism regarding the number of markers is quite common from those that have not had the experience to work with powerful Ancestry Informative Markers. Most forensic scientists are accustomed to working with markers such as the CODIS markers, which have very little Ancestry Information content, and from them, as we have discussed, inferring genetic ancestry is hazardous. Until one has filtered through tens of thousands of markers to find the best AIMs, and through validation trials (or reading this book) learned what type of power they provide when used together in concert, it is quite natural to be skeptical and/or uninformed about what can and cannot be accomplished. Though more markers usually provide greater ancestry information content, typing a forensic sample for 10,000 or 100,000 SNPs is not always possible due to sample quantity limitations. In addition, it is far more expensive than typing 171 carefully selected AIMs.

Consider the quotation: ''How far are we from having (ancestry tests) used to justify taking DNA from any black man on a street corner, because we think a sub-Saharan African committed the crime?'' asks … a Chicago lawyer who has lectured on ancestry testing at the American Academy of Forensic Sciences.

It seem this argument is somewhat shallow, since law enforcement already uses race and other human constructs to shape their investigations, yet we rarely see indiscriminate dragnets focused on minority groups. There have been instances of targeted dragnets, but it would seem that their application is a sociological and judicial issue, not directly a function of where the knowledge about race or ancestry came from. In other words, the problem some have with them is a problem with the dragnets themselves—not the eyewitness testimony, or the legitimacy of the idea that an eyewitness can contribute meaningfully to a case. Using information obtained from DNA is no different fundamentally than using it when obtained from a human eyewitness, except it is more accurate, quantifiable, and falsifiable. It seems that the legitimacy of what happens afterward, whether a local dragnet is established for example, is a separate issue from that of what came before.

A defense lawyer is quoted in the article as saying: "... It's (ancestry testing technology) supposed to be new and foolproof, but that's traditionally what they say about all new" crime-fighting innovations. "By the time we find out there are serious questions ... a whole bunch of innocent people have had their DNA searched."

This argument is saying that if problems with the technology are discovered some years off in the future, some bad thing such as X will have happened in the interim. Here it is notable that the worst consequence X suggested here is that innocent people will have had their DNA searched (at ancestry informative markers, or possibly at CODIS STR loci). Of all the horrible things that could happen to a person—to be murdered, raped, illegally imprisoned, battered, and so on—having had a scientist attempt and fail to match the DNA fingerprint of an innocent person with that present at a crime scene could hardly be considered one of them. Nonetheless, many people feel, no matter how unreasonably, that their DNA is somehow a taboo territory and that all their other DNA-encoded features, such as how they look, behave, what diseases they express in an outwardly manifested way, and which can be "read" by eye in an instant by anyone on the street, are somehow different. The insinuation is that having someone read your DNA fingerprint is far more disagreeable than having your face searched by an investigator's eye, or your psyche searched through questioning tactics used by an investigator, or your photograph searched by witnesses looking at a photo album. It also suggests that unnecessary DNA reading to identify a culprit is far more harmful to the innocent people part of that project than not doing it would be to the future victims of the unapprehended criminal (who may remain free because the DNA technology wasn't applied). Although this may seem to be an illogical fear to many, it is one held by many nonetheless, and so we would like to understand its origin.

How you feel about DNA evidence, and molecular photofitting specifically, may depend on whether you are defense- or prosecution-minded. Let us look at the differing viewpoints that tend to be experienced among defense and prosecution-minded individuals—not to pontificate on how a criminal justice system should function, but rather to gain some insight into where people are coming from who criticize and defend DNA methods.

CONCERNS OF THE DEFENSE-MINDED

The basic premise underlying this fear is that DNA information carries with it great potential for misuse. This is not an unfounded fear. For example, the more frequently your DNA is inspected for match to a crime scene, the more likely it is that a laboratory error (such as mixing up your sample with the crime scene sample) would produce a false positive match. DNA thus carries a disproportional weight in an investigation and later in court. Few of us would feel comfortable knowing that our future is one laboratory error away from being ended. On the other hand, false accusation and unjustified imprisonment are already possibilities for all of us—all that is needed are one, preferably two eyewitnesses with an axe to grind. The impact that DNA technology has had on public safety is indisputable—through September 6, 2006, 39,291 criminal investigations have been aided (added value to the investigative process) in 49 U.S. states (http://www.fbi.gov/hq/lab/codis/aidedmap.htm). It would seem foolish to discard the entire CODIS system because of the risk of false positive linkages, for in so doing we exchange tens of thousands of true positives and the safety their identification provides to avoid the possibility of a few false positives. Our courts have recognized that there needs to be a balance in how DNA data is used. For example, though DNA carries disproportionate weight in many investigations, other evidence usually is required and convictions usually require a preponderance of evidence, not just a DNA match.

Some would argue that health information might be inadvertently or purposefully obtained by scanning a person's DNA for AIMs or CODIS markers. It is true that one might use the direct method of inference discussed herein to estimate the likelihood that an individual exhibits a particular disease of unequal distribution among the world's populations (pending the demonstration of this unequal distribution, and the construction of the proper database). For known persons, such information is already available through physician records, yet we have laws in place that regulate invasion of medical privacy to protect this information. For unknown persons, such information if extracted from the DNA might constitute an invasion of this unknown person's privacy, but so too might the description of what this unknown person looks like from a human eyewitness.

The argument could be made that when a person leaves a sample at a crime scene, they abdicate their right to that sample and any information it contains. Aside from this, the fact is that most DNA labs have neither the time nor the budget to search through evidence for medical information. Such an exercise would be very expensive, and as of the writing of this text, simply not possible for most types of medical information. For this likelihood of medical misuse of DNA evidence to be put into proper context, it will have to be explained to the general public that markers for multifactorial and quantitative diseases (e.g., most human diseases) are not only distinct from AIMs and CODIS markers but very difficult and sometimes impossible to identify with modern technology (because their numbers are so great, and the penetrance of each so low we often lack the ability to separate environment from genetics). Even if known, their measurement is expensive, time consuming, and generally most professionals have little incentive to invade our privacy (how frequently do our doctors conduct unauthorized diagnostics tests out of curiosity?).

Though AIMs impart good information for certain physical characteristics, and can sometimes impart information about certain disease risks if the disease was disparately distributed among parental groups, the fact remains that diseases are usually far more complicated than simple physical traits. Since they involve so many genetic (and environmental) factors, prediction of non-Mendelian (most) disease status with suitable predictive power, sensitivity, and specificity from DNA is impossible at present and probably will remain so for most such diseases in the foreseeable future. For example, though prostate cancer rates are greater for African Americans than Caucasians, the relationship is not strong enough to provide useful guidance to an investigation. The loci underlying the expression of prostate cancer are not yet clear, and may never be, yet because of the frequency of the disease and the weakness of the ancestry correlation, markers at these functional loci would be needed to infer the existence of such a complex phenotype with any precision. Nonetheless, the correlation between genetic ancestry and some diseases creates some justification to this concern and so, we cannot simply dismiss the possibility that, technical limitations notwithstanding, there exists a risk that medical information could be gleaned from DNA of a known or unknown person.

It would seem then that the debate on this risk needs to be focused on concepts of law and public policy. We trust doctors with much more directly relevant data concerning disease and health—the type contained in medical files, which actually indicate disease status, not just likelihood to develop a disease. Of course, much of this data is genetic in nature, and could be used to genetically discriminate. Though we trust doctors with our private information, we have still created laws governing the use of such information to prevent abuses, but such laws apply to all people who are legally in possession of such information, not just doctors. It would seem to be not any more egregious for a law enforcement officer to hold data only remotely related to disease risk (even if not discernible, due to technical limitations) than a doctor to hold data more directly related to disease expression or risk. Many distinguish between doctors and police—that the former can be trusted but the latter cannot. However, in many respects law enforcement personnel are not unlike doctors—doctors whose patient is society and with the goal to ameliorate aspects of social disease (crime) rather than a doctor who treats biological disease for an individual patient.

A defense-minded individual might point out that cases of police abuse make this equivalence illusory, however there have been many cases of doctors and nurses convicted of serial murder and it is grossly unfair to say that one type of professional is more trustworthy than another (just as it is unfair to say that a person from one population is more criminally minded, or less intelligent than those from another). Nonetheless, in some people's minds these arguments will not impress, and the double standard between DNA and other types of evidence, and between doctor versus police access, will still live on. Ultimately, these will be matters for voters and legislators to decide, not scientists.

CONCERNS OF THE PROSECUTION-MINDED

We have discussed many of the types of complaints people have against the application of DNA technology in forensics, but many people feel the opposite. They would argue that the system has a large number of more serious problems to worry about than whether a cop might learn about a murder suspect's likelihood to develop asthma, or mistakenly assign an innocent person's DNA profile to a sample from a murder scene. They would point to problems that actually do affect life and death on a daily basis, such as catch-and-release policies, early release/parole programs, and poor conviction rates. Many U.S. citizens feel that the criminal justice system is too lenient.

For many lay people, it is very annoying to have to listen to a suspect's or criminal's side of the story. There is no doubt a good reason for why we do so —our system of criminal justice is set up the way it is to make sure we get things right. Unfortunately, in affording protection against unjust conviction, our system is sometimes (maybe even often times) inefficient. In general, compared to other parts of the world, it is hard to argue that the U.S. criminal justice system is not relatively lenient. On the one hand, it is a good thing that our system is based on the presumption of innocence and we err on the side of freedom. On the other hand, when prison overcrowding becomes an issue, rather than make life more uncomfortable for felons by packing them into cells more densely, or rather than build more prisons, we tend to release them early. Many of these early releases return to their former lifestyles, murdering or raping again, and sometimes they return to jail in what many cynics refer to as the revolving door of American justice. In this case, we as a society may be saying that the life of the person killed by the recidivist is not worth as much as the comfort of the convicted murderer, or his right to live in a prison that is not overcrowded. Many argue that it is worse to over-punish a felon than it is to subject innocent society to the recidivist but for many, prisons do not exist to punish or rehabilitate—they exist to protect innocent society from criminals and so for them, this causes a great deal of frustration. The fact apparent to many is that the U.S. criminal justice system is very liberal compared to those in other parts of the world.

For example, whereas in the United Kingdom, society accepts the use of cameras in public places to make crime fighting more efficient, in the United States it is considered a taboo violation of civil liberties. Why is this? Part of the answer might be found in the anti-establishment nature of how the United States was founded—by individuals who desired to leave what they viewed as a repressive social structure in order to gain the liberty to overcome legacy limitations in a new land. This overarching theme of freedom seems to bleed over into our criminal justice system, where in erring on the side of freedom, it seems sometimes that a good part of the system is designed to protect criminals from the long arm of the law, even at the expense of meting out freedom to those who common sense would dictate do not deserve it. For example, it sometimes seems that only in America could it happen that a rapist against whom an overwhelming amount of evidence exists be released or have his or her conviction overturned because some of the evidence was obtained from a wire-tap for which the court order had expired (i.e., a technicality). A defense attorney would argue that although this causes the uncomfortable situation where a rapist is set free, overall it is a good thing for the system, since any capital case should be strong enough to withstand any devil's advocate assault (not to mention also, that enforcing our standards and laws to the letter is good for an efficient system).

The technical problem seen by many who are prosecution-minded is that for many cases where there exists a preponderance of evidence, the threshold of reasonable doubt needed to convict has not been or cannot be met, and that as a result, innocent people are harmed by the inability to sequester individuals for whom ample evidence exists who pose a threat to society. In most phases of a jurist's life, common sense is the rule by which evidence is judged. Whether one's son deserves punishment usually is decided based on a preponderance of the evidence rather than whether or not there is a chance, no matter how remote, of innocence. When the consequences of conviction are more devastating, such as in a capital murder case, the standard is understandably higher.

To the prosecution-minded, the justice system often seems rife with double standards and cynicism is sometimes the order of the day. Why is it that testimony from a supposed eyewitness can be accepted in a court of law even though there is no way to test the validity of that testimony, but scientific methods (which are falsifiable) have to pass through vigorous and sometimes unreasonable periods of validation and study before being used? Many find it frustrating that our society sees little wrong with laws designed to allow the Internal Revenue Service to monitor bank records to make sure business-people are paying the proper taxes (even those that have never committed or been charged with a crime), but everything wrong with a homicide detective invading the privacy of a convicted felon by recording his conversations or sampling his DNA outside of a specific court order. Indeed, for many white-collar crimes, it sometimes seems that the accused must prove his innocence rather than the prosecution prove guilt, but for more serious crimes the opposite is true, and society bends over backward to provide protections to these accused.

Many would complain that this is a dangerous double standard, which would lend itself to higher conviction rates for those who commit white-collar crime than those who commit far more serious crimes, such as rape or murder. For example in the federal court in the District of Massachusetts in 1998, the conviction rate for white-collar crime was nearly 91.7%. The guidelines are complicated and often require mandatory sentences, though in many U.S. states, mandatory sentences do not exist even for child rapists, which one could argue do far more and longer lasting damage than the average white-collar criminal. In contrast, the national average for felony convictions is only about 55%, and most serious felonies are never even tried in a court of law. For example, Boston police failed to make arrests in more than 70% of the city's 75 killings in 2005 and in 2004, and area (Suffolk) prosecutors failed to win convictions (on any charge, not just murder) in more than half of the 17 homicide cases they brought to trial. A prosecution-minded person might point out that this is not due to poor police work, but a byproduct of a system that coddles miscreants, murders, rapists, and thieves while expanding capacity to pursue white collar criminals. (A cynic might also point out that the government makes no money by putting a murderer in jail, but stands to make a lot by fingering a white collar criminal.) A defense attorney would argue here that the potential consequences of conviction are less damaging for the white collar accused than the accused of murder/rape and so such a double standard is justified, though this is debatable given that white collar sentences are becoming increasingly long (e.g., reference, the Sarbanes-Oxley Act of 2002 and increased prosecution and sentence lengths in the wake of scandals like that at former Enron Corporation). Others would counter that the consequences for society of mistakenly acquitting a tax cheat are less onerous than those of mistakenly acquitting a murderer. For many, since the job of the criminal justice system (and indeed the government itself) is to protect society's pursuit of life, liberty, and happiness rather than to punish or rehabilitate criminals, or to maximize tax revenues, the double standard seems backward in application.

Regardless of where you fall on these issues, the fact is that in the United States, and indeed in much of western society, the standard prosecution that law enforcement must meet is higher than in many other parts of the world. New technology that has the potential to impact the law enforcement system is studied very carefully, sometimes with overkill, harshly criticized and forced to defend itself against such criticism. We should try to separate the process that DNA tests are put through from the warts and deficiencies of our criminal justice system and come to accept that DNA is subject to a double standard because many people (especially criminals, prospective criminals, the accused, and the defense-minded) are afraid of it. Viewed in this light, the criticism of genetic ancestry testing methods becomes more clear, and reveals itself as a natural, healthy process. Charges, even those that seem personal to the scientist who studies in the field of molecular photofitting, can be considered devils-advocate challenges intended to vet new technology with potential to significantly change the way aspects of our criminal justice system operate. If the technology is useful, it should be of a strong enough foundation to withstand the charges/challenges, and if the scientist is a good one, with an objective and healthy mind, he or she should be able to live with them as well. So, scientists who work in the field of molecular photofitting should take care not to adopt a cynical or overly defensive mind set when it comes to dealing with criticisms, even those that seem ideologically based. Challenge and criticism are part of the scientific method and they help us make sure we get things right.

RESISTANCE IN THE SCIENTIFIC COMMUNITY

Lawyers, ethicists, and other social advocates are not the only critics of molecular photofitting—the academic research community has shunned research in the area of individual admixture estimation for many years. As we have mentioned, part of the reason for this is that until recently, these estimates could not be practically and properly obtained, but part of it also relates to concerns about ethics and potential social consequences. A *Popular Science* article (Jan. 2004 edition) illustrates perhaps most clearly the nature of the academic community's resistance to admixture testing. The magazine contained an article describing the use of genetics testing for the purposes of genealogy and focused on the use of admixture (DNAPrint Genomics, Inc.'s 171 AIM panel discussed in this text, referred to as AncestryBy DNA) and single chromosome haplotype (Family Tree DNA's Y-chromosome and mtDNA) tests for genealogical purposes. Though some of the technical description provided in the article was incorrect, the nongeneticist author did a reasonable job of distilling what is certainly a difficult topic for a lay reader, but in "surveying the academic crowd," she unwittingly provided a glimpse into the mind of the seemingly ideologically averse. In fact, it was more the tone of the comments than their content that was interesting from a sociological, perhaps even a political angle.

Academic scientists quoted in the article criticized genetic testing on three main grounds:

1. The tests attempt to provide a definitive answer for what cannot be definitively determined.
2. The tests overpromise what they can do, destroying public trust of genetics tests, potentially an ominous development on the eve of personalized medicine, which will use such tests.
3. The nonmedical application of data developed for medical reasons is unethical.

We will address these one by one.

THE NONMEDICAL APPLICATION OF DATA DEVELOPED FOR MEDICAL REASONS IS UNETHICAL.

In the *Popular Science* article, one geneticist criticized admixture tests as an unintended extension of the human genome project, which was data "collected ... for medical research." "What these companies are doing is, in my opinion, an unfortunate consequence of that research." The tone of this comment underlies a stiff resistance that is very interesting in a sociological/geopolitical context.

Unintended consequences are not always *bad* consequences. For example, magnetic resonance imaging and the Internet were not intended consequences of NASA's space program, yet they were derived from this program and so clearly benefit humanity. The critic is stating, however, that the nonmedical use of admixture panels is specifically unfortunate. Why, according to the critic, is not entirely clear from the statement.

In the same *Popular Science* article, one prominent scientist previously associated with this work complained that a test for admixture " ... cannot be very precise."

Recall that the very first admixture panels preceded the human genome project, having been developed by several groups for the purposes of Admixture Mapping and Mapping by Admixture Linkage Disequilibrium. One of the first notable efforts in this area was for the inference of Native American/European admixture proportions in individuals for the purpose of mapping risk loci for diabetes and gallbladder disease among Mexican-Americans (Hanis et al. 1986). These studies were initiated on the basis that knowledge of the ancestry of alleles enables more efficient genome screening in admixed populations, and MLE accuracies reported in the range of 5 to 10% were accepted with enthusiasm for this purpose. The critic quoted earlier was actually part of the team that conducted the work of Hanis et al. (1986), so their opinion carries particular weight. What the critic has said essentially is that accuracy is more important for a genealogist than a medical researcher, the latter of whom could tolerate a reasonable ± 5 to 10% error but the former of whom cannot.

What is interesting about these comments is that this person holds an admixture test being sold to a human customer (detective, genealogist) to a different standard than an admixture test applied to biomedical research. It would not be very difficult to fashion an argument that sometimes the reverse should be true—that accuracy in biomedical research is more important than accuracy for a genealogist. The common suggestion that the inferences of genetic ancestry are illegitimate because they cannot pinpoint the precise set of percentages with 100% has been discussed already in this chapter.

One scientist enunciated his criticism in the *Popular Science* article an interesting way stating, "In no way am I saying their product is totally bogus." Of course, this statement seems to imply that the product/method is partially bogus, which as we have seen from the data presented up to now in this text is not a fair implication. There are always more experiments to do, and we can look forward to an even more detailed characterization of various AIM panels in the future, but having not yet done all of the possible experiments for a given panel is not a logical basis for concluding that the panel is not useful. For example, we might use different models of admixture in future simulations, and use these alternative models to retest our hypotheses on the RMSE and bias of the 171 AIM panel. That only one simple model tested so far is not an indication of lack of panel value—we need to look at the preponderance of the data accumulated so far, which is reasonably encouraging.

THE TESTS ATTEMPT TO PROVIDE A DEFINITIVE ANSWER FOR WHAT CANNOT BE DEFINITIVELY DETERMINED

This topic was dealt with in detail in Chapters 3 and 5, and is a complex topic since, as with any ambitious statistical exercise, there are many sources of error in our estimates of ancestry. This error usually is handled in a Likelihood or Bayesian framework, based on the laws of probability, and/or with carefully thought-out procedures and communicated using confidence intervals. Unfortunately, the *Popular Science* article in 2004 incorrectly portrayed the determination of admixture as an exercise in defining a single set of values, rather than a statistical exercise of estimating a most likely value along with its confidence intervals (that is, producing a set of likely values, one of which is the most likely). No mention was made in this article about the confidence contours and likelihood intervals that are so integral for interpreting genetic ancestry admixture results for any one particular individual (Frudakis 2005; Shriver & Kittles 2004), and instead the test was criticized for producing any values at all since "every expert agreed that no DNA test can make definitive statements about ancestry (such as 90 percent European and 10 percent Native American)." Though this particular fact is true—we cannot define with 100% certainty what the correct proportions are—the implication that statistical methods for the inference of genetic ancestry are illegitimate of course is misleading and unfounded. If applied for genetic ancestry, such an unfair accusation would extend to many other areas of physical and biological science, which rely on statistical methods.

For example, it is also true that though a single hurricane projection is almost never accurate, if one looks at historical data it is clear that collections of them created for a given storm (a statistical sampling) make projection cones that are impressively accurate. It is not uncommon to form estimates with wider confidence intervals in many other areas of work, such as in the social sciences and liberal arts, and even in epidemiology. This was not the first article that held a genetics test for ancestry or admixture to different standards than other statistical tests. The fact that many scientifically trained critics so readily make the assumption that admixture tests attempt to produce a single definitive answer, in contrast to how statistical methods work in general, highlights the ingrained resistance to genetics tests of ancestry. To be fair, it also may simply be a lack of familiarity with the methods described in this text.

One prominent scientist later said in this article,

> It is possible to learn something about a person's ancestry with autosomal (nuclear) DNA testing, but you have to be very careful. When they say you're 90% European and 10% Native American, that's based on samples they collected from people they decided were representative of Europeans and Native Americans. So really what they've told you isn't what ethnicity you are, but how similar you are to the people in that database. If you change the database you change the results.

This was a very useful statement. We are attempting to make inferences based on characteristics of populations that no longer exist (the parental populations). Although this can be accomplished in modern-day populations using advanced statistical methods (i.e., Bayesian methods), these methods are not easily grasped by many lay people and even scientists, and it is difficult to assess how well these methods work. As discussed in Chapter 3, strictly speaking we can never define the ancestral groups we wish to report affiliation with; we can only estimate their characteristics and existence through an appreciation of modern-day diversity. The ancestral groups lived many thousands of years ago, and modern day descendents cannot be assumed to be perfect representatives of them. However, as we discussed in Chapters 2 and 3, there are statistical methods (Bayesian methods) that can be used to approximate the ancestry group character quite well.

There are other ways of handling this type of uncertainty, such as at the sample collection step (large samples collected with reference to the body of anthropological knowledge, a sort of Bayesian approach in and of itself). In fact, because for an especially powerful AIM, the between-group variation in allele frequency is greater than the within-group variation, ancestry information content can be harnessed for accurate admixture estimation using surprisingly small reference samples for allele frequency estimation, as long as there is no sampling bias (see Chapter 3). However it is very true that the quality of our estimates depends on the care with which the database was constructed, what samples were used, and indeed, what AIMs are used. If we built a database exclusively composed of African populations and attempted to infer admixture among these African populations for a European, we would get a misleading impression. The results would be good indications of genetic relatedness between the European and African populations, and could be interpreted with reference to a phylogenetic diagram, but would not mean necessarily that the European "came from" one or more, or any of the African populations. Note however that even with such a defective population model, if our database is of good size, the indirect (empirical) method of phenotype inference would illustrate a huge variation in phenotypes corresponding to a European type of profile—one that would undoubtedly not be useful and would preclude any useful inference of phenotype.

The application of statistical methods of course is not unique for genetic ancestry estimation, and indeed all statistical methods suffer from sampling effects. The entire basis for the field of statistics is to form generalizations based on observations in samples. The concept of Bayesian logic is based on the use of historical, empirical observations to guide future interpretations. What we are doing when inferring admixture is not fundamentally different from what we do in many other aspects of our lives. A diagnostics test for detecting strep throat works the same way as any other statistical exercise—it wasn't tested on every bug or person in the world prior to implementation, just a reasonable sample, and statistical generalizations were employed. A given strep throat test therefore may not pick up every strain, or have equal sensitivity among all the various strains. This doesn't mean that strep throat tests have no value, or shouldn't be used.

In our case, we are estimating affiliation with modern-day populations as a means by which to estimate proportional ancestry among parental populations that lived long ago. More specifically, we are estimating affiliation among various elements of modern-day population stratification, wherever and however that stratification arose. We give arbitrary names to these elements, as we have discussed in Chapters 3, 4, 5, and 6, and we cannot prove precisely from whence and when these elements came, but the body of anthropological knowledge provides good guidance. What we can prove is that there exists a very real pattern in the apportionment of modern genetic diversity, and we know that parental populations of some form, geography, and time gave rise to modern populations that exist today. Rather than rely on the body of anthropological knowledge, perhaps alternative assays of the future may report affiliation with numerically indexed elements of modern population stratification. Then, only anthropologists would need to be concerned with the origin of those elements, and what affiliation with them from an admixture test really means.

THE TESTS OVERPROMISE WHAT THEY CAN DO, DESTROYING PUBLIC TRUST OF GENETICS TESTS, POTENTIALLY AN OMINOUS DEVELOPMENT ON THE EVE OF PERSONALIZED MEDICINE THAT WILL USE SUCH TESTS

One geneticist was quoted in the *Popular Science* article as having made this complaint, and it is indeed an understandable one. Genomics tests for the purpose of inferring the likelihood a trait will exist at some point in the future will always be subject to statistical errors, but we repeat that all statistical tests are subject to statistical error and this does not render them useless or worse, malignant. It would seem a stretch to say that statistical error associated with genetic ancestry estimation will frighten patients away from taking pharmacogenomics tests in the future, but traditional diagnostics tests (which are also based on statistical samples as we have discussed) or other types of statistical tests will not or do not similarly frighten them. Based on the complexity of most human phenotypes, one could argue that the public needs to become accustomed to managing the complexity of predicting complex conditions using statistics if they are to truly benefit from the human genome project.

Many in the academic community are conflicted by the notion that genetic ancestry could be useful for their epidemiology or genetic research. These same scientists already commonly use methods for defining population structure in their sample sets, but somehow when this structure is characterized in terms of ancestry, conclusions seem less appetizing. One of the reasons for this might be the arbitrary nature of the parental group nomenclature, which we discussed in Chapters 3 through 6. Some of the resistance takes the patina of politics. Sometimes we find with critics that negative results become proof for postulates, though in science, negative results are never considered to be informative. Consider the following article, which discusses data that is factually correct but leaves a misleading impression.

Soon after the human genome project was finished, an October 27, 2004 article in the Times Online entitled "Gene tests prove that we are all the same under the skin" by Mark Henderson was written, proclaiming that inspection of the genome has revealed the fallacy of the concept of "race."

The main theme of this article is that individuals should be considered as individuals rather than as members of socially defined (e.g., racial) groups, whether it be for medical purposes or any other purpose. Note that many social scientists and politicians who advocate various social programs would seemingly sometimes disagree with this statement, but for the most part this idea is consistent with everything we have learned in this book up to now. When we infer phenotype indirectly we are considering the subject as an individual with a unique genetic history (of origin residing somewhere along a continuum of ancestry admixture) relevant to that phenotype. Were we to not know a subject's unique ancestry admixture proportions we would not be able to infer the phenotype as accurately (or in some cases, at all). The article states that variation among human populations is continuous rather than discrete, due to our relatively recent history as a species, and that race is a substandard variable for use in medical research due to its "poor" correlation with biology.

Notwithstanding the overall correctness of the take-home message—that individual data is more informative than data based on group membership—there are a number of intimations from the article that are misleading. Note the following passage from the article:

"It is impossible to look at people's genetic code and deduce whether they are black, Caucasian or Asian, and there is no human population that fits the biological definition of a race, the (genome sequencing) study found. Ethnicity is almost entirely socially and culturally constructed, and even the trait used most commonly to define it—skin color—varies widely among people of similar ancestry."

From the passage it could be concluded by the lay reader that there exist no information about "race" in the DNA. While it is true that "race" is a social concept, and not perfectly correlated with the concept of genetic ancestry or "genetic populations," it is not responsible to suggest that there is no correlation at all. As we have seen throughout this text, there are indeed such correlations as assumption free clustering methods are able to group individuals largely along continental lines and marker panels derived from these methods can even partition populations along self-reported "racial" lines as we discussed in Chapters 5 through 7. Although it is correct in stating that we are over 99% identical at the level of our DNA, it seems to conveniently side-step the fact that a few percent of the remaining 1% are extremely informative for ancestry as the bulk of this book and decades of research has clearly shown.

Let us focus on the statement from this passage: **"Research has also shown that when scientists try to guess a person's ethnic origin by looking at genes, they get it wrong between a third and two thirds of the time."**

The reader is left with the impression that there is simply no way to measure race or genetic ancestry from DNA—that the tools/markers required to do this don't exist. However, the implication of the statement is not entirely correct as we have discussed in detail in earlier chapters. Depending on the samples, "binning" can achieve accuracies in the 90% range if good AIMs are used (Chapter 2, where we discussed binning, and Chapter 3, where we discussed the binning of samples based on genetic ancestry admixture proportions). The article doesn't distinguish between "binning" Hispanics, which is difficult using any marker set, and Europeans, which is much easier and instead makes a sweeping generalization. In addition, the statement cleverly uses the word genes, which makes it not incorrect in most cases (depending on the genes, and again, the samples), since genes are subject to selective evolutionary forces and do not usually make good AIMs. Unfortunately, most of the readers of this article do not appreciate these subtleties and they walk away with a somewhat biased presentation of the topic. In fact it is not unusual to find this statement, in one of several forms, in many media articles written on the topics of DNA, race, and ancestry in the past few years. The companion to this type of implication is often found in the statement that "race cannot be read from the DNA," which, since race is not caused by DNA, is of course true if one defines "reading" as information achieved only through direct inference, and depending on the accuracy one requires for an inference to qualify as "read." But it is misleading if we consider that race is correlated with DNA variation and that indirect methods may be used to provide solid (quantifiable, testable) guidance. No mention of this point usually is made in these stories.

Consider a second passage from the article:

"Standard concepts of race, indeed, are so misleading that they are undermining efforts to untangle the true contribution that genetics make to individuality, and ought to be abandoned by science, the researchers said. In medicine, for example, race is often used to predict whether patients will respond to particular drugs. While this can be true on average, it leads to generalizations that deny useful medicines to millions who do not meet ethnic stereotypes."

The theme of this passage is true—that if possible for a given medical problem, individuals are best considered as individuals rather than as members of groups. However, the concept that "race" has no value in biomedical science is one not shared by many scientists. For example, see the U.S. Food and Drug Administration mandated collection of "race" data for all subjects of human clinical trials (see "FDA Issues Guidance on Race and Ethnicity Data";

www.fda.gov/fdac/features/2003/303_race.html). Indeed, the passage from the article admits that race-based conclusions may be true on average, but laments that certain individuals are not best served with the use of the non-scientific concept. There are two points a reader may graduate to next—use no population or race data, or use better data (such as measures of genomic ancestry, which are indeed better correlated with phenotypes). The main problem with the article is one of omission—as the latter option is not discussed at all. One could appreciate from reading the entire article that there is no discussion about the concept of genetic ancestry or population affiliation. Thus, the lay-reader unfortunately extrapolates from passages such as this that the inference of genetic ancestry (in addition to "race") serves no legitimate scientific end for a medical researcher. Placing aside the concept of indirect phenotype inference, which we have already discussed in this book as legitimate in certain cases (depending on the markers and phenotype), there are other more fundamental applications of genetic ancestry in biomedical as well as forensic science not mentioned in the article. As we have discussed, techniques that rely on measurement of admixture such as Admixture Mapping represent powerful tools for honing in on disease or drug-response genes in addition to those genes controlling common anthropometric phenotypes. There are other applications that are of even more broad interest. For example, genetic screens commonly identify false positives that are caused by ancestral stratification in study samples and an appreciation of genetic ancestry admixture is probably the most powerful method by which to correct for this stratification.

This last application is crucial to understand, as you probably appreciate already if you have read Chapter 9. We can consider a brief example here to review the use of genomic ancestry estimates to help control type I and II error. Consider a hypothetical study for the discovery of functionally relevant loci underlying variable skin pigmentation (such as that described in Chapter 8, which like many drug-response, disease and forensically relevant phenotypes is distributed to a certain extent as a function of genetic ancestry). Were we to study an eclectic group of European, Africans, and European/African mixed subjects, we would incorrectly identify thousands of markers for skin pigmentation that have no direct role in melanin production. In this case, we would be identifying a large number of European/African AIMs, since population stratification in this sample is correlated with phenotype. That is, the AIMs are correlated with the relevant loci, not linked to them in a genetic sense, and therefore not informative about biology directly (only indirectly). Indeed, as we have discussed in Chapter 9—using multiple regression (Table 9–6), and Boyesian analyses (Tables 9–11 A,B)—it is especially useful to condition genetic association results on ancestry admixture to ensure that associations identified in a genetic screen are not false positives—that they speak to biology, not correlations with biology. In the field of forensics, our goals are sometimes different and we can take advantage of the correlations to learn something about an individual who left a sample at a crime scene. This is practical even though we are all 99.9% identical at the level of our DNA.

The author of the article we are discussing here suggests or implies that there are no functional or biological differences between population groups because we are all 99.9% identical at the level of our DNA, but then contradicts this theme when admitting that self-reported race can be useful in predicting drug response "on average." Let's focus on this particular point in a bit more detail. From the preceding passage note the following two sentences:

> In medicine, for example, race is often used to predict whether patients will respond to particular drugs. While this can be true on average, it leads to generalizations that deny useful medicines to millions who do not meet ethnic stereotypes.

According to this statement, race can be used as a prognosticator of drug response, but it should nevertheless not be used because it is not a perfect prognosticator. In other words, race would not work as well as knowing the genetic basis for response (obviously) and using it as a proxy until such a knowledge is in hand is a bad thing because it may do more harm than good, such as keep certain people from being treated. This would seem to be a very weak argument because science is replete with examples of the utility of intermediate-stage technologies. Overlooked are the facts that

- There are usually many drugs for the same indication, and not all of them would be expected to have the same ethnic component to their response, meaning that viable alternatives usually would be available.
- An appreciation for race or genetic ancestry could help specific patients get the right drug before the genetic basis for response is formally elucidated (so that direct methods of inference can be practiced).
- For many drugs, using race or genetic ancestry as a proxy or prediction variable provides a tangible net improvement in medical condition for the population as a whole, not just the odd individual. That is to say, many more individuals are helped than hurt by using the information, if any are hurt at all (see the first point). The fact that the focus is on those for whom treatment would be withheld or recommended against is somewhat unfortunate.

For example, no mention was made of the NitroMed/BiDil case that indicated the pharmacogenomic relevance of tailoring the use of certain drugs for individuals of certain ancestry. BiDil was first rejected due to poor efficacy in a cosmopolitan sample, but worked well in the African-American segment of this population, so the company executing the trial redesigned it to focus on African Americans (Franciosa et al. 2002; Taylor et al. 2002). In the redesigned trial, the drug showed such a clear-cut efficacy that the FDA stopped the trial so that the placebo arm could benefit from therapy as well. This resulted in FDA approval for the first "ethnic" drug—and many media and journals (Bloche 2006; Branca 2005; Duster 2005; Duster 2007; Echols et al. 2006; Flack 2007; Kahn 2005; Temple and Stockbridge 2007; Wadman 2005) editorials were since drafted. It is not the results of this study that are surprising or rare, but the fact that it has taken so long for investigators to gain the courage to actually use race in the design of a clinical trial for a drug with unequal population response rates. Though this study would have benefited from the use of more objective measures of ancestry and admixture, it could be argued that using race in this case was better than not using anything at all.

In fact, many drugs show different response proclivities in different racial or ethnic groups, so much so that the FDA requires an assessment of race for every clinical trial, and epidemiologists, pharmacologists, and population geneticists usually correct their observations for racial covariates in an attempt to minimize the contribution of type I or II error. So it seems obvious that race is not a useless concept in science and medicine. The position from the previous statement seems to be that it should not be used anyway. If one is advocating the use of more objective measures (which perhaps this author meant to do, since they had discussed using measures that account more for individuality), then the statement is a good one. From the article, which admittedly has to abide by space limitations, this nuance does not seem to be apparent. One of the implicit goals of this article—of teaching people to consider other people as individuals rather than members of groups—is a noble one, and as forensics (or other) scientists we always prefer to use direct rather than indirect methods for inferring phenotypes. However, the major defect associated with the article is one of omission, and though it would benefit from a bit more balance and disclosure, 99% of the readers would likely not appreciate this. It is natural to disfavor indirect methods of phenotype inference in favor of more direct methods, but improper to conceal through omission the feasibility of indirect methods altogether.

We have considered this article in an attempt to illustrate the way the "No Race—DNA" position is promoted and discuss why, through omission, it is promoted so vigorously by so many writers. There are many possible reasons. In this particular article, much of the story is left out and the reader is thus left with a series of misimpressions. In a short article it is not possible to provide every angle for consideration and up until recently, there has been a dearth of data demonstrating the feasibility for indirect methods (the article was published right after the first draft of the human genome sequence was released). This is probably a direct result of a funding bias by governmental agencies, which have traditionally eschewed research aimed at elucidating the relationship between DNA, race, and phenotypes. In addition, some of the points omitted in the article may be a bit too arcane for most lay readers. Further, it is possible that the way questions are posed by lay people (journalists) to scientists quoted in such stories promote discussion of only part of the story.

However it would be naive to think that the omissions are always due to the complexity of the topic or the lack of reference material. As for many topics of social interest, journalists bring their own biases to the table and the crafting of terms in many articles such as this one lead to the inescapable conclusion that an agenda is being pursued. However, the incomplete information provided by many of the quoted scientists is more disturbing. It would be unfortunate if some scientists intentionally obfuscate reality in loyalty to some sort of political/social agenda, making the common mistake of believing that noble ends justify dishonest means, or that actions should be judged solely on their intentions (which may be noble), rather than their consequences (which may be regrettable, as when misinformation or propaganda is supplied). Since scientists are trained to be objective and to separate politics from their work, we would think intentional obfuscation is (hopefully) a rare event. We expect biased or incomplete reporting by journalists on social issues, which are political by their very nature, but it is unacceptable for scientists to engage in ''social engineering'' through omission or half-truthing. As we pointed out earlier in this chapter, science is supposed to be the opposite of politics—based on objectively determined data rather than subjective bias and/or advocacy. Scientists are supposed to recognize the truth, regardless of whether or not it is the truth they want to see. Given its power to contaminate the scientific method with subjectivity, advocacy has no place in science. It is presumably due to this objectivity that a person's title as a scientist lends a certain amount of credibility that is not normally afforded to others, like politicians or lawyers who are well known to say what they need to in order to get what they desire.

BiDil

We have already broached the BiDil example in analyzing the last article but we should consider other interesting points brought up in the articles written in response to the NitroMed/BiDil study (African-American Heart Failure Trial (A-HeFT)). Consider the *Nature Genetics* statement in the editorial:

> The use of race as a proxy is inhibiting scientists from doing their job of separating and identifying the real environmental and genetic causes of disease.

It certainly would not be a good thing if scientists, having found that indirect methods work for a given trait, stopped there in satisfaction because as we have discussed, direct methods are almost always expected to be more accurate. However, this is not to say that indirect methods have no role in any scientific process. As we have discussed throughout this text, an appreciation of population structure is actually very helpful when attempting to locate or infer from phenotypically active loci because they potentially give us information about loci we have not yet discovered (if they are distributed as a function of genetic ancestry) as well as helping us qualify those we have discovered as bonafide. Indeed, the indirect method of inference is inherent to the fundamentals with which admixture mapping approaches operate—where we recognize and use correlations between ancestry and genetic state as a mathematical fuel by which to find new disease genes. This editorial statement assumes that all underlying loci can be easily discovered by scientists as long as they make the effort, and thus there should be no need for proxies. However, the reality is that many traits are underlain by a large number of interacting variants, each with a relatively low penetrance.

Sample sizes of millions and budgets of billions would be required to completely solve many medically relevant traits. In some cases, there will be a pressing need to predict phenotype (such as response to a desperately needed but potentially dangerous drug), which is unevenly distributed as a function of ancestry, but the genetic research defining the specific variants that cause the phenotype will not yet have been finished. If race is useful for some of these traits, are we to eschew its use and accept the consequences of a relatively inefficient post-market performance of the drug? This would seem similar to "not reading" a "book" describing a donor who left DNA at a crime scene because a better book could have conceivably been available if our state of knowledge was more advanced. Our argument in this textbook is that in these cases, genetic ancestry serves as a useful proxy and that it is socially irresponsible to *not* use it. Indeed we have little problem using race for the distribution of social resources and in formulating governmental policies, and it is unfortunate that a consensus of a double standard exists when it comes to its use in medicine and forensics.

Many might argue that it is OK to use social constructs (such as race) for social purposes (such as the distribution of governmental resources), but not for scientific purposes (like drug development or use, or forensics). Put another way, social constructs (as opposed to scientific variables) are legitimately applied to address social (as opposed to medical) problems, and scientific variables are useful for medical or scientific purposes, but cannot be legitimately applied to social problems. We would agree with this logic, and suggest that optimally more scientific genetic ancestry, discussed throughout this text, be used for clinical research in preference to the social construct of race. However in lieu of access to such data, we would argue that if the application of a social construct to a medical problem helps save lives or ameliorate morbidity, it is irresponsible to not use it. Whether this use has larger implications on the health of societies would seem to be a function of human and institutional behavior, which can and should be regulated with the rule of law. Tools that can save lives should be used to save lives, and protection against misuse should be provided through the development of coherent and logical social structures, not the obfuscation of the utility of the tool.

DNA IS "DIFFERENT"

We have discussed treating DNA differently from other forms of evidence in a scientific context. Since DNA evidence carries (or should carry) more weight than an eye-witness account, it should be subject to more certainty and to higher quality control standards. However, placing quality control issues aside, there are many people that feel the application of DNA should also be subject to double standards (one for DNA, another for other forms of evidence). Consider a CBS News story from April 5, 2007 titled "A Not So Perfect Match—How Near-DNA Matches Can Incriminate Relatives of Criminals" (produced by Shari Finkelstein, interview by Leslie Stahl). This balanced story discusses the pros and cons of using partial DNA matches from CODIS searches as a means of identifying relatives for unidentified crime scene DNA donors. For example, if a sample from an alleged rapist does not match anyone in CODIS exactly, but shows an unusual partial match we might be able to infer that, while the donor could not be identified directly (their sample has never been loaded into CODIS), the database contained data for a relative of the donor. Obviously, this knowledge could lead to identification of the donor him/herself. It turns out that the statistical power inherent to the CODIS STR maker panel is adequate for the detection of brother-brother, brother-sister, or sister-sister relationships and the likelihood values of such are called the sibship index. The pros of using such methods include, obviously, getting criminals off the street—but as the story points out could also help clarify cases involving degraded DNA which sometimes result in the imprisonment of siblings of the guilty. In the story, the cons of using such methods were provided by Stephen Mercer, who is a Maryland attorney specializing in DNA cases. Mr. Mercer was opposed to the idea because he felt that enhancing the efficiency of the criminal justice system would unfairly target individuals from certain ethnic backgrounds, and because it amounted to unreasonable search (a violation of privacy) for innocent siblings of crime scene DNA donors. Denver, CO District Attorney Mitch Morrissey rebutted with the analogy of a license plate.

"Say we have a hit and run accident where someone was killed. And we have witnesses that say, 'This is the color of the car. This is the style of the car. But I only got the first three license plate numbers' . . . Only a partial. What should the police do? Just say, 'Oh no, it's only partial, so we're not gonna do anything?' "

Mercer's response was illuminating for our discussion in this chapter, "Of course they're gonna come up with analogies that seem to do away with any sense of wrongdoing or violation of privacy by the government. So they say, 'Oh, well this is like a partial plate, and we're just following up on these leads.' . . . "

When the interviewer Leslie Stahl asked "And what's wrong with that?" Mercer replied, "Because its not a partial plate. We're talking about DNA. DNA is different. DNA contains a vast amount of intensely personal information". This response represents the basis for many arguments that come from DNA critics, and it is this belief that we want to study in this section of Chapter 11. Why is it that DNA is different? What makes it so different that a double standard should be applied on its application (rather than merely its evaluation)?

Recall that the CODIS system for human identification (or in this case, sibship analysis) uses Short Tandem Repeat (STR) markers as a sort of barcode. Each individual has a unique combination of STR variants inherited from their parents, who also had unique combinations. Unlike the marker panels we have focused on throughout this text, and as we have discussed earlier, STRs are derived from the so-called "junk DNA." The reason it's called "junk DNA" is because it is not part of any human genes. Genes are required for, and directly cause phenotypes (upon which "intensely personal information" would be based), and the "junk DNA" in between genes (actually between gene regulatory and transcription domains) serves no overt purpose. "Intensely personal information" is not possible to obtain with CODIS barcodes because STRs are purely "junk DNA." They provide no reliable information about "race" or ancestry (because they are average markers not specially selected ancestry informative markers), and absolutely no information at all about disease status, phenotype, personality, religion, or any other type of proclivity. So, while DNA itself may contain this information (and we can develop assays to get some of it), the CODIS data specifically being discussed here does not. Indeed, screening a Department of Motor Vehicles (DMV) database with a partial license plate is not only no fundamentally different than screening CODIS with a partial STR profile, but even worse. Consider that from the DMV database, we can obtain personal information for people who match the partial license plate (certain diseases, eyesight limitations, weight, sex, age, prior driving or other offenses, some of which may be embarrassing). Then, consider that from a partial DNA match, we can obtain nothing other than the bar code of identity. That DNA itself, from which the STR data comes, contains more information is not relevant to this particular argument.

Mercer's comments would have seemed more logically applied to the technology discussed in this text. Let us place aside one of the questions that would then arise—given that a criminal left a book behind describing him/herself, should we read it or not? Rather, let's focus here on the question of whether a partial physical profile, or a partial STR match constitute probable cause justifying the intrusion of government and law enforcement into a person's life? Both the physical and partial STR profiles represent elements of identity—the physical profile constituting a combination of low-order variables and the partial STR profile constituting a combination of high-order variables. These variable combinations are highly characteristic of (if not completely informative for) identity. A partial license plate is very similar—a combination of high-order variables that is also characteristic of (if not completely informative for) identity. It is the fundamental relationship with identity that renders physical profiles, partial STR matches and partial license plate matches (even partial eyewitness accounts) adequate bases for probable cause. It is the use of data that is completely de-coupled with identity that would cause a problem.

RACISM AND GENETIC ANCESTRY TESTING

Colonization, genocide, slavery, legalized segregation, apartheid, Jim Crow laws, and concentration camps are but a few of the atrocities that are the history of our civilized world, and every culture has its own list to be ashamed of. Given the enormity of these events, their long-term consequences will take generations to overcome. Modern conceptions of race, racism, and racialization are some of the fallout of these events. Part of our responsibility as population geneticists is to work toward the abolition of these misconceptions and the social injustice that result from racism and racialization, not necessarily because they are undesirable (which as a human being, this author must admit they are), but because they are scientifically untenable, as we will discuss in this section of the chapter. In this light, it is imperative that proprietors of genomic ancestry admixture methods and products devote considerable internal resources to education regarding the different perspectives (sociocultural, political, and biological) on race and the meaning of populations in light of genomic science and biomedical research. It should be taught that:

1. Race is not a biological concept. Genes don't cause race, and even though population structure exists, there is not enough genetic differentiation among

human populations to consider them as separate zoological entities that differ among qualities that are innately human.

2. Race is a social construct. This means that these classifications (black, white, Hispanic, Jewish) are defined (and redefined) by the prevailing sociopolitical structure.
   a. Race is often a great amalgamation of many diverse populations and ethnicities.
   b. Race is often ascribed only to the minority populations.
   c. In the United States, any minority population ancestry is dominant and the person is completely of the minority group (e.g., ''the one-drop rule'').

3. Despite the veracity of the first two points, since there is a correspondence among broad racial categories and populations, the conclusion that there are no average biological differences among any racially described groups may not be true.

4. Racism continues. ''In some places, and for some people, overt racism has given way to implicit racialization and 'Colorblind Racism,' a term coined by Dr. Eduardo Bonilla-Silva (Stanford University).''

5. Race is a poor surrogate for population—a theme we have come across many times in this text up until this point (see Chapter 6). Population usually refers to a unit of evolution and refers to a group of persons who generally select mates from within the group. For certain scientists, population may refer to individuals united by genetic history, phenotype, and/or proclivity (such as in expressing a disease or a phenotype). Needless to say, these things need not necessarily correlate with race.

6. Being respectful is the first important step in not having a racialized perspective.
   a. Each person is a human being first and foremost. It is disrespectful, at any level, on the street, in the lab, or in the clinic, to consider his or her population group first. When lives are involved and the data support, the consideration of ancestry may be justified, but only as a subjugate to individuality.
   b. Populations should be described (not defined) in precise language that members of the community would use.

7. We believe that the physical and cultural diversity of the world's peoples should be embraced as a valuable and even sacred resource. Indeed, the genomic variation both within and among populations is in many ways our human biodiversity and will provide important clues as to the origins, our physiological construction, and the possible futures of our fragile species.

RACISM AND THE COMMON RACIST MANTRA

It is still an ugly fact of life that many human beings believe, whether or not they admit or vocalize it, that there are qualitative differences among the world's various populations—heritable differences that translate into substantive and fundamental disparities in those qualities that make us human. Stated more directly, it is not uncommonly held that individual human beings are not of equal quality. Such a conclusion would be difficult enough to defend scientifically, since quality is in the eye of the beholder, but taking this one step further and claiming that quality is a function of heritage makes the idea even more untenable. To many people, the quality of a human being is a function of their soul, and therefore is intangible and distinctly unique to the individual; it is best indicated by who they are and what they do with their life rather than who their ancestors were—as if a soul were heritable.

The racist, however, forms conclusions on individual quality simply through inference, based on who a person's ancestors were. Of course, the groups of the highest quality often are anointed by these people to be their own. To a European racist, the average European is of superior quality compared to non-Europeans, because the original Europeans were more intelligent, better looking, or better athletes than non-Europeans. This specifically was Adolph Hitler's argument. To an East Asian racist, the criteria met for determining superiority might overlap (e.g., Asians are more/most intelligent and best-looking) or be different than those of an African racist. Racists that vocalize such opinions point to standardized testing results, racial proportions on professional sporting teams, or differences in socioeconomic status between populations as proof of their belief. Most of the people who still believe this are not themselves what most would call thinkers, though many thinkers share these beliefs as well, whether they speak them or not. Indeed, racism seems to be imbedded into the natural history of human development. The very notion of assortative (sexually selective) mating, from which genetic drift produces allele frequency differentials among human populations is based not just on accessibility but on active selection for phenotypes. That is, selection of mates has been and still is often based on the assessment of qualitative differences between potential mates based largely on overt phenotypes. Phenotypes used by Europeans to judge ''health'' and ''suitability'' (attractiveness) as potential mates—such as square jaws, blue penetrating eyes, light colored and straight hair etc.—are characteristically European in nature. Lest the reader doubt this bias, note how frequently European (American, Australian, etc.) television programming involves individuals with crisp, clearly defined features such as square jaws, sharp noses, etc. (which are probably very much unadmixed European, probably northern European phenotypes) compared to ambiguously defined features such as round jaws and bulbous noses, etc. (which seem non-northern European in prevalence—which one might appreciate for example, from scanning nose shape in digital photographs of 100 West Africans or East Asians against those of 100 Northern Europeans). One could argue that our choice of mates, dictated by what we think is attractive, is the ultimate and most fundamental expression of a seemingly innate racism—since the choice of mates is viewed by most as one of the most important choices they make in their lives. For example, it is one thing to say you don't have any bias against people of group X and a different thing altogether to have children with someone from group X. When all of our mates are of one racial background, what are we saying really? When one population's choice of mates are almost always other members of that population, what can we say really about that population? Is not assortative mating, which is one basis for today's population structure, a by product of racism in and of itself? In the past, travel was difficult and choice of mates was necessarily restricted to populations. Today intercontinental travel is trivial, yet Europeans still tend to be attracted to other Europeans—blonds attracted to other blonds, brunettes to other brunettes. East Asians find other East Asians most attractive. I would maintain that our expression of what is and is not physically attractive is the expression of a basal, almost innate form of racisim from which all other racism flourishes. Yet it seems to be something we are born with. When you find someone you think is beautiful, it's difficult to imagine that you think this simply because you were taught to. Rather, it seems that to each of us, beauty is a function of certain immutable laws associated with the expression of phenotypes, which tend to be population or group-specific, and that the ability to recognize adherence to these laws is innate. Perhaps it is a function of narcisim, which is definitely innate and common to us all. In other words, it may be that we are all born racists—that our animal baseline is segregation, clanism, and racism. Perhaps as we become more educated and wise, and depending on the level of cultural advance within which we live, we gradually grow from this baseline and lose the racism. Racism is indeed quite a problem—most wars fought on our planet over the past 100,000 years were based on the concept of "population", "clan," or "race." But then again, nature demands diversity (as we will discuss later), and without "population," "clan," or "race" our ancestors would have lacked diversity and we probably would never have survived to this point. Perhaps you can see the conundrum, and the irony. However, most of us recognize that what may have been necessary a millennia ago for our primitive ancestors may no longer be necessary today. As our species has learned to think, we have acquired an ability to control our own destiny and are no longer totally in nature's hands. We take it today, as thinking, evolved human beings, that while racism may have been useful for our primitive animal-like ancestors it is of no use in today's civilization because we are no longer animals but thinking souls. Viewed in this light, racism is something that is ugly, harsh, and unfair. Indeed, when practiced it causes pain, death, and regret. Racism is indeed a problem, and to this day it is still endemic across nations and regions and seemingly an innate, baseline function of what it is, or rather used to be a human being. However most would agree that we as a species are improving—and that racism is gradually dissipating as human society and indeed as we humans ourselves continue to evolve. Thus, it would seem that the way to correct the problem of racism must be with logical thought—and the succession of ideas over animal reflexes. It is upon this thought we will focus on in the rest of this chapter.

In addition to overt phenotypes, it is most probable that population differences in certain allele frequencies among our ancestors conferred disease susceptibility (such as the Duffy Null polymorphism and malaria resistance). However, while segregation of ancestral populations might have once been useful for dealing with diverse geographical climes, how our ancestors deal with those climes is largely a function of overt, rather than innate phenotypes—such as pigmentation of hair, eyes, skin, xenobiotic metabolism, shape of external features, etc. Though we might expect differences in skin color, or even nose shape among parental populations, we would not necessarily expect differences in cranial capacity or kidney function. Most modern-day racism involves equilibrating all human phenotypes—that because there are differences in skin color there must be differences in brain function, for example. Aside from the fact that our ancestors were probably racists, and we today are all a little bit racist too, it appears that much of the foundation upon which racism lies is not structurally sound.

We will look a bit more deeply into the roots of European racism to illustrate the problems with the racist point of view. (Why European racism? Because the author of this text is European and most familiar with viewpoints held by other Europeans. An African author focused on African-centric racism would likely be able to weave similar examples into arguments.) It is an unspoken notion held by many Europeans, some of them even scientists, that Europeans are more adept at manipulating their environments than non-Europeans. Whether this is true is debatable, but even if it were, whether it is due to heredity and whether the ability to manipulate equates with intelligence is a completely different issue. The thinking European racist will point to the development of agrarian economies some 30 to 50 KYA to the present, where non-African peoples tended to reside in colder climes, which they would claim forced the development of greater cerebral ability in an evolutionary sense—maybe to deal with cold in a pure Darwinian sense, or maybe due to greater cerebral heat exchange with the environment (reasoning that thinking generates considerable heat, and a warm brain exchanges heat better with cooler air). The development of farming practices enabled more food to be produced than needed at the present time by those who did the farming, and populations expanded rapidly within which labor was divided among specialists.

In the new agrarian economy adopted by these populations that left Africa, not every citizen had to worry about hunting down their next meal. Individuals acquired the opportunity to specialize and grow in different intellectual directions—lawyers, bureaucrats, ironworkers, carpenters, and such were occupations that were born by agrarian civilizations. Most of these agrarian populations resided in northerly climes where sedentary, patrilocal populations were the rule rather than the exception and, so the thinking goes, they expanded to develop in the colder corners of the globe because it was these regions that were devoid of competition. It is commonly proposed that living in such a hostile environment required the development of societal cooperation, specialization, and structure. In Africa, these things were not as crucial for the hunter–gatherer means of subsistence and so, the thinking goes, there was little evolutionary pressure to develop the brain capacity to invent and manage them.

To the European racist, it was evolutionary pressure, applied by the hostile northerly climes and enabled by the development of agriculture, that forced certain humans to develop the ability to manipulate their environments, and it is an ability of a human being to manipulate their environment that is the foundation of modern intelligence. The ancestors of the intelligent races would have had to develop this superior intellectual ability in order to survive, to develop tools for ensuring temperature homeostasis, dwelling structures capable of withstanding a harsher climate, methods and systems for ensuring the availability of food year round—even during the dead of winter. Internal combustion engines, heaters, or any other tool that humans currently use in order to manipulate their environment are presumed by them to trace their genesis eventually to this evolutionary pressure placed on the agrarians who migrated to inhabit the colder regions of the globe. The European racist will point to the construction of airplanes, computers, space shuttles, factories, and medicines in today's world as a byproduct of a rapid development of intelligence in those African-derived populations that went north (mainly focusing on Europeans, of course).

THE DATA DOES NOT AND PROBABLY CANNOT SUPPORT THE RACIST VIEWPOINT

It may seem plausible that given the difference in environmental history experienced among the worlds various peoples that there exist some adaptive, substantial and fundamental differences in physiological constitution, but as scientists we are skeptical of whether differences in behavioral emphasis are of genetic or social base. Some believe that they are one and the same, that social structure is imbedded in the genes, but a belief is not the same thing as a fact. As scientists, we require more than plausibility before drawing conclusions. We require evidence. It so happens that for most complex behavioral traits, or those of complex physiological or social basis, we can never really acquire this evidence.

There is no doubt that there are sociological and cultural differences among the world's various populations, and of course cultural differences almost certainly owe at least some of their genesis to differences in geography and history that is correlated with ancestry, but it has never been shown to be true that intellectual or physical differences among today's modern populations are genetic in base rather than environmental and/or sociocultural. Assume we could agree what intelligence is and is not. It would seem to be impossible to prove genes play any role at all in intellectual differences among populations, because the skills possessed by any human being are a function not only of their genes but on the cultural and social forces that guided their development, as well as that of their ancestors from whom they obtained their customs.

Simply put, it is impossible to extricate environment and culture from genetics and in order for our example European racist to prove this theory, he or she must be able to do so. Why can't we extricate genes from environment to answer the question once and for all? To do so we would have to reduce ourselves to savages and conduct controlled experiments on our own kind where we hold certain variables constant from birth in some subjects, and others in other subjects. Geneticists have devised methods to separate environmental and genetic components of phenotype variance. These methods rely on observations within and between pedigrees, but they cannot separate environment and genetics completely and it cannot even be verified to what extent they are able to separate them. Because the separation of environment and genetics in human populations is not falsifiable, we simply do not know whether genes and environment, or just one or the other underlie any differences in the expression of extremely complex phenotypes between the worlds peoples. Man cannot be studied like a laboratory organism and of course, experiments cannot be undertaken on human beings with controlled environments.

Why is this an issue with intelligence and behavior but not for skin pigmentation? The difficulty relating genes to phenotype independent of environment increases geometrically with the complexity of the phenotype. Whereas it is easy to measure skin pigmentation in a manner that controls for environmental variability (we can control UV exposure quite easily), it is impossible to measure neurobehavioral traits in this manner, or even many complex physiological processes—there are just too many variables to consider. For example, we cannot control an individual's intellectual environment during childhood, or that of his parents when they were being raised. Indeed, epidemiologists even struggle with this problem when studying diseases such as cancer and heart disease. For most diseases it is virtually impossible to tease environmental from genetic etiology, even for diseases that show vast differences in prevalence among continental population groups.

Some diseases, such as prostate cancer show dramatic differences among populations, and are expressed in a manner that is independent of socioeconomic status or geography. For prostate cancer, there is undoubtedly an ancestry component to the genetics of expression, and for traits like this, we can conclude that genes unequally distributed among populations explain most extant variation. Admixture mapping is applied to better understand such diseases, mapping the genes at work. However, intelligence or physical ability is far more complex than prostate cancer, which has a binary expression (either you have it or don't have it), is easily diagnosed (are cells dividing uncontrollably, causing illness or not?), and probably is caused by a handful of gene variants rather than thousands of variants in hundreds of genes.

That does not mean that the European racist's hypothesis is not true, only that we can never know whether or not it is. The European (or other) racist's beliefs are based on a certain type of faith, which cannot be falsified and is therefore not scientific. To a scientist, a hypothesis that cannot be falsified is not worth considering. As scientists, we are devoted to the truth, even if it is an ugly truth, but we are also wary of attempting to answer questions that cannot be answered because attempts to do so increase the probability of producing false knowledge. However, this is a different thing than saying we do not want to know the truth if it is ugly. If ugly things about ancestry and phenotype are true, we want to be able to discuss them objectively, and not deny them because they have undesirable nonscientific consequences. We control the science downstream of its discovery, with laws and ethical cannons.

Were scientists only to perform "good" types of experiments but not potentially "bad" ones, they would be functioning as advocates, and advocacy is to science what cancer is to an organism—it undermines the scientific process through the destruction of objectivity, which supports the foundation of what science is all about. However, in this case, the truth is there is no way to know whether the European racist viewpoint is correct or incorrect. It is possible that this viewpoint is correct, and as scientists we have to be honest about this possibility, but we can never know for certain and concluding that it is because it feels correct is inherently unscientific and more an exercise in faith. Forming likelihood statements on whether the theory is true, given the huge number of environmental variables at work, would be futile. We require evidence to test hypotheses, and in this area it seems to most reasonable scientists that it is impossible to acquire the evidence required. Until an idea can be proved, it must be treated skeptically and so scientists in general are correct to debunk ideas linking ancestry and qualitative characteristics that make us intrinsically human.

Thus, it is not necessarily that scientists are functioning as politicians when they claim that racist viewpoints are untenable; they are merely acting like scientists in the face of unproven assertions and an untestable hypothesis.

Aside from the fact that we cannot prove the concepts that support modern (and historical) notions of racism, that these notions are therefore scientifically untenable, and likely to remain so for a very long time if not forever, what is it about the agrarian intelligence theory that makes it racist and offensive from a sociopolitical perspective? Many things: For one thing, the racist viewpoint on agrarian links to the development of intelligence would seem to obviously have been generated by a European, used to hearing the prejudices and biases common to others of his culture who probably value technological advancement over other kinds of progress or signs of advancement. A slightly different theory could have been propounded by East Asians, or Africans, or Native Americans, who point to other features they feel are characteristic to their populations. In fact, most of the readers of this text will unfortunately be Europeans, not a small part due to social imbalances and cultural differences that persist to this day. Many of the world's cultures have different values—values that persist even in immigrant communities—and it is important that when we find ourselves passing judgment on a given aspect of a culture or even a population, that we realize that our own lens is colored with bias. Jobling et al. (2004) treat this subject admirably, showing how each culture has historically painted glorious pictures of their own kind, defining beauty in terms of the norms of their limited experience. Of course beauty as well as cultural superiority is in the eye of the beholder.

SUBJECTIVE NATURE OF THE WORD INTELLIGENCE

The term intelligence deserves some special attention here, because it has been invoked as a differentially distributed trait by so many racists over modern history. Let us assume that nature has indeed built into the human population qualitative and/or quantitative differences in neurological function. There is still a problem the racist faces—choosing objectively which type is better than the other. This is a completely arbitrary decision, without scientific base. What one population considers intelligence is not often what another presumes it to be. Piano playing is considered by many, and actually is to this author, one form of intelligence. Is it not challenging to perceive and execute complex thought patterns and then translate these into complex finger and foot movements like a master pianist? What about athletic ability—is not a certain type of intellectual ability required to plan and execute complex musculoskeletal movements with precision, often on extremely short notice? The concept of muscle memory is crucial for sports such as basketball where a very precise number and quality of movement is required to be able to make 95% of one's free throws.

Are these actions not controlled by the brain, and if so, why is one type of neural competency or memory better than another type? Both musical ability and sporting prowess are thought of as talents for a good reason—talent comes from the brain and both are controlled by neurons residing in the brain, so why do we (mostly Europeans) consider the ability to translate abstract thought onto paper in a standardized test as the only measure of intelligence? What about the ability to weave complex basket and matting topologies possessed by a South Pacific Islander (I have always had difficulty visualizing chromosomal topologies, which are similar in structure)? If a European child were compared to a South Pacific Islander child in the ability to perceive complex three-dimensional topology and pattern required for this practice, and if such an ability is indeed a sign of intelligence (which, of course it is), would not the European child from civilization almost certainly be measured as less intelligent than the savage Islander child? What about multilinguality? If we use this as a measure of intelligence (which again, it is), many Middle Eastern, Central Asian, and African populations would be considered the most intelligent, and U.S. European Americans the least intelligent. Indeed, mathematics itself is a type of language that some of us in later years seem better able to grasp than others, much like languages. If one were to use the ability to solve an algebraic equation, the opposite classification might result; of course, in this case, due to societal emphasis. Why is one of these a better indication of intelligence than another? The answer would seem to be—for cultural reasons.

ACCORDING TO NATURE, DIVERSITY IS A GOOD THING

On the flipside, many scientists, guided by good intentions (such as to foster harmony among the world's peoples) take upon these arguments to project that there are no differences at all among the world's various populations. Many scientists hold on to this position even though evidence has accumulated that many of the overt phenotypes vary as a function of phylogeography. Such scientists are equally biased, just in the opposite direction and for a more benign reason than the racist is. Just as the racist cannot gather the data to support his or her viewpoint, neither can the well-meaning agnostics support their position that *no* differences exist among the world's various populations. The agnostics typically point to the low F_{ST} between populations for randomly selected loci and the fact that there exists far more genetic diversity between individuals of populations than between the populations themselves as proof that there are no substantial biological differences between human populations. On the surface, this type of statement is not logical, however, for as we can easily prove, in fact average value differences among populations do exist for many overt traits. They would be correct if they focused their discussion on qualities that make us intrinsically human, but often they don't.

Some geneticists take steps in this direction, but then leave a misleading impression. For example, they might argue that the only differences between populations are superficial—those traits whose expression is under extreme selective pressure such as cutaneous pigmentation (which as we have explained earlier in Chapters 8 and 9 is related to vitamin D production and UV protection), a few anthropometric phenotypes, and maybe a few token complex diseases as well—but that evolution drew a line or an invisible boundary beyond which it has not progressed to produce interpopulation differences for traits related to intelligence, behavior, or physical abilities. These scientists would deny—almost as if they have conducted the experiments necessary to know—that there is any genetic basis for physical or intellectual differences among populations. They are correct to point out that we cannot know the answer to these questions, but not correct to imply that the answer is known. Those who do are speaking while wearing their political, rather than their scientific hat, and it may be no surprise to the reader that scientists often wear political hats when giving interviews.

The basis for this is of course a desire for peace among the world's peoples, and the belief that differences cause conflict. Though most differences between populations are likely to be rooted in both genes and environment, we know from study of a variety of organisms that nature builds in diversity as a means by which to hedge its bets. To nature, qualitative and quantitative diversity is a beautiful thing, the fuel upon which natural selection operates. It is even true that to nature competition between subpopulations is a healthy force for ensuring fitness and longevity of populations (just as we have discussed, a devil's advocate position in the criminal justice system is a good thing overall, though it often seems frustrating). Though we might suspect that evolutionary forces have acted on our ancestors just as they have every other species of animal to create or begin to create significant biological differences between human subpopulations, we cannot prove that any of the extant differences extend to complex qualities that are intimately related to what it is that makes us human, such as intellect or behavior. In this respect, the last article we discussed, ''Gene tests prove that we are all the same under the skin,'' left one impression that was very much correct—that our common root as human beings was probably far too recent for extremely complex phenotypes, those that make us human, like intelligence and behavior, to diverge substantially (in contrast to overt phenotypes, and even certain physical abilities, which are controlled by circuitry far less complex than intellect).

Our difficulty in addressing hypotheses on innate differences is therefore not merely a function of our inability to separate genes from environment for the most complex phenotypes, but also a byproduct of the lack of speciation among human populations. These two complicating factors are probably inherently related to one another, intrinsically entwined. The fact that these sorts of hypotheses are so difficult to test should be telling us something. Not that diversity is bad and needs to be obfuscated, but that there exists little purchase for compelling arguments in favor or against them. Indeed, if it were possible to conduct a controlled human experiment to measure populational differences in intelligence, a racist may be very surprised at the outcome.

Assume that, no matter how unlikely, thousands of years of selection, perhaps statistical chance had created differences among populations in these innate, human phenotypes—it's just that we couldn't prove them. Given that the phenotypes we are discussing are so complex, so easily modifiable by environment and so learned, it would seem that classifying humans into behavioral or intelligence classes based on group membership would be poor practice. Recall the information we lost about ancestry (and the outright mistakes we make) when binning samples in Chapter 2, or the penalty we incur when we infer iris color from genomic ancestry admixture in Chapter 8 (recall, we could not make inferences for most individuals of high European admixture; see Figure 8-21). Recall the improvement we experience by considering individual OCA2 genotypes in Chapter 9 rather than relying on ancestry admixture.

Clearly, to report ancestry properly we must treat human beings as individuals, and to infer iris color with the highest accuracy, we must report from the genes that underlie variable iris color. These qualities (ancestry, iris color) are relatively simple, more easily discussed in terms of binning and less influenced by environment than an innate human phenotype like intelligence, yet even for these easy characteristics, we can see that generalization sacrifices considerable information. Our ability to infer phenotype indirectly is proportional to the simplicity of the phenotype; though it may work well for something as simple as iris color or cutaneous melanin index values, it would be foolhardy to believe it could work for something as complex and difficult to measure and define as intelligence or behavior. Even if it were possible to unambiguously define intelligence, and to parameterize it in populations, in all likelihood the indirect method would probably just not work with such phenotypes—they are far too complex both genetically and environmentally.

So, for most phenotypes other than the overt, we simply cannot generalize based on ancestry and in my opinion, it is unlikely we will ever be able to do so. Fortunately for the forensic scientist, most of the phenotypes that can be addressed with statistical tools (the so-called anthropometric phenotypes) play a role in physical appearance; that is, they are overt and descriptive of an element (though a small element) of individuality. This is probably no accident! Not just in terms of selective advantage in various environments but because sexual selection can operate more efficiently on outwardly appearing phenotypes of relatively simple inheritance, rather than the extremely complex behavioral/intellectual phenotypes.

Though there may indeed be qualitative and/or quantitative differences in innate human characteristics (the nonovert phenotypes) among the human subpopulations:

- We can never know for sure
- Using such hypothetical differences to judge individuals is not very wise
- Determining which characteristics themselves are superior is a completely subjective exercise

For these reasons, aside from the fact that racism leads to disharmony and increased human suffering, the racist's mantra will forever likely be viewed as a fringe line of thought, not well steeped in objective thought and useful for little but misguided self-promotion.

BIBLIOGRAPHY

Abbott, C., Jackson, I.J., Carritt, B., Povey, S. (1991). The human homolog of the mouse brown gene maps to the short arm of chromosome 9 and extends the known region of homology with mouse chromosome 4. *Genomics* **11**:471–473.

Adams, D., Ward, R. (1973). Admixture studies and detection of selection. *Science* **180**:1137–1143.

Akey, J.M., Sosnoski, D., Parra, E., Dios, S., Hiester, K., Su, B. et al. (2001). Melting curve analysis of SNPs (McSNP): A simple gel-free low-cost approach to SNP genotyping and DNA fragment analysis. *Biotechniques* **30**:358–367.

Akey, J., Wang, H., Xiong, M., Wu, H., Liu, W. et al. (2001). Interaction between the melanocortin-1 receptor and P genes contributes to inter-individual variation in skin pigmentation phenotypes in a Tibetan population. *Hum Genet* **108**: 516–520.

Akey, J.M., Zhang, G., Zhang, K., Jin, L., Shriver, M.D. (2002). Interrogating a high-density SNP map for signatures of natural selection. *Genome Res* **12**(12): 1805–1814.

Aldridge, K., Boyadjiev, S.A., Capone, G.T., DeLeon, V.B., Richtsmeier, J.T. (2005). Precision and error of three-dimensional phenotypic measures acquired from 3dMD photogrammetric images. *Am J Med Genet A* **138**(3):247–253.

Allison, D.B., Edlen-Nezin, L., Clay-Williams, G. (1997). Obesity among African-American women: Prevalence, consequences, causes, and developing research. *Women's Health* **3**:243–274.

Ancans, J., Hoogduijn, M.J., Thody, A.J. (2001). Melanosomal pH, pink locus protein and their roles in melanogenesis. *J Invest Dermatol* **117**:158–159.

Anderson, E., Thompson, E. (2002). A model-based method for identifying species hybrids using multilocus genetic data. *Genetics* **160**:1217–1229.

Balding, D., Bishop, M., Cannings, C. (2000). *Handbook of Statistical Genetics.* John Wiley & Sons, LTD., Chichester, England.

Bamshad, M.J., Wooding, S., Watkins, W.S., Ostler, C.T., Batzer, M.A., Jorde, L.B. (2003). Human population genetic structure and inference of group membership. *Am J Hum Genet* **72**(3):578–589.

Bansal, A., van den Boom, D., Kammerer, S., Honisch, C., Adams, G., Cantor, C. (2002). Association testing by DNA pooling: An effective initial screen. *PNAS* **99**(26):16871–16874.

Barac, L., Pericie, M., Klarie, I.M., Rootsi, S., Janicijevic, B., Kivisild, T. et al. (2003). Y chromosomal heritage of Croation population and its island isolates. *Eur J Hum Genet* **11**(7):535–542.

Barbujani, G., Magagni, A., Minch, E., Cavalli-Sforza, L. (1997). An apportionment of human DNA diversity. *Proc Natl Acad Sci USA* **94**:4516–4519.

Barsh, G.S. (1996). The genetics of pigmentation: From fancy genes to complex traits. *Trends Genet* **12**(8):299–305.

Bastiaens, M., ter Huurne, J., Gruis, N., Bergman, W., Westendorp, R., Vermeer, B.J. et al. (2001a). The melanocortin-1-receptor gene is the major freckle gene. *Hum Mol Genet* **10**(16):1701–1708.

Bastiaens, M.T., ter Huurne, J.A., Kielich, C., Gruis, N.A., Westendorp, R.G., Vermeer, B.J. et al. (2001b). Melanocortin-1 receptor gene variants determine the risk of nonmelanoma skin cancer independently of fair skin and red hair. *Am J Hum Genet* **68**(4):884–894.

Basu, A., Mukherjee, N., Roy, S., Sengupta, S., Banerjee, S., Chakraborty, M. et al. (2003). Ethnic India: A genomic view, with special reference to peopling and structure. *Genome Res* **13**(10):2277–2290.

Bauchet, M., McEvoy, B., Pearson, L., Quillen, E., Sarkisian, T., Hovhannesyan, K., Deka, R., Bradley, D., Shriver, M. (2007). Measuring European population stratification using microarray genotype data. In press, *Am J Hum Genet.*

Behar, D., Garrigan, D., Kaplan, M., Mobasher, Z., Rosengarten, D., Karafet, T. *et al.* (2004). Contrasting patterns of Y chromosome variation in Ashkenazi Jewish and host non-Jewish European populations. *Hum Genet* **114**:354–365.

Behar, D., Hammer, M., Bonne-Tamir, B., Richards, M., Villems, R., Rosengarten, D. (2004). Differential bottleneck effects in the mtDNA gene pool of Ashkenazi Jewish populations. In press.

Belkhir, K., Borsa, P., Chikhi, L., Raufaste, N., Bonhomme, F. (2001). GENETIX, software under windows for the genetic of populations. 4.02 ed. Montpellier, France: Laboratory Genome, Populations, Interactions CNRS UMR 5000, University of Montpellier II (Montpellier, France).

Bermisheva, M., Tambets, K., Villems, R., Khusnutdinova, E. (2002). Diversity of mitochondrial DNA haplotypes in ethnic populations of the Volga-Ural region of Russia. *Mol Biol (Mosk)* **36**(6):990–1001.

Bernstein, F. (1931). Die geographische Verteilung der blutgruppen und ihre anthropologische bedeutung. In: Comitato italiano per lo studio dei problemi della populazione. Roma: Instituto Poligrafico dello Stato. 227–243.

Bertorelle, G., Excoffier, L. (1998). Inferring admixture proportions from molecular data. *Mol Biol Evol* **15**(10):1298–1311.

Bhasin M., Singh, I., Sudhakar, K., Bhardwaj, V., Chahal, S., Walter, H., Dannewitz, A. (1985). Genetic studies in four tribal populations of the Surat district, Gujarat (India). *Ann Hum Biol* **12**:27–39.

Biasutti, R. (1959). Razze e popoli della terra, Torino.

Bild, A., Yao, G., Chang, J., Wang, Q., Potti, A., Chasse, D. et al. (2006). Oncogenic pathway signatures in human cancers as a guide to targeted therapies. *Nature* **439**:353–357.

Bito, L.Z., Matheny, A., Cruickshanks, K.J., Nondahl, D.M., Carino, O.B. (1997). Iris color changes past early childhood. The Louisville Twin Study. *Arch Ophthalmol* **115**(5):659–663.

Blangero, J. (1986). Admixture estimation during multivariate quantitative traits. *Am J Phys Anthropol* **69**:177 (Abstract).

Bloche M. (2006). Race, money and medicines. *J Law Med Ethics* **34**(3):555–558.

Blumberg, B., Hesser, J. (1971). Loci differentially affected by selection in two American Black populations. *Proc Natl Acad Sci USA* **68**:2554–2558.

Blumberg, B.S., Workman, P.L., Hirschfeld, J. (1964). Gamma-globulin, group specific, and lipoprotein groups in a U.S. white and negro population. *Nature* **202**:561–563.

Boissy, R.E., Zhao, H., Og, W.S., Austin, L.M., Wildenberg, S.C., Boissy, Y.L. et al. (1996). Mutation in and lack of expression of tyrosinase-related protein-1 (TRP-1) in melanocytes from an individual with brown oculocutaneous albinism: A new subtype of albinism classified as "OCA3." *Am J Hum Genet* **58**:1145–1156.

Bonilla, C., Shriver, M.D., Parra, E.J., Jones, A., Ferna?ndez, J.R. (2004a). Ancestral proportions and their association with skin pigmentation and bone mineral density in Puerto Rican women from New York City. *Hum Genet* **115**:57–68.

Bonilla, C., Parra, E.J., Pfaff, C.L., Dios, S., Marshall, J.A., Hamman, R.F. et al. (2004b). Admixture in the Hispanics of the San Luis Valley, Colorado and its implications for complex trait gene mapping. *Ann Hum Genet* **68**:139–153.

Bonilla, C., Boxill, L.-A., McDonald, S.A., Williams, T., Sylvester, N., Parra, E.J. et al. (2005). The g.8818G variant of the Agouti Signaling Protein (ASIP) gene is the ancestral allele and is associated with darker skin color in African Americans. *Human Genetics* **116**:402–406.

Bonilla, C., Gutierrez, G., Parra, E.J., Kline, C., Shriver, M.D. Admixture analysis of a rural population of the state of Guerrero, Mexico. *Am J Phys Anthro.* In press.

Bowcock, A., Ruiz-Linares, A., Tomfohrde, J., Minch, E., Kidd, J., Cavalli-Sforza, L.L. (1994). High resolution of human evolutionary trees with polymorphic micro-satellites. *Nature* **368**:455–457.

Bowman, B.H., Barnett, D.R., Hite, R. (1967). Hemoglobin G-Coushatta: A beta variant with a delta-like substitution. *Biochem Biophys Res Commun* **26**(4):466–470.

Box, N.F., Wyeth, J.R., O'Gorman, L.E., Martin, N.G., Sturm, R.A. (1997). Characterization of melanocyte stimulating hormone variant alleles in twins with red hair. *Hum Mol Genet* **6**:1891–1897.

Box, N.F., Duffy, D.L., Irving, R.E., Russell, A., Chen, W., Griffyths, L.R. et al. (2001). Melanocortin-1 receptor genotype is a risk factor for basal and squamous cell carcinoma. *J Invest Dermatol* **116**:224–229.

Branca M. (2005). BiDil raises questions about race as a marker. *Nat Rev Drug Discov* **4**(8):615–616.

Brauer, G., Chopra, V.P. (1978). Estimation of the heritability of hair and iris color. *Anthropol Anz* **36**(2):109–120.

Brenner, C. (1998). Difficulties in the estimation of ethnic affiliation. *Am J Hum Genet* **62**:1558–1560.

Brues, A.M. (1975). Rethinking human pigmentation. *Am J Phys Anthropol* **43**(3):387–391.

Budowle, B., Smith, J., Moretti, T. (2000). DNA typing protocols: Molecular biology and forensic analysis. Eaton Publishing Company/Biotechniques Books.

Burroughs, V.J., Maxey, R.W., Levy, R.A. (2002). Racial and ethnic differences in response to medicines: Towards individualized pharmaceutical treatment. *J Natl Med Assoc* **94**(10 Suppl):1–26.

Butler, J.M., Devaney, J.M., Marino, M.A., Vallone, P.M. (2001). Quality control of PCR primers used in multiplex STR amplification reactions. *Forensic Sci Int* **119**(1):87–96.

Butler, J.M., Shen, Y., McCord, B.R. (2003). The development of reduced size STR amplicons as tools for analysis of degraded DNA. *J Forensic Sci* **48**(5):1054–1064.

Carlin, J.B., Wolfe, R., Brown, C.H., Gelman, A. (2001). A case study on the choice, interpretation and checking of multilevel models for longitudinal binary outcomes. *Biostatistics* **2**(4):397–416.

Cavalli-Sforza, L.L. (1966a). Some old and new data on the genetics of human populations. *Ala J Med Sci* **3**(4):376–381.

Cavalli-Sforza, L.L. (1966b). Population structure and human evolution. *Proc R Soc Lond B Biol Sci* **164**(995):362–379.

Cavalli-Sforza, L., Menozzi, P., Piazza, A. (1994). The History and Geography of Human Genes. Princeton University Press, Princeton, NJ.

Cavalli-Sforza, L., Minch, E. (1997). Paleolithic and Neolithic lineages in the European mitochondrial gene pool. *Am J Hum Genet* **61**:247–251.

Cavalli-Sforza, L.L., Bodmer, F. (1999). Genetics of human populations. Dover Publications, Mineola, NY, 713.

Chakraborty, R. (1975). Estimation of race admixture—A new method. *Am J Phys Anthropol* **42**(3):507–511.

Chakraborty, R. (1985). Gene identity in racial hybrids and estimation of admixture rates. In J.V. Neel, Y. Ahuja (Eds.), Genetic microdifferentiation in man and other animals. Delhi: Indian anthropological association, Delhi University, Anthropology Department, 171–180.

Chakraborty, R. (1986). Gene admixture in human populations: Models and predictions. *Yearbook of Phys Anthropol* **29**:1–43.

Chakraborty, R. (1990). Mitochondrial DNA polymorphism reveals hidden heterogeneity within some Asian populations. *Am J Hum Genet* **47**(1):87–94.

Chakraborty, R. (1992). Multiple alleles and estimation of genetic parametric computational equations showing involvement of all alleles. *Genetics* **130**:231–243.

Chakraborty, R., Ferrell, R., Stern, M., Haffner, S., Hazuda, H., Rosenthal, M. (1986a). Relationship of prevalence of non-insulin-dependent diabetes mellitus with Amerindian admixture in the Mexican Americans of San Antonio, Texas. *Genet Epidemiol* **3**(6):435–454.

Chakraborty, R., Walter, H., Mukherjee, B., Sauber, P., Malhorta, K., Banerjee, S., Roy, M. (1986b). Immunoglobulin (Gm and Km) allotypes in nine endogamous groups of West Bengal, India. *Ann Hum Biol* **14**(2):155–167.

Chakraborty, R., Weiss, K.M. (1988). Admixture as a tool for finding linked genes and detecting that difference from allelic associations between loci. *Proc Natl Acad Sci USA* **85**:9119–9123.

Chakraborty, R., Kamboh, M.I., Ferrell, R.E. (1991). Unique alleles in admixed populations: A strategy for determining hereditary population differences of disease frequencies. *Ethnicity and Disease* **1**:245–256.

Chakraborty, R., Kamboh, M.I., Nwankwo, Ferrell, R.E. (1992). Caucasian genes in American blacks: New data. *Am J Hum Genet* **50**:145–155.

Chance, J.E., Turner, A.L., Goldstein, A.G. (1982). Development of differential recognition for own- and other-race faces. *Psychol* **112**:29–37.

Chikhi, L., Nichols, R., Barbujani, G., Beaumont, M. (2002). Y genetic data support the Neolithic demic diffusion model. *Proc Natl Acad Sci USA* **99**:11008–11013.

Chikhi, L., Bruford, M.W., Beaumont, M.A. (2001). Estimation of admixture proportions: A likelihood-based approach using Markov chain Monte Carlo. *Genetics* **158**(3):1347–1362.

Chintamaneni, C.D., Ramsay, M., Colman, M.-A., Fox, M.F., Pickard, R.T., Kwon, B.S. (1991). Mapping the human CAS2 gene, the homologue of the mouse brown (b) locus, to human chromosome 9p22-pter. *Biochem Biophys Res Commun* **178**:227–235.

Chopra, V. (1970). Studies on serum groups in the Kumaon region, India. *Humangenetik* **10**:35–43.

Clark, A., Hubisz, M., Bustamante, C., Williamson, S., Nielsen, R. (2005). Ascertainment bias in studies of human genome-wide polymorphism. *Genome Research* **15**:1496–1502.

Cockerham, C.C., Weir, B.S. (1984). Covariances of relatives stemming from a population undergoing mixed self and random mating. *Biometrics* **40**(1):157–164.

Cockerham, C.C., Weir, B.S. (1986). Estimation of inbreeding parameters in stratified populations. *Ann Hum Genet* **50**(Pt 3):271–281.

Collins, R. (1986). The Basques. Blackwell, Oxford.

Collins-Schramm, H.E., Phillips, C.M., Operario, D.J., Lee, J.S., Weber, J.L., Hanson, R.L. et al. (2002). Ethnic-difference markers for use in mapping by admixture linkage disequilibrium. *Am J Hum Genet* **70**:737–750.

Corander, J., Waldmann, P., Sillanpaa, M.J. (2003). Bayesian analysis of genetic differentiation between populations. *Genetics* **163**(1):367–374.

Corander, J., Waldmann, P., Marttinen, P., Sillanpaa, M.J. (2004). BAPS 2: enhanced possibilities for the analysis of genetic population structure. *Bioinformatics* **20**(15):2363–2369.

Davenport and Davenport. (1906). Davenport's Station for Experimental Genetics, at Cold Spring Harbor on Long Island, New York.

Dawson, K.J., Belkhir, K. (2001). A Bayesian approach to the identification of panmitic populations and the assignment of individuals. *Genet Res* **78**(1):59–77.

Dean, M., Stephens, J.C., Winkler, C., Lomb, D.A, Ramsburg, M., Boaze, R. et al. (1994). Polymorphic admixture typing in human ethnic populations. *Am J Hum Genet* **55**:788–808.

Deka, R., Shriver, M.D., Yu, L.M., Ferrell, R.E., Chakraborty, R. (1995). Intra- and inter-population diversity at short tandem repeat loci in diverse populations of the world. *Electrophoresis* **16**:1659–1664.

Derenko, M., Maliarchuk, B., Denisova, G., Dambueva, I., Kakpako, D., Luzina, F. et al. (2002). Molecular genetic differentiation of ethnic populations in southern and eastern Siberia based on mitochondrial DNA polymorphism. *Genetika* **38**(10):1409–1416.

Derenko, M., Grzybowski, T., Malyarcuk, B., Dambueva, I., Denisov, G., Czarny, J., Dorzu, C. et al. (2003). Diversity of mitochondrial DNA lineages in South Siberia. *Ann Hum Genet* **67**(5):391–411.

Devlin, B., Roeder, K. (1999). Genomic control for association studies. *Biometrics* **55**:997–1004.

Diffey, B.L., Oliver, R.J., Farr, P.M. (1984). A portable instrument for quantifying erythema induced by ultraviolet radiation. *Br J Dermatol* **111**(6):663–672.

Duffy, D.L., Box, N.F., Chen, W., Palmer, J.S., Montgomery, G.W., James, M.R. et al. (2004). Interactive effects of MC1R and OCA2 on melanoma risk phenotypes. *Hum Mol Genet* **13**(4):447–461.

Durham-Pierre, D., Gardner, J.M., Nakatsu, Y., King, R.A., Francke, U., Ching, A. et al. (1994). African origin of an intragenic deletion of the human P gene in tyrosinase positive oculocutaneous albinism. *Nature Genet* **7**:176–179.

Durham-Pierre, D., King, R.A., Naber, J.M., Laken, S., Brilliant, M.H. (1996). Estimation of carrier frequency of a 2.7 kb deletion allele of the P gene associated with OCA2 in African-Americans. *Hum Mutat* **7**:370–373.

Duster T. (2005). Medicine. Race and reification in science. *Science* **307**(5712):1050–1051.

Duster T. (2007). Medicalisation of race. *Lancet* **369**(9562):702–704.

Echols M., Yancy C. (2006). Isosorbide dinitrate-hydralazine combination therapy in African Americans with heart failure. *Vasc Health Risk Manag* **2**(4): 423–431.

Edwards, E.A., Duntley, S.Q. (1939). The pigments and color of living human skin. *Am J Anat* **65**:1–33.

Eiberg, H., Mohr, J. (1987). Major locus for red hair color linked to MNS blood groups on chromosome 4. *Clin Genet* **32**(2):125–128.

Eiberg, H., Mohr, J. (1996). Assignment of genes coding for brown iris colour (BEY2) and brown hair colour (HCL3) on chromosome 15q. *Eur J Hum Genet* **4**(4):237–241.

Ewens, W. (1969). *Population Genetics*. London, Methuen.

Falush, D., Stephens, M., Pritchard, J.K. (2003). Inference of population structure using multilocus genotype data: Linked loci and correlated allele frequencies. *Genetics* **164**(4):1567–1587.

Farr, P.M., Diffey, B.L. (1984). Quantitative studies on cutaneous erythema induced by ultraviolet radiation. *Br J Dermatol* **111**(6):673–682.

Ferna?ndez, J.R., Shriver, M.D., Beasley, T.M., Rafla-Demetrious, N., Parra, E., Albu, J. et al. (2003). Association of African genetic admixture with resting metabolic rate and obesity among African American women. *Obesity Research* **11**:904–911.

Fisher, R.A. (1936). The use of multiple measurements in taxonomic problems. *Annals of Eug* **7**:179–188.

Flack J. (2007). Editorial commentary on fixed combination isosorbide dinitrate/hydralazine for nitric-oxide-enhancing therapy in heart failure. *Expert Opin Pharmacother.* **8**(3):275–277.

Flanagan, N., Healy, E., Ray, A., Philips, S., Todd, C., Jackson, I.J. (2000). Pleiotropic effects of the melanocortin 1 receptor (MC1R) gene on human pigmentation. *Hum Molec Genet* **9**:2531–2537.

Flesch, P., Rothman, S. (1945). Isolation of an iron pigment from human red hair. *J Invest Dermatol* **6**:257–270.

Franciosa, J.A., Taylor, A.L., Cohn, J.N., Yancy, C.W., Ziesche, S., Olukotun, A. et al. (2002). African-American Heart Failure Trial (A-HeFT): Rationale, design, and methodology. *J Card Fail* **8**(3):128–135.

Frandberg, P.A., Doufexis, M., Kapas, S., Chhajlani, V. (1998). Amino acid residues in third intracellular loop of melanocortin 1 receptor are involved in G-protein coupling. *Biochem Mol Biol Int* **46**(5):913–922.

Frudakis, T. (2005). Powerful but requiring caution: Genetic tests of ancestral origins. *Natl Geneal Soc Qtly* **93**:260–268.

Frudakis, T., Venkateswarlu, K., Thomas, M., Gaskin, Z., Ginjupalli, S., Gunturi, S. et al. (2003a). A classifier for the SNP-based inference of ancestry. *J Forensic Sci* **48**(4):771–782. Erratum in: *J Forensic Sci* (2004). **49**(5):1145–1146.

Frudakis, T., Thomas, M., Gaskn, Z., Venkateswarlu, K., Chandra, K., Ginjupalli, S. et al. (2003b). Sequences associated with human iris pigmentation. *Genetics* **165**:2071–2083.

Frudakis, T., Terravainen, T., Thomas, T. (2006). Multilocus OCA2 genotypes specify human iris colors. *Genetics*, In review.

Gardner, J.M., Nakatsu, Y., Gondo, Y., Lee, S., Lyon, M.F., King, R.A., Brilliant, M.H. (1992). The mouse pink-iris dilution gene: Association with human Prader-Willi and Angelman syndromes. *Science* **257**:1121–1124.

Gardner, B., MacAdam, D. (1934). Colorimetric analysis of hair color. *Am J Phys Anthropol* **19**(2):187–201.

Gelman, A., Rubin, D. (1992). Inference from iterative simulation using multiple sequences. *Stat Sci* **7**:457–511.

Gelman, A., Carlin, D., Stern, H., Rubin, D. (1995). *Bayesian Data Analysis.* London: Chapman & Hall.

Ghosh, A., Kirk, R., Joshi, S., Bhatia, H. (1977). A population genetic study of the Kota in the Nilgiri hills, South India. *Hum Hered* **27**:225–241.

Glass, B. (1955). On the unlikelihood of significant admixture of genes from the North American Indians in the present composition of the Negroes of the United States. *Am J Hum Genet* **7**:368–385.

Glass, B., Li, C. (1953). The dynamics of racial intermixture—An analysis based on the American Negro. *Am J Hum Genet* **5**:1–19.

Gohler, W. (1966). Genetische untersuchungen zum Gm system. Doctoral dissertation, Karl-Marx Institat, Leipzig.

Gomulkiewicz, R., Brodziak, J., Mangel, M. (1990). Ranking loci for genetic stock identification by curvature methods. *Can J Fish Aquat Sci* **47**:611–619.

Grant (2002). An estimate of premature cancer mortality in the U.S. due to inadequate doses of solar UV-B radiation. *Cancer* **94**:1867–1875.

Guillot, G., Mortier, F., Estoup, A. (2005) Geneland: A computer package for landscape genetics. *Mol Ecol* Notes **5**(3):708–711.

Gyske, L.I., Ivanov, P.L. (1996). A method for differential cytolysis in the molecular genetic expert identification of material evidence: Problems in optimizing the procedure. *Sud Med Ekspert* **39**(1):16–21.

Haldane, J.B. (1949). The association of characters as a result of inbreeding and linkage. *Ann Eugen* **15**(1):15–23.

Halder, I. (2005). Doctoral Dissertation. The Pennsylvania State University, PA.

Halder, I., Shriver, M.D. (2003). Measuring and using admixture to study the genetics of complex diseases. *Hum Genom* **1**:52–62.

Halder, I., Shriver, M., Thomas, M., Fernandez, J., Frudakis, T. (2006). A panel of ancestry informative markers for estimating individual biogeographical ancestry and admixture from four continents: utility and applications. *Am J Hum Genet*. In review.

Hamabe, J., Fukushima, Y., Harada, N., Abe, K., Matsuo, N., Nagai, T. et al. (1991). Molecular study of the Prader-Willi syndrome: Deletion, RFLP, and phenotype analyses of 50 patients. *Am J Med Genet* **41**:54–63.

Hamman, R.F., Mayer, E.J., Moo-Young, G.A., Hildebrandt, W., Marshall, J.A., Baxter, J. (1989). Prevalence and risk factors of diabetic retinopathy in non-Hispanic whites and Hispanics with NIDDM. San Luis Valley Diabetes Study. *Diabetes* **38**(10):1231–1237.

Hammer, M., Redd, A., Wood, E., Bonner, M., Jarjanazi, H., Karafet, T. et al. (2000). Jewish and Middle Eastern non-Jewish populations share a common pool of Y-chromosome biallelic haplotypes. *Proc Natl Acad Sci U S A* **97**:6769–6774.

Hammer, M.F., Zegura, S. (2002). The human Y chromosome haplogroup tree: Nomenclature and phylogeography of its major subdivisions. *Annu Rev Anthropol* **31**:303–321.

Hanis, C.L., Chakraborty, R., Ferrell, R.E., Schull, W.J. (1986). Individual admixture estimates: Disease associations and individual risk of diabetes and gallbladder disease among Mexican-Americans in Starr County, Texas. *Am J Phys Anthropol* **70**(4):433–441.

Harding, R., Healy, E., Ray, A., Ellis, N., Flanagan, N., Todd, C. et al. (2000). Evidence for variable selective pressures at MC1R. *Am J Hum Genet* **66**(4):1351–1364.

Harrison, G.A., Owen, J. (1964). Studies on the inheritance of human skin colour. *Ann Hum Genet* **28**:27–37.

Hartl, D., Clark, A. (1998). *Principles of Population Genetics*. Sinauer Associates Inc.

Helgason, A., Siguroardottir, S., Nicholson, J. et al. (2000). Estimating Scandinavian and Gaelic ancestry in the male settlers of Iceland. *Am J Hum Genet* **67**:697–717.

Helgason, A., Hickey, E., Goodacre, S., Bosnes, V., Stefansson, K., Ward, R., Sykes, B. (2001). mtDna and the islands of the North Atlantic: Estimating the proportions of Norse and Gaelic ancestry. *Am J Hum Genet* **68**(3):723–737. Epub 2001 Feb 01.

Helleman, J., Jansen, M., Span, P., van Staveren, I., Massuger, L., Meijer-van Gelder, M. et al. (2006). Molecular profiling of platinum resistant ovarian cancer. *Int J Cancer* **118**:1963–1971.

Hertzog, K., Johnston, F. (1968). Selection and the Rh polymorphism. *Hum Biol* **40**:85–97.

Hoggart, C.J., Parra, E.J., Shriver, M.D., Kittles, R.A., Clayton, D.G., McKeigue, P.M. (2003). Control of confounding of genetic associations in stratified populations. *Am J Hum Genet* **72**:1492–1504.

Hoggart, C.J., Shriver, M.D., Kittles, R.A., Clayton, D.G., McKeigue, P.M. (2004). Design and analysis of admixture mapping studies. *Am J Hum Genet* **74**(5):965–978. Epub 2004 Apr 14.

Howells, W.W. (1989). Skull shapes and the map: Craniometric analyses in the dispersion of modern Homo. *Peabody Museum Papers* **79**:1–189.

Howells, W.W. (1995). Who's who in skulls: Ethnic identification of crania from measurements. *Peabody Museum Papers* **82**:1–108.

Hunter Labs (1996). Hunter Lab Color Scale. Insight on Color Vol. 8 No. 9 (August 1–15, 1996). Reston, VA, USA: Hunter Associates Laboratories.

Imesch, P.D., Wallow, I.H., Albert, D.M. (1997). The color of the human iris: A review of morphologic correlates and of some conditions that affect iridial pigmentation. *Surv Ophthalmol* **41**(Suppl 2):S117–123.

Inman, K., Rudin, N. (2002). The origin of evidence. *Forensic Sci Int* **126**(1):11–16.

Ito, S., Prota, G. (1977). Novel reaction of cysteine with phenolic amino acids in hydrobromic acid: Reversible formation of 3-cysteinyl-S-yltyrosine and cystein-S-yldopas. *JCS Chem Comm* (13)4:251–252.

Jablonski, N.G., Chaplin, G. (2000). The evolution of human skin coloration. *J Hum Evol* **39**:57–106.

Jeffreys, A.J., Wilson, V., Thein, S.L. (1985). Individual-specific "fingerprints" of human DNA. *Nature* **316**(6023):76–79.

Jeffries, A. (2005). Genetic fingerprinting. *Nature Medicine* **11**(10):xiv–xvii.

Jimbow, K., Salopek, T.G., Dixon, W.T., Searles, G.E., Yamada, K. (1991). The epidermal melanin unit in the pathophysiology of malignant melanoma. *Am J Dermatopathol* **13**(2):179–188.

Jimbow, K., Alena, F., Dixon, W., Hara, H. (1992). Regulatory factors of pheo- and eumelanogenesis in melanogenic compartments. *Pigment Cell Res* Suppl **2**:36–42.

Jobling, M., Hurles, M., Tyler-Smith, C. (2004). *Human Evolutionary Genetics. Origins, Peoples & Disease.* Garland Publishing, New York, NY.

Jorde, L., Watkins, W., Bamshad, M., Dixon, M., Ricker, C., Seielstad, M., Batzer, M. (2000). The distribution of human genetic diversity: A comparison of mitochondrial, autosomal, and Y-chromosome data. *Am J Hum Genet* **66**:979–988.

Jorgensen, T.H., Buttenschon, H.N., Wang, A.G., Als, T.D., Borglum, A.D., Ewald, H. (2004). The origin of the isolated population of the Faroe Islands investigated using Y chromosomal markers. *Hum Genet* **115**(1):19–28. Epub 2004 Apr 09.

Juberg, R.C., Sholte, F.G., Touchstone, W.J. (1975). Normal values for intercanthal distances of 5- to 11-year-old American blacks. *Pediatrics* **55**(3):431–436.

Jurmain, R., Nelson, H., Kilgore, L., Trevathan, W. (2000). *Introduction to Physical Anthropology.* Wadsworth/Thomson Learning, Belmont, CA.

Kahn J. (2005). Misreading race and genomics after BiDil. *Nat Genet* **37**(7):655–656.

Kalaydjieva, L., Gresham, D., Gooding, R. (2000). Genetic studies of the Roma (Gypsies): A review. *BMC Med Genet* **2**:5.

Kalow, W. (Ed). (1992). *Pharmacogenetics of Drug Metabolism.* Pergamon Press. Elmsford, NY.

Kanetsky, P., Swoyer, J., Panossian, S., Holmes, R., Guerry, D., Rebbeck, T. (2002). A polymorphism in the Agouti signaling protein gene is associated with human pigmentation. *Am J Hum Gen* **70**:770–775.

Karafet, T.M., Zegura, S.L., Posukh, O., Osipova, L., Bergen, A., Long, J et al. (1999). Ancestral Asian source(s) of new world Y-chromosome founder haplotypes. *Am J Hum Genet* **64**(3):817–831.

Kashyap, V.K., Ashma, R., Gaikwad, S., Sarkar, B.N., Trivedi, R. (2004). Deciphering diversity in populations of various linguistic and ethnic affiliations of different geographical regions of India: Analysis based on 15 microsatellite markers. *J Genet* **83**(1):49–63.

Ke, Y., Su, B., Song, X., Lu, D., Chen, L., Li, H. et al. (2001). African origin of modern humans in East Asia: A tale of 12,000 Y chromosomes. *Science* **292**:1151–1153.

Kendall, M., Stuart, A. (1958). *The Advanced Theory of Statistics*. London: Charles Griffin.

Kennedy, C., ter Huurne, J., Berkhout, M., Gruis, N., Bastiaens, M., Bergman, W. et al. (2001). Melanocortin 1 receptor (MC1R) gene variants are associated with an increased risk for cutaneous melanoma which is largely independent of skin type and hair color. *J Invest Dermatol* **117**(2):294–300.

Kimura, M. (1955). Stochastic processes and distribution of gene frequencies under natural selection. Cold Spring Harbor Symp. *Quant Biol* **20**:33–53.

Kimura, M., Crow, J. (1964). The number of alleles that can be maintained in a finite population. *Genetics* **49**:725–738.

Kittles, R., Perola, M., Peltonen, L., Bergen, A., Aragon, R., Virkkunen, M., Linnoila, M. et al. (1998). Dual origins of Finns revealed by Y chromosome haplotype variation. *Am J Hum Genet* **62**(5):1171–1179.

Kittles, R.A., Chen, W., Panguluri, R.K., Ahaghotu, C., Jackson, A., Adebamowo, C.A. et al. (2002). CYP3A4-V and prostate cancer in African Americans: Causal or confounding association because of population stratification? *Hum Genet* **110**:553–560.

Kollias, N., Baqer, A.H. (1987). Absorption mechanisms of human melanin in the visible, 400–720 nm. *J Invest Dermatol* **89**(4):384–388.

Kollias, N., Sayre, R.M., Zeise, L., Chedekel, M.R. (1991). Photoprotection by melanin. *J Photochem Photobiol B* **9**(2):135–160.

Kollias, N., Baqer, A., Sadiq, I. (1994). Minimum erythema dose determination in individuals of skin type V and VI with diffuse reflectance spectroscopy. *Photodermatol Photoimmunol Photomed* **10**(6):249–254.

Koppula, S.V., Robbins, L.S., Lu, D., Baack, E., White, C.R. Jr., Swanson, N.A., Cone, R.D. (1997). Identification of common polymorphisms in the coding sequence of the human MSH receptor (MC1R) with possible biological effects. *Hum Mutat* **9**(1):30–36.

Krude, H., Biebermann, H., Luck, W., Horn, R., Brabant, G., Gruters, A. (1998). Severe early-onset obesity, adrenal insufficiency and red hair pigmentation caused by POMC mutations in humans. *Nat Genet* **19**(2):155–157.

Kurczynski, T. (1970). Generalized distance and discrete variables. *Biometrics* **26**:525–534.

Lamason, R.L., Mohideen, M.A., Mest, J.R., Wong, A.C., Norton, H.L., Aros, M.C. et al. (2005). SLC24A5, a putative cation exchanger, affects pigmentation in zebrafish and humans. *Science* **310**(5755):1782–1786.

Larsson, M., Pedersen, N.L., Stattin, H. (2003). Importance of genetic effects for characteristics of the human iris. *Twin Res* **6**(3):192–200.

Lee, S.-T., Nicholls, R.D., Schnur, R.E., Guida, L.C., Lu-Kuo, J., Spinner, N.B. et al. (1994). Diverse mutations of the P gene among African-Americans with type II (tyrosinase-positive) oculocutaneous albinism (OCA2). *Hum Molec Genet* **3**:2047–2051.

Lele, S., Richtsmeier, J. (2001). *An Invariant Approach to Statistical Analysis of Shapes*. CRC Press Boca Raton, FL.

Lell, J.T., Sukernik, R.I., Starikovskaya, Y.B., Su, B., Jin, L., Schurr, T.G. et al. (2002). The dual origin and Siberian affinities of Indigenous American Y chromosomes. *Am J Hum Genet* **70**(1):192–206.

Lewontin, R. (1972). The apportionment of human diversity. *Evol Biol* **6**:381–398.

Lindsay, D.S., Jack, P.C., Christian, M.A. (1991). Other-race face perception. *J Appl Psychol* **76**:587–589.

Lindsey, J.D., Jones, H.L., Hewitt, E.G., Angert, M., Weinreb, R.N. (2001). Induction of tyrosinase gene transcription in human iris organ cultures exposed to latanoprost. *Arch Ophthalmol J* **119**(6):853–860.

Little, M.A., Wolff, M.E. (1981). Skin and hair reflectance in women with red hair. *Ann Hum Biol* **8**(3):231–241.

Lloyd, V., Ramaswami, M., Kramer, H. (1998). Not just pretty iris: Drosophila iris-colour mutations and lysosomal delivery. *Trends Cell Biol* **8**(7):257–259.

Long, J., Smouse, P. (1983). Intertribal gene flow between the Ye'cuana and Yanomama: Genetic analysis of an admixed village. *Am J Phys Anthropol* **61**:411–422.

Long, J.C. (1991). The genetic structure of admixed populations. *Genetics* **127**:417–428.

Loomis, W.F. (1967). Skin-pigment regulation of vitamin-D biosynthesis in man. *Science* **157**(788):501–506.

MacLean, C., Workman, P. (1973a). Genetic studies on hybrid populations. I. Individual estimates of ancestry and their relation to quantitative traits. *Ann Hum Genet* **36**:341–351.

MacLean, C., Workman, P. (1973b). Genetic studies on hybrid populations. II. Estimation of distribution of ancestry. *Ann Hum Gent* **36**:459–465.

Mahli, R., Eshleman, J. (In press). What can and can't be determined about Indigenous American (American Indian) ancestry through the analysis of DNA or the uses and limitations of DNA based ancestry tests for Indigenous Americans.

Mason, H.S. (1967). The structure of melanin. In Montagna, W., Hu, F., *The Pigmentary System: Advanced in Biology of Skin, vol. 8.* Oxford: Pergamon Press, 293–312.

McKeigue, P.M. (1998). Mapping genes that underlie ethnic differences in disease risk: Methods for detecting linkage in admixed populations, by conditioning on parental admixture. *Am J Hum Genet* **63**:241–251.

McKeigue, P.M., Carpenter, J.R., Parra, E.J., Shriver, M.D. (2000). Estimation of admixture and detection of linkage in admixed populations by a Bayesian approach: Application to African-American populations. *Ann Hum Genet* **64**:171–186.

Meinhold-Heerlein, I., Bauerschlag, D., Hilpert, F., Dimitrov, P., Sapinoso, L., Orlowska-Volk, M. et al. (2005). Molecular and prognostic distinction between serous ovarian carcinomas of varying grade and malignant potential. *Oncogene* **34**:1053–1065.

Mendel, G. (1865). Experiments in Plant Hybridization. Read at the February 8th and March 8th, 1865 meetings of the Brunn Natural History Society.

Menon, I.A., Basu, P.K., Persad, S., Avaria, M., Felix, C.C., Kalyanaraman, B. (1987). Is there any difference in the photobiological properties of melanins isolated from human blue and brown eyes? *Br J Ophthalmol* **71**(7):549–552.

Millar, R. (1991). Selecting loci for genetic stock identification using maximum likelihood, and the connection with curvature methods. *Can J Fish Aquat Sci* **48**:2173–2179.

Molokhia, M., Hoggart, C., Patrick, A.L., Shriver, M., Parra, E., Ye, J. et al. (2003). Relation of risk of systemic lupus erythematosus to west African admixture in a Caribbean population. *Hum Genet* **112**(3):310–318. Epub 2003 Jan 24.

Mourant, A., Kopec, A., Domaniewska-Sobczak, K. (1976). *The Distribution of the Human Blood Groups and Other Polymorphisms.* Oxford University Press, London.

National Research Council. (1996). Commission on DNA forensic science: An update. National Academy Press, Washington, D.C.

Napolitano, A., Vincensi, M.R., Di Donato, P., Monfrecola, G., Prota, G. (2000). Microanalysis of melanins in mammalian hair by alkaline hydrogen peroxide degradation: Identification of a new structural marker of pheomelanins. *J Invest Dermatol* **114**(6):1141–1147.

Naysmith, L., Waterston, K., Ha, T., Flanagan, N., Bisset, Y., Ray, A. et al. (2004). Quantitative measures of the effect of the melanocortin 1 receptor on human pigmentary status. *J Invest Dermatol* **122**(2):423–428.

Neel, J.V. (1974). Developments in monitoring human populations for mutation rates. *Mutation Res* **26**:319–328.

Nei, M. (1978). Estimation of average heterozygosity and genetic distance from a small number of individuals. *Genetics* **89**:583–590.

Nei, M. (1987). *Molecular Population Genetics.* Columbia University Press, New York.

Nei, M., Chakraborty, R. (1973). Genetic distance and electrophoretic identity of proteins between taxa. *J Mol Evol* **2**(4):323–328.

Nordlund, J.J. (Ed.). (1998). *The Pigmentary System: Physiology and Pathophysiology.* Oxford University Press.

Nordlund, J.J., Abdel-Malek, Z.A., Boissy, R.E., Rheins, L.A. (1989). Pigment cell biology: An historical review. *J Invest Dermatol* **92**(4 Suppl):53S–60S.

Norton, H. (2005). Dissertation Thesis. The Pennsylvania State University, PA.

Norton, H.L., Friedlaender, J.S., Merriwether, D.A., Koki, G., Mgone, C.S., Shriver, M.D. (2006). Skin and hair pigmentation variation in Island Melanesia. *Am J Phys Anthropol* **130**(2):254–268.

Oetting, W.S., King, R.A. (1991). Mutations within the promoter region of the tyrosinase gene in type I (tyrosinase-related) oculocutaneous albinism. (Abstract) *Clin Res* **39**:267A.

Oetting, W.S., King, R.A. (1992). Molecular analysis of type I-A (tyrosine negative) oculocutaneous albinism. *Hum Genet* **90**:258–262.

Oetting, W.S., King, R.A. (1993). Molecular basis of type I (tyrosinase-related) oculocutaneous albinism: Mutations and polymorphisms of the human tyrosinase gene. *Hum Mutat* **2**:1–6.

Oetting, W.S., King, R.A. (1999). Molecular basis of albinism: Mutations and polymorphisms of pigmentation genes associated with albinism. *Hum Mutat* **13**:99–115.

Ooi, C.E., Moreira, J.E., Dell'Angelica, E.C., Poy, G., Wassarman, D.A., Bonifacino, J.S. (1997). Altered expression of a novel adaptin leads to defective pigment granule biogenesis in the Drosophila iris color mutant garnet. *EMBO J* **16**(15):4508–4518.

Ortonne, J.P., Prota, G. (1993). Hair melanins and hair color: Ultrastructural and biochemical aspects. *J Invest Dermatol* **101**(1 Suppl):82S–89S.

Ouellet, V., Provencher, D., Maugard, C., Le Page, C., Ren, F., Lussier, C. et al. (2005). Discrimination between serous low malignant potential and invasive epithelial ovarian tumors using molecular profiling. *Oncogene* **24**:4672–4687.

Palmer, J.S., Duffy, D.L., Box, N.F., Aitken, J.F., O'Gorman, L.E., Green, A.C. et al. (2000). Melanocortin-1 receptor polymorphisms and risk of melanoma: Is the association explained solely by pigmentation phenotype? *Am J Hum Genet* **66**(1):176–186.

Parker, W., Bearn, A. (1961). Haptoglobin and transferring variation in humans and primates: Two new variants in Chinese and Japanese populations. *Ann Hum Genet* **25**:227–241.

Parra, E., Kittles, R., Shriver, M. 2004. Implications of correlations between skin color and genetic ancestry for biomedical research. *Nature Genetics* **36**(11):S54–S60.

Parra, E.J., Marcini, A., Akey, J., Martinson, J., Batzer, M.A., Cooper, R. et al. (1998). Estimating African-American admixture proportions by use of population-specific alleles. *Am J Hum Genet* **63**:1839–1851.

Parra, E.J., Kittles, R.A., Argyropoulos, G., Pfaff, C.L., Hiester, K., Bonilla, C. et al. (2001). Ancestral proportions and admixture dynamics in geographically defined African Americans living in South Carolina. *Am J Phys Anthropol* **114**(1):18–29.

Parra, E.J., Hoggart, C.J., Bonilla, C., Dios, S., Norris, J.M., Marshall, J.A. et al. (2004). Relation of type 2 diabetes to individual admixture and candidate gene polymorphisms in the Hispanic American population of San Luis Valley, Colorado. *J Med Genet* **41**(11):e116.

Parra, F.C., Amado, R.C., Lambertucci, J.R., Rocha, J., Antunes, C.M., Pena, S.D. (2003). Color and genomic ancestry in Brazilians. *Proc Natl Acad Sci U S A* **100**(1):177–182. Epub 2002 Dec 30.

Pastural, E., Ersoy, F., Yalman, N., Wulffraat, N., Grillo, E., Ozkinay, F. et al. (2000). Two genes are responsible for Griscelli syndrome at the same 15q21 locus. *Genomics* **63**(3):299–306.

Patterson, N., Hattangadi, N., Lane, B., Lohmueller, K.E., Hafler, D.A., Oksenberg, J.R. et al. (2004). Methods for high-density admixture mapping of disease genes. *Am J Hum Genet* **74**:979–1000.

Pearson, E. S., Hartley, H.O. (Eds.). (1962). Biometrika Tables for Statisticians, 1, second ed. Cambridge University Press, London.

Pfaff, C., Parra, E., Bonilla, C., Hiester, K., McKeigue, P., Kamboh, M. et al. (2001). Population structure in admixed populations: Effect of admixture dynamics on the pattern of linkage disequilibrium. *Am J Hum Genet* **68**(1):198–207.

Pfaff, C.L., Kittles, R.A., Shriver, M.D. (2002). Adjusting for population structure in admixed populations. *Genet Epidemiol* **22**:196–201.

Pfaff, C.L., Barnholtz-Sloan, J., Wagner, J.K., Long, J.C. (2004). Information on ancestry from genetic markers. *Genet Epidemiol* **26**(4):305–315.

Pritchard, J.K., Stephens, M., Donelly, P. (2000). Inference of population structure using multilocus genotype data. *Genetics* **155**:945–959.

Pritchard, J.K., Donelly, P. (2001). Case-control studies of association in structured or admixed populations. *Theor Popul Biol* **60**:227–237.

Prota, G. (1993). Regulatory mechanism of melanogenesis: Beyond the tyrosinase concept. *J Invest Dermatol* **100**:156S–161S.

Prota, G. (2000). Melanins, melanogenesis and melanocytes: Looking at their functional significance from the chemist's viewpoint. *Pigment Cell Res* **13**:283–293.

Prota, G., R. Nicolaus, R. (1967). On the biogenesis of phaeomelanins. In Montagna, W., Hu, F. *Advances in biology of skin. The pigmentary system.* Vol. VIII. New York: Pergamon Press, 323–328.

Puri, N., Gardner, J.M., Brilliant, M.H. (2000). Aberrant pH of melanosomes in pink-eyed dilution (p) mutant melanocytes. *J Invest Dermatol* **115**(4):607–613.

Rajkumar, R., Kashyap, V.K. (2004). Genetic structure of four socio-culturally diversified caste populations of southwest India and their affinity with related Indian and global groups. *BMC Genet* **5**(1):23.

Ramana, G.V., Su, B., Jin, L., Singh, L., Wang, N., Underhill, P., Chakraborty, R. (2001). Y-chromosome SNP haplotypes suggest evidence of gene flow among caste, tribe, and the migrant Siddi populations of Andhra Pradesh, South India. *Eur J Hum Genet* **9**(9):695–700.

Ramaswamy, S., Ross, K., Lander, E., Golub, T. (2003). A molecular signature of metastasis in primary solid tumors. *Nature Gen* **33**:49–54.

Rao, D.C., Chakraborty, R. (1974). The generalized Wright's model and population structure with special reference to the ABO blood group system. *Am J Hum Genet* **26**(4):444–453.

Rebbeck, T.R., Kanetsky, P.A., Walker, A.H., Holmes, R., Halpern, A.C., Schuchter, L.M. et al. (2002). P gene as an inherited biomarker of human eye color. *Cancer Epidemiol Biomarkers Prev* **11**(8):782–784.

Reed, T. (1969). Caucasian genes in American Negroes. *Science* **165**:762–768.

Reed, T. (1973). Number of gene loci required for accurate estimation of ancestral population proportions in individual human hybrids. *Nature* **244**:575–576.

Reed, T.E. (1952). Red hair colour as a genetical character. *Ann Eugen* **17**(2):115–139.

Reidla, M., Kivisild, T., Metspalu, E., Kaldma, K., Tambets, K., Tolk, H.V. et al. (2003). Origin and diffusion of mtDNA haplogroup X. *Am J Hum Genet* **73**(6):1178–1190.

Relethford, J.H. (1992). Cross-cultural analysis of migration rates: Effects of geographic distance and population size. *Am J Phys Anthropol* **89**(4):459–466.

Relethford, J.H. (1997). Hemispheric differences in human skin color. *Am J Phys Anthropol* **104**:449–457.

Relethford, J.H. (2002). Apportionment of global human genetic diversity based on craneometrics and skin color. *Am J Phys Anthropol* **118**:393–398.

Risch, N. (2001). Implications of multilocus inheritance for gene-disease association studies. *Theor Popul Biol* **60**(3):215–220.

Risch, N., Burchard, E., Ziv, E., Tang, H. (2002). Categorization of humans in biomedical research: Genes, race and disease. *Genome Biol* **3**. Comment 2007. Epub 2002 Jul. 1.

Roberts, D. (1955). The dynamics of racial admixture in the American Negro: Some anthropological considerations. *Am J Hum Genet* **7**:361–367.

Roberts, D., Hirons, R. (1962). The dynamics of racial admixture. *Am J Hum Genet* **14**:261–277.

Roberts, D., Hirons, R. (1965). Methods of analysis of the genetic composition of a hybrid population. *Hum Biol* **37**:38–43.

Robbins, J.H., Brumback, R.A., Mendiones, M., Barrett, S.F., Carl, J.R., Cho, S. et al. (1991). Neurological disease in xeroderma pigmentosum: Documentation of a late onset type of the juvenile onset form. *Brain* **114**(Pt 3):1335–1361.

Robbins, L.S., Nadeau, J.H., Johnson, K.R., Kelly, M.A., Roselli-Rehfuss, L., Baack, E. et al. (1993). Pigmentation phenotypes of variant extension locus alleles result from point mutations that alter MSH receptor function. *Cell* **72**:827–834.

Robins, A.H. (1991). *Biological Perspectives on Human Pigmentation*. Cambridge University Press, Cambridge.

Romauldi, C., Balding, D., Nasidze, I. et al. (2002). Patterns of human diversity, within and among continents, inferred from biallelic DNA polymorphisms. *Genome Res* **12**:602–612.

Rosenberg, N., Pritchard, J., Weber, J., Cann, H., Kidd, K., Zhivotovsky, L., Feldman, M. (2002). Genetic structure of human populations. *Science* **298**:2381–2385.

Rosenberg, N., Lei, L., Ward, R., Pritchard, J. (2003). Informativeness of genetic markers for inference of ancestry. *Am J Hum Genet* **73**:1402–1422.

Roychoudhury, A.K., Nei, M. (1988). *Human Polymorphic Genes: World Distribution*. Oxford University Press, New York.

Ruhlen, M. (1998). The origin of the Na-Dene. *Proc Natl Acad Sci U S A* **95**(23):13994–13996.

Rybicki, B.A., Iyengar, S.K., Harris, T., Liptak, R., Elston, R.C., Sheffer, R. et al. (2003). *Hum Hered* **53**:187–196.

Saha, N., Kirk, R., Shanbhag, S., Joshi, S., Bhatia, H. (1974). Genetic studies among the Kadar of Kerala. *Hum Hered* **24**:198–218.

Saha, N., Kirl, R., Shanbhag, S., Joshi, S., Bhatia, H. (1976). Population genetic studies in Kerala and the Nilgirls (South West India). *Hum Hered* **26**:175–197.

Saitou, N., Nei, M. (1987). The neighbor-joining method: A new method for reconstructing phylogenetic trees. *Mol Biol Evol* **4**:406–425.

Salas, A., Richards, M., De la Fe, T. et al. (2002). The making of the African mtDNA landscape. *Am J Hum Genet* **71**:1082–1111.

Saldanha, P. (1957). Gene flow from white into Negro populations in Brazil. *Am J Hum Genet* **9**:299–309.

Schioth, H.B., Phillips, S.R., Rudzish, R., Birch-Machin, M.A., Wikberg, J.E., Rees, J.L. (1999). Loss of function mutations of the human melanocortin 1 receptor are common and are associated with red hair. *Biochem Biophys Res Commun* **260**:488–491.

Schlotterer, C. (2000). Evolutionary dynamics of microsatellite DNA. *Chromosoma* **109**(6):365–371.

Schmidt, A., Beermann, F. (1994). Molecular basis of dark-irised albinism in the mouse. *Proc Natl Acad Sci U S A* **91**(11):4756–4760.

Schurr, T.G., Sherry, S.T. (2004). Mitochondrial DNA and Y chromosome diversity and the peopling of the Americas: Evolutionary and demographic evidence. *Am J Hum Biol* **16**(4):420–439.

Seldin, M.F., Shigeta, R., Villoslada, P., Selmi, C., Tuomilehto, J., Silva, G., Belmont, J.W., Klareskog, L., Gregersen, P.K. (2006). European population substructure: clustering of northern and southern populations. *PloS Genet* 15;**2**(9):e143.

Shriver, M.D., Smith, M., Jin, L. (1997). Forensic ethnic affiliation estimation by use of population specific allele DNA markers. *Am J Hum Genet* **60**:957–964.

Shriver, M.D., Parra, E.J. (2000). Comparison of narrow-band reflectance spectroscopy and tristimulus colorimetry for measurements of skin and hair color in persons of different biological ancestry. *Am J Phys Anthropol* **112**:17–27.

Shriver, M.D., Parra, E.J., Dios, S., Bonilla, C., Norton, H., Jovel, C. et al. (2003). Skin pigmentation, biogeographical ancestry and admixture mapping. *Hum Genet* **112**:387–399. Epub 2003 Feb 11.

Shriver, M.D., Kennedy, G.C., Parra, E.J., Lawson, H.A., Sonpar, V., Huang, J. et al. (2004). The genomic distribution of population substructure in four populations using 8,525 autosomal SNPs. *Hum Genomics* **1**(4):274–286.

Shriver, M.D., Kittles, R.A. (2004). Genetic ancestry and the search for personalized genetic histories. *Nat Rev Genet* **5**(8):611–618.

Shriver, M.D., Mei, R., Parra, E.J., Sonpar, V., Halder, I., Tishkoff, S.A. et al. (2005). Large-scale SNP analysis reveals clustered and continuous patterns of human genetic variation. *Hum Genomics* **2**(2):81–89.

Smith, R., Healy, E., Siddiqui, S., Flanagan, N., Steijlen, P.M., Rosdahl, I. et al. (1998). Melanocortin 1 receptor variants in an Irish population. *J Invest Derm* **111**:119–122.

Smith, M., Lautenberger, J., Shin, H., Chretien, J., Shrestha, S., Gilbert, D., O'Brien, S. (2001). Markers for mapping by admixture linkage disequilibrium in African American and Hispanic populations. *Am J Hum Genet* **69**:1080–1094.

Smith, M., Patterson, N., Lautenberger, J., Truelove, A., McDonald, G., Waliszewska, A. et al. (2004). A high density admixture map for disease gene discovery in African Americans. *Am J Hum Genet* **74**(5):1001–1013.

Smouse, P., Neel, J. (1977). Multivariate analysis of gametic disequilibrium in the Yanomama. *Genetics* **85**(4):733–753.

Sneath, P., Snokal, R. (1973). *Numerical Taxonomy*. W.H. Freeman and Company, San Francisco, 230–234.

Sorby, H.C. (1878). On the colouring matters found in human hair. *J Anthropol Inst Lond* **8**:1–14.

Spitsyn, V., Kuchheuser, W., Makarov, S., Bychkovskaia, L., Pai, G., Balanovskii, O., Afanas'eva, I. (2001). The Russian gene pool. Frequency of genetic markers. *Genetika* **37**(3):386–401.

Spritz, R.A. (1995). A study in scarlet. *Nature Genet* **11**:225–226.

Stamatas, G.N., Zmudzka, B.Z., Kollias, N., Beer, J.Z. (2004). Non-invasive measurements of skin pigmentation in situ. *Pigment Cell Res* **17**(6):618–626.

Stanley, S. (2006). *An Invisible Man; the Hunt for a Serial Killer Who Got Away with a Decade of Murder*. Berkley Books, New York.

Staricco, R.G. (1963). Amelanotic melanocytes in the outer sheath of the human hair follicle and their role in the repigmentation of regenerated epidermis. *Ann N Y Acad Sci* **100**:239–255.

Steinberg, A. (1974). The Gm and Inv immunoglobulin allotypes in Indian populations: A review. In L.D. Sanghvi, V. Balakrishnan, H. Bhatia, P. Sukumaran, J.V. Undevia (Eds), *Human Population Genetics in India*. New Delhi, Orient Longman, 112–126.

Steinberg, R., Giles, A., Stauffer, R. (1960). A Gm-like factor present in Negroes and rare or absent in whites: Its relation to Gma and Gmx. *Am J Hum Genet* **12**:44–51.

Stephens, J.C., Briscoe, D., O'Brien, S.J. (1994). Mapping by admixture linkage disequilibrium in human populations: Limits and guidelines. *Am J Hum Genet* **55**:809–824.

Stephens, M., Smith, N.J., Donnelly, P. (2001). A new statistical method for haplotype reconstruction from population data. *Am J Hum Genet* **68**(4):978–989.

Strachan, R. (1999). *Human Molecular Genetics 2, 2nd ed.* John Wiley & Sons, New York.

Sturm, R.A. (2002). Skin colour and skin cancer—MC1R, the genetic link. *Melanoma Res* **12**(5):405–416.

Sturm, R. (2006). A golden age of human pigmentation genetics. *Trends Genet* **22**(9):464–468. Epub 2006 Jul 18.

Sturm, R.A., Box, N.F., Ramsay, M. (1998). Human pigmentation genetics: The difference is only skin deep. *Bioessays* **20**(9):712–721.

Sturm, R., Teasdale, R., Box, N. (2001). Human pigmentation genes: Identification, structure and consequences of polymorphic variation. *Gene* **277**:49–62.

Sturm, R.A., Frudakis, T.N. (2004). Eye colour: Portals into pigmentation genes and ancestry. *Trends Genet* **20**(8):327–332.

Su, B., Jin, L., Underhill, P. et al. (2000). Polynesian origins: Insights from the Y chromosome. *Proc Natl Acad Sci U S A* **97**:8225–8228.

Sun, G., McGarvey, S.T., Bayoumi, R., Mulligan, C.J., Barrantes, R., Raskin, S. et al. (2003). Global genetic variation at nine short tandem repeat loci and implications on forensic genetics. *Eur J Hum Genet* **11**(1):39–49.

Sunderland, E. (1956). Hair-colour variation in the United Kingdom. *Ann Hum Genet* **20**(4):312–333.

Tang, H., Quertermous, T., Rodriguez, B., Kardia, S.L., Zhu, X., Brown, A. et al. (2005). Genetic structure, self-identified race/ethnicity, and confounding in case-control association studies. *Am J Hum Genet* **76**(2):268–275.

Taylor, A.L., Cohn, J.N., Worcel, M., Franciosa, J.A.., A-HeFT Investigators. (2002). The African-American Heart Failure Trial: Background, rationale and significance. *J Natl Med Assoc* **94**(9):762–769.

Temple R., Stockbridge, N. (2007). BiDil for heart failure in black patients: The U.S. Food and Drug Administration perspective. *Ann Intern Med* **146**(1):57–62.

Templeton, A. (1998). Nested clade analyses of phylogeographic data: Testing hypotheses about gene flow and population history. *Mol Ecol* **7**:381–397.

Templeton, A.R. (2002). Out of Africa again and again. *Nature* **416**:45–51.

Templeton, A.R. (2004). Statistical phylogeography: Methods of evaluating and minimizing inference errors. *Mol Ecol* **13**(4):789–809.

Terwilliger, J.D., Goring, H.H. (2000). Gene mapping in the 20th and 21st centuries: Statistical methods, data analysis, and experimental design. *Hum Biol* **72**(1): 63–132.

Thody, A.J., Higgins, E.M., Wakamatsu, K., Ito, S., Burchill, S.A., Marks, J.M. (1991). Pheomelanin as well as eumelanin is present in human epidermis. *J Invest Dermatol* **97**(2):340–344.

Thomas, M., Weale, M., Jones, A., Richards, M., Smith, A., Rodhead, N. et al. (2002). Founding mothers of Jewish communities: Geographically separated Jewish groups were independently founded by very few female ancestors. *Am J Hum Genet* **70**:1411–1420.

Thompson, E.A. (1973). The Icelandic admixture problem. *Ann Hum Genet* **37**(1):69–80.

Underhill, P., Passarino, G., Lin, A., Shen, P., Mirazon Lahr, M., Foley, R. et al. (2001). The phylogeography of Y chromosome binary haplotypes and the origins of modern human populations. *Ann Hum Genet* **65**:43–62.

Valverde, P., Healy, E., Jackson, I., Rees, J.L., Thody, A.J. (1995). Variants of the melanocyte-stimulating hormone receptor gene are associated with red hair and fair skin in humans. *Nature Genet* **11**:328–330.

Verrelli, B.C., Tishkoff, S.A. (2004). Signatures of selection and gene conversion associated with human color vision variation *Am J Hum Genet* **75**:363–375.

Vos, G., Rirk, R., Steinberg, A. (1963). The distribution of the gamma globulin types Gm(a), Gm(h), Gm(x) and Gm-like in South and Southeast Asia and Australia. *Am J Hum Genet* **15**:44–52.

Wadman, M. (2005). Drug targeting: Is race enough? Nature **435**(7045):1008–1009.

Wagner, J., Jovel, C., Norton, H., Parra, E., Shriver, M. (2002). Comparing quantitative measures of erythema, pigmentation and skin response using reflectometry. *Pigment Cell Res* **15**:379–384.

Wahlund, S. (1928). Zusammensetzung von populationen and korrelationserscheinungen von standpunkt der vererbungslehre aus betrachtet. *Hereditas* **11**:65–106.

Walter, H., Hilling, M., Singh, I., Bhasin, M., Goud, J., Veerraju, P. (1980). Gm and Iny phenotypes and gene frequencies in Indian populations. *S Asian Anthropol* **1**:69–75.

Wang, J. (2003). Maximum-likelihood estimation of admixture proportions from genetic data. *Genetics* **164**:747–765.

Weir, B. (1996). The second national research council report on forensic DNA evidence. *Am J Hum Genet* **59**(3):497–500.

Weir, B.S., Cockerham, C.C. (1969). Group inbreeding with two linked loci. *Genetics* **63**(3):711–742.

Weir, B.S., Cardon, L.R., Anderson, A.D., Nielsen, D.M., Hill, W.G. (2005). Measures of human population structure show heterogenity among genomic regions. *Genome Research* **15**:1468–1476.

Weiss, K.M., Fullerton, S.M. (2000). Phenogenetic drift and the evolution of genotype-phenotype relationships. *Theor Popul Biol* **57**(3):187–195.

Wijsman, E. (1984). Techniques for estimating genetic admixture and applications to the problems of the origin of the Icelanders and the Ashkenazi Jews. *Hum Genet* **67**:441–448.

Wilson, J., Weale, M., Smith, A. et al. (2001). Population genetic structure of variable drug response. *Nature Genet* **29**:265–269.

Workman, P. (1968). Gene flow and the search for natural selection in man. *Hum Biol* **40**:260–279.

Workman, P., Blumberg, B., Cooper, A. (1963). Selection, gene migration, and polymorphic stability in US White and Negro populations. *Am J Hum Genet* **15**:429–437.

Wright, S. (1967). *Evolution and the Genetics of Populations, Vol. 1*. Chicago: Genetic and Biometric Foundation.

Wu, B., Liu, N., Zhao, H. (2006). PSMIX: An R package for population structure inference via maximum likelihood method. *Bioinformatics* **7**:317–326.

Ye, J., Parra, E.J., Sosnoski, D.M., Hiester, K., Underhill, P.A., Shriver, M.D. (2002). Melting curve SNP (McSNP) genotyping: a useful approach for diallelic genotyping in forensic science. *J Forensic Sci* **47**:593–600.

Zaykin, D., Zhivotovsky, L., Weir, B.S. (1995). Exact tests for association between alleles at arbitrary numbers of loci. *Genetica* **96**(1–2):169–178.

Zanetti, R., Prota, G., Napolitano, A., Martinez, C., Sancho-Garnier, H., Osterlind, A. et al. (2001). Development of an integrated method of skin phenotype measurement using the melanins. *Melanoma Res* **11**(6):551–557.

Zegura, S.L., Karafet, T.M., Zhivotovsky, L.A., Hammer, M.F. (2004). High-resolution SNPs and microsatellite haplotypes point to a single, recent entry of Indigenous American Y chromosomes into the Americas. *Mol Biol Evol* **21**(1):164–175. Epub 2003 Oct 31.

Zhu, G., Evans, D.M., Duffy, D.L., Montgomery, G.W., Medland, S.E., Gillespie, N.A. et al. (2004). A genome scan for eye color in 502 twin families: Most variation is due to a QTL on chromosome 15q. *Twin Res* **7**(2):197–210.

Page numbers followed by ''f'' denote figures; those followed by ''t'' denote tables; and those followed by ''b'' denote boxes